DEFINITE INTEGRAL MADE EASY

OTHER BOOKS IN MADE EASY SERIES BY THE SAME AUTHOR

1. Limits and Derivatives Made Easy
2. Applications of Derivatives Made Easy
3. Indefinite Integral Made Easy
4. Co-ordinate Geometry Made Easy
5. Complex Numbers Made Easy

CONTENTS

PREFACE

This book on Definite Integral is designed for students of schools and colleges. The aim of the book is to present the subject in such a way that even an average student may find no difficulty in understanding it.

There are, of course, a number of books available on the subject , which are too concise to be intelligible to the students. There is hardly any single book which covers the entire syllabus of Definite Integral prescribed in various state boards.

Every effort has been made to present the subject matter in the most lucid style. The following are some of the salient features of this book :

(*i*) The book serves the purpose of text as well as a help book.

(*ii*) A large number of typical and important solved examples have been provided to enable the student to have a clear grasp of the subject and to equip him for attempting problems in the board examination without any difficulty.

(*iii*) Care has been taken to grade the problems in such a way that students move from basic to intricate problems with ease.

(*iv*) The 'Remarks and Notes' have been added quite often in the book so that they should help in understanding the ideas better.

(*v*) At end of each chapter, a 'Revision Exercise' has been incorporated. The purpose of this exercise is to give an opportunity to students to revise the entire chapter in the minimum possible time.

It is hoped with all the above characteristics the book will found really useful by teachers and students alike.

A serious effort has been made to keep the book free from errors, but even then some errors might have crept in.

Suggestions from the teachers as well as students for further improvement will be warmly received.

— AUTHOR

SYMBOLS

Greek Alphabets

Α	α	Alpha	Ι	ι	Iota	Ρ	ρ	Rho
Β	β	Beta	Κ	κ	Kappa	Σ	σ	Sigma
Γ	γ	Gamma	Λ	λ	Lambda	Τ	τ	Tau
D	δ	Delta	Μ	μ	Mu	Υ	υ	Upsilon
Ε	ε	Epsilon	Ν	ν	Nu	Φ	φ	Phi
Ζ	ζ	Zeta	Ξ	ξ	Xi	Χ	χ	Chi
Η	η	Eta	Ο	ο	Omicron	Ψ	ψ	Psi
Θ	θ	Theta	Π	π	Pi	Ω	ω	Omega
	∃	there exists		∀	for all			

Metric Weights and Measures

LENGTH

10 millimetres	= 1 centimetre
10 centimetres	= 1 decimetre
10 decimetres	= 1 metre
10 metres	= 1 decametre
10 decametres	= 1 hectometre
10 hectometres	= 1 kilometre

CAPACITY

10 millilitres	= 1 centilitre
10 centilitres	= 1 decilitre
10 decilitres	= 1 litre
10 litre	= 1 dekalitre
10 decalitres	= 1 hectolitre
10 hectolitres	= 1 kilolitre

VOLUME

1000 cubic centimetres	= 1 centigram
1000 cubic decimetres	= 1 cubic metre

AREA

100 square metres	= 1 are
100 ares	= 1 hectare
100 hectares	= 1 square Kilometre

WEIGHT

10 milligrams	= 1 centigram
10 centigrams	= 1 decigram
10 decigrams	= 1 gram
10 grams	= 1 decagram
10 decagrams	= 1 hectogram
10 hectograms	= 1 kilogram
100 kilograms	= 1 quintal
10 quintals	= 1 metric ton (tonne)

ABBREVIATIONS

kilometre	km	tonne	t
metre	m	quintal	q
centimetre	cm	kilogram	kg
millimetre	mm	gram	g
kilolitre	kl	are	a
litre	l	hectare	ha
millilitre	ml	centiare	ca

1

Definite Integral as the Limit of a Sum

1.1. DEFINITION : DEFINITE INTEGRAL AS THE LIMIT OF A SUM

If $f(x)$ is a continuous and single-valued function in the closed interval $[a, b]$, then

$$\lim_{h \to 0} h[f(a) + f(a + h) + f(a + 2h) + \ldots\ldots + f(a + \overline{n-1}\, h)]$$

where : $$h = \frac{b-a}{n} \text{ or } nh = b - a$$

is called the definite integral of $f(x)$ between the limits a and b and is written as :

$$\int_a^b f(x)\, dx$$

Thus : $$\int_a^b f(x)\, dx = \lim_{h \to 0} h[f(a) + f(a + h) + f(a + 2h) + \ldots\ldots + f(a + \overline{n-1}\, h)]$$

where : $nh = b - a$

Or $$\int_a^b f(x)\, dx = \lim \sum_{r=0}^{n-1} hf(a + rh)$$

where : $h \to 0$, $n \to \infty$, $nh = b - a$

Note 1. The above method of evaluating $\int_a^b f(x)\, dx$

i.e., $$\int_a^b f(x)\, dx = \lim_{h \to 0} h[f(a) + f(a + h) + f(a + 2h) + \ldots\ldots + f(a + \overline{n-1}\, h)]$$

where : $nh = b - a$

is called integration by summation or integration from definition or integration from first principles or integration by *ab*-nitio.

Note 2. Proceed to the limit as $h \to 0$ only after putting $nh = b - a$.

1.2. DEFINITION : WALLI'S FORMULA

If $f(x)$ be a function such that :

(*i*) $f(x)$ is continuous function defined on $[a, b]$ and

(*ii*) $f(x)$ does not change its sign on $[a, b]$.

Then, $$\lim_{r \to 1} (r - 1)\, [f(a) \,.\, a + f(ar) \,.\, ar + f(ar^2) \,.\, ar^2 + \ldots\ldots + f(ar^{n-1})\, ar^{n-1}]$$

where : $r^n = \dfrac{b}{a}$, $(a \neq 0)$

This definition is useful to integrate the functions of the form $\int_a^b x^n \, dx$.

where : n is a positive integer.

Remark : The value of the definite integral of a function over any particular interval depends on the function and the interval, but not on the variable of integration that we choose to represent the independent variable.

If the independent variable is denoted by t or u or y instead of x, we simply write the integral as $\int_a^b f(t)\, dt$ or $\int_a^b f(u)\, du$ or $\int_a^b f(y)\, dy$ instead of $\int_a^b f(x)\, dx$. Hence, the value of integration is called a dummy variable.

1.3. SOME USEFUL RESULTS HELPFUL FOR THE EVALUATION OF INTEGRALS BY SUMMATION

1. $1 + 2 + 3 + \ldots\ldots + (n-1) = \dfrac{n(n-1)}{2}$
2. $1 + 2 + 3 + \ldots\ldots + n = \dfrac{n(n+1)}{2}$
3. $1^2 + 2^2 + 3^2 + \ldots\ldots + (n-1)^2 = \dfrac{n(n-1)(2n-1)}{6}$
4. $1^2 + 2^2 + 3^2 + \ldots\ldots + n^2 = \dfrac{n(n+1)(2n+1)}{6}$
5. $1^3 + 2^3 + 3^3 + \ldots\ldots + (n-1)^3 = \dfrac{n^2(n-1)^2}{4} = \left[\dfrac{n(n-1)}{2}\right]^2$
6. $1^3 + 2^3 + 3^3 + \ldots\ldots + n^3 = \dfrac{n^2(n+1)^2}{4} = \left[\dfrac{n(n+1)}{2}\right]^2$
7. $a + ar + ar^2 + \ldots\ldots + ar^{n-1} = \dfrac{a(r^n - 1)}{(r-1)}, \; r \neq 1, r > 1.$
8. $\sin a + \sin (a + h) + \sin (a + 2h) + \ldots\ldots + \sin (a + \overline{n-1}\, h)$

$$= \frac{\sin\left[a + \left(\frac{n-1}{2}\right)h\right] \sin\left(\frac{nh}{2}\right)}{\sin\left(\frac{h}{2}\right)}$$

9. $\cos a + \cos(a + h) + \cos(a + 2h) + \ldots\ldots + \cos (a + \overline{n-1}\, h)$

$$= \frac{\cos\left[a + \left(\frac{n-1}{2}\right)h\right] \sin\left(\frac{nh}{2}\right)}{\sin\left(\frac{h}{2}\right)}$$

1.4. IMPORTANT NOTE

A check on the answer :

If $f(x)$ is positive in the interval (a, b), then $\int_a^b f(x)\, dx$ is also positive.

$$\therefore \qquad \int_a^b f(x)\,dx = \lim_{h \to 0} h[f(a) + f(a+h) + f(a+2h) + \ldots\ldots + f\{a + (n-1)\,h\}]$$

where : $nh = b - a$.

Now, each term of the series on the R.H.S. of above equation is positive.

Therefore their sum and therefore the limit of their sum is also positive.

Similarly, if $f(x)$ is negative in the interval (a, b), then $\int_a^b f(x)\,dx$ is also negative.

The above note enables us to say whether the value of a definite integral should be positive or negative and, therefore, serves as a check on the answer.

SOME SOLVED EXAMPLES

Example 1. *Evaluate the following definite integrals as limit of sums :*

(i) $\int_1^2 x\,dx$ (ii) $\int_0^3 (x+4)\,dx$

(iii) $\int_a^b x\,dx$ (iv) $\int_0^5 (x+1)\,dx$

(v) $\int_0^2 (2x+1)\,dx$ (vi) $\int_{-1}^1 (x+3)\,dx$

(vii) $\int_2^7 (5x-9)\,dx$ (viii) $\int_0^4 2x\,dx$.

Solution. (i) Let $I = \int_1^2 x\,dx$

Comparing I with $\int_a^b f(x)\,dx$, we get : $a = 1,\ b = 2,\ f(x) = x$

Let $\qquad h = \dfrac{b-a}{n} = \dfrac{2-1}{n} = \dfrac{1}{n} \Rightarrow nh = 1 \;;\; n \in N$

$$\therefore \quad I = \lim_{h \to 0} h[f(a) + f(a+h) + f(a+2h) + \ldots\ldots + f\{a + (n-1)\,h\}]$$

$$= \lim_{h \to 0} h[f(1) + f(1+h) + f(1+2h) + \ldots\ldots + f\{1 + (n-1)\,h\}] \qquad [\because \ a = 1]$$

$$= \lim_{h \to 0} h[1 + (1+h) + (1+2h) + \ldots\ldots + \{1 + (n-1)\,h\}] \qquad [\because \ f(x) = x]$$

$$= \lim_{h \to 0} h[(1 + 1 + 1 + \ldots\ldots + 1) + h\{1 + 2 + 3 + \ldots\ldots + (n-1)\}]$$

$$= \lim_{h \to 0} h[n + h\{1 + 2 + 3 + \ldots\ldots + (n-1)\}]$$

$$= \lim_{h \to 0} h\left[n + h \cdot \frac{n(n-1)}{2}\right] \qquad \left[\because \ [1 + 2 + 3 + \ldots\ldots + (n-1)] = \frac{n(n-1)}{2}\right]$$

$$= \lim_{h \to 0}\left[nh + \frac{nh(nh-h)}{2}\right]$$

$$= \lim_{h \to 0}\left[1 + \frac{1(1-h)}{2}\right] \qquad [\because \ nh = 1]$$

$$= 1 + \frac{1}{2} = \frac{3}{2}.$$

(*ii*) Let $\quad \text{I} = \int_0^3 (x + 4)\, dx$

Comparing I with $\int_a^b f(x)\, dx$, we get : $a = 0,\ b = 3,\ f(x) = x + 4$

Let $\quad h = \dfrac{b-a}{n} = \dfrac{3-0}{n} = \dfrac{3}{n} \quad \Rightarrow \quad nh = 3,\ n \in \text{N}.$

$$\therefore \quad \text{I} = \lim_{h\to 0} h[f(a) + f(a+h) + f(a+2h) + \ldots\ldots + f\{a + (n-1)h\}]$$

$$= \lim_{h\to 0} h[f(0) + f(0+h) + f(0+2h) + \ldots\ldots + f\{0 + (n-1)h\}] \qquad [\because \quad a = 0]$$

$$= \lim_{h\to 0} h[f(0) + f(h) + f(2h) + \ldots\ldots + f\{(n-1)h\}]$$

$$= \lim_{h\to 0} h[(0+4) + (h+4) + (2h+4) + \ldots\ldots + \{(n-1)h + 4\}] \qquad [\because \quad f(x) = (x+4)]$$

$$= \lim_{h\to 0} h[(4 + 4 + 4 + \ldots\ldots + 4) + \{h + 2h + \ldots\ldots + (n-1)h\}]$$

$$= \lim_{h\to 0} h[4n + h\{1 + 2 + 3 + \ldots\ldots + (n-1)\}]$$

$$= \lim_{h\to 0} h\left[4n + h \cdot \frac{n(n-1)}{2}\right] \qquad \left[\because \quad [1 + 2 + 3 + \ldots\ldots + (n-1)] = \frac{n(n-1)}{2}\right]$$

$$= \lim_{h\to 0}\left[4nh + \frac{nh(nh-h)}{2}\right] = \lim_{h\to 0}\left[4(3) + \frac{3(3-h)}{2}\right] \qquad [\because \quad nh = 3]$$

$$= 12 + \frac{3(3-0)}{2} = 12 + \frac{9}{2} = \frac{33}{2}.$$

(*iii*) Let $\quad \text{I} = \int_a^b x\, dx$

Comparing I with $\int_a^b f(x)\, dx$, we get : $a = a,\ b = b,\ f(x) = x$

Let $\quad h = \dfrac{b-a}{n} \quad \Rightarrow \quad nh = b - a,\ n \in \text{N}.$

$$\therefore \quad \text{I} = \lim_{h\to 0} h[f(a) + f(a+h) + f(a+2h) + \ldots\ldots + f\{a + (n-1)h\}]$$

$$= \lim_{h\to 0} h[a + (a+h) + (a+2h) + \ldots\ldots + \{a + (n-1)h\}] \qquad [\because \quad f(x) = x]$$

$$= \lim_{h\to 0} h[(a + a + a + \ldots\ldots + a) + h(1 + 2 + 3 + \ldots\ldots + (n-1))]$$

$$= \lim_{h\to 0} h\left[na + h \cdot \frac{n(n-1)}{2}\right] \qquad \left[\because \quad [1 + 2 + 3 + \ldots\ldots + (n-1)] = \frac{n(n-1)}{2}\right]$$

$$= \lim_{h\to 0}\left[nha + \frac{nh(nh-h)}{2}\right]$$

$$= \lim_{h\to 0}\left[(b-a)a + \frac{(b-a)(b-a-h)}{2}\right] \qquad [\because \quad nh = (b-a)]$$

$$= a(b-a) + \frac{(b-a)(b-a)}{2}$$

$$= (b-a)\left[a + \frac{b-a}{2}\right] = \frac{(b-a)(b+a)}{2} = \frac{b^2-a^2}{2}.$$

(*iv*) Let $\quad I = \int_0^5 (x+1)\,dx$

Comparing I with $\int_a^b f(x)\,dx$, we get : $a = 0$, $b = 5$, $f(x) = x + 1$.

Let $\quad h = \frac{b-a}{n} = \frac{5-0}{n} = \frac{5}{n} \quad \Rightarrow \quad nh = 5,\ n \in N.$

$\therefore \quad I = \lim_{h\to 0} h[f(a) + f(a+h) + f(a+2h) + \ldots\ldots + f\{a+(n-1)h\}]$

$= \lim_{h\to 0} h[f(0) + f(0+h) + f(0+2h) + \ldots\ldots + f\{0+(n-1)h\}] \qquad [\because\ a = 0]$

$= \lim_{h\to 0} h[f(0) + f(h) + f(2h) + \ldots\ldots + f\{(n-1)h\}]$

$= \lim_{h\to 0} h[(0+1) + (h+1) + (2h+1) + \ldots\ldots + \{(n-1)h+1\}] \qquad [\because\ f(x) = (x+1)]$

$= \lim_{h\to 0} h[(1+1+1+\ldots\ldots+1) + \{h + 2h + \ldots\ldots + (n-1)h\}]$

$= \lim_{h\to 0} h[n + h\{1+2+3+\ldots\ldots+(n-1)\}]$

$= \lim_{h\to 0} h\left[n + h\,.\,\frac{n(n-1)}{2}\right] \qquad \left[\because\ [1+2+3+\ldots\ldots+(n-1)] = \frac{n(n-1)}{2}\right]$

$= \lim_{h\to 0}\left[nh + \frac{nh(nh-h)}{2}\right] = \lim_{h\to 0}\left[5 + \frac{5(5-h)}{2}\right] \qquad [\because\ nh = 5]$

$= 5 + \frac{25}{2} = \frac{35}{2}.$

(*v*) Let $\quad I = \int_0^2 (2x+1)\,dx$

Comparing I with $\int_a^b f(x)\,dx$, we get : $a = 0$, $b = 2$, $f(x) = (2x+1)$.

Let $\quad h = \frac{b-a}{n} = \frac{2-0}{n} = \frac{2}{n} \quad \Rightarrow \quad nh = 2.$

$\therefore \quad I = \lim_{h\to 0} h[f(a) + f(a+h) + f(a+2h) + \ldots\ldots + f\{a+(n-1)h\}]$

$= \lim_{h\to 0} h[f(0) + f(0+h) + f(0+2h) + \ldots\ldots + f\{0+(n-1)h\}] \qquad [\because\ a = 0]$

$= \lim_{h\to 0} h[f(0) + f(h) + f(2h) + \ldots\ldots + f\{(n-1)h\}]$

$= \lim_{h\to 0} h[\{2(0)+1\} + (2h+1) + (2(2h)+1\} + \ldots\ldots + \{2(n-1)h+1\}]$

$[\because\ f(x) = (2x+1)]$

$= \lim_{h\to 0} h[1 + (2h+1) + (4h+1) + 2h(n-1) + 1]$

$= \lim_{h\to 0} h[(1+1+1+\ldots\ldots+1) + 2h[1+2+3+\ldots\ldots(n-1)\}]$

$$= \lim_{h\to 0} h\left[n + 2h\,\frac{n(n-1)}{2}\right] \qquad \left[\because\ [1+2+3+\ldots\ldots+(n-1)] = \frac{n(n-1)}{2}\right]$$

$$= \lim_{h\to 0}\ [nh + nh(nh - h]$$

$$= \lim_{h\to 0}\ [2 + 2(2 - h)] = 2 + 4 = 6. \qquad [\because\ nh = 2]$$

(*vi*) Let $\quad I = \int_{-1}^{1} (x + 3)\, dx$

Comparing I with $\int_a^b f(x)\, dx$, we get : $a = -1,\ b = 1,\ f(x) = x + 3$.

Let $\qquad h = \dfrac{b-a}{n} = \dfrac{1-(-1)}{n} = \dfrac{2}{n} \quad \Rightarrow \quad nh = 2,\ n \in N.$

$$\therefore \qquad I = \lim_{h\to 0}\ h[f(a) + f(a + h) + f(a + 2h) + \ldots\ldots + f\{a + (n-1)\,h\}]$$

$$= \lim_{h\to 0}\ h[f(-1) + f(-1 + h) + f(-1 + 2h) + \ldots\ldots + f\{-1 + (n-1)\,h\}]$$

$$[\because\ a = -1]$$

$$= \lim_{h\to 0}\ h[(-1 + 3) + (-1 + h + 3) + (-1 + 2h + 3) + \ldots\ldots + \{-1 + (n-1)\,h + 3\}]$$

$$[\because\ f(x) = x + 3]$$

$$= \lim_{h\to 0}\ h[2 + (h + 2) + (2h + 2) + \ldots\ldots + \{(n-1)\,h + 2\}]$$

$$= \lim_{h\to 0}\ h[(2 + 2 + 2 + \ldots\ldots + 2) + h\{1 + 2 + 3 + \ldots\ldots + (n-1)\}]$$

$$= \lim_{h\to 0} h\left[2n + h\,.\,\frac{n(n-1)}{2}\right] \qquad \left[\because\ [1+2+3+\ldots\ldots+(n-1)] = \frac{n(n-1)}{2}\right]$$

$$= \lim_{h\to 0}\left[2nh + \frac{nh(nh-h)}{2}\right]$$

$$= \lim_{h\to 0}\left[2(2) + \frac{2(2-h)}{2}\right] \qquad [\because\ nh = 2]$$

$$= 4 + (2 - 0) = 6.$$

(*vii*) Let $I = \int_{2}^{7} (5x - 9)\, dx$

Comparing I with $\int_a^b f(x)\, dx$, we get : $\quad a = 2,\ b = 7,\ f(x) = 5x - 9$.

Let $\qquad h = \dfrac{b-a}{n} = \dfrac{7-2}{n} = \dfrac{5}{n} \quad \Rightarrow \quad nh = 5,\ n \in N.$

$$\therefore \quad I = \lim_{h\to 0}\ h[f(a) + f(a + h) + f(a + 2h) + \ldots\ldots + f\{a + (n-1)\,h\}]$$

$$= \lim_{h\to 0}\ h[f(2) + f(2 + h) + f(2 + 2h) + \ldots\ldots + f\{2 + (n-1)\,h\}] \qquad [\because\ a = 2]$$

$$= \lim_{h\to 0}\ h[\{5(2) - 9\} + \{5(2 + h) - 9\} + \{5(2 + 2h) - 9\} + \ldots\ldots + \{5(2 + (n-1)\,h) - 9\}]$$

$$[\because\ f(x) = (5x - 9)]$$

$$= \lim_{h\to 0} h[1 + (5h + 1) + (10h + 1) + ... + \{5(n-1)h + 1\}]$$

$$= \lim_{h\to 0} h[(1 + 1 + 1 + ... + 1) + 5h\{1 + 2 + 3 + ... + (n-1)\}]$$

$$= \lim_{h\to 0} h\left[n + 5h \cdot \frac{n(n-1)}{2}\right] \qquad \left[\because \ [1+2+3+......+(n-1)] = \frac{n(n-1)}{2}\right]$$

$$= \lim_{h\to 0}\left[nh + \frac{5nh(nh-h)}{2}\right] = \lim_{h\to 0}\left[5 + \frac{5.5(5-h)}{2}\right] \qquad [\because \ nh = 5]$$

$$= 5 + \frac{25(5-0)}{2} = 5 + \frac{125}{2} = \frac{135}{2}.$$

(*viii*) Let $I = \int_0^4 2x\, dx$

Comparing I with $\int_a^b f(x)\, dx$, we get : $a = 0$, $b = 4$, $f(x) = 2x$

Let $h = \frac{b-a}{n} = \frac{4-0}{n} \Rightarrow nh = 4, n \in N.$

$$\therefore \quad I = \lim_{h\to 0} h[f(a) + f(a+h) + f(a+2h) + ... + f\{a + (n-1)h\}]$$

$$= \lim_{h\to 0} h[f(0) + f(0+h) + f(0+2h) + ... + f\{0 + (n-1)h\}] \qquad [\because \ a = 0]$$

$$= \lim_{h\to 0} h[f(0) + f(h) + f(2h) + ... + f\{(n-1)h\}]$$

$$= \lim_{h\to 0} h[\{2(0) + 2(h) + 2(2h) + ... + (n-1)2h] \qquad [\because \ f(x) = 2x]$$

$$= \lim_{h\to 0} h[2h(1 + 2 + 3 + ... + (n-1))]$$

$$= \lim_{h\to 0} h\left[2h \cdot \frac{n(n-1)}{2}\right] \qquad \left[\because \ [1+2+3+......+(n-1)] = \frac{n(n-1)}{2}\right]$$

$$= \lim_{h\to 0} [nh(nh-h]$$

$$= 4(4-0)] = 16. \qquad [\because \ nh = 4]$$

Example 2. *Evaluate the following integrals as limit of sums :*

(*i*) $\int_0^2 (x + 3)\, dx$ (*ii*) $\int_3^5 (2 - x)\, dx$

(*iii*) $\int_1^2 (2x + 3)\, dx$ (*iv*) $\int_1^3 (2x + 1)\, dx$

(*v*) $\int_0^2 (x + 4)\, dx$ (*vi*) $\int_2^4 (2x - 1)\, dx.$

Solution. (*i*) Let $I = \int_0^2 (x + 3)\, dx$

Comparing I with $\int_a^b f(x)\, dx$, we get : $a = 0$, $b = 2$, $f(x) = x + 3$.

Let $h = \frac{b-a}{n} = \frac{2-0}{n} = \frac{2}{n} \Rightarrow nh = 2, n \in N.$

$\therefore \quad I = \lim_{h\to 0} h[f(a) + f(a+h) + f(a+2h) + \ldots\ldots + f\{a + (n-1)h\}]$

$= \lim_{h\to 0} h[f(0) + f(0+h) + f(0+2h) + \ldots\ldots + f\{0 + (n-1)h\}]$ $\quad [\because \; a = 0]$

$= \lim_{h\to 0} h[f(0) + f(h) + f(2h) + \ldots\ldots + f\{(n-1)h\}]$

$= \lim_{h\to 0} h[3 + (h+3) + (2h+3) + \ldots\ldots + \{(n-1)h + 3\}]$ $\quad [\because \; f(x) = (x+3)]$

$= \lim_{h\to 0} h[(3 + 3 + 3 + \ldots.. + 3) + \{h + 2h + \ldots\ldots + (n-1)h\}]$

$= \lim_{h\to 0} h[3n + h\{[1 + 2 + 3 + \ldots\ldots + (n-1)\}]$

$= \lim_{h\to 0} h\left[3n + h \cdot \frac{n(n-1)}{2}\right]$ $\quad \left[\because \; 1 + 2 + 3 + \ldots\ldots + (n-1) = \frac{n(n-1)}{2}\right]$

$= \lim_{h\to 0}\left[3nh + \frac{nh(nh-h)}{2}\right] = \lim_{h\to 0}\left[3(2) + \frac{2(2-h)}{2}\right]$ $\quad [\because \; nh = 2]$

$= 6 + 2 = 8.$

(*ii*) Let $I = \int_3^5 (2-x)\,dx$

Comparing I with $\int_a^b f(x)\,dx$, we get : $a = 3,\ b = 5,\ f(x) = (2-x)$.

Let $h = \frac{b-a}{n} = \frac{5-3}{n} = \frac{2}{n} \quad \Rightarrow \quad nh = 2,\ n \in \mathrm{N}$

$\therefore \quad I = \lim_{h\to 0} h[f(a) + f(a+h) + f(a+2h) + \ldots\ldots + f\{a + (n-1)h\}]$

$= \lim_{h\to 0} h[f(3) + f(3+h) + f(3+2h) + \ldots\ldots + f\{3 + (n-1)h\}]$ $\quad [\because \; a = 3]$

$= \lim_{h\to 0} h[(2-3) + \{2 - (3+h)\} + \{2 - (3+2h)\} + \ldots\ldots + \{2 - (3 + (n-1)h)\}]$

$[\because \; f(x) = (2-x)]$

$= \lim_{h\to 0} h[-1 + (-1-h) + (-1-2h) + \ldots\ldots + \{-1 - (n-1)h\}]$

$= \lim_{h\to 0} h[(-1-1-1 \ldots. -1) - \{h + 2h + 3h + \ldots\ldots + (n-1)h\}]$

$= \lim_{h\to 0} h[-n - h\{1 + 2 + 3 + \ldots\ldots + (n-1)h\}]$

$= \lim_{h\to 0} h\left[-n - h \cdot \frac{n(n-1)}{2}\right]$ $\quad \left[\because \; 1 + 2 + 3 + \ldots\ldots + (n-1) = \frac{n(n-1)}{2}\right]$

$= \lim_{h\to 0}\left[-nh - \frac{nh(nh-h)}{2}\right] = \lim_{h\to 0}\left[-2 - \frac{2(2-h)}{2}\right]$ $\quad [\because \; nh = 2]$

$= -2 - 2 = -4.$

(*iii*) Let $I = \int_1^2 (2x+3)\,dx$

Comparing I with $\int_a^b f(x)\,dx$, we get : $a = 1,\ b = 2,\ f(x) = (2x+3)$.

Let $h = \dfrac{b-a}{n} = \dfrac{2-1}{n} = \dfrac{1}{n} \Rightarrow nh = 1, n \in \text{N}$

$\therefore \quad \text{I} = \lim\limits_{h\to 0} h[f(a) + f(a+h) + f(a+2h) + \ldots\ldots + f\{a+(n-1)h\}]$

$= \lim\limits_{h\to 0} h[f(1) + f(1+h) + f(1+2h) + \ldots\ldots + f\{1+(n-1)h\}]$ $[\because \ a = 1]$

$= \lim\limits_{h\to 0} h[\{2(1)+3\} + \{2(1+h)+3\} + \{2(1+2h)+3\} + \ldots\ldots + f\{2(1+(n-1)h)+3\}]$

$[\because \ f(x) = (2x+3)]$

$= \lim\limits_{h\to 0} h[5 + (5+2h) + (5+4h) + \ldots\ldots + \{5+2(n-1)h\}]$

$= \lim\limits_{h\to 0} h[(5+5+5+\ldots\ldots+5) + \{2h+4h+\ldots\ldots+2(n-1)h\}]$

$= \lim\limits_{h\to 0} h[5n + 2h[1+2+3+\ldots\ldots+(n-1)\}]$

$= \lim\limits_{h\to 0} h\left[5n + 2h \cdot \dfrac{n(n-1)}{2}\right]$ $\left[\because \ 1+2+3+\ldots\ldots+(n-1) = \dfrac{n(n-1)}{2}\right]$

$= \lim\limits_{h\to 0} [5nh + nh(nh-h)] = \lim\limits_{h\to 0} [5 + 1(1-h)]$ $[\because \ nh = 1]$

$= 5 + 1 = 6.$

(*iv*) Let $\text{I} = \displaystyle\int_1^3 (2x+1)\, dx$

Comparing I with $\displaystyle\int_a^b f(x)\, dx$, we get : $a = 1$, $b = 3$, $f(x) = (2x+1)$.

Let $h = \dfrac{b-a}{n} = \dfrac{3-1}{n} = \dfrac{2}{n} \Rightarrow nh = 2, n \in \text{N}$

$\therefore \quad \text{I} = \lim\limits_{h\to 0} h[f(a) + f(a+h) + f(a+2h) + \ldots\ldots + f\{a+(n-1)h\}]$

$= \lim\limits_{h\to 0} h[f(1) + f(1+h) + f(1+2h) + \ldots\ldots + f\{1+(n-1)h\}]$ $[\because \ a = 1]$

$= \lim\limits_{h\to 0} h[\{2(1)+1\} + \{2(1+h)+1\} + \{2(1+2h)+1\} + \ldots\ldots + \{2(1+(n-1)h)+1\}]$

$[\because \ f(x) = (2x+1)]$

$= \lim\limits_{h\to 0} h[3 + (3+2h) + (3+4h) + \ldots\ldots + \{3+2(n-1)h\}]$

$= \lim\limits_{h\to 0} h[(3+3+3+\ldots\ldots+3) + \{2h+4h+\ldots\ldots+2(n-1)h\}]$

$= \lim\limits_{h\to 0} h[3n + 2h[\{1+2+3+\ldots\ldots+(n-1)\}]$

$\left[\because \ [1+2+3+\ldots\ldots+(n-1)] = \dfrac{n(n-1)}{2}\right]$

$= \lim\limits_{h\to 0} h\left[3n + 2h \cdot \dfrac{n(n-1)}{2}\right]$

$= \lim\limits_{h\to 0} [3nh + nh(nh-h)] = \lim\limits_{h\to 0} [3(2) + 2(2-h)]$ $[\because \ nh = 2]$

$= 6 + 4 = 10.$

(*v*) Let $I = \int_0^2 (x + 4)\, dx$.

Comparing I with $\int_a^b f(x)\, dx$, we get : $a = 0$, $b = 2$, $f(x) = (x + 4)$.

Let $h = \frac{b-a}{n} = \frac{2-0}{n} = \frac{2}{n} \Rightarrow nh = 2, n \in N$.

$\therefore \quad I = \lim_{h \to 0} h[f(a) + f(a + h) + f(a + 2h) + \ldots\ldots + f\{a + (n - 1)\, h\}]$

$= \lim_{h \to 0} h[f(0) + f(0 + h) + f(0 + 2h) + \ldots\ldots + f\{0 + (n - 1)h\}]$ $\quad [\because \ a = 0]$

$= \lim_{h \to 0} h[f(0) + f(h) + f(2h) + \ldots\ldots + f\{(n - 1)h\}]$

$= \lim_{h \to 0} h[(0 + 4) + (h + 4) + (2h + 4) + \ldots\ldots + \{(n - 1)h + 4\}]$ $\quad [\because \ f(x) = (x + 4)]$

$= \lim_{h \to 0} h[(4 + 4 + 4 + \ldots\ldots + 4) + \{h + 2h + 3h + \ldots\ldots + (n - 1)\, h\}]$

$= \lim_{h \to 0} h[4n + h\{1 + 2 + 3 + \ldots\ldots + (n - 1)\}]$

$= \lim_{h \to 0} h\left[4n + h \cdot \frac{n(n-1)}{2}\right]$ $\quad \left[\because \ [1 + 2 + 3 + \ldots\ldots + (n-1)] = \frac{n(n-1)}{2}\right]$

$= \lim_{h \to 0} \left[4nh + \frac{nh(nh - h)}{2}\right] = \lim_{h \to 0} \left[4(2) + \frac{2(2-h)}{2}\right]$ $\quad [\because \ nh = 2]$

$= 8 + 2 = 10.$

(*vi*) Let $I = \int_2^4 (2x - 1)\, dx$

Comparing I with $\int_a^b f(x)\, dx$, we get : $a = 2$, $b = 4$, $f(x) = (2x - 1)$.

Let $h = \frac{b-a}{n} = \frac{4-2}{n} = \frac{2}{n} \Rightarrow nh = 2, n \in N$.

$\therefore \quad I = \lim_{h \to 0} h[f(a) + f(a + h) + f(a + 2h) + \ldots\ldots + f\{a + (n - 1)\, h\}]$

$= \lim_{h \to 0} h[f(2) + f(2 + h) + f(2 + 2h) + \ldots\ldots + f\{2 + (n - 1)h\}]$ $\quad [\because \ a = 2]$

$= \lim_{h \to 0} h[\{2(2) - 1\} + \{2(2 + h) - 1\} + \{2(2 + 2h) - 1\} + \ldots\ldots + \{2(2 + (n - 1)h) - 1\}]$

$[\because \ f(x) = (2x - 1)]$

$= \lim_{h \to 0} h[3 + (3 + 2h) + (3 + 4h) + \ldots\ldots + \{3 + 2(n - 1)h\}]$

$= \lim_{h \to 0} h[(3 + 3 + 3 + \ldots\ldots + 3) + \{2h + 4h + \ldots\ldots + 2(n - 1)\, h\}]$

$= \lim_{h \to 0} h[3n + 2h[1 + 2 + 3 + \ldots\ldots + (n - 1)h\}]$

$= \lim_{h \to 0} h\left[3n + 2h \cdot \frac{n(n-1)}{2}\right]$ $\quad \left[\because \ 1 + 2 + 3 + \ldots\ldots + (n-1) = \frac{n(n-1)}{2}\right]$

$= \lim_{h \to 0} [3nh + nh(nh - h)] = \lim_{h \to 0} [3(2) + 2(2 - h)]$ $\quad [\because \ nh = 2]$

$= 6 + 4 = 10.$

Example 3. *Evaluate the following integrals as the limit of sums :*

(i) $\int_a^b x^2\,dx$ (ii) $\int_1^2 x^2\,dx$

(iii) $\int_0^2 (x^2+3)\,dx$ (iv) $\int_1^3 (2x^2+5x)\,dx$

(v) $\int_1^3 (x^2+5x)\,dx$.

Solution. (i) Let $I = \int_a^b x^2\,dx$

Comparing I with $\int_a^b f(x)\,dx$, we get : $a = a$, $b = b$, $f(x) = x^2$.

Let $h = \dfrac{b-a}{n} \Rightarrow nh = b - a,\ n \in N$

$$\therefore\quad I = \lim_{h\to 0} h[f(a) + f(a+h) + f(a+2h) + \ldots\ldots + f\{a+(n-1)h\}]$$

$$= \lim_{h\to 0} h[a^2 + (a+h)^2 + (a+2h)^2 + \ldots\ldots + \{a+(n-1)h\}^2] \qquad [\because\ f(x) = x^2]$$

$$= \lim_{h\to 0} h[a^2 + (a^2+h^2+2ah) + (a^2+(2h)^2+4ah) + \ldots\ldots + \{a^2+(n-1)^2h^2+2a(n-1)h\}]$$

$$= \lim_{h\to 0} h[(a^2+a^2+a^2+\ldots\ldots+a^2) + \{h^2+(2h)^2+\ldots\ldots+(n-1)^2h^2\} + \{2ah+4ah+\ldots\ldots+2a(n-1)h\}]$$

$$= \lim_{h\to 0} h[na^2 + h^2\{1^2+2^2+\ldots\ldots+(n-1)^2\} + 2ah\{1+2+\ldots\ldots+(n-1)\}]$$

$$= \lim_{h\to 0} h\left[na^2 + h^2 . \frac{n(n-1)(2n-1)}{6} + 2ah . \frac{n(n-1)}{2}\right]$$

$$\left[\begin{aligned} \because\ & [1+2+3+\ldots\ldots+(n-1)] = \frac{n(n-1)}{2} \\ & [1^2+2^2+3^2+\ldots\ldots+(n-1)^2] = \frac{n(n-1)(2n-1)}{6} \end{aligned}\right]$$

$$= \lim_{h\to 0}\left[(nh)a^2 + \frac{nh(nh-h)(2nh-h)}{6} + anh(nh-h)\right]$$

$$= \lim_{h\to 0}\left[(b-a)a^2 + \frac{(b-a)(b-a-h)\{2(b-a)-h\}}{6} + a(b-a)(b-a-h)\right] \qquad [\because\ nh = (b-a)]$$

$$= \left[(b-a)a^2 + \frac{(b-a)(b-a)\,2(b-a)}{6} + a(b-a)(b-a)\right]$$

$$= \left[(b-a)a^2 + \frac{1}{3}(b-a)^3 + a(b-a)^2\right] = \frac{(b-a)}{3}[3a^2 + (b-a)^2 + 3a(b-a)]$$

$$= \frac{b-a}{3}[3a^2 + b^2 + a^2 - 2ab + 3ab - 3a^2] = \frac{(b-a)}{3}(a^2+ab+b^2)$$

$$= \frac{1}{3}(b^3 - a^3). \qquad [\because\ (a^3-b^3) = (a-b)(a^2+ab+b^2)]$$

(*ii*) Let $I = \int_1^2 x^2\,dx$

Comparing I with $\int_a^b f(x)\,dx$, we get : $a = 1$, $b = 2$, $f(x) = x^2$

Let $h = \frac{b-a}{n} = \frac{2-1}{n} = \frac{1}{n} \Rightarrow nh = 1,\ n \in N.$

$$\therefore\ I = \lim_{h\to 0} h[f(a) + f(a+h) + f(a+2h) + \ldots\ldots + f\{a + (n-1)h\}]$$

$$= \lim_{h\to 0} h[f(1) + f(1+h) + f(1+2h) + \ldots\ldots + f\{1 + (n-1)h\}] \qquad [\because\ a = 1]$$

$$= \lim_{h\to 0} h[1^2 + (1+h)^2 + (1+2h)^2 + \ldots\ldots + \{1 + (n-1)h\}^2] \qquad [\because\ f(x) = x^2]$$

$$= \lim_{h\to 0} h[1 + (1 + h^2 + 2h) + (1 + (2h)^2 + 4h) + \ldots\ldots + \{1 + (n-1)^2h^2 + 2(n-1)h\}]$$

$$= \lim_{h\to 0} h[(1 + 1 + 1 + \ldots\ldots + 1) + \{h^2 + (2h)^2 + \ldots\ldots + (n-1)^2 h^2\} + \{2h + 4h + \ldots\ldots + 2(n-1)h\}]$$

$$= \lim_{h\to 0} h[n + h^2(1^2 + 2^2 + 3^2 + \ldots\ldots(n-1)^2) + \{2h + 4h + \ldots\ldots + 2(n-1)h\}]$$

$$= \lim_{h\to 0} h[n + h^2(1^2 + 2^2 + 3^2 + \ldots\ldots(n-1)^2) + 2h\{1 + 2 + 3 + \ldots\ldots + (n-1)\}]$$

$$= \lim_{h\to 0} h\left[n + h^2 \cdot \frac{n(n-1)(2n-1)}{6} + 2h \cdot \frac{n(n-1)}{2}\right]$$

$$\left[\begin{array}{l}\because\ [1+2+3+\ldots\ldots+(n-1)] = \dfrac{n(n-1)}{2} \\ \ \ \ [1^2+2^2+3^2+\ldots\ldots+(n-1)^2] = \dfrac{n(n-1)(2n-1)}{6}\end{array}\right]$$

$$= \lim_{h\to 0}\left[nh + \frac{nh(nh-h)(2nh-h)}{6} + nh(nh-h)\right] \qquad [\because\ nh = 1]$$

$$= \lim_{h\to 0}\left[1 + \frac{1(1-h)(2-h)}{6} + 1(1-h)\right]$$

$$= 1 + \frac{1}{3} + 1 = \frac{3+1+3}{3} = \frac{7}{3}.$$

(*iii*) Let $I = \int_0^2 (x^2 + 3)\,dx$

Comparing I with $\int_a^b f(x)\,dx$, we get : $a = 0$, $b = 2$, $f(x) = (x^2 + 3)$.

Let $\qquad h = \frac{b-a}{n} = \frac{2-0}{n} = \frac{2}{n} \Rightarrow nh = 2,\ n \in N.$

$$\therefore\ I = \lim_{h\to 0} h[f(a) + f(a+h) + f(a+2h) + \ldots\ldots + f\{a + (n-1)h\}]$$

$$= \lim_{h\to 0} h[f(0) + f(0+h) + f(0+2h) + \ldots\ldots + f\{0 + (n-1)h\}] \qquad [\because\ a = 0]$$

$$= \lim_{h\to 0} h[f(0) + f(h) + f(2h) + \ldots\ldots + f\{(n-1)h\}]$$

$$= \lim_{h\to 0} h[(0+3)+(h^2+3)+\{(2h)^2+3\}+\ldots\ldots+\{((n-1)h)^2+3\}] \qquad [\because \; f(x)=(x^2+3)]$$

$$= \lim_{h\to 0} h[(3+3+3+\ldots\ldots+3)+\{h^2+(2h)^2+\ldots\ldots+(n-1)^2h^2\}]$$

$$= \lim_{h\to 0} h[3n+h^2(1^2+2^2+3^2+\ldots\ldots+(n-1)^2]$$

$$= \lim_{h\to 0} h\left[3n+h^2 \cdot \frac{n(n-1)(2n-1)}{6}\right]$$

$$\left[\because \; [1^2+2^2+3^2+\ldots\ldots+(n-1)^2] = \frac{n(n-1)(2n-1)}{6}\right]$$

$$= \lim_{h\to 0}\left[3nh+\frac{nh(nh-h)(2nh-h)}{6}\right] = \lim_{h\to 0}\left[3(2)+\frac{2(2-h)(2.2-h)}{6}\right] \qquad [\because \; nh=2]$$

$$= 6+\frac{2(2-0)(4-0)}{6} = 6+\frac{16}{6} = 6+\frac{8}{3} = \frac{26}{3}.$$

(*iv*) Let $I = \int_1^3 (2x^2+5x)\,dx$

Comparing I with $\int_a^b f(x)\,dx$, we get : $a=1$, $b=3$, $f(x)=(2x^2+5x)$.

Let $\qquad h = \frac{b-a}{n} = \frac{3-1}{n} = \frac{2}{n} \;\Rightarrow\; nh = 2,\; n \in N.$

$$\therefore \quad I = \lim_{h\to 0} h[f(a)+f(a+h)+f(a+2h)+\ldots\ldots+f\{a+(n-1)h\}]$$

$$= \lim_{h\to 0} h[f(1)+f(1+h)+f(1+2h)+\ldots\ldots+f\{1+(n-1)h\}] \qquad [\because \; a=1]$$

$$= \lim_{h\to 0} h[\{2(1)^2+5(1)\}+\{2(1+h)^2+5(1+h)\}+\{2(1+2h)^2+5(1+2h)\}$$
$$+\ldots\ldots+\{2(1+(n-1)h)^2+5(1+(n-1)h)\}]$$
$$[\because \; f(x)=(2x^2+5x)]$$

$$= \lim_{h\to 0} h[7+\{2(1+h^2+2h)+5+5h\}+\{2(1+(2h)^2+2(2h)+5+10h\}+\ldots\ldots$$
$$+\{2(1+(n-1)^2h^2+2(n-1)h+5+5(n-1)h\}]$$

$$= \lim_{h\to 0} h[7+(7+2h^2+9h)+\{7+2(2h)^2+18h\}+\ldots\ldots+\{7+2(n-1)^2h^2+9(n-1)h\}]$$

$$= \lim_{h\to 0} h[\{7+7+7+\ldots\ldots+7)+2h^2\{1^2+2^2+\ldots\ldots+(n-1)^2\}$$
$$+9h\{1+2+\ldots\ldots+(n-1)\}]$$

$$\left[\begin{aligned} \because \; [1+2+3+\ldots\ldots+(n-1)] &= \frac{n(n-1)}{2} \\ [1^2+2^2+3^2+\ldots\ldots+(n-1)^2] &= \frac{n(n-1)(2n-1)}{6} \end{aligned}\right]$$

$$= \lim_{h\to 0} h\left[7n+2h^2 \cdot \frac{n(n-1)(2n-1)}{6}+9h \cdot \frac{n(n-1)}{2}\right]$$

$$= \lim_{h\to 0}\left[7nh+\frac{nh(nh-h)(2nh-h)}{3}+\frac{9}{2}nh(nh-h)\right]$$

$$= \lim_{h\to 0}\left[7(2) + \frac{2(2-h)(2.2-h)}{3} + \frac{9}{2}.2(2-h)\right] \qquad [\because \quad nh = 2]$$

$$= 14 + \frac{16}{3} + 18 = \frac{42+16+54}{3} = \frac{112}{3}.$$

(*v*) Let I $= \int_1^3 (x^2 + 5x)\, dx.$

Comparing I with $\int_a^b f(x)\, dx$, we get : $a = 1$, $b = 3$, $f(x) = (x^2 + 5x)$

Let $\qquad h = \frac{b-a}{n} = \frac{3-1}{n} = \frac{2}{n} \quad \Rightarrow \quad nh = 2,\ n \in$ N.

$$\therefore \quad \text{I} = \lim_{h\to 0} h[f(a) + f(a+h) + f(a+2h) + \ldots\ldots + f\{a+(n-1)h\}]$$

$$= \lim_{h\to 0} h[f(1) + f(1+h) + f(1+2h) + \ldots\ldots + f\{1+(n-1)h\}] \qquad [\because \quad a = 1]$$

$$= \lim_{h\to 0} h[\{1+5(1)\} + \{(1+h)^2 + 5(1+h)\} + \{(1+2h)^2 + 5(1+2h)\} + \ldots\ldots + \{(1+(n-1)h)^2 + 5(1+(n-1)h)\}]$$

$$[\because \quad f(x) = (x^2 + 5x)]$$

$$= \lim_{h\to 0} h[6 + (1+h^2+2h+5+5h) + (1+(2h)^2+2(2h)+5+10h) + \ldots\ldots + \{1+(n-1)^2h^2 + 2(n-1)h + 5 + 5(n-1)h\}]$$

$$= \lim_{h\to 0} h[6 + (6+h^2+7h) + (6+(2h)^2+14h) + \ldots\ldots + \{6+(n-1)^2h^2 + 7(n-1)h\}]$$

$$= \lim_{h\to 0} h[\{6+6+6+\ldots\ldots+6) + (h^2 + (2h)^2 + \ldots\ldots + (n-1)^2h^2) + \{7h + 14h + \ldots\ldots + 7(n-1)h\}]$$

$$= \lim_{h\to 0} h[6n + h^2\{1^2 + 2^2 + \ldots\ldots + (n-1)^2\} + 7h\{1 + 2 + \ldots\ldots + (n-1)\}]$$

$$= \lim_{h\to 0} h\left[6n + h^2 . \frac{n(n-1)(2n-1)}{6} + 7h . \frac{n(n-1)}{2}\right]$$

$$\left[\begin{array}{l}\because \quad [1+2+3+\ldots\ldots+(n-1)] = \dfrac{n(n-1)}{2}\\ \quad\ [1^2+2^2+3^2+\ldots\ldots+(n-1)^2] = \dfrac{n(n-1)(2n-1)}{6}\end{array}\right]$$

$$= \lim_{h\to 0}\left[6nh + \frac{nh(nh-h)(2nh-h)}{6} + \frac{7nh(nh-h)}{2}\right]$$

$$= \lim_{h\to 0}\left[6(2) + \frac{2(2-h)(2.2-h)}{6} + \frac{7.2(2-h)}{2}\right] \qquad [\because \quad nh = 2]$$

$$= 12 + \frac{8}{3} + 14 = 26 + \frac{8}{3} = \frac{78+8}{3} = \frac{86}{3}.$$

Example 4. *Evaluate the following integrals as the limit of sums :*

(*i*) $\int_0^2 (x^2 + 2)\, dx$ (*ii*) $\int_1^2 (x^2 - 1)\, dx$

(*iii*) $\int_1^3 (2x^2 + 5)\, dx$ (*iv*) $\int_1^4 (x^2 + 3x)\, dx$

(*v*) $\int_0^3 (2x^2 + 3)\, dx$ (*vi*) $\int_1^3 (x^2 + x)\, dx.$

Solution. (*i*) Let $I = \int_0^2 (x^2 + 2)\, dx$

Comparing I with $\int_a^b f(x)\, dx$, we get : $a = 0,\ b = 2,\ f(x) = (x^2 + 2)$.

Let $\quad h = \dfrac{b-a}{n} = \dfrac{2-0}{n} = \dfrac{2}{n} \quad \Rightarrow \quad nh = 2\ ;\ n \in N.$

$$\therefore \quad I = \lim_{h\to 0} h[f(a) + f(a+h) + f(a+2h) + \ldots\ldots + f\{a + (n-1)h\}]$$

$$= \lim_{h\to 0} h[f(0) + f(0+h) + f(0+2h) + \ldots\ldots + f\{0 + (n-1)h\}] \qquad [\because\ a = 0]$$

$$= \lim_{h\to 0} h[f(0) + f(h) + f(2h) + \ldots\ldots + f\{(n-1)h\}]$$

$$= \lim_{h\to 0} h[(0+2) + (h^2+2) + \{(2h)^2 + 2\} + \ldots\ldots + \{(n-1)^2 h^2 + 2\}]$$

$$[\because\ f(x) = (x^2 + 2)]$$

$$= \lim_{h\to 0} h[(2 + 2 + 2 + \ldots\ldots + 2) + \{h^2 + (2h)^2 + \ldots\ldots + (n-1)^2 h^2\}]$$

$$= \lim_{h\to 0} h[2n + h^2(1^2 + 2^2 + 3^2 + \ldots\ldots + (n-1)^2)]$$

$$= \lim_{h\to 0} h\left[2n + h^2 \cdot \frac{n(n-1)(2n-1)}{6}\right]$$

$$\left[\because\ [1^2 + 2^2 + 3^2 + \ldots\ldots + (n-1)^2] = \frac{n(n-1)(2n-1)}{6}\right]$$

$$= \lim_{h\to 0}\left[2nh + \frac{nh(nh-h)(2nh-h)}{6}\right]$$

$$= \lim_{h\to 0}\left[2(2) + \frac{2(2-h)(2.2-h)}{6}\right] \qquad [\because\ nh = 2]$$

$$= 4 + \frac{(2-0)(4-0)}{3} = 4 + \frac{8}{3} = \frac{12+8}{3} = \frac{20}{3}.$$

(*ii*) Let $I = \int_1^2 (x^2 - 1)\, dx$

Comparing I with $\int_a^b f(x)\, dx$, we get : $a = 1,\ b = 2,\ f(x) = (x^2 - 1)$.

Let $\quad h = \dfrac{b-a}{n} = \dfrac{2-1}{n} = \dfrac{1}{n} \quad \Rightarrow \quad nh = 1,\ n \in N.$

$$\therefore \quad I = \lim_{h\to 0} h[f(a) + f(a+h) + f(a+2h) + \ldots\ldots + f\{a + (n-1)h\}]$$

$$= \lim_{h\to 0} h[f(1) + f(1+h) + f(1+2h) + \ldots\ldots + f\{1 + (n-1)h\}] \qquad [\because\ a = 1]$$

$$= \lim_{h\to 0} h[(1^2 - 1) + \{(1+h)^2 - 1\} + \{(1+2h)^2 - 1) + \ldots\ldots + \{(1 + (n-1)h)^2 - 1\}]$$

$$[\because\ f(x) = (x^2 - 1)]$$

$$= \lim_{h\to 0} h[0 + (1 + h^2 + 2h - 1) + \{1 + (2h)^2 + 2(2h) - 1\} + \ldots\ldots + \{1 + (n-1)^2 h^2 + 2(n-1)h - 1\}]$$

$$= \lim_{h\to 0} h[(h^2 + 2h) + \{(2h)^2 + 2(2h)\} + \ldots\ldots + \{(n-1)^2 h^2 + 2(n-1)h\}]$$

$$= \lim_{h\to 0} h[\{h^2 + 4h^2 + \ldots\ldots + (n-1)^2 h^2\} + \{2h + 2(2h) + \ldots\ldots + 2(n-1)h\}]$$

$$= \lim_{h\to 0} h\{h^2 (1^2 + 2^2 + \ldots\ldots + (n-1)^2) + 2h\{1 + 2 + \ldots\ldots + (n-1)\}]$$

$$= \lim_{h\to 0} h\left[h^2 . \frac{n(n-1)(2n-1)}{6} + \frac{2h . n(n-1)}{2}\right]$$

$$\left[\begin{array}{l} \because\ [1+2+3+\ldots\ldots+(n-1)] = \dfrac{n(n-1)}{2} \\ [1^2+2^2+3^2+\ldots\ldots+(n-1)^2] = \dfrac{n(n-1)(2n-1)}{6} \end{array}\right]$$

$$= \lim_{h\to 0}\left[\frac{nh(nh-h)(2nh-h)}{6} + nh(nh-h)\right]$$

$$= \lim_{h\to 0}\left[\frac{1(1-h)(2-h)}{6} + 1(1-h)\right] \qquad [\because\ nh = 1]$$

$$= \frac{1}{3} + 1 = \frac{4}{3}.$$

(*iii*) Let $I = \int_1^3 (2x^2 + 5)\,dx$

Comparing I with $\int_a^b f(x)\,dx$, we get : $a = 1$, $b = 3$, $f(x) = (2x^2 + 5)$

Let $h = \frac{b-a}{n} = \frac{3-1}{n} = \frac{2}{n} \Rightarrow nh = 2, n \in N.$

$$\therefore\quad I = \lim_{h\to 0} h[f(a) + f(a+h) + f(a+2h) + \ldots\ldots + f\{a + (n-1)h\}]$$

$$= \lim_{h\to 0} h[f(1) + f(1+h) + f(1+2h) + \ldots\ldots + f\{1 + (n-1)h\}] \qquad [\because\ a = 1]$$

$$= \lim_{h\to 0} h[\{2(1)^2 + 5\} + \{2(1+h)^2 + 5\} + \{2(1+2h)^2 + 5\} + \ldots\ldots + \{2(1 + (n-1)h)^2 + 5\}]$$

$$[\because\ f(x) = (2x^2 + 5)]$$

$$= \lim_{h\to 0} h[7 + \{2(1 + h^2 + 2h) + 5\} + \{2(1 + (2h)^2 + 2(2h)) + 5\} + \ldots\ldots + \{2(1 + (n-1)^2 h^2 + 2(n-1)h) + 5\}]$$

$$= \lim_{h\to 0} h[7 + (7 + 2h^2 + 4h) + (7 + 2(2h)^2 + 8h) + \ldots\ldots + \{7 + 2(n-1)^2 h^2 + 4(n-1)h\}]$$

$$= \lim_{h\to 0} h[(7 + 7 + 7 + \ldots\ldots + 7) + \{2h^2 + 2(2h)^2 + \ldots\ldots + 2(n-1)^2 h^2\} + \{4h + 8h + \ldots\ldots + 4(n-1)h\}]$$

$$= \lim_{h\to 0} h[7n + 2h^2 \{1^2 + 2^2 + \ldots\ldots + (n-1)^2\} + 4h\{1 + 2 + \ldots\ldots + (n-1)\}]$$

$$= \lim_{h\to 0} h\left[7n + 2h^2 . \frac{n(n-1)(2n-1)}{6} + 4h . \frac{n(n-1)}{2}\right]$$

$$\left[\begin{array}{l} \because\ [1+2+3+\ldots\ldots+(n-1)] = \dfrac{n(n-1)}{2} \\ [1^2+2^2+3^2+\ldots\ldots+(n-1)^2] = \dfrac{n(n-1)(2n-1)}{6} \end{array}\right]$$

$$= \lim_{h\to 0}\left[7nh + \frac{nh(nh-h)(2nh-h)}{3} + 2nh(nh-h)\right]$$

$$= \lim_{h \to 0} \left[7(2) + \frac{2(2-h)(2.2-h)}{3} + 2(2)(2-h) \right] \qquad [\because \; nh = 2]$$

$$= 14 + \frac{16}{3} + 8 = 22 + \frac{16}{3} = \frac{66+16}{3} = \frac{82}{3}.$$

(*iv*) Let $I = \int_1^4 (x^2 + 3x)\, dx$

Comparing I with $\int_a^b f(x)\, dx$, we get : $a = 1$, $b = 4$, $f(x) = (x^2 + 3x)$.

Let $h = \frac{b-a}{n} = \frac{4-1}{n} = \frac{3}{n} \Rightarrow nh = 3, n \in \text{N}.$

$$\therefore \quad I = \lim_{h \to 0} h[f(a) + f(a+h) + f(a+2h) + \ldots\ldots + f\{a + (n-1)h\}]$$

$$= \lim_{h \to 0} h[f(1) + f(1+h) + f(1+2h) + \ldots\ldots + f\{1 + (n-1)h\}] \qquad [\because \; a = 1]$$

$$= \lim_{h \to 0} h[\{1^2 + 3(1)\} + \{(1+h)^2 + 3(1+h)\} + \{(1+2h)^2 + 3(1+2h)\} + \ldots\ldots + \{(1+(n-1)h)^2 + 3(1+(n-1)h)\}]$$

$$[\because \; f(x) = (x^2 + 3x)]$$

$$= \lim_{h \to 0} h[4 + (1 + h^2 + 2h + 3 + 3h) + \{1 + (2h)^2 + 2(2h) + 3 + 6h\} + \ldots\ldots + \{1 + (n-1)^2h^2 + 2(n-1)h + 3 + 3(n-1)h\}]$$

$$= \lim_{h \to 0} h[4 + (4 + h^2 + 5h) + (4 + (2h)^2 + 10h) + \ldots\ldots + \{4 + (n-1)^2h^2 + 5(n-1)h\}]$$

$$= \lim_{h \to 0} h[(4 + 4 + 4 + \ldots\ldots + 4) + \{h^2 + (2h)^2 + \ldots\ldots + (n-1)^2 h^2\} + \ldots\ldots + \{5h + 10h + \ldots\ldots + 5(n-1)h\}]$$

$$= \lim_{h \to 0} h[4n + h^2 \{1^2 + 2^2 + \ldots\ldots + (n-1)^2\} + 5h\{1 + 2 + \ldots\ldots + (n-1)\}]$$

$$= \lim_{h \to 0} h \left[4n + h^2 . \frac{n(n-1)(2n-1)}{6} + 5h . \frac{n(n-1)}{2} \right]$$

$$\left[\begin{array}{l} \because \; [1 + 2 + 3 + \ldots\ldots + (n-1)] = \dfrac{n(n-1)}{2} \\ [1^2 + 2^2 + 3^2 + \ldots\ldots + (n-1)^2] = \dfrac{n(n-1)(2n-1)}{6} \end{array} \right]$$

$$= \lim_{h \to 0} \left[4nh + \frac{nh(nh-h)(2nh-h)}{6} + \frac{5nh(nh-h)}{2} \right]$$

$$= \lim_{h \to 0} \left[4(3) + \frac{3(3-h)(2.3-h)}{6} + \frac{5(3)(3-h)}{2} \right] \qquad [\because \; nh = 3]$$

$$= 12 + 9 + \frac{45}{2} = 21 + \frac{45}{2} = \frac{42+45}{2} = \frac{87}{2}.$$

(*v*) Let $I = \int_0^3 (2x^2 + 3)\, dx$

Comparing I with $\int_a^b f(x)\, dx$, we get : $a = 0$, $b = 3$, $f(x) = (2x^2 + 3)$.

Let $h = \frac{b-a}{n} = \frac{3-0}{n} = \frac{3}{n} \Rightarrow nh = 3, n \in \text{N}.$

$\therefore \quad \mathrm{I} = \lim\limits_{h\to 0} h[f(a) + f(a+h) + f(a+2h) + \ldots\ldots + f\{a + (n-1)h\}]$

$= \lim\limits_{h\to 0} h[f(0) + f(0+h) + f(0+2h) + \ldots\ldots + f\{0 + (n-1)h\}]$ $\quad [\because \ a = 0]$

$= \lim\limits_{h\to 0} h[f(0) + f(h) + f(2h) + \ldots\ldots + f\{(n-1)h\}]$

$= \lim\limits_{h\to 0} h[\{2(0) + 3\} + (2h^2 + 3) + \{2(2h)^2 + 3\} + \ldots\ldots + \{2((n-1)h)^2 + 3\}]$

$[\because \ f(x) = (2x^2 + 3)]$

$= \lim\limits_{h\to 0} h[3 + (2h^2 + 3) + \{2(4h^2) + 3\} + \ldots\ldots + \{2(n-1)^2 h^2 + 3\}]$

$= \lim\limits_{h\to 0} h[(3 + 3 + 3 + \ldots\ldots + 3) + 2h^2\{1^2 + 2^2 + \ldots\ldots + (n-1)^2\}]$

$= \lim\limits_{h\to 0} h\left[3n + 2h^2 \cdot \frac{n(n-1)(2n-1)}{6}\right]$

$\left[\because \ [1^2 + 2^2 + 3^2 + \ldots\ldots + (n-1)^2] = \frac{n(n-1)(2n-1)}{6}\right]$

$= \lim\limits_{h\to 0} \left[3nh + \frac{nh(nh-h)(2nh-h)}{3}\right]$

$= \lim\limits_{h\to 0} \left[3(3) + \frac{3(3-h)(2.3-h)}{3}\right]$ $\quad [\because \ nh = 3]$

$= 9 + 18 = 27.$

(*vi*) Let $\quad \mathrm{I} = \int_1^3 (x^2 + x)\, dx$

Comparing I with $\int_a^b f(x)\, dx$, we get : $a = 1$, $b = 3$, $f(x) = (x^2 + x)$.

Let $\quad h = \frac{b-a}{n} = \frac{3-1}{n} = \frac{2}{n} \quad \Rightarrow \quad nh = 2, n \in \mathrm{N}.$

$\therefore \quad \mathrm{I} = \lim\limits_{h\to 0} h[f(a) + f(a+h) + f(a+2h) + \ldots\ldots + f\{a + (n-1)h\}]$

$= \lim\limits_{h\to 0} h[f(1) + f(1+h) + f(1+2h) + \ldots\ldots + f\{1 + (n-1)h\}]$ $\quad [\because \ a = 1]$

$= \lim\limits_{h\to 0} h[(1^2 + 1) + \{(1+h)^2 + (1+h)\} + \{(1+2h)^2 + (1+2h)\} +$

$\ldots\ldots + \{(1 + (n-1)h)^2 + 1 + (n-1)h\}]$

$[\because \ f(x) = (x^2 + x)]$

$= \lim\limits_{h\to 0} h[2 + (1 + h^2 + 2h + 1 + h) + \{1 + 4h^2 + 2(2h) + 1 + 2h\} +$

$\ldots\ldots + \{1 + (n-1)^2h^2 + 2(n-1)h + 1 + (n-1)h\}]$

$= \lim\limits_{h\to 0} h[2 + (2 + h^2 + 3h) + (2 + 4h^2 + 6h) + \ldots\ldots + \{2 + (n-1)^2h^2 + 3(n-1)h\}]$

$= \lim\limits_{h\to 0} h[(2 + 2 + 2 + \ldots\ldots + 2) + \{h^2 + (2h)^2 + \ldots\ldots + (n-1)^2\}$

$+ \{3h + 6h + \ldots\ldots + 3(n-1)h\}]$

$= \lim\limits_{h\to 0} h[2n + h^2\{1^2 + 2^2 + 3^2 + \ldots\ldots + (n-1)^2\} + 3h\{1 + 2 + 3 + \ldots\ldots + (n-1)\}]$

$$= \lim_{h \to 0} h\left[2n + h^2 \cdot \frac{n(n-1)(2n-1)}{6} + 3h \cdot \frac{n(n-1)}{2}\right]$$

$$\left[\begin{array}{l} \because \ [1+2+3+\ldots\ldots+(n-1)] = \dfrac{n(n-1)}{2} \\ \ \ \ [1^2+2^2+3^2+\ldots\ldots+(n-1)^2] = \dfrac{n(n-1)(2n-1)}{6} \end{array}\right]$$

$$= \lim_{h \to 0}\left[2nh + \frac{nh(nh-h)(2nh-h)}{6} + \frac{3nh(nh-h)}{2}\right]$$

$$= \lim_{h \to 0}\left[2(2) + \frac{2(2-h)(2.2-h)}{6} + \frac{3(2)(2-h)}{2}\right] \qquad [\because \ nh = 2]$$

$$= 4 + \frac{2(2-0)(4-0)}{6} + 3(2-0)$$

$$= 4 + \frac{8}{3} + 6 = 10 + \frac{8}{3} = \frac{30+8}{3} = \frac{38}{3}.$$

Example 5. *Evaluate the following integrals as the limit of sums :*

(*i*) $\int_1^4 (x^2 - x)\, dx$ (*ii*) $\int_1^3 (2x^2 + x + 9)\, dx$

(*iii*) $\int_0^2 (x^2 + 1)\, dx$ (*iv*) $\int_1^2 x^3\, dx$

(*v*) $\int_0^1 (3x^2 + 2x + 1)\, dx.$

Solution. (*i*) Let $I = \int_1^4 (x^2 - x)\, dx$

Comparing I with $\int_a^b f(x)\, dx$, we get : $a = 1$, $b = 4$, $f(x) = (x^2 - x)$.

Let $h = \frac{b-a}{n} = \frac{4-1}{n} = \frac{3}{n} \Rightarrow nh = 3, n \in N$.

$$\therefore \quad I = \lim_{h \to 0} h[f(a) + f(a+h) + f(a+2h) + \ldots\ldots + f\{a+(n-1)h\}]$$

$$= \lim_{h \to 0} h[f(1) + f(1+h) + f(1+2h) + \ldots\ldots + f\{1+(n-1)h\}] \qquad [\because \ a = 1]$$

$$= \lim_{h \to 0} h[\{(1)^2 - 1\} + \{(1+h)^2 - (1+h)\} + \{(1+2h)^2 - (1+2h)\} + \ldots\ldots + \{(1+(n-1)h)^2 - (1+(n-1)h)\}]$$

$$[\because \ f(x) = (x^2 - x)]$$

$$= \lim_{h \to 0} h[0 + (1 + h^2 + 2h - 1 - h) + (1 + (2h)^2 + 2(2h) - 1 - 2h) + \ldots\ldots + \{1 + (n-1)^2h^2 + 2(n-1)h - 1 - (n-1)h\}]$$

$$= \lim_{h \to 0} h[(h^2 + h) + \{(2h)^2 + (2h)\} + \ldots\ldots + \{(n-1)^2 h^2 + (n-1) h\}]$$

$$= \lim_{h \to 0} h[\{h^2 + (2h)^2 + \ldots\ldots + (n-1)^2 h^2\} + \{h + 2h + \ldots\ldots + (n-1)h\}]$$

$$= \lim_{h \to 0} h[h^2 \{1^2 + 2^2 + \ldots\ldots + (n-1)^2\} + h\{1 + 2 + \ldots\ldots + (n-1)\}]$$

$$= \lim_{h\to 0} h\left[h^2 \cdot \frac{n(n-1)(2n-1)}{6} + h \cdot \frac{n(n-1)}{2}\right]$$

$$\left[\begin{aligned} \because\ [1+2+3+\ldots\ldots+(n-1)] &= \frac{n(n-1)}{2} \\ [1^2+2^2+3^2+\ldots\ldots+(n-1)^2] &= \frac{n(n-1)(2n-1)}{6}\end{aligned}\right]$$

$$= \lim_{h\to 0}\left[\frac{nh(nh-h)(2nh-h)}{6} + \frac{nh(nh-h)}{2}\right]$$

$$= \lim_{h\to 0}\left[\frac{3(3-h)(2.3-h)}{6} + \frac{3(3-h)}{2}\right] \qquad [\because\ nh = 3]$$

$$= 9 + \frac{9}{2} = \frac{18+9}{2} = \frac{27}{2}.$$

(*ii*) Let $I = \int_1^3 (2x^2 + x + 9)\, dx$

Comparing I with $\int_a^b f(x)\, dx$, we get : $a = 1$, $b = 3$, $f(x) = (2x^2 + x + 9)$.

Let $h = \frac{b-a}{n} = \frac{3-1}{n} = \frac{2}{n} \Rightarrow nh = 2, n \in N.$

$$\therefore\ I = \lim_{h\to 0} h[f(a) + f(a+h) + f(a+2h) + \ldots\ldots + f\{a+(n-1)h\}]$$

$$= \lim_{h\to 0} h[f(1) + f(1+h) + f(1+2h) + \ldots\ldots + f\{1+(n-1)h\}] \qquad [\because\ a = 1]$$

$$= \lim_{h\to 0} h[\{2(1)^2 + 1 + 9\} + \{2(1+h)^2 + (1+h) + 9\} + \{2(1+2h)^2 + (1+2h) + 9\} + \ldots\ldots + \{2(1+(n-1)h)^2 + (1+(n-1)h) + 9\}]$$

$$[\because\ f(x) = (2x^2 + x + 9)]$$

$$= \lim_{h\to 0} h[12 + \{2(1+h^2+2h) + 1 + h + 9\} + \{2(1+4h^2+2(2h)) + 1 + 2h + 9\} + \ldots\ldots + \{2(1+(n-1)^2h^2 + 2(n-1)h) + 1 + (n-1)h + 9\}]$$

$$= \lim_{h\to 0} h[12 + (12 + 2h^2 + 5h) + (12 + 2(2h)^2 + 10h) + \ldots\ldots + \{12 + 2(n-1)^2 h^2 + 5(n-1)h\}]$$

$$= \lim_{h\to 0} h[(12 + 12 + 12 + \ldots\ldots + 12) + \{2h^2 + 2(2h)^2 + \ldots\ldots + 2(n-1)^2 h^2\} + \{5h + 10h + \ldots\ldots + 5(n-1)h\}]$$

$$= \lim_{h\to 0} h[12n + 2h^2\{1^2 + 2^2 + \ldots\ldots + (n-1)^2\} + 5h\{1 + 2 + \ldots\ldots + (n-1)\}]$$

$$= \lim_{h\to 0} h\left[12n + 2h^2 \cdot \frac{n(n-1)(2n-1)}{6} + 5h \cdot \frac{n(n-1)}{2}\right]$$

$$\left[\begin{aligned} \because\ [1+2+3+\ldots\ldots+(n-1)] &= \frac{n(n-1)}{2} \\ [1^2+2^2+3^2+\ldots\ldots+(n-1)^2] &= \frac{n(n-1)(2n-1)}{6}\end{aligned}\right]$$

$$= \lim_{h\to 0}\left[12nh + \frac{nh(nh-h)(2nh-h)}{3} + \frac{5nh(nh-h)}{2}\right]$$

$$= \lim_{h\to 0}\left[12(2) + \frac{2(2-h)(2.2-h)}{3} + \frac{5(2)(2-h)}{2}\right] \qquad [\because \ nh = 2]$$

$$= 24 + \frac{16}{3} + 10 = 34 + \frac{16}{3} = \frac{102+16}{3} = \frac{118}{3}.$$

(*iii*) Let $I = \int_0^2 (x^2 + 1)\,dx$

Comparing I with $\int_a^b f(x)\,dx$, we get : $a = 0$, $b = 2$, $f(x) = (x^2 + 1)$.

Let $\qquad h = \frac{b-a}{n} = \frac{2-0}{n} = \frac{2}{n} \quad \Rightarrow \quad nh = 2, n \in N.$

$$\therefore \quad I = \lim_{h\to 0} h[f(a) + f(a+h) + f(a+2h) + \ldots\ldots + f\{a + (n-1)h\}]$$

$$= \lim_{h\to 0} h[f(0) + f(0+h) + f(0+2h) + \ldots\ldots + f\{0 + (n-1)h\}] \qquad [\because \ a = 0]$$

$$= \lim_{h\to 0} h[f(0) + f(h) + f(2h) + \ldots\ldots + f\{(n-1)h\}]$$

$$= \lim_{h\to 0} h[(0+1) + (h^2+1) + \{(2h)^2 + 1\} + \ldots\ldots + \{(n-1)^2 h^2 + 1\}]$$

$$[\because \ f(x) = (x^2+1)]$$

$$= \lim_{h\to 0} h[(1 + 1 + 1 + \ldots\ldots + 1) + \{h^2 + (2h)^2 + \ldots\ldots + (n-1)^2 h^2\}]$$

$$= \lim_{h\to 0} h[n + h^2\{1^2 + 2^2 + \ldots\ldots + (n-1)^2\}]$$

$$= \lim_{h\to 0} h\left[n + h^2 \cdot \frac{n(n-1)(2n-1)}{6}\right]$$

$$\left[\because \ [1^2 + 2^2 + 3^2 + \ldots\ldots + (n-1)^2] = \frac{n(n-1)(2n-1)}{6}\right]$$

$$= \lim_{h\to 0}\left[nh + \frac{nh(nh-h)(2nh-h)}{6}\right]$$

$$= \lim_{h\to 0}\left[2 + \frac{2(2-h)(2.2-h)}{6}\right] \qquad [\because \ nh = 2]$$

$$= 2 + \frac{(2-0)(4-0)}{3} = 2 + \frac{8}{3} = \frac{6+8}{3} = \frac{14}{3}.$$

(*iv*) Let $I = \int_1^2 x^3\,dx$

Comparing I with $\int_a^b f(x)\,dx$, we get : $a = 1$, $b = 2$, $f(x) = x^3$.

Let $\quad h = \frac{b-a}{n} = \frac{2-1}{n} = \frac{1}{n} \quad \Rightarrow \quad nh = 1, n \in N.$

$$\therefore \quad I = \lim_{h\to 0} h[f(a) + f(a+h) + f(a+2h) + \ldots\ldots + f\{a + (n-1)h\}]$$

$$= \lim_{h\to 0} h[f(1) + f(1+h) + f(1+2h) + \ldots\ldots + f\{1 + (n-1)h\}] \qquad [\because \ a = 1]$$

$$= \lim_{h\to 0} h[1^3 + (1+h)^3 + (1+2h)^3 + \ldots\ldots + \{1 + (n-1)h\}^3] \qquad [\because \ f(x) = x^3]$$

$$= \lim_{h\to 0} h[1 + (1 + 3h + 3h^2 + h^3) + (1 + 6h + 12h^2 + 8h^3) +$$

$$\ldots\ldots + \{1 + 3(n-1)h + 3(n-1)^2 h^2 + (n-1)^3 h^3\}]$$

$$= \lim_{h \to 0} h\,[(1 + 1 + 1 + \ldots\ldots + 1) + \{3h + 6h + \ldots\ldots + 3(n-1)h\}$$
$$+ \{3h^2 + 12h^2 + \ldots\ldots + 3(n-1)^2 h^2\} + \{h^3 + 8h^3 + \ldots\ldots + (n-1)^3 h^3\}]$$

$$= \lim_{h \to 0} h[n + 3h\{1 + 2 + \ldots\ldots + (n-1)\} + 3h^2\{1^2 + 2^2 + \ldots\ldots + (n-1)^2\}$$
$$+ h^3\{1^3 + 2^3 + \ldots\ldots + (n-1)^3\}]$$

$$= \lim_{h \to 0} h\left[n + 3h \cdot \frac{n(n-1)}{2} + 3h^2 \cdot \frac{n(n-1)(2n-1)}{6} + h^3 \cdot \frac{n^2(n-1)^2}{4}\right]$$

$$\left[\begin{array}{l} \because\ [1+2+3+\ldots\ldots+(n-1)] = \dfrac{n(n-1)}{2} \\ [1^2+2^2+3^2+\ldots\ldots+(n-1)^2] = \dfrac{n(n-1)(2n-1)}{6} \\ [1^3+2^3+3^3+\ldots\ldots+(n-1)^3] = \dfrac{n^2(n-1)^2}{4} \end{array}\right]$$

$$= \lim_{h \to 0}\left[nh + \frac{3nh(nh-h)}{2} + \frac{nh(nh-h)(2nh-h)}{2} + \frac{n^2h^2(nh-h)^2}{4}\right]$$

$$= \lim_{h \to 0}\left[1 + \frac{3(1-h)}{2} + \frac{1(1-h)(2-h)}{2} + \frac{1(1-h)^2}{4}\right] \qquad [\because\ nh = 1]$$

$$= 1 + \frac{3(1-0)}{2} + \frac{(1-0)(2-0)}{2} + \frac{(1-0)^2}{4}$$

$$= 1 + \frac{3}{2} + 1 + \frac{1}{4} = 2 + \frac{3}{2} + \frac{1}{4} = \frac{8+6+1}{4} = \frac{15}{4}.$$

(*v*) Let $I = \int_0^1 (3x^2 + 2x + 1)\, dx$

Comparing I with $\int_a^b f(x)\, dx$, we get : $a = 0$, $b = 1$, $f(x) = (3x^2 + 2x + 1)$.

Let $h = \frac{b-a}{n} = \frac{1-0}{n} = \frac{1}{n} \Rightarrow nh = 1, n \in N.$

$$\therefore\quad I = \lim_{h \to 0} h[f(a) + f(a+h) + f(a+2h) + \ldots\ldots + f\{a + (n-1)h\}]$$

$$= \lim_{h \to 0} h[f(0) + f(0+h) + f(0+2h) + \ldots\ldots + f\{0 + (n-1)h\}] \qquad [\because\ a = 0]$$

$$= \lim_{h \to 0} h[f(0) + f(h) + f(2h) + \ldots\ldots + f\{(n-1)h\}]$$

$$= \lim_{h \to 0} h[\{3(0)^2 + 2(0) + 1\} + \{3h^2 + 2h + 1\} + \{3(2h)^2 + 2(2h) + 1\} +$$
$$\ldots\ldots + \{3(n-1)^2h^2 + 2(n-1)h + 1\}]$$
$$[\because\ f(x) = (3x^2 + 2x + 1)]$$

$$= \lim_{h \to 0} h[(1 + 1 + 1 + \ldots\ldots + 1) + \{3h^2 + 3(2h)^2 + \ldots\ldots + 3(n-1)^2 h^2\} +$$
$$+ \{2h + 2(2h) + \ldots\ldots + 2(n-1)h\}]$$

$$= \lim_{h \to 0} h[n + 3h^2\{1^2 + 2^2 + \ldots\ldots + (n-1)^2\} + 2h\{1 + 2 + \ldots\ldots + (n-1)\}]$$

$$= \lim_{h \to 0} h\left[n + 3h^2 \cdot \frac{n(n-1)(2n-1)}{6} + 2h \cdot \frac{n(n-1)}{2}\right]$$

$$\left[\because \; 1 + 2 + 3 + \ldots\ldots + (n-1) = \frac{n(n-1)}{2} \\ 1^2 + 2^2 + 3^2 + \ldots\ldots + (n-1)^2 = \frac{n(n-1)(2n-1)}{6}\right]$$

$$= \lim_{h \to 0}\left[nh + \frac{nh(nh-h)(2nh-h)}{2} + nh(nh-h)\right]$$

$$= \lim_{h \to 0}\left[1 + \frac{1(1-h)(2-h)}{2} + 1(1-h)\right] \qquad [\because \; nh = 1]$$

$= 1 + 1 + 1 = 3.$

Example 6. *Evaluate by summation, the definite integral :*

(*i*) $\int_a^b e^x\, dx$ (*ii*) $\int_0^4 e^x\, dx$

(*iii*) $\int_0^5 e^{-2x}\, dx$ (*iv*) $\int_0^2 e^x\, dx.$

Solution. (*i*) Let $I = \int_a^b e^x\, dx.$

Comparing I with $\int_a^b f(x)\, dx$, we get : $a = a$, $b = b$, $f(x) = e^x$.

Let $h = \frac{b-a}{n} = \frac{b-a}{n} \Rightarrow nh = b - a, n \in N.$

$\therefore \qquad I = \lim_{h \to 0} h[f(a) + f(a+h) + f(a+2h) + \ldots\ldots + f\{a + (n-1)h\}]$

$$= \lim_{h \to 0} h[e^a + e^{a+h} + e^{a+2h} + \ldots\ldots + e^{a+(n-1)h}] \qquad [\because \; f(x) = e^x]$$

Since $e^a + e^{a+h} + e^{a+2h} + \ldots\ldots + e^{a+(n-1)h}$ is a G.P. (of n-terms), where first term is e^a and common ratio is e^h.

$$\therefore \qquad \text{Sum} = \frac{e^a(e^{nh}-1)}{e^h - 1} \qquad \left[\because \; \frac{a(r^n-1)}{r-1}\right]$$

$$= \lim_{h \to 0} h\left[\frac{e^a(e^{nh}-1)}{e^h-1}\right] = \lim_{h \to 0}\left[\frac{he^a(e^{b-a}-1)}{e^h-1}\right] \qquad [\because \; nh = (b-a)]$$

$$= \lim_{h \to 0}\left(\frac{h}{e^h-1}\right)(e^b - e^a) = \lim_{h \to 0}\frac{1}{\left(\frac{e^h-1}{h}\right)} \cdot (e^b - e^a)$$

$$= e^b - e^a. \qquad \left[\because \; \lim_{h \to 0}\left(\frac{e^h-1}{h}\right) = 1\right]$$

(*ii*) Let $\quad I = \int_0^4 e^x\, dx$

Comparing I with $\int_a^b f(x)\,dx$, we get : $a = 0$, $b = 4$, $f(x) = e^x$.

Let $\quad h = \dfrac{b-a}{n} = \dfrac{4-0}{n} = \dfrac{4}{n} \quad \Rightarrow \quad nh = 4,\ n \in \text{N}.$

$$\therefore \quad \text{I} = \lim_{h\to 0} h[f(a) + f(a+h) + f(a+2h) + \ldots\ldots + f\{a+(n-1)h\}]$$

$$= \lim_{h\to 0} h[f(0) + f(0+h) + f(0+2h) + \ldots\ldots + f\{0+(n-1)h\}] \qquad [\because \ a = 0]$$

$$= \lim_{h\to 0} h[f(0) + f(h) + f(2h) + \ldots\ldots + f\{(n-1)h\}]$$

$$= \lim_{h\to 0} h[e^0 + e^h + e^{2h} + \ldots\ldots + e^{(n-1)h}] \qquad [\because \ f(x) = e^x]$$

$$= \lim_{h\to 0} h\left[\frac{1\{1-(e^h)^n\}}{1-e^h}\right] \qquad \left[\because \ \text{S}_n = \frac{a(1-r^n)}{1-r}\right]$$

$$= \lim_{h\to 0} \frac{h(1-e^{nh})}{1-e^h}$$

$$= \lim_{h\to 0} \frac{h(1-e^4)}{1-e^h} = (1-e^4) \,.\, \lim_{h\to 0} \frac{h}{(1-e^h)} \qquad [\because \ nh = 4]$$

$$= (1-e^4) \,.\, \lim_{h\to 0} \frac{1}{\left(\dfrac{e^h-1}{h}\right)} = (e^4-1) \times \frac{1}{1} = (e^4-1). \qquad \left[\because \ \lim_{h\to 0} \frac{e^h-1}{h} = 1\right]$$

(*iii*) Let $\quad \text{I} = \int_0^5 e^{-2x}\,dx$

Comparing I with $\int_a^b f(x)\,dx$, we get : $a = 0$, $b = 5$, $f(x) = e^{-2x}$

Let $\quad h = \dfrac{b-a}{n} = \dfrac{5-0}{n} = \dfrac{5}{n} \quad \Rightarrow \quad nh = 5,\ n \in \text{N}.$

$$\therefore \quad \text{I} = \lim_{h\to 0} h[f(a) + f(a+h) + f(a+2h) + \ldots\ldots + f\{a+(n-1)h\}]$$

$$= \lim_{h\to 0} h[f(0) + f(0+h) + f(0+2h) + \ldots\ldots + f\{0+(n-1)h\}] \qquad [\because \ a = 0]$$

$$= \lim_{h\to 0} h[f(0) + f(h) + f(2h) + \ldots\ldots + f\{(n-1)h\}]$$

$$= \lim_{h\to 0} h[e^0 + e^{-2h} + e^{-4h} + \ldots\ldots + e^{-2(n-1)h}] \qquad [\because \ f(x) = e^{-2x}]$$

$$= \lim_{h\to 0} h[1 + e^{-2h} + (e^{-2h})^2 + \ldots\ldots + (e^{-2h})^{n-1}]$$

$$= \lim_{h\to 0} h\left[\frac{1\,(1-(e^{-2h})^n}{1-e^{-2h}}\right] \qquad \left[\because \ \text{S}_n = \frac{a(1-r^n)}{1-r}\right]$$

$$= \lim_{h\to 0} \frac{h(1-e^{-2nh})}{1-e^{-2h}} = \lim_{h\to 0} \frac{h(1-e^{-2(5)})}{1-e^{-2h}} \qquad [\because \ nh = 5]$$

$$= \lim_{h\to 0} \frac{h(e^{-10} - 1)}{(e^{-2h} - 1)} = (e^{-10} - 1) \cdot \lim_{h\to 0} \frac{1}{\left(\frac{e^{-2h} - 1}{h}\right)}$$

$$= (e^{-10} - 1) \cdot \lim_{h\to 0} \frac{1}{\frac{e^{-2h} - 1}{-2h} \times (-2)}$$ [Multiply and divided by (– 2)]

$$= (e^{-10} - 1) \frac{1}{1(-2)} = -\frac{1}{2} (e^{-10} - 1).$$ $\left[\because \lim_{h\to 0} \frac{e^h - 1}{h} = 1\right]$

(*iv*) Let $I = \int_0^2 e^x \, dx$

Comparing I with $\int_a^b f(x)\, dx$, we get : $a = 0$, $b = 2$, $f(x) = e^x$.

Let $h = \frac{b-a}{n} = \frac{2-0}{n} = \frac{2}{n} \Rightarrow nh = 2, n \in N.$

$\therefore$ $I = \lim_{h\to 0} h[f(a) + f(a + h) + f(a + 2h) + ... + f\{a + (n - 1) h\}]$

$= \lim_{h\to 0} h[f(0) + f(0 + h) + f(0 + 2h) + ... + f\{0 + (n - 1) h\}]$ $[\because a = 0]$

$= \lim_{h\to 0} h[f(0) + f(h) + f(2h) + ... + f\{(n - 1) h\}]$

$= \lim_{h\to 0} h[e^0 + e^h + e^{2h} + ... + e^{(n-1)h}]$ $[\because f(x) = e^x]$

$$= \lim_{h\to 0} h\left[e^0 \left\{\frac{(e^h)^n - 1}{e^h - 1}\right\}\right]$$ $\left[\because a + ar + ar^2 + ... + ar^{n-1} = a\left(\frac{r^n - 1}{r - 1}\right)\right]$

$$= \lim_{h\to 0}\left[\frac{h(e^{nh} - 1)}{e^h - 1}\right] = \lim_{h\to 0}\left[\frac{h(e^2 - 1)}{e^h - 1}\right]$$ $[\because nh = 2]$

$$= (e^2 - 1) \lim_{h\to 0} \left(\frac{1}{\frac{e^h - 1}{h}}\right)$$

$$= (e^2 - 1) \times \frac{1}{1} = (e^2 - 1).$$ $\left[\because \lim_{h\to 0} \frac{e^h - 1}{h} = 1\right]$

Example 7. *Evaluate by summation, the definite integral :*

(*i*) $\int_{-1}^{1} e^x \, dx$ (*ii*) $\int_a^b e^{\lambda x} \, dx$

(*iii*) $\int_0^1 e^{2-3x} \, dx$ (*iv*) $\int_0^4 (x + e^{2x}) \, dx.$

Solution. (*i*) Let $I = \int_{-1}^{1} e^x \, dx$

Comparing I with $\int_a^b f(x)\, dx$, we get : $a = -1$, $b = 1$, $f(x) = e^x$.

Let $h = \dfrac{b-a}{n} = \dfrac{1-(-1)}{n} = \dfrac{2}{n} \Rightarrow nh = 2, n \in N.$

$\therefore \quad I = \lim_{h\to 0} h[f(a) + f(a+h) + f(a+2h) + \ldots\ldots + f\{a+(n-1)h\}]$

$= \lim_{h\to 0} h[f(-1) + f(-1+h) + f(-1+2h) + \ldots\ldots + f\{-1+(n-1)h\}]$ $\quad [\because \ a = -1]$

$= \lim_{h\to 0} h[e^{-1} + (e^{-1+h}) + (e^{-1+2h}) + \ldots\ldots + (e^{-1+(n-1)h})]$ $\quad [\because \ f(x) = e^x]$

$= \lim_{h\to 0} h[e^{-1} + e^{-1}.e^{h} + e^{-1}.e^{2h} + \ldots\ldots + e^{-1}.e^{(n-1)h}]$

$= \lim_{h\to 0} he^{-1}[1 + e^{h} + e^{2h} + \ldots.. + e^{(n-1)h}]$

$$= \lim_{h\to 0} \frac{h}{e}\left[\frac{(e^h)^n - 1}{e^h - 1}\right] = \lim_{h\to 0} \frac{h}{e}\left[\frac{e^{nh} - 1}{e^h - 1}\right] \qquad \left[\because \text{ For a G.P., } S_n = \frac{a(r^n - 1)}{r-1}\right]$$

$$= \lim_{h\to 0} \frac{h}{e}\left[\frac{e^2 - 1}{e^h - 1}\right] = \frac{e^2-1}{e} . \lim_{h\to 0}\left(\frac{h}{e^h - 1}\right) \qquad [\because \ nh = 2]$$

$$= \frac{e^2-1}{e} \lim_{h\to 0}\left(\frac{1}{\frac{e^h - 1}{h}}\right) = \frac{e^2-1}{e} \times \frac{1}{1} = e - \frac{1}{e}.$$

(*ii*) Let $I = \int_a^b e^{\lambda x}\, dx$

Comparing I with $\int_a^b f(x)\, dx$, we get : $a = a$, $b = b$, $f(x) = e^{\lambda x}$.

Let $h = \dfrac{b-a}{n} \Rightarrow nh = b - a, n \in N.$

$\therefore \quad I = \lim_{h\to 0} h[f(a) + f(a+h) + f(a+2h) + \ldots\ldots + f\{a+(n-1)h\}]$

$= \lim_{h\to 0} h[e^{\lambda a} + e^{\lambda(a+h)} + e^{\lambda(a+2h)} + \ldots\ldots + e^{\lambda\{a+(n-1)h\}}]$ $\quad [\because \ f(x) = e^{\lambda x}]$

$= \lim_{h\to 0} h[e^{\lambda a} + e^{\lambda a} . e^{\lambda h} + e^{\lambda a} . e^{2\lambda h} + \ldots\ldots + e^{\lambda a} . e^{\lambda(n-1)h}]$

$= \lim_{h\to 0} he^{\lambda a}[1 + e^{\lambda h} + e^{2\lambda h} + \ldots\ldots + e^{\lambda(n-1)h}]$

$$= \lim_{h\to 0} he^{\lambda a}\left[\frac{1(1-(e^{\lambda h})^n}{1-e^{\lambda h}}\right] \qquad \left[\because \text{ For a G.P., } S_n = \frac{a(1-r^n)}{1-r}\right]$$

$$= e^{\lambda a} \lim_{h\to 0} h\left[\frac{1-e^{\lambda nh}}{1-e^{\lambda h}}\right] = e^{\lambda a} \lim_{h\to 0} h\left[\frac{1-e^{\lambda(b-a)}}{1-e^{\lambda h}}\right] \qquad [\because \ nh = (b-a)]$$

$$= e^{\lambda a} (e^{\lambda(b-a)} - 1) \lim_{h\to 0}\left(\frac{h}{e^{\lambda h} - 1}\right)$$

$$= e^{\lambda a} [e^{\lambda(b-a)} - 1] . \lim_{h\to 0}\left(\frac{1}{\frac{e^{\lambda h}-1}{\lambda h} \times \lambda}\right) \qquad \text{[Multiply and divided by } \lambda]$$

$$\left[\because \lim_{h\to 0}\left(\frac{e^h - 1}{h}\right) = 1\right]$$

$$= (e^{\lambda b} - e^{\lambda a}) \cdot \frac{1}{1.\lambda} = \frac{1}{\lambda}(e^{\lambda b} - e^{\lambda a}).$$

(*iii*) Let $I = \int_0^1 e^{2-3x}\,dx$

Comparing I with $\int_a^b f(x)\,dx$, we get : $a = 0$, $b = 1$, $f(x) = e^{2-3x}$.

Let $\quad h = \frac{b-a}{n} = \frac{1-0}{n} = \frac{1}{n} \quad \Rightarrow \quad nh = 1,\ n \in N.$

$\therefore \quad I = \lim_{h\to 0} h[f(a) + f(a+h) + f(a+2h) + \ldots\ldots + f\{a+(n-1)h\}]$

$= \lim_{h\to 0} h[f(0) + f(0+h) + f(0+2h) + \ldots\ldots + f\{0+(n-1)h\}] \qquad [\because \ a = 0]$

$= \lim_{h\to 0} h[f(0) + f(h) + f(2h) + \ldots\ldots + f\{(n-1)h\}]$

$= \lim_{h\to 0} h[e^{2-3(0)} + e^{2-3h} + e^{2-3(2h)} + \ldots\ldots + e^{2-3(n-1)h}] \qquad [\because \ f(x) = e^{2-3x}]$

$= \lim_{h\to 0} h[e^2 + e^2 \cdot e^{-3h} + e^2 \cdot e^{-3(2h)} + \ldots\ldots + e^2 \cdot e^{-3(n-1)h}]$

$= \lim_{h\to 0} h\, e^2[1 + e^{-3h} + e^{-3(2h)} + \ldots\ldots + e^{-3(n-1)h}]$

$= \lim_{h\to 0} he^2\left[\frac{(e^{-3h})^n - 1}{e^{-3h} - 1}\right] \qquad \left[\because \text{ For a G.P., } S_n = \frac{a(r^n - 1)}{r - 1}\right]$

$= \lim_{h\to 0} he^2\left[\frac{e^{-3nh} - 1}{e^{-3h} - 1}\right]$

$= e^2 \lim_{h\to 0} h\left[\frac{e^{-3} - 1}{e^{-3h} - 1}\right] = e^2(e^{-3} - 1) \cdot \lim_{h\to 0}\left(\frac{h}{e^{-3h} - 1}\right) \qquad [\because \ nh = 1]$

$= (e^{-1} - e^2) \cdot \lim_{h\to 0}\left[\frac{1}{\left(\frac{e^{-3h} - 1}{-3h}\right) \times (-3)}\right] \qquad$ [Multiply and divided by (-3)]

$= (e^{-1} - e^2) \cdot \frac{1}{1(-3)} = \frac{1}{3}(e^2 - e^{-1}).$

(*iv*) Let $\quad I = \int_0^4 (x + e^{2x})\,dx$

Comparing I with $\int_0^4 (x + e^{2x})\,dx$, we get : $a = 0$, $b = 4$, $f(x) = x + e^{2x}$.

Let $\quad h = \frac{b-a}{n} = \frac{4-0}{n} = \frac{4}{n} \quad \Rightarrow \quad nh = 4,\ n \in N.$

$\therefore \quad I = \lim_{h\to 0} h[f(a) + f(a+h) + f(a+2h) + \ldots\ldots + f\{a+(n-1)h\}]$

$= \lim_{h\to 0} h[f(0) + f(0+h) + f(0+2h) + \ldots\ldots + f\{0+(n-1)h\}] \qquad [\because \ a = 0]$

$= \lim_{h\to 0} h[f(0) + f(h) + f(2h) + \ldots\ldots + f\{(n-1)h\}]$

$= \lim_{h\to 0} h[(0 + e^0) + (h + e^{2h}) + \{2h + e^{2(2h)}\} + \ldots\ldots + \{(n-1)h + e^{2(n-1)h}\}]$

$[\because \ f(x) = (x + e^{2x})]$

$$= \lim_{h\to 0} h[\{h + 2h + ... + (n-1)h\} + \{1 + e^{2h} + e^{2(2h)} + ... + e^{2(n-1)h}\}]$$

$$= \lim_{h\to 0} h[h\{1 + 2 + ... + (n-1)\} + \left\{1.\frac{(e^{2h})^n - 1}{e^{2h} - 1}\right\}$$

$$= \lim_{h\to 0} h\left[h.\frac{n(n-1)}{2} + \frac{e^{2nh} - 1}{e^{2h} - 1}\right] \qquad \left[\because [1+2+3+......+(n-1)] = \frac{n(n-1)}{2}\right]$$

$$= \lim_{h\to 0} \frac{nh(nh-h)}{2} + \lim_{h\to 0} \frac{h(e^{2nh} - 1)}{e^{2h} - 1}$$

$$= \lim_{h\to 0} \frac{4(4-h)}{2} + \lim_{h\to 0} \frac{h(e^{2(4)} - 1)}{e^{2h} - 1} \qquad [\because nh = 4]$$

$$= \frac{16}{2} + (e^8 - 1) \lim_{h\to 0} \left(\frac{1}{\frac{e^{2h} - 1}{2h} \times 2}\right) \qquad \text{[Multiply and divided by 2]}$$

$$= 8 + (e^8 - 1) . \frac{1}{1(2)} \qquad \left[\because \lim_{h\to 0}\left(\frac{e^h - 1}{h}\right) = 1\right]$$

$$= 8 + \frac{1}{2}(e^8 - 1) = \frac{16 + e^8 - 1}{2} = \frac{15 + e^8}{2}.$$

Example 8. *Evaluate the following integrals as the limit of a sum :*

(*i*) $\int_a^b e^{-x}\, dx$ (*ii*) $\int_2^4 2^x\, dx$ (*iii*) $\int_0^1 (e^x + 1)\, dx$.

Solution. (*i*) Let $I = \int_a^b e^{-x}\, dx$

Comparing I with $\int_a^b f(x)\, dx$, we get : $a = a$, $b = b$, $f(x) = e^{-x}$.

Let $h = \frac{b-a}{n} \Rightarrow nh = b - a, n \in N$.

$$\therefore \quad I = \lim_{h\to 0} h[f(a) + f(a+h) + f(a+2h) + ... + f\{a + (n-1)h\}]$$

$$= \lim_{h\to 0} h[e^{-a} + e^{-(a+h)} + e^{-(a+2h)} + ... + e^{-\{a+(n-1)h\}}] \qquad [\because f(x) = e^{-x}]$$

$$= \lim_{h\to 0} h[e^{-a} + e^{-a} . e^{-h} + e^{-a} . e^{-2h} + ... + e^{-a} . e^{-(n-1)h}]$$

$$= \lim_{h\to 0} he^{-a}[1 + e^{-h} + e^{-2h} + ... + e^{-(n-1)h}]$$

$$= \lim_{h\to 0} he^{-a}\left[\frac{1-(e^{-h})^n}{1-e^{-h}}\right] = \lim_{h\to 0} he^{-a}\left[\frac{1-e^{-nh}}{1-e^{-h}}\right]$$

$$= e^{-a} . \lim_{h\to 0} h\left[\frac{1-e^{-nh}}{1-e^{-h}}\right] = e^{-a} . \lim_{h\to 0} h\left[\frac{1-e^{-(b-a)}}{1-e^{-h}}\right] \qquad [\because nh = (b-a)]$$

$$= e^{-a}(1 - e^{-(b-a)}) . \lim_{h\to 0}\left(\frac{h}{1-e^{-h}}\right) = (e^{-a} - e^{-b})(1) \qquad \left[\because \lim_{h\to 0}\left(\frac{e^h - 1}{h}\right) = 1\right]$$

$$= e^{-a} - e^{-b}.$$

(*ii*) Let $\quad I = \int_2^4 2^x \, dx$

Comparing I with $\int_a^b f(x)\, dx$, we get : $a = 2$, $b = 4$, $f(x) = 2^x$.

Let $\quad h = \dfrac{b-a}{n} = \dfrac{4-2}{n} = \dfrac{2}{n} \quad \Rightarrow \quad nh = 2,\ n \in N.$

$$\therefore \quad I = \lim_{h \to 0} h[f(a) + f(a+h) + f(a+2h) + \ldots\ldots + f\{a + (n-1)h\}]$$

$$= \lim_{h \to 0} h[f(2) + f(2+h) + f(2+2h) + \ldots\ldots + f\{2 + (n-1)h\}] \qquad [\because \ a = 2]$$

$$= \lim_{h \to 0} h[2^2 + 2^{2+h} + 2^{2+2h} + \ldots\ldots + 2^{2+(n-1)h}] \qquad [\because \ f(x) = 2^x]$$

$$= \lim_{h \to 0} h[2^2 + 2^2 \,.\, 2^h + 2^2 \,.\, 2^{2h} + \ldots\ldots + 2^2 \,.\, 2^{(n-1)h}]$$

$$= \lim_{h \to 0} 4h[1 + 2^h + 2^{2h} + \ldots\ldots + 2^{(n-1)h}]$$

$$= \lim_{h \to 0} 4h\left[\frac{(2^h)^n - 1}{2^h - 1}\right] = \lim_{h \to 0} 4h\left[\frac{2^{nh} - 1}{2^h - 1}\right]$$

$$= \lim_{h \to 0} 4h\left[\frac{2^2 - 1}{2^h - 1}\right] = \lim_{h \to 0} 4\left[\frac{4-1}{\dfrac{2^h - 1}{h}}\right] \qquad [\because \ nh = 2]$$

$$= \frac{12}{\log 2}\,. \qquad \left[\because \ \lim_{h \to 0} \frac{2^h - 1}{h} = \log 2\right]$$

(*iii*) Let $\quad I = \int_0^1 (e^x + 1)\, dx$

Comparing I with $\int_a^b f(x)\, dx$, we get : $a = 0$, $b = 1$, $f(x) = (e^x + 1)$.

Let $\quad h = \dfrac{b-a}{n} = \dfrac{1-0}{n} = \dfrac{1}{n} \quad \Rightarrow \quad nh = 1,\ n \in N.$

$$\therefore \quad I = \lim_{h \to 0} h[f(a) + f(a+h) + f(a+2h) + \ldots\ldots + f\{a + (n-1)h\}]$$

$$= \lim_{h \to 0} h[f(0) + f(0+h) + f(0+2h) + \ldots\ldots + f\{0 + (n-1)h\}] \qquad [\because \ a = 0]$$

$$= \lim_{h \to 0} h[f(0) + f(h) + f(2h) + \ldots\ldots + f\{(n-1)h\}]$$

$$= \lim_{h \to 0} h[(e^0 + 1) + (e^h + 1) + (e^{2h} + 1) + \ldots\ldots + \{e^{(n-1)h} + 1\}] \qquad [\because \ f(x) = (e^x + 1)]$$

$$= \lim_{h \to 0} h[(1 + 1 + 1 + \ldots\ldots + 1) + \{e^0 + e^h + e^{2h} + \ldots\ldots + e^{(n-1)h}\}]$$

$$= \lim_{h \to 0} h\left[n + \frac{1 - (e^h)^n}{1 - e^h}\right] = \lim_{h \to 0}\left[nh + \frac{h(1 - e^{nh})}{1 - e^h}\right]$$

$$= \lim_{h \to 0}\left[1 + \frac{h(1-e)}{1 - e^h}\right] \qquad [\because \ nh = 1]$$

$$= 1 + (1-e) \lim_{h \to 0} \left(\frac{1}{\frac{1-e^h}{h}} \right) = 1 + (e-1) \lim_{h \to 0} \left(\frac{1}{\frac{e^h - 1}{h}} \right)$$

$$= [1 + (e-1) \cdot 1] = e. \qquad \left[\because \lim_{h \to 0} \left(\frac{e^h - 1}{h} \right) = 1 \right]$$

Example 9. *Evaluate the following as limit of sum :*

(*i*) $\int_a^b \sin x \, dx$ (*ii*) $\int_a^b \cos x \, dx$

(*iii*) $\int_0^{\pi/2} \cos x \, dx$ (*iv*) $\int_0^{\pi/2} \sin x \, dx$

(*v*) $\int_a^b \frac{1}{x} \, dx$ (*vi*) $\int_a^b \frac{1}{x^2} \, dx$

(*vii*) $\int_a^b \frac{1}{\sqrt{x}} \, dx$ (*viii*) $\int_a^b \sin^2 x \, dx.$

Solution. (*i*) Let $I = \int_a^b \sin x \, dx$

Comparing I with $\int_a^b f(x) \, dx$, we get : $a = a$, $b = b$, $f(x) = \sin x$.

Let $h = \frac{b-a}{n} \Rightarrow nh = b - a, n \in N.$

$$\therefore \quad I = \lim_{h \to 0} h[f(a) + f(a+h) + f(a+2h) + \ldots\ldots + f\{a + (n-1)h\}]$$

$$= \lim_{h \to 0} h[\sin a + \sin(a+h) + \sin(a+2h) + \ldots\ldots + \sin\{a + (n-1)h\}]$$

$$[\because \quad f(x) = \sin x]$$

$$= \lim_{h \to 0} h \left[\frac{\sin\left[a + \left(\frac{n-1}{2}\right)h\right] . \sin\left(\frac{nh}{2}\right)}{\sin\left(\frac{h}{2}\right)} \right]$$

$$\left[\because \; [\sin a + \sin(a+h) + \sin(a+2h) + \ldots\ldots + \sin(a + \overline{n-1}\,h)] = \frac{\sin\left[a + \left(\frac{n-1}{2}\right)h\right] . \sin\left(\frac{nh}{2}\right)}{\sin\left(\frac{h}{2}\right)} \right]$$

$$= \lim_{h \to 0} \left[2 \left(\frac{\frac{h}{2}}{\sin \frac{h}{2}} \right) . \sin\left\{a + \frac{1}{2}(n-1)h\right\} . \sin\left(\frac{nh}{2}\right) \right] \qquad \text{[Multiply and divided by 2]}$$

$$= \lim_{h \to 0} \left[\left(\frac{\frac{h}{2}}{\sin \frac{h}{2}} \right) . 2 \sin\left\{a + \frac{nh}{2} - \frac{h}{2}\right\} \sin\left(\frac{nh}{2}\right) \right]$$

$$= \lim_{h\to 0}\left[\left(\frac{\left(\frac{h}{2}\right)}{\sin\frac{h}{2}}\right).2\sin\left\{a+\frac{b-a}{2}-\frac{h}{2}\right\}\sin\left(\frac{b-a}{2}\right)\right] \qquad [\because \quad nh=(b-a)]$$

$$= \lim_{h\to 0}\left(\frac{\frac{h}{2}}{\sin\frac{h}{2}}\right).\lim_{h\to 0}\left[2\sin\left(\frac{a+b}{2}-\frac{h}{2}\right)\right].\sin\left(\frac{b-a}{2}\right)$$

$$= 1\,.\,2\;\sin\left(\frac{a+b}{2}\right)\sin\left(\frac{b-a}{2}\right) = 2\sin\left(\frac{b+a}{2}\right)\sin\left(\frac{b-a}{2}\right) \qquad \left[\because \lim_{x\to 0}\frac{\sin x}{x}=1\right]$$

$$= \cos\left(\frac{b+a-b+a}{2}\right)-\cos\left(\frac{b+a+b-a}{2}\right)$$

$$[\because \quad 2\sin A\sin B=\cos(A-B)-\cos(A+B)]$$

$= \cos a - \cos b.$

(*ii*) Let $I = \int_a^b \cos x\, dx$

Comparing I with $\int_a^b f(x)\,dx$, we get : $a = a$, $b = b$, $f(x) = \cos x$.

Let $h = \frac{b-a}{n} \quad \Rightarrow \quad nh = (b-a),\ n \in N.$

$$\therefore \quad I = \lim_{h\to 0} h[f(a)+f(a+h)+f(a+2h)+......+f\{a+(n-1)\,h\}]$$

$$= \lim_{h\to 0} h[\cos a+\cos(a+h)+\cos(a+2h)+......+\cos\{a+(n-1)h\}]$$

$$[\because \quad f(x)=\cos x]$$

$$= \lim_{h\to 0} h\left[\frac{\cos\left[a+\left(\frac{n-1}{2}\right)h\right].\sin\left(\frac{nh}{2}\right)}{\sin\left(\frac{h}{2}\right)}\right]$$

$$\left[\because \quad [\cos a+\cos(a+h)+\cos(a+2h)+......+\cos(a+\overline{n-1}\,h)] = \frac{\cos\left[a+\left(\frac{n-1}{2}\right)h\right].\sin\left(\frac{nh}{2}\right)}{\sin\left(\frac{h}{2}\right)}\right]$$

$$= \lim_{h\to 0}\left[\left(\frac{\frac{h}{2}}{\sin\left(\frac{h}{2}\right)}\right)2\cos\left\{a+\frac{nh}{2}-\frac{h}{2}\right\}\sin\left(\frac{nh}{2}\right)\right] \qquad \text{[Multiply and divided by 2]}$$

$$= \lim_{h\to 0}\left[\left(\frac{\frac{h}{2}}{\sin\left(\frac{h}{2}\right)}\right)2\cos\left\{a+\frac{b-a}{2}-\frac{h}{2}\right\}\sin\left(\frac{b-a}{2}\right)\right] \qquad [\because \quad nh=(b-a)]$$

$$= \lim_{h\to 0}\left(\frac{1}{\dfrac{\sin\left(\dfrac{h}{2}\right)}{h/2}}\right) . 2\lim_{h\to 0}\left[\cos\left(\frac{b+a}{2}-\frac{h}{2}\right)\right].\sin\left(\frac{b-a}{2}\right)$$

$$= 2\cos\left(\frac{b+a}{2}\right).\sin\left(\frac{b-a}{2}\right)$$

$$= \sin\left(\frac{b+a+b-a}{2}\right)-\sin\left(\frac{b+a-b+a}{2}\right)$$

[$\because$ 2 cos A sin B = sin (A + B) – sin (A – B)]

$$= \sin b - \sin a.$$

(*iii*) Let $\quad I = \int_0^{\pi/2} \cos x\, dx$

Comparing I with $\int_a^b f(x)\,dx$, we get : $a = 0,\ b = \dfrac{\pi}{2}$, $f(x) = \cos x$.

Proceed as part (*ii*), we get

$$I = \sin b - \sin a$$

Put $\quad a = 0,\ b = \dfrac{\pi}{2}$

$$I = \sin\frac{\pi}{2} - \sin 0 = 1 - 0 = 1.$$

(*iv*) Let $I = \int_0^{\pi/2} \sin x\, dx$

Comparing I with $\int_a^b f(x)\,dx$, we get : $a = 0,\ b = \dfrac{\pi}{2}$, $f(x) = \sin x$.

Proceed as part (*i*), we get

$$I = \cos a - \cos b$$

Put $\quad a = 0,\ b = \dfrac{\pi}{2}$

$$I = \cos 0 - \cos\frac{\pi}{2}$$
$$= 1 - 0 = 1.$$

(*v*) Let $\quad I = \int_a^b \frac{1}{x}\,dx \qquad \text{...(1)}$

Comparing I with $\int_a^b f(x)\,dx$, we get : $a = a,\ b = b,\ f(x) = \dfrac{1}{x}$ and $r^n = \dfrac{b}{a}$, $a \neq 0$

$\therefore$ By Walli's Formula,

$$\int_a^b f(x)\,dx = \lim_{r\to 1}(r-1)\,[af(a) + arf(ar) + ar^2f(ar^2) + \ldots\ldots + ar^{n-1}f(ar^{n-1})]$$

$$\therefore \qquad I = \lim_{r\to 1}(r-1)\left[a.\frac{1}{a} + ar.\frac{1}{ar} + (ar^2).\frac{1}{(ar^2)} + \ldots\ldots + (ar^{n-1}).\frac{1}{(ar^{n-1})}\right]$$

$$\left[\because\ f(x) = \frac{1}{x}\right]$$

$$= \lim_{r \to 1} (r-1)[1+1+1+ \ldots\ldots + 1]$$

$$\Rightarrow \qquad I = \lim_{r \to 1} (r-1) \, . \, n \qquad \ldots(2)$$

Now $\quad r^n = \dfrac{b}{a}$

Taking log on both sides, $\log r^n = \log \dfrac{b}{a} \quad \Rightarrow \quad n \log r = \log \left(\dfrac{b}{a}\right)$

$$\Rightarrow \qquad n = \frac{\log\left(\frac{b}{a}\right)}{\log r} \qquad \ldots(3)$$

Substituting this value of n in equation (2), we get

$$I = \lim_{r \to 1} \frac{(r-1)\log\left(\frac{b}{a}\right)}{\log r} = \log\left(\frac{b}{a}\right) . \lim_{r \to 1}\left(\frac{r-1}{\log r}\right)$$

Put $\quad r = 1 + h \quad \Rightarrow \quad h \to 0$ as $r \to 1$

$$\therefore \qquad I = \log\left(\frac{b}{a}\right) . \lim_{h \to 0} \frac{(1+h-1)}{\log(1+h)} = \log\left(\frac{b}{a}\right) . \lim_{h \to 0} \frac{h}{\log(1+h)}$$

$$= \log\left(\frac{b}{a}\right) . \lim_{h \to 0} \left[\frac{1}{\frac{1}{h}\log(1+h)}\right]$$

$$= \log\left(\frac{b}{a}\right) . 1 = \log b - \log a.$$

(*vi*) Let $I = \displaystyle\int_a^b \frac{1}{x^2}\, dx$

Comparing I with $\displaystyle\int_a^b f(x)\, dx$, we get : $a = a$, $b = b$, $f(x) = \dfrac{1}{x^2}$ and $r^n = \dfrac{b}{a}$

$\therefore$ By Walli's Formula,

$$\int_a^b f(x)\, dx = \lim_{r \to 1} (r-1)[af(a) + arf(ar) + ar^2 f(ar^2) + \ldots\ldots + ar^{n-1} . f(ar^{n-1})]$$

$$\therefore \qquad I = \lim_{r \to 1} (r-1)\left[a . \frac{1}{a^2} + ar . \frac{1}{(ar)^2} + ar^2 . \frac{1}{(ar^2)^2} + \ldots\ldots + ar^{n-1} . \frac{1}{(ar^{n-1})^2}\right]$$

$$\left[\because \ f(x) = \frac{1}{x^2}\right]$$

$$= \lim_{r \to 1} (r-1)\left[\frac{1}{a} + \frac{1}{ar} + \frac{1}{ar^2} + \ldots\ldots + \frac{1}{ar^{n-1}}\right]$$

$$= \lim_{r \to 1} (r-1)\left[\frac{\frac{1}{a}\left[\frac{1}{r^n} - 1\right]}{\frac{1}{r} - 1}\right] \qquad \left[\because \text{ In a G.P., sum } S_n = \frac{a(r^n - 1)}{r-1}\right]$$

$$= \lim_{r\to 1}(r-1)\left[\frac{\frac{1}{a}\left[\frac{1-r^n}{r^n}\right]}{\frac{1-r}{r}}\right] = \lim_{r\to 1}\frac{(r-1).r}{(1-r)}\left[\frac{1}{a}\left(\frac{1-\frac{b}{a}}{\frac{b}{a}}\right)\right] \qquad \left[\because\ r^n = \frac{b}{a}\right]$$

$$= \lim_{r\to 1}\frac{r(r-1)}{1-r}\cdot\frac{1}{a}\left[\frac{a}{b}-1\right]$$

$$= \lim_{r\to 1} -r\left[\frac{1}{b}-\frac{1}{a}\right] = -1\left(\frac{1}{b}-\frac{1}{a}\right) = \frac{1}{a}-\frac{1}{b}.$$

(*vii*) Let $I = \int_a^b \frac{1}{\sqrt{x}}\,dx$

Comparing I with $\int_a^b f(x)\,dx$, we get : $a = a$, $b = b$, $f(x) = \frac{1}{\sqrt{x}}$, and $r^n = \frac{b}{a}$

$\therefore$ By Walli's Formula,

$$\int_a^b f(x)\,dx = \lim_{r\to 1}(r-1)\,[af(a) + arf(ar) + ar^2f(ar^2) + \ldots\ldots + ar^{n-1}f(ar^{n-1})]$$

$$\therefore \qquad I = \lim_{r\to 1}(r-1)\left[a.\frac{1}{\sqrt{a}} + ar.\frac{1}{\sqrt{ar}} + ar^2.\frac{1}{\sqrt{ar^2}} + \ldots\ldots + ar^{n-1}.\frac{1}{\sqrt{ar^{n-1}}}\right]$$

$$\left[\because\ f(x) = \frac{1}{\sqrt{x}}\right]$$

$$= \lim_{r\to 1}(r-1)\left[\sqrt{a}+\sqrt{ar}+\sqrt{ar^2}+\ldots\ldots+\sqrt{ar^{n-1}}\right]$$

$$= \lim_{r\to 1}(r-1)\left[\frac{\sqrt{a}\left\{\left(\sqrt{r}\right)^n-1\right\}}{\sqrt{r}-1}\right] \qquad \left[\because\ \text{In a G.P., sum } S_n = \frac{a(r^n-1)}{r-1}\right]$$

$$= \lim_{r\to 1}\left[\frac{\left(\sqrt{r}+1\right)\left(\sqrt{r}-1\right)}{\sqrt{r}-1}\right].\left[\sqrt{a}.\left(\sqrt{r}\right)^n-1\right]$$

$$= \lim_{r\to 1}\left(\sqrt{r}+1\right).\sqrt{a}.\left[\sqrt{\frac{b}{a}}-1\right] \qquad \left[\because\ r^n = \frac{b}{a}\right]$$

$$= (1+1)\,\sqrt{a}.\left[\frac{\sqrt{b}-\sqrt{a}}{\sqrt{a}}\right] = 2\left(\sqrt{b}-\sqrt{a}\right).$$

(*viii*) Let $I = \int_a^b \sin^2 x\,dx = \frac{1}{2}\int_a^b (1-\cos 2x)\,dx$ $\qquad [\because\ 1-\cos 2A = 2\sin^2 A]$

Comparing I with $\int_a^b f(x)\,dx$, we get : $a = a$, $b = b$, $f(x) = (1-\cos 2x)$.

Let $h = \dfrac{b-a}{n} \quad \Rightarrow \quad nh = (b-a),\ n \in \mathrm{N}.$

$$\therefore \quad \mathrm{I} = \lim_{h \to 0} \frac{h}{2} \left[f(a) + f(a+h) + f(a+2h) + \ldots\ldots + f\{a + (n-1)h\}\right]$$

$$= \lim_{h \to 0} \frac{h}{2} \left[1\,(1 - \cos 2a) + \{1 - \cos 2(a+h)\} + \ldots\ldots + \left\{1 - \cos 2\left(a + \overline{n-1}\,h\right)\right\}\right]$$

$$[\because \quad f(x) = (1 - \cos 2x)]$$

$$= \lim_{h \to 0} \frac{h}{2} \left[(1 + 1 + 1 + \ldots\ldots + 1) - \left\{\cos 2a + \cos 2\,(a+h) + \ldots\ldots + \cos 2\left(a + \overline{n-1}\,h\right)\right\}\right]$$

$$= \lim_{h \to 0} \frac{h}{2} \left[n - \frac{\cos\left(2a + \overline{n-1}\,h\right) \sin nh}{\sin h}\right]$$

$$= \lim_{h \to 0} \frac{1}{2} \left[nh - \frac{h \cos\left(2a + \overline{n-1}\,h\right) \sin nh}{\sin h}\right]$$

$$= \lim_{h \to 0} \frac{1}{2} \left[(b-a) - \frac{\cos\,[2a + b - a - h]\,\sin\,(b-a)}{\left(\dfrac{\sin h}{h}\right)}\right] \qquad [\because \quad nh = (b-a)$$

$$= \frac{1}{2} \left[(b-a) - \frac{\cos\,(b+a)\,\sin\,(b-a)}{1}\right]$$

$$= \frac{1}{2} \left[(b-a) - \frac{1}{2} \cdot 2 \cos\,(b+a)\,\sin\,(b-a)\right] \qquad \text{[Multiply and divided by 2]}$$

$$= \frac{1}{2} \left[(b-a) - \frac{1}{2} \{\sin\,(b+a+b-a) - \sin\,(b+a-b+a)\}\right]$$

$$[\because \quad 2 \cos \mathrm{A} \sin \mathrm{B} = \sin\,(\mathrm{A}+\mathrm{B}) - \sin\,(\mathrm{A}-\mathrm{B})]$$

$$= \frac{1}{2} \left[(b-a) - \frac{1}{2} \{\sin 2b - \sin 2a\}\right]$$

$$= \frac{1}{2} \left[(b-a) - \frac{1}{2} \{2 \sin b \cos b - 2 \sin a \cos a\}\right] \qquad [\because \quad \sin 2\mathrm{A} = 2 \sin \mathrm{A} \cos \mathrm{A}]$$

$$= \frac{1}{2}\,(b-a) - \frac{1}{2}\,(\sin a \cos a - \sin b \cos b).$$

EXERCISE FOR PRACTICE

Evaluate the following integrals as the limit of a sum :

1. $\displaystyle\int_0^5 (x+1)\,dx$

2. $\displaystyle\int_3^5 (2-x)\,dx$

3. $\displaystyle\int_2^3 (2x^2+1)\,dx$

4. $\displaystyle\int_a^b e^{-x}\,dx$

5. $\displaystyle\int_1^3 (x^2+x)\,dx$

6. $\displaystyle\int_0^2 e^{3x+1}\,dx$

7. $\int_0^4 (x + e^{2x})\, dx$

8. $\int_0^{\pi/2} \sin x\, dx$

9. $\int_0^3 (x^2 + 1)\, dx$

10. $\int_2^3 e^{-x}\, dx$

11. $\int_0^5 (x + 1)\, dx$

12. $\int_1^2 (x^2 - 1)\, dx$

Answers

1. $\frac{35}{2}$
2. -4
3. 8
4. $e^{-a} - e^{-b}$
5. $\frac{38}{3}$
6. $\frac{e(e^6 - 1)}{3}$
7. $8 + \frac{e^8}{2}$
8. 1
9. 12
10. $e^{-2} - e^{-3}$
11. $\frac{35}{2}$
12. $\frac{4}{3}$.

2

Definite Integral by Using Indefinite Integral

2.1. INTRODUCTION

The concept of definite integrals links the area to other important concepts such as lengths, graphs of the different curves, volume, density, probability and work etc. Definite integral is very useful in calculating areas bounded by curves, arc length, volumes, velocity, length, work, moment of inertia etc.

We will also see that the Indefinite Integral and the Definite Integral are closely related to each other.

2.2. FUNDAMENTAL THEOREM OF INTEGRAL CALCULUS

Statement. If $f(x)$ is a continuous function defined on closed interval $[a, b]$, and if $F(x)$ be the primitive or antiderivative of $f(x)$ *i.e.*, $\frac{d}{dx}\{F(x)\} = f(x)$, then the definite integral of $f(x)$ over $[a, b]$ is denoted by $\int_a^b f(x)\,dx$ and is equal to $[F(b) - F(a)]$.

i.e.,
$$\int_a^b f(x)\,dx = [F(x)]_a^b = [F(b) - F(a)].$$

Here, a and b are called the limits of integration, 'a' is called the lower limit and 'b' is called the upper limit.

The interval $[a, b]$ is called the interval of integration.

Theorem 1. *If $f(x)$ is a continuous and single valued function of x in the interval $[a, b]$ where a and b are finite and $a < b$, then*

$$\lim_{h \to 0} h[f(a) + f(a + h) + f(a + 2h) + \ldots\ldots + f\{a + (n - 1)h\}] = F(b) - F(a).$$

i.e.,
$$\int_a^b f(x)\,dx = F(b) - F(a).$$

Proof. Let AB be the graph of the curve $y = f(x)$. LA and MB are the ordinates $x = a$, $x = b$.

Then, $\quad S = \text{area ALMB}$

$$= F(b) - F(a) \qquad \ldots(1)$$

where $\quad F'(x) = f(x)$.

Divide LM into n equal parts, each equal to h such that

$$nh = \text{LM} = \text{OM} - \text{OL}$$
$$= b - a. \qquad ...(2)$$

by means of points $N_1, N_2, ..., N_{n-1}$ and through these points draw $N_1P_1, N_2P_2, ..., N_{n-1}P_{n-1}$ perpendiculars to x-axis meeting the curve in points $P_1, P_2, ..., P_{n-1}$.

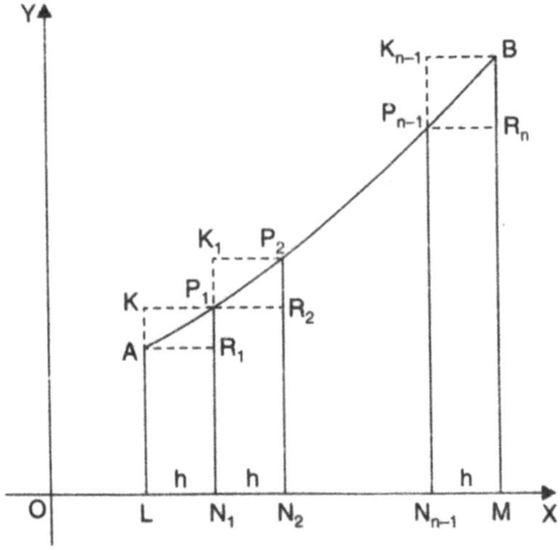

Complete the two sets of rectangles $AN_1, P_1N_2, ..., P_{n-1}M$ and $LP_1, N_1P_2, ..., N_{n-1}B$.

Abscissae of the points A, $P_1, P_2, ..., P_{n-1}$, B are :

$$a, a + h, a + 2h, ..., a + (n-1)h, a + nh = b \qquad \text{[By using equation (2)]}$$

Their respective ordinates are :

$$\text{LA} = f(a), N_1P_1 = f(a + h), N_2P_2 = f(a + 2h), ...,$$
$$......, N_{n-1}P_{n-1} = f(a + \overline{n-1}\, h), \text{MB} = f(a + nh)$$

Now sum of the areas of inscribed rectangles $AN_1, P_1N_2, ..., P_{n-1}M$.

$$= h\, f(a) + h\, f(a + h) + h\, f(a + 2h) + ... + h\, f(a + \overline{n-1}\, h)$$
$$= h\, [f(a) + f(a + h) + f(a + 2h) + ... + f(a + \overline{n-1}\, h)]$$
$$= S_n \quad \text{(say)} \qquad ...(3)$$

Sum of the areas of circumscribed rectangles $LP_1, N_1P_2, ..., N_{n-1}B$.

$$= h\, f(a + h) + h\, f(a + 2h) + ... + h\, f(a + nh)$$
$$= h\, [f(a + h) + f(a + 2h) + ... + f(a + nh)]$$
$$= h\, [f(a + h) + f(a + 2h) + ... + f(a + \overline{n-1}\, h)] + h\, f(a + nh)$$
$$= h\, [f(a) + f(a + h) + f(a + 2h) + ... + f(a + \overline{n-1}\, h)] + h\, f(a + nh) - h\, f(a)$$

[Add and subtract $h\, f(a)$]

$$= h\, [f(a) + f(a + h) + f(a + 2h) + ... + f(a + \overline{n-1}\, h) + h\, [f(b) - f(a)]$$

[$\because \quad nh = b - a \quad \Rightarrow \quad a + nh = b$]

$$= S_n + h\, [f(b) - f(a)] \qquad ...(4) \quad \text{[By using equation (3)]}$$

From the fig, it is obvious that area ALMB lies between the sum of the areas of inscribed and circumscribed rectangles.

$\therefore$ S lies between S_n and $S_n + h\,[f(b) - f(a)]$

As $h \to 0$ $\qquad S = \lim_{h\to 0} S_n = \int_a^b f(x)\,dx.$

$$= \lim_{h\to 0} h\,[f(a) + f(a+h) + f(a+2h) + \ldots\ldots + f(a + \overline{n-1}\,h)] \qquad \ldots(5)$$

$\therefore$ From equations (1) and (5) ;

Equating the two values of S, *i.e.*, area ALMB, we get

$$\lim_{h\to 0} h\,[f(a) + f(a+h) + f(a+2h) + \ldots\ldots + f(a + \overline{n-1}\,h)] = F(b) - F(a).$$

i.e., $$\int_a^b f(x)\,dx = F(b) - F(a).$$

where $\qquad F'(x) = f(x).$

Remark 1. In the $\int_a^b f(x)\,dx$, it does not matter which antiderivative is used to evaluate definite integral, because :

Let us take $F(x) + c$ instead of $F(x)$ as the antiderivative of $f(x)$.

$$\therefore \qquad \int_a^b f(x)\,dx = [F(x) + c]_a^b$$
$$= [F(b) + c] - [F(a) + c]$$
$$= F(b) - F(a).$$

As the constant of integration disappears so $\int_a^b f(x)\,dx$ has a definite value $[F(b) - F(a)]$.

In other words, to evaluate the definite integral there is no need to keep constant of integration.

Remark 2. This theorem is very useful, because it gives us a method of calculating the definite integral more easily, without calculating the limit of a sum. Though the fundamental theorem of integral calculus has eased the process of evaluating definite integrals, but this has involved the search of a function, whose derivative is the integrand in the given definite integral. In other words, we shall find the definite integral of a function by first finding its indefinite integral.

2.3. EVALUATION OF DEFINITE INTEGRAL BY CHANGING LIMITS AFTER SUBSTITUTION

Method to evaluate $\int_a^b f(x)\,dx$ by the substitution $x = g(z)$.

In the integrand put $x = g(z) \Rightarrow dx = g'(z)\,dz$

$\therefore$ when $x = a$ and $x = b$. Then find the corresponding values of z say $g(a)$ and $g(b)$, respectively.

Note. While evaluating the definite integral by substitution, the limits of integration must also be changed.

Remember 1. It is a common mistake of not changing the limits of integration while evaluating the integral in z between the old limits a and b which are the values of x and not of y.

Remember 2. The new limits $g(a)$ and $g(b)$ in evaluating the definite integral $\int_a^b f(x)\,dx$ by substituting $x = g(z)$ must be taken in such a manner that as x increases continuously from a to b, z increases from $g(a)$ to $g(b)$ or decreases from $g(a)$ to $g(b)$ continuously.

SOME SOLVED EXAMPLES

Example 1. *Evaluate the following definite integrals :*

(*i*) $\int_{-4}^{-1} \frac{1}{x}\, dx$ (*ii*) $\int_{1}^{5} x^2\, dx$

(*iii*) $\int_{2}^{3} \frac{1}{x}\, dx$ (*iv*) $\int_{1}^{2} x^2\, dx$

(*v*) $\int_{-5}^{5} x\, dx$ (*vi*) $\int_{2}^{5} (x^3 + x)\, dx.$

Solution. (*i*) Let $\mathrm{I} = \int_{-4}^{-1} \frac{1}{x}\, dx = \int_{-4}^{-1} \frac{1}{x}\, dx = \Big[\log |x|\Big]_{-4}^{-1}$

$= \log |-1| - \log |-4| = \log 1 - \log 4$

$= 0 - \log 4 = -\log 4.$ $[\because \ \log 1 = 0]$

(*ii*) Let $\mathrm{I} = \int_{1}^{5} x^2\, dx = \int_{1}^{5} x^2\, dx = \left[\frac{x^3}{3}\right]_1^5$

$= \frac{(5)^3}{3} - \frac{(1)^3}{3} = \frac{125}{3} - \frac{1}{3}$

$= \frac{124}{3}.$

(*iii*) Let $\mathrm{I} = \int_{2}^{3} \frac{1}{x}\, dx$

$= \int_{2}^{3} \frac{1}{x}\, dx = \Big[\log |x|\Big]_2^3 = \log |3| - \log |2| = \log 3 - \log 2$

$= \log \frac{3}{2}.$ $\left[\because \ \log m - \log n = \log\left(\frac{m}{n}\right)\right]$

(*iv*) Let $\mathrm{I} = \int_{1}^{2} x^2\, dx$

$= \int_{1}^{2} x^2\, dx = \left[\frac{x^3}{3}\right]_1^2 = \frac{(2)^3}{3} - \frac{(1)^3}{3}$

$= \frac{8}{3} - \frac{1}{3} = \frac{7}{3}.$

(*v*) Let $\mathrm{I} = \int_{-5}^{5} x\, dx$

$= \int_{-5}^{5} x\, dx = \left[\frac{x^2}{2}\right]_{-5}^5 = \frac{(5)^2}{2} - \frac{(-5)^2}{2}$

$= \frac{25}{2} - \frac{25}{2} = 0.$

(*vi*) Let $$I = \int_2^5 (x^3 + x)\, dx$$

$$= \int_2^5 (x^3 + x)\, dx = \left[\frac{x^4}{4} + \frac{x^2}{2}\right]_2^5 = \left[\frac{(5)^4}{4} + \frac{(5)^2}{2}\right] - \left[\frac{(2)^4}{4} + \frac{(2)^2}{2}\right]$$

$$= \frac{625}{4} + \frac{25}{2} - \frac{16}{4} - \frac{4}{2} = \frac{625}{4} + \frac{25}{2} - 4 - 2$$

$$= \frac{625}{4} + \frac{25}{2} - 6 = \frac{625 + 50 - 24}{4} = \frac{651}{4}.$$

Example 2. *Evaluate the following definite integrals :*

(*i*) $\int_0^1 \frac{1}{2x-3}\, dx$ (*ii*) $\int_{-1}^1 (x+1)\, dx$

(*iii*) $\int_0^8 x^{5/3}\, dx$ (*iv*) $\int_0^4 (x + x^{3/2})\, dx$

(*v*) $\int_1^2 \frac{x^2 - 3x + 2}{x^4}\, dx$ (*vi*) $\int_0^4 \left(\sqrt{x} - 2x + x^2\right) dx.$

Solution. (*i*) Let $I = \int_0^1 \frac{1}{2x-3}\, dx$

$$= \int_0^1 \frac{1}{2x-3}\, dx = \left[\frac{\log|(2x-3)|}{2}\right]_0^1 = \frac{1}{2}\left[\log|(2x-3)|\right]_0^1$$

$$= \frac{1}{2}\,[\log|2(1)-3| - \log|2(0)-3|]$$

$$= \frac{1}{2}\,[\log|-1| - \log|-3|] = \frac{1}{2}\,[\log 1 - \log 3]$$

$$= \frac{1}{2}\,(0 - \log 3) = -\frac{1}{2}\log 3. \qquad [\because \log 1 = 0]$$

(*ii*) Let $$I = \int_{-1}^1 (1+x)\, dx$$

$$= \int_{-1}^1 (1+x)\, dx = \left[x + \frac{x^2}{2}\right]_{-1}^1 = \left[1 + \frac{(1)^2}{2}\right] - \left[(-1) + \frac{(-1)^2}{2}\right]$$

$$= \left(1 + \frac{1}{2}\right) - \left(-1 + \frac{1}{2}\right) = \frac{3}{2} - \left(-\frac{1}{2}\right)$$

$$= \frac{3}{2} + \frac{1}{2} = 2.$$

(*iii*) Let $$I = \int_0^8 x^{5/3}\, dx$$

$$= \int_0^8 x^{5/3}\, dx = \left[\frac{x^{\frac{5}{3}+1}}{\frac{5}{3}+1}\right]_0^8 = \left[\frac{x^{\frac{8}{3}}}{\frac{8}{3}}\right]_0^8 = \frac{3}{8}\,[(8)^{8/3} - (0)^{8/3}]$$

$$= \frac{3}{8}\,[(2^3)^{8/3} - 0] = \frac{3}{8}\,[2^8] = \frac{3}{8} \times 256$$

$$= 96.$$

(*iv*) Let $\quad I = \int_0^4 (x + x^{3/2})\, dx$

$$= \int_0^4 (x + x^{3/2})\, dx = \left[\frac{x^2}{2} + \frac{x^{\frac{3}{2}+1}}{\frac{3}{2}+1}\right]_0^4 = \left[\frac{x^2}{2} + \frac{x^{\frac{5}{2}}}{\frac{5}{2}}\right]_0^4$$

$$= \left[\frac{(4)^2}{2} + \frac{2}{5}\,(4)^{5/2}\right] - [0 + 0]$$

$$= \frac{16}{2} + \frac{2}{5}\,(32) = 8 + \frac{64}{5} = \frac{40 + 64}{5} \qquad [\because \quad 4^{5/2} = (2^2)^{5/2} = 2^5 = 32]$$

$$= \frac{104}{5}.$$

(*v*) Let $\quad I = \int_1^2 \frac{x^2 - 3x + 2}{x^4}\, dx$

$$= \int_1^2 \left(\frac{x^2}{x^4} - \frac{3x}{x^4} + \frac{2}{x^4}\right) dx = \int_1^2 \left(\frac{1}{x^2} - \frac{3}{x^3} + \frac{2}{x^4}\right) dx$$

$$= \int_1^2 (x^{-2} - 3x^{-3} + 2x^{-4})\, dx = \left[\frac{x^{-2+1}}{-2+1} - 3\,\frac{x^{-3+1}}{-3+1} + 2\,\frac{x^{-4+1}}{-4+1}\right]_1^2$$

$$= \left[\frac{x^{-1}}{-1} - 3\,\frac{x^{-2}}{-2} + 2\,\frac{x^{-3}}{-3}\right]_1^2 = \left[-\frac{1}{x} + \frac{3}{2x^2} - \frac{2}{3x^3}\right]_1^2$$

$$= \left[-\frac{1}{2} + \frac{3}{2(2)^2} - \frac{2}{3(2)^3}\right] - \left[-\frac{1}{1} + \frac{3}{2(1)^2} - \frac{2}{3(1)^3}\right]$$

$$= \left[-\frac{1}{2} + \frac{3}{8} - \frac{1}{12}\right] - \left[-1 + \frac{3}{2} - \frac{2}{3}\right] = \left[\frac{-12 + 9 - 2}{24}\right] - \left(-\frac{1}{6}\right)$$

$$= -\frac{5}{24} + \frac{1}{6} = -\frac{1}{24}.$$

(*vi*) Let $\quad I = \int_0^4 \left(\sqrt{x} - 2x + x^2\right) dx$

$$= \int_0^4 \left(\sqrt{x} - 2x + x^2\right) dx = \left[\frac{x^{\frac{1}{2}+1}}{\frac{1}{2}+1} - 2\,\frac{x^2}{2} + \frac{x^3}{3}\right]_0^4$$

$$= \left[\frac{x^{3/2}}{3/2} - x^2 + \frac{x^3}{3}\right]_0^4 = \left[\frac{2}{3}(4)^{3/2} - (4)^2 + \frac{(4)^3}{3}\right] - [0 - 0 + 0]$$

$$= \left[\frac{2(8)}{3} - \frac{16}{1} + \frac{64}{3}\right] = \frac{16 - 48 + 64}{3}$$

$$= \frac{32}{3}.$$

Example 3. *Evaluate the following definite integrals :*

(*i*) $\int_1^2 (4x^3 - 5x^2 + 6x + 9)\,dx$ (*ii*) $\int_0^\pi \sin\frac{x}{2}\,dx$

(*iii*) $\int_0^1 \frac{1}{\sqrt{1+x}+\sqrt{x}}\,dx$ (*iv*) $\int_0^{\pi/2} (\sin x + \cos x)\,dx$

(*v*) $\int_0^\pi \cos x\,dx$ (*vi*) $\int_0^{\pi/2} \cos 2x\,dx.$

Solution. (*i*) Let $I = \int_1^2 (4x^3 - 5x^2 + 6x + 9)\,dx$

$$= \int_1^2 (4x^3 - 5x^2 + 6x + 9)\,dx = \left[\frac{4x^4}{4} - \frac{5x^3}{3} + \frac{6x^2}{2} + 9x\right]_1^2$$

$$= \left[x^4 - \frac{5}{3}x^3 + 3x^2 + 9x\right]_1^2$$

$$= \left[(2)^4 - \frac{5}{3}(2)^3 + 3(2)^2 + 9(2)\right] - \left[(1)^4 - \frac{5}{3}(1)^3 + 3(1)^2 + 9(1)\right]$$

$$= \left[16 - \frac{40}{3} + 12 + 18\right] - \left[1 - \frac{5}{3} + 3 + 9\right] = 46 - \frac{40}{3} - 13 + \frac{5}{3} = 33 - \frac{35}{3}$$

$$= \frac{99 - 35}{3} = \frac{64}{3}.$$

(*ii*) Let $I = \int_0^\pi \sin\frac{x}{2}\,dx$

$$= \int_0^\pi \sin\frac{x}{2}\,dx = \left[-\frac{\cos x/2}{1/2}\right]_0^\pi = -2\left[\cos\frac{x}{2}\right]_0^\pi$$

$$= -2\left[\cos\frac{\pi}{2} - \cos 0\right] = -2(0 - 1)$$

$$= 2.$$

(*iii*) Let $I = \int_0^1 \frac{1}{\sqrt{1+x}+\sqrt{x}}\,dx$

$$= \int_0^1 \frac{1}{\sqrt{1+x}+\sqrt{x}} \times \frac{\sqrt{1+x}-\sqrt{x}}{\sqrt{1+x}-\sqrt{x}}\,dx$$ [Rationalization]

$$= \int_0^1 \frac{\sqrt{1+x} - \sqrt{x}}{(1+x-x)}\, dx = \int_0^1 \left(\sqrt{1+x} - \sqrt{x}\right) dx$$

$$= \left[\frac{(1+x)^{\frac{1}{2}+1}}{\frac{1}{2}+1} - \frac{(x)^{\frac{1}{2}+1}}{\frac{1}{2}+1} \right]_0^1 = \left[\frac{2}{3}(1+x)^{3/2} - \frac{2}{3}x^{3/2} \right]_0^1$$

$$= \left[\frac{2}{3}(1+1)^{3/2} - \frac{2}{3}(1)^{3/2} \right] - \left[\frac{2}{3}(1+0)^{3/2} - \frac{2}{3}(0)^{3/2} \right]$$

$$= \frac{2}{3}[(2)^{3/2} - 1] - \frac{2}{3}[1-0] = \frac{2}{3}\left(2\sqrt{2}\right) - \frac{2}{3} - \frac{2}{3} = \frac{4\sqrt{2}}{3} - \frac{4}{3}$$

$$= \frac{4}{3}\left(\sqrt{2} - 1\right).$$

(*iv*) Let $\quad I = \int_0^{\pi/2} (\sin x + \cos x)\, dx$

$$= \int_0^{\pi/2} (\sin x + \cos x)\, dx = [-\cos x + \sin x]_0^{\pi/2}$$

$$= \left[-\cos\frac{\pi}{2} + \sin\frac{\pi}{2} \right] - [-\cos 0 + \sin 0] = (0+1) - (-1+0)$$

$$= 1 + 1 = 2.$$

(*v*) Let $\quad I = \int_0^{\pi} \cos x\, dx$

$$= \int_0^{\pi} \cos x\, dx = \Big[\sin x \Big]_0^{\pi} = \sin \pi - \sin 0 \qquad [\because \quad \sin 180° = 0]$$

$$= 0.$$

(*vi*) Let $\quad I = \int_0^{\pi/2} \cos 2x\; dx$

$$= \int_0^{\pi/2} \cos 2x\, dx = \left[\frac{\sin 2x}{2} \right]_0^{\pi/2} = \frac{1}{2}\left[\sin 2\left(\frac{\pi}{2}\right) - \sin 0 \right]$$

$$= \frac{1}{2}[\sin \pi - \sin 0] = 0.$$

Example 4. *Evaluate the following definite integrals :*

(*i*) $\int_0^{\pi/2} \sin^2 x\, dx$ (*ii*) $\int_{\pi/6}^{\pi/2} \frac{\cos x}{\sin^2 x}\, dx$

(*iii*) $\int_0^{\pi/2} \sin 2x\, dx$ (*iv*) $\int_0^{\pi/4} \frac{1}{\cos^2 x}\, dx$

(*v*) $\int_4^5 e^x\, dx$ (*vi*) $\int_0^{\pi/2} \frac{\sin x}{\sqrt{1+\cos x}}\, dx.$

Solution. (*i*) Let $I = \int_0^{\pi/2} \sin^2 x \, dx$

$$= \int_0^{\pi/2} \left(\frac{1 - \cos 2x}{2}\right) dx \qquad \left[\begin{array}{l} \because \ 1 - \cos 2A = 2 \sin^2 A \\ \Rightarrow \dfrac{1 - \cos 2A}{2} = \sin^2 A \end{array}\right]$$

$$= \frac{1}{2} \int_0^{\pi/2} (1 - \cos 2x) \, dx = \frac{1}{2} \left[x - \frac{\sin 2x}{2}\right]_0^{\pi/2}$$

$$= \frac{1}{2} \left[\frac{\pi}{2} - \frac{\sin 2\left(\frac{\pi}{2}\right)}{2}\right] - \frac{1}{2}\left[0 - \frac{\sin 0}{2}\right] = \frac{1}{2}\left[\frac{\pi}{2} - \frac{\sin \pi}{2}\right] - 0$$

$$= \frac{1}{2}\left[\frac{\pi}{2} - 0\right] = \frac{\pi}{4}. \qquad [\because \ \sin \pi = \sin 180^\circ = 0]$$

(*ii*) Let $I = \int_{\pi/6}^{\pi/2} \frac{\cos x}{\sin^2 x} dx$

$$= \int_{\pi/6}^{\pi/2} \frac{1}{\sin x} \cdot \frac{\cos x}{\sin x} dx = \int_{\pi/6}^{\pi/2} \text{cosec } x \cot x \, dx$$

$$= \left[- \text{cosec } x\right]_{\pi/6}^{\pi/2} = - \left[\text{cosec } \frac{\pi}{2} - \text{cosec } \frac{\pi}{6}\right]$$

$$= - [1 - 2] = 1.$$

(*iii*) Let $I = \int_0^{\pi/2} \sin 2x \, dx$

$$= \int_0^{\pi/2} \sin 2x \, dx = \left[- \frac{\cos 2x}{2}\right]_0^{\pi/2} = - \frac{1}{2}\left[\cos 2\left(\frac{\pi}{2}\right) - \cos 2(0)\right]$$

$$= - \frac{1}{2} [\cos \pi - \cos 0]$$

$$= - \frac{1}{2} (- 1 - 1) = 1. \qquad [\because \ \cos \pi = \cos 180^\circ = - 1]$$

(*iv*) Let $I = \int_0^{\pi/4} \frac{1}{\cos^2 x} dx$

$$= \int_0^{\pi/4} \sec^2 x \, dx = [\tan x]_0^{\pi/4} = \left[\tan \frac{\pi}{4} - \tan 0\right]$$

$$= 1 - 0 = 1. \qquad \left[\because \ \tan \frac{\pi}{4} = \tan 45^\circ = 1\right]$$

(*v*) Let $\qquad I = \int_4^5 e^x\, dx$

$$= \int_4^5 e^x\, dx = \left[e^x\right]_4^5 = e^5 - e^4$$

$$= e^4\,(e - 1).$$

(*vi*) Let $\qquad I = \int_0^{\pi/2} \frac{\sin x}{\sqrt{1+\cos x}}\, dx$

$$= \int_0^{\pi/2} \frac{\sin x}{\sqrt{1+\cos x}}\, dx = \int_0^{\pi/2} \frac{\sin x}{\sqrt{2\cos^2 \frac{x}{2}}}\, dx \qquad \left[\begin{array}{l}\because\ 1+\cos 2A = 2\cos^2 A \\ \Rightarrow\ 1+\cos A = 2\cos^2 \frac{A}{2}\end{array}\right]$$

$$= \frac{1}{\sqrt{2}} \int_0^{\pi/2} \frac{2\sin\frac{x}{2}\cos\frac{x}{2}}{\cos\frac{x}{2}}\, dx \qquad \left[\begin{array}{l}\because\ \sin 2A = 2\sin A\cos A \\ \Rightarrow\ \sin A = 2\sin\frac{A}{2}\cos\frac{A}{2}\end{array}\right]$$

$$= \frac{2}{\sqrt{2}} \int_0^{\pi/2} \sin\frac{x}{2}\, dx = \sqrt{2}\left[-\frac{\cos x/2}{1/2}\right]_0^{\pi/2} = -2\sqrt{2}\left[\cos\frac{x}{2}\right]_0^{\pi/2}$$

$$= -2\sqrt{2}\left[\cos\frac{\pi}{4} - \cos 0°\right] = -2\sqrt{2}\left[\frac{1}{\sqrt{2}} - 1\right] \qquad \left[\because\ \cos\frac{\pi}{4} = \cos 45° = \frac{1}{\sqrt{2}}\right]$$

$$= -2 + 2\sqrt{2}\,.$$

Example 5. *Evaluate the following definite integrals :*

(*i*) $\int_0^{\pi/2} \sin x \sin 2x\, dx$ (*ii*) $\int_0^1 \frac{1}{1+x^2}\, dx$

(*iii*) $\int_2^3 \frac{1}{x^2-1}\, dx$ (*iv*) $\int_0^1 \frac{1}{\sqrt{1-x^2}}\, dx$

(*v*) $\int_3^4 \frac{1}{x^2-4}\, dx$ (*vi*) $\int_0^{\pi/2} \frac{\sin^2 x}{(1+\cos x)^2}\, dx.$

Solution. (*i*) Let $I = \int_0^{\pi/2} \sin x \sin 2x\, dx$ [Multiply and divided by 2]

$$= \frac{1}{2}\int_0^{\pi/2} 2\sin 2x \sin x\, dx$$

$$= \frac{1}{2}\int_0^{\pi/2} [\cos(2x - x) - \cos(2x + x)]\, dx$$

$$[\because\ 2\sin A \sin B = \cos(A - B) - \cos(A + B)]$$

$$= \frac{1}{2}\int_0^{\pi/2} (\cos x - \cos 3x)\, dx = \frac{1}{2}\left[\sin x - \frac{\sin 3x}{3}\right]_0^{\pi/2}$$

$$= \frac{1}{2}\left[\sin\frac{\pi}{2} - \frac{\sin 3(\pi/2)}{3}\right] - \frac{1}{2}\left[\sin 0° - \frac{\sin 0°}{3}\right]$$

$$= \frac{1}{2}\left[1 - \frac{1}{3}(-1)\right] - \frac{1}{2}[0-0] \qquad \left[\because \sin\frac{3\pi}{2} = \sin\left(\pi + \frac{\pi}{2}\right) = -\sin\frac{\pi}{2} = -1\right]$$

$$= \frac{1}{2}\left[\frac{4}{3}\right] = \frac{2}{3}.$$

(*ii*) Let $\quad I = \int_0^1 \frac{1}{1+x^2}\,dx$

$$= \int_0^1 \frac{1}{1+x^2}\,dx = \left[\tan^{-1} x\right]_0^1 \qquad \left[\because \int \frac{1}{a^2+x^2}\,dx = \frac{1}{a}\tan^{-1}\frac{x}{a} + c\right]$$

$$= [\tan^{-1}(1) - \tan^{-1}(0)] = \frac{\pi}{4} - 0 = \frac{\pi}{4}.$$

(*iii*) Let $\quad I = \int_2^3 \frac{1}{x^2-1}\,dx$

$$= \int_2^3 \frac{1}{x^2-1}\,dx = \left[\frac{1}{2}\log\left|\frac{x-1}{x+1}\right|\right]_2^3 \qquad \left[\because \int \frac{1}{x^2-a^2}\,dx = \frac{1}{2a}\log\left|\frac{x-a}{x+a}\right| + c\right]$$

$$= \frac{1}{2}\left[\log\left(\frac{3-1}{3+1}\right) - \log\left(\frac{2-1}{2+1}\right)\right] = \frac{1}{2}\left[\log\left(\frac{2}{4}\right) - \log\left(\frac{1}{3}\right)\right]$$

$$= \frac{1}{2}\left[\log\left(\frac{1}{2}\right) - \log\left(\frac{1}{3}\right)\right] = \frac{1}{2}\log\left(\frac{1}{2}\times\frac{3}{1}\right) \qquad \left[\because \log m - \log n = \log\left(\frac{m}{n}\right)\right]$$

$$= \frac{1}{2}\log\frac{3}{2}.$$

(*iv*) Let $\quad I = \int_0^1 \frac{1}{\sqrt{1-x^2}}\,dx$

$$= \int_0^1 \frac{1}{\sqrt{1-x^2}}\,dx = \left[\sin^{-1} x\right]_0^1 \qquad \left[\because \int \frac{1}{\sqrt{a^2-x^2}}\,dx = \sin^{-1}\frac{x}{a} + c\right]$$

$$= \sin^{-1}(1) - \sin^{-1}(0) = \frac{\pi}{2} - 0$$

$$= \frac{\pi}{2}.$$

(*v*) Let $\quad I = \int_3^4 \frac{1}{x^2-4}\,dx$

$$= \int_3^4 \frac{1}{x^2-2^2}\,dx = \frac{1}{2(2)}\left[\log\left|\frac{x-2}{x+2}\right|\right]_3^4 \qquad \left[\because \int \frac{1}{x^2-a^2}\,dx = \frac{1}{2a}\log\left|\frac{x-a}{x+a}\right| + c\right]$$

$$= \frac{1}{4}\left[\log\left(\frac{4-2}{4+2}\right) - \log\left(\frac{3-2}{3+2}\right)\right] = \frac{1}{4}\left[\log\frac{1}{3} - \log\frac{1}{5}\right]$$

$$= \frac{1}{4}\log\left[\frac{1}{3}\times\frac{5}{1}\right] \qquad \left[\because \log m - \log n = \log\left(\frac{m}{n}\right)\right]$$

$$= \frac{1}{4}\log\frac{5}{3}.$$

(*vi*) Let $\quad I = \int_0^{\pi/2} \frac{\sin^2 x}{(1+\cos x)^2}\,dx = \int_0^{\pi/2}\left(\frac{\sin x}{1+\cos x}\right)^2 dx$

$$= \int_0^{\pi/2}\left(\frac{2\sin\frac{x}{2}\cos\frac{x}{2}}{2\cos^2\frac{x}{2}}\right)^2 dx \qquad \left[\begin{array}{l}\because \sin 2A = 2\sin A\cos A \\ \Rightarrow \sin A = 2\sin\frac{A}{2}\cos\frac{A}{2} \\ \because 1+\cos 2A = 2\cos^2 A \\ \Rightarrow 1+\cos A = 2\cos^2\frac{A}{2}\end{array}\right]$$

$$= \int_0^{\pi/2} \tan^2\frac{x}{2}\,dx = \int_0^{\pi/2}\left(\sec^2\frac{x}{2} - 1\right)dx \qquad [\because \sec^2 A - \tan^2 A = 1]$$

$$= \left[\frac{\tan\frac{x}{2}}{1/2} - x\right]_0^{\pi/2} = \left[2\tan\frac{x}{2} - x\right]_0^{\pi/2} = \left[2\tan\left(\frac{\pi/2}{2}\right) - \frac{\pi}{2}\right] - [2\tan 0^\circ - 0^\circ]$$

$$= \left[2\tan\frac{\pi}{4} - \frac{\pi}{2}\right] - 0$$

$$= 2 - \frac{\pi}{2}. \qquad \left[\because \tan\frac{\pi}{4} = \tan 45^\circ = 1\right]$$

Example 6. *Evaluate the following definite integrals :*

(*i*) $\int_0^{\pi/4} \tan^2 x\,dx$ (*ii*) $\int_0^{\pi/2} \cos^2 x\,dx$

(*iii*) $\int_0^{\pi/2} \cos^4 x\,dx$ (*iv*) $\int_0^{\pi/4} \sin^4 x\,dx$

(*v*) $\int_0^{\infty} \frac{1}{a^2+b^2x^2}\,dx$ (*vi*) $\int_0^{\pi/6} \cos x\cos 2x\,dx.$

Solution. (*i*) Let $I = \int_0^{\pi/4} \tan^2 x\,dx$

$$= \int_0^{\pi/4} (\sec^2 x - 1)\,dx \qquad [\because \sec^2 A - \tan^2 A = 1]$$

$$= [\tan x - x]_0^{\pi/4} = \left[\tan\frac{\pi}{4} - \frac{\pi}{4}\right] - [\tan 0^\circ - 0^\circ]$$

$$= \left(1 - \frac{\pi}{4}\right). \qquad \left[\because \tan\frac{\pi}{4} = \tan 45^\circ = 1\right]$$

(*ii*) Let $I = \int_0^{\pi/2} \cos^2 x \, dx = \int_0^{\pi/2} \left(\frac{1 + \cos 2x}{2} \right) dx$ $\left[\because 1 + \cos 2A = 2\cos^2 A \Rightarrow \frac{1 + \cos 2A}{2} = \cos^2 A \right]$

$= \frac{1}{2} \int_0^{\pi/2} (1 + \cos 2x) \, dx = \frac{1}{2} \left[x + \frac{\sin 2x}{2} \right]_0^{\pi/2}$

$= \frac{1}{2} \left[\left(\frac{\pi}{2} + \frac{\sin \pi}{2} \right) - \left(0 + \frac{\sin 0^\circ}{2} \right) \right]$

$= \frac{1}{2} \left[\frac{\pi}{2} + 0 - 0 \right] = \frac{\pi}{4}$. $[\because \sin \pi = \sin 180^\circ = 0]$

(*iii*) Let $I = \int_0^{\pi/2} \cos^4 x \, dx$ $\left[\because 1 + \cos 2A = 2\cos^2 A \Rightarrow \frac{1 + \cos 2A}{2} = \cos^2 A \right]$

$= \int_0^{\pi/2} (\cos^2 x)^2 \, dx = \int_0^{\pi/2} \left(\frac{1 + \cos 2x}{2} \right)^2 dx$

$= \frac{1}{4} \int_0^{\pi/2} (1 + \cos^2 2x + 2 \cos 2x) \, dx$

$= \frac{1}{4} \int_0^{\pi/2} \left(1 + \frac{1 + \cos 4x}{2} + 2 \cos 2x \right) dx$ $\left[\because 1 + \cos 2A = 2\cos^2 A \Rightarrow 1 + \cos 4A = 2\cos^2 2A \Rightarrow \left(\frac{1 + \cos 4A}{2} \right) = \cos^2 2A \right]$

$= \frac{1}{4} \int_0^{\pi/2} \left(1 + \frac{1}{2} + \frac{1}{2} \cos 4x + 2 \cos 2x \right) dx = \frac{1}{4} \int_0^{\pi/2} \left(\frac{3}{2} + \frac{1}{2} \cos 4x + 2 \cos 2x \right) dx$

$= \frac{3}{8} \int_0^{\pi/2} 1 \, . \, dx + \frac{1}{8} \int_0^{\pi/2} \cos 4x \, dx + \frac{1}{2} \int_0^{\pi/2} \cos 2x \, dx$

$= \frac{3}{8} \Big[x \Big]_0^{\pi/2} + \frac{1}{8} \left[\frac{\sin 4x}{4} \right]_0^{\pi/2} + \frac{1}{2} \left[\frac{\sin 2x}{2} \right]_0^{\pi/2}$

$= \frac{3}{8} \left(\frac{\pi}{2} - 0 \right) + \frac{1}{32} [\sin 2\pi - \sin 0^\circ] + \frac{1}{4} [\sin \pi - \sin 0^\circ]$

$= \frac{3\pi}{16} + \frac{1}{32} (0 - 0) + \frac{1}{4} (0 - 0)$

$= \frac{3\pi}{16}$.

(*iv*) Let $I = \int_0^{\pi/4} \sin^4 x \, dx = \int_0^{\pi/4} (\sin^2 x)^2 \, dx$

$= \int_0^{\pi/4} \left(\frac{1 - \cos 2x}{2} \right)^2 dx$ $\left[\because 1 - \cos 2A = 2\sin^2 A \Rightarrow \left(\frac{1 - \cos 2A}{2} \right) = \sin^2 A \right]$

$$= \frac{1}{4}\int_0^{\pi/4} (1 + \cos^2 2x - 2\cos 2x)\, dx$$

$$= \frac{1}{4}\int_0^{\pi/4}\left[1 + \frac{1+\cos 4x}{2} - 2\cos 2x\right] dx \qquad \left[\begin{array}{l} \because\ 1+\cos 2A = 2\cos^2 A \\ \Rightarrow 1+\cos 4A = 2\cos^2 2A \\ \Rightarrow \left(\dfrac{1+\cos 4A}{2}\right) = \cos^2 2A \end{array}\right]$$

$$= \frac{1}{4}\int_0^{\pi/4}\left[1 + \frac{1}{2} + \frac{1}{2}\cos 4x - 2\cos 2x\right] dx = \frac{1}{4}\int_0^{\pi/4}\left(\frac{3}{2} + \frac{1}{2}\cos 4x - 2\cos 2x\right) dx$$

$$= \frac{3}{8}\int_0^{\pi/4} 1.\, dx + \frac{1}{8}\int_0^{\pi/4} \cos 4x\, dx - \frac{1}{2}\int_0^{\pi/4} \cos 2x\, dx$$

$$= \frac{3}{8}[x]_0^{\pi/4} - \frac{1}{8}\left[\frac{\sin 4x}{4}\right]_0^{\pi/4} - \frac{1}{2}\left[\frac{\sin 2x}{2}\right]_0^{\pi/4}$$

$$= \frac{3}{8}\left(\frac{\pi}{4} - 0\right) - \frac{1}{32}[\sin \pi - \sin 0°] - \frac{1}{4}\left[\sin\frac{\pi}{2} - \sin 0°\right]$$

$$= \frac{3\pi}{32} - \frac{1}{32}(0 - 0) - \frac{1}{4}(1 - 0)$$

$$= \frac{3\pi}{32} - \frac{1}{4}.$$

(*v*) Let $\quad I = \int_0^{\infty} \frac{1}{a^2 + b^2 x^2}\, dx = \frac{1}{b^2}\int_0^{\infty} \frac{1}{\left(\frac{a}{b}\right)^2 + x^2}\, dx$

$$= \frac{1}{b^2}\cdot\left[\frac{1}{\left(\frac{a}{b}\right)} \tan^{-1}\frac{x}{\frac{a}{b}}\right]_0^{\infty} \qquad \left[\because \int \frac{1}{a^2+x^2}\, dx = \frac{1}{a}\tan^{-1}\frac{x}{a} + c\right]$$

$$= \frac{1}{b^2} \times \frac{b}{a}\left[\tan^{-1}\frac{bx}{a}\right]_0^{\infty} = \frac{1}{ab}[\tan^{-1}\infty - \tan^{-1} 0] = \frac{1}{ab}\left[\frac{\pi}{2} - 0\right]$$

$$= \frac{\pi}{2ab}.$$

(*vi*) Let $\quad I = \int_0^{\pi/6} \cos x \cos 2x\, dx$

$$= \frac{1}{2}\int_0^{\pi/6} 2\cos 2x \cos x\, dx \qquad \text{[Multiply and divided by 2]}$$

$$= \frac{1}{2}\int_0^{\pi/6} [\cos(2x + x) + \cos(2x - x)]\, dx$$

[$\because$ 2 cos A cos B = cos (A + B) + cos (A – B)]

$$= \frac{1}{2}\int_0^{\pi/6} (\cos 3x + \cos x)\, dx$$

$$= \frac{1}{2}\left[\frac{\sin 3x}{3}\right]_0^{\pi/6} + \frac{1}{2}[\sin x]_0^{\pi/6}$$

$$= \frac{1}{6}\left[\sin 3\left(\frac{\pi}{6}\right) - \sin 3(0°)\right] + \frac{1}{2}\left[\sin\frac{\pi}{6} - \sin 0°\right]$$

$$= \frac{1}{6}[1-0] + \frac{1}{2}\left[\frac{1}{2} - 0\right] = \frac{1}{6} + \frac{1}{4} \qquad \left[\because \sin\frac{3\pi}{6} = \sin\frac{\pi}{2} = 1,\ \sin\frac{\pi}{6} = \sin 30° = \frac{1}{2}\right]$$

$$= \frac{2+3}{12} = \frac{5}{12}.$$

Example 7. *Evaluate the following definite integrals :*

(*i*) $\int_0^{\pi} \sin^3 x\, dx$ (*ii*) $\int_0^{\pi/2} \cos^3 x\, dx$

(*iii*) $\int_0^{\pi/2} \sin^4 x\, dx$ (*iv*) $\int_0^{\pi}\left(\sin^2\frac{x}{2} - \cos^2\frac{x}{2}\right) dx.$

Solution. (*i*) Let $I = \int_0^{\pi} \sin^3 x\, dx$

$$= \int_0^{\pi} \frac{1}{4}[3\sin x - \sin 3x]\, dx \qquad \left[\begin{array}{l}\because \sin 3A = 3\sin A - 4\sin^3 A \\ \Rightarrow 4\sin^3 A = 3\sin A - \sin 3A \\ \Rightarrow \sin^3 A = \frac{1}{4}[3\sin A - \sin 3A]\end{array}\right]$$

$$= \frac{1}{4}\int_0^{\pi} (3\sin x - \sin 3x)\, dx = \frac{1}{4}\left[-3\cos x + \frac{\cos 3x}{3}\right]_0^{\pi}$$

$$= \frac{1}{4}\left[\left(-3\cos\pi + \frac{\cos 3\pi}{3}\right) - \left(-3\cos 0° + \frac{\cos 0°}{3}\right)\right]$$

$$= \frac{1}{4}\left[-3(-1) + \frac{(-1)}{3}\right] - \frac{1}{4}\left[-3 + \frac{1}{3}\right]$$

$$= \frac{1}{4}\left[3 - \frac{1}{3}\right] - \frac{1}{4}\left[\frac{-9+1}{3}\right] = \frac{1}{4}\left[\frac{9-1}{3}\right] - \frac{1}{4}\left[-\frac{8}{3}\right]$$

$$= \frac{2}{3} + \frac{2}{3} = \frac{4}{3}.$$

(*ii*) Let $I = \int_0^{\pi/2} \cos^3 x\, dx$

$$= \int_0^{\pi/2} \frac{1}{4}[\cos 3x + 3\cos x]\, dx \qquad \left[\begin{array}{l}\because \cos 3A = 4\cos^3 A - 3\cos A \\ \Rightarrow 4\cos^3 A = \cos 3A + 3\cos A \\ \Rightarrow \cos^3 A = \frac{1}{4}[\cos 3A + 3\cos A]\end{array}\right]$$

$$= \frac{1}{4}\int_0^{\pi/2} (\cos 3x + 3\cos x)\, dx$$

$$= \frac{1}{4}\left[\frac{\sin 3x}{3} + 3\sin x\right]_0^{\pi/2}$$

$$= \frac{1}{4}\left[\left(\frac{\sin 3\left(\frac{\pi}{2}\right)}{3} + 3\sin \pi\right) - \left(\frac{\sin 0^\circ}{3} + 3\sin 0^\circ\right)\right]$$

$$\left[\because \ \sin\frac{3\pi}{2} = \sin\left(\pi + \frac{\pi}{2}\right) = -1\right]$$

$$= \frac{1}{4}\left[\left(-\frac{1}{3} + 3\right) - (0+0)\right] = \frac{1}{4}\left[\frac{-1+9}{3}\right]$$

$$= \frac{2}{3}.$$

(*iii*) Let $\quad I = \int_0^{\pi/2} \sin^4 x\, dx = \int_0^{\pi/2} (\sin^2 x)^2\, dx$

$$= \int_0^{\pi/2} \left(\frac{1-\cos 2x}{2}\right)^2 dx \qquad \left[\begin{array}{l}\because \ 1-\cos 2A = 2\sin^2 A \\ \Rightarrow \left(\frac{1-\cos 2A}{2}\right) = \sin^2 A\end{array}\right]$$

$$= \frac{1}{4}\int_0^{\pi/2} (1 + \cos^2 2x - 2\cos 2x)\, dx$$

$$= \frac{1}{4}\int_0^{\pi/2} \left(1 + \frac{1+\cos 4x}{2} - 2\cos 2x\right) dx \qquad \left[\begin{array}{l}\because \ 1+\cos 2A = 2\cos^2 A \\ \Rightarrow 1+\cos 4A = 2\cos^2 2A \\ \Rightarrow \left(\frac{1+\cos 4A}{2}\right) = \cos^2 2A\end{array}\right]$$

$$= \frac{1}{4}\int_0^{\pi/2} \left(1 + \frac{1}{2} + \frac{1}{2}\cos 4x - 2\cos 2x\right) dx = \frac{1}{4}\int_0^{\pi/2} \left(\frac{3}{2} + \frac{1}{2}\cos 4x - 2\cos 2x\right) dx$$

$$= \frac{3}{8}\int_0^{\pi/2} 1 \,.\, dx + \frac{1}{8}\int_0^{\pi/2} \cos 4x\, dx - \frac{1}{2}\int_0^{\pi/2} \cos 2x\, dx$$

$$= \frac{3}{8}\Big[x\Big]_0^{\pi/2} + \frac{1}{8}\left[\frac{\sin 4x}{4}\right]_0^{\pi/2} - \frac{1}{2}\left[\frac{\sin 2x}{2}\right]_0^{\pi/2}$$

$$= \frac{3}{8}\left[\frac{\pi}{2} - 0\right] + \frac{1}{32}\left[\sin 4\left(\frac{\pi}{2}\right) - \sin 0^\circ\right] - \frac{1}{4}\left[\sin 2\left(\frac{\pi}{2}\right) - \sin 0^\circ\right]$$

$$= \frac{3\pi}{16} + \frac{1}{32}[0-0] - \frac{1}{4}[0-0]$$

$$= \frac{3\pi}{16}.$$

(*iv*) Let $\quad I = \int_0^{\pi} \left(\sin^2 \frac{x}{2} - \cos^2 \frac{x}{2}\right) dx$

$$= \int_0^{\pi} \left(\sin^2 \frac{x}{2} - \cos^2 \frac{x}{2}\right) dx = \int_0^{\pi} (-\cos x)\, dx$$

$$= -\int_0^{\pi} \cos x \, dx \qquad \left[\begin{array}{l} \because \cos 2A = \cos^2 A - \sin^2 A \\ \Rightarrow \cos A = \cos^2 \frac{A}{2} - \sin^2 \frac{A}{2} \\ \Rightarrow (-\cos A) = \left(\sin^2 \frac{A}{2} - \cos^2 \frac{A}{2}\right) \end{array}\right]$$

$$= -[\sin x]_0^{\pi} = -[\sin \pi - \sin 0°] = -[0 - 0]$$

$$= 0.$$

Example 8. *Evaluate the following definite integrals :*

(*i*) $\int_0^{\pi/2} sin^3 x\, dx$ (*ii*) $\int_0^{\pi/4} 2\, tan^3 x\, dx$

(*iii*) $\int_0^{\pi/4} \sqrt{1 - sin\, 2x}\, dx$ (*iv*) $\int_0^{\pi/2} \sqrt{1 + sin\, x}\, dx$

(*v*) $\int_{\pi/6}^{\pi/4} cosec\, x\, dx$ (*vi*) $\int_0^1 \frac{1-x}{1+x}\, dx.$

Solution. (*i*) Let $I = \int_0^{\pi/2} \sin^3 x \, dx$

$$= \int_0^{\pi/2} \frac{1}{4} [3 \sin x - \sin 3x]\, dx$$

$$= \frac{1}{4}\int_0^{\pi/2} (3 \sin x - \sin 3x)\, dx \qquad \left[\begin{array}{l} \because \sin 3A = 3 \sin A - 4 \sin^3 A \\ \Rightarrow 4 \sin^3 A = 3 \sin A - \sin 3A \\ \Rightarrow \sin^3 A = \frac{1}{4}[3 \sin A - \sin 3A] \end{array}\right]$$

$$= \frac{3}{4}\int_0^{\pi/2} \sin x \, dx - \frac{1}{4}\int_0^{\pi/2} \sin 3x \, dx = \frac{3}{4}[-\cos x]_0^{\pi/2} - \frac{1}{4}\left[-\frac{\cos 3x}{3}\right]_0^{\pi/2}$$

$$= -\frac{3}{4}\left[\cos \frac{\pi}{2} - \cos 0°\right] + \frac{1}{12}\left[\cos 3\left(\frac{\pi}{2}\right) - \cos 0°\right]$$

$$= -\frac{3}{4}[0 - 1] + \frac{1}{12}[0 - 1] = \frac{3}{4} - \frac{1}{12} = \frac{9-1}{12} = \frac{8}{12}$$

$$= \frac{2}{3}.$$

(*ii*) Let $\quad I = \int_0^{\pi/4} 2\tan^3 x\,dx = 2\int_0^{\pi/4} \tan x \,.\, \tan^2 x\,dx$

$= 2\int_0^{\pi/4} \tan x\,(\sec^2 x - 1)\,dx \qquad [\because \sec^2 A - \tan^2 A = 1]$

$= 2\int_0^{\pi/4} \tan x \sec^2 x\,dx - 2\int_0^{\pi/4} \tan x\,dx$

$= 2\left[\frac{\tan^2 x}{2}\right]_0^{\pi/4} - 2\Big[\log|\sec x|\Big]_0^{\pi/4} \qquad \left[\because \int [f(x)]^n f'(x)\,dx = \frac{[f(x)]^{n+1}}{n+1} + c\right]$

$= \left[\tan^2\left(\frac{\pi}{4}\right) - \tan^2 0^\circ\right] - 2\left[\log\left|\sec\left(\frac{\pi}{4}\right)\right| - \log|\sec 0^\circ|\right]$

$= [(1)^2 - 0] - 2\,[\log(\sqrt{2}) - \log 1]$

$= 1 - 2\log\sqrt{2} + 0 = 1 - 2\log\sqrt{2} = 1 - \log\left(\sqrt{2}\right)^2 \qquad [\because m\log n = \log n^m]$

$= 1 - \log 2.$

(*iii*) Let $\quad I = \int_0^{\pi/4} \sqrt{1 - \sin 2x}\;dx \qquad \left[\because \begin{array}{l}\cos^2 A + \sin^2 A = 1\\ \sin 2A = 2\sin A\cos A\end{array}\right]$

$= \int_0^{\pi/4} \sqrt{1 - \sin 2x}\;dx = \int_0^{\pi/4} \sqrt{\cos^2 x + \sin^2 x - 2\sin x\cos x}\;.\,dx$

$= \int_0^{\pi/4} \sqrt{(\cos x - \sin x)^2}\;dx$

$= \int_0^{\pi/4} (\cos x - \sin x)\,dx$

$= \Big[(\sin x + \cos x)\Big]_0^{\pi/4} \qquad \left[\because \begin{array}{l}\sin\frac{\pi}{4} = \sin 45^\circ = \frac{1}{\sqrt{2}}\\ \cos\frac{\pi}{4} = \cos 45^\circ = \frac{1}{\sqrt{2}}\end{array}\right]$

$= \left[\left(\sin\frac{\pi}{4} + \cos\frac{\pi}{4}\right) - (\sin 0^\circ + \cos 0^\circ)\right]$

$= \left(\frac{1}{\sqrt{2}} + \frac{1}{\sqrt{2}}\right) - (0 + 1) = \frac{2}{\sqrt{2}} - 1 \;=\; \sqrt{2} - 1.$

(*iv*) Let $\quad I = \int_0^{\pi/2} \sqrt{1 + \sin x}\;dx = \int_0^{\pi/2} \sqrt{1 + \sin x}\;dx$

$\left[\begin{array}{l}\because \cos^2 A + \sin^2 A = 1\\ \Rightarrow \cos^2\frac{A}{2} + \sin^2\frac{A}{2} = 1\\ \quad \sin 2A = 2\sin A\cos A\\ \Rightarrow \sin A = 2\sin\frac{A}{2}\cos\frac{A}{2}\end{array}\right]$

$= \int_0^{\pi/2} \sqrt{\left(\cos^2\frac{x}{2} + \sin^2\frac{x}{2} + 2\sin\frac{x}{2}\cos\frac{x}{2}\right)}\;dx$

$$= \int_0^{\pi/2} \sqrt{\left(\cos\frac{x}{2} + \sin\frac{x}{2}\right)^2}\, dx = \int_0^{\pi/2} \left(\cos\frac{x}{2} + \sin\frac{x}{2}\right) . dx$$

$$= \int_0^{\pi/2} \cos\frac{x}{2}\, dx + \int_0^{\pi/2} \sin\frac{x}{2}\, dx = \left[\frac{\sin\frac{x}{2}}{\frac{1}{2}}\right]_0^{\pi/2} + \left[\frac{-\cos\frac{x}{2}}{\frac{1}{2}}\right]_0^{\pi/2}$$

$$= 2\left[\sin\frac{\pi}{4} - \sin 0^\circ\right] - 2\left[\cos\frac{\pi}{4} - \cos 0^\circ\right]$$

$$= 2\left[\frac{1}{\sqrt{2}} - 0\right] - 2\left[\frac{1}{\sqrt{2}} - 1\right] \qquad \left[\because \begin{array}{l} \sin\frac{\pi}{4} = \sin 45^\circ = \frac{1}{\sqrt{2}} \\ \cos\frac{\pi}{4} = \cos 45^\circ = \frac{1}{\sqrt{2}} \end{array}\right]$$

$$= \sqrt{2} - \sqrt{2} + 2 = 2.$$

(*v*) Let $\quad I = \int_{\pi/6}^{\pi/4} \operatorname{cosec} x\, dx$

$$= \int_{\pi/6}^{\pi/4} \operatorname{cosec} x\, dx = \Big[\log|\operatorname{cosec} x - \cot x|\Big]_{\pi/6}^{\pi/4}$$

$$= \left[\log\left|\operatorname{cosec}\frac{\pi}{4} - \cot\frac{\pi}{4}\right|\right] - \left[\log\left|\operatorname{cosec}\frac{\pi}{6} - \cot\frac{\pi}{6}\right|\right]$$

$$= \log|\sqrt{2} - 1| - \log|2 - \sqrt{3}|$$

$$= \log\left|\frac{\sqrt{2}-1}{2-\sqrt{3}}\right|. \qquad \left[\because \log m - \log n = \log\left(\frac{m}{n}\right)\right]$$

Alternative Method :

$$\int_{\pi/6}^{\pi/4} \operatorname{cosec} x\, dx = \Big[\log|\operatorname{cosec} x - \cot x|\Big]_{\pi/6}^{\pi/4}$$

$$= \left[\log\left|\frac{1}{\sin x} - \frac{\cos x}{\sin x}\right|\right]_{\pi/6}^{\pi/4} = \left[\log\left|\frac{1-\cos x}{\sin x}\right|\right]_{\pi/6}^{\pi/4}$$

$$= \log\left|\frac{1-\cos\frac{\pi}{4}}{\sin\frac{\pi}{4}}\right| - \log\left|\frac{1-\cos\frac{\pi}{6}}{\sin\frac{\pi}{6}}\right|$$

$$= \log\left|\frac{1-\frac{1}{\sqrt{2}}}{\frac{1}{\sqrt{2}}}\right| - \log\left|\frac{1-\frac{\sqrt{3}}{2}}{\frac{1}{2}}\right| = \log|\sqrt{2}-1| - \log|2-\sqrt{3}|$$

$$= \log\left|\frac{\sqrt{2}-1}{2-\sqrt{3}}\right|. \qquad \left[\because \log m - \log n = \log\left(\frac{m}{n}\right)\right]$$

(*vi*) Let $\quad I = \int_0^1 \frac{1-x}{1+x}\, dx = -\int_0^1 \frac{x-1}{x+1}\, dx$

$= -\int_0^1 \left(\frac{x+1-2}{x+1}\right) dx$ [Add and subtract 1 to the numerator]

$= -\int_0^1 \left[\frac{x+1}{x+1} - \frac{2}{x+1}\right] dx = -\int_0^1 1\, dx + 2\int_0^1 \frac{1}{x+1}\, dx$

$= -\left[x\right]_0^1 + 2\left[\log|x+1|\right]_0^1$

$= -(1-0) + 2\,[\log(1+1) - \log(0+1)]$

$= -1 + 2\log 2 - 2\log 1$ $\quad [\because \log 1 = 0]$

$= -1 + 2\log 2 = -1 + \log(2)^2$ $\quad [\because m\log n = \log n^m]$

$= -1 + \log 4.$

Example 9. *Evaluate the following definite integrals :*

(*i*) $\int_{\pi/4}^{\pi/2} \sqrt{1+\sin 2x}\, dx$ (*ii*) $\int_0^{\pi/2} \sqrt{1+\cos x}\, dx$

(*iii*) $\int_{\pi/4}^{\pi/2} \cot x\, dx$ (*iv*) $\int_{\pi/4}^{\pi/2} \sqrt{1-\sin 2x}\, dx$

(*v*) $\int_0^{\pi/4} \sqrt{1+\sin 2x}\, dx$ (*vi*) $\int_0^{\pi/2} \sqrt{1-\cos 2x}\, dx.$

Solution. (*i*) Let $I = \int_{\pi/4}^{\pi/2} \sqrt{1+\sin 2x}\, dx$

$= \int_{\pi/4}^{\pi/2} \sqrt{\cos^2 x + \sin^2 x + 2\sin x\cos x}\, dx$ $\quad \left[\begin{array}{l}\because \cos^2 A + \sin^2 A = 1 \\ \sin 2A = 2\sin A\cos A\end{array}\right]$

$= \int_{\pi/4}^{\pi/2} \sqrt{(\cos x + \sin x)^2}\, dx = \int_{\pi/4}^{\pi/2} (\cos x + \sin x)\, dx$

$= \int_{\pi/4}^{\pi/2} \cos x\, dx + \int_{\pi/4}^{\pi/2} \sin x\, dx$

$= \left[\sin x\right]_{\pi/4}^{\pi/2} + \left[-\cos x\right]_{\pi/4}^{\pi/2}$

$= \left[\sin\frac{\pi}{2} - \sin\frac{\pi}{4}\right] - \left[\cos\frac{\pi}{2} - \cos\frac{\pi}{4}\right] = \left(1 - \frac{1}{\sqrt{2}}\right) - \left(0 - \frac{1}{\sqrt{2}}\right)$

$= 1 - \frac{1}{\sqrt{2}} + \frac{1}{\sqrt{2}} = 1.$

(*ii*) Let $\quad I = \int_0^{\pi/2} \sqrt{1+\cos x}\, dx$

$= \int_0^{\pi/2} \sqrt{2\cos^2\frac{x}{2}}\, dx$ $\quad \left[\begin{array}{l}\because 1+\cos 2A = 2\cos^2 A \\ \Rightarrow 1+\cos A = 2\cos^2\frac{A}{2}\end{array}\right]$

$$= \sqrt{2}\int_0^{\pi/2} \cos\frac{x}{2}\, dx = \sqrt{2}\left[\frac{\sin\frac{x}{2}}{\frac{1}{2}}\right]_0^{\pi/2} = 2\sqrt{2}\left[\sin\left(\frac{\pi/2}{2}\right) - \sin 0^\circ\right]$$

$$= 2\sqrt{2}\left[\sin\frac{\pi}{4} - \sin 0^\circ\right] = 2\sqrt{2}\left[\frac{1}{\sqrt{2}} - 0\right]$$

$= 2.$

(*iii*) Let $\quad I = \int_{\pi/4}^{\pi/2} \cot x\, dx$

$$= \int_{\pi/4}^{\pi/2} \cot x\, dx = \Big[\log|(\sin x)|\Big]_{\pi/4}^{\pi/2} = \left[\log\left(\sin\frac{\pi}{2}\right) - \log\left(\sin\frac{\pi}{4}\right)\right]$$

$$= \left[\log 1 - \log\frac{1}{\sqrt{2}}\right] = 0 - \log(2)^{-1/2} = -\left(-\frac{1}{2}\right)\log 2 \qquad [\because\ n\log m = \log m^n]$$

$$= \frac{1}{2}\log 2.$$

(*iv*) Let $\quad I = \int_{\pi/4}^{\pi/2} \sqrt{1 - \sin 2x}\; dx \qquad \left[\begin{array}{l}\because\ \cos^2 A + \sin^2 A = 1\\ \quad \sin 2A = 2\sin A\cos A\end{array}\right]$

$$= \int_{\pi/4}^{\pi/2} \sqrt{\cos^2 x + \sin^2 x - 2\sin x\cos x}\; dx$$

$$= \int_{\pi/4}^{\pi/2} \sqrt{(\cos x - \sin x)^2}\; dx = \int_{\pi/4}^{\pi/2} |\cos x - \sin x|\; dx \qquad \left[\because\ \sqrt{x^2} = |x|\right]$$

$$= \int_{\pi/4}^{\pi/2} -(\cos x - \sin x)\, dx \qquad \left[\begin{array}{l}\because\ \cos x < \sin x \quad \text{for } \frac{\pi}{4} < x < \frac{\pi}{2}\\ \therefore\ \cos x - \sin x < 0\\ \Rightarrow |\cos x - \sin x| = -(\cos x - \sin x)\end{array}\right]$$

$$= \int_{\pi/4}^{\pi/2} \sin x\, dx - \int_{\pi/4}^{\pi/2} \cos x\, dx$$

$$= \Big[-\cos x\Big]_{\pi/4}^{\pi/2} - \Big[\sin x\Big]_{\pi/4}^{\pi/2} = -\left[\cos\frac{\pi}{2} - \cos\frac{\pi}{4}\right] - \left[\sin\frac{\pi}{2} - \sin\frac{\pi}{4}\right]$$

$$= -\left[0 - \frac{1}{\sqrt{2}}\right] - \left[1 - \frac{1}{\sqrt{2}}\right] = \frac{1}{\sqrt{2}} - 1 + \frac{1}{\sqrt{2}} = \frac{2}{\sqrt{2}} - 1$$

$$= \sqrt{2} - 1.$$

(*v*) Let $\quad I = \int_0^{\pi/4} \sqrt{1 + \sin 2x}\; dx$

$$= \int_0^{\pi/4} \sqrt{\cos^2 x + \sin^2 x + 2\sin x\cos x}\; dx \qquad \left[\begin{array}{l}\because\ \cos^2 A + \sin^2 A = 1\\ \quad \sin 2A = 2\sin A\cos A\end{array}\right]$$

$$= \int_0^{\pi/4} \sqrt{(\cos x + \sin x)^2}\, dx = \int_0^{\pi/4} (\cos x + \sin x)\, dx$$

$$= \int_0^{\pi/4} \cos x\, dx + \int_0^{\pi/4} \sin x\, dx = \left[\sin x\right]_0^{\pi/4} + \left[-\cos x\right]_0^{\pi/4}$$

$$= \left[\sin\frac{\pi}{4} - \sin 0^\circ\right] - \left[\cos\frac{\pi}{4} - \cos 0^\circ\right] = \left(\frac{1}{\sqrt{2}} - 0\right) - \left(\frac{1}{\sqrt{2}} - 1\right)$$

$$= \frac{1}{\sqrt{2}} - \frac{1}{\sqrt{2}} + 1 = 1.$$

(*vi*) Let $\quad I = \int_0^{\pi/2} \sqrt{1 - \cos 2x}\, dx$

$$= \int_0^{\pi/2} \sqrt{2\sin^2 x}\, dx = \sqrt{2}\int_0^{\pi/2} \sin x\, dx \qquad [\because \; 1 - \cos 2A = 2\sin^2 A]$$

$$= \sqrt{2}\left[-\cos x\right]_0^{\pi/2} = -\sqrt{2}\left[\cos x\right]_0^{\pi/2} = -\sqrt{2}\left[\cos\frac{\pi}{2} - \cos 0^\circ\right]$$

$$= -\sqrt{2}\,[0 - 1] = \sqrt{2}.$$

Example 10. *Evaluate the following definite integrals :*

(*i*) $\int_0^{\pi} \frac{1}{1 + \sin x}\, dx$ $\qquad$ (*ii*) $\int_0^{\pi/4} \sin 2x \sin 3x\, dx$

(*iii*) $\int_{-\pi/4}^{+\pi/4} \frac{1}{1 + \sin x}\, dx.$

Solution. (*i*) Let $I = \int_0^{\pi} \frac{1}{1 + \sin x}\, dx$

$$= \int_0^{\pi} \frac{1}{1 + \sin x} \times \frac{1 - \sin x}{1 - \sin x}\, dx \qquad \text{[Rationalization]}$$

$$= \int_0^{\pi} \frac{1 - \sin x}{1 - \sin^2 x}\, dx = \int_0^{\pi} \frac{1 - \sin x}{\cos^2 x}\, dx \qquad [\because \; \cos^2 A + \sin^2 A = 1]$$

$$= \int_0^{\pi} \left(\frac{1}{\cos^2 x} - \frac{\sin x}{\cos x} \cdot \frac{1}{\cos x}\right) dx = \int_0^{\pi} (\sec^2 x - \tan x \sec x)\, dx$$

$$= \int_0^{\pi} \sec^2 x\, dx - \int_0^{\pi} \tan x \sec x\, dx$$

$$= \left[\tan x\right]_0^{\pi} - \left[\sec x\right]_0^{\pi} = (\tan \pi - \tan 0^\circ) - (\sec \pi - \sec 0^\circ)$$

$$= (0 - 0) - (-1 - 1) = -(-2)$$

$$= 2.$$

(*ii*) Let $\quad I = \int_0^{\pi/4} \sin 2x \sin 3x\, dx$

$$= \frac{1}{2}\int_0^{\pi/4} 2\sin 3x \sin 2x\, dx \qquad \text{[Multiply and divided by 2]}$$

$$= \frac{1}{2}\int_0^{\pi/4} [\cos(3x-2x) - \cos(3x+2x)]\,dx \qquad \left[\because\ 2\sin A\sin B = \cos(A-B) - \cos(A+B)\right]$$

$$= \frac{1}{2}\int_0^{\pi/4} \cos x\,dx - \frac{1}{2}\int_0^{\pi/4}\cos 5x\,dx = \frac{1}{2}[\sin x]_0^{\pi/4} - \frac{1}{2}\left[\frac{\sin 5x}{5}\right]_0^{\pi/4}$$

$$= \frac{1}{2}\left[\sin\frac{\pi}{4} - \sin 0^\circ\right] - \frac{1}{10}\left[\sin\frac{5\pi}{4} - \sin 0^\circ\right]$$

$$= \frac{1}{2}\left[\frac{1}{\sqrt{2}} - 0\right] - \frac{1}{10}\left[-\frac{1}{\sqrt{2}} - 0\right] \qquad \left[\because\ \sin\frac{5\pi}{4} = \sin\left(\pi + \frac{\pi}{4}\right) = -\frac{1}{\sqrt{2}}\right]$$

$$= \frac{1}{2\sqrt{2}} + \frac{1}{10\sqrt{2}} = \frac{5+1}{10\sqrt{2}} = \frac{6}{10\sqrt{2}} = \frac{3}{5\sqrt{2}}.$$

(*iii*) Let $I = \int_{-\pi/4}^{\pi/4} \frac{1}{1+\sin x}\,dx$

$$= \int_{-\pi/4}^{\pi/4} \frac{1}{1+\sin x} \times \frac{1-\sin x}{1-\sin x}\,dx \qquad \text{[Rationalization]}$$

$$= \int_{-\pi/4}^{\pi/4} \frac{1-\sin x}{1-\sin^2 x}\,dx = \int_{-\pi/4}^{\pi/4} \frac{1-\sin x}{\cos^2 x}\,dx$$

$$= \int_{-\pi/4}^{\pi/4}\left(\frac{1}{\cos^2 x} - \frac{\sin x}{\cos^2 x}\right)dx = \int_{-\pi/4}^{\pi/4}\sec^2 x\,dx - \int_{-\pi/4}^{\pi/4}\sec x\tan x\,dx$$

$$= \Big[\tan x\Big]_{-\pi/4}^{\pi/4} - \Big[\sec x\Big]_{-\pi/4}^{\pi/4} = \left[\tan\frac{\pi}{4} - \tan\left(-\frac{\pi}{4}\right)\right] - \left[\sec\frac{\pi}{4} - \sec\left(-\frac{\pi}{4}\right)\right]$$

$$= [1-(-1)] - \left[\sqrt{2} - \sqrt{2}\right] = 1 + 1 - \sqrt{2} + \sqrt{2}$$

$$= 2.$$

Example 11. (*i*) *If* $\int_0^a \frac{1}{2+8x^2}\,dx = \frac{\pi}{16}$, *find the value of a.*

(*ii*) *If* $\int_a^b x^3\,dx = 0$ *and* $\int_a^b x^2\,dx = \frac{2}{3}$, *then find the values of a and b.*

(*iii*) *If* $\int_1^m (3x^2 + 2x + 1)\,dx = 11$, *find the value of m.*

(*iv*) *Evaluate* : $\int_0^4 \frac{2x+3}{5x^2+1}\,dx$.

Solution. (*i*) Let $I = \int_0^a \frac{1}{2+8x^2}\,dx$

$$= \frac{1}{8}\int_0^a \frac{1}{\left(\frac{1}{4}+x^2\right)}\,dx = \frac{1}{8}\int_0^a \frac{1}{\left(\frac{1}{2}\right)^2 + x^2}\,dx$$

$$\left[\because \int \frac{1}{a^2+x^2}\,dx = \frac{1}{a}\tan^{-1}\frac{x}{a} + c\right]$$

$$= \frac{1}{8}\cdot\left[\frac{1}{1/2}\tan^{-1}\frac{x}{1/2}\right]_0^a = \frac{2}{8}\left[\tan^{-1}2x\right]_0^a$$

$$= \frac{1}{4}\left[\tan^{-1}2a - \tan^{-1}0\right] = \frac{1}{4}\tan^{-1}2a.$$

Since it is given that $I = \frac{\pi}{16}$

$$\therefore \quad \frac{1}{4}\tan^{-1}2a = \frac{\pi}{16}$$

$$\Rightarrow \quad \tan^{-1}2a = \frac{\pi}{4}$$

$$\Rightarrow \quad 2a = \tan\left(\frac{\pi}{4}\right) \Rightarrow 2a = 1 \Rightarrow a = \frac{1}{2}.$$

(*ii*) Let $\quad I_1 = \int_a^b x^3\,dx$

$$= \left[\frac{x^4}{4}\right]_a^b = \frac{1}{4}\left[x^4\right]_a^b = \frac{1}{4}[b^4 - a^4]$$

Since, it is given that, $I_1 = 0$

$$\therefore \quad \frac{1}{4}[b^4 - a^4] = 0 \Rightarrow \quad b^4 - a^4 = 0$$

$$\Rightarrow \quad b^4 - a^4 = 0$$

$$\Rightarrow \quad (b^2 - a^2)(b^2 + a^2) = 0$$

$$\Rightarrow \quad (b-a)(b+a)(b^2+a^2) = 0 \qquad \text{...(1)}$$

Let $\quad I_2 = \int_a^b x^2\,dx$

$$= \int_a^b x^2\,dx = \left[\frac{x^3}{3}\right]_a^b = \frac{1}{3}\left[x^3\right]_a^b = \frac{1}{3}[b^3 - a^3]$$

Also, it is given that : $I_2 = \frac{2}{3}$.

$$\therefore \quad \frac{1}{3}[b^3 - a^3] = \frac{2}{3} \Rightarrow b^3 - a^3 = 2. \qquad \text{...(2)}$$

Now, from equation (1), we have

$$(b-a)(b+a)(b^2+a^2) = 0 \qquad \left[\begin{array}{l}\because (b-a)=0 \Rightarrow b^3-a^3 \neq 2\\ \because (b^2+a^2 \neq 0)\end{array}\right]$$

$$\Rightarrow \quad (b + a) = 0$$
$$\Rightarrow \quad b = -a$$

$\therefore$ From equation (2), we get

$$(-a)^3 - a^3 = 2 \;\Rightarrow\; -2a^3 = 2 \;\Rightarrow\; a^3 = -1 \;\Rightarrow\; a = -1.$$

$$\therefore \quad b = -a$$
$$\Rightarrow \quad b = -(-1) = 1.$$

(*iii*) It is given that $\int_1^m (3x^2 + 2x + 1)\, dx = 11$

$$\Rightarrow \quad \left[\frac{3x^3}{3} + \frac{2x^2}{2} + x\right]_1^m = 11$$

$$\Rightarrow \quad \left[x^3 + x^2 + x\right]_1^m = 11$$

$$\Rightarrow \quad [(m^3 + m^2 + m) - ((1)^3 + (1)^2 + 1)] = 11$$
$$\Rightarrow \quad m^3 + m^2 + m - 3 = 11$$
$$\Rightarrow \quad m^3 + m^2 + m - 14 = 0$$
$$\Rightarrow \quad (m - 2)(m^2 + 3m + 7) = 0$$
$$\Rightarrow \quad m - 2 = 0, \quad m^2 + 3m + 7 = 0$$

$$\Rightarrow \quad m = 2, \quad m = \frac{-b \pm \sqrt{b^2 - 4ac}}{2a}$$

$$\Rightarrow \quad m = \frac{-3 \pm \sqrt{9 - 28}}{2}$$

$$\Rightarrow \quad m = \frac{-3 \pm \sqrt{-19}}{2}$$

$$\Rightarrow \quad m = -\frac{3}{2} \pm \frac{\sqrt{19}}{2}\, i\,.$$

(*iv*) Let $\quad I = \int_0^4 \frac{2x + 3}{5x^2 + 1}\, dx$

$$= \int_0^4 \left(\frac{2x}{5x^2 + 1} + \frac{3}{5x^2 + 1}\right) dx$$

$$= \int_0^4 \frac{2x}{5x^2 + 1}\, dx + 3\int_0^4 \frac{1}{5x^2 + 1}\, dx$$

$$= \frac{1}{5}\int_0^4 \frac{10x}{5x^2 + 1}\, dx + \frac{3}{5}\int_0^4 \frac{1}{x^2 + \left(\frac{1}{\sqrt{5}}\right)^2}\, dx$$

[Multiply and divide the first integral by 5]

$$= \frac{1}{5}\left[\log|5x^2+1|\right]_0^4 + \frac{3}{5}\left[\frac{1}{1/\sqrt{5}}\tan^{-1}\frac{x}{1/\sqrt{5}}\right]_0^4$$

$$\left[\because \int \frac{f'(x)}{f(x)}\,dx = \log|f(x)|+c \quad \int \frac{1}{x^2+a^2}\,dx = \frac{1}{a}\tan^{-1}\frac{x}{a}+c\right]$$

$$= \frac{1}{5}[\log(5(4)^2+1) - \log(5(0)+1)] + \frac{3}{5}\sqrt{5}\left[\tan^{-1}\sqrt{5}x\right]_0^4$$

$$= \frac{1}{5}[\log 81 - \log 1] + \frac{3}{\sqrt{5}}[\tan^{-1}4\sqrt{5} - \tan^{-1}0]$$

$$= \frac{1}{5}\log 81 + \frac{3}{\sqrt{5}}\tan^{-1}4\sqrt{5}. \qquad [\because \log 1 = 0]$$

Example 12. *Evaluate the following definite integrals :*

(*i*) $\int_0^{\pi/2} \frac{\sin^2 x}{1+\cos x}\,dx$ (*ii*) $\int_{\pi/2}^{\pi} \frac{1+\cos x}{1-\cos x}\,dx$

(*iii*) $\int_0^{\sqrt{2}} \sqrt{2-x^2}\,dx$ (*iv*) $\int_0^{\infty} e^{-x}\,dx$

(*v*) $\int_3^8 \frac{x}{\sqrt{1+x}}\,dx.$

Solution. (*i*) Let $I = \int_0^{\pi/2} \frac{\sin^2 x}{1+\cos x}\,dx$

$$= \int_0^{\pi/2} \frac{\sin^2 x}{1+\cos x}\,dx = \int_0^{\pi/2} \frac{1-\cos^2 x}{1+\cos x}\,dx \qquad [\because \cos^2 A + \sin^2 A = 1]$$

$$= \int_0^{\pi/2} \frac{(1-\cos x)(1+\cos x)}{(1+\cos x)}\,dx = \int_0^{\pi/2} (1-\cos x)\,dx$$

$$= \int_0^{\pi/2} 1.\,dx - \int_0^{\pi/2} \cos x\,dx$$

$$= \left[x\right]_0^{\pi/2} - \left[\sin x\right]_0^{\pi/2} = \left[\frac{\pi}{2} - 0\right] - \left[\sin\frac{\pi}{2} - \sin 0\right]$$

$$= \frac{\pi}{2} - 1.$$

(*ii*) Let $I = \int_{\pi/2}^{\pi} \frac{1+\cos x}{1-\cos x}\,dx$

$$\left[\begin{array}{l}\because\ 1+\cos 2A = 2\cos^2 A \\ \quad 1+\cos A = 2\cos^2\frac{A}{2} \\ \quad 1-\cos 2A = 2\sin^2 A \\ \Rightarrow 1-\cos 2A = 2\sin^2\frac{A}{2}\end{array}\right]$$

$$= \int_{\pi/2}^{\pi} \frac{2\cos^2\frac{x}{2}}{2\sin^2\frac{x}{2}}\,dx$$

$$= \int_{\pi/2}^{\pi} \cot^2 \frac{x}{2}\, dx$$

$$= \int_{\pi/2}^{\pi} \left(\operatorname{cosec}^2 \frac{x}{2} - 1\right) dx \qquad [\because \quad \operatorname{cosec}^2 A - \cot^2 A = 1]$$

$$= \int_{\pi/2}^{\pi} \operatorname{cosec}^2 \frac{x}{2}\, dx - \int_{\pi/2}^{\pi} 1 \,.\, dx = \left[\frac{-\cot \frac{x}{2}}{\frac{1}{2}}\right]_{\pi/2}^{\pi} - \left[x\right]_{\pi/2}^{\pi}$$

$$= -2\left[\cot \frac{\pi}{2} - \cot \frac{\pi}{4}\right] - \left[\pi - \frac{\pi}{2}\right] = -2\,[0 - 1] - \left[\pi - \frac{\pi}{2}\right]$$

$$= 2 - \frac{\pi}{2}\,.$$

(*iii*) Let $$I = \int_0^{\sqrt{2}} \sqrt{2 - x^2}\, dx = \int_0^{\sqrt{2}} \sqrt{\left(\sqrt{2}\right)^2 - x^2}\, dx$$

$$= \left[\frac{x}{2}\sqrt{2 - x^2} + \frac{2}{2}\sin^{-1}\frac{x}{\sqrt{2}}\right]_0^{\sqrt{2}}$$

$$\left[\because \int \sqrt{a^2 - x^2}\, dx = \frac{x\sqrt{a^2 - x^2}}{2} + \frac{a^2}{2}\sin^{-1}\frac{x}{a} + c\right]$$

$$= \left(\frac{\sqrt{2}}{2}\sqrt{2 - \left(\sqrt{2}\right)^2} + \sin^{-1}\frac{\sqrt{2}}{\sqrt{2}}\right) - \left(\frac{0}{2}\sqrt{2 - 0} + \sin^{-1} 0\right)$$

$$= (0 + \sin^{-1} 1) - (0)$$

$$= \sin^{-1} 1 = \frac{\pi}{2}\,.$$

(*iv*) Let $$I = \int_0^{\infty} e^{-x}\, dx$$

$$= \left[\frac{e^{-x}}{-1}\right]_0^{\infty} = -\left[\frac{1}{e^x}\right]_0^{\infty} = -\left[\frac{1}{e^{\infty}} - \frac{1}{e^0}\right]$$

$$= -[0 - 1] = 1.$$

(*v*) Let $$I = \int_3^8 \frac{x}{\sqrt{1 + x}}\, dx$$

$$= \int_3^8 \frac{1 + x - 1}{\sqrt{1 + x}}\, dx \qquad \text{[Add and subtract 1 to the numerator]}$$

$$= \int_3^8 \left(\frac{1 + x}{\sqrt{1 + x}} - \frac{1}{\sqrt{1 + x}}\right) dx = \int_3^8 \sqrt{1 + x}\, dx - \int_3^8 (1 + x)^{-1/2}\, dx$$

$$= \left[\frac{(1 + x)^{\frac{1}{2}+1}}{\frac{1}{2} + 1}\right]_3^8 - \left[\frac{(1 + x)^{-\frac{1}{2}+1}}{-\frac{1}{2} + 1}\right]_3^8 = \frac{2}{3}\left[(1 + x)^{3/2}\right]_3^8 - 2\left[(1 + x)^{1/2}\right]_3^8$$

$$= \frac{2}{3}[(1+8)^{3/2} - (1+3)^{3/2}] - 2[(1+8)^{1/2} - (1+3)^{1/2}]$$

$$= \frac{2}{3}[27 - 8] - 2[3 - 2] \qquad \left[\because \begin{array}{l} (9)^{3/2} = 9\sqrt{9} = 9(3) = 27 \\ (4)^{3/2} = 4\sqrt{4} = 4(2) = 8 \end{array}\right]$$

$$= \frac{2}{3}(19) - 2(1) = \frac{38}{3} - 2 = \frac{38-6}{3}$$

$$= \frac{32}{3}.$$

Example 13. *Evaluate the following definite integrals :*

(i) $\int_0^1 \frac{1}{[ax + b(1-x)]^2}\, dx$ (ii) $\int_0^1 \frac{1}{1 + x + x^2}\, dx$

(iii) $\int_2^3 \frac{x}{x^2 + 1}\, dx$ (iv) $\int_0^1 \frac{2x + 3}{5x^2 + 1}\, dx$

(v) $\int_0^2 \frac{5x + 1}{x^2 + 4}\, dx$ (vi) $\int_0^{\pi/4} (2\sec^2 x + x^3 + 2)\, dx.$

Solution. (*i*) Let $I = \int_0^1 \frac{1}{[ax + b(1-x)]^2}\, dx$

$$= \int_0^1 \frac{1}{(ax + b - bx)^2}\, dx = \int_0^1 \frac{1}{[(a-b)x + b]^2}\, dx$$

$$= \int_0^1 [(a-b)x + b]^{-2}\, dx = \left[\frac{[(a-b)x + b]^{-2+1}}{(-2+1)(a-b)}\right]_0^1$$

$$= \frac{1}{b-a}\left[\frac{1}{(a-b)x + b}\right]_0^1 = \frac{1}{b-a}\left[\frac{1}{a-b+b} - \frac{1}{(a-b)0 + b}\right] = \frac{1}{b-a}\left[\frac{1}{a} - \frac{1}{b}\right]$$

$$= \frac{1}{b-a}\left(\frac{b-a}{ab}\right) = \frac{1}{ab}.$$

(*ii*) Let $I = \int_0^1 \frac{1}{1 + x + x^2}\, dx = \int_0^1 \frac{1}{x^2 + x + 1}\, dx$

$$= \int_0^1 \frac{1}{\left(x^2 + x + \frac{1}{4}\right) + \left(1 - \frac{1}{4}\right)}\, dx \qquad \left[\begin{array}{l} \text{Add and subtract } \frac{1}{4} \text{ to the denom.} \\ \because \left(\frac{1}{2}\text{ co-eff. of } x\right)^2 = \frac{1}{4} \end{array}\right]$$

$$= \int_0^1 \frac{1}{\left(x + \frac{1}{2}\right)^2 + \frac{3}{4}}\, dx = \int_0^1 \frac{1}{\left(x + \frac{1}{2}\right)^2 + \left(\frac{\sqrt{3}}{2}\right)^2}\, dx$$

$$= \left[\frac{1}{\frac{\sqrt{3}}{2}} \tan^{-1}\left(\frac{\left(x+\frac{1}{2}\right)}{\left(\frac{\sqrt{3}}{2}\right)}\right)\right]_0^1 \qquad \left[\because \int \frac{1}{x^2+a^2}\,dx = \frac{1}{a}\tan^{-1}\frac{x}{a}+c\right]$$

$$= \frac{2}{\sqrt{3}}\left[\tan^{-1}\left(\frac{2x+1}{\sqrt{3}}\right)\right]_0^1 = \frac{2}{\sqrt{3}}\left[\tan^{-1}\left(\frac{2+1}{\sqrt{3}}\right) - \tan^{-1}\left(\frac{0+1}{\sqrt{3}}\right)\right]$$

$$= \frac{2}{\sqrt{3}}\left[\tan^{-1}\sqrt{3} - \tan^{-1}\frac{1}{\sqrt{3}}\right]$$

$$= \frac{2}{\sqrt{3}}\left[\frac{\pi}{3} - \frac{\pi}{6}\right] = \frac{2}{\sqrt{3}}\left(\frac{\pi}{6}\right) \qquad \left[\begin{array}{l}\because \tan\frac{\pi}{6} = \tan 30^\circ = \frac{1}{\sqrt{3}} \\ \tan\frac{\pi}{3} = \tan 60^\circ = \sqrt{3}\end{array}\right]$$

$$= \frac{\pi}{3\sqrt{3}}.$$

(*iii*) Let $\quad I = \int_2^3 \frac{x}{x^2+1}\,dx$

$$= \frac{1}{2}\int_2^3 \frac{2x}{x^2+1}\,dx \qquad \text{[Multiply and divided by 2]}$$

$$= \frac{1}{2}\left[\log|(x^2+1)|\right]_2^3 \qquad \left[\because \int \frac{f'(x)}{f(x)}\,dx = \log|f(x)|+c\right]$$

$$= \frac{1}{2}\left[\log(3^2+1) - \log(2^2+1)\right] = \frac{1}{2}\,[\log 10 - \log 5]$$

$$= \frac{1}{2}\log\left(\frac{10}{5}\right) = \frac{1}{2}\log 2. \qquad \left[\because \log m - \log n = \log\left(\frac{m}{n}\right)\right]$$

(*iv*) Let $\quad I = \int_0^1 \frac{2x+3}{5x^2+1}\,dx$

$$= \int_0^1 \left(\frac{2x}{5x^2+1} + \frac{3}{5x^2+1}\right) dx = \int_0^1 \frac{2x}{5x^2+1}\,dx + 3\int_0^1 \frac{1}{5x^2+1}\,dx$$

$$= \frac{1}{5}\int_0^1 \frac{10x}{5x^2+1}\,dx + \frac{3}{5}\int_0^1 \frac{1}{x^2+\left(\frac{1}{\sqrt{5}}\right)^2}\,dx \qquad \left[\begin{array}{l}\text{Multiply and divide} \\ \text{the first integral by 5}\end{array}\right]$$

$$= \frac{1}{5}\left[\log|5x^2+1|\right]_0^1 + \frac{3}{5}\left[\frac{1}{1/\sqrt{5}}\tan^{-1}\frac{x}{1/\sqrt{5}}\right]_0^1$$

$$\left[\begin{array}{l}\because \int \frac{f'(x)}{f(x)}\,dx = \log|f(x)|+c \\ \int \frac{1}{a^2+x^2}\,dx = \frac{1}{a}\tan^{-1}\frac{x}{a}+c\end{array}\right]$$

$$= \frac{1}{5}\,[\log(5+1) - \log(0+1)] + \frac{3}{5}\,.\,\sqrt{5}\left[\tan^{-1}\sqrt{5}x\right]_0^1$$

$$= \frac{1}{5}[\log 6 - \log 1] + \frac{3}{\sqrt{5}}[\tan^{-1}\sqrt{5} - \tan^{-1} 0]$$

$$= \frac{1}{5}\log 6 + \frac{3}{\sqrt{5}}\tan^{-1}\sqrt{5}. \qquad [\because \log 1 = 0]$$

(*v*) Let $\quad I = \int_0^2 \frac{5x+1}{x^2+4}\,dx$

$$= \int_0^2 \left(\frac{5x}{x^2+4} + \frac{1}{x^2+4}\right) dx = 5\int_0^2 \frac{x}{x^2+4}\,dx + \int_0^2 \frac{1}{x^2+4}\,dx$$

$$= \frac{5}{2}\int_0^2 \frac{2x}{x^2+4}\,dx + \int_0^2 \frac{1}{x^2+2^2}\,dx$$ [Multiply and divide the first integral by 2]

$$= \frac{5}{2}\left[\log|x^2+4|\right]_0^2 + \left[\frac{1}{2}\tan^{-1}\frac{x}{2}\right]_0^2$$

$$\left[\because \int \frac{f'(x)}{f(x)}\,dx = \log|f(x)| + c,\quad \int \frac{1}{a^2+x^2}\,dx = \frac{1}{a}\tan^{-1}\frac{x}{a} + c\right]$$

$$= \frac{5}{2}[\log|(2)^2+4| - \log|0+4|] + \frac{1}{2}\left[\tan^{-1}\frac{2}{2} - \tan^{-1} 0\right]$$

$$= \frac{5}{2}\,[\log 8 - \log 4] + \frac{1}{2}\,[\tan^{-1} 1 - 0]$$

$$= \frac{5}{2}\log\left(\frac{8}{4}\right) + \frac{1}{2}\left(\frac{\pi}{4}\right)$$

$$\left[\because \log m - \log n = \log\frac{m}{n},\quad \tan\frac{\pi}{4} = \tan 45^\circ = 1\right]$$

$$= \frac{5}{2}\log 2 + \frac{\pi}{8}.$$

(*vi*) Let $\quad I = \int_0^{\pi/4} (2\sec^2 x + x^3 + 2)\,dx$

$$= 2\int_0^{\pi/4} \sec^2 x\,dx + \int_0^{\pi/4} x^3\,dx + 2\int_0^{\pi/4} 1\,.\,dx$$

$$= 2\left[\tan x\right]_0^{\pi/4} + \left[\frac{x^4}{4}\right]_0^{\pi/4} + 2\left[x\right]_0^{\pi/4}$$

$$= 2\left[\tan\frac{\pi}{4} - \tan 0^\circ\right] + \frac{1}{4}\left[\left(\frac{\pi}{4}\right)^4 - 0\right] + 2\left[\frac{\pi}{4} - 0\right]$$

$$= 2\,(1-0) + \frac{1}{4}\left[\frac{\pi^4}{256} - 0\right] + \frac{\pi}{2} = 2 + \frac{\pi^4}{1024} + \frac{\pi}{2}$$

$$= \frac{\pi^4}{1024} + \frac{\pi}{2} + 2.$$

Example 14. *Evaluate the following :*

(i) $\int_0^2 \frac{1}{4+x-x^2}\,dx$ (ii) $\int_0^2 \frac{1}{\sqrt{3+2x-x^2}}\,dx$

(iii) $\int_1^2 \frac{1}{\sqrt{x^2+4x+3}}\,dx$ (iv) $\int_0^4 \frac{1}{\sqrt{x^2+2x+3}}\,dx$

(v) $\int_{1/4}^{1/2} \frac{1}{\sqrt{x-x^2}}\,dx$ (vi) $\int_0^a \frac{1}{\sqrt{ax-x^2}}\,dx.$

Solution. (i) Let $I = \int_0^2 \frac{1}{4+x-x^2}\,dx = \int_0^2 \frac{1}{4-(x^2-x)}\,dx$

$$= \int_0^2 \frac{1}{\left(4+\frac{1}{4}\right)-\left(x^2-x+\frac{1}{4}\right)}\,dx \qquad \left[\begin{array}{l}\text{Add and subtract } \frac{1}{4} \text{ to the denominator} \\ \because \left(\frac{1}{2}\text{ co-eff. of } x\right)^2 = \frac{1}{4}\end{array}\right]$$

$$= \int_0^2 \frac{1}{\frac{17}{4}-\left(x-\frac{1}{2}\right)^2}\,dx = \int_0^2 \frac{1}{\left(\frac{\sqrt{17}}{2}\right)^2-\left(x-\frac{1}{2}\right)^2}\,dx$$

$$= \left[\frac{1}{2.\left(\frac{\sqrt{17}}{2}\right)} . \log\left|\frac{\frac{\sqrt{17}}{2}+\left(x-\frac{1}{2}\right)}{\frac{\sqrt{17}}{2}-\left(x-\frac{1}{2}\right)}\right|\right]_0^2 \qquad \left[\because \int \frac{1}{a^2-x^2}\,dx = \frac{1}{2a}\log\left|\frac{a+x}{a-x}\right| + c\right]$$

$$= \frac{1}{\sqrt{17}}\left[\log\left|\frac{\frac{\sqrt{17}}{2}+\left(2-\frac{1}{2}\right)}{\frac{\sqrt{17}}{2}-\left(2-\frac{1}{2}\right)}\right| - \log\left|\frac{\frac{\sqrt{17}}{2}+\left(0-\frac{1}{2}\right)}{\frac{\sqrt{17}}{2}-\left(0-\frac{1}{2}\right)}\right|\right]$$

$$= \frac{1}{\sqrt{17}}\left[\log\left|\frac{\frac{\sqrt{17}}{2}+\frac{3}{2}}{\frac{\sqrt{17}}{2}-\frac{3}{2}}\right| - \log\left|\frac{\frac{\sqrt{17}}{2}-\frac{1}{2}}{\frac{\sqrt{17}}{2}+\frac{1}{2}}\right|\right]$$

$$= \frac{1}{\sqrt{17}}\left[\log\left(\frac{\sqrt{17}+3}{\sqrt{17}-3}\right) - \log\left(\frac{\sqrt{17}-1}{\sqrt{17}+1}\right)\right]$$

$$= \frac{1}{\sqrt{17}}\left[\log\left(\frac{\frac{\sqrt{17}+3}{\sqrt{17}-3}}{\frac{\sqrt{17}-1}{\sqrt{17}+1}}\right)\right] \qquad \left[\because \log m - \log n = \log\left(\frac{m}{n}\right)\right]$$

$$= \frac{1}{\sqrt{17}} \log\left(\frac{\sqrt{17}+3}{\sqrt{17}-3} \times \frac{\sqrt{17}+1}{\sqrt{17}-1}\right) = \frac{1}{\sqrt{17}} \log\left[\frac{17+3\sqrt{17}+\sqrt{17}+3}{17-3\sqrt{17}-\sqrt{17}+3}\right]$$

$$= \frac{1}{\sqrt{17}} \log\left(\frac{20+4\sqrt{17}}{20-4\sqrt{17}}\right) = \frac{1}{\sqrt{17}} \log\left[\frac{4(5+\sqrt{17})}{4(5-\sqrt{17})}\right]$$

$$= \frac{1}{\sqrt{17}} \log\left(\frac{5+\sqrt{17}}{5-\sqrt{17}}\right).$$

(*ii*) Let $I = \int_0^2 \frac{1}{\sqrt{3+2x-x^2}}\, dx = \int_0^2 \frac{1}{\sqrt{-(x^2-2x-3)}}\, dx$

$$= \int_0^2 \frac{1}{\sqrt{-[x^2-2x+1-3-1]}}\, dx \qquad \left[\begin{array}{l}\text{Add and subtract 1 to the denom.} \\ \because \left(\frac{1}{2}\text{ co-eff. of } x\right)^2 = 1\end{array}\right]$$

$$= \int_0^2 \frac{1}{\sqrt{-[(x-1)^2-4]}}\, dx = \int_0^2 \frac{1}{\sqrt{(2)^2-(x-1)^2}}\, dx$$

$$= \left[\sin^{-1}\left(\frac{x-1}{2}\right)\right]_0^2 \qquad \left[\because \int \frac{1}{\sqrt{a^2-x^2}}\, dx = \sin^{-1}\frac{x}{a} + c\right]$$

$$= \left[\sin^{-1}\left(\frac{2-1}{2}\right) - \sin^{-1}\left(\frac{0-1}{2}\right)\right]$$

$$= \sin^{-1}\frac{1}{2} - \sin^{-1}\left(-\frac{1}{2}\right) = \frac{\pi}{6} - \left(-\frac{\pi}{6}\right) = \frac{\pi}{6} + \frac{\pi}{6} \qquad \left[\because \sin\frac{\pi}{6} = \sin 30^\circ = \frac{1}{2}\right]$$

$$= \frac{2\pi}{6} = \frac{\pi}{3}.$$

(*iii*) Let $I = \int_1^2 \frac{1}{\sqrt{x^2+4x+3}}\, dx$

$$= \int_1^2 \frac{1}{\sqrt{(x^2+4x+4)+(3-4)}}\, dx \qquad \left[\begin{array}{l}\text{Add and subtract 4 to the denom.} \\ \because \left(\frac{1}{2}\text{ co-eff. of } x\right)^2 = 4\end{array}\right]$$

$$= \int_1^2 \frac{1}{\sqrt{(x+2)^2-1}}\, dx \qquad \left[\because \int \frac{1}{\sqrt{x^2-a^2}}\, dx = \log\left|x+\sqrt{x^2-a^2}\right| + c\right]$$

$$= \left[\log\left|(x+2)+\sqrt{(x+2)^2-1}\right|\right]_1^2 = \left[\log\left|(x+2)+\sqrt{x^2+4x+3}\right|\right]_1^2$$

$$= [\log|(2+2)+\sqrt{4+8+3}| - \log|(1+2)+\sqrt{1+4+3}|]$$

$$= \log\left(4+\sqrt{15}\right) - \log\left(3+\sqrt{8}\right)$$

$$= \log\left(\frac{4+\sqrt{15}}{3+\sqrt{8}}\right). \qquad \left[\because \log m - \log n = \log\left(\frac{m}{n}\right)\right]$$

(*iv*) Let $\quad I = \int_0^4 \frac{1}{\sqrt{x^2+2x+3}}\,dx$

$$= \int_0^4 \frac{1}{\sqrt{(x^2+2x+1)+(3-1)}}\,dx \qquad \left[\begin{array}{l}\text{Add and subtract 1 to the denom.}\\ \because \left(\frac{1}{2}\text{ co-eff. of } x\right)^2 = 1\end{array}\right]$$

$$= \int_0^4 \frac{1}{\sqrt{(x+1)^2+\left(\sqrt{2}\right)^2}}\,dx \qquad \left[\because \int \frac{1}{\sqrt{x^2+a^2}}\,dx = \log\left|\, x+\sqrt{x^2+a^2}\,\right| + c\right]$$

$$= \left[\log\left|\,(x+1)+\sqrt{(x+1)^2+\left(\sqrt{2}\right)^2}\,\right|\right]_0^4 = \log\left|\,(x+1)+\sqrt{x^2+2x+3}\,\right|_0^4$$

$$= [\log|(4+1)+\sqrt{16+8+3}| - \log|(0+1)+\sqrt{0+0+3}|]$$

$$= [\log(5+\sqrt{27}) - \log(1+\sqrt{3})]$$

$$= \log\left(\frac{5+3\sqrt{3}}{1+\sqrt{3}}\right) \qquad \left[\because \log m - \log n = \log\left(\frac{m}{n}\right)\right]$$

$$= \log\left(\frac{5+3\sqrt{3}}{1+\sqrt{3}} \times \frac{1-\sqrt{3}}{1-\sqrt{3}}\right) \qquad \text{(Rationalization)}$$

$$= \log\left(\frac{5+3\sqrt{3}-5\sqrt{3}-9}{1-3}\right) = \log\left(\frac{-4-2\sqrt{3}}{-2}\right)$$

$$= \log(2+\sqrt{3}).$$

(*v*) Let $\quad I = \int_{1/4}^{1/2} \frac{1}{\sqrt{x-x^2}}\,dx = \int_{1/4}^{1/2} \frac{1}{\sqrt{-(x^2-x)}}\,dx$

$$= \int_{1/4}^{1/2} \frac{1}{\sqrt{-\left[x^2-x+\frac{1}{4}-\frac{1}{4}\right]}}\,dx \qquad \left[\begin{array}{l}\text{Add and subtract } \frac{1}{4} \text{ to the denom.}\\ \because \left(\frac{1}{2}\text{ co-eff. of } x\right)^2 = \frac{1}{4}\end{array}\right]$$

$$= \int_{1/4}^{1/2} \frac{1}{\sqrt{-\left[\left(x-\frac{1}{2}\right)^2-\left(\frac{1}{2}\right)^2\right]}}\,dx = \int_{1/4}^{1/2} \frac{1}{\sqrt{\left(\frac{1}{2}\right)^2-\left(x-\frac{1}{2}\right)^2}}\,dx$$

$$= \left[\sin^{-1}\left(\frac{x-\frac{1}{2}}{\frac{1}{2}}\right)\right]_{1/4}^{1/2} \qquad \left[\because \int \frac{1}{\sqrt{a^2-x^2}}\,dx = \sin^{-1}\frac{x}{a}+c\right]$$

$$= \left[\sin^{-1}\left(\frac{2x-1}{1}\right)\right]_{1/4}^{1/2} = \left[\sin^{-1}\left\{2\left(\frac{1}{2}\right)-1\right\} - \sin^{-1}\left\{2\left(\frac{1}{4}\right)-1\right\}\right]$$

$$= \sin^{-1} 0 - \sin^{-1}\left(-\frac{1}{2}\right) \qquad [\because \ \sin(-\theta) = -\sin\theta]$$

$$= 0 + \sin^{-1}\frac{1}{2} = \frac{\pi}{6}. \qquad \left[\because \ \sin\frac{\pi}{6} = \sin 30^\circ = \frac{1}{2}\right]$$

(*vi*) Let $\quad I = \int_0^a \frac{1}{\sqrt{ax-x^2}}\,dx = \int_0^a \frac{1}{\sqrt{-[x^2-ax]}}\,dx$

$$= \int_0^a \frac{1}{\sqrt{-\left[x^2-ax+\frac{a^2}{4}-\frac{a^2}{4}\right]}}\,dx \qquad \left[\begin{array}{l}\text{Add and subtract } \frac{a^2}{4} \text{ to the denom.}\\ \because \left(\frac{1}{2}\text{ co-eff. of } x\right)^2 = \frac{a^2}{4}\end{array}\right]$$

$$= \int_0^a \frac{1}{\sqrt{-\left[\left(x-\frac{a}{2}\right)^2-\left(\frac{a}{2}\right)^2\right]}}\,dx = \int_0^a \frac{1}{\sqrt{\left(\frac{a}{2}\right)^2-\left(x-\frac{a}{2}\right)^2}}\,dx$$

$$= \left[\sin^{-1}\frac{\left(x-\frac{a}{2}\right)}{\frac{a}{2}}\right]_0^a \qquad \left[\because \int \frac{1}{\sqrt{a^2-x^2}}\,dx = \sin^{-1}\frac{x}{a}+c\right]$$

$$= \left[\sin^{-1}\left(\frac{2x-a}{a}\right)\right]_0^a = \left[\sin^{-1}\left(\frac{2a-a}{a}\right) - \sin^{-1}\left(\frac{0-a}{a}\right)\right]$$

$$= [\sin^{-1}(1) - \sin^{-1}(-1)] \qquad [\because \ \sin(-\theta) = -\sin\theta]$$

$$= \sin^{-1} 1 + \sin^{-1} 1 = \frac{\pi}{2}+\frac{\pi}{2} \qquad \left[\because \ \sin\frac{\pi}{2} = \sin 90^\circ = 1\right]$$

$$= \frac{2\pi}{2} = \pi.$$

Example 15. *Evaluate the following :*

(*i*) $\int_0^{\pi/2} \frac{1}{5+3\cos x}\,dx$ (*ii*) $\int_1^3 \frac{1}{x^2(x+1)}\,dx$

(*iii*) $\int_0^\infty \frac{1}{(x^2+a^2)(x^2+b^2)}\,dx$ (*iv*) $\int_1^2 \frac{1}{(x+1)(x+2)}\,dx$

(*v*) $\int_1^2 \frac{5x^2}{x^2+4x+3}\,dx.$

Solution. (*i*) Let $I = \int_0^{\pi/2} \frac{1}{5+3\cos x}\, dx$

$$= \int_0^{\pi/2} \frac{1}{5+3\left(\frac{1-\tan^2\frac{x}{2}}{1+\tan^2\frac{x}{2}}\right)}\, dx \qquad \left[\because \cos 2A = \frac{1-\tan^2 A}{1+\tan^2 A} \Rightarrow \cos A = \frac{1-\tan^2\frac{A}{2}}{1+\tan^2\frac{A}{2}}\right]$$

$$= \int_0^{\pi/2} \frac{\left(1+\tan^2\frac{x}{2}\right)}{5\left(1+\tan^2\frac{x}{2}\right)+3\left(1-\tan^2\frac{x}{2}\right)}\, dx$$

$$= \int_0^{\pi/2} \frac{\left(1+\tan^2\frac{x}{2}\right)}{8+2\tan^2\frac{x}{2}}\, dx$$

Put $\tan\frac{x}{2} = z \Rightarrow \frac{1}{2}\sec^2\frac{x}{2}\, dx = dz \Rightarrow \sec^2\frac{x}{2}\, dx = 2dz$

$$\Rightarrow \left(1+\tan^2\frac{x}{2}\right) dx = 2dz \qquad [\because \sec^2 A - \tan^2 A = 1]$$

When $x = 0 \Rightarrow z = \tan 0^\circ \Rightarrow z = 0$

and when $x = \frac{\pi}{2} \Rightarrow z = \tan\left(\frac{\pi}{4}\right) \Rightarrow z = 1$

$\therefore$
$$I = \int_0^1 \frac{2}{8+2z^2}\, dz$$

$$= \int_0^1 \frac{1}{4+z^2}\, dz = \int_0^1 \frac{1}{(2)^2+z^2}\, dz$$

$$= \left[\frac{1}{2}\tan^{-1}\left(\frac{z}{2}\right)\right]_0^1 \qquad \left[\because \int \frac{1}{a^2+x^2}\, dx = \frac{1}{a}\tan^{-1}\frac{x}{a} + c\right]$$

$$= \frac{1}{2}\left[\tan^{-1}\left(\frac{1}{2}\right) - \tan^{-1} 0^\circ\right] = \frac{1}{2}\left[\tan^{-1}\frac{1}{2} - 0\right]$$

$$= \frac{1}{2}\tan^{-1}\left(\frac{1}{2}\right).$$

(*ii*) Let
$$I = \int_1^3 \frac{1}{x^2(x+1)}\, dx$$

Let
$$\frac{1}{x^2(x+1)} = \frac{A}{x} + \frac{B}{x^2} + \frac{C}{x+1} \qquad ...(1)$$

Multiplying both sides by $x^2(x + 1)$, we get

$$1 = Ax(x+1) + B(x+1) + Cx^2 \qquad ...(2)$$

Put $x = 0$ in (2), we get

$$1 = B(0 + 1) \Rightarrow B = 1$$

Put $x = -1$ in (2), we get

$$1 = A(-1)(-1 + 1) + B(-1 + 1) + C(-1)^2$$

$$\Rightarrow \quad 1 = C \Rightarrow C = 1$$

Comparing the co-efficients of x on both sides of equation (2), we have

$$0 = A + B \Rightarrow A = -B \Rightarrow A = -1.$$

Substituting the values of A, B and C in equation (1), we have

$$\frac{1}{x^2(x+1)} = \frac{-1}{x} + \frac{1}{x^2} + \frac{1}{x+1}$$

$$\therefore \quad I = \int_1^3 \frac{1}{x^2(x+1)}\,dx = \int_1^3 \left[\frac{-1}{x} + \frac{1}{x^2} + \frac{1}{x+1}\right] dx$$

$$= -\int_1^3 \frac{1}{x}\,dx + \int_1^3 x^{-2}\,dx + \int_1^3 \frac{1}{x+1}\,dx$$

$$= -\Big[\log|x|\Big]_1^3 + \left[\frac{x^{-2+1}}{-2+1}\right]_1^3 + \Big[\log|x+1|\Big]_1^3$$

$$= -[\log 3 - \log 1] - 1\left[\frac{1}{x}\right]_1^3 + [\log|3+1| - \log|1+1|]$$

$$= -\log 3 - \left[\frac{1}{3} - \frac{1}{1}\right] + [\log 4 - \log 2]$$

$$= -\log 3 + \frac{2}{3} + \log\left(\frac{4}{2}\right) \qquad \left[\because \log m - \log n = \log\left(\frac{m}{n}\right)\right]$$

$$= -\log 3 + \frac{2}{3} + \log 2$$

$$= \frac{2}{3} + \log\frac{2}{3}.$$

(*iii*) Let $I = \int_0^\infty \frac{1}{(x^2 + a^2)(x^2 + b^2)}\,dx$

Let $\frac{1}{(x^2 + a^2)(x^2 + b^2)} = \frac{1}{(z + a^2)(z + b^2)} = \frac{A}{(z + a^2)} + \frac{B}{(z + b^2)}$ ...(1) [Put $(x^2 = z)$]

Multiplying both sides by $(z + a^2)(z + b^2)$, we get

$$1 = A(z + b^2) + B(z + a^2) \qquad ...(2)$$

Put $z = -a^2$ in (2), we get

$$1 = A(-a^2 + b^2) + B(-a^2 + a^2) \Rightarrow 1 = A(b^2 - a^2) \Rightarrow A = \frac{1}{b^2 - a^2}$$

Put $z = -b^2$ in (2), we get

$$1 = A(-b^2 + b^2) + B(-b^2 + a^2) \Rightarrow 1 = B(a^2 - b^2) \Rightarrow B = \frac{1}{a^2 - b^2}$$

Substituting the values of A and B in (1), we have

$$\frac{1}{(z+a^2)(z+b^2)} = \frac{\frac{1}{b^2-a^2}}{(z+a^2)} + \frac{\frac{1}{a^2-b^2}}{(z+b^2)} = \frac{1}{(b^2-a^2)(z+a^2)} - \frac{1}{(b^2-a^2)(z+b^2)}$$

or
$$\frac{1}{(x^2+a^2)(x^2+b^2)} = \frac{1}{b^2-a^2}\left[\frac{1}{x^2+a^2} - \frac{1}{x^2+b^2}\right] \qquad [\because \; z = x^2]$$

$$\therefore \quad I = \int_0^\infty \frac{1}{(x^2+a^2)(x^2+b^2)}\,.\,dx = \int_0^\infty \left[\frac{1}{b^2-a^2}\left(\frac{1}{x^2+a^2} - \frac{1}{x^2+b^2}\right)\right] dx$$

$$= \frac{1}{b^2-a^2}\left[\int_0^\infty \frac{1}{x^2+a^2}\,dx - \int_0^\infty \frac{1}{x^2+b^2}\,dx\right]$$

$$= \frac{1}{b^2-a^2}\left[\left(\frac{1}{a}\tan^{-1}\left(\frac{x}{a}\right)\right)_0^\infty - \left(\frac{1}{b}\tan^{-1}\left(\frac{x}{b}\right)\right)_0^\infty\right]$$

$$\left[\because \int \frac{1}{x^2+a^2}\,dx = \frac{1}{a}\tan^{-1}\frac{x}{a} + c\right]$$

$$= \frac{1}{b^2-a^2}\left[\left(\frac{1}{a}\tan^{-1}\infty - \frac{1}{a}\tan^{-1}0^\circ\right) - \left(\frac{1}{b}\tan^{-1}\infty - \frac{1}{b}\tan^{-1}0^\circ\right)\right]$$

$$= \frac{1}{b^2-a^2}\left[\left(\frac{\pi}{2a} - 0\right) - \left(\frac{\pi}{2b} - 0\right)\right] = \frac{1}{b^2-a^2}\left[\frac{\pi}{2a} - \frac{\pi}{2b}\right]$$

$$= \frac{\pi}{b^2-a^2}\left(\frac{b-a}{2ab}\right) = \frac{\pi(b-a)}{2ab(b-a)(b+a)}$$

$$= \frac{\pi}{2ab(a+b)}.$$

(*iv*) Let $\quad I = \int_1^2 \frac{1}{(x+1)(x+2)}\,dx$

Let
$$\frac{1}{(x+1)(x+2)} = \frac{A}{x+1} + \frac{B}{x+2} \qquad ...(1)$$

Multiplying both sides by $(x + 1)(x + 2)$, we get

$$1 = A(x+2) + B(x+1) \qquad ...(2)$$

Put $x = -1$ in (2), we get

$$1 = A(-1+2) + B(-1+1) \Rightarrow 1 = A \Rightarrow A = 1$$

Put $x = -2$ in (2), we get

$$1 = A(-2+2) + B(-2+1) \Rightarrow 1 = -B \Rightarrow B = -1$$

Substituting the values of A and B in (1), we have

$$\frac{1}{(x+1)(x+2)} = \frac{1}{x+1} + \frac{-1}{x+2}$$

$$\therefore \qquad I = \int_1^2 \frac{1}{(x+1)(x+2)}\,dx = \int_1^2 \left[\frac{1}{x+1} - \frac{1}{x+2}\right] dx$$

$$= \int_1^2 \frac{1}{x+1}\,dx - \int_1^2 \frac{1}{x+2}\,dx$$

$$= \left[\log|x+1|\right]_1^2 - \left[\log|x+2|\right]_1^2$$

$$= [\log|2+1| - \log|1+1|] - [\log|2+2| - \log|1+2|]$$

$$= (\log 3 - \log 2) - (\log 4 - \log 3)$$

$$= \log\frac{3}{2} - \log\frac{4}{3} = \log\left(\frac{3/2}{4/3}\right) = \log\left(\frac{3}{2}\times\frac{3}{4}\right) \qquad \left[\because \log m - \log n = \log\left(\frac{m}{n}\right)\right]$$

$$= \log\frac{9}{8}.$$

(*v*) Let $\quad I = \int_1^2 \frac{5x^2}{x^2+4x+3}\,dx$

Since the integrand is an improper fraction *i.e.*, (the degree of numerator is equal to the degree of denominator), therefore, by actual division, we have

$$\therefore \quad \int_1^2 \frac{5x^2}{x^2+4x+3}\,dx = \int_1^2 \left[5 - \frac{20x+15}{x^2+4x+3}\right] dx$$

$$= \int_1^2 5\,dx - \int_1^2 \frac{5(4x+3)}{(x+1)(x+3)}\,dx$$

$$x^2+4x+3 \overline{)\,5x^2\,}(5$$
$$\underline{\;5x^2 + 20x + 15\;}$$
$$\;\;-\;\;-\;\;-$$
$$\underline{\;-20x - 15\;}$$

Let $\quad \dfrac{4x+3}{(x+1)(x+3)} = \dfrac{A}{x+1} + \dfrac{B}{x+3}$ ...(1)

Multiplying both sides by $(x + 1)(x + 3)$, we get

$$(4x + 3) = A(x + 3) + B(x + 1) \qquad ...(2)$$

Put $x = -1$ in (2), we get

$$4(-1) + 3 = A(-1 + 3) + B(-1 + 1) \;\Rightarrow\; -4 + 3 = 2A \;\Rightarrow\; A = \frac{-1}{2}$$

Put $x = -3$ in (2), we get

$$4(-3) + 3 = A(-3 + 3) + B(-3 + 1)$$

$$\Rightarrow \qquad -12 + 3 = -2B \;\Rightarrow\; -9 = -2B \;\Rightarrow\; B = \frac{9}{2}$$

Substituting the values of A and B in (1), we have

$$\frac{4x+3}{(x+1)(x+3)} = \frac{\frac{-1}{2}}{x+1} + \frac{\frac{9}{2}}{x+3} = \frac{-1}{2(x+1)} + \frac{9}{2(x+3)}$$

$$\therefore \qquad I = 5\int_1^2 1\,.\,dx - 5\int_1^2 \left[\frac{-1}{2(x+1)} + \frac{9}{2(x+3)}\right] dx$$

$$= 5\int_1^2 1 \, . \, dx + \frac{5}{2}\int_1^2 \frac{1}{x+1}\, dx - \frac{45}{2}\int_1^2 \frac{1}{x+3}\, dx$$

$$= 5\left[x \right]_1^2 + \frac{5}{2}\left[\log|x+1| \right]_1^2 - \frac{45}{2}\left[\log|x+3| \right]_1^2$$

$$= 5\,[2-1] + \frac{5}{2}\,[\log|2+1| - \log|1+1|] - \frac{45}{2}\,[\log|2+3| - \log|1+3|]$$

$$= 5 + \frac{5}{2}\,[\log 3 - \log 2] - \frac{45}{2}\,[\log 5 - \log 4] \qquad \left[\because \log m - \log n = \log\left(\frac{m}{n}\right)\right]$$

$$= 5 + \frac{5}{2}\log\frac{3}{2} - \frac{45}{2}\log\frac{5}{4}.$$

Example 16. *Evaluate the following :*

(*i*) $\displaystyle\int_0^{\pi/2} \frac{\cos x}{(1+\sin x)(2+\sin x)}\, dx$ (*ii*) $\displaystyle\int_2^3 \frac{x^2+1}{(2x+1)(x^2-1)}\, dx$

(*iii*) $\displaystyle\int_0^1 \frac{x}{x^2+4x+3}\, dx$ (*iv*) $\displaystyle\int_0^1 \sqrt{x(1-x)}\, dx$

(*v*) $\displaystyle\int_1^2 \frac{1}{x(1+x^2)}\, dx.$

Solution. (*i*) Let $I = \displaystyle\int_0^{\pi/2} \frac{\cos x}{(1+\sin x)(2+\sin x)}\, dx$

Put $\sin x = z \Rightarrow \cos x\, dx = dz$

When $x = 0 \Rightarrow z = \sin 0^\circ = 0$ and when $x = \pi/2 \Rightarrow z = \sin \pi/2 = 1.$

$$\therefore \qquad I = \int_0^1 \frac{1}{(1+z)(2+z)}\, dz$$

Let $\displaystyle\frac{1}{(1+z)(2+z)} = \frac{A}{1+z} + \frac{B}{2+z}$ ...(1)

Multiplying both sides by $(1 + z)(2 + z)$, we get

$1 = A(2 + z) + B(1 + z)$ ...(2)

Put $z = -1$ in (2), we get

$1 = A[2 + (-1)] + B[1 + (-1)] \Rightarrow 1 = A \Rightarrow A = 1$

Put $z = -2$ in (2), we get

$1 = A[2 + (-2)] + B[1 + (-2)] \Rightarrow 1 = -B \Rightarrow B = -1.$

Substituting the values of A and B in equation (1), we have

$$\frac{1}{(1+z)(2+z)} = \frac{1}{1+z} + \frac{-1}{2+z}$$

$$\therefore \qquad I = \int_0^1 \frac{1}{(1+z)(2+z)}\, dz = \int_0^1 \left[\frac{1}{1+z} - \frac{1}{2+z}\right] dz$$

$$= \int_0^1 \frac{1}{1+z}\, dz - \int_0^1 \frac{1}{2+z}\, dz$$

$$= \Big[\log|1+z|\Big]_0^1 - \Big[\log|2+z|\Big]_0^1$$

$= [\log|1+1| - \log|1+0|] - [\log|2+1| + \log|2+0|]$

$= (\log 2 - \log 1) - (\log 3 - \log 2)$ $[\because \log 1 = 0]$

$= \log 2 - \log 3 + \log 2 = 2\log 2 - \log 3$

$= \log (2)^2 - \log 3$ $[\because n\log m = \log m^n]$

$$= \log 4 - \log 3 = \log \frac{4}{3}. \qquad \left[\because \log m - \log n = \log\left(\frac{m}{n}\right)\right]$$

(*ii*) Let $$I = \int_2^3 \frac{x^2+1}{(2x+1)(x^2-1)}\,dx = \int_2^3 \frac{x^2+1}{(2x+1)(x-1)(x+1)}\,dx$$

Let $$\frac{x^2+1}{(2x+1)(x-1)(x+1)} = \frac{A}{(2x+1)} + \frac{B}{(x-1)} + \frac{C}{(x+1)} \qquad ...(1)$$

Multiplying both sides by $(2x + 1)(x - 1)(x + 1)$, we get

$$x^2 + 1 = A(x-1)(x+1) + B(2x+1)(x+1) + C(2x+1)(x-1) \qquad ...(2)$$

Put $x = 1$ in (2), we get

$$1 + 1 = A(1-1)(1+1) + B[2+1][1+1] + C[2+1][1-1]$$

$$\Rightarrow \qquad 2 = 6B \Rightarrow B = \frac{1}{3}$$

Put $x = -1$ in (2), we get

$$[(-1)^2 + 1] = A[-1-1][-1+1] + B[2(-1)+1][-1+1] + C[2(-1)+1][-1-1]$$

$$\Rightarrow \qquad 2 = C(-1)(-2) \Rightarrow 2 = 2C \Rightarrow C = 1$$

Comparing the co-efficients of x^2 on both sides of equation (2), we have

$$1 = A + 2B + 2C \Rightarrow 1 = A + 2\left(\frac{1}{3}\right) + 2(1)$$

$$\Rightarrow \qquad 1 - 2 - \frac{2}{3} = A \Rightarrow -1 - \frac{2}{3} = A \Rightarrow A = -\frac{5}{3}$$

Substituting the values of A, B and C in equation (1), we have

$$\frac{x^2+1}{(2x+1)(x-1)(x+1)} = \frac{-5/3}{(2x+1)} + \frac{1/3}{x-1} + \frac{1}{x+1} = -\frac{5}{3(2x+1)} + \frac{1}{3(x-1)} + \frac{1}{x+1}$$

$$\therefore \qquad I = \int_2^3 \frac{x^2+1}{(2x+1)(x^2-1)}\,dx = \int_2^3 \left[-\frac{5}{3(2x+1)} + \frac{1}{3(x-1)} + \frac{1}{x+1}\right] dx$$

$$= -\frac{5}{3}\int_2^3 \frac{1}{2x+1}\,dx + \frac{1}{3}\int_2^3 \frac{1}{x-1}\,dx + \int_2^3 \frac{1}{x+1}\,dx$$

$$= -\frac{5}{3}\left[\frac{\log|2x+1|}{2}\right]_2^3 + \frac{1}{3}\Big[\log|x-1|\Big]_2^3 + \Big[\log|x+1|\Big]_2^3$$

$$= -\frac{5}{6}[\log|2(3)+1| - \log|2(2)+1|] + \frac{1}{3}[\log|3-1| - \log|2-1|]$$
$$+ [\log|3+1| - \log|2+1|]$$

$$= -\frac{5}{6}[\log 7 - \log 5] + \frac{1}{3}[\log 2 - \log 1] + [\log 4 - \log 3]$$

$$= -\frac{5}{6}\log\frac{7}{5} + \frac{1}{3}\log 2 + \log\frac{4}{3}. \qquad \left[\because \log m - \log n = \log\left(\frac{m}{n}\right)\right]$$

(*iii*) Let $\quad I = \int_0^1 \frac{x}{x^2+4x+3}\,dx = \int_0^1 \frac{x}{(x+1)(x+3)}\,dx$

Let $\qquad \frac{x}{(x+1)(x+3)} = \frac{A}{x+1} + \frac{B}{x+3} \qquad ...(1)$

Multiplying both sides by $(x + 1)(x + 3)$, we get

$$x = A(x+3) + B(x+1) \qquad ...(2)$$

Put $x = -1$ in equation (2), we get

$$-1 = A(-1+3) + B(-1+1) \;\Rightarrow\; -1 = 2A \;\Rightarrow\; A = -\frac{1}{2}$$

Put $x = -3$ in equation (2), we get

$$-3 = A(-3+3) + B(-3+1) \;\Rightarrow\; -3 = -2B \;\Rightarrow\; B = \frac{3}{2}$$

Substituting the values of A and B in equation (1), we have

$$\frac{x}{(x+1)(x+3)} = \frac{-1/2}{x+1} + \frac{3/2}{x+3}$$

or $$\frac{x}{x^2+4x+3} = \frac{-1}{2(x+1)} + \frac{3}{2(x+3)}$$

$\therefore$ $$I = \int_0^1 \frac{x}{x^2+4x+3}\,dx = \int_0^1 \left[-\frac{1}{2(x+1)} + \frac{3}{2(x+3)}\right]dx$$

$$= -\frac{1}{2}\int_0^1 \frac{1}{x+1}\,dx + \frac{3}{2}\int_0^1 \frac{1}{x+3}\,dx$$

$$= -\frac{1}{2}\Big[\log|x+1|\Big]_0^1 + \frac{3}{2}\Big[\log|x+3|\Big]_0^1$$

$$= -\frac{1}{2}[\log|1+1| - \log|0+1|] + \frac{3}{2}[\log|1+3| - \log|0+3|]$$

$$= -\frac{1}{2}[\log 2 - \log 1] + \frac{3}{2}[\log 4 - \log 3] \qquad \left[\because \log m - \log n = \log\left(\frac{m}{n}\right)\right]$$

$$= -\frac{1}{2}\log 2 + \frac{3}{2}\log\left(\frac{4}{3}\right) = \frac{1}{2}\left[3\log\left(\frac{4}{3}\right) - \log 2\right]$$

$$= \frac{1}{2}\left[\log\left(\frac{4}{3}\right)^3 - \log 2\right] \qquad [\because n\log m = \log m^n]$$

$$= \frac{1}{2}\left[\log\frac{64}{27} - \log 2\right] = \frac{1}{2}\left[\log\frac{64/27}{2/1}\right] = \frac{1}{2}\log\left(\frac{64}{27}\times\frac{1}{2}\right)$$

$$= \frac{1}{2}\log\frac{32}{27}.$$

(*iv*) Let $\quad I = \int_0^1 \sqrt{x(1-x)}\, dx$

$$= \int_0^1 \sqrt{x - x^2}\, dx = \int_0^1 \sqrt{-[x^2 - x]}\, dx$$

$$= \int_0^1 \sqrt{-\left[x^2 - x + \frac{1}{4} - \frac{1}{4}\right]}\, dx \qquad \left[\text{Add and subtract } \frac{1}{4},\ \because \left(\frac{1}{2}\text{ co-eff. of } x\right)^2 = \frac{1}{4}\right]$$

$$= \int_0^1 \sqrt{-\left[\left(x - \frac{1}{2}\right)^2 - \left(\frac{1}{2}\right)^2\right]}\, dx = \int_0^1 \sqrt{\left(\frac{1}{2}\right)^2 - \left(x - \frac{1}{2}\right)^2}\, dx$$

$$= \left[\frac{\left(x - \frac{1}{2}\right)}{2}\sqrt{\left(\frac{1}{2}\right)^2 - \left(x - \frac{1}{2}\right)^2} + \frac{\left(\frac{1}{2}\right)^2}{2}\sin^{-1}\frac{\left(x - \frac{1}{2}\right)}{\frac{1}{2}}\right]_0^1$$

$$\left[\because \int \sqrt{a^2 - x^2}\, dx = \frac{x}{2}\sqrt{a^2 - x^2} + \frac{a^2}{2}\sin^{-1}\frac{x}{a} + c\right]$$

$$= \left[\frac{(2x-1)}{4}\sqrt{x - x^2} + \frac{1}{8}\sin^{-1}(2x - 1)\right]_0^1$$

$$= \left[\frac{2(1) - 1}{4}\sqrt{1 - (1)^2} + \frac{1}{8}\sin^{-1}(2(1) - 1)\right] - \left[\frac{2(0) - 1}{4}\sqrt{0 - 0} + \frac{1}{8}\sin^{-1}(0 - 1)\right]$$

$$= 0 + \frac{1}{8}\sin^{-1}(1) - \frac{1}{8}\sin^{-1}(-1) \qquad [\because \sin(-\theta) = -\sin\theta]$$

$$= \frac{1}{8}\left(\frac{\pi}{2}\right) + \frac{1}{8}\left(\frac{\pi}{2}\right) = \frac{\pi}{16} + \frac{\pi}{16} = \frac{2\pi}{16}$$

$$= \frac{\pi}{8}.$$

(*v*) Let $I = \int_1^2 \frac{1}{x(1 + x^2)}\, dx$

Let $\quad \frac{1}{x(1+x^2)} = \frac{A}{x} + \frac{Bx + C}{(1+x^2)}$ ...(1)

Multiplying both sides by $x(1 + x^2)$, we have

$$1 = A(1 + x^2) + (Bx + C)(x) \qquad ...(2)$$

Put $x = 0$ in equation (2), we get

$$1 = A(1 + 0) + (B.0 + C) . 0 \Rightarrow A = 1$$

Comparing co-efficients of x^2 on both sides of equation (2), we have

$$0 = A + B \Rightarrow B = - A \Rightarrow B = - 1$$

Comparing co-efficients of x on both sides of equation (2), we have

$$0 = C \Rightarrow C = 0$$

Substituting the values of A, B and C in equation (1), we have

$$\frac{1}{x(1+x^2)} = \frac{1}{x} + \frac{-x}{1+x^2}$$

$$\therefore \quad I = \int_1^2 \frac{1}{x(1+x^2)}\,dx = \int_1^2 \left[\frac{1}{x} - \frac{x}{1+x^2}\right] dx = \int_1^2 \frac{1}{x}\,dx - \int_1^2 \frac{x}{1+x^2}\,dx$$

$$= \int_1^2 \frac{1}{x}\,dx - \frac{1}{2}\int_1^2 \frac{2x}{1+x^2}\,dx \qquad \left[\text{Multiply and divide the second integral by 2}\right]$$

$$= \Big[\log|x|\Big]_1^2 - \frac{1}{2}\Big[\log|1+x^2|\Big]_1^2 \qquad \left[\because \int \frac{f'(x)}{f(x)}\,dx = \log|f(x)| + c\right]$$

$$= [\log 2 - \log 1] - \frac{1}{2}[\log |1 + 2^2| - \log |1 + 1|]$$

$$= \log 2 - \frac{1}{2}[\log 5 - \log 2]$$

$$= \log 2 - \frac{1}{2}\log\frac{5}{2}. \qquad \left[\because \log m - \log n = \log\left(\frac{m}{n}\right)\right]$$

Example 17. *Evaluate the following* :

(*i*) $\int_{\pi/2}^{\pi} e^x \left(\frac{1-\sin x}{1-\cos x}\right) dx$ (*ii*) $\int_0^1 x\,e^x\,dx$

(*iii*) $\int_0^1 \tan^{-1} x\,dx$ (*iv*) $\int_1^2 \frac{\log x}{x^2}\,dx$

(*v*) $\int_0^{\pi/2} x^2 \cos 2x\,dx.$

Solution. (*i*) Let $I = \int_{\pi/2}^{\pi} e^x \left(\frac{1-\sin x}{1-\cos x}\right) dx$

$$\left[\begin{array}{l}\because \sin 2A = 2\sin A\cos A \\ \Rightarrow \sin A = 2\sin\frac{A}{2}\cos\frac{A}{2} \\ 1-\cos 2A = 2\sin^2 A \\ \Rightarrow 1-\cos A = 2\sin^2\frac{A}{2}\end{array}\right]$$

$$= \int_{\pi/2}^{\pi} e^x \left[\frac{1 - 2\sin\frac{x}{2}\cos\frac{x}{2}}{2\sin^2\frac{x}{2}}\right] dx$$

$$= \int_{\pi/2}^{\pi} e^x \left[\frac{1}{2\sin^2\frac{x}{2}} - \frac{2\sin\frac{x}{2}\cos\frac{x}{2}}{2\sin^2\frac{x}{2}}\right] dx$$

$$= \int_{\pi/2}^{\pi} e^x \left[\frac{1}{2}\operatorname{cosec}^2 \frac{x}{2} - \cot \frac{x}{2}\right] dx$$

$$\left[\because \int e^x [f(x) + f'(x)]\, dx = e^x f(x) + c\right]$$

$$= \left[e^x \left(-\cot \frac{x}{2}\right)\right]_{\pi/2}^{\pi}$$

$$= \left[-e^{\pi} \cot \frac{\pi}{2} + e^{\pi/2} \cot \frac{\pi}{4}\right] \qquad \left[\begin{array}{l}\because \cot \frac{\pi}{2} = \cot 90^\circ = 0 \\ \cot \frac{\pi}{4} = \cot 45^\circ = 1\end{array}\right]$$

$$= [0 + e^{\pi/2}] = e^{\pi/2}.$$

(*ii*) Let $\quad I = \int_0^1 x\, e^x\, dx$

Let $\quad \int x\, e^x\, dx = \int \underset{\text{I}}{x}\, \underset{\text{II}}{e^x}\, dx$

Integrating by parts, we have

$$\int x.e^x\, dx = x \,.\int e^x\, dx - \int \left\{\frac{d}{dx}(x) \,.\int e^x\, dx\right\} dx$$

$$= x\, e^x - \int 1 \,.\, e^x\, dx$$

$$= x\, e^x - e^x$$

$$\therefore \quad \int_0^1 x\, e^x\, dx = \left[x\, e^x - e^x\right]_0^1 = [(1 \,.\, e^1 - e^1) - (0 - e^0)]$$

$$= 1.$$

(*iii*) Let $\quad I = \int_0^1 \tan^{-1} x\, dx$

Let $\quad \int \tan^{-1} x\, dx = \int \underset{\text{II}}{1} . \underset{\text{I}}{\tan^{-1} x}\, dx$ [Taking unity as the second function]

Integrating by parts, we have

$$\int \tan^{-1} x\, dx = \tan^{-1} x \,.\int 1 . dx - \int \left\{\frac{d}{dx}(\tan^{-1} x) \,.\int 1 . dx\right\} dx$$

$$= \tan^{-1} x \,.\, x - \int \frac{1}{1 + x^2} \,.\, x\, dx$$

$$= x \tan^{-1} x - \frac{1}{2}\int \frac{2x}{1 + x^2}\, dx \qquad \text{[Multiply and divided by 2]}$$

$$= x \tan^{-1} x - \frac{1}{2}\,[\log |(1 + x^2)|] \qquad \left[\because \int \frac{f'(x)}{f(x)}\, dx = \log |f(x)| + c\right]$$

$$\therefore \quad I = \int_0^1 \tan^{-1} x\, dx = \left[x \tan^{-1} x - \frac{1}{2} \log |(1 + x^2)|\right]_0^1$$

$$= \left[1 \tan^{-1} 1 - \frac{1}{2} \log (1 + 1)\right] - \left[0 - \frac{1}{2} \log (1 + 0)\right]$$

$$= \frac{\pi}{4} - \frac{1}{2} \log 2 + \frac{1}{2} \log 1 \qquad \left[\because \tan\frac{\pi}{4} = \tan 45^\circ = 1\right]$$

$$= \frac{\pi}{4} - \frac{1}{2} [\log 2 - \log 1]$$

$$= \frac{\pi}{4} - \frac{1}{2} \log 2. \qquad \left[\because \log m - \log n = \log\left(\frac{m}{n}\right)\right]$$

(*iv*) Let $\quad I = \int_1^2 \frac{\log x}{x^2}\, dx$

Let $\int \log x \cdot \frac{1}{x^2} \cdot dx = \int \underset{\text{I}}{\log x} \cdot \underset{\text{II}}{(x^{-2})}\, dx$

Integrating by parts, we have

$$\int \log x \cdot \frac{1}{x^2} \cdot dx = \log x \cdot \int x^{-2}\, dx - \int \left\{\frac{d}{dx}(\log x) \cdot \int x^{-2}\, dx\right\} dx$$

$$= \log x \cdot \frac{x^{-2+1}}{-2+1} - \int \frac{1}{x} \cdot \frac{x^{-2+1}}{(-2+1)}\, dx$$

$$= -\frac{1}{x}\log x + \int \frac{1}{x^2}\, dx = -\frac{1}{x}\log x + \frac{x^{-2+1}}{-2+1}$$

$$= -\frac{1}{x}\log x - \frac{1}{x}$$

$$\therefore \quad \int_1^2 \frac{\log x}{x^2}\, dx = \left[-\frac{1}{x}\log x - \frac{1}{x}\right]_1^2 = \left[\left(-\frac{1}{2}\log 2 - \frac{1}{2}\right) - (-1 \log 1 - 1)\right]$$

$$= -\frac{1}{2}\log 2 - \frac{1}{2} + 1 = \frac{1}{2} - \frac{1}{2}\log 2$$

$$= \frac{1}{2}[1 - \log 2].$$

(*v*) Let $\quad I = \int_0^{\pi/2} x^2 \cos 2x\, dx$

Let $\quad I_1 = \int \underset{\text{I}}{x^2} \cdot \underset{\text{II}}{\cos 2x}\, dx$

Integrating by parts, we have

$$I_1 = x^2 \cdot \int \cos 2x\, dx - \int \left\{\frac{d}{dx}(x^2) \cdot \int \cos 2x\, dx\right\} dx$$

$$= x^2 \frac{\sin 2x}{2} - \int 2x \cdot \frac{\sin 2x}{2}\, dx$$

$$= \frac{1}{2} x^2 \sin 2x - \int \underset{\text{I}}{x}\, \underset{\text{II}}{\sin 2x}\, dx$$

Integrating again by parts, we have

$$I_1 = \frac{1}{2} x^2 \sin 2x - \left[x \cdot \int \sin 2x\, dx - \int \left\{\frac{d}{dx}(x) \cdot \int \sin 2x\, dx\right\} dx\right]$$

$$= \frac{1}{2} x^2 \sin 2x - \left[x \cdot \frac{(-\cos 2x)}{2} - \int 1 \cdot \frac{(-\cos 2x)}{2}\, dx \right]$$

$$= \frac{1}{2} x^2 \sin 2x + \frac{1}{2} x \cos 2x - \frac{1}{2} \int \cos 2x\, dx$$

$$= \frac{1}{2} x^2 \sin 2x + \frac{1}{2} x \cos 2x - \frac{1}{2} \frac{\sin 2x}{2}$$

$$= \frac{1}{2} x^2 \sin 2x + \frac{1}{2} x \cos 2x - \frac{1}{4} \sin 2x$$

$$\therefore \quad I = \int_0^{\pi/2} x^2 \cos 2x\, dx = \left[\frac{1}{2} x^2 \sin 2x + \frac{1}{2} x \cos 2x - \frac{1}{4} \sin 2x \right]_0^{\pi/2}$$

$$= \frac{1}{4} \left[2x^2 \sin 2x + 2x \cos 2x - \sin 2x \right]_0^{\pi/2}$$

$$= \frac{1}{4} \left[2\left(\frac{\pi}{2}\right)^2 \sin\left(2 \cdot \frac{\pi}{2}\right) + \frac{2\pi}{2} \cos\left(2 \cdot \frac{\pi}{2}\right) - \sin\left(2 \cdot \frac{\pi}{2}\right) \right] - \frac{1}{4} [0 + 0 - \sin 0^\circ]$$

$$= \frac{1}{4} \left[2 \frac{\pi^2}{4} \sin \pi + \pi \cos \pi - \sin \pi \right] - 0$$

$$= \frac{1}{4} [0 + \pi(-1) - 0] = -\frac{\pi}{4}.$$

Example 18. *If* $\int_0^a \sqrt{x}\, dx = 2a \int_0^{\pi/2} \sin^3 x\, dx$, *find the value of* $\int_a^{a+1} x \cdot dx$.

Solution. Let $I_1 = \int_0^a \sqrt{x}\, dx = \left[\frac{x^{\frac{1}{2}+1}}{\frac{1}{2}+1} \right]_0^a = \frac{2}{3} \left[x^{3/2} \right]_0^a = \frac{2}{3} [a^{3/2} - 0]$

$$\Rightarrow \quad I_1 = \frac{2}{3} a^{3/2} \qquad ...(2)$$

$$\text{Let} \quad I_2 = 2a \int_0^{\pi/2} \sin^3 x\, dx \qquad ...(3)$$

$$= 2a \int_0^{\pi/2} \frac{1}{4} [3 \sin x - \sin 3x]\, dx$$

$$\left[\begin{aligned} &\because \sin 3A = 3 \sin A - 4 \sin^3 A \\ &\Rightarrow 4 \sin^3 A = 3 \sin A - \sin 3A \\ &\Rightarrow \sin^3 A = \frac{1}{4} [3 \sin A - \sin 3A] \end{aligned} \right]$$

$$= \frac{6a}{4} \int_0^{\pi/2} \sin x\, dx - \frac{2a}{4} \int_0^{\pi/2} \sin 3x\, dx$$

$$= \frac{3a}{2} [-\cos x]_0^{\pi/2} - \frac{a}{2} \left[-\frac{\cos 3x}{3} \right]_0^{\pi/2} = -\frac{3a}{2} \left[\cos \frac{\pi}{2} - \cos 0^\circ \right] + \frac{a}{6} \left[\cos \frac{3\pi}{2} - \cos 0^\circ \right]$$

$$= -\frac{3a}{2} [0 - 1] + \frac{a}{6} [0 - 1]$$

$$= \frac{3a}{2} - \frac{a}{6} = \frac{9a - a}{6} = \frac{8a}{6} = \frac{4a}{3} \qquad ...(4)$$

Since, it is given that

$$I_1 = I_2 \qquad \text{[Using equations (3) and (4)]}$$

$$\therefore \quad \frac{2}{3} a^{3/2} = \frac{4a}{3}$$

$$\Rightarrow \quad a^{3/2} = 2a$$

$$\Rightarrow \quad a^{3/2} - 2a = 0$$

$$\Rightarrow \quad a(a^{1/2} - 2) = 0$$

$$\Rightarrow \quad a = 0 \quad \text{or} \quad a^{1/2} = 2$$

$$\Rightarrow \quad a = 0 \quad \text{or} \quad a = 4.$$

Now, two case arises :

Case I : When $a = 0$

$$\int_a^{a+1} x\,dx = \int_0^1 x\,dx = \left[\frac{x^2}{2}\right]_0^1 = \left[\frac{1}{2} - 0\right] = \frac{1}{2}.$$

Case II : When $a = 4$.

$$\int_a^{a+1} x\,dx = \int_4^5 x\,dx = \left[\frac{x^2}{2}\right]_4^5 = \frac{1}{2}[(5)^2 - (4)^2] = \frac{1}{2}[25 - 16]$$

$$= \frac{9}{2}.$$

Example 19. *Evaluate the following definite integrals :*

(i) $\int_0^1 \left(x\,e^x + \sin\frac{\pi x}{4}\right) dx$ (ii) $\int_0^1 \left(x\,e^x + \cos\frac{\pi x}{4}\right) dx$

(iii) $\int_0^{\pi/4} (1 - x^2) \sin 2x\,dx$ (iv) $\int_0^{1/\sqrt{2}} \frac{\sin^{-1} x}{(1 - x^2)^{3/2}}\,dx.$

Solution. (i) Let $I = \int_0^1 \left(x\,e^x + \sin\frac{\pi x}{4}\right) dx = \int_0^1 x\,e^x\,dx + \int_0^1 \sin\frac{\pi x}{4}\,dx$

Let $\quad I_1 = \int \underset{\text{I}}{x}\,\underset{\text{II}}{e^x}\,dx$

Integrating by parts, we have

$$I_1 = \left[x\,.\int e^x\,dx - \int\left\{\frac{d}{dx}(x)\,.\int e^x\,dx\right\}dx\right]$$

$$= \left[x\,e^x - \int 1\,.\,e^x\,dx\right] = [x\,e^x - e^x]$$

$$\therefore \quad \int_0^1 x\,e^x\,dx + \int_0^1 \sin\frac{\pi x}{4}\,dx = \left[x\,e^x - e^x\right]_0^1 + \left[\frac{-\cos\frac{\pi x}{4}}{\frac{\pi}{4}}\right]_0^1$$

$$= [(1\,e^1 - e^1) - (0 - e^0)] - \frac{4}{\pi}\left[\cos\frac{\pi}{4} - \cos 0^\circ\right]$$

$$= [0 - (-1)] - \frac{4}{\pi}\left[\frac{1}{\sqrt{2}} - 1\right]$$

$$= 1 - \frac{4}{\pi}\left(\frac{1-\sqrt{2}}{\sqrt{2}}\right).$$

(*ii*) Let $\quad I = \int_0^1 \left(x\, e^x + \cos \frac{\pi x}{4} \right) dx$

$$= \int_0^1 x\, e^x\, dx + \int_0^1 \cos \frac{\pi x}{4}\, dx \qquad [\because \text{ See Part } (i)]$$

$$= \left[x\, e^x - e^x \right]_0^1 + \left[\frac{\sin \frac{\pi x}{4}}{\frac{\pi}{4}} \right]_0^1 = [(1 \,.\, e^1 - e^1) - (0 - e^0)] + \frac{4}{\pi}\left[\sin \frac{\pi}{4} - \sin 0^\circ \right]$$

$$= [0 - (-1)] + \frac{4}{\pi}\left[\frac{1}{\sqrt{2}} - 0 \right]$$

$$= 1 + \frac{2\sqrt{2}}{\pi}.$$

(*iii*) Let $\quad I = \int_0^{\pi/4} (1 - x^2) \sin 2x\, dx$

Let $I_1 = \int \underset{\text{I}}{(1 - x^2)}\, \underset{\text{II}}{\sin 2x}\, dx$

Integrating by parts, we have

$$I_1 = (1 - x^2) \,.\, \int \sin 2x\, dx - \int \left\{ \frac{d}{dx}(1 - x^2) \,.\, \int \sin 2x\, dx \right\} dx$$

$$= (1 - x^2)\left(-\frac{\cos 2x}{2} \right) - \int (-2x)\left(-\frac{\cos 2x}{2} \right) dx$$

$$= -\frac{1}{2}(1 - x^2) \cos 2x - \int \underset{\text{I}}{x}\, \underset{\text{II}}{\cos 2x}\, dx$$

Integrating by parts again, we have

$$I_1 = \frac{1}{2}(x^2 - 1) \cos 2x - \left[x \,.\, \int \cos 2x\, dx - \int \left\{ \frac{d}{dx}(x) \,.\, \int \cos 2x\, dx \right\} dx \right]$$

$$= \frac{1}{2}(x^2 - 1) \cos 2x - \left[x \,.\, \frac{\sin 2x}{2} - \int \frac{1 . \sin 2x}{2}\, dx \right]$$

$$= \frac{1}{2}(x^2 - 1) \cos 2x - \frac{1}{2} x \sin 2x + \frac{1}{2}\left[-\frac{\cos 2x}{2} \right]$$

$$= \frac{1}{2}(x^2 - 1) \cos 2x - \frac{1}{2} x \sin 2x - \frac{1}{4} \cos 2x$$

$$= \frac{1}{2} x^2 \cos 2x - \frac{1}{2} \cos 2x - \frac{1}{2} x \sin 2x - \frac{1}{4} \cos 2x$$

$$= \frac{1}{2} x^2 \cos 2x - \frac{1}{2} x \sin 2x - \frac{3}{4} \cos 2x$$

$$= \frac{1}{4}(2x^2 - 3) \cos 2x - \frac{1}{2} x \sin 2x.$$

$\therefore \quad I = \int_0^{\pi/4} (1 - x^2) \sin 2x \, dx = \left[\frac{1}{4}(2x^2 - 3)\cos 2x - \frac{1}{2} x \sin 2x\right]_0^{\pi/4}$

$= \frac{1}{4}\left[(2x^2 - 3)\cos 2x - 2x \sin 2x\right]_0^{\pi/4}$

$= \frac{1}{4}\left[\left\{2\left(\frac{\pi}{4}\right)^2 - 3\right\}\cos\frac{\pi}{2} - 2 . \frac{\pi}{4}\sin\frac{\pi}{2}\right] - \frac{1}{4}\left[(0-3)\cos 0° - 0\right]$

$= \frac{1}{4}\left[\left(2 . \frac{\pi^2}{16} - 3\right) . 0 - \frac{\pi}{2}\right] - \frac{1}{4}\left[-3 - 0\right]$

$= -\frac{\pi}{8} + \frac{3}{4} = \frac{-\pi + 6}{8} = \frac{6 - \pi}{8}.$

(*iv*) Let $\quad I = \int_0^{1/\sqrt{2}} \frac{\sin^{-1} x}{(1 - x^2)^{3/2}} dx = \int_0^{1/\sqrt{2}} \frac{\sin^{-1} x}{(1 - x^2)\sqrt{1 - x^2}} dx$

Put $\sin^{-1} x = z \Rightarrow x = \sin z \Rightarrow \frac{1}{\sqrt{1 - x^2}} dx = dz$

When $\quad x = 0, \sin z = 0 \Rightarrow z = 0$ and when $x = \frac{1}{\sqrt{2}}, \sin z = \frac{1}{\sqrt{2}} \Rightarrow z = \frac{\pi}{4}$

$\therefore \quad I = \int_0^{\pi/4} \frac{z}{(1 - \sin^2 z)} dz = \int_0^{\pi/4} \frac{z}{\cos^2 z} dz = \int_0^{\pi/4} z \sec^2 z \, dz$

Let $\quad I_1 = \int \underset{\text{I}}{z} \, \underset{\text{II}}{\sec^2 z} \, dz$

Integrating by parts, we get

$I_1 = z . \int \sec^2 z \, dz - \int \left\{\frac{d}{dz}(z) . \int \sec^2 z \, dz\right\} dz$

$= z \tan z - \int 1 . \tan z \, dz$

$= z \tan z - \log |\sec z|$

$\therefore \quad I = \int_0^{\pi/4} z \sec^2 z \, dz = \left[z \tan z - \log|\sec z|\right]_0^{\pi/4}$

$= \left[\frac{\pi}{4}\tan\frac{\pi}{4} - \log\left(\sec\frac{\pi}{4}\right)\right] - \left[0 - \log(\sec 0°)\right]$

$= \left(\frac{\pi}{4}\tan\frac{\pi}{4} - \log\sqrt{2}\right) - (-\log 1) \qquad [\because \log 1 = 0]$

$= \frac{\pi}{4}\tan\frac{\pi}{4} - \frac{1}{2}\log 2. \qquad [\because m \log n = \log n^m]$

Example 20. *Evaluate the following :*

(i) $\int_0^{2\pi} e^{x/2} \sin\left(\frac{x}{2}+\frac{\pi}{4}\right) dx$ (ii) $\int_2^4 \frac{x^2+x}{\sqrt{2x+1}}\, dx$

(iii) $\int_0^{\pi/2} x \sin x \, dx$ (iv) $\int_0^{\pi/6} (2+3x^2) \cos 3x \, dx$

(v) $\int_{\pi/4}^{\pi/2} \cos 2x \log \sin x \, dx$ (vi) $\int_0^{\pi/2} x^2 \sin 3x \, dx$

(vii) $\int_1^3 \frac{\log x}{(x+1)^2}\, dx$ (viii) $\int_1^2 \log x \, dx.$

Solution. (i) Let $I = \int_0^{2\pi} e^{x/2} \sin\left(\frac{x}{2}+\frac{\pi}{4}\right) dx$

Let $I_1 = \int \underset{\text{I}}{e^{x/2}} \underset{\text{II}}{\sin\left(\frac{x}{2}+\frac{\pi}{4}\right)} dx$

Integrating by parts, we have

$$I_1 = e^{x/2} . \int \sin\left(\frac{x}{2}+\frac{\pi}{4}\right) dx - \int \left\{\frac{d}{dx}(e^{x/2}) . \int \sin\left(\frac{x}{2}+\frac{\pi}{4}\right) dx\right\} dx$$

$$= e^{x/2} . \left[\frac{-\cos\left(\frac{x}{2}+\frac{\pi}{4}\right)}{\frac{1}{2}}\right] - \int \frac{1}{2} e^{x/2} . \left[\frac{-\cos\left(\frac{x}{2}+\frac{\pi}{4}\right)}{\frac{1}{2}}\right] dx$$

$$= -2e^{x/2} \cos\left(\frac{x}{2}+\frac{\pi}{4}\right) + \int \underset{\text{I}}{e^{x/2}} \underset{\text{II}}{\cos\left(\frac{x}{2}+\frac{\pi}{4}\right)} dx$$

Integrating by parts again, we have

$$I_1 = -2e^{x/2} \cos\left(\frac{x}{2}+\frac{\pi}{4}\right) + \left[e^{x/2} . \int \cos\left(\frac{x}{2}+\frac{\pi}{4}\right) dx - \int \left\{\frac{d}{dx}(e^{x/2}) . \int \cos\left(\frac{x}{2}+\frac{\pi}{4}\right) dx\right\} dx\right]$$

$$= -2e^{x/2} \cos\left(\frac{x}{2}+\frac{\pi}{4}\right) + \left[e^{x/2} \left\{\frac{\sin\left(\frac{x}{2}+\frac{\pi}{4}\right)}{\frac{1}{2}}\right\} - \int \frac{1}{2} e^{x/2} . \frac{\sin\left(\frac{x}{2}+\frac{\pi}{4}\right)}{\frac{1}{2}} . dx\right]$$

$$= -2e^{x/2} \cos\left(\frac{x}{2}+\frac{\pi}{4}\right) + 2e^{x/2} \sin\left(\frac{x}{2}+\frac{\pi}{4}\right) - \int e^{x/2} \sin\left(\frac{x}{2}+\frac{\pi}{4}\right) dx$$

$$= -2e^{x/2} \cos\left(\frac{x}{2}+\frac{\pi}{4}\right) + 2e^{x/2} \sin\left(\frac{x}{2}+\frac{\pi}{4}\right) - I_1$$

$$\Rightarrow \quad 2I_1 = -2e^{x/2} \cos\left(\frac{x}{2}+\frac{\pi}{4}\right) + 2e^{x/2} \sin\left(\frac{x}{2}+\frac{\pi}{4}\right)$$

$$\Rightarrow \quad I_1 = -e^{x/2}\left[\cos\left(\frac{x}{2}+\frac{\pi}{4}\right) - \sin\left(\frac{x}{2}+\frac{\pi}{4}\right)\right]$$

$$\therefore \quad I = \int_0^{2\pi} e^{x/2} \sin\left(\frac{x}{2}+\frac{\pi}{4}\right) dx = -\left[e^{x/2}\left\{\cos\left(\frac{x}{2}+\frac{\pi}{4}\right) - \sin\left(\frac{x}{2}+\frac{\pi}{4}\right)\right\}\right]_0^{2\pi}$$

$$= -\left[e^{2\pi/2}\left\{\cos\left(\frac{2\pi}{2}+\frac{\pi}{4}\right) - \sin\left(\frac{2\pi}{2}+\frac{\pi}{4}\right)\right\}\right] - \left[e^0\left\{\cos\left(0+\frac{\pi}{4}\right) - \sin\left(0+\frac{\pi}{4}\right)\right\}\right]$$

$$= -\left[e^{\pi}\left\{\cos\frac{5\pi}{4} - \sin\frac{5\pi}{4}\right\}\right] - \left[\cos\frac{\pi}{4} - \sin\frac{\pi}{4}\right]$$

$$= -\left[e^{\pi}\left\{-\frac{1}{\sqrt{2}} - \left(-\frac{1}{\sqrt{2}}\right)\right\}\right] - \left[\frac{1}{\sqrt{2}} - \frac{1}{\sqrt{2}}\right] \qquad \left[\because \begin{array}{l} \cos\frac{5\pi}{4} = \cos\left(\pi+\frac{\pi}{4}\right) = -\frac{1}{\sqrt{2}} \\ \sin\frac{5\pi}{4} = \sin\left(\pi+\frac{\pi}{4}\right) = -\frac{1}{\sqrt{2}} \end{array}\right]$$

$$= -e^{\pi}\left(-\frac{1}{\sqrt{2}} + \frac{1}{\sqrt{2}}\right) - \left(\frac{1}{\sqrt{2}} - \frac{1}{\sqrt{2}}\right)$$

$$= -e^{\pi}(0) - (0) = 0.$$

(ii) Let $I = \int_2^4 \frac{x^2+x}{\sqrt{2x+1}}\, dx$

Let $I_1 = \int \underset{\text{I}}{(x^2+x)} \,.\, \underset{\text{II}}{(2x+1)^{-1/2}}\, dx$

Integrating by parts, we have

$$I_1 = (x^2+x) \,.\, \int (2x+1)^{-1/2}\, dx - \int\left\{\frac{d}{dx}(x^2+x) \,.\, \int (2x+1)^{-1/2}\, dx\right\} dx$$

$$= (x^2+x) \,.\, \frac{(2x+1)^{-\frac{1}{2}+1}}{2.\left(-\frac{1}{2}+1\right)} - \int \frac{(2x+1)\,.\,(2x+1)^{-\frac{1}{2}+1}}{2.\left(-\frac{1}{2}+1\right)}\, dx$$

$$\left[\because \int (ax+b)^n\, dx = \frac{(ax+b)^{n+1}}{a(n+1)} + c\right]$$

$$= (x^2+x)\sqrt{2x+1} - \int (2x+1)(2x+1)^{1/2}\, dx$$

$$= (x^2+x)\sqrt{2x+1} - \int (2x+1)^{3/2}\, dx$$

$$= (x^2+x)\sqrt{2x+1} - \frac{(2x+1)^{\frac{3}{2}+1}}{2.\left(\frac{3}{2}+1\right)} = (x^2+x)\sqrt{2x+1} - \frac{1}{5}(2x+1)^{5/2}$$

$$\therefore \quad \int_2^4 \frac{x^2+x}{\sqrt{2x+1}}\, dx = \left[(x^2+x)\sqrt{2x+1} - \frac{1}{5}(2x+1)^{5/2}\right]_2^4$$

$$= \left[(4^2 + 4)\sqrt{2(4)+1} - \frac{1}{5}[2(4)+1]^{5/2}\right] - \left[(2^2 + 2)\sqrt{2(2)+1} - \frac{1}{5}[2(2)+1]^{5/2}\right]$$

$$= \left[20\sqrt{9} - \frac{1}{5}(9)^{5/2}\right] - \left[6\sqrt{5} - \frac{1}{5}(5)^{5/2}\right]$$

$$= \left(20(3) - \frac{243}{5}\right) - 6\sqrt{5} + \frac{25\sqrt{5}}{5} \qquad \left[\because \begin{array}{l} (9)^{5/2} = (9^{1/2})^5 = (3)^5 = 243 \\ (5)^{5/2} = (5^5)^{1/2} = \sqrt{3125} = \sqrt{625 \times 5} = 25\sqrt{5} \end{array}\right]$$

$$= 60 - \frac{243}{5} - 6\sqrt{5} + 5\sqrt{5}$$

$$= \frac{300 - 243}{5} - \sqrt{5} = \frac{57}{5} - \sqrt{5}\,.$$

(*iii*) Let $\quad I = \int_0^{\pi/2} x \sin x \, dx$

Let $\quad I_1 = \int \underset{\text{I}}{x} \, . \underset{\text{II}}{\sin x} \, dx$

Integrating by parts, we have

$$I_1 = x \, . \int \sin x \, dx - \int \left\{\frac{d}{dx}(x) \, . \int \sin x \, dx\right\} dx$$

$$= x(-\cos x) - \int 1 \, . (-\cos x) \, dx = -x\cos x + \int \cos x \, dx = -x \cos x + \sin x$$

$\therefore \quad I = \int_0^{\pi/2} x \sin x \, dx = [-x \cos x + \sin x]_0^{\pi/2}$

$$= \left[-\frac{\pi}{2}\cos\frac{\pi}{2} + \sin\frac{\pi}{2}\right] - [-0 + \sin 0°] = \left[-\frac{\pi}{2}(0) + 1\right] - [0 + 0]$$

$$= 1.$$

(*iv*) Let $\quad I = \int_0^{\pi/6} (2 + 3x^2) \cos 3x \, dx$

Let $\quad I_1 = \int \underset{\text{I}}{(2 + 3x^2)} . \underset{\text{II}}{\cos 3x} \, dx$

Integrating by parts, we have

$$I_1 = (2 + 3x^2) \, . \int \cos 3x \, dx - \int \left\{\frac{d}{dx}(2 + 3x^2) \, . \int \cos 3x \, dx\right\} dx$$

$$= (2 + 3x^2) \, . \frac{\sin 3x}{3} - \int 6x \, . \frac{\sin 3x}{3} \, dx$$

$$= \frac{1}{3}(2 + 3x^2) \sin 3x - 2\int \underset{\text{I}}{x} \, \underset{\text{II}}{\sin 3x} \, dx$$

Integrating again by parts, we have

$$I_1 = \frac{1}{3}(2 + 3x^2)\sin 3x - 2\left[x \, . \int \sin 3x \, dx - \int \left\{\frac{d}{dx}(x) \, . \int \sin 3x \, dx\right\} dx\right]$$

$$= \frac{1}{3}(2 + 3x^2)\sin 3x - 2\left[x \, . \left(-\frac{\cos 3x}{3}\right) - \int 1 \, . \left(-\frac{\cos 3x}{3}\right) dx\right]$$

$$= \frac{1}{3}(2+3x^2)\sin 3x + \frac{2}{3}x\cos 3x - \frac{2}{3}\int \cos 3x\, dx$$

$$= \frac{1}{3}(2+3x^2)\sin 3x + \frac{2}{3}x\cos 3x - \frac{2}{3}\left(\frac{\sin 3x}{3}\right)$$

$$= \frac{1}{3}(2+3x^2)\sin 3x + \frac{2}{3}x\cos 3x - \frac{2}{9}\sin 3x$$

$$= \frac{1}{9}(6+9x^2-2)\sin 3x + \frac{2}{3}x\cos 3x$$

$$= \frac{1}{9}(9x^2+4)\sin 3x + \frac{2}{3}x\cos 3x$$

$$\therefore \quad \int_0^{\pi/6}(2+3x^2)\cos 3x\, dx = \left[\frac{1}{9}(9x^2+4)\sin 3x + \frac{2}{3}x\cos 3x\right]_0^{\pi/6}$$

$$= \left[\frac{1}{9}\left\{9\left(\frac{\pi}{6}\right)^2+4\right\}\sin\frac{3\pi}{6} + \frac{2}{3}\left(\frac{\pi}{6}\right)\cos\frac{3\pi}{6}\right] - \left[\frac{1}{9}(0+4)\sin 0° + 0\right]$$

$$= \left(\frac{1}{9}\left[\frac{9\pi^2}{36}+4\right]\sin\frac{\pi}{2} + \frac{\pi}{9}\cos\frac{\pi}{2}\right) - 0$$

$$= \left[\frac{\pi^2}{36}+\frac{4}{9}+0\right] = \frac{1}{36}(\pi^2+16).$$

(*v*) Let $\quad I = \int_{\pi/4}^{\pi/2} \cos 2x \log(\sin x)\, dx$

Let $\quad I_1 = \int \underset{\text{I}}{\cos 2x}\ \underset{\text{II}}{\log(\sin x)}\, dx$

Integrating by parts, we have

$$I_1 = \log(\sin x) . \int \cos 2x\, dx - \int \left\{\frac{d}{dx}\{\log(\sin x)\} . \int \cos 2x\, dx\right\} dx$$

$$= \log(\sin x) . \frac{\sin 2x}{2} - \int \frac{1}{\sin x} . \cos x . \frac{\sin 2x}{2}\, dx$$

$[\because \quad \sin 2A = 2\sin A\cos A]$

$$= \frac{1}{2}\log(\sin x) . \sin 2x - \frac{1}{2}\int \frac{\cos x}{\sin x} . 2\sin x\cos x\, dx$$

$$= \frac{1}{2}\log(\sin x) . \sin 2x - \int \cos^2 x\, dx \qquad \left[\begin{array}{l}\because \ 1+\cos 2A = 2\cos^2 A \\ \Rightarrow \frac{1}{2}(1+\cos 2A) = \cos^2 A\end{array}\right]$$

$$= \frac{1}{2}\log(\sin x) . \sin 2x - \frac{1}{2}\int (1+\cos 2x)\, dx$$

$$= \frac{1}{2}\log(\sin x) . \sin 2x - \frac{1}{2}\left[x + \frac{\sin 2x}{2}\right]$$

$$= \frac{1}{2}\left[\log(\sin x) . \sin 2x - x - \frac{\sin 2x}{2}\right]$$

$$\therefore \quad I = \int_{\pi/4}^{\pi/2} \cos 2x \log(\sin x)\, dx = \frac{1}{2}\left[\log(\sin x) . \sin 2x - x - \frac{\sin 2x}{2}\right]_{\pi/4}^{\pi/2}$$

$$= \frac{1}{2}\left[\log\left(\sin\frac{\pi}{2}\right)\sin 2\left(\frac{\pi}{2}\right) - \frac{\pi}{2} - \frac{\sin^2\left(\frac{\pi}{2}\right)}{2}\right]$$

$$- \frac{1}{2}\left[\log\left(\sin\frac{\pi}{4}\right)\sin 2\left(\frac{\pi}{4}\right) - \frac{\pi}{4} - \frac{\sin^2\left(\frac{\pi}{4}\right)}{2}\right]$$

$$= \frac{1}{2}\left[(\log 1)\sin\pi - \frac{\pi}{2} - \frac{1}{2}\sin\pi\right] - \frac{1}{2}\left[\log\left(\frac{1}{\sqrt{2}}\right)\sin\frac{\pi}{2} - \frac{\pi}{4} - \frac{1}{2}\sin\frac{\pi}{2}\right]$$

$$= \frac{1}{2}\left[0 - \frac{\pi}{2} - 0\right] - \frac{1}{2}\left[\log\left(\frac{1}{\sqrt{2}}\right) - \frac{\pi}{4} - \frac{1}{2}\right]$$

$$= -\frac{\pi}{4} - \frac{1}{2}\log\left(\frac{1}{\sqrt{2}}\right) + \frac{\pi}{8} + \frac{1}{4}$$

$$= -\frac{\pi}{8} + \frac{1}{4} - \frac{1}{2}\log\left(\frac{1}{\sqrt{2}}\right).$$

(*vi*) Let $I = \int_0^{\pi/2} x^2 \sin 3x\, dx$

Let $I_1 = \int \underset{\text{I}}{x^2} \underset{\text{II}}{\sin 3x}\, dx$

Integrating by parts, we have

$$I_1 = x^2 . \int \sin 3x\, dx - \int \left\{\frac{d}{dx}(x^2) . \int \sin 3x\, dx\right\} dx$$

$$= x^2 . \frac{(-\cos 3x)}{3} - \int 2x \frac{(-\cos 3x)}{3}\, dx$$

$$= -\frac{1}{3}x^2 \cos 3x + \frac{2}{3}\int \underset{\text{I}}{x} \underset{\text{II}}{\cos 3x}\, dx$$

Integrating by parts again, we have

$$I_1 = -\frac{1}{3}x^2\cos 3x + \frac{2}{3}\left[x . \int \cos 3x\, dx - \int \left\{\frac{d}{dx}(x) . \int \cos 3x\, dx\right\} dx\right]$$

$$= -\frac{1}{3}x^2\cos 3x + \frac{2}{3}\left[x . \frac{\sin 3x}{3} - \int \frac{1 . \sin 3x}{3}\, dx\right]$$

$$= -\frac{1}{3}x^2\cos 3x + \frac{2}{9}x\sin 3x - \frac{2}{9}\int \sin 3x\, dx$$

$$= -\frac{1}{3}x^2\cos 3x + \frac{2}{9}x\sin 3x - \frac{2}{9}\left(-\frac{\cos 3x}{3}\right)$$

$$= -\frac{1}{3}x^2\cos 3x + \frac{2}{9}x\sin 3x + \frac{2}{27}\cos 3x$$

$$= \frac{1}{27}(-9x^2+2)\cos 3x + \frac{2}{9}x\sin 3x$$

$$\therefore \qquad I = \int_0^{\pi/2} x^2 \sin 3x\, dx = \left[\frac{1}{27}(-9x^2+2)\cos 3x + \frac{2}{9}x\sin 3x\right]_0^{\pi/2}$$

$$= \left[\frac{1}{27}\left\{-9\left(\frac{\pi}{2}\right)^2+2\right\}\cos\frac{3\pi}{2} + \frac{2}{9}\left(\frac{\pi}{2}\right)\sin\frac{3\pi}{2}\right] - \left[\frac{1}{27}\{0+2\}\cos 0^\circ + 0\right]$$

$$= \left[0 + \frac{\pi}{9}(-1)\right] - \frac{2}{27} \qquad \left[\because \cos\left(\frac{3\pi}{2}\right) = \cos\left(\pi+\frac{\pi}{2}\right) = 0 \quad \sin\left(\frac{3\pi}{2}\right) = \sin\left(\pi+\frac{\pi}{2}\right) = -1\right]$$

$$= -\frac{\pi}{9} - \frac{2}{27} = -\frac{1}{27}(3\pi+2).$$

(*vii*) Let $I = \int_1^3 \frac{\log x}{(x+1)^2}\, dx$ ...(1)

Let $I_1 = \int \underset{\text{I}}{\log x} \,.\, \underset{\text{II}}{(x+1)^{-2}}\, dx$ ...(2)

Integrating by parts, we get

$$I_1 = \log x \,.\, \int (x+1)^{-2}\, dx - \int \left\{\frac{d}{dx}(\log x) \,.\, \int (x+1)^{-2}\, dx\right\} dx$$

$$= \log x \,.\, \frac{(x+1)^{-2+1}}{(-2+1)} - \int \frac{1}{x} \,.\, \frac{(x+1)^{-2+1}}{(-2+1)}\, dx$$

$$= -\frac{\log x}{x+1} + \int \frac{1}{x(x+1)}\, dx \qquad \text{...(3)}$$

Let $\frac{1}{x(x+1)} = \frac{A}{x} + \frac{B}{x+1}$ ...(4)

Muliplying both sides by $x(x + 1)$, we have

$$1 = A(x+1) + Bx \qquad \text{...(5)}$$

Put $x = 0$ in equation (5), we get

$$1 = A(0+1) + 0 \quad \Rightarrow \quad A = 1$$

Put $x = -1$ in equation (5), we get

$$1 = A(-1+1) + B(-1) \quad \Rightarrow \quad 1 = -B \quad \Rightarrow \quad B = -1$$

Substituting the values of A and B in equation (4), we have

$$\frac{1}{x(x+1)} = \frac{1}{x} + \frac{-1}{x+1}$$

$\therefore$ From equation (3), we get

$$I_1 = -\frac{\log x}{x+1} + \int \left(\frac{1}{x} - \frac{1}{x+1}\right) dx$$

$$= -\frac{\log x}{x+1} + \int \frac{1}{x}\, dx - \int \frac{1}{x+1}\, dx = -\frac{\log x}{x+1} + \log|x| - \log|x+1|$$

$$= -\frac{\log x}{x+1} + \log\left|\frac{x}{x+1}\right| \qquad \left[\because \log m - \log n = \log\left(\frac{m}{n}\right)\right]$$

$$\therefore \quad \int_1^3 \frac{\log x}{(x+1)^2}\,dx = \left[-\frac{\log (x)}{(x+1)} + \log\left|\frac{x}{x+1}\right|\right]_1^3$$

$$= \left[-\frac{\log 3}{3+1} + \log\left(\frac{3}{3+1}\right)\right] - \left[-\frac{\log 1}{1+1} + \log\left(\frac{1}{1+1}\right)\right]$$

$$= -\frac{\log 3}{4} + \log\left(\frac{3}{4}\right) - \left[0 + \log\left(\frac{1}{2}\right)\right]$$

$$= -\frac{1}{4}\log 3 + \log 3 - \log 4 - [\log 1 - \log 2]$$

$$= \frac{3}{4}\log 3 - \log 4 + \log 2 \qquad \left[\because \ \log m - \log n = \log\left(\frac{m}{n}\right)\right]$$

$$= \frac{3}{4}\log 3 - 2\log 2 + \log 2 \qquad [\because \ \log 4 = \log 2^2 = 2\log 2]$$

$$= \frac{3}{4}\log 3 - \log 2.$$

(*viii*) Let $\qquad I = \int_1^2 \log x \, . \, dx$

Let $\qquad I_1 = \int \underset{II}{1} \, . \underset{I}{\log x} \qquad$ [Taking unity as the second function]

Integrating by parts, we have

$$= \log x \, . \int 1 \, . \, dx - \int \left\{\frac{d}{dx}(\log x) \, . \int 1 \, . \, dx\right\} dx$$

$$= \log x \, . \, x - \int \frac{1}{x} \, . \, x\,dx = x\log x - \int 1 \, . \, dx$$

$$= x\log x - x$$

$$\therefore \quad \int_1^2 \log x\,dx = \Big[x\log x - x\Big]_1^2$$

$$= [2\log 2 - 2] - [1\log 1 - 1]$$

$$= 2\log 2 - 2 - 0 + 1 = 2\log 2 - 1 \qquad [\because \ n\log m = \log m^n]$$

$$= \log 2^2 - 1 = \log 4 - 1$$

$$= -1 + \log 4.$$

Example 21. *If f(x) is of the form :*

$$f(x) = a + bx + cx^2$$

Show that : $\displaystyle \int_0^1 f(x)\,dx = \frac{1}{6}\left[f(0) + 4f\left(\frac{1}{2}\right) + f(1)\right].$

Solution. It is given that

$$f(x) = cx^2 + bx + a \qquad ...(1)$$

$$\therefore \quad \int_0^1 f(x)\,dx = \int_0^1 (cx^2 + bx + a)\,dx$$

$$= c\int_0^1 x^2\,dx + b\int_0^1 x\,dx + a\int_0^1 1 \, . \, dx$$

$$= c\left[\frac{x^3}{3}\right]_0^1 + b\left[\frac{x^2}{2}\right]_0^1 + a\left[x\right]_0^1$$

$$= \frac{c}{3}[1-0] + \frac{b}{2}[1-0] + a[1-0]$$

$$\int_0^1 f(x)\,dx = \frac{c}{3} + \frac{b}{2} + a \qquad ...(2)$$

Now $\quad f(x) = a + bx + cx^2$

For $x = 0$; $\quad f(0) = a + b(0) + c(0) = a$

For $x = \frac{1}{2}$; $f\left(\frac{1}{2}\right) = a + b\left(\frac{1}{2}\right) + c\left(\frac{1}{2}\right)^2 = a + \frac{b}{2} + \frac{c}{4}$

For $x = 1$; $\quad f(1) = a + b(1) + c(1)^2 = a + b + c$

$$\therefore \quad \frac{1}{6}\left[f(0) + 4f\left(\frac{1}{2}\right) + f(1)\right] = \frac{1}{6}\left[a + 4 + \left(a + \frac{b}{2} + \frac{c}{4}\right) + a + b + c\right]$$

$$= \frac{1}{6}[6a + 3b + 2c] = a + \frac{b}{2} + \frac{c}{3} \qquad ...(3)$$

$\therefore$ From equations (2) and (3), we have

$$\int_0^1 f(x)\,dx = \frac{1}{6}\left[f(0) + 4f\left(\frac{1}{2}\right) + f(1)\right].$$

Example 22. *Evaluate the following integrals :*

(*i*) $\int_0^1 \frac{5x}{(4+x^2)^2}\,dx$ (*ii*) $\int_0^1 x\,e^{x^2}\,dx$

(*iii*) $\int_0^2 \frac{5x+1}{x^2+4}\,dx$ (*iv*) $\int_2^4 \frac{x}{x^2+1}\,dx$

(*v*) $\int_1^2 3x\sqrt{5-x^2}\,dx$ (*vi*) $\int_0^1 \frac{x^2}{1+x^6}\,dx.$

Solution. (*i*) Let $I = \int_0^1 \frac{5x}{(4+x^2)^2}\,dx$

Put $\quad (4 + x^2) = z \;\Rightarrow\; 2x\,dx = dz \;\Rightarrow\; x\,dx = \frac{1}{2}\,dz$

When $x = 0$, $z = 4$ and when $x = 1$, $z = 4 + 1^2 = 5$

$$\therefore \qquad I = 5\int_4^5 \frac{1}{z^2}\left(\frac{1}{2}\,dz\right) = \frac{5}{2}\int_4^5 z^{-2}\,dz = \frac{5}{2}\left[\frac{z^{-2+1}}{-2+1}\right]_4^5$$

$$= -\frac{5}{2}\left[\frac{1}{z}\right]_4^5 = -\frac{5}{2}\left[\frac{1}{5} - \frac{1}{4}\right] = -\frac{5}{2}\left[\frac{4-5}{20}\right]$$

$$= \left(-\frac{5}{2}\right)\left(-\frac{1}{20}\right) = \frac{1}{8}.$$

(*ii*) Let $\quad I = \int_0^1 x\, e^{x^2}\, dx$

Put $\quad x^2 = z \Rightarrow 2x\, dx = dz \Rightarrow x\, dx = \frac{1}{2}\, dz$

When $\quad x = 0,\ z = 0$ and when $x = 1,\ z = (1)^2 = 1$

$$\therefore \quad I = \int_0^1 e^z \left(\frac{1}{2}\right) dz = \frac{1}{2}\int_0^1 e^z\, dz = \frac{1}{2}\left[e^z\right]_0^1$$

$$= \frac{1}{2}\,[e^1 - e^0] = \frac{1}{2}\,[e - 1].$$

(*iii*) Let $\quad I = \int_0^2 \frac{5x+1}{x^2+4}\,.\,dx$

$$= \int_0^2 \left[\frac{5x}{x^2+4} + \frac{1}{x^2+4}\right] dx$$

$$= 5\int_0^2 \frac{x}{x^2+4}\, dx + \int_0^2 \frac{1}{x^2+2^2}\, dx$$

$$= \frac{5}{2}\int_0^2 \frac{2x}{x^2+4}\, dx + \int_0^2 \frac{1}{x^2+2^2}\, dx \qquad \left[\begin{array}{l}\text{Multiply and divide}\\ \text{the first integral by 2}\end{array}\right]$$

$$= \frac{5}{2}\left[\log|x^2+4|\right]_0^2 + \frac{1}{2}\left[\tan^{-1}\frac{x}{2}\right]_0^2 \qquad \left[\begin{array}{l}\because \int \frac{f'(x)}{f(x)}\, dx = \log|f(x)| + c\\ \int \frac{1}{a^2+x^2}\, dx = \frac{1}{a}\tan^{-1}\frac{x}{a} + c\end{array}\right]$$

$$= \frac{5}{2}\,[\log(2^2+4) - \log(0+4)] + \frac{1}{2}\left[\tan^{-1}\frac{2}{2} - \tan^{-1} 0\right]$$

$$= \frac{5}{2}\,[\log 8 - \log 4] + \frac{1}{2}\,[\tan^{-1} 1 - \tan^{-1} 0]$$

$$= \frac{5}{2}\left[\log\frac{8}{4}\right] + \frac{1}{2}\left[\frac{\pi}{4} - 0\right] \qquad \left[\begin{array}{l}\because \log m - \log n = \log\left(\frac{m}{n}\right)\\ \tan\frac{\pi}{4} = \tan 45° = 1\end{array}\right]$$

$$= \frac{5}{2}\log 2 + \frac{\pi}{8}\,.$$

(*iv*) Let $\quad I = \int_2^4 \frac{x}{x^2+1}\, dx$

Put $\quad x^2 + 1 = z \Rightarrow 2x\, dx = dz \Rightarrow x\, dx = \frac{1}{2}\, dz$

When $\quad x = 2,\ z = (2)^2 + 1 = 4 + 1 = 5$ and when $x = 4,\ z = (4)^2 + 1 = 16 + 1 = 17$

$\therefore \quad I = \int_5^{17} \frac{1}{z}\left(\frac{1}{2}\, dz\right) = \frac{1}{2}\int_5^{17} \frac{1}{z}\, dz$

$= \frac{1}{2}\Big[\log|z|\Big]_5^{17}$

$= \frac{1}{2}\,[\log|17| - \log|5|]$ $\quad \left[\because \log m - \log n = \log\left(\frac{m}{n}\right)\right]$

$= \frac{1}{2}\log\frac{17}{5}$.

(*v*) Let $\quad I = \int_1^2 3x\sqrt{5-x^2}\, dx$

Put $\quad 5 - x^2 = z \Rightarrow -2x\, dx = dz \Rightarrow x\, dx = -\frac{1}{2}\, dz$

When $\quad x = 1,\ z = 5 - (1)^2 = 5 - 1 = 4$ and when $x = 2,\ z = 5 - (2)^2 = 5 - 4 = 1$

$\therefore \quad I = 3\int_4^1 \sqrt{z}\,.\left(-\frac{1}{2}\, dz\right) = -\frac{3}{2}\int_4^1 z^{1/2}\, dz$

$= -\frac{3}{2}\left[\frac{z^{\frac{1}{2}+1}}{\frac{1}{2}+1}\right]_4^1 = \left(-\frac{3}{2}\right)\left(\frac{2}{3}\right)\Big[z^{3/2}\Big]_4^1$ $\quad [\because (4)^{3/2} = (2^2)^{3/2} = (2)^3 = 8]$

$= -[(1)^{3/2} - (4)^{3/2}] = -[1 - 8] = 7$.

(*vi*) Let $\quad I = \int_0^1 \frac{x^2}{1+x^6}\, dx$

Put $\quad x^3 = z \Rightarrow 3x^2\, dx = dz \Rightarrow x^2\, dx = \frac{1}{3}\, dz$

When $\quad x = 0,\ z = 0$ and when $x = 1,\ z = (1)^3 = 1$

$\therefore \quad I = \int_0^1 \frac{1}{1+z^2}\left(\frac{1}{3}\, dz\right) = \frac{1}{3}\int_0^1 \frac{1}{1+z^2}\, dz$ $\quad \left[\because \int \frac{1}{a^2+x^2}\, dx = \frac{1}{a}\tan^{-1}\frac{x}{a} + c\right]$

$= \frac{1}{3}\Big[\tan^{-1} z\Big]_0^1 = \frac{1}{3}\,[\tan^{-1} 1 - \tan^{-1} 0]$ $\quad \left[\because \tan\frac{\pi}{4} = \tan 45° = 1\right]$

$= \frac{1}{3}\left(\frac{\pi}{4} - 0\right) = \frac{\pi}{12}$.

Example 23. *Evaluate the following integrals :*

(*i*) $\int_0^1 \frac{x}{1+\sqrt{x}}\, dx$ $\qquad$ (*ii*) $\int_0^1 \frac{e^x}{1+e^{2x}}\, dx$

(*iii*) $\int_0^1 \frac{24x^3}{(1+x^2)^4}\, dx$ $\qquad$ (*iv*) $\int_1^2 \frac{3x}{9x^2-1}\, dx$

(*v*) $\int_0^1 x\sqrt{1-x^2}\, dx$ $\qquad$ (*vi*) $\int_0^1 x(1-x)^5\, dx$.

Solution. (*i*) Let $I = \int_0^1 \frac{x}{1+\sqrt{x}}\, dx$

Put $\quad \sqrt{x} = z \Rightarrow x = z^2 \Rightarrow dx = 2z\, dz$

When $\quad x = 0, z = 0$ and when $x = 1, z = \sqrt{1} = 1$

$$\therefore \quad I = \int_0^1 \frac{z^2}{1+z}(2z\, dz) = 2\int_0^1 \frac{z^3}{(1+z)}\, dz$$

$$= 2\int_0^1 \frac{z^3+1-1}{z+1}\, dz \qquad \text{[Add and subtract 1 to the numerator]}$$

$$= 2\int_0^1 \left[\frac{z^3+1}{z+1} - \frac{1}{z+1}\right] dz = 2\int_0^1 \frac{z^3+1}{z+1}\, dz - 2\int_0^1 \frac{1}{z+1}\, dz$$

Remember. If the integrand is a rational function whose numerator and denominator are polynomials and degree of numerator is greater than or equal to that of the denominator, first divide the numerator by the denominator and then use the result :

$$\left[\frac{\text{Numerator}}{\text{Denominator}} = \text{Quatient} + \frac{\text{Remainder}}{\text{Denominator}}\right]$$

$\therefore$ By actual division ;

$$= 2\int_0^1 (z^2 - z + 1)\, dz - 2\int_0^1 \frac{1}{z+1}\, dz$$

$$= 2\left[\frac{z^3}{3} - \frac{z^2}{2} + z\right]_0^1 - 2\left[\log|z+1|\right]_0^1$$

$$= 2\left[\frac{1}{3} - \frac{1}{2} + 1 - 0\right] - 2\,[\log(1+1) - \log(0+1)]$$

$$= 2\left[\frac{2-3+6}{6}\right] - 2\,[\log 2 - \log 1]$$

$$= \frac{5}{3} - 2\log 2.$$

$$\begin{array}{r|l}
z+1) & z^3 + 1 \;(z^2 - z + 1 \\
 & z^3 + z^2 \\
 & -\quad - \\ \hline
 & -z^2 + 1 \\
 & -z^2 - z \\
 & +\quad + \\ \hline
 & +z + 1 \\
 & z + 1 \\
 & -\quad - \\ \hline
 & \times
\end{array}$$

(*ii*) Let $\quad I = \int_0^1 \frac{e^x}{1+e^{2x}}\, dx$

Put $\quad e^x = z \Rightarrow e^x\, dx = dz$

When $\quad x = 0 \Rightarrow z = e^0 = 1$ and when $x = 1 \Rightarrow z = e^1 = e$

$$\therefore \quad I = \int_1^e \frac{1}{1+z^2}\, dz$$

$$= \left[\tan^{-1} z\right]_1^e \qquad \left[\because \int \frac{1}{a^2+x^2}\, dx = \frac{1}{a}\tan^{-1}\frac{x}{a} + c\right]$$

$$= [\tan^{-1} e - \tan^{-1} 1] = \tan^{-1} e - \frac{\pi}{4}.$$

(*iii*) Let $\quad I = \int_0^1 \frac{24x^3}{(1+x^2)^4}\, dx = \int_0^1 \frac{12x^2 \,.\, 2x}{(1+x^2)^4}\, dx$ [Note this step]

Put $(1 + x^2) = z \;\Rightarrow\; 2x\, dx = dz$

When $x = 0,\ z = (0 + 1) = 1$ and when $x = 1,\ z = (1 + 1^2) = 2$

$$\therefore \quad I = \int_1^2 \frac{12(z-1)}{z^4}\, dz$$

$$= 12 \int_1^2 \left(\frac{z}{z^4} - \frac{1}{z^4}\right) dz = 12 \int_1^2 \frac{1}{z^3}\, dz - 12 \int_1^2 \frac{1}{z^4}\, dz$$

$$= 12 \left[\frac{z^{-3+1}}{-3+1}\right]_1^2 - 12 \left[\frac{z^{-4+1}}{-4+1}\right]_1^2 = -6\left[\frac{1}{z^2}\right]_1^2 + 4\left[\frac{1}{z^3}\right]_1^2$$

$$= -6\left[\frac{1}{2^2} - \frac{1}{1^2}\right] + 4\left[\frac{1}{2^3} - \frac{1}{1^3}\right]$$

$$= -6\left[\frac{1}{4} - 1\right] + 4\left[\frac{1}{8} - 1\right] = -6\left(-\frac{3}{4}\right) + 4\left(-\frac{7}{8}\right)$$

$$= \frac{9}{2} - \frac{7}{2} = \frac{2}{2} = 1.$$

(*iv*) Let $\quad I = \int_1^2 \frac{3x}{9x^2 - 1}\, dx$

Put $9x^2 - 1 = z \;\Rightarrow\; 18x\, dx = dz \;\Rightarrow\; 3x\, dx = \frac{1}{6}\, dz$

When $x = 1,\ z = 9(1)^2 - 1 = 9 - 1 = 8$ and when $x = 2,\ z = 9(2)^2 - 1 = 9(4) - 1 = 36 - 1 = 35$

$$\therefore \quad I = \int_8^{35} \frac{1}{z}\left(\frac{1}{6}\, dz\right) = \frac{1}{6}\int_8^{35} \frac{1}{z}\, dz$$

$$= \frac{1}{6}\Big[\log|z|\Big]_8^{35} = \frac{1}{6}\,[\log 35 - \log 8]$$

$$= \frac{1}{6} \log \frac{35}{8}. \qquad \left[\because \log m - \log n = \log\left(\frac{m}{n}\right)\right]$$

(*v*) Let $I = \int_0^1 x\sqrt{1-x^2}\, dx$

Put $1 - x^2 = z \;\Rightarrow\; -2x\, dx = dz \;\Rightarrow\; x\, dx = -\frac{1}{2}\, dz$

When $x = 0,\ z = 1 - 0 = 1$ and when $x = 1,\ z = 1 - (1)^2 = 1 - 1 = 0$

$$\therefore \quad I = \int_1^0 \sqrt{z}\,.\left(-\frac{1}{2}\right) dz = -\frac{1}{2}\int_1^0 z^{1/2}\, dz$$

$$= -\frac{1}{2}\left[\frac{z^{\frac{1}{2}+1}}{\frac{1}{2}+1}\right]_1^0 = -\frac{1}{2} \times \frac{2}{3}\left[z^{3/2}\right]_1^0$$

$$= -\frac{1}{3}[0-1] = \frac{1}{3}.$$

(*vi*) Let $\quad I = \int_0^1 x(1-x)^5\, dx$

Put $\quad (1-x) = z \Rightarrow 1 - z = x \Rightarrow -dx = dz \Rightarrow dx = -dz$

When $\quad x = 0,\ z = 1 - 0 = 1$ and when $x = 1,\ z = 1 - 1 = 0$

$$\therefore \quad I = \int_1^0 (1-z)\, z^5\, (-dz) = -\int_1^0 (z^5 - z^6)\, dz$$

$$= \int_1^0 (z^6 - z^5)\, dz = \left[\frac{z^7}{7} - \frac{z^6}{6}\right]_1^0$$

$$= (0-0) - \left(\frac{1}{7} - \frac{1}{6}\right) = -\left(\frac{6-7}{42}\right)$$

$$= -\left(-\frac{1}{42}\right) = \frac{1}{42}.$$

Example 24. *Evaluate the following integrals :*

(*i*) $\int_0^1 \sin^{-1} x\, dx$ (*ii*) $\int_a^b \frac{\log x}{x}\, dx$

(*iii*) $\int_0^{\pi/2} \frac{\sin x \cos x}{1 + \sin^4 x}\, dx$ (*iv*) $\int_1^2 x\sqrt{3x-2}\, dx$

(*v*) $\int_0^1 \cos^{-1} x\, dx$ (*vi*) $\int_0^{\pi/2} \frac{\sin x}{1 + \cos^2 x}\, dx.$

Solution. (*i*) Let $I = \int_0^1 \sin^{-1} x\, dx$

Put $\quad \sin^{-1} x = z \Rightarrow x = \sin z \Rightarrow dx = \cos z\, dz$

When $\quad x = 0,\ \sin z = 0 \Rightarrow z = 0$ and when $x = 1,\ \sin z = 1 \Rightarrow z = \frac{\pi}{2}$

$$\therefore \quad I = \int_0^{\pi/2} z \cos z\, dz$$

Let $\quad I_1 = \int \underset{\text{I}}{z} \underset{\text{II}}{\cos z}\, dz$

Integrating by parts, we get

$$I_1 = z \cdot \int \cos z\, dz - \int \left\{\frac{d}{dz}(z) \cdot \int \cos z\, dz\right\} dz$$

$$= z \sin z - \int 1 \cdot \sin z\, dz$$

$$= z \sin z - (-\cos z) = (z \sin z + \cos z)$$

$$\therefore \quad \int_0^{\pi/2} z \cos z\, dz = \Big[z \sin z + \cos z\Big]_0^{\pi/2}$$

$$= \left[\frac{\pi}{2} \sin \frac{\pi}{2} + \cos \frac{\pi}{2}\right] - [0 + \cos 0°]$$

$$= \left(\frac{\pi}{2} + 0\right) - (0 + 1) = \frac{\pi}{2} - 1.$$

(*ii*) Let $\quad I = \int_a^b \frac{\log x}{x}\, dx$

Put $\quad \log x = z \Rightarrow \frac{1}{x}\, dx = dz$

When $\quad x = a,\ z = \log a$ and when $x = b,\ z = \log b$

$$\therefore \quad I = \int_{\log a}^{\log b} z\, dz = \left[\frac{z^2}{2}\right]_{\log a}^{\log b} = \frac{1}{2}\, [(\log b)^2 - (\log a)^2]$$

$$= \frac{1}{2}\, [(\log b + \log a)(\log b - \log a)] \qquad \left[\begin{array}{l} \because \log m + \log n = \log m.n \\ \log m - \log n = \log\left(\frac{m}{n}\right) \end{array}\right]$$

$$= \frac{1}{2} \log (ba) \log\left(\frac{b}{a}\right).$$

(*iii*) Let $\quad I = \int_0^{\pi/2} \frac{\sin x \cos x}{1 + \sin^4 x}\, dx$

Put $\quad \sin^2 x = z \Rightarrow 2 \sin x \cos x\, dx = dz \Rightarrow \sin x \cos x = \frac{1}{2}\, dz$

When $\quad x = 0,\ z = \sin^2 0 = 0$ and when $x = \frac{\pi}{2},\ z = \sin^2 \frac{\pi}{2} = (1)^2 = 1$

$$\therefore \quad I = \int_0^1 \frac{1}{1 + z^2}\left(\frac{1}{2}\, dz\right) = \frac{1}{2}\int_0^1 \frac{1}{1 + z^2}\, dz$$

$$= \frac{1}{2}\left[\tan^{-1} z\right]_0^1 \qquad \left[\because \int \frac{1}{a^2 + x^2}\, dx = \frac{1}{a} \tan^{-1} \frac{x}{a} + c\right]$$

$$= \frac{1}{2}\, [\tan^{-1} 1 - \tan^{-1} 0] = \frac{1}{2}\left(\frac{\pi}{4} - 0\right)$$

$$= \frac{\pi}{8}.$$

(*iv*) Let $\quad I = \int_1^2 x\sqrt{3x - 2}\, dx$

Put $\sqrt{3x - 2} = z \Rightarrow (3x - 2) = z^2 \Rightarrow 3x = z^2 + 2 \Rightarrow 3\, dx = 2z\, dz \Rightarrow dx = \frac{2}{3}\, z\, dz$

When $\quad x = 1,\ z = \sqrt{3(1) - 2} = \sqrt{1} = 1$ and when $x = 2,\ z = \sqrt{3(2) - 2} = \sqrt{6 - 2} = \sqrt{4} = 2.$

$$\therefore \quad I = \int_1^2 \left(\frac{z^2 + 2}{3}\right).\, z.\left(\frac{2}{3}\, z\, dz\right)$$

$$= \frac{2}{9}\int_1^2 (z^4 + 2z^2)\, dz = \frac{2}{9}\left[\frac{z^5}{5} + \frac{2z^3}{3}\right]_1^2$$

$$= \frac{2}{9}\left[\frac{(2)^5}{5} + \frac{2(2)^3}{3}\right] - \frac{2}{9}\left[\frac{(1)^5}{5} + \frac{2(1)^3}{3}\right]$$

$$= \frac{2}{9}\left[\frac{32}{5} + \frac{16}{3}\right] - \frac{2}{9}\left[\frac{1}{5} + \frac{2}{3}\right] = \frac{2}{9}\left(\frac{96+80}{15}\right) - \frac{2}{9}\left(\frac{3+10}{15}\right) = \frac{2}{9}\left[\frac{176}{15} - \frac{13}{15}\right] = \frac{2}{9}\left[\frac{176-13}{15}\right]$$

$$= \frac{2}{9} \times \frac{163}{15} = \frac{326}{135}.$$

(*v*) Let $\quad I = \int_0^1 \cos^{-1} x \, dx$

Put $\quad \cos^{-1} x = z \Rightarrow x = \cos z \Rightarrow dx = -\sin z \, dz$

When $\quad x = 0, \cos z = 0 \Rightarrow z = \pi/2$ and when $x = 1, \cos z = 1 \Rightarrow z = 0$

$\therefore \quad I = \int_{\pi/2}^0 z(-\sin z \, dz) = -\int_{\pi/2}^0 z \sin z \, dz$

Let $\quad I_1 = \int \underset{\text{I}}{z} \underset{\text{II}}{\sin z} \, dz$

Integrating by parts, we have

$$I_1 = z \cdot \int \sin z \, dz - \int \left\{\frac{d}{dz}(z) \cdot \int \sin z \, dz\right\} dz$$

$$= z(-\cos z) - \int 1 \cdot (-\cos z) \, dz = -z \cos z + \int \cos z \, dz$$

$$= -z \cos z + \sin z$$

$$\therefore \quad -\int_{\pi/2}^0 z \sin z \, dz = -[-z \cos z + \sin z]_{\pi/2}^0$$

$$= \Big[z \cos z - \sin z\Big]_{\pi/2}^0$$

$$= [0 - \sin 0°] - \left[\frac{\pi}{2}\cos\frac{\pi}{2} - \sin\frac{\pi}{2}\right] = 0 - (0 - 1) = 1.$$

(*vi*) Let $\quad I = \int_0^{\pi/2} \frac{\sin x}{1 + \cos^2 x} \, dx$

Put $\quad \cos x = z \Rightarrow -\sin x \, dx = dz \Rightarrow \sin x \, dx = -dz$

When $\quad x = 0, z = \cos 0° = 1$ and when $x = \frac{\pi}{2}, z = \cos\frac{\pi}{2} = 0$

$$\therefore \quad I = \int_1^0 \frac{1}{1+z^2}(-dz) = -\int_1^0 \frac{1}{1+z^2} \, dz$$

$$= -\Big[\tan^{-1} z\Big]_1^0 \qquad \left[\because \int \frac{1}{a^2 + x^2} dx = \frac{1}{a}\tan^{-1}\frac{x}{a} + c\right]$$

$$= -[\tan^{-1} 0 - \tan^{-1} 1] = -\left(0 - \frac{\pi}{4}\right) = \frac{\pi}{4}.$$

Example 25. *Evaluate the following integrals :*

(i) $\int_0^{\pi/2} \sin 2x \tan^{-1} (\sin x)\, dx$ (ii) $\int_0^{\pi/2} \sqrt{\cos\theta}\, \sin^3\theta\, d\theta$

(iii) $\int_0^{\pi/2} \sqrt{\sin\theta}\, \cos^5\theta\, d\theta$ (iv) $\int_0^{\pi/2} \cos^3\theta\, (\sin\theta)^{1/4}\, d\theta$

(v) $\int_0^1 \sin^{-1}\left(\frac{2x}{1+x^2}\right) dx.$

Solution. (i) Let $I = \int_0^{\pi/2} \sin 2x \tan^{-1} (\sin x)\, dx$

$\Rightarrow \quad I = \int_0^{\pi/2} 2 \sin x \cos x \tan^{-1} (\sin x)\, dx$ [$\because$ $\sin 2A = 2 \sin A \cos A$]

Put $\sin x = z \Rightarrow \cos x\, dx = dz$

When $x = 0$, $z = \sin 0° = 0$ and when $x = \frac{\pi}{2}$, $z = \sin\frac{\pi}{2} = 1$

$\therefore \quad I = \int_0^1 2z (\tan^{-1} z)\, dz = 2\int_0^1 z \tan^{-1} z\, dz$

Let $I_1 = \int \underset{\text{II}}{z} \underset{\text{I}}{\tan^{-1} z}\, dz$

Integrating by parts, we get

$$I_1 = \tan^{-1} z . \int z\,dz - \int \left\{\frac{d}{dz}(\tan^{-1} z) . \int z\,dz\right\} dz$$

$$= \tan^{-1} z . \frac{z^2}{2} - \int \frac{1}{1+z^2} . \frac{z^2}{2}\, dz = \frac{1}{2} z^2 \tan^{-1} z - \frac{1}{2}\int \frac{z^2}{z^2+1}\, dz$$

$$= \frac{1}{2} z^2 \tan^{-1} z - \frac{1}{2}\int \left(\frac{z^2+1-1}{z^2+1}\right) dz$$ [Add and subtract 1 to the numerator]

$$= \frac{1}{2} z^2 \tan^{-1} z - \frac{1}{2}\int 1 . dz + \frac{1}{2}\int \frac{1}{1+z^2}\, dz$$

$$= \frac{1}{2} z^2 \tan^{-1} z - \frac{1}{2} z + \frac{1}{2}\tan^{-1} z$$ $\left[\because \int \frac{1}{a^2+x^2}\, dx = \frac{1}{a}\tan^{-1}\frac{x}{a} + c\right]$

$$= \frac{1}{2}[(z^2+1)\tan^{-1} z - z]$$

$$\therefore \quad I = 2\int_0^1 z \tan^{-1} z\, dz = 2\left[\frac{1}{2}\{(z^2+1)\tan^{-1} z - z\}\right]_0^1 = \left[(z^2+1)\tan^{-1} z - z\right]_0^1$$

$$= [(1+1)\tan^{-1} 1 - 1] - [(0+1)\tan^{-1} 0 - 0]$$

$$= [2\tan^{-1} 1 - 1] - \tan^{-1} 0 = 2\,\frac{\pi}{4} - 1 - 0$$

$$= \frac{\pi}{2} - 1.$$

(ii) Let $\quad I = \int_0^{\pi/2} \sqrt{\cos\theta} . \sin^3\theta\, d\theta$

$$= \int_0^{\pi/2} \sqrt{\cos\theta} . \sin^2\theta \sin\theta\, d\theta$$ [Note this step]

$$= \int_0^{\pi/2} \sqrt{\cos\theta}\,.\,(1-\cos^2\theta)\,.\,\sin\theta\,d\theta \qquad [\because \sin^2 A + \cos^2 A = 1]$$

Put $\cos\theta = z \Rightarrow -\sin\theta\,d\theta = dz \Rightarrow \sin\theta\,d\theta = -dz$

When $\theta = 0,\ z = \cos 0° = 1$ and when $\theta = \frac{\pi}{2},\ z = \cos\frac{\pi}{2} = 0$

$$\therefore \quad I = \int_1^0 \sqrt{z}\,(1-z^2)(-dz) = \int_1^0 \sqrt{z}\,(z^2-1)\,dz = \int_1^0 (z^{5/2} - z^{1/2})\,dz$$

$$= \left[\frac{z^{\frac{5}{2}+1}}{\frac{5}{2}+1} - \frac{z^{\frac{1}{2}+1}}{\frac{1}{2}+1}\right]_1^0 = \left[\frac{2}{7}z^{7/2} - \frac{2}{3}z^{3/2}\right]_1^0 = \left[(0-0) - \left(\frac{2}{7} - \frac{2}{3}\right)\right]$$

$$= -\left(\frac{6-14}{21}\right) = \frac{8}{21}.$$

(*iii*) Let $I = \int_0^{\pi/2} \sqrt{\sin\theta}\,.\,\cos^5\theta\,d\theta$

$$= \int_0^{\pi/2} \sqrt{\sin\theta}\,.\,(\cos^2\theta)^2\,.\,\cos\theta\,d\theta \qquad \text{[Note this step]}$$

$$= \int_0^{\pi/2} \sqrt{\sin\theta}\,.\,(1-\sin^2\theta)^2\cos\theta\,d\theta \qquad [\because \cos^2 A + \sin^2 A = 1]$$

Put $\sin\theta = z \Rightarrow \cos\theta\,d\theta = dz$

When $\theta = 0,\ z = \sin 0° = 0$ and when $\theta = \frac{\pi}{2},\ z = \sin\frac{\pi}{2} = 1$

$$\therefore \quad I = \int_0^1 \sqrt{z}\,(1-z^2)^2\,dz$$

$$= \int_0^1 \sqrt{z}\,(1 + z^4 - 2z^2)\,dz \qquad [\because (a-b)^2 = a^2 + b^2 - 2ab]$$

$$= \int_0^1 (z^{1/2} + z^{9/2} - 2z^{5/2})\,dz$$

$$= \left[\frac{z^{\frac{1}{2}+1}}{\frac{1}{2}+1} + \frac{z^{\frac{9}{2}+1}}{\frac{9}{2}+1} - \frac{2z^{\frac{5}{2}+1}}{\frac{5}{2}+1}\right]_0^1 = \left[\frac{2}{3}z^{3/2} + \frac{2}{11}z^{11/2} - \frac{4}{7}z^{7/2}\right]_0^1$$

$$= \left[\frac{2}{3}(1)^{3/2} + \frac{2}{11}(1)^{11/2} - \frac{4}{7}(1)^{7/2}\right] - [0+0-0] = \frac{2}{3} + \frac{2}{11} - \frac{4}{7}$$

$$= \frac{154 + 42 - 132}{231} = \frac{64}{231}.$$

(*iv*) Let $I = \int_0^{\pi/2} \cos^3\theta\,(\sin\theta)^{1/4}\,d\theta$

$$= \int_0^{\pi/2} (\sin\theta)^{1/4}\,.\,\cos^2\theta\,.\,\cos\theta\,d\theta \qquad \text{[Note this step]}$$

$$= \int_0^{\pi/2} (\sin \theta)^{1/4} . (1 - \sin^2 \theta) . \cos \theta \, d\theta \qquad [\because \sin^2 A + \cos^2 A = 1]$$

Put $\sin \theta = z \Rightarrow \cos \theta \, d\theta = dz$

When $\theta = 0, z = \sin 0° = 0$ and when $\theta = \frac{\pi}{2}, z = \sin \frac{\pi}{2} = 1$

$$\therefore \quad I = \int_0^1 (z)^{1/4} (1 - z^2) \, dz = \int_0^1 (z^{1/4} - z^{9/4}) \, dz$$

$$= \left[\frac{z^{\frac{1}{4}+1}}{\frac{1}{4}+1} - \frac{z^{\frac{9}{4}+1}}{\frac{9}{4}+1} \right]_0^1 = \left[\frac{4}{5} z^{5/4} - \frac{4}{13} z^{13/4} \right]_0^1$$

$$= \left[\frac{4}{5} (1)^{5/4} - \frac{4}{13} (1)^{13/4} \right] - [0 - 0] = \left(\frac{4}{5} - \frac{4}{13} \right) - 0$$

$$= \frac{52 - 20}{65} = \frac{32}{65}.$$

(*v*) Let $I = \int_0^1 \sin^{-1} \left(\frac{2x}{1 + x^2} \right) dx$

Put $x = \tan \theta \Rightarrow dx = \sec^2 \theta \, d\theta$

When $x = 0, \tan \theta = 0 \Rightarrow \theta = 0$ and when $x = 1, \tan \theta = 1 \Rightarrow \theta = \frac{\pi}{4}$

$$\therefore \quad I = \int_0^{\pi/4} \sin^{-1} \left(\frac{2 \tan \theta}{1 + \tan^2 \theta} \right) . \sec^2 \theta \, d\theta \qquad \left[\because \sin 2A = \frac{2 \tan A}{1 + \tan^2 A} \right]$$

$$= \int_0^{\pi/4} \sin^{-1} (\sin 2\theta) \sec^2 \theta \, d\theta$$

$$\Rightarrow \quad I = \int_0^{\pi/4} 2\theta \sec^2 \theta \, d\theta = 2 \int_0^{\pi/4} \theta \sec^2 \theta \, d\theta$$

Let $I_1 = \int \underset{\text{I}}{\theta} \; \underset{\text{II}}{\sec^2 \theta} \, d\theta$

Integrating by parts, we get

$$= \theta . \int \sec^2 \theta \, d\theta - \int \left\{ \frac{d}{d\theta} (\theta) . \int \sec^2 \theta \, d\theta \right\} d\theta$$

$$= \theta \tan \theta - \int 1 . \tan \theta \, d\theta = \theta \tan \theta - \log |\sec \theta|$$

$$\therefore \quad I = 2 \int_0^{\pi/4} \theta \sec^2 \theta \, d\theta = 2 \Big[\theta \tan \theta - \log |\sec \theta| \Big]_0^{\pi/4}$$

$$= 2 \left[\frac{\pi}{4} \tan \frac{\pi}{4} - \log \left(\sec \frac{\pi}{4} \right) \right] - 2 [0 - \log (\sec 0°)]$$

$$= 2 \left[\frac{\pi}{4} - \log \sqrt{2} \right] - 2(0 - \log 1)$$

$$= \frac{\pi}{2} - 2\log\sqrt{2} = \frac{\pi}{2} - \log\left(\sqrt{2}\right)^2 \qquad \begin{bmatrix} \because \log 1 = 0 \\ m\log n = \log n^m \end{bmatrix}$$

$$= \frac{\pi}{2} - \log 2.$$

Example 26. *Evaluate the following integrals :*

(*i*) $\int_0^a \frac{x}{\sqrt{a^2 + x^2}}\, dx$ (*ii*) $\int_0^a \frac{x^4}{\sqrt{a^2 - x^2}}\, dx$

(*iii*) $\int_1^3 \frac{\cos(\log x)}{x}\, dx$ (*iv*) $\int_0^1 \frac{x \tan^{-1} x}{(1 + x^2)^{3/2}}\, dx$

(*v*) $\int_0^{\pi/6} (1 - \cos 3\theta) \sin 3\theta\, d\theta$ (*vi*) $\int_{\pi/3}^{\pi/2} \frac{\sqrt{1 + \cos x}}{(1 - \cos x)^{5/2}}\, dx$

Solution. (*i*) Let $I = \int_0^a \frac{x}{\sqrt{a^2 + x^2}}\, dx$

Put $\sqrt{a^2 + x^2} = z \Rightarrow a^2 + x^2 = z^2 \Rightarrow 2x\,dx = 2z\,dz \Rightarrow x\,dx = z\,dz$

When $x = 0, z = \sqrt{a^2 + 0} = \sqrt{a^2} = a$ and when $x = a, z = \sqrt{a^2 + a^2} = \sqrt{2a^2} = \sqrt{2}\,a$

$$\therefore \quad I = \int_a^{\sqrt{2}a} \frac{1}{z} . (z . dz) = \int_a^{\sqrt{2}a} 1 . dz = \Big[z\Big]_a^{\sqrt{2}a} = [\sqrt{2}a - a]$$

$$= a(\sqrt{2} - 1) .$$

(*ii*) Let $I = \int_0^a \frac{x^4}{\sqrt{a^2 - x^2}}\, dx$

Put $x = a \sin\theta \Rightarrow dx = a\cos\theta\, d\theta$

When $x = 0, a\sin\theta = 0 \Rightarrow \sin\theta = 0 \Rightarrow \theta = 0$

and when $x = a, a\sin\theta = a \Rightarrow \sin\theta = 1 \Rightarrow \theta = \frac{\pi}{2}$.

$$\therefore \quad I = \int_0^{\pi/2} \frac{a^4 \sin^4\theta}{\sqrt{a^2 - a^2\sin^2\theta}} . a\cos\theta\, d\theta$$

$$= \int_0^{\pi/2} \frac{a^5 \sin^4\theta\cos\theta}{\sqrt{a^2(1 - \sin^2\theta)}}\, d\theta = \int_0^{\pi/2} \frac{a^5\sin^4\theta\cos\theta}{a\sqrt{\cos^2\theta}}\, d\theta \qquad [\because \sin^2 A + \cos^2 A = 1]$$

$$= a\int_0^{\pi/2} \sin^4\theta\, d\theta = a\int_0^{\pi/2} (\sin^2\theta)^2\, d\theta$$

$$= a^4 \int_0^{\pi/2} \left(\frac{1 - \cos 2\theta}{2}\right)^2 d\theta \qquad \begin{bmatrix} \because 1 - \cos 2A = 2\sin^2 A \\ \frac{1 - \cos 2A}{2} = \sin^2 A \end{bmatrix}$$

$$= \frac{a^4}{4} \int_0^{\pi/2} (1 + \cos^2 2\theta - 2\cos 2\theta)\, d\theta$$

$$= \frac{a^4}{4}\int_0^{\pi/2}\left[1+\frac{1}{2}(1+\cos 4\theta) - 2\cos 2\theta\right] d\theta \qquad \left[\begin{array}{l}\because\ 1+\cos 2A = 2\cos^2 A \\ \Rightarrow 1+\cos 4A = 2\cos^2 2A \\ \Rightarrow \frac{1}{2}(1+\cos 4A) = \cos^2 2A\end{array}\right]$$

$$= \frac{a^4}{8}\int_0^{\pi/2}(2+1+\cos 4\theta - 4\cos 2\theta)\, d\theta$$

$$= \frac{a^4}{8}\left[3\int_0^{\pi/2} 1.\, d\theta + \int_0^{\pi/2}\cos 4\theta\, d\theta - 4\int_0^{\pi/2}\cos 2\theta\, d\theta\right]$$

$$= \frac{a^4}{8}\left(3.\left[\theta\right]_0^{\pi/2} + \left[\frac{\sin 4\theta}{4}\right]_0^{\pi/2} - 4\left[\frac{\sin 2\theta}{2}\right]_0^{\pi/2}\right)$$

$$= \frac{a^4}{8}\left[3\left(\frac{\pi}{2}-0\right)+\frac{1}{4}\left(\sin\frac{4\pi}{2} - \sin 0°\right) - 2\left(\sin\frac{2\pi}{2} - \sin 0°\right)\right]$$

$$= \frac{a^4}{8}\left[\frac{3\pi}{2}+\frac{1}{4}(0-0) - 2(0-0)\right]$$

$$= \frac{3a^4\pi}{16}.$$

(*iii*) Let $\quad I = \int_1^3 \frac{\cos(\log x)}{x}\, dx$

Put $\quad \log x = z \quad \Rightarrow \quad \frac{1}{x}\, dx = dz$

When $\quad x = 1,\ z = \log 1 \quad \Rightarrow \quad z = 0$ and when $x = 3,\ z = \log 3 \quad \Rightarrow \quad z = \log 3.$

$\therefore \qquad I = \int_0^{\log 3} \cos z\, dz$

$$= \left[\sin z\right]_0^{\log 3} = [\sin(\log 3) \quad \sin 0°] = \sin(\log 3) - 0$$

$$= \sin(\log 3).$$

(*iv*) Let $\quad I = \int_0^1 \frac{x\tan^{-1} x}{(1+x^2)^{3/2}}\, dx = \int_0^1 \frac{x\tan^{-1} x}{(1+x^2)\sqrt{1+x^2}}\, dx$

Put $\quad \tan^{-1} x = z \quad \Rightarrow \quad x = \tan z \quad \Rightarrow \quad \frac{1}{1+x^2}\, dx = dz$

When $\quad x = 0,\ \tan z = 0 \quad \Rightarrow \quad z = 0$ and when $x = 1,\ \tan z = 1 \quad \Rightarrow \quad z = \frac{\pi}{4}.$

$\therefore \qquad I = \int_0^{\pi/4} \frac{z\tan z}{\sqrt{1+\tan^2 z}}\, dz$

$$= \int_0^{\pi/4} \frac{z\tan z}{\sqrt{\sec^2 z}}\, dz = \int_0^{\pi/4} \frac{z\tan z}{\sec z}\, dz \qquad [\because \sec^2 A - \tan^2 A = 1]$$

$$= \int_0^{\pi/4} z \,.\, \frac{\sin z}{\cos x} \,.\, \frac{1}{\frac{1}{\cos z}}\, dz = \int_0^{\pi/4} z \sin z \, dz$$

Let $\quad I_1 = \int \underset{I}{z} \underset{II}{\sin z} \, dz$

Integrating by parts, we get

$$I_1 = z \,.\, \int \sin z \, dz - \int \left\{ \frac{d}{dz}(z) \,.\, \int \sin z \, dz \right\} dz$$

$$= z(-\cos z) - \int 1 \,.\, (-\cos z) \, dz$$

$$= -z \cos z + \int \cos z \, dz = -z \cos z + \sin z$$

$\therefore$
$$I = \int_0^{\pi/4} z \sin z \, dz = \Big[-z \cos z + \sin z \Big]_0^{\pi/4}$$

$$= \left[-\frac{\pi}{4} \cos \frac{\pi}{4} + \sin \frac{\pi}{4} \right] - [0 + \sin 0°] = \left(-\frac{\pi}{4} \,.\, \left(\frac{1}{\sqrt{2}} \right) + \frac{1}{\sqrt{2}} \right) - 0$$

$$= \frac{1}{\sqrt{2}} - \frac{\pi}{4\sqrt{2}} = \left(\frac{4 - \pi}{4\sqrt{2}} \right).$$

(*v*) Let $\quad I = \int_0^{\pi/6} (1 - \cos 3\theta) \sin 3\theta \, d\theta$

Put $\cos 3\theta = z \;\Rightarrow\; -3 \sin 3\theta \, d\theta = dz \;\Rightarrow\; \sin 3\theta \, d\theta = -\frac{1}{3} dz$

When $\quad \theta = 0,\ z = \cos 0 \;\Rightarrow\; z = 1$ and when $\theta = \frac{\pi}{6}$, $z = \cos \frac{3\pi}{6} \;\Rightarrow\; z = \cos \frac{\pi}{2} = 0$

$\therefore$
$$I = \int_1^0 (1 - z) \left(-\frac{1}{3} dz \right) = -\frac{1}{3} \int_1^0 (1 - z) \, dz$$

$$= -\frac{1}{3} \left[z - \frac{z^2}{2} \right]_1^0 = -\frac{1}{3} \left[(0 - 0) - \left(1 - \frac{1}{2} \right) \right] = -\frac{1}{3} \left[-\frac{1}{2} \right]$$

$$= \frac{1}{6}.$$

(*vi*) Let $\quad I = \int_{\pi/3}^{\pi/2} \frac{\sqrt{1 + \cos x}}{(1 - \cos x)^{5/2}} \, dx = \int_{\pi/3}^{\pi/2} \frac{\sqrt{1 + \cos x}}{(1 - \cos x)^{5/2}} \times \frac{\sqrt{1 - \cos x}}{\sqrt{1 - \cos x}} \, dx$

[Rationalization]

$$= \int_{\pi/3}^{\pi/2} \frac{\sqrt{1 - \cos^2 x}}{(1 - \cos x)^3} \, dx = \int_{\pi/3}^{\pi/2} \frac{\sqrt{\sin^2 x}}{(1 - \cos x)^3} \, dx \qquad [\because \sin^2 A + \cos^2 A = 1]$$

$$= \int_{\pi/3}^{\pi/2} \frac{\sin x}{(1 - \cos x)^3} \, dx$$

Put $(1-\cos x)=z \Rightarrow -(-\sin x)\,dx=dz \Rightarrow \sin x\,dx=dz$

When $x=\frac{\pi}{3}$, $z=1-\cos\frac{\pi}{3}=1-\frac{1}{2}=\frac{1}{2}$ and when $x=\frac{\pi}{2}$, $z=1-\cos\frac{\pi}{2}=1-0=1$

$$\therefore \quad I=\int_{1/2}^{1}\frac{1}{z^3}\,dz=\int_{1/2}^{1}z^{-3}\,dz=\left[\frac{z^{-3+1}}{-3+1}\right]_{1/2}^{1}=-\frac{1}{2}\left[\frac{1}{z^2}\right]_{1/2}^{1}$$

$$=-\frac{1}{2}\left[\frac{1}{(1)^2}-\frac{1}{(1/2)^2}\right]=-\frac{1}{2}\left[1-\frac{1}{1/4}\right]=-\frac{1}{2}[1-4]$$

$$=-\frac{1}{2}(-3)=\frac{3}{2}.$$

Example 27. *Evaluate the following integrals :*

(i) $\int_0^{2\pi}\frac{\cos\theta}{\sqrt{4+3\sin\theta}}\,d\theta$ (ii) $\int_0^{a}\sin^{-1}\sqrt{\frac{x}{a+x}}\,dx$

(iii) $\int_0^{\left(\frac{\pi}{2}\right)^{1/3}} x^2\sin x^3\,dx$ (iv) $\int_0^{\pi/4}\tan^3 x\,dx$

(v) $\int_0^{1}\frac{\sin(\tan^{-1}x)}{1+x^2}\,dx$ (vi) $\int_0^{1}\frac{\sqrt{\tan^{-1}x}}{1+x^2}\,dx.$

Solution. (i) Let $I=\int_0^{2\pi}\frac{\cos\theta}{\sqrt{4+3\sin\theta}}\,d\theta$

Put $4+3\sin\theta=z \Rightarrow 3\cos\theta\,d\theta=dz \Rightarrow \cos\theta\,d\theta=\frac{1}{3}\,dz$

When $\theta=0$, $z=4+3\sin 0°=4+3(0)=4$

and when $\theta=2\pi$, $z=4+3\sin 2\pi=4+3(0)=4$

$$\therefore \quad I=\int_4^4\frac{1}{\sqrt{z}}\left(\frac{1}{3}\,dz\right)=\frac{1}{3}\int_4^4 z^{-1/2}\,dz$$

$$=\frac{1}{3}\left[\frac{z^{-\frac{1}{2}+1}}{-\frac{1}{2}+1}\right]_4^4=\frac{2}{3}\left[z^{1/2}\right]_4^4=\frac{2}{3}[(4)^{1/2}-(4)^{1/2}]=\frac{2}{3}[2-2]$$

$$=\frac{2}{3}(0)=0.$$

(ii) Let $I=\int_0^{a}\sin^{-1}\sqrt{\frac{x}{a+x}}\,dx$

Put $x=a\tan^2\theta \Rightarrow dx=2a\tan\theta\sec^2\theta\,d\theta$

When $x=0,\ a\tan^2\theta=0 \Rightarrow \tan^2\theta=0 \Rightarrow \theta=0$

and when $x=a,\ a\tan^2\theta=a \Rightarrow \tan^2\theta=1 \Rightarrow \tan\theta=1 \Rightarrow \theta=\frac{\pi}{4}.$

$$\therefore \quad I=\int_0^{\pi/4}\sin^{-1}\sqrt{\frac{a\tan^2\theta}{a+a\tan^2\theta}}\,(2a\tan\theta\sec^2\theta\,d\theta)$$

$$= 2a \int_0^{\pi/4} \sin^{-1} \sqrt{\frac{\tan^2 \theta}{1+\tan^2 \theta}} \tan \theta \sec^2 \theta \, d\theta$$

$$= 2a \int_0^{\pi/4} \sin^{-1} \sqrt{\frac{\tan^2 \theta}{\sec^2 \theta}} \tan \theta \sec^2 \theta \, d\theta \qquad [\because \sec^2 A - \tan^2 A = 1]$$

$$= 2a \int_0^{\pi/4} \sin^{-1} \left(\frac{\tan \theta}{\sec \theta}\right) \tan \theta \sec^2 \theta \, d\theta = 2a \int_0^{\pi/4} \sin^{-1} \left[\frac{\frac{\sin \theta}{\cos \theta}}{\frac{1}{\cos \theta}}\right] \tan \theta \sec^2 \theta \, d\theta$$

$$= 2a \int_0^{\pi/4} \sin^{-1} (\sin \theta) \,.\, \tan \theta \sec^2 \theta \, d\theta = 2a \int_0^{\pi/4} \theta \,.\, \tan \theta \sec^2 \theta \, d\theta$$

Let $I_1 = \int \underset{I}{\theta} \, \underset{II}{\tan \theta \sec^2 \theta} \, d\theta$

Integrating by parts, we get

$$I_1 = \theta \,.\, \int \tan \theta \sec^2 \theta \, d\theta - \int \left\{\frac{d}{d\theta}(\theta) \,.\, \int \tan \theta \sec^2 \theta \, d\theta\right\} d\theta$$

$$= \theta \,.\, \frac{\tan^2 \theta}{2} - \int 1 . \frac{\tan^2 \theta}{2} \, d\theta \qquad \left[\because \int [f(x)]^n f'(x) \, dx = \frac{f(x)^{n+1}}{n+1} + c\right]$$

$$= \frac{1}{2} \theta \tan^2 \theta - \frac{1}{2} \int (\sec^2 \theta - 1) \, d\theta \qquad [\because \sec^2 A - \tan^2 A = 1]$$

$$= \frac{1}{2} \theta \tan^2 \theta - \frac{1}{2} \int \sec^2 \theta \, d\theta + \frac{1}{2} \int 1 \,.\, d\theta = \frac{1}{2} \theta \tan^2 \theta - \frac{1}{2} \tan \theta + \frac{\theta}{2}$$

$$\therefore \qquad I = 2a \int_0^{\pi/4} \theta \,.\, \tan \theta \sec^2 \theta \, d\theta = 2a \left[\frac{1}{2} \theta \tan^2 \theta - \frac{1}{2} \tan \theta + \frac{\theta}{2}\right]_0^{\pi/4}$$

$$= 2a \,.\, \frac{1}{2} \left[\theta \tan^2 \theta - \tan \theta + \theta\right]_0^{\pi/4} = a \left[\left(\frac{\pi}{4} \tan^2 \frac{\pi}{4} - \tan \frac{\pi}{4} + \frac{\pi}{4}\right) - (0 - \tan 0^\circ + 0)\right]$$

$$= a \left[\left(\frac{\pi}{4} - 1 + \frac{\pi}{4}\right) - (0)\right] = a \left[\frac{2\pi}{4} - 1\right]$$

$$= \frac{a\pi}{2} - a.$$

(*iii*) Let $I = \int_0^{\left(\frac{\pi}{2}\right)^{1/3}} x^2 \sin x^3 \, dx$

Put $\quad x^3 = z \Rightarrow 3x^2 \, dx = dz \Rightarrow x^2 \, dx = \frac{1}{3} dz$

When $\quad x = 0, z = (0)^3 = 0$ and when $x = \left(\frac{\pi}{2}\right)^{1/3}, z = \left[\left(\frac{\pi}{2}\right)^{1/3}\right]^3 \Rightarrow z = \frac{\pi}{2}$

$$\therefore \qquad I = \int_0^{\pi/2} \sin z \,.\left(\frac{1}{3} dz\right) = \frac{1}{3} \int_0^{\pi/2} \sin z \, dz$$

$$= \frac{1}{3}\left[-\cos z\right]_0^{\pi/2} = -\frac{1}{3}\left[\cos\frac{\pi}{2} - \cos 0°\right]$$

$$= -\frac{1}{3}[0-1] = \frac{1}{3}.$$

(*iv*) Let $\quad I = \int_0^{\pi/4} \tan^3 x\,dx = \int_0^{\pi/4} \tan^2 x \,.\, \tan x\,dx$

$$= \int_0^{\pi/4} (\sec^2 x - 1)\,.\,\tan x\,dx \qquad [\because \quad \sec^2 A - \tan^2 A = 1]$$

$$= \int_0^{\pi/4} \tan x\,.\,\sec^2 x\,dx - \int_0^{\pi/4} \tan x\,dx$$

$$= \left[\frac{\tan^2 x}{2}\right]_0^{\pi/4} - \left[\log|\sec x|\right]_0^{\pi/4} \qquad \left[\because \int [f(x)]^n f'(x)\,dx = \frac{[f(x)]^{n+1}}{n+1} + c\right]$$

$$= \frac{1}{2}\left[\tan^2\frac{\pi}{4} - \tan^2 0°\right] - \left[\log\left|\sec\frac{\pi}{4}\right| - \log|\sec 0°|\right]$$

$$= \frac{1}{2}[(1)^2 - 0] - [\log(\sqrt{2}) - \log 1]$$

$$= \frac{1}{2} - \log\sqrt{2} \qquad \left[\because \log m - \log n = \log\left(\frac{m}{n}\right)\right]$$

$$= \frac{1}{2} - \log(2)^{1/2} = \frac{1}{2} - \frac{1}{2}\log 2 \qquad [\because \quad m\log n = \log n^m]$$

$$= \frac{1}{2}(1 - \log 2).$$

(*v*) Let $\quad I = \int_0^1 \frac{\sin(\tan^{-1} x)}{1+x^2}\,dx$

Put $\tan^{-1} x = z \;\Rightarrow\; x = \tan z \;\Rightarrow\; \frac{1}{1+x^2}\,dx = dz$

When $\quad x = 0, \tan z = 0 \;\Rightarrow\; z = 0$ and when $x = 1, \tan z = 1 \;\Rightarrow\; z = \frac{\pi}{4}$

$\therefore \quad I = \int_0^{\pi/4} \sin z\,dz$

$$= \left[-\cos z\right]_0^{\pi/4} = -\left[\cos\frac{\pi}{4} - \cos 0^c\right] = -\left[\frac{1}{\sqrt{2}} - 1\right]$$

$$= 1 - \frac{1}{\sqrt{2}} = \frac{\sqrt{2}-1}{\sqrt{2}}.$$

(*vi*) Let $\quad I = \int_0^1 \frac{\sqrt{\tan^{-1} x}}{1+x^2}\,dx$

Put $\quad \tan^{-1} x = z \;\Rightarrow\; x = \tan z \;\Rightarrow\; \frac{1}{1+x^2}\,dx = dz$

When $x = 0, \tan z = 0 \Rightarrow z = 0°$ and when $x = 1, \tan z = 1 \Rightarrow z = \frac{\pi}{4}$.

$$\therefore \quad I = \int_0^{\pi/4} \sqrt{z} \, . \, dz$$

$$= \left[\frac{z^{\frac{1}{2}+1}}{\frac{1}{2}+1}\right]_0^{\pi/4} = \frac{2}{3}\left[z^{3/2}\right]_0^{\pi/4} = \frac{2}{3}\left[\left(\frac{\pi}{4}\right)^{3/2} - (0)^{3/2}\right]$$

$$= \frac{2}{3} . \frac{\pi^{3/2}}{8} = \frac{\pi^{3/2}}{12} . \qquad [\because \; 4^{3/2} = (2^2)^{3/2} = 2^3 = 8]$$

Example 28. *Evaluate the following integrals :*

(i) $\int_0^{\pi/4} \frac{(3\tan^2 x + 2)\sec^2 x}{(\tan^3 x + 2\tan x + 9)^2} dx$ (ii) $\int_0^{\pi} 5(5 - 4\cos\theta)^{1/4} . \sin\theta \, d\theta$

(iii) $\int_0^1 2x \cos^{-1} x^2 \, dx$ (iv) $\int_0^{\pi/6} \cos^{-3} 2\theta \sin 2\theta \, d\theta$

(v) $\int_0^{\pi/2} \frac{\cos x}{1 + \sin^2 x} dx$ (vi) $\int_0^{(\pi^2)^{1/3}} \sqrt{x} \cos^2 (x^{3/2}) \, dx.$

Solution. (*i*) Let $I = \int_0^{\pi/4} \frac{(3\tan^2 x + 2)\sec^2 x}{(\tan^3 x + 2\tan x + 9)^2} dx$

Put $\tan^3 x + 2\tan x + 9 = z$

$\Rightarrow (3\tan^2 x \sec^2 x + 2\sec^2 x) \, dx = dz \Rightarrow (3\tan^2 x + 2)\sec^2 x \, dx = dz$

When $x = 0, z = \tan^3 0° + 2\tan 0° + 9 \Rightarrow z = 0 + 0 + 9 \Rightarrow z = 9.$

and when $x = \frac{\pi}{4}, z = \tan^3 \frac{\pi}{4} + 2\tan\frac{\pi}{4} + 9 = (1)^3 + 2(1) + 9 = 1 + 2 + 9 = 12.$

$$\therefore \quad I = \int_9^{12} \frac{1}{z^2} dz = \int_9^{12} z^{-2} \, dz = \left[\frac{z^{-2+1}}{-2+1}\right]_9^{12} = -\left[\frac{1}{z}\right]_9^{12} = -\left[\frac{1}{12} - \frac{1}{9}\right]$$

$$= -\left[\frac{3-4}{36}\right] = \frac{1}{36} .$$

(*ii*) Let $I = \int_0^{\pi} 5(5 - 4\cos\theta)^{1/4} . \sin\theta \, d\theta$

Put $(5 - 4\cos\theta) = z \Rightarrow 4\sin\theta \, d\theta = dz \Rightarrow \sin\theta \, d\theta = \frac{1}{4} dz$

When $\theta = 0, \quad z = 5 - 4\cos 0° \Rightarrow z = 5 - 4(1) \Rightarrow z = 1$

and when $\theta = \pi, \quad z = 5 - 4\cos\pi \Rightarrow z = 5 - 4(-1) \Rightarrow z = 5 + 4 = 9.$

$$\therefore \quad I = \int_1^9 5z^{1/4}\left(\frac{1}{4} dz\right) = \frac{5}{4}\int_1^9 z^{1/4} \, dz$$

$$= \frac{5}{4}\left[\frac{z^{\frac{1}{4}+1}}{\frac{1}{4}+1}\right]_1^9 = \frac{5}{4} . \frac{4}{5}\left[z^{5/4}\right]_1^9$$

$$= [(9)^{5/4} - (1)^{5/4}] \qquad \left[\because \; 9^{5/4} = (3^2)^{5/4} = (3)^{5/2} = \left[(3)^{1/2}\right]^5 = \left(\sqrt{3}\right)^5 = 9\sqrt{3}\right]$$

$$= \left(9\sqrt{3} - 1\right).$$

(*iii*) Let $\quad I = \int_0^1 2x \cos^{-1} x^2 \, dx$

Put $\quad x^2 = z \;\Rightarrow\; 2x\, dx = dz$

When $\quad x = 0, \quad z = (0)^2 = 0$ and when $x = 1, \quad z = (1)^2 = 1$

$\therefore \quad I = \int_0^1 \cos^{-1} z \, dz$

$$= \left[-\frac{1}{\sqrt{1-z^2}}\right]_0^1 = -\left[\frac{1}{\sqrt{1-(1)^2}} - \frac{1}{\sqrt{1-0}}\right]$$

$$= -[0 - 1] = 1.$$

(*iv*) Let $\quad I = \int_0^{\pi/6} \cos^{-3} 2\theta \sin 2\theta \, d\theta$

Put $\quad \cos 2\theta = z \;\Rightarrow\; -2 \sin 2\theta \, d\theta = dz \;\Rightarrow\; \sin 2\theta \, d\theta = -\frac{1}{2} dz$

When $\quad \theta = 0, \quad z = \cos 0° = 1$ and when $\theta = \frac{\pi}{6}, \quad z = \cos 2\left(\frac{\pi}{6}\right) = \cos\frac{\pi}{3} = \frac{1}{2}$

$\therefore \quad I = \int_1^{1/2} (z)^{-3} \cdot \left(-\frac{1}{2} dz\right) = -\frac{1}{2}\int_1^{1/2} z^{-3} \, dz$

$$= -\frac{1}{2}\left[\frac{z^{-3+1}}{-3+1}\right]_1^{1/2} = \left(-\frac{1}{2}\right)\left(-\frac{1}{2}\right)\left[\frac{1}{z^2}\right]_1^{1/2}$$

$$= \frac{1}{4}\left[\frac{1}{(1/2)^2} - \frac{1}{(1)^2}\right] = \frac{1}{4}[4 - 1] = \frac{3}{4}.$$

(*v*) Let $\quad I = \int_0^{\pi/2} \frac{\cos x}{1 + \sin^2 x} \, dx$

Put $\quad \sin x = z \;\Rightarrow\; \cos x \, dx = dz$

When $\quad x = 0, \quad z = \sin 0° = 0$ and when $x = \frac{\pi}{2}, \quad z = \sin\frac{\pi}{2} = 1$

$\therefore \quad I = \int_0^1 \frac{1}{1+z^2} \, dz$

$$= \left[\tan^{-1} z\right]_0^1 = [\tan^{-1} 1 - \tan^{-1} 0] \qquad \left[\because \int \frac{1}{a^2 + x^2} dx = \frac{1}{a}\tan^{-1}\frac{x}{a} + c\right]$$

$$= \left(\frac{\pi}{4} - 0\right) = \frac{\pi}{4}.$$

(*vi*) Let $\quad I = \int_0^{(\pi^2)^{1/3}} \sqrt{x} \cos^2 (x^{3/2})\, dx$

Put $\quad x^{3/2} = z \;\Rightarrow\; \frac{3}{2} x^{1/2}\, dx = dz \;\Rightarrow\; \sqrt{x}\, dx = \frac{2}{3}\, dz$

When $\quad x = 0,\; z = (0)^{3/2} = 0$ and when $x = (\pi^2)^{1/3}, \quad z = (\pi^{2/3})^{3/2} = \pi$

$$\therefore \quad I = \int_0^{\pi} \cos^2 z \left(\frac{2}{3}\, dz\right) = \frac{2}{3}\int_0^{\pi} \cos^2 z\, dz$$

$$= \frac{2}{3}\int_0^{\pi}\left(\frac{1+\cos 2z}{2}\right) dz = \frac{1}{3}\int_0^{\pi} (1 + \cos 2z)\, dz \qquad \left[\begin{array}{l}\because\; 1+\cos 2A = 2\cos^2 A \\ \Rightarrow\; \dfrac{1+\cos 2A}{2} = \cos^2 A\end{array}\right]$$

$$= \frac{1}{3}\left[z + \frac{\sin 2z}{2}\right]_0^{\pi} = \frac{1}{3}\left[\left(\pi + \frac{\sin 2\pi}{2}\right) - \left(0 + \frac{\sin 0^\circ}{2}\right)\right] = \frac{1}{3}[(\pi - 0) - (0 + 0)]$$

$$= \frac{\pi}{3}.$$

Example 29. *Evaluate the following integrals :*

(*i*) $\int_0^1 (\cos^{-1} x)^2\, dx$ (*ii*) $\int_2^4 \frac{x^2 + x}{\sqrt{2x+1}}\, dx$

(*iii*) $\int_0^1 \tan^{-1}\left(\frac{2x}{1-x^2}\right) dx$ (*iv*) $\int_{\pi/6}^{\pi/2} \frac{\operatorname{cosec} x \cot x}{1 + \operatorname{cosec}^2 x}\, dx$

(*v*) $\int_0^{(\pi/2)^{1/3}} x^5 \sin x^3\, dx$ (*vi*) $\int_0^{\pi/4} (\tan^2 \theta + \tan^4 \theta)\, d\theta.$

Solution. (*i*) Let $I = \int_0^1 (\cos^{-1} x)^2\, dx$

Put $\cos^{-1} x = \theta \;\Rightarrow\; x = \cos\theta \;\Rightarrow\; dx = -\sin\theta\, d\theta$

When $x = 0,\; \cos\theta = 0 \;\Rightarrow\; \theta = \frac{\pi}{2}$ and when $x = 1,\; \cos\theta = 1 \;\Rightarrow\; \theta = 0.$

$$\therefore \quad I = \int_{\pi/2}^{0} \theta^2 (-\sin\theta\, d\theta) = -\int_{\pi/2}^{0} \theta^2 \sin\theta\, d\theta$$

Let $\quad I_1 = \int \underset{\text{I}}{\theta^2}\; \underset{\text{II}}{\sin\theta}\, d\theta$

Integrating by parts, we have

$$I_1 = \theta^2 . \int \sin\theta . d\theta - \int \left\{\frac{d}{d\theta}(\theta^2) . \int \sin\theta\, d\theta\right\} d\theta$$

$$= \theta^2 (-\cos\theta) - \int 2\theta . (-\cos\theta)\, d\theta$$

$$= -\theta^2 \cos\theta + 2\int \underset{\text{I}}{\theta}\; \underset{\text{II}}{\cos\theta}\, d\theta$$

Integrating by parts again, we have

$$I_1 = -\theta^2 \cos\theta + 2\left[\theta . \int \cos\theta\, d\theta - \int \left\{\frac{d}{d\theta}(\theta) . \int \cos\theta\, d\theta\right\} d\theta\right]$$

$$= -\theta^2 \cos\theta + 2\left[\theta \sin\theta - \int 1 . \sin\theta \, d\theta\right]$$

$$= -\theta^2 \cos\theta + 2\theta \sin\theta + 2\cos\theta = (2-\theta^2)\cos\theta + 2\theta\sin\theta$$

$$\therefore \qquad I = -\int_{\pi/2}^{0} \theta^2 \sin\theta \, d\theta = -\left[(2-\theta^2)\cos\theta + 2\theta\sin\theta\right]_{\pi/2}^{0}$$

$$= -\left[\{(2-0)\cos 0^\circ + 0\} - \left\{\left(2 - \left(\frac{\pi}{2}\right)^2\right)\cos\frac{\pi}{2} + 2 . \frac{\pi}{2}\sin\frac{\pi}{2}\right\}\right]$$

$$= -[2-(0+\pi)] = -2+\pi$$

$$= \pi - 2.$$

(*ii*) Let $I = \displaystyle\int_2^4 \frac{x^2+x}{\sqrt{2x+1}}\,dx$

Put $\sqrt{2x+1} = z \;\Rightarrow\; (2x+1) = z^2 \;\Rightarrow\; x = \left(\dfrac{z^2-1}{2}\right)$

$$\Rightarrow \; 2\,dx = 2z\,dz \;\Rightarrow\; dx = z\,dz$$

When $\quad x = 2,\; z = \sqrt{2(2)+1} = \sqrt{4+1} = \sqrt{5}$

and when $\quad x = 4,\; z = \sqrt{2(4)+1} = \sqrt{8+1} = \sqrt{9} = 3.$

$$\therefore \quad I = \int_{\sqrt{5}}^{3} \frac{\left(\frac{z^2-1}{2}\right)^2 + \left(\frac{z^2-1}{2}\right)}{z}(z\,dz) = \int_{\sqrt{5}}^{3}\left(\frac{z^4+1-2z^2}{4} + \frac{z^2-1}{2}\right)dz$$

$$= \int_{\sqrt{5}}^{3}\left(\frac{z^4+1-2z^2+2z^2-2}{4}\right)dz = \frac{1}{4}\int_{\sqrt{5}}^{3}(z^4-1)\,dz$$

$$= \frac{1}{4}\left[\frac{z^5}{5} - z\right]_{\sqrt{5}}^{3} = \frac{1}{4}\left[\left(\frac{(3)^5}{5} - 3\right) - \left(\frac{(\sqrt{5})^5}{5} - \sqrt{5}\right)\right]$$

$$= \frac{1}{4}\left[\left(\frac{243}{5} - 3\right) - \left(\frac{25\sqrt{5}}{5} - \sqrt{5}\right)\right] = \frac{1}{4}\left[\left(\frac{243-15}{5}\right) - \left(5\sqrt{5} - \sqrt{5}\right)\right] = \frac{1}{4}\left[\frac{228}{5} - \frac{4\sqrt{5}}{1}\right]$$

$$= \left(\frac{57}{5} - \sqrt{5}\right).$$

(*iii*) Let $\qquad I = \displaystyle\int_0^1 \tan^{-1}\left(\frac{2x}{1-x^2}\right)dx$

Put $\qquad x = \tan\theta \;\Rightarrow\; dx = \sec^2\theta\,d\theta$

When $\qquad x = 0,\; \tan\theta = 0 \;\Rightarrow\; \theta = 0$ and when $x = 1,\; \tan\theta = 1 \;\Rightarrow\; \theta = \dfrac{\pi}{4}$

$$\therefore \qquad I = \int_0^{\pi/4} \tan^{-1}\left(\frac{2\tan\theta}{1-\tan^2\theta}\right)\sec^2\theta\,d\theta$$

$$= \int_0^{\pi/4} \tan^{-1} (\tan 2\theta) \sec^2 \theta \, d\theta \qquad \left[\because \tan 2A = \frac{2 \tan A}{1 - \tan^2 A}\right]$$

$$= \int_0^{\pi/4} 2\theta \sec^2 \theta \, d\theta = 2 \int_0^{\pi/4} \theta \sec^2 \theta \, d\theta$$

Let $$I_1 = \int \underset{I}{\theta} \underset{II}{\sec^2 \theta} \, d\theta$$

Integrating by parts, we have

$$I_1 = \theta \, . \int \sec^2 \theta \, d\theta - \int \left\{\frac{d}{d\theta}(\theta) . \int \sec^2 \theta \, d\theta\right\} d\theta$$

$$= \theta \tan \theta - \int 1 . \tan \theta \, d\theta = \theta \tan \theta - \log |\sec \theta|$$

$$\therefore \qquad I = 2 \int_0^{\pi/4} \theta \sec^2 \theta \, d\theta = 2 \Big[\theta \tan \theta - \log |\sec \theta|\Big]_0^{\pi/4}$$

$$= 2\left[\left(\frac{\pi}{4} \tan \frac{\pi}{4} - \log\left|\sec \frac{\pi}{4}\right|\right) - (0 - \log |\sec 0°|)\right]$$

$$= 2\left[\left(\frac{\pi}{4} - \log \sqrt{2}\right) - (0 - \log 1)\right] = \frac{\pi}{2} - 2 \log \sqrt{2}$$

$$= \frac{\pi}{2} - 2 \log (2)^{1/2} = \frac{\pi}{2} - 2 \frac{1}{2} \log 2 \qquad [\because \; m \log n = \log n^m]$$

$$= \frac{\pi}{2} - \log 2.$$

(*iv*) Let $$I = \int_{\pi/6}^{\pi/2} \frac{\text{cosec } x \cot x}{1 + \text{cosec}^2 x} \, dx$$

Put $\text{cosec } x = z \Rightarrow -\text{cosec } x \cot x \, dx = dz \Rightarrow \text{cosec } x \cot x \, dx = -dz$

When $x = \frac{\pi}{2}$, $z = \text{cosec } \frac{\pi}{2} \Rightarrow z = 1$ and when $x = \frac{\pi}{6}$, $z = \text{cosec } \frac{\pi}{6} \Rightarrow z = 2.$

$$\therefore \quad I = \int_2^1 \frac{1}{1 + z^2} (-dz) = -\int_2^1 \frac{1}{1 + z^2} \, dz$$

$$= -\Big[\tan^{-1} z\Big]_2^1 \qquad \left[\because \int \frac{1}{a^2 + x^2} dx = \frac{1}{a} \tan^{-1} \frac{x}{a} + c\right]$$

$$= -[\tan^{-1}(1) - \tan^{-1}(2)] = -\left[\frac{\pi}{4} - \tan^{-1} 2\right]$$

$$= \tan^{-1} 2 - \frac{\pi}{4}.$$

(*v*) Let $$I = \int_0^{\left(\frac{\pi}{2}\right)^{1/3}} x^5 \sin x^3 \, dx = \int_0^{\left(\frac{\pi}{2}\right)^{1/3}} x^3 . x^2 \sin x^3 \, dx$$

Put $x^3 = z \Rightarrow 3x^2 \, dx = dz \Rightarrow x^2 \, dx = \frac{1}{3} dz$

When $x = 0$, $z = (0)^3 = 0$ and when $x = \left(\frac{\pi}{2}\right)^{1/3}$, $z = \left[\left(\frac{\pi}{2}\right)^{1/3}\right]^3 = \frac{\pi}{2}$.

$$\therefore \qquad I = \int_0^{\pi/2} z \sin z \left(\frac{1}{3}\, dz\right) = \frac{1}{3}\int_0^{\pi/2} z \sin z\, dz$$

Let $\quad I_1 = \int \underset{\text{I}}{z} \underset{\text{II}}{\sin z}\, dz$

Integrating by parts, we have

$$I_1 = z \,.\int \sin z\, dz - \int \left\{\frac{d}{dz}(z)\,.\int \sin z\, dz\right\} dz$$

$$= z(-\cos z) - \int 1\,.(-\cos z)\, dz = -z\cos z + \int \cos z\, dz$$

$$= -z\cos z + \sin z$$

$$\therefore \qquad I = \frac{1}{3}\int_0^{\pi/2} z\sin z\, dz = \frac{1}{3}\Big[-z\cos z + \sin z\Big]_0^{\pi/2}$$

$$= \frac{1}{3}\left[\left(-\frac{\pi}{2}\cos\frac{\pi}{2} + \sin\frac{\pi}{2}\right) - (0 + \sin 0°)\right] = \frac{1}{3}[(0+1) - (0+0)]$$

$$= \frac{1}{3}.$$

(*vi*) Let $\quad I = \int_0^{\pi/4} (\tan^2\theta + \tan^4\theta)\, d\theta = \int_0^{\pi/4} \tan^2\theta\,(1 + \tan^2\theta)\, d\theta$

$$I = \int_0^{\pi/4} \tan^2\theta \sec^2\theta\, d\theta \qquad [\because \sec^2 A - \tan^2 A = 1]$$

Put $\quad \tan\theta = z \Rightarrow \sec^2\theta\, d\theta = dz$

When $\theta = 0$, $z = \tan 0° = 0$ and when $\theta = \frac{\pi}{4}$, $z = \tan\frac{\pi}{4} = 1$

$$\therefore \qquad I = \int_0^1 z^2\, dz = \left[\frac{z^3}{3}\right]_0^1 = \frac{1}{3}\Big[z^3\Big]_0^1 = \frac{1}{3}[(1)^3 - (0)^3]$$

$$= \frac{1}{3}(1 - 0) = \frac{1}{3}.$$

Example 30. *Evaluate the following definite integrals :*

(*i*) $\int_0^{\pi/6} (2 + 3x^2) \cos 3x\, dx$ (*ii*) $\int_0^1 x \log\left(1 + \frac{x}{2}\right) dx$

(*iii*) $\int_0^1 x \tan^{-1} x\, dx$ (*iv*) $\int_0^{\infty} \frac{x \tan^{-1} x}{(1 + x^2)^2}\, dx$

(*v*) $\int_0^{\pi/2} \left(\sqrt{\tan x} + \sqrt{\cot x}\right) dx$ (*vi*) $\int_0^{\pi/2} \frac{1}{4\sin^2 x + 5\cos^2 x}\, dx.$

Solution. (*i*) Let $I = \int_0^{\pi/6} \underset{\text{I}}{(2 + 3x^2)} \underset{\text{II}}{\cos 3x}\, dx$

Integrating by parts, we have

$$I = \left[(2+3x^2)\cdot\frac{\sin 3x}{3}\right]_0^{\pi/6} - \int_0^{\pi/6} \frac{d}{dx}(2+3x^2)\cdot\frac{\sin 3x}{3}\,dx$$

$$= \frac{1}{3}\left[(2+3x^2)\sin 3x\right]_0^{\pi/6} - \frac{1}{3}\int_0^{\pi/6} 6x\sin 3x\,dx$$

$$= \frac{1}{3}\left[\left(2+3\left(\frac{\pi}{6}\right)^2\right)\sin\frac{3\pi}{6} - (2+0)\sin 0^\circ\right] - 2\int_0^{\pi/6} \underset{\text{I}}{x}\,\underset{\text{II}}{\sin 3x}\,dx$$

Integrating again by parts, we have

$$I = \frac{1}{3}\left[\left(2+\frac{\pi^2}{12}\right)-0\right] - 2\left[x\cdot\left(-\frac{\cos 3x}{3}\right)\right]_0^{\pi/6} + 2\int_0^{\pi/6}\frac{d}{dx}(x)\left(-\frac{\cos 3x}{3}\right)dx$$

$$= \frac{2}{3}+\frac{\pi^2}{36}+\frac{2}{3}\left[x\cos 3x\right]_0^{\pi/6} - \frac{2}{3}\int_0^{\pi/6} 1\,.\cos 3x\,dx$$

$$= \frac{2}{3}+\frac{\pi^2}{36}+\frac{2}{3}\left[\frac{\pi}{6}\cos\frac{3\pi}{6}-0\right]-\frac{2}{3}\left[\frac{\sin 3x}{3}\right]_0^{\pi/6}$$

$$= \frac{2}{3}+\frac{\pi^2}{36}+\frac{2}{3}[0]-\frac{2}{9}\left[\sin\frac{3\pi}{6}-\sin 0^\circ\right]$$

$$= \frac{2}{3}+\frac{\pi^2}{36}-\frac{2}{9} = \frac{24+\pi^2-8}{36} = \frac{\pi^2+16}{36}.$$

(*ii*) Let $\quad I = \int_0^1 \underset{\text{II}}{x}\,\underset{\text{I}}{\log\left(1+\frac{x}{2}\right)}\,dx$

Integrating by parts, we have

$$I = \left[\log\left(1+\frac{x}{2}\right)\cdot\frac{x^2}{2}\right]_0^1 - \int_0^1 \frac{d}{dx}\left\{\log\left(1+\frac{x}{2}\right)\right\}\cdot\frac{x^2}{2}\,dx$$

$$= \left[\log\left(1+\frac{1}{2}\right)\cdot\frac{1}{2} - \log(1+0)\,.\,0\right] - \int_0^1 \frac{1}{\left(1+\frac{x}{2}\right)}\cdot\left(\frac{1}{2}\right)\cdot\frac{x^2}{2}\,dx$$

$$= \frac{1}{2}\log\frac{3}{2} - \frac{1}{4}\int_0^1 \frac{x^2}{\left(\frac{2+x}{2}\right)}\,dx$$

$$= \frac{1}{2}\log\frac{3}{2} - \frac{1}{2}\int_0^1 \frac{x^2}{x+2}\,dx$$

$$= \frac{1}{2}\log\frac{3}{2} - \frac{1}{2}\int_0^1\left[(x-2)+\frac{4}{x+2}\right]dx$$

$$= \frac{1}{2}\log\frac{3}{2} - \frac{1}{2}\int_0^1 x-2+\frac{4}{x+2}\,dx$$

$$\begin{array}{r|l} x+2 & x^2 \quad (x-2 \\ & \underline{\overset{-}{x^2} + \overset{-}{2x}} \\ & -2x \\ & \underline{\overset{+}{-2x} \overset{+}{-4}} \\ & 4 \end{array}$$

$$= \frac{1}{2}\log\frac{3}{2} - \frac{1}{2}\left[\frac{x^2}{2} - 2x + 4\log|x+2|\right]_0^1$$

$$= \frac{1}{2}\log\frac{3}{2} - \frac{1}{2}\left[\frac{(1)^2}{2} - 2(1) + 4\log|1+2| - 0 - 0 + 4\log|0+2|\right]$$

$$= \frac{1}{2}\log\frac{3}{2} - \frac{1}{2}\left[\frac{1}{2} - 2 + 4\log 3 - 4\log 2\right]$$

$$= \frac{1}{2}\log\frac{3}{2} - \frac{1}{2}\left[-\frac{3}{2} + 4\log\frac{3}{2}\right] = \frac{1}{2}\log\frac{3}{2} + \frac{3}{4} - 2\log\frac{3}{2}$$

$$\left[\because \log m - \log n = \log\left(\frac{m}{n}\right)\right]$$

$$= \frac{3}{4} - \frac{3}{2}\log\frac{3}{2}.$$

(*iii*) Let $\quad I = \int_0^1 \underset{\text{II}}{x}\ \underset{\text{I}}{\tan^{-1} x}\, dx$

Integrating by parts, we have

$$I = \left[\tan^{-1} x \,.\, \frac{x^2}{2}\right]_0^1 - \int_0^1 \frac{d}{dx}(\tan^{-1} x)\,.\,\frac{x^2}{2}\,dx$$

$$= \frac{1}{2}\left[x^2 \tan^{-1} x\right]_0^1 - \frac{1}{2}\int_0^1 \frac{x^2}{1+x^2}\,dx$$

$$= \frac{1}{2}[1\tan^{-1} 1 - 0] - \frac{1}{2}\int_0^1 \frac{1+x^2-1}{1+x^2}\,dx \qquad \left[\because \text{Add and subtract 1 to the numerator}\right]$$

$$= \frac{1}{2}\left[\frac{\pi}{4} - 0\right] - \frac{1}{2}\int_0^1\left(\frac{1+x^2}{1+x^2} - \frac{1}{1+x^2}\right)dx = \frac{\pi}{8} - \frac{1}{2}\int_0^1 1\,dx + \frac{1}{2}\int_0^1 \frac{1}{1+x^2}\,dx$$

$$= \frac{\pi}{8} - \frac{1}{2}\left[x\right]_0^1 + \frac{1}{2}\left[\tan^{-1} x\right]_0^1 \qquad \left[\because \int \frac{1}{a^2+x^2}\,dx = \frac{1}{a}\tan^{-1}\frac{x}{a} + c\right]$$

$$= \frac{\pi}{8} - \frac{1}{2}[1-0] + \frac{1}{2}[\tan^{-1} 1 - \tan^{-1} 0] = \frac{\pi}{8} - \frac{1}{2} + \frac{1}{2}\left[\frac{\pi}{4} - 0\right] = \frac{\pi}{8} - \frac{1}{2} + \frac{\pi}{8}$$

$$= \frac{\pi}{4} - \frac{1}{2} = \frac{\pi - 2}{4}.$$

(*iv*) Let $\quad I = \int_0^\infty \frac{x\tan^{-1} x}{(1+x^2)^2}\,dx$

Put $\tan^{-1} x = z \;\Rightarrow\; x = \tan z \;\Rightarrow\; \frac{1}{1+x^2}\,dx = dz$

When $x = 0$, $\tan z = 0 \;\Rightarrow\; z = 0$ and when $x = \infty$, $\tan z = \infty \;\Rightarrow\; z = \frac{\pi}{2}$

$$\therefore \quad I = \int_0^{\pi/2} \frac{z \tan z}{(1+\tan^2 z)}\, dz = \int_0^{\pi/2} \frac{z \tan z}{\sec^2 z}\, dz \qquad [\because \ \sec^2 A - \tan^2 A = 1]$$

$$= \int_0^{\pi/2} z \tan z \,.\, \cos^2 z\, dz = \int_0^{\pi/2} z \sin z \cos z\, dz$$

$$= \frac{1}{2}\int_0^{\pi/2} z\,(2 \sin z \cos z)\, dz \qquad \text{[Multiply and divided by 2]}$$

$$= \frac{1}{2}\int_0^{\pi/2} \underset{\text{I}}{z}\ \underset{\text{II}}{\sin 2z}\, dz \qquad [\because \ 2 \sin A \cos A = \sin 2A]$$

Integrating by parts, we have

$$I = \frac{1}{2}\left[z \,.\, \frac{(-\cos 2z)}{2}\right]_0^{\pi/2} - \frac{1}{2}\int_0^{\pi/2} \frac{d}{dz}(z)\,.\,\frac{(-\cos 2z)}{2}\, dz$$

$$= -\frac{1}{4}\left[z \cos 2z\right]_0^{\pi/2} + \frac{1}{4}\int_0^{\pi/2} 1 \,.\, \cos 2z\, dz$$

$$= -\frac{1}{4}\left[\frac{\pi}{2}\cos\frac{2\pi}{2} - 0\right] + \frac{1}{4}\left[\frac{\sin 2z}{2}\right]_0^{\pi/2} = -\frac{1}{4}\left[\frac{\pi}{2}(-1)\right] + \frac{1}{8}\left[\sin\frac{2\pi}{2} - \sin 0\right]$$

$$= \frac{\pi}{8} + \frac{1}{8}(0-0) = \frac{\pi}{8}.$$

(*v*) Let $\quad I = \int_0^{\pi/2} \left(\sqrt{\tan x} + \sqrt{\cot x}\right) dx = \int_0^{\pi/2} \left(\sqrt{\frac{\sin x}{\cos x}} + \sqrt{\frac{\cos x}{\sin x}}\right) dx$

$$= \int_0^{\pi/2} \left(\frac{\sqrt{\sin x}}{\sqrt{\cos x}} + \frac{\sqrt{\cos x}}{\sqrt{\sin x}}\right) dx = \int_0^{\pi/2} \frac{(\sin x + \cos x)}{\sqrt{\sin x \cos x}}\, dx$$

$$= \sqrt{2}\int_0^{\pi/2} \frac{(\sin x + \cos x)}{\sqrt{2 \sin x \cos x}}\, dx \qquad [\text{Multiply and divided by } \sqrt{2}]$$

$$= \sqrt{2}\int_0^{\pi/2} \frac{(\sin x + \cos x)}{\sqrt{\sin 2x}}\, dx \qquad [\because \ \sin 2A = 2 \sin A \cos A]$$

$$= \sqrt{2}\int_0^{\pi/2} \frac{(\sin x + \cos x)}{\sqrt{1-(1-\sin 2x)}}\, dx \qquad \text{[Note this step. Add and subtract 1]}$$

$$= \sqrt{2}\int_0^{\pi/2} \frac{(\sin x + \cos x)}{\sqrt{1-[\sin^2 x + \cos^2 x - 2 \sin x \cos x]}}\, dx \qquad \left[\begin{array}{l}\because \ \sin 2A = 2 \sin A \cos A \\ \sin^2 A + \cos^2 A = 1\end{array}\right]$$

$$= \sqrt{2}\int_0^{\pi/2} \frac{(\sin x + \cos x)}{\sqrt{1-(\sin x - \cos x)^2}}\, dx$$

Put $\quad \sin x - \cos x = z \ \Rightarrow \ (\cos x + \sin x)\, dx = dz$

When $\quad x = 0, \quad z = \sin 0° - \cos 0° = 0 - 1 = -1$

and when $\quad x = \frac{\pi}{2}, \quad z = \sin\frac{\pi}{2} - \cos\frac{\pi}{2} = 1 - 0 = 1$

$\therefore \quad I = \sqrt{2}\int_{-1}^{1} \frac{1}{\sqrt{1-z^2}}\, dz$

$= \sqrt{2}\left[\sin^{-1} z\right]_{-1}^{1}$ $\quad \left[\because \int \frac{1}{\sqrt{a^2 - x^2}}\, dx = \sin^{-1}\frac{x}{a} + c\right]$

$= \sqrt{2}\ [\sin^{-1}(1) - \sin^{-1}(-1)] = \sqrt{2}\left[\frac{\pi}{2} - \left(-\frac{\pi}{2}\right)\right] = \sqrt{2}\left[\frac{\pi}{2} + \frac{\pi}{2}\right]$

$= \sqrt{2}\pi.$

(*vi*) Let $\quad I = \int_0^{\pi/2} \frac{1}{4\sin^2 x + 5\cos^2 x}\, dx$

$= \int_0^{\pi/2} \left(\frac{\frac{1}{\cos^2 x}}{\frac{4\sin^2 x}{\cos^2 x} + \frac{5\cos^2 x}{\cos^2 x}}\right) dx$ $\quad$ [Dividing the numerator and the denominator by $\cos^2 x$]

$= \int_0^{\pi/2} \frac{\sec^2 x}{4\tan^2 x + 5}\, dx$

Put $\quad \tan x = z \Rightarrow \sec^2 x\, dx = dz$

When $\quad x = 0, \quad z = \tan 0° = 0$ and when $x = \frac{\pi}{2}, \quad z = \tan\frac{\pi}{2} = \infty.$

$\therefore \quad I = \int_0^{\infty} \frac{1}{4z^2 + 5}\, dz = \frac{1}{4}\int_0^{\infty} \frac{1}{z^2 + \frac{5}{4}}\, dz = \frac{1}{4}\int_0^{\infty} \frac{1}{z^2 + \left(\frac{\sqrt{5}}{2}\right)^2}\, dz$

$= \frac{1}{4}\left[\frac{1}{\frac{\sqrt{5}}{2}} \tan^{-1}\frac{z}{\frac{\sqrt{5}}{2}}\right]_0^{\infty}$ $\quad \left[\because \int \frac{1}{x^2 + a^2}\, dx = \frac{1}{a}\tan^{-1}\frac{x}{a} + c\right]$

$= \frac{1}{4}\cdot\frac{2}{\sqrt{5}}\left[\tan^{-1}\frac{2z}{\sqrt{5}}\right]_0^{\infty} = \frac{1}{2\sqrt{5}}\ [\tan^{-1}\infty - \tan^{-1} 0]$

$= \frac{1}{2\sqrt{5}}\left[\frac{\pi}{2} - 0\right] = \frac{\pi}{4\sqrt{5}}.$

Example 31. *Evaluate the following integrals :*

(*i*) $\int_0^{\pi/2} \frac{1}{5 + 4\sin x}\, dx$ $\qquad$ (*ii*) $\int_0^{\pi/2} \frac{1}{3 + 2\cos x}\, dx$

(*iii*) $\int_0^{\pi/2} \frac{1}{6 - \cos x}\, dx$ $\qquad$ (*iv*) $\int_0^{\pi} \frac{1}{3 + 2\sin x + \cos x}\, dx$

(*v*) $\int_0^{\pi/2} \frac{1}{1 - 2\sin x}\, dx$ $\qquad$ (*vi*) $\int_0^{\pi/2} \frac{1}{7 + 2\sin x}\, dx$

(*vii*) $\int_0^{\pi/2} \frac{1}{5\cos x + 3\sin x}\,dx$ (*viii*) $\int_0^{\pi} \frac{\sin x}{\sin x + \cos x}\,dx$

(*ix*) $\int_0^{\pi/2} \frac{1}{4 + 5\sin x}\,dx$ (*x*) $\int_0^{\pi/2} \frac{1}{2\cos x + 4\sin x}\,dx.$

Solution. (*i*) Let $I = \int_0^{\pi/2} \frac{1}{5 + 4\sin x}\,dx$

Put $z = \tan\frac{x}{2} \Rightarrow dz = \frac{1}{2}\sec^2\frac{x}{2}\,dx$ $[\because \sec^2 A - \tan^2 A = 1]$

$$\Rightarrow dz = \frac{1}{2}\left(1 + \tan^2\frac{x}{2}\right)dx \Rightarrow 2dz = (1 + z^2)\,dx \Rightarrow \frac{2}{1+z^2}\,dz = dx$$

Now, $\sin x = \dfrac{2\tan\frac{x}{2}}{1 + \tan^2\frac{x}{2}} \Rightarrow \sin x = \left(\dfrac{2z}{1+z^2}\right)$

When $x = 0$, $z = \tan 0° = 0$ and when $x = \frac{\pi}{2}$, $z = \tan\frac{\pi}{4} = 1$

$$\therefore \quad I = \int_0^1 \frac{1}{5 + 4\left(\frac{2z}{1+z^2}\right)} \cdot \frac{2}{1+z^2} \cdot dz$$

$$= \int_0^1 \frac{1+z^2}{5 + 5z^2 + 8z} \cdot \left(\frac{2}{1+z^2}\right) dz = 2\int_0^1 \frac{1}{5z^2 + 8z + 5}\,dz = \frac{2}{5}\int_0^1 \frac{1}{z^2 + \frac{8}{5}z + 1}\,dz$$

$$= \frac{2}{5}\int_0^1 \frac{1}{\left(z^2 + \frac{8}{5}z + \frac{16}{25}\right) + \left(1 - \frac{16}{25}\right)}\,dz$$

$\left[\text{Add and subtract } \frac{16}{25} \text{ to the denom. } \because \left(\frac{1}{2}\text{ co-eff. of } z\right)^2 = \frac{16}{25}\right]$

$$= \frac{2}{5}\int_0^1 \frac{1}{\left(z + \frac{4}{5}\right)^2 + \frac{9}{25}}\,dz = \frac{2}{5}\int_0^1 \frac{1}{\left(z + \frac{4}{5}\right)^2 + \left(\frac{3}{5}\right)^2}\,dz$$

$$= \frac{2}{5}\left[\frac{1}{\left(\frac{3}{5}\right)} \tan^{-1}\frac{\left(z + \frac{4}{5}\right)}{\frac{3}{5}}\right]_0^1$$

$\left[\because \int \frac{1}{a^2 + x^2}\,dx = \frac{1}{a}\tan^{-1}\frac{x}{a} + c\right]$

$$= \frac{2}{5} \cdot \frac{5}{3}\left[\tan^{-1}\left(\frac{5z+4}{3}\right)\right]_0^1 = \frac{2}{3}\left[\tan^{-1}\left(\frac{5+4}{3}\right) - \tan^{-1}\left(\frac{0+4}{3}\right)\right]$$

$$= \frac{2}{3}\left[\tan^{-1}\frac{9}{3} - \tan^{-1}\frac{4}{3}\right] = \frac{2}{3}\left[\tan^{-1} 3 - \tan^{-1}\frac{4}{3}\right]$$

$$= \frac{2}{3}\tan^{-1}\left[\frac{\left(3-\frac{4}{3}\right)}{1+3\left(\frac{4}{3}\right)}\right] == \frac{2}{3}\tan^{-1}\left[\frac{\frac{5}{3}}{1+4}\right]$$

$$\left[\because \tan^{-1}A - \tan^{-1}B = \tan^{-1}\left(\frac{A-B}{1+AB}\right)\right]$$

$$= \frac{2}{3}\tan^{-1}\frac{1}{3}.$$

(*ii*) Let $I = \int_0^{\pi/2} \frac{1}{3+2\cos x}\, dx$

Put $z = \tan\frac{x}{2} \Rightarrow dz = \frac{1}{2}\sec^2\frac{x}{2}\, dx$

$\Rightarrow \quad 2dz = \left(1+\tan^2\frac{x}{2}\right) dx \Rightarrow \frac{2}{1+z^2}\, dz = dx \qquad [\because \sec^2 A - \tan^2 A = 1]$

Now $\cos x = \dfrac{1-\tan^2\frac{x}{2}}{1+\tan^2\frac{x}{2}} \Rightarrow \cos x = \dfrac{1-z^2}{1+z^2}$

When $x = 0$, $z = \tan 0° = 0$ and when $x = \frac{\pi}{2}$, $z = \tan\frac{\pi}{4} = 1$

$$\therefore \quad I = \int_0^1 \frac{1}{3+2\left(\frac{1-z^2}{1+z^2}\right)} \cdot \frac{2}{1+z^2}\, dz = \int_0^1 \frac{1+z^2}{(3+3z^2+2-2z^2)} \cdot \frac{2}{1+z^2}\, dz$$

$$= 2\int_0^1 \frac{1}{z^2+5}\, dz = 2\int_0^1 \frac{1}{z^2+\left(\sqrt{5}\right)^2}\, dz$$

$$= 2\left[\frac{1}{\sqrt{5}}\tan^{-1}\frac{z}{\sqrt{5}}\right]_0^1 \qquad \left[\because \int \frac{1}{x^2+a^2}\, dx = \frac{1}{a}\tan^{-1}\frac{x}{a} + c\right]$$

$$= \frac{2}{\sqrt{5}}\left[\tan^{-1}\frac{1}{\sqrt{5}} - \tan^{-1} 0°\right]$$

$$= \frac{2}{\sqrt{5}}\tan^{-1}\left(\frac{1}{\sqrt{5}}\right).$$

(*iii*) Let $I = \int_0^{\pi/2} \frac{1}{6-\cos x}\, dx$

Put $z = \tan\frac{x}{2} \Rightarrow dz = \frac{1}{2}\sec^2\frac{x}{2}\, dx$

$\Rightarrow \quad 2dz = \left(1+\tan^2\frac{x}{2}\right) dx \Rightarrow \frac{2}{1+z^2}\, dz = dx \qquad [\because \sec^2 A - \tan^2 A = 1]$

Now $\cos x = \dfrac{1-\tan^2 \frac{x}{2}}{1+\tan^2 \frac{x}{2}} \Rightarrow \cos x = \dfrac{1-z^2}{1+z^2}$

When $x = 0$, $z = \tan 0° = 0$ and when $x = \frac{\pi}{2}$, $z = \tan \frac{\pi}{4} = 1$.

$$\therefore \quad I = \int_0^1 \frac{1}{6-\left(\frac{1-z^2}{1+z^2}\right)} \cdot \frac{2}{1+z^2}\, dz = \int_0^1 \frac{1+z^2}{6+6z^2-1+z^2} \cdot \frac{2}{1+z^2}\, dz$$

$$= 2\int_0^1 \frac{1}{7z^2+5}\, dz = \frac{2}{7}\int_0^1 \frac{1}{z^2+\frac{5}{7}}\, dz = \frac{2}{7}\int_0^1 \frac{1}{z^2+\left(\sqrt{\frac{5}{7}}\right)^2}\, dz$$

$$= \frac{2}{7}\left[\frac{1}{\sqrt{\frac{5}{7}}} \tan^{-1} \frac{z}{\sqrt{\frac{5}{7}}}\right]_0^1 \qquad \left[\because \int \frac{1}{x^2+a^2}\, dx = \frac{1}{a} \tan^{-1} \frac{x}{a} + c\right]$$

$$= \frac{2}{7} \cdot \frac{\sqrt{7}}{\sqrt{5}}\left[\tan^{-1} \frac{\sqrt{7}z}{\sqrt{5}}\right]_0^1 = \frac{2}{\sqrt{7}\sqrt{5}}\left[\tan^{-1} \frac{\sqrt{7}}{\sqrt{5}} - \tan^{-1} 0\right]$$

$$= \frac{2}{\sqrt{35}} \tan^{-1}\left(\sqrt{\frac{7}{5}}\right).$$

(*iv*) Let $I = \displaystyle\int_0^{\pi} \frac{1}{(3+2\sin x+\cos x)}\, dx$

Put $z = \tan \frac{x}{2} \Rightarrow dz = \frac{1}{2} \sec^2 \frac{x}{2}\, dx$

$\Rightarrow \quad 2dz = \left(1+\tan^2 \frac{x}{2}\right) dx \Rightarrow \dfrac{2}{1+z^2}\, dz = dx \qquad [\because \sec^2 A - \tan^2 A = 1]$

Now $\sin x = \dfrac{2\tan \frac{x}{2}}{1+\tan^2 \frac{x}{2}}, \cos x = \dfrac{1-\tan^2 \frac{x}{2}}{1+\tan^2 \frac{x}{2}}$

$\Rightarrow \quad \sin x = \dfrac{2z}{1+z^2}, \cos x = \dfrac{1-z^2}{1+z^2}$

When $x = 0$, $z = \tan 0° = 0$ and when $x = \pi$, $z = \tan \frac{\pi}{2} = \infty$

$$\therefore \quad I = \int_0^{\infty} \frac{1}{\left[3+2\left(\frac{2z}{1+z^2}\right)+\frac{1-z^2}{1+z^2}\right]} \cdot \frac{2}{1+z^2}\, dz$$

$$= 2\int_0^\infty \frac{1+z^2}{(3+3z^2+4z+1-z^2)} \cdot \frac{1}{1+z^2}\, dz$$

$$= 2\int_0^\infty \frac{1}{2z^2+4z+4}\, dz = \int_0^\infty \frac{1}{z^2+2z+2}\, dz$$

$$= \int_0^\infty \frac{1}{(z^2+2z+1)+(2-1)}\, dz \qquad \left[\begin{array}{l}\text{Add and subtract 1 to the denom.}\\ \because \left(\frac{1}{2}\text{co-eff. of } z\right)^2 = 1\end{array}\right]$$

$$= \int_0^\infty \frac{1}{(z+1)^2+1}\, dz$$

$$= \left[\tan^{-1}(z+1)\right]_0^\infty \qquad \left[\because \int \frac{1}{x^2+a^2}\, dx = \frac{1}{a}\tan^{-1}\frac{x}{a} + c\right]$$

$$= [\tan^{-1}\infty - \tan^{-1}(0+1)] = [\tan^{-1}\infty - \tan^{-1} 1]$$

$$= \frac{\pi}{2} - \frac{\pi}{4} = \frac{\pi}{4}.$$

(*v*) Let $\quad I = \displaystyle\int_0^{\pi/2} \frac{1}{1-2\sin x}\, dx$

Put $\quad z = \tan\frac{x}{2} \Rightarrow dz = \frac{1}{2}\sec^2\frac{x}{2}\, dx$

$\Rightarrow \quad 2dz = \sec^2\frac{x}{2}\, dx \Rightarrow 2dz = \left(1+\tan^2\frac{x}{2}\right) dx \Rightarrow \dfrac{2}{1+z^2}\, dz = dx$

$[\because \sec^2 A - \tan^2 A = 1]$

Now $\quad \sin x = \dfrac{2\tan\frac{x}{2}}{1+\tan^2\frac{x}{2}} \Rightarrow \sin x = \dfrac{2z}{1+z^2}$

When $\quad x = 0, \; z = \tan 0° = 0$ and when $x = \frac{\pi}{2}, \; z = \tan\frac{\pi}{4} = 1$

$\therefore \quad I = \displaystyle\int_0^1 \frac{1}{1-2\left(\frac{2z}{1+z^2}\right)} \cdot \frac{2}{1+z^2}\, dz$

$$= 2\int_0^1 \frac{1+z^2}{1+z^2-4z} \cdot \frac{1}{1+z^2}\, dz = 2\int_0^1 \frac{1}{z^2-4z+1}\, dz$$

$$= 2\int_0^1 \frac{1}{(z^2-4z+4)+(1-4)}\, dz \qquad \left[\begin{array}{l}\text{Add and subtract 4 to the denom.}\\ \because \left(\frac{1}{2}\text{co-eff. of } z\right)^2 = 4\end{array}\right]$$

$$= 2\int_0^1 \frac{1}{(z-2)^2-3}\, dz = 2\int_0^1 \frac{1}{(z-2)^2-\left(\sqrt{3}\right)^2}\, dz$$

$$= 2\left[\frac{1}{2\sqrt{3}}\cdot\log\left|\frac{z-2-\sqrt{3}}{z-2+\sqrt{3}}\right|\right]_0^1 \qquad \left[\because \int \frac{1}{x^2-a^2}\,dx = \frac{1}{2a}\log\left|\frac{x-a}{x+a}\right| + c\right]$$

$$= \frac{1}{\sqrt{3}}\left[\log\left|\frac{1-2-\sqrt{3}}{1-2+\sqrt{3}}\right| - \log\left|\frac{0-2-\sqrt{3}}{0-2+\sqrt{3}}\right|\right]$$

$$= \frac{1}{\sqrt{3}}\left[\log\left|\frac{-1-\sqrt{3}}{-1+\sqrt{3}}\right| - \log\left|\frac{-2-\sqrt{3}}{-2+\sqrt{3}}\right|\right] = \frac{1}{\sqrt{3}}\left[\log\left(\frac{\sqrt{3}+1}{\sqrt{3}-1}\right) - \log\left(\frac{\sqrt{3}+2}{\sqrt{3}-2}\right)\right]$$

$$= \frac{1}{\sqrt{3}}\left[\log\left(\frac{\dfrac{\sqrt{3}+1}{\sqrt{3}-1}}{\dfrac{\sqrt{3}+2}{\sqrt{3}-2}}\right)\right] \qquad \left[\because \log m - \log n = \log\left(\frac{m}{n}\right)\right]$$

$$= \frac{1}{\sqrt{3}}\left[\log\left(\frac{\sqrt{3}+1}{\sqrt{3}-1}\times\frac{\sqrt{3}-2}{\sqrt{3}+2}\right)\right] = \frac{1}{\sqrt{3}}\left[\log\left(\frac{3+\sqrt{3}-2\sqrt{3}-2}{3-\sqrt{3}+2\sqrt{3}-2}\right)\right]$$

$$= \frac{1}{\sqrt{3}}\log\left(\frac{1-\sqrt{3}}{1+\sqrt{3}}\right)$$

$$= \frac{1}{\sqrt{3}}\log\left(\frac{1-\sqrt{3}}{1+\sqrt{3}}\times\frac{1-\sqrt{3}}{1-\sqrt{3}}\right) \qquad \text{[Rationalization]}$$

$$= \frac{1}{\sqrt{3}}\log\left(\frac{1+3-2\sqrt{3}}{1-3}\right) = \frac{1}{\sqrt{3}}\log\left(\frac{4-2\sqrt{3}}{-2}\right) = \frac{1}{\sqrt{3}}\log[-(2-\sqrt{3})]$$

$$= \frac{1}{\sqrt{3}}\log(\sqrt{3}-2)\,.$$

(*vi*) Let $\quad \mathrm{I} = \displaystyle\int_0^{\pi/2} \frac{1}{7+2\sin x}\,dx$

Put $\quad z = \tan\dfrac{x}{2} \;\Rightarrow\; dz = \dfrac{1}{2}\sec^2\dfrac{x}{2}\,dx$

$\Rightarrow \quad 2dz = \left(1+\tan^2\dfrac{x}{2}\right)dx \;\Rightarrow\; \dfrac{2}{1+z^2}\,dz = dx \qquad [\because \sec^2 \mathrm{A} - \tan^2 \mathrm{A} = 1]$

Now $\quad \sin x = \dfrac{2\tan\dfrac{x}{2}}{1+\tan^2\dfrac{x}{2}} = \dfrac{2z}{1+z^2}$

When $x = 0$, $\;z = \tan 0° = 0$ and when $x = \dfrac{\pi}{2}$, $\;z = \tan\dfrac{\pi}{4} = 1$

$$\therefore \quad \mathrm{I} = \int_0^1 \frac{1}{7+2\left(\dfrac{2z}{1+z^2}\right)}\cdot\frac{2}{1+z^2}\,dz = 2\int_0^1 \frac{1+z^2}{7+7z^2+4z}\cdot\frac{1}{1+z^2}\,dz$$

$$= 2\int_0^1 \frac{1}{7z^2+4z+7}\,dz = \frac{2}{7}\int_0^1 \frac{1}{\left(z^2+\frac{4}{7}z+1\right)}\,dz$$

$$\therefore \quad I = \frac{2}{7}\int_0^1 \frac{1}{\left(z^2+\frac{4}{7}z+\frac{4}{49}\right)+\left(1-\frac{4}{49}\right)}\,dz \qquad \left[\begin{array}{l}\text{Add and subtract } \frac{4}{49} \text{ to the denom.}\\ \because \left(\frac{1}{2}\text{ co-eff. of } z\right)^2 = \frac{4}{49}\end{array}\right]$$

$$= \frac{2}{7}\int_0^1 \frac{1}{\left(z+\frac{2}{7}\right)^2+\frac{45}{49}}\,dz = \frac{2}{7}\int_0^1 \frac{1}{\left(z+\frac{2}{7}\right)^2+\left(\frac{\sqrt{45}}{7}\right)^2}\,dz$$

$$= \frac{2}{7}\left[\frac{1}{\frac{\sqrt{45}}{7}}\cdot\tan^{-1}\left(\frac{z+\frac{2}{7}}{\frac{\sqrt{45}}{7}}\right)\right]_0^1 \qquad \left[\because \int\frac{1}{x^2+a^2}\,dx = \frac{1}{a}\tan^{-1}\frac{x}{a}+c\right]$$

$$= \frac{2}{7}\cdot\frac{7}{\sqrt{45}}\left[\tan^{-1}\frac{7z+2}{\sqrt{45}}\right]_0^1 = \frac{2}{\sqrt{45}}\left[\tan^{-1}\left(\frac{7+2}{3\sqrt{5}}\right)-\tan^{-1}\left(\frac{0+2}{3\sqrt{5}}\right)\right]$$

$$= \frac{2}{\sqrt{45}}\left[\tan^{-1}\left(\frac{9}{3\sqrt{5}}\right)-\tan^{-1}\left(\frac{2}{3\sqrt{5}}\right)\right] = \frac{2}{\sqrt{45}}\left[\tan^{-1}\frac{3}{\sqrt{5}}-\tan^{-1}\frac{2}{3\sqrt{5}}\right]$$

$$= \frac{2}{\sqrt{45}}\left[\tan^{-1}\left(\frac{\frac{3}{\sqrt{5}}-\frac{2}{3\sqrt{5}}}{1+\frac{3}{\sqrt{5}}\cdot\frac{2}{3\sqrt{5}}}\right)\right] \qquad \left[\because \tan^{-1}A-\tan^{-1}B = \tan^{-1}\left(\frac{A-B}{1+AB}\right)\right]$$

$$= \frac{2}{\sqrt{45}}\left[\tan^{-1}\left(\frac{\frac{9-2}{3\sqrt{5}}}{1+\frac{2}{5}}\right)\right] = \frac{2}{3\sqrt{5}}\left[\tan^{-1}\left(\frac{\frac{7}{3\sqrt{5}}}{\frac{7}{5}}\right)\right] = \frac{2}{3\sqrt{5}}\tan^{-1}\left(\frac{7}{3\sqrt{5}}\cdot\frac{5}{7}\right)$$

$$= \frac{2}{3\sqrt{5}}\tan^{-1}\left(\frac{\sqrt{5}}{3}\right).$$

(*vii*) Let $\quad I = \int_0^{\pi/2} \frac{1}{5\cos x + 3\sin x}\,dx$

Put $\quad z = \tan\frac{x}{2} \quad \Rightarrow \quad dz = \frac{1}{2}\sec^2\frac{x}{2}\,dx$

$\Rightarrow \quad 2dz = \left(1+\tan^2\frac{x}{2}\right)dx \quad \Rightarrow \quad \frac{2}{1+z^2}\,dz = dx \qquad [\because \sec^2 A - \tan^2 A = 1]$

Now $\sin x = \dfrac{2\tan\frac{x}{2}}{1+\tan^2\frac{x}{2}}$, $\cos x = \dfrac{1-\tan^2\frac{x}{2}}{1+\tan^2\frac{x}{2}}$

$\Rightarrow$ $\sin x = \dfrac{2z}{1+z^2}$, $\cos x = \dfrac{1-z^2}{1+z^2}$

When $x = 0$, $z = \tan 0° = 0$ and when $x = \dfrac{\pi}{2}$, $z = \tan\dfrac{\pi}{4} = 1$

$$\therefore \quad I = \int_0^1 \frac{1}{5\left(\frac{1-z^2}{1+z^2}\right)+3\left(\frac{2z}{1+z^2}\right)} \cdot \frac{2}{1+z^2}\,dz = 2\int_0^1 \frac{1+z^2}{5-5z^2+6z} \cdot \frac{1}{1+z^2}\,dz$$

$$= 2\int_0^1 \frac{1}{5-5z^2+6z}\,.\,dz = -\frac{2}{5}\int_0^1 \frac{1}{z^2-\frac{6}{5}z-1}\,dz$$

$$= -\frac{2}{5}\int_0^1 \frac{1}{\left(z^2-\frac{6}{5}z+\frac{9}{25}\right)-\left(1+\frac{9}{25}\right)}\,dz$$

$$\left[\begin{array}{l}\text{Add and subtract } \frac{9}{25} \text{ to the denom.} \\ \because \left(\frac{1}{2}\text{ co-eff. of } z\right)^2 = \frac{9}{25}\end{array}\right]$$

$$= -\frac{2}{5}\int_0^1 \frac{1}{\left(z-\frac{3}{5}\right)^2-\frac{34}{25}}\,dz = -\frac{2}{5}\int_0^1 \frac{1}{\left(z-\frac{3}{5}\right)^2-\left(\frac{\sqrt{34}}{5}\right)^2}\,dz$$

$$= -\frac{2}{5}\cdot\left[\frac{1}{2\,.\,\frac{\sqrt{34}}{5}}\log\left|\frac{z-\frac{3}{5}-\frac{\sqrt{34}}{5}}{z-\frac{3}{5}+\frac{\sqrt{34}}{5}}\right|\right]_0^1 \quad \left[\because \int \frac{1}{x^2-a^2}\,dx = \frac{1}{2a}\log\left|\frac{x-a}{x+a}\right|+c\right]$$

$$= -\frac{2}{5}\cdot\frac{5}{2\sqrt{34}}\left[\log\left|\frac{5z-3-\sqrt{34}}{5z-3+\sqrt{34}}\right|\right]_0^1$$

$$= -\frac{1}{\sqrt{34}}\left[\log\left|\frac{5-3-\sqrt{34}}{5-3+\sqrt{34}}\right| - \log\left|\frac{0-3-\sqrt{34}}{0-3+\sqrt{34}}\right|\right]$$

$$= -\frac{1}{\sqrt{34}}\left[\log\left|\frac{2-\sqrt{34}}{2+\sqrt{34}}\right| - \log\left|\frac{\sqrt{34}+3}{\sqrt{34}-3}\right|\right]$$

$$= -\frac{1}{\sqrt{34}}\left[\log\left|\frac{\sqrt{34}-2}{\sqrt{34}+2}\right| - \log\left|\frac{\sqrt{34}+3}{\sqrt{34}-3}\right|\right]$$

$$= -\frac{1}{\sqrt{34}}\left[\log\left|\frac{\sqrt{34}-2}{\sqrt{34}+2}\times\frac{\sqrt{34}-3}{\sqrt{34}+3}\right|\right] \qquad \left[\because \ \log m - \log n = \log\left(\frac{m}{n}\right)\right]$$

$$= -\frac{1}{\sqrt{34}}\left[\log\left|\frac{34-2\sqrt{34}-3\sqrt{34}+6}{34+2\sqrt{34}+3\sqrt{34}+6}\right|\right] = -\frac{1}{\sqrt{34}}\left[\log\left|\frac{40-5\sqrt{34}}{40+5\sqrt{34}}\right|\right]$$

$$= -\frac{1}{\sqrt{34}}\log\left(\frac{8-\sqrt{34}}{8+\sqrt{34}}\right)$$

$$= -\frac{1}{\sqrt{34}}\log\left[\frac{8-\sqrt{34}}{8+\sqrt{34}}\times\frac{8-\sqrt{34}}{8-\sqrt{34}}\right] \qquad \text{[Rationalization]}$$

$$= -\frac{1}{\sqrt{34}}\log\left[\frac{64+34-16\sqrt{34}}{64-34}\right] = -\frac{1}{\sqrt{34}}\log\left[\frac{98-16\sqrt{34}}{30}\right]$$

$$= -\frac{1}{\sqrt{34}}\log\left[\frac{49-8\sqrt{34}}{15}\right].$$

(*viii*) Let $I = \int_0^{\pi}\frac{\sin x}{\sin x+\cos x}\,dx$

$$= \frac{1}{2}\int_0^{\pi}\frac{2\sin x}{\sin x+\cos x}\,dx \qquad \text{[Multiply and divided by 2]}$$

$$= \frac{1}{2}\int_0^{\pi}\frac{\sin x+\sin x}{\sin x+\cos x}\,dx \qquad \text{[Note this step]}$$

$$= \frac{1}{2}\int_0^{\pi}\frac{\sin x+\cos x+\sin x-\cos x}{\sin x+\cos x}\,dx \qquad \text{[Add and subtract } \cos x\text{]}$$

$$= \frac{1}{2}\int_0^{\pi}\left(\frac{\sin x+\cos x}{\sin x+\cos x}+\frac{\sin x-\cos x}{\sin x+\cos x}\right)dx$$

$$= \frac{1}{2}\int_0^{\pi}1.\,dx+\frac{1}{2}\int_0^{\pi}\frac{\sin x-\cos x}{\sin x+\cos x}\,dx = \frac{1}{2}\Big[x\Big]_0^{\pi}-\frac{1}{2}\int_0^{\pi}\frac{\cos x-\sin x}{\sin x+\cos x}\,dx$$

$$= \frac{1}{2}[\pi-0]-\frac{1}{2}\Big[\log|\sin x+\cos x|\Big]_0^{\pi} \qquad \left[\because \int\frac{f'(x)}{f(x)}\,dx = \log|f(x)|+c\right]$$

$$= \frac{\pi}{2}-\frac{1}{2}\,[\log\,|\sin\pi+\cos\pi| - \log\,|\sin 0°+\cos 0°|]$$

$$= \frac{\pi}{2}-\frac{1}{2}\,[\log\,|0-1| - \log\,|0+1|] = \frac{\pi}{2}-\frac{1}{2}[\log 1-\log 1]$$

$$= \frac{\pi}{2}.$$

(*ix*) Let $I = \int_0^{\pi/2}\frac{1}{4+5\sin x}\,dx$

Put $\quad z = \tan\frac{x}{2} \Rightarrow dz = \frac{1}{2}\sec^2\frac{x}{2}\,dx$

$\Rightarrow \quad 2dz = \left(1+\tan^2\frac{x}{2}\right)dx \Rightarrow \frac{2}{1+z^2}\,dz = dx \qquad [\because \sec^2 A - \tan^2 A = 1]$

Now $\sin x = \dfrac{2\tan\frac{x}{2}}{1+\tan^2\frac{x}{2}} = \dfrac{2z}{1+z^2}$

When $\quad x = 0,\ z = \tan 0° = 0$ and when $x = \frac{\pi}{2},\ z = \tan\frac{\pi}{4} = 1$

$$\therefore \quad I = \int_0^1 \frac{1}{4+5\left(\frac{2z}{1+z^2}\right)}\cdot\frac{2}{1+z^2}\,dz$$

$$= 2\int_0^1 \frac{1+z^2}{4+4z^2+10z}\cdot\frac{1}{1+z^2}\,dz = 2\int \frac{1}{4z^2+10z+4}\,dz = \frac{1}{2}\int \frac{1}{z^2+\frac{5}{2}z+1}\,dz$$

$$\therefore \quad I = \frac{1}{2}\int_0^1 \frac{1}{\left(z^2+\frac{5}{2}z+\frac{25}{16}\right)+\left(1-\frac{25}{16}\right)}\,dz \qquad \left[\begin{array}{l}\text{Add and subtract } \frac{25}{16} \text{ to the denom.}\\ \because \left(\frac{1}{2}\text{ co-eff. of } z\right)^2 = \frac{25}{16}\end{array}\right]$$

$$= \frac{1}{2}\int_0^1 \frac{1}{\left(z+\frac{5}{4}\right)^2-\frac{9}{16}}\,dz = \frac{1}{2}\int_0^1 \frac{1}{\left(z+\frac{5}{4}\right)^2-\left(\frac{3}{4}\right)^2}\,dz$$

$$= \frac{1}{2}\left[\frac{1}{2\left(\frac{3}{4}\right)}\cdot\log\left|\frac{z+\frac{5}{4}-\frac{3}{4}}{z+\frac{5}{4}+\frac{3}{4}}\right|\right]_0^1 \qquad \left[\because \int\frac{1}{x^2-a^2}\,dx = \frac{1}{2a}\log\left|\frac{x-a}{x+a}\right|+c\right]$$

$$= \frac{1}{2}\cdot\frac{2}{3}\left[\log\left|\frac{4z+2}{4z+8}\right|\right]_0^1 = \frac{1}{3}\left[\log\left|\frac{4+2}{4+8}\right| - \log\left|\frac{0+2}{0+8}\right|\right]$$

$$= \frac{1}{3}\left[\log\left(\frac{6}{12}\right)-\log\left(\frac{2}{8}\right)\right] = \frac{1}{3}\left[\log\frac{1}{2}-\log\frac{1}{4}\right]$$

$$= \frac{1}{3}\log\left[\frac{1}{2}\times\frac{4}{1}\right] = \frac{1}{3}\log 2. \qquad \left[\because \log m - \log n = \log\left(\frac{m}{n}\right)\right]$$

(x) Let $\quad I = \displaystyle\int_0^{\pi/2} \frac{1}{(2\cos x + 4\sin x)}\,dx$

Put $\quad z = \tan \frac{x}{2} \quad \Rightarrow \quad dz = \frac{1}{2} \sec^2 \frac{x}{2}\, dx$

$$\Rightarrow \quad 2dz = \left(1 + \tan^2 \frac{x}{2}\right) dx \quad \Rightarrow \quad \frac{2}{1+z^2}\, dz = dx \qquad [\because \sec^2 A - \tan^2 A = 1]$$

Now $\quad \sin x = \dfrac{2 \tan \frac{x}{2}}{1 + \tan^2 \frac{x}{2}}, \quad \cos x = \dfrac{1 - \tan^2 \frac{x}{2}}{1 + \tan^2 \frac{x}{2}}$

$$\Rightarrow \quad \sin x = \frac{2z}{1+z^2}, \quad \cos x = \frac{1-z^2}{1+z^2}$$

When $\quad x = 0, \quad z = \tan 0° = 0$ and when $x = \frac{\pi}{2}, \quad z = \tan \frac{\pi}{4} = 1$

$$\therefore \quad I = \int_0^1 \frac{1}{2\left(\frac{1-z^2}{1+z^2}\right) + 4\left(\frac{2z}{1+z^2}\right)} \cdot \frac{2}{1+z^2}\, dz$$

$$= 2\int_0^1 \frac{1+z^2}{2 - 2z^2 + 8z} \cdot \frac{1}{1+z^2}\, dz = 2\int_0^1 \frac{1}{-2z^2 + 8z + 2}\, dz = -\int_0^1 \frac{1}{z^2 - 4z - 1}\, dz$$

$$= -\int_0^1 \frac{1}{(z^2 - 4z + 4) + (-1 - 4)}\, dz \qquad \left[\begin{array}{l}\text{Add and subtract 4 to the denom.}\\ \because \left(\frac{1}{2}\text{ co-eff. of } z\right)^2 = 4\end{array}\right]$$

$$= -\int_0^1 \frac{1}{(z-2)^2 - 5}\, dz = -\int_0^1 \frac{1}{(z-2)^2 - \left(\sqrt{5}\right)^2}\, dz$$

$$= \int_0^1 \frac{1}{\left(\sqrt{5}\right)^2 - (z-2)^2}\, dz$$

$$= \frac{1}{2\sqrt{5}}\left[\log\left|\frac{\sqrt{5} + z - 2}{\sqrt{5} - z + 2}\right|\right]_0^1 \qquad \left[\because \int \frac{1}{a^2 - x^2}\, dx = \frac{1}{2a}\log\left|\frac{a+x}{a-x}\right| + c\right]$$

$$= \frac{1}{2\sqrt{5}}\left[\log\left|\frac{\sqrt{5} + 1 - 2}{\sqrt{5} - 1 + 2}\right| - \log\left|\frac{\sqrt{5} + 0 - 2}{\sqrt{5} - 0 + 2}\right|\right]$$

$$= \frac{1}{2\sqrt{5}}\left[\log\left|\frac{\sqrt{5} - 1}{\sqrt{5} + 1}\right| - \log\left|\frac{\sqrt{5} - 2}{\sqrt{5} + 2}\right|\right]$$

$$= \frac{1}{2\sqrt{5}}\log\left|\frac{\frac{\sqrt{5}-1}{\sqrt{5}+1}}{\frac{\sqrt{5}-2}{\sqrt{5}+2}}\right| = \frac{1}{2\sqrt{5}}\log\left[\frac{\sqrt{5}-1}{\sqrt{5}+1} \times \frac{\sqrt{5}+2}{\sqrt{5}-2}\right]$$

$$= \frac{1}{2\sqrt{5}} \log\left[\frac{5-\sqrt{5}+2\sqrt{5}-2}{5+\sqrt{5}-2\sqrt{5}-2}\right] = \frac{1}{2\sqrt{5}} \log\left[\frac{3+\sqrt{5}}{3-\sqrt{5}}\right]$$

$$= \frac{1}{2\sqrt{5}} \log\left[\frac{3+\sqrt{5}}{3-\sqrt{5}} \times \frac{3+\sqrt{5}}{3+\sqrt{5}}\right] \qquad \text{[Rationalization]}$$

$$= \frac{1}{2\sqrt{5}} \log\left[\frac{9+3\sqrt{5}+3\sqrt{5}+5}{9-5}\right] = \frac{1}{2\sqrt{5}} \log\left[\frac{14+6\sqrt{5}}{4}\right]$$

$$= \frac{1}{2\sqrt{5}} \log\left(\frac{7+3\sqrt{5}}{2}\right).$$

Example 32. *Evaluate the following integrals :*

(*i*) $\int_0^{\pi} \frac{1}{5+4\cos x}\,dx$ (*ii*) $\int_0^{\pi/2} \frac{\cos x}{3\cos x+\sin x}\,dx$

(*iii*) $\int_0^{\pi/2} \frac{\cos x}{\left(\cos\frac{x}{2}+\sin\frac{x}{2}\right)^3}\,dx$ (*iv*) $\int_0^{\pi/2} \frac{\sin 2x}{\sin^4 x+\cos^4 x}\,dx$

(*v*) $\int_0^{\pi/4} \sec^4 x\,dx.$

Solution. (*i*) Let $I = \int_0^{\pi} \frac{1}{5+4\cos x}\,dx$

Put $z = \tan\frac{x}{2} \Rightarrow dz = \frac{1}{2}\sec^2\frac{x}{2}\,dx$

$\Rightarrow$ $2dz = \left(1+\tan^2\frac{x}{2}\right)dx \Rightarrow \frac{2}{1+z^2}\,dz = dx$ [$\because \sec^2 A - \tan^2 A = 1$]

Now $\cos x = \dfrac{1-\tan^2\frac{x}{2}}{1+\tan^2\frac{x}{2}} = \dfrac{1-z^2}{1+z^2}$

When $x = 0,\ z = \tan 0° = 0$ and when $x = \pi,\ z = \tan\frac{\pi}{2} = \infty.$

$$\therefore \quad I = \int_0^{\infty} \frac{1}{5+4\left(\frac{1-z^2}{1+z^2}\right)} \cdot \frac{2}{1+z^2}\,dz = 2\int_0^{\infty} \frac{1+z^2}{5+5z^2+4-4z^2} \cdot \frac{1}{1+z^2}\,dz$$

$$= 2\int_0^{\infty} \frac{1}{z^2+9}\,dz = 2\int_0^{\infty} \frac{1}{z^2+3^2}\,dz$$

$$= 2\,.\left[\frac{1}{3}\tan^{-1}\frac{z}{3}\right]_0^{\infty} \qquad \left[\because \int \frac{1}{a^2+x^2}\,dx = \frac{1}{a}\tan^{-1}\frac{x}{a}+c\right]$$

$$= \frac{2}{3}[\tan^{-1}\infty - \tan^{-1}0] = \frac{2}{3}\left[\frac{\pi}{2}-0\right]$$

$$= \frac{\pi}{3}.$$

(*ii*) Let $\quad I = \int_0^{\pi/2} \frac{\cos x}{(3\cos x + \sin x)}\, dx$

Let $\quad \cos x = \lambda(3\cos x + \sin x) + \mu \frac{d}{dx}(3\cos x + \sin x)$

$\Rightarrow \quad \cos x = \lambda(3\cos x + \sin x) + \mu(-3\sin x + \cos x)$ ...(1)

$\Rightarrow \quad \cos x = (3\lambda + \mu)\cos x + (\lambda - 3\mu)\sin x$ ...(2)

Comparing the co-efficients of $\cos x$ and $\sin x$ on both sides of equation (2), we have

$1 = 3\lambda + \mu$ ...(3)

$0 = \lambda - 3\mu$ ...(4)

Now, multiplying equation (3) by 3 and adding to equation (4), we have

$$3\mu + 9\lambda = 3$$
$$-3\mu + \lambda = 0$$

$$10\lambda = 3 \Rightarrow \lambda = \frac{3}{10}$$

Put $\lambda = \frac{3}{10}$ in (4), we have

$$\frac{3}{10} - 3\mu = 0 \Rightarrow -3\mu = -\frac{3}{10} \Rightarrow \mu = \frac{1}{10}$$

Substitituting the values of λ and μ in equation (1), we get

$$\cos x = \frac{3}{10}(3\cos x + \sin x) + \frac{1}{10}(-3\sin x + \cos x)$$

$$\therefore \quad I = \int_0^{\pi/2} \frac{\cos x}{(3\cos x + \sin x)}\, dx$$

$$= \int_0^{\pi/2} \frac{\frac{3}{10}(3\cos x + \sin x) + \frac{1}{10}(-3\sin x + \cos x)}{(3\cos x + \sin x)}\, dx$$

$$= \frac{3}{10}\int_0^{\pi/2} \frac{3\cos x + \sin x}{3\cos x + \sin x}\, dx + \frac{1}{10}\int_0^{\pi/2} \frac{-3\sin x + \cos x}{3\cos x + \sin x}\, dx$$

$$= \frac{3}{10}\int_0^{\pi/2} 1 . dx + \frac{1}{10}\int_0^{\pi/2} \frac{-3\sin x + \cos x}{3\cos x + \sin x}\, dx$$

$$= \frac{3}{10}\Big[x\Big]_0^{\pi/2} + \frac{1}{10}\Big[\log|3\cos x + \sin x|\Big]_0^{\pi/2} \qquad \left[\because \int \frac{f'(x)}{f(x)}\, dx = \log|f(x)| + c\right]$$

$$= \frac{3}{10}\left[\frac{\pi}{2} - 0\right] + \frac{1}{10}\left[\log\left|3\cos\frac{\pi}{2} + \sin\frac{\pi}{2}\right| - \log|3\cos 0° + \sin 0°|\right]$$

$$= \frac{3\pi}{20} + \frac{1}{10}[\log 1 - \log 3]$$

$$= \frac{3\pi}{20} + \frac{1}{10}\log\left(\frac{1}{3}\right). \qquad \left[\because \log m - \log n = \log\left(\frac{m}{n}\right)\right]$$

(iii) Let $I = \int_0^{\pi/2} \frac{\cos x}{\left(\cos\frac{x}{2} + \sin\frac{x}{2}\right)^3} dx$

$$\left[\because \cos 2A = \cos^2 A - \sin^2 A \Rightarrow \cos A = \cos^2 \frac{A}{2} - \sin^2 \frac{A}{2}\right]$$

$$= \int_0^{\pi/2} \frac{\left(\cos^2\frac{x}{2} - \sin^2\frac{x}{2}\right)}{\left(\cos\frac{x}{2} + \sin\frac{x}{2}\right)^3} dx$$

$$= \int_0^{\pi/2} \frac{\left(\cos\frac{x}{2} + \sin\frac{x}{2}\right)\left(\cos\frac{x}{2} - \sin\frac{x}{2}\right)}{\left(\cos\frac{x}{2} + \sin\frac{x}{2}\right)^3} dx \qquad [\because (a^2 - b^2) = (a + b)(a - b)]$$

$$= \int_0^{\pi/2} \frac{\left(\cos\frac{x}{2} - \sin\frac{x}{2}\right)}{\left(\cos\frac{x}{2} + \sin\frac{x}{2}\right)^2} dx$$

Put $\left(\cos\frac{x}{2} + \sin\frac{x}{2}\right) = z \Rightarrow \left(-\frac{1}{2}\sin\frac{x}{2} + \frac{1}{2}\cos\frac{x}{2}\right) dx = dz$

$\Rightarrow \frac{1}{2}\left(\cos\frac{x}{2} - \sin\frac{x}{2}\right) dx = dz \Rightarrow \left(\cos\frac{x}{2} - \sin\frac{x}{2}\right) dx = 2dz$

When $x = 0, \quad z = \cos 0° + \sin 0° = 1 + 0 = 1$

and when $x = \frac{\pi}{2}, \quad z = \cos\frac{\pi}{4} + \sin\frac{\pi}{4} = \frac{1}{\sqrt{2}} + \frac{1}{\sqrt{2}} = \frac{2}{\sqrt{2}} = \sqrt{2}$

$\therefore \quad I = \int_1^{\sqrt{2}} \frac{1}{z^2}(2dz) = 2\int_1^{\sqrt{2}} z^{-2}\, dz$

$$= 2\left[\frac{z^{-2+1}}{-2+1}\right]_1^{\sqrt{2}} - 2\left[\frac{1}{z}\right]_1^{\sqrt{2}} = -2\left(\frac{1}{\sqrt{2}} - \frac{1}{1}\right)$$

$$= -\sqrt{2} + 2.$$

(iv) Let $I = \int_0^{\pi/2} \frac{\sin 2x}{\sin^4 x + \cos^4 x} dx = \int_0^{\pi/2} \frac{\sin 2x}{\sin^4 x + (\cos^2 x)^2} dx$

$$= \int_0^{\pi/2} \frac{\sin 2x}{\sin^4 x + (1 - \sin^2 x)^2} dx$$

Put $\sin^2 x = z \Rightarrow 2\sin x \cos x\, dx = dz \Rightarrow \sin 2x\, dx = dz$

When $x = 0, \quad z = \sin^2 0° = 0$ and when $x = \frac{\pi}{2}, \quad z = \sin^2\frac{\pi}{2} = (1)^2 = 1$

$\therefore \quad I = \int_0^1 \frac{1}{z^2 + (1 - z)^2} dz = \int_0^1 \frac{1}{z^2 + 1 + z^2 - 2z} dz$

$$= \int_0^1 \frac{1}{2z^2 - 2z + 1}\, dz = \frac{1}{2}\int_0^1 \frac{1}{z^2 - z + \frac{1}{2}}\, dz$$

$$= \frac{1}{2}\int_0^1 \frac{1}{\left(z^2 - z + \frac{1}{4}\right) + \left(\frac{1}{2} - \frac{1}{4}\right)}\, dz \qquad \left[\begin{array}{l}\text{Add and subtract } \frac{1}{4} \text{ to the denom.}\\ \because \left(\frac{1}{2}\text{ co-eff. of } z\right)^2 = \frac{1}{4}\end{array}\right]$$

$$= \frac{1}{2}\int_0^1 \frac{1}{\left(z - \frac{1}{2}\right)^2 + \frac{1}{4}}\, dz = \frac{1}{2}\int_0^1 \frac{1}{\left(z - \frac{1}{2}\right)^2 + \left(\frac{1}{2}\right)^2}\, dz$$

$$= \frac{1}{2}\left[\frac{1}{\frac{1}{2}} \tan^{-1}\left(\frac{z - \frac{1}{2}}{\frac{1}{2}}\right)\right]_0^1 \qquad \left[\because \int \frac{1}{x^2 + a^2}\, dx = \frac{1}{a}\tan^{-1}\frac{x}{a} + c\right]$$

$$= \frac{1}{2} . 2\left[\tan^{-1}(2z - 1)\right]_0^1 = [\tan^{-1}(2 - 1) - \tan^{-1}(0 - 1)] = \tan^{-1} 1 - \tan^{-1}(-1)$$

$$= \frac{\pi}{4} - \left(-\frac{\pi}{4}\right) = \frac{\pi}{2}.$$

(*v*) Let $\quad I = \int_0^{\pi/4} \sec^4 x\, dx = \int_0^{\pi/4} \sec^2 x . \sec^2 x\, dx$

$$= \int_0^{\pi/4} (1 + \tan^2 x) \sec^2 x\, dx \qquad [\because \sec^2 A - \tan^2 A = 1]$$

Put $\tan x = z \Rightarrow \sec^2 x\, dx = dz$

When $\quad x = 0, \quad z = \tan 0° = 0$ and when $x = \frac{\pi}{4}, \quad z = \tan\frac{\pi}{4} = 1$

$\therefore \qquad I = \int_0^1 (1 + z^2)\, dz$

$$= \left[z + \frac{z^3}{3}\right]_0^1 = \left[1 + \frac{(1)^3}{3}\right] - [0 + 0] = \left(1 + \frac{1}{3}\right) - 0$$

$$= \frac{4}{3}.$$

Example 33. *Evaluate the following integrals :*

(*i*) $\int_0^1 x(\tan^{-1} x)^2\, dx$ (*ii*) $\int_0^a \sqrt{a^2 - x^2}\, dx$

(*iii*) $\int_0^{\pi/4} \frac{\tan^3 x}{1 + \cos 2x}\, dx$ (*iv*) $\int_{-a}^{+a} \sqrt{\frac{a - x}{a + x}}\, dx$

(*v*) $\int_0^{\pi/2} \frac{\sin x}{(1 + \cos x)(2 + \cos x)}\, dx$ (*vi*) $\int_0^{\pi/2} \frac{1}{a^2 \sin^2 x + b^2 \cos^2 x}\, dx.$

Solution. (*i*) Let $\mathrm{I} = \int_0^1 \underset{\text{II}}{x}\ (\underset{\text{I}}{\tan^{-1} x})^2\ dx$

Integrating by parts, we have

$$= (\tan^{-1} x)^2 \cdot \frac{x^2}{2} - \int_0^1 \frac{d}{dx} (\tan^{-1} x)^2 \cdot \frac{x^2}{2}\ dx$$

$$= \left[\frac{x^2}{2} (\tan^{-1} x)^2\right]_0^1 - \int_0^1 \left(2 \tan^{-1} x \cdot \frac{1}{1+x^2}\right) \cdot \frac{x^2}{2}\ dx$$

$$= \left[\frac{1}{2} (\tan^{-1} 1)^2 - 0\right] - \int_0^1 \frac{x^2}{1+x^2} \tan^{-1} x\ dx$$

$$= \frac{1}{2}\left(\frac{\pi}{4}\right)^2 - \int_0^1 \frac{1+x^2-1}{1+x^2} \tan^{-1} x\ dx \qquad \text{[Note this step]}$$

$$= \frac{1}{2}\left(\frac{\pi^2}{16}\right) - \int_0^1 \left(1 - \frac{1}{1+x^2}\right) \tan^{-1} x\ dx$$

$$= \frac{\pi^2}{32} - \int_0^1 \tan^{-1} x\ dx + \int_0^1 \frac{1}{1+x^2} \tan^{-1} x\ dx$$

$$= \frac{\pi^2}{32} + \int_0^1 \frac{1}{1+x^2} \tan^{-1} x\ dx - \int_0^1 1 \cdot (\tan^{-1} x)\ dx$$

$$\left[\because \int [f(x)]^n f'(x)\ dx = \frac{[f(x)]^{n+1}}{n+1} + c\right]$$

$$= \frac{\pi^2}{32} + \left[\frac{(\tan^{-1} x)^2}{2}\right]_0^1 - \int_0^1 \underset{\text{II}}{1} \cdot (\underset{\text{I}}{\tan^{-1} x})\ dx \qquad \text{[Integrating by parts]}$$

$$= \frac{\pi^2}{32} + \frac{1}{2} [(\tan^{-1} 1)^2 - (\tan^{-1} 0)^2] - \left[\left| \tan^{-1} x \cdot x \right|_0^1 - \int_0^1 \frac{d}{dx} (\tan^{-1} x) \cdot x\ dx\right]$$

$$= \frac{\pi^2}{32} + \frac{1}{2}\left[\left(\frac{\pi}{4}\right)^2 - 0\right] - \left[(1 \tan^{-1} 1 - 0) - \int_0^1 \frac{1}{1+x^2}\, x\ dx\right]$$

$$= \frac{\pi^2}{32} + \frac{1}{2}\left(\frac{\pi^2}{16}\right) - \frac{\pi}{4} - \int_0^1 \frac{x}{1+x^2}\ dx$$

$$= \frac{\pi^2}{32} + \frac{\pi^2}{32} - \frac{\pi}{4} - \frac{1}{2}\int_0^1 \frac{2x}{1+x^2}\ dx \qquad \text{[Multiply and divided by 2]}$$

$$= \frac{2\pi^2}{32} - \frac{\pi}{4} - \frac{1}{2}\left[\log|1+x^2|\right]_0^1 \qquad \left[\because \int \frac{f'(x)}{f(x)}\ dx = \log|f(x)| + c\right]$$

$$= \frac{\pi^2}{16} - \frac{\pi}{4} - \frac{1}{2} [\log|1+1| - \log|1+0|] = \frac{\pi^2}{16} - \frac{\pi}{4} - \frac{1}{2} [\log 2 - \log 1]$$

$$= \frac{\pi^2}{16} - \frac{\pi}{4} - \frac{1}{2} \log 2.$$

(*ii*) Let $\quad I = \int_0^a \sqrt{a^2 - x^2}\, dx$

Put $\quad x = a \sin\theta \quad \Rightarrow \quad dx = a\cos\theta\, d\theta$

When $\quad x = 0, \quad a\sin\theta = 0 \quad \Rightarrow \quad \sin\theta = 0 \quad \Rightarrow \quad \theta = 0$

and when $\quad x = a, \quad a\sin\theta = a \quad \Rightarrow \quad \sin\theta = 1 \quad \Rightarrow \quad \theta = \frac{\pi}{2}$

$\therefore$
$$I = \int_0^{\pi/2} \sqrt{a^2 - a^2\sin^2\theta}\ .\ a\cos\theta\, d\theta$$

$$= \int_0^{\pi/2} a\sqrt{1-\sin^2\theta}\ .\ a\cos\theta\, d\theta = a^2\int_0^{\pi/2}\sqrt{\cos^2\theta}\ .\ \cos\theta\, d\theta$$

$$= a^2\int_0^{\pi/2}\cos\theta\ .\ \cos\theta\, d\theta = a^2\int_0^{\pi/2}\cos^2\theta\, d\theta$$

$$= a^2\int_0^{\pi/2}\left(\frac{1+\cos 2\theta}{2}\right) d\theta \qquad \left[\begin{array}{l}\because\ (1+\cos 2A) = 2\cos^2 A \\ \Rightarrow\ \dfrac{1+\cos 2A}{2} = \cos^2 A\end{array}\right]$$

$$= \frac{a^2}{2}\int_0^{\pi/2}(1+\cos 2\theta)\, d\theta$$

$$= \frac{a^2}{2}\left[\theta + \frac{\sin 2\theta}{2}\right]_0^{\pi/2} = \frac{a^2}{2}\left[\left(\frac{\pi}{2} + \frac{\sin\pi}{2}\right) - (0+0)\right] = \frac{a^2}{2}\left(\frac{\pi}{2} + 0\right)$$

$$= \frac{\pi a^2}{4}.$$

(*iii*) Let $\quad I = \int_0^{\pi/4} \frac{\tan^3 x}{1+\cos 2x}\, dx = \int_0^{\pi/4}\frac{\tan^3 x}{2\cos^2 x}\, dx \qquad [\because\ 1 + \cos 2A = 2\cos^2 A]$

$$= \frac{1}{2}\int_0^{\pi/4}\tan^3 x\ .\ \sec^2 x\, dx$$

Put $\quad \tan x = z \quad \Rightarrow \quad \sec^2 x\, dx = dz$

When $\quad x = 0, \quad z = \tan 0° = 0$ and when $x = \frac{\pi}{4}, \quad z = \tan\frac{\pi}{4} = 1$

$\therefore$
$$I = \frac{1}{2}\int_0^1 z^3\, dz = \frac{1}{2}\left[\frac{z^4}{4}\right]_0^1 = \frac{1}{8}\left[z^4\right]_0^1 = \frac{1}{8}\,[(1)^4 - (0)^4]$$

$$= \frac{1}{8}[1-0] = \frac{1}{8}.$$

(*iv*) Let $\quad I = \int_{-a}^a \sqrt{\frac{a-x}{a+x}}\, dx$

Put $\quad x = a\cos\theta \quad \Rightarrow \quad dx = -\, a\sin\theta\, d\theta$

When $x = -a$, $a \cos\theta = -a \Rightarrow \cos\theta = -1 \Rightarrow \theta = \pi$
and when $x = a$, $a\cos\theta = a \Rightarrow \cos\theta = 1 \Rightarrow \theta = 0°$.

$$\therefore \quad I = \int_{\pi}^{0} \sqrt{\frac{a - a\cos\theta}{a + a\cos\theta}}\,(-a\sin\theta\, d\theta)$$

$$= -a\int_{\pi}^{0} \sqrt{\frac{a(1-\cos\theta)}{a(1+\cos\theta)}}\sin\theta\, d\theta = -a\int_{\pi}^{0}\sqrt{\frac{1-\cos\theta}{1+\cos\theta}}\sin\theta\, d\theta$$

$$= -a\int_{\pi}^{0}\sqrt{\frac{2\sin^2\frac{\theta}{2}}{2\cos^2\frac{\theta}{2}}}\,.\sin\theta\, d\theta$$

$$\left[\begin{aligned} \because\ & 1+\cos 2A = 2\cos^2 A \\ \Rightarrow\ & 1+\cos A = 2\cos^2\frac{A}{2} \\ & 1-\cos 2A = 2\sin^2 A \\ \Rightarrow\ & 1-\cos A = 2\sin^2\frac{A}{2}\end{aligned}\right]$$

$$= -a\int_{\pi}^{0}\frac{\sin\frac{\theta}{2}}{\cos\frac{\theta}{2}}\,.\,2\sin\frac{\theta}{2}\cos\frac{\theta}{2}\, d\theta$$

$$\left[\begin{aligned} \because\ & \sin 2A = 2\sin A\cos A \\ \Rightarrow\ & \sin A = 2\sin\frac{A}{2}\cos\frac{A}{2}\end{aligned}\right]$$

$$= -2a\int_{\pi}^{0}\sin^2\frac{\theta}{2}\, d\theta = -2a\int_{\pi}^{0}(1-\cos\theta)\, d\theta$$

$$= -2a\Big[\theta - \sin\theta\Big]_{\pi}^{0} = -2a\,[(0-\sin 0°) - (\pi - \sin\pi)] = -2a\,[0 - \pi + 0]$$

$$= 2a\pi.$$

(*v*) Let $$I = \int_{0}^{\pi/2} \frac{\sin x}{(1+\cos x)(2+\cos x)}\, dx$$

Put $\cos x = z \Rightarrow -\sin x\, dx = dz \Rightarrow \sin x\, dx = -dz$

When $x = 0$, $z = \cos 0° = 1$ and when $x = \frac{\pi}{2}$, $z = \cos\frac{\pi}{2} = 0$

$$\therefore \quad I = \int_{1}^{0}\frac{1}{(1+z)(2+z)}(-dz) = -\int_{1}^{0}\frac{1}{(z+1)(z+2)}\, dz \qquad ...(1)$$

Let $$\frac{1}{(z+1)(z+2)} = \frac{A}{(z+1)} + \frac{B}{(z+2)} \qquad ...(2)$$

Multiplying both sides by $(z+1)(z+2)$, we get

$$1 = A(z+2) + B(z+1) \qquad ...(3)$$

Put $z = -1$ in (3), we get

$$1 = A(-1+2) + B(-1+1) \Rightarrow 1 = A \Rightarrow A = 1$$

Put $z = -2$ in (3), we get

$$1 = A(-2+2) + B(-2+1) \Rightarrow 1 = -B \Rightarrow B = -1$$

Substituting the values of A and B in equation (2), we have

$$\frac{1}{(z+1)(z+2)}=\frac{1}{(z+1)}+\frac{-1}{(z+2)}$$

$$\therefore \qquad I=-\int_1^0 \frac{1}{(z+1)(z+2)}\,dz=-\int_1^0\left(\frac{1}{z+1}-\frac{1}{z+2}\right)dz$$

$$=-\int_1^0 \frac{1}{z+1}\,dz+\int_1^0 \frac{1}{z+2}\,dz$$

$$=-\Big[\log|z+1|\Big]_1^0+\Big[\log|z+2|\Big]_1^0$$

$$=-[\log(0+1)-\log(1+1)]+[\log(0+2)-\log(1+2)]$$

$$=-[\log 1-\log 2]+[\log 2-\log 3] \qquad [\because \log 1=0]$$

$$=\log 2+\log\frac{2}{3} \qquad \left[\because \log m-\log n=\log\left(\frac{m}{n}\right)\right]$$

$$=\log\left[2\left(\frac{2}{3}\right)\right]=\log\frac{4}{3}. \qquad [\because \log(m.n)=\log m+\log n]$$

(*vi*) Let $$I=\int_0^{\pi/2}\frac{1}{a^2\sin^2 x+b^2\cos^2 x}\,dx$$

$$=\int_0^{\pi/2}\frac{\frac{1}{\cos^2 x}}{a^2\frac{\sin^2 x}{\cos^2 x}+b^2\frac{\cos^2 x}{\cos^2 x}}\,dx \qquad \left[\begin{array}{l}\text{Dividing the numerator and}\\ \text{the denominator by } \cos^2 x\end{array}\right]$$

$$=\int_0^{\pi/2}\frac{\sec^2 x}{a^2\tan^2 x+b^2}\,dx$$

Put $\tan x=z \Rightarrow \sec^2 x\,dx=dz$

When $x=0$, $z=\tan 0°=0$ and when $x=\frac{\pi}{2}$, $z=\tan\frac{\pi}{2}=\infty$

$$\therefore \qquad I=\int_0^{\infty}\frac{1}{a^2z^2+b^2}\,.\,dz=\frac{1}{a^2}\int_0^{\infty}\frac{1}{z^2+\frac{b^2}{a^2}}\,dz$$

$$=\frac{1}{a^2}\cdot\left[\frac{1}{\left(\frac{b}{a}\right)}\tan^{-1}\left(\frac{z}{\frac{b}{a}}\right)\right]_0^{\infty} \qquad \left[\because \int\frac{1}{x^2+a^2}\,dx=\frac{1}{a}\tan^{-1}\frac{x}{a}+c\right]$$

$$=\frac{1}{a^2}\cdot\frac{a}{b}\left[\tan^{-1}\frac{az}{b}\right]_0^{\infty}=\frac{1}{ab}\,[\tan^{-1}\infty-\tan^{-1}0]=\frac{1}{ab}\left[\frac{\pi}{2}-0\right]$$

$$=\frac{\pi}{2ab}.$$

Example 34. *Evaluate the following integrals :*

(*i*) $\displaystyle\int_0^{\pi/3} \frac{\cos x}{3+4\sin x}\,dx$ (*ii*) $\displaystyle\int_0^{\pi/2} \frac{\cos x}{1+\cos x+\sin x}\,dx$

(*iii*) $\displaystyle\int_1^{2} \frac{1}{x(1+\log x)^2}\,dx$ (*iv*) $\displaystyle\int_0^{\pi/2} \frac{\cos^2\theta}{\cos^2\theta+4\sin^2\theta}\,d\theta.$

Solution. (*i*) Let $\displaystyle \text{I} = \int_0^{\pi/3} \frac{\cos x}{3+4\sin x}\,dx$

$$= \frac{1}{4}\int_0^{\pi/3} \frac{4\cos x}{3+4\sin x}\,dx \qquad \text{[Multiply and divided by 4]}$$

$$= \frac{1}{4}\Big[\log|3+4\sin x|\Big]_0^{\pi/3} \qquad \left[\because \int \frac{f'(x)}{f(x)}\,dx = \log|f(x)|+c\right]$$

$$= \frac{1}{4}\left[\log\left|3+4\sin\frac{\pi}{3}\right| - \log|3+4\sin 0°|\right]$$

$$= \frac{1}{4}\left[\log\left(3+4\,.\,\frac{\sqrt{3}}{2}\right) - \log(3+0)\right] = \frac{1}{4}\left[\log\left(3+2\sqrt{3}\right)-\log 3\right]$$

$$= \frac{1}{4}\log\left(\frac{3+2\sqrt{3}}{3}\right). \qquad \left[\because \log m - \log n = \log\left(\frac{m}{n}\right)\right]$$

(*ii*) Let $\displaystyle \text{I} = \int_0^{\pi/2} \frac{\cos x}{1+\cos x+\sin x}\,dx$

Put $\displaystyle z = \tan\frac{x}{2} \Rightarrow dz = \frac{1}{2}\sec^2\frac{x}{2}\,dx$

$$\Rightarrow \quad 2dz = \left(1+\tan^2\frac{x}{2}\right)dx \Rightarrow \frac{2}{1+z^2}\,dz = dx \qquad [\because \sec^2 A - \tan^2 A = 1]$$

Now $\displaystyle \sin x = \frac{2\tan\frac{x}{2}}{1+\tan^2\frac{x}{2}}, \quad \cos x = \frac{1-\tan^2\frac{x}{2}}{1+\tan^2\frac{x}{2}}$

$$\Rightarrow \quad \sin x = \frac{2z}{1+z^2}, \quad \cos x = \frac{1-z^2}{1+z^2}$$

When $x = 0$, $z = \tan 0° = 0$ and when $x = \dfrac{\pi}{2}$, $z = \tan\dfrac{\pi}{4} = 1$

$$\therefore \quad \text{I} = \int_0^1 \frac{\left(\frac{1-z^2}{1+z^2}\right)}{1+\left(\frac{1-z^2}{1+z^2}\right)+\left(\frac{2z}{1+z^2}\right)}\cdot\frac{2}{1+z^2}\,dz = 2\int_0^1 \frac{1-z^2}{1+z^2+1-z^2+2z}\cdot\frac{1}{1+z^2}\,dz$$

$$= 2\int_0^1 \frac{1-z^2}{(2z+2)(1+z^2)}\,dz = 2\int_0^1 \frac{(1-z)(1+z)}{2(z+1)(1+z^2)}\,dz$$

$$= \int_0^1 \frac{(1-z)}{1+z^2}\,dz = -\int_0^1 \frac{z-1}{z^2+1}\,dz$$

$$= -\int_0^1 \frac{z}{z^2+1}\,dz + \int_0^1 \frac{1}{z^2+1}\,dz$$

$$= -\frac{1}{2}\int_0^1 \frac{2z}{z^2+1}\,dz + \int_0^1 \frac{1}{z^2+1}\,dz$$ [Multiply and divide the first integral by 2]

$$= -\frac{1}{2}\Big[\log|z^2+1|\Big]_0^1 + \Big[\tan^{-1} z\Big]_0^1$$ $\left[\because \int \frac{f'(x)}{f(x)}\,dx = \log|f(x)| + c \quad \int \frac{1}{x^2+a^2}\,dx = \frac{1}{a}\tan^{-1}\frac{x}{a} + c\right]$

$$= -\frac{1}{2}[\log(1+1) - \log(0+1)] + [\tan^{-1} 1 - \tan^{-1} 0]$$

$$= -\frac{1}{2}[\log 2 - \log 1] + \left[\frac{\pi}{4} - 0\right]$$

$$= -\frac{1}{2}\log 2 + \frac{\pi}{4}$$

$$= \frac{\pi}{4} - \frac{1}{2}\log 2.$$

(*iii*) Let $I = \int_1^2 \frac{1}{x(1+\log x)^2}\,dx$

Put $(1 + \log x) = z \Rightarrow \frac{1}{x}\,dx = dz$

When $x = 1$, $z = 1 + \log 1 = 1$ and when $x = 2$, $z = 1 + \log 2 = (1 + \log 2)$

$$\therefore \quad I = \int_1^{1+\log 2} \frac{1}{z^2}\,dz = \int_1^{1+\log 2} z^{-2}\,dz$$

$$= \left[\frac{z^{-2+1}}{-2+1}\right]_1^{1+\log 2} = -\left[\frac{1}{z}\right]_1^{1+\log 2} = -\left[\frac{1}{1+\log 2} - 1\right]$$

$$= 1 - \frac{1}{1+\log 2} = \frac{1+\log 2 - 1}{1+\log 2} = \frac{\log 2}{1+\log 2}$$

(*iv*) Let $I = \int_0^{\pi/2} \frac{\cos^2\theta}{\cos^2\theta + 4\sin^2\theta}\,d\theta$

$$= \int_0^{\pi/2} \frac{\cos^2\theta}{\cos^2\theta + 4(1-\cos^2\theta)}\,d\theta$$ [$\because$ $\sin^2 A + \cos^2 A = 1$]

$$= \int_0^{\pi/2} \frac{\cos^2\theta}{\cos^2\theta + 4 - 4\cos^2\theta}\,d\theta = \int_0^{\pi/2} \frac{\cos^2\theta}{4 - 3\cos^2\theta}\,d\theta$$

$$= -\frac{1}{3}\int_0^{\pi/2} \frac{-3\cos^2\theta}{4 - 3\cos^2\theta}\,d\theta$$ [Multiply and divided by – 3]

$$= -\frac{1}{3}\int_0^{\pi/2} \frac{4-3\cos^2\theta-4}{4-3\cos^2\theta}\,d\theta \qquad \text{[Add and subtract 4 to the numerator]}$$

$$= -\frac{1}{3}\int_0^{\pi/2}\left(\frac{4-3\cos^2\theta}{4-3\cos^2\theta}-\frac{4}{4-3\cos^2\theta}\right)d\theta$$

$$= -\frac{1}{3}\int_0^{\pi/2}\left(1-\frac{4}{4-3\cos^2\theta}\right)d\theta = -\frac{1}{3}\int_0^{\pi/2} 1\,.\,d\theta + \frac{4}{3}\int_0^{\pi/2}\frac{1}{4-3\cos^2\theta}\,d\theta$$

$$= -\frac{1}{3}\Big[\theta\Big]_0^{\pi/2} + \frac{4}{3}\int_0^{\pi/2}\frac{1}{4-3\cos^2\theta}\,d\theta = -\frac{1}{3}\left[\frac{\pi}{2}-0\right]+\frac{4}{3}\int_0^{\pi/2}\frac{1}{4-3\cos^2\theta}\,d\theta$$

$$= -\frac{\pi}{6}+\frac{4}{3}\int_0^{\pi/2}\frac{\sec^2\theta}{4\sec^2\theta-3}\,d\theta \qquad \left[\begin{array}{l}\text{Dividing the numerator and}\\ \text{the denominator by } \cos^2\theta\end{array}\right]$$

$$= -\frac{\pi}{6}+\frac{4}{3}\int_0^{\pi/2}\frac{\sec^2\theta}{4(1+\tan^2\theta)-3}\,d\theta \qquad [\because \sec^2 A - \tan^2 A = 1]$$

$$= -\frac{\pi}{6}+\frac{4}{3}\int_0^{\pi/2}\frac{\sec^2\theta}{4\tan^2\theta+1}\,d\theta$$

Put $\tan\theta = z \Rightarrow \sec^2\theta\, d\theta = dz$

When $\theta = 0,\ z = \tan 0° = 0$ and when $\theta = \frac{\pi}{2},\ z = \tan\frac{\pi}{2} = \infty$

$$\therefore \qquad I = -\frac{\pi}{6}+\frac{4}{3}\int_0^{\infty}\frac{1}{4z^2+1}\,dz = -\frac{\pi}{6}+\frac{1}{3}\int_0^{\infty}\frac{1}{z^2+\frac{1}{4}}\,dz$$

$$= -\frac{\pi}{6}+\frac{1}{3}\left[\frac{1}{\frac{1}{2}}\tan^{-1}\frac{z}{\frac{1}{2}}\right]_0^{\infty} \qquad \left[\because \int\frac{1}{x^2+a^2}\,dx = \frac{1}{a}\tan^{-1}\frac{x}{a}+c\right]$$

$$= -\frac{\pi}{6}+\frac{2}{3}\Big[\tan^{-1}2z\Big]_0^{\infty} = -\frac{\pi}{6}+\frac{2}{3}\,[\tan^{-1}\infty-\tan^{-1}0] = -\frac{\pi}{6}+\frac{2}{3}\left[\frac{\pi}{2}-0\right]$$

$$= -\frac{\pi}{6}+\frac{\pi}{3} = \frac{\pi}{6}.$$

Example 35. *Show that* $\int_0^{\pi/2} \sin^6\theta\, d\theta = \frac{5}{6}\int_0^{\pi/2}\sin^4\theta\, d\theta.$

Solution. Let $I = \int_0^{\pi/2}\sin^6\theta\, d\theta$ ...(1)

$$= \int_0^{\pi/2} \underset{\text{I}}{\sin^5\theta}\ \underset{\text{II}}{\sin\theta}\ d\theta$$

Integrating by parts, we have

$$I = \Big[\sin^5\theta\,.\,(-\cos\theta)\Big]_0^{\pi/2} - \int_0^{\pi/2}\frac{d}{d\theta}(\sin^5\theta)\,.\,(-\cos\theta)\,d\theta$$

$$= -\left[\sin^5\frac{\pi}{2}\cos\frac{\pi}{2} - \sin^5 0° \cos 0°\right] + \int_0^{\pi/2} 5\sin^4\theta\cos\theta\,.\,\cos\theta\, d\theta$$

$$= -[0 - 0] + 5\int_0^{\pi/2} \sin^4\theta\cos^2\theta\, d\theta$$

$$= 5\int_0^{\pi/2} \sin^4\theta\,.\,(1 - \sin^2\theta)\, d\theta \qquad [\because \sin^2 A + \cos^2 A = 1]$$

$$= 5\int_0^{\pi/2} \sin^4\theta\, d\theta - 5\int_0^{\pi/2} \sin^6\theta\, d\theta$$

$$= 5\int_0^{\pi/2} \sin^4\theta\, d\theta - 5I \qquad [\because \text{ By using equation (1)}]$$

$$\Rightarrow \qquad 6I = 5\int_0^{\pi/2} \sin^4\theta\, d\theta$$

$$\therefore \qquad I = \frac{5}{6}\int_0^{\pi/2} \sin^4\theta\, d\theta.$$

Example 36. *Evaluate the following integrals :*

(*i*) $\displaystyle\int_0^{\pi/2} \frac{1}{a\cos x + b\sin x}\, dx$ (*ii*) $\displaystyle\int_{\pi/3}^{\pi/2} \frac{\sqrt{1+\cos x}}{(1-\cos x)^{5/2}}\, dx$

(*iii*) $\displaystyle\int_0^1 x\sqrt{\frac{1-x^2}{1+x^2}}\, dx$ (*iv*) $\displaystyle\int_0^{\pi/2} \frac{1}{9 + 16\cos^2 x}\, dx$

(*v*) $\displaystyle\int_0^{\pi/2} \frac{1}{4 + 9\cos^2 x}\, dx$ (*vi*) $\displaystyle\int_0^{\pi/4} \frac{\sin x + \cos x}{9 + 16\sin 2x}\, dx.$

Solution. (*i*) Let $I = \displaystyle\int_0^{\pi/2} \frac{1}{a\cos x + b\sin x}\, dx$

Put $a = r\cos\theta,\ b = r\sin\theta$

Squaring and adding, we get

$$a^2 + b^2 = r^2(\cos^2\theta + \sin^2\theta)$$

$$\Rightarrow \quad a^2 + b^2 = r^2 \quad \Rightarrow \quad r = \sqrt{a^2 + b^2}$$

Also $\displaystyle\frac{b}{a} = \frac{r\sin\theta}{r\cos\theta} = \tan\theta \quad \Rightarrow \quad \theta = \tan^{-1}\frac{b}{a}$

$$\therefore \qquad I = \int_0^{\pi/2} \frac{1}{r\cos\theta\cos x + r\sin\theta\sin x}\, dx$$

$$= \frac{1}{r}\int_0^{\pi/2} \frac{1}{\cos(x-\theta)}\, dx \qquad [\because \cos(A - B) = \cos A\cos B + \sin A\sin B]$$

$$= \frac{1}{\sqrt{a^2+b^2}}\int_0^{\pi/2} \sec(x-\theta)\, d\theta$$

$$= \frac{1}{\sqrt{a^2+b^2}}\Big[\log|\sec(x-\theta) + \tan(x-\theta)|\Big]_0^{\pi/2}$$

$$= \frac{1}{\sqrt{a^2+b^2}}\left[\log\left|\sec\left(\frac{\pi}{2}-\theta\right) + \tan\left(\frac{\pi}{2}-\theta\right)\right| - \log|\sec(0-\theta) + \tan(0-\theta)|\right]$$

$$= \frac{1}{\sqrt{a^2+b^2}} [\log|\text{cosec}\,\theta + \cot\theta| - \log|\sec\theta - \tan\theta|]$$

$$\left[\begin{array}{l} \because \ \sec\left(\frac{\pi}{2} - A\right) = \text{cosec}\,A \\ \tan\left(\frac{\pi}{2} - A\right) = \cot A \\ \sec(-\theta) = \sec\theta,\ \tan(-\theta) = -\tan\theta \end{array}\right]$$

$$= \frac{1}{\sqrt{a^2+b^2}} [\log|\text{cosec}\,\theta + \cot\theta| - \log|\sec\theta - \tan\theta|]$$

$$= \frac{1}{\sqrt{a^2+b^2}} \left[\log\left|\sqrt{1+\cot^2\theta} + \cot\theta\right| - \log\left|\sqrt{1+\tan^2\theta} - \tan\theta\right|\right]$$

$$\left[\begin{array}{l} \because \ \text{cosec}^2 A - \cot^2 A = 1 \\ \Rightarrow \text{cosec}^2 A = 1 + \cot^2 A \\ \Rightarrow \text{cosec}\,A = \sqrt{1+\cot^2 A} \\ \quad \sec^2 A - \tan^2 A = 1 \\ \Rightarrow \sec^2 A = 1 + \tan^2 A \\ \Rightarrow \sec A = \sqrt{1+\tan^2 A} \end{array}\right]$$

$$= \frac{1}{\sqrt{a^2+b^2}} \left[\log\left|\sqrt{1+\frac{a^2}{b^2}} + \frac{a}{b}\right| - \log\left|\sqrt{1+\frac{b^2}{a^2}} - \frac{b}{a}\right|\right]$$

$$\left[\because \ \frac{b}{a} = \tan\theta \ \Rightarrow \ \frac{a}{b} = \cot\theta\right]$$

$$= \frac{1}{\sqrt{a^2+b^2}} \left[\log\left|\frac{\sqrt{b^2+a^2}+a}{b}\right| - \log\left|\frac{\sqrt{a^2+b^2}-b}{a}\right|\right]$$

$$= \frac{1}{\sqrt{a^2+b^2}} \left[\log\left|\frac{\sqrt{a^2+b^2}+a}{b} \times \frac{a}{\sqrt{a^2+b^2}-b}\right|\right]$$

$$\left[\because \ \log m - \log n = \log\left(\frac{m}{n}\right)\right]$$

$$= \frac{1}{\sqrt{a^2+b^2}} \left[\log\left|\frac{a}{b} \cdot \frac{\left(\sqrt{a^2+b^2}+a\right)}{\left(\sqrt{a^2+b^2}-b\right)}\right|\right].$$

(*ii*) Let $\quad I = \int_{\pi/3}^{\pi/2} \frac{\sqrt{1+\cos x}}{(1-\cos x)^{5/2}}\, dx$

$$= \int_{\pi/3}^{\pi/2} \frac{\sqrt{2\cos^2 \frac{x}{2}}}{\left(2\sin^2 \frac{x}{2}\right)^{5/2}}\, dx$$

$$\left[\begin{array}{l} \because\ 1+\cos 2A = 2\cos^2 A \\ \Rightarrow 1+\cos A = 2\cos^2 \frac{A}{2} \\ \quad 1-\cos 2A = 2\sin^2 A \\ \Rightarrow 1-\cos A = 2\sin^2 \frac{A}{2} \end{array}\right]$$

$$= \int_{\pi/3}^{\pi/2} \frac{\sqrt{2}\cos \frac{x}{2}}{4\sqrt{2}\sin^5 \frac{x}{2}}\, dx = \frac{1}{4}\int_{\pi/3}^{\pi/2} \frac{\cos \frac{x}{2}}{\sin^5 \frac{x}{2}}\, dx$$

Put $\sin \frac{x}{2} = z \Rightarrow \frac{1}{2}\cos \frac{x}{2}\, dx = dz \Rightarrow \cos \frac{x}{2}\, dx = 2dz$

When $x = \frac{\pi}{3}$, $z = \sin \frac{\pi}{6} = \frac{1}{2}$ and when $x = \frac{\pi}{2}$, $z = \sin \frac{\pi}{4} = \frac{1}{\sqrt{2}}$

$$\therefore \quad I = \frac{1}{4}\int_{1/2}^{1/\sqrt{2}} \frac{1}{z^5}(2dz) = \frac{1}{2}\int_{1/2}^{1/\sqrt{2}} z^{-5}\, dz$$

$$= \frac{1}{2}\left[\frac{z^{-5+1}}{-5+1}\right]_{1/2}^{1/\sqrt{2}} = -\frac{1}{8}\left[\frac{1}{z^4}\right]_{1/2}^{1/\sqrt{2}}$$

$$= -\frac{1}{8}\left[\frac{1}{\left(\frac{1}{\sqrt{2}}\right)^4} - \frac{1}{\left(\frac{1}{2}\right)^4}\right] = -\frac{1}{8}\left[\frac{1}{\frac{1}{4}} - \frac{1}{\frac{1}{16}}\right] = -\frac{1}{8}[4-16]$$

$$= -\frac{1}{8}(-12) = \frac{3}{2}.$$

(*iii*) Let $I = \int_0^1 x\sqrt{\frac{1-x^2}{1+x^2}}\, dx$

Put $x^2 = z \Rightarrow 2x\, dx = dz \Rightarrow x\, dx = \frac{1}{2}\, dz$

When $x = 0$, $z = (0)^2 = 0$ and when $x = 1$, $z = (1)^2 = 1$

$$\therefore \quad I = \int_0^1 \sqrt{\frac{1-z}{1+z}}\cdot\left(\frac{1}{2}\, dz\right) = \frac{1}{2}\int_0^1 \sqrt{\frac{1-z}{1+z}}\, dz$$

$$= \frac{1}{2}\int_0^1 \sqrt{\frac{1-z}{1+z}\times\frac{1-z}{1-z}}\, dz \qquad \text{[Rationalization]}$$

$$= \frac{1}{2}\int_0^1 \frac{\sqrt{(1-z)^2}}{\sqrt{(1+z)(1-z)}}\, dz = \frac{1}{2}\int_0^1 \frac{1-z}{\sqrt{1-z^2}}\, dz$$

$$= \frac{1}{2}\int_0^1 \frac{1}{\sqrt{1-z^2}}\, dz - \frac{1}{2}\int_0^1 \frac{z}{\sqrt{1-z^2}}\, dz$$

$$= \frac{1}{2}\int_0^1 \frac{1}{\sqrt{1-z^2}}\,dz + \frac{1}{4}\int_0^1 (-2z)\,(1-z^2)^{-1/2}\,dz$$

[Multiply and divide the second integral by 2]

$$= \frac{1}{2}\Big[\sin^{-1} z\Big]_0^1 + \frac{1}{4}\left[\frac{(1-z^2)^{-\frac{1}{2}+1}}{-\frac{1}{2}+1}\right]_0^1 \qquad \left[\because \int [f(x)]^n f'(x)\,dx = \frac{[f(x)]^{n+1}}{n+1} + c \quad \int \frac{1}{\sqrt{a^2-x^2}}\,dx = \sin^{-1}\frac{x}{a} + c\right]$$

$$= \frac{1}{2}\,[\sin^{-1} 1 - \sin^{-1} 0] + \frac{1}{4}\cdot\frac{2}{1}\Big[(1-z^2)^{1/2}\Big]_0^1$$

$$= \frac{1}{2}\left[\frac{\pi}{2} - 0\right] + \frac{1}{2}\,[(1-1)^{1/2} - (1-0)^{1/2}] = \frac{\pi}{4} + \frac{1}{2}\,[0-1]$$

$$= \frac{\pi}{4} - \frac{1}{2}\,.$$

(*iv*) Let $\quad \mathrm{I} = \displaystyle\int_0^{\pi/2} \frac{1}{9 + 16\cos^2 x}\,dx$

$$= \int_0^{\pi/2} \frac{\frac{1}{\cos^2 x}}{\frac{9}{\cos^2 x} + \frac{16\cos^2 x}{\cos^2 x}}\,dx \qquad \left[\text{Dividing the numerator and the denominator by } \cos^2 x\right]$$

$$= \int_0^{\pi/2} \frac{\sec^2 x}{9\sec^2 x + 16}\,dx$$

$$= \int_0^{\pi/2} \frac{\sec^2 x}{9(1+\tan^2 x) + 16}\,dx \qquad [\because \sec^2 A - \tan^2 A = 1]$$

$$= \int_0^{\pi/2} \frac{\sec^2 x}{9\tan^2 x + 25}\,dx$$

Put $\quad \tan x = z \;\Rightarrow\; \sec^2 x\,dx = dz$

When $\quad x = 0,\;\; z = \tan 0° = 0$ and when $x = \frac{\pi}{2},\;\; z = \tan\frac{\pi}{2} = \infty$

$\therefore$

$$\mathrm{I} = \int_0^{\infty} \frac{1}{9z^2 + 25}\,dz = \frac{1}{9}\int_0^{\infty} \frac{1}{z^2 + \frac{25}{9}}\,dz = \frac{1}{9}\int_0^{\infty} \frac{1}{z^2 + \left(\frac{5}{3}\right)^2}\,dz$$

$$= \frac{1}{9}\left[\frac{1}{5/3}\tan^{-1}\frac{z}{5/3}\right]_0^{\infty} \qquad \left[\because \int \frac{1}{x^2+a^2}\,dx = \frac{1}{a}\tan^{-1}\frac{x}{a} + c\right]$$

$$= \frac{1}{9}\cdot\frac{3}{5}\left[\tan^{-1}\frac{3z}{5}\right]_0^{\infty} = \frac{1}{15}\,[\tan^{-1}\infty - \tan^{-1} 0] = \frac{1}{15}\left[\frac{\pi}{2} - 0\right]$$

$$= \frac{\pi}{30}\,.$$

(*v*) Let $$I = \int_0^{\pi/2} \frac{1}{4 + 9\cos^2 x}\, dx$$

Please try yourself as part (*iv*).

[**Hint.** Dividing the numerator and the denominator by $\cos^2 x$.] [**Ans.** $\frac{\pi}{4\sqrt{13}}$.]

(*vi*) Let $$I = \int_0^{\pi/4} \frac{\sin x + \cos x}{9 + 16\sin 2x}\, dx$$

Put $\sin x - \cos x = z \Rightarrow (\cos x + \sin x)\, dx = dz$

Squaring on both sides

$$(\sin x - \cos x)^2 = z^2$$

$$\Rightarrow \sin^2 x + \cos^2 x - 2\sin x \cos x = z^2$$

$$\Rightarrow 1 - \sin 2x = z^2 \Rightarrow \sin 2x = (1 - z^2) \qquad \left[\begin{matrix}\because \sin^2 A + \cos^2 A = 1 \\ \sin 2A = 2\sin A \cos A\end{matrix}\right]$$

When $x = 0,\quad z = \sin 0° - \cos 0° = (0 - 1) = -1$

and when $x = \frac{\pi}{4},\quad z = \sin\frac{\pi}{4} - \cos\frac{\pi}{4} = \left(\frac{1}{\sqrt{2}} - \frac{1}{\sqrt{2}}\right) = 0$

$$\therefore \quad I = \int_{-1}^{0} \frac{1}{9 + 16(1 - z^2)}\, dz$$

$$= \int_{-1}^{0} \frac{1}{25 - 16z^2}\, dz = \frac{1}{16}\int_{-1}^{0} \frac{1}{\frac{25}{16} - z^2}\, dz = \frac{1}{16}\int_{-1}^{0} \frac{1}{\left(\frac{5}{4}\right)^2 - z^2}\, dz$$

$$= \frac{1}{16}\left[\frac{1}{2 \cdot \frac{5}{4}} \log\left|\frac{\frac{5}{4} + z}{\frac{5}{4} - z}\right|\right]_{-1}^{0} \qquad \left[\because \int \frac{1}{a^2 - x^2}\, dx = \frac{1}{2a}\log\left|\frac{a + x}{a - x}\right| + c\right]$$

$$= \frac{1}{16} \cdot \frac{2}{5}\left[\log\left|\frac{5 + 4z}{5 - 4z}\right|\right]_{-1}^{0} = \frac{1}{40}\left[\log\left|\frac{5 + 0}{5 - 0}\right| - \log\left|\frac{5 + 4(-1)}{5 - 4(-1)}\right|\right]$$

$$= \frac{1}{40}\left[\log 1 - \log\left(\frac{1}{9}\right)\right] = \frac{1}{40}\,[0 - \log (3)^{-2}] = \frac{1}{40}\,[-(-2)\log 3]$$

$$= \frac{2}{40}\log 3 = \frac{1}{20}\log 3.$$

Example 37. *Evaluate the following integrals :*

(*i*) $\int_{\pi/6}^{\pi/3} \frac{\sin x + \cos x}{\sqrt{\sin 2x}}\, dx$ (*ii*) $\int_0^a x\sqrt{\frac{a^2 - x^2}{a^2 + x^2}}\, dx$

(*iii*) $\int_1^2 \frac{x + 1}{(x + 2)^2}\, e^x\, dx$ (*iv*) $\int_{\pi/4}^{\pi/2} e^x (\log \sin x + \cot x)\, dx$

(*v*) $\int_0^{\pi/4} \sqrt{\tan x}\, dx.$

Solution. (*i*) Let $I = \int_{\pi/6}^{\pi/3} \frac{\sin x + \cos x}{\sqrt{\sin 2x}}\, dx$

Put $\sin x - \cos x = z \Rightarrow (\cos x + \sin x)\, dx = dz$

Squaring on both sides,

$$(\sin x - \cos x)^2 = z^2 \Rightarrow \sin^2 x + \cos^2 x - 2 \sin x \cos x = z^2$$

$$\Rightarrow \quad 1 - \sin 2x = z^2 \Rightarrow 1 - z^2 = \sin 2x \qquad \left[\because \begin{array}{l} \sin^2 A + \cos^2 A = 1 \\ \sin 2A = 2 \sin A \cos A \end{array}\right]$$

When $x = \frac{\pi}{6}$, $z = \sin \frac{\pi}{6} - \cos \frac{\pi}{6} = \frac{1}{2} - \frac{\sqrt{3}}{2} = \frac{1-\sqrt{3}}{2}$

and when $x = \frac{\pi}{3}$, $z = \sin \frac{\pi}{3} - \cos \frac{\pi}{3} = \frac{\sqrt{3}}{2} - \frac{1}{2} = \frac{\sqrt{3}-1}{2}$

$$\therefore \quad I = \int_{\frac{1-\sqrt{3}}{2}}^{\frac{\sqrt{3}-1}{2}} \frac{1}{\sqrt{1-z^2}}\, dz = \left[\sin^{-1} z\right]_{\frac{1-\sqrt{3}}{2}}^{\frac{\sqrt{3}-1}{2}} \qquad \left[\because \int \frac{1}{a^2 - x^2}\, dx = \sin^{-1} \frac{x}{a} + c\right]$$

$$= \left[\sin^{-1}\left(\frac{\sqrt{3}-1}{2}\right) - \sin^{-1}\left(\frac{1-\sqrt{3}}{2}\right)\right]$$

$$= \sin^{-1}\left(\frac{\sqrt{3}-1}{2}\right) + \sin^{-1}\left(\frac{\sqrt{3}-1}{2}\right) \qquad [\because \sin(-\theta) = -\sin\theta]$$

$$= 2 \sin^{-1}\left(\frac{\sqrt{3}-1}{2}\right).$$

(*ii*) Let $I = \int_0^a x \sqrt{\frac{a^2 - x^2}{a^2 + x^2}}\, dx$

Put $x^2 = a^2 \cos\theta$

$$\Rightarrow \quad 2x\, dx = -a^2 \sin\theta\, d\theta \Rightarrow x\, dx = -\frac{a^2}{2} \sin\theta\, d\theta$$

When $x = 0$, $a^2 \cos\theta = 0 \Rightarrow \cos\theta = 0 \Rightarrow \theta = \frac{\pi}{2}$

and when $x = a$, $a^2 \cos\theta = a^2 \Rightarrow \cos\theta = 1 \Rightarrow \theta = 0°$

$$\therefore \quad I = \int_{\pi/2}^{0} \frac{\sqrt{a^2 - a^2 \cos\theta}}{\sqrt{a^2 + a^2\cos\theta}} \cdot \left(-\frac{a^2}{2} \sin\theta\, d\theta\right)$$

$$= \int_{\pi/2}^{0} \frac{a\sqrt{1-\cos\theta}}{a\sqrt{1+\cos\theta}} \left(-\frac{a^2}{2} \sin\theta\right) d\theta$$

$$= -\frac{a^2}{2}\int_{\pi/2}^{0} \frac{\sqrt{2\sin^2\frac{\theta}{2}}}{\sqrt{2\cos^2\frac{\theta}{2}}} \sin\theta \, d\theta \qquad \begin{bmatrix} \because\ 1-\cos 2A = 2\sin^2 A \\ \Rightarrow\ 1-\cos A = 2\sin^2 \frac{A}{2} \\ 1+\cos 2A = 2\cos^2 A \\ \Rightarrow\ 1+\cos A = 2\cos^2\frac{A}{2} \end{bmatrix}$$

$$= -\frac{a^2}{2}\int_{\pi/2}^{0} \frac{\sin\frac{\theta}{2}}{\cos\frac{\theta}{2}} . 2\sin\frac{\theta}{2}\cos\frac{\theta}{2}\, d\theta \qquad \begin{bmatrix} \because\ \sin 2A = 2\sin A\cos A \\ \Rightarrow\ \sin A = 2\sin\frac{A}{2}\cos\frac{A}{2} \end{bmatrix}$$

$$= -\frac{a^2}{2}\int_{\pi/2}^{0} 2\sin^2\frac{\theta}{2}\, d\theta = -\frac{a^2}{2}\int_{\pi/2}^{0} (1-\cos\theta)\, d\theta = -\frac{a^2}{2}\int_{\pi/2}^{0} 1\, d\theta + \frac{a^2}{2}\int_{\pi/2}^{0} \cos\theta\, d\theta$$

$$= -\frac{a^2}{2}\Big[\theta\Big]_{\pi/2}^{0} + \frac{a^2}{2}\Big[\sin\theta\Big]_{\pi/2}^{0} = -\frac{a^2}{2}\left[0-\frac{\pi}{2}\right] + \frac{a^2}{2}\left[\sin 0° - \sin\frac{\pi}{2}\right] = \frac{a^2\pi}{4} + \frac{a^2}{2}(0-1)$$

$$= \frac{\pi a^2}{4} - \frac{a^2}{2} = \frac{a^2}{2}\left(\frac{\pi}{2}-1\right).$$

(*iii*) Let $\quad I = \int_1^2 \frac{x+1}{(x+2)^2} e^x \, dx$

$$= \int_1^2 \left[\frac{x+2-1}{(x+2)^2}\right] e^x \, dx \qquad \text{[Add and subtract 1 to the numerator]}$$

$$= \int_1^2 \left[\frac{x+2}{(x+2)^2} - \frac{1}{(x+2)^2}\right] e^x\, dx = \int_1^2 \left[\frac{1}{x+2} + \frac{-1}{(x+2)^2}\right] e^x \, dx$$

$$= \int_1^2 \underset{\text{I}}{\frac{1}{x+2}}\ \underset{\text{II}}{e^x}\, dx - \int_1^2 \frac{1}{(x+2)^2} e^x\, dx$$

Integrating the first integral by parts, we get

$$I = \left[\frac{1}{x+2} . e^x\right]_1^2 - \int_1^2 \frac{d}{dx}\{(x+2)^{-1}\} e^x\, dx - \int_1^2 \frac{1}{(x+2)^2} e^x \, dx$$

$$= \left[\frac{1}{2+2} . e^2 - \frac{1}{1+2} e^1\right] + \int_1^2 \frac{1}{(x+2)^2} e^x \, dx - \int_1^2 \frac{1}{(x+2)^2} e^x\, dx$$

$$= \frac{e^2}{4} - \frac{e}{3}.$$

(*iv*) Let $\quad I = \int_{\pi/4}^{\pi/2} e^x (\log \sin x + \cot x)\, dx$

$$= \int_{\pi/4}^{\pi/2} \underset{\text{II}}{e^x}\ \underset{\text{I}}{\log\sin x}\, dx + \int_{\pi/4}^{\pi/2} e^x \cot x\, dx$$

Integrating the first integral by parts, we have

$$= \Big[\log(\sin x) . e^x\Big]_{\pi/4}^{\pi/2} - \int_{\pi/4}^{\pi/2} \frac{d}{dx}\{\log(\sin x)\} . e^x\, dx + \int_{\pi/4}^{\pi/2} e^x \cot x\, dx$$

$$= \left[\log\left(\sin\frac{\pi}{2}\right)e^{\pi/2} - \log\left(\sin\frac{\pi}{4}\right)e^{\pi/4}\right] - \int_{\pi/4}^{\pi/2}\frac{1}{\sin x}\,.\cos x\,.\,e^x\,dx + \int_{\pi/4}^{\pi/2} e^x\cot x\,dx$$

$$= \left[(\log 1)\,e^{\pi/2} - \left[\log\left(\frac{1}{\sqrt{2}}\right)\right]e^{\pi/4}\right] - \int_{\pi/4}^{\pi/2} e^x\cot x\,dx + \int_{\pi/4}^{\pi/2} e^x\cot x\,dx$$

$$= 0 - e^{\pi/4}\,.\log\frac{1}{\sqrt{2}} = -\,e^{\pi/4}\log(2)^{-1/2} = -\,e^{\pi/4}\,.\left(-\frac{1}{2}\right)\log 2$$

$$= \frac{1}{2}e^{\pi/4}\log 2.$$

(*v*) Let $I = \int_0^{\pi/4}\sqrt{\tan x}\,dx$ ...(1)

Put $\sqrt{\tan x} = z \Rightarrow \tan x = z^2 \Rightarrow \sec^2 x\,dx = 2z\,dz$

$\Rightarrow (1+\tan^2 x)\,dx = 2z\,dz$ [$\because \sec^2 A - \tan^2 A = 1$]

$\Rightarrow (1+z^4)\,dx = 2z\,dz \Rightarrow dx = \dfrac{2z}{1+z^4}\,dz$

When $x = 0,\; z = \sqrt{\tan 0^\circ} = \sqrt{0} = 0$ and when $x = \dfrac{\pi}{4},\; z = \sqrt{\tan\dfrac{\pi}{4}} = \sqrt{1} = 1.$

$$\therefore \quad I = \int_0^1 z\,.\frac{2z}{1+z^4}\,dz$$

$$= \int_0^1 \frac{2z^2}{z^4+1}\,dz = \int_0^1 \frac{z^2+z^2}{z^4+1}\,dz \quad \text{[Note this step]}$$

$$= \int_0^1 \frac{z^2+1+z^2-1}{z^4+1}\,dz \quad \text{[Add and subtract 1 to the numerator]}$$

$$= \int_0^1 \frac{z^2+1}{z^4+1}\,dz + \int_0^1 \frac{z^2-1}{z^4+1}\,dz$$

$$\Rightarrow \quad I = I_1 + I_2 \quad \text{(say)} \qquad ...(2)$$

Now $$I_1 = \int \frac{z^2+1}{z^4+1}\,dz = \int \frac{1+\frac{1}{z^2}}{z^2+\frac{1}{z^2}}\,dz$$ [Dividing the numerator and the denominator by z^2]

Put $\left(z-\dfrac{1}{z}\right) = y \Rightarrow \left(1+\dfrac{1}{z^2}\right)dz = dy$

$$\left[\because \left(z-\frac{1}{z}\right) = y \Rightarrow \left(z-\frac{1}{z}\right)^2 = y^2 \Rightarrow z^2+\frac{1}{z^2}-2 = y^2 \Rightarrow z^2+\frac{1}{z^2} = y^2+2\right]$$

$$\therefore \quad I_1 = \int \frac{1}{y^2+2}\,dy$$

$$= \int \frac{1}{y^2+\left(\sqrt{2}\right)^2}\,dy$$

$$= \left[\frac{1}{\sqrt{2}} \tan^{-1} \frac{y}{\sqrt{2}}\right] \qquad \left[\because \int \frac{1}{x^2 + a^2}\, dx = \frac{1}{a} \tan^{-1} \frac{x}{a} + c\right]$$

$$= \left[\frac{1}{\sqrt{2}} \tan^{-1} \left(\frac{z - \frac{1}{z}}{\sqrt{2}}\right)\right] \qquad \left[\because\ y = \left(z - \frac{1}{z}\right)\right]$$

$$\Rightarrow \quad I_1 = \left[\frac{1}{\sqrt{2}} \tan^{-1} \left(\frac{z^2 - 1}{\sqrt{2}z}\right)\right] \qquad \ldots(3)$$

and

$$I_2 = \int \frac{z^2 - 1}{z^4 + 1}\, dz = \int \frac{1 - \frac{1}{z^2}}{z^2 + \frac{1}{z^2}}\, dz \qquad \left[\text{Dividing the numerator and the denominator by } z^2\right]$$

Put $\left(z + \frac{1}{z}\right) = y \Rightarrow \left(1 - \frac{1}{z^2}\right) dz = dy$

$$\left[\begin{array}{l}\because \text{ Put} \left(z + \frac{1}{z}\right) = y \\ \Rightarrow \left(z + \frac{1}{z}\right)^2 = y^2 \\ \Rightarrow z^2 + \frac{1}{z^2} + 2 = y^2 \\ \Rightarrow z^2 + \frac{1}{z^2} = y^2 - 2\end{array}\right]$$

$$\therefore \quad I_2 = \int \frac{1}{y^2 - 2}\, dy = \int \frac{1}{y^2 - \left(\sqrt{2}\right)^2}\, dy$$

$$= \frac{1}{2\sqrt{2}} \log \left|\frac{y - \sqrt{2}}{y + \sqrt{2}}\right| \qquad \left[\because \int \frac{1}{x^2 - a^2}\, dx = \frac{1}{2a} \log \left|\frac{x - a}{x + a}\right| + c\right]$$

$$= \frac{1}{2\sqrt{2}} \log \left|\frac{z + \frac{1}{z} - \sqrt{2}}{z + \frac{1}{z} + \sqrt{2}}\right| \qquad \left[\because\ y = \left(z + \frac{1}{z}\right)\right]$$

$$\Rightarrow \quad I_2 = \frac{1}{2\sqrt{2}} \log \left|\frac{z^2 - \sqrt{2}z + 1}{z^2 + \sqrt{2}z + 1}\right| \qquad \ldots(4)$$

$\therefore$ From equation (2), we have

$I = I_1 + I_2$

$$= \left[\frac{1}{\sqrt{2}} \tan^{-1} \frac{z^2 - 1}{\sqrt{2}z}\right]_0^1 + \left[\frac{1}{2\sqrt{2}} \log \left|\frac{z^2 - \sqrt{2}z + 1}{z^2 + \sqrt{2}z + 1}\right|\right]_0^1 \qquad \text{[Using equations (3) and (4)]}$$

$$= \frac{1}{\sqrt{2}}\left[\tan^{-1}\left(\frac{1 - 1}{\sqrt{2}}\right) - \tan^{-1}\left(\frac{0 - 1}{0}\right)\right] + \frac{1}{2\sqrt{2}}\left[\log\left|\frac{1 - \sqrt{2} + 1}{1 + \sqrt{2} + 1}\right| - \log\left|\frac{0 - 0 + 1}{0 + 0 + 1}\right|\right]$$

$$= \frac{1}{\sqrt{2}}\left[\tan^{-1} 0 - \tan^{-1}(-\infty)\right] + \frac{1}{2\sqrt{2}}\left[\log\left|\frac{2 - \sqrt{2}}{2 + \sqrt{2}}\right| - \log 1\right]$$

$$= \frac{1}{\sqrt{2}}\left[0-\left(-\frac{\pi}{2}\right)\right]+\frac{1}{2\sqrt{2}}\log\left(\frac{\sqrt{2}-1}{\sqrt{2}+1}\right) \qquad [\because \log 1 = 0]$$

$$= \frac{\pi}{2\sqrt{2}}+\frac{1}{2\sqrt{2}}\log\left(\frac{\sqrt{2}-1}{\sqrt{2}+1}\times\frac{\sqrt{2}-1}{\sqrt{2}-1}\right) \qquad \text{(Rationalization)}$$

$$= \frac{\pi}{2\sqrt{2}}+\frac{1}{2\sqrt{2}}\log\left(\frac{\left(\sqrt{2}-1\right)^2}{1}\right)$$

$$= \frac{\pi}{2\sqrt{2}}+\frac{1}{2\sqrt{2}}\,.\,2\log\left(\sqrt{2}-1\right) \qquad [\because m\log n = \log n^m]$$

$$= \frac{\pi}{2\sqrt{2}}+\frac{1}{\sqrt{2}}\log\left(\sqrt{2}-1\right).$$

Example 38. *If p and q are non-zero integers, prove that :*

(i) $\int_0^\pi \cos px \cos qx\, dx = 0 \quad \text{if } p \neq q$
$\qquad = \frac{\pi}{2} \quad \text{if } p = q.$

(ii) $\int_0^\pi \sin px \sin qx\, dx = 0 \quad \text{if } p \neq q$
$\qquad = \frac{\pi}{2} \quad \text{if } p = q.$

Solution. (*i*) If $p \neq q$; Let $\mathrm{I} = \int_0^\pi \cos px \cos qx\, dx$

$$= \frac{1}{2}\int_0^\pi 2\cos px \cos qx\, dx \qquad \text{[Multiply and divided by 2]}$$

$$= \frac{1}{2}\int_0^\pi [\cos (p+q)x + \cos (p-q)x]\, dx$$

$$[\because 2\cos A \cos B = \cos (A+B) + \cos (A-B)]$$

$$= \frac{1}{2}\left[\int_0^\pi \cos (p+q)x\, dx + \int_0^\pi \cos (p-q)\, x\, dx\right] = \frac{1}{2}\left\{\left[\frac{\sin (p+q)x}{p+q}\right]_0^\pi + \left[\frac{\sin (p-q)x}{p-q}\right]_0^\pi\right\}$$

$$= \frac{1}{2(p+q)}[\sin (p+q)\pi - \sin 0°] + \frac{1}{2(p-q)}[\sin (p-q)\pi - \sin 0°]$$

$$\left[\begin{array}{l}\because \sin n\pi = 0, n \in Z \\ \quad (p+q), (p-q) \in Z\end{array}\right]$$

$$= \frac{1}{2(p+q)}[0-0] + \frac{1}{2(p-q)}[0-0] = 0.$$

If $p = q$;

$$\therefore \quad \mathrm{I} = \int_0^\pi \cos^2 px\, dx$$

$$= \int_0^\pi \left(\frac{1+\cos 2px}{2}\right) dx \qquad \left[\begin{array}{l}(1+\cos 2A = 2\cos^2 A \\ \Rightarrow \left(\frac{1+\cos 2A}{2}\right) = \cos^2 A\end{array}\right]$$

$$= \frac{1}{2}\int_0^{\pi}(1+\cos 2px)\,dx = \frac{1}{2}\left[x+\frac{\sin 2px}{2p}\right]_0^{\pi} = \frac{1}{2}\left[\left(\pi+\frac{\sin 2p\pi}{2p}\right)-\left(0+\frac{\sin 0^\circ}{2p}\right)\right]$$

$$= \frac{1}{2}[(\pi+0)-0] = \frac{\pi}{2}.$$

(*ii*) If $p \neq q$;

Let $\mathrm{I} = \int_0^{\pi} \sin px \sin qx\,dx$

$$= \frac{1}{2}\int_0^{\pi} 2\sin px \sin qx\,dx \quad \text{[Multiply and divided by 2]}$$

$$= \frac{1}{2}\int_0^{\pi}[\cos(p-q)x - \cos(p+q)x]\,dx$$

$$[\because \ 2\sin \mathrm{A}\sin \mathrm{B} = \cos(\mathrm{A}-\mathrm{B}) - \cos(\mathrm{A}+\mathrm{B})]$$

$$= \frac{1}{2}\int_0^{\pi}\cos(p-q)\,x\,dx - \frac{1}{2}\int_0^{\pi}\cos(p+q)\,x\,dx = \frac{1}{2}\left[\frac{\sin(p-q)x}{p-q}\right]_0^{\pi} - \frac{1}{2}\left[\frac{\sin(p+q)x}{p+q}\right]_0^{\pi}$$

$$= \frac{1}{2(p-q)}[\sin(p-q)\pi - \sin 0^\circ] - \frac{1}{2(p+q)}[\sin(p+q)\pi - \sin 0^\circ]$$

$$= \frac{1}{2(p-q)}[0-0] - \frac{1}{2(p+q)}[0-0] = 0.$$

If $p = q$;

$$\therefore \quad \mathrm{I} = \int_0^{\pi}\sin^2 px\,dx = \frac{1}{2}\int_0^{\pi}(1-\cos 2px)\,dx \quad \left[\begin{array}{l}\because \ 1-\cos 2\mathrm{A} = 2\sin^2 \mathrm{A}\\ \Rightarrow \left(\dfrac{1-\cos 2\mathrm{A}}{2}\right) = \sin^2 \mathrm{A}\end{array}\right]$$

$$= \frac{1}{2}\left[x-\frac{\sin 2px}{2p}\right]_0^{\pi} = \frac{1}{2}\left[\left(\pi-\frac{\sin 2p\pi}{2p}\right)-\left(0-\frac{\sin 0^\circ}{2p}\right)\right]$$

$$= \frac{1}{2}[(\pi-0)-(0-0)]$$

$$= \frac{\pi}{2}.$$

EXERCISE FOR PRACTICE

Evaluate the following definite integrals : (1—16)

1. $\int_0^1 \frac{1}{2x-3}\,dx$

2. $\int_{-\pi/2}^{\pi/2} \cos x\,dx$

3. $\int_0^{\pi/2} \frac{1}{1+\cos\theta}\,d\theta$

4. $\int_0^{\pi/2} \sqrt{1-\cos 2x}\,dx$

5. $\int_0^{\pi/2} (x+6\cos 2x)\,dx$

6. $\int_0^{\pi/4} \sin^4 x.dx$

7. $\int_0^3 \sqrt{9-x^2}\,dx$

8. $\int_1^3 \frac{\log x}{(1+x)^2}\,dx$

9. $\int_0^{\pi/2} x \cos x \, dx$

10. $\int_1^2 \frac{x+3}{x(x+2)} \, dx$

11. $\int_0^1 \frac{x^2}{1+x^6} \, dx$

12. $\int_0^{\pi/2} \frac{\sin x}{1+\cos^2 x} \, dx$

13. $\int_0^1 \sin^{-1} x \, dx$

14. $\int_0^{\pi/2} \frac{1}{5+4\cos x} \, dx$

15. $\int_0^{\pi/4} \frac{1}{4+9\cos^2 x} \, dx$

16. $\int_0^1 \frac{x \sin^{-1} x}{\sqrt{1-x^2}} \, dx$

Answers

1. $-\frac{1}{2}\log 3$	**2.** 2	**3.** 1	**4.** $\sqrt{2}$
5. $\frac{\pi^2}{8}$	**6.** $\frac{3\pi}{32} - \frac{1}{4}$	**7.** $\frac{9\pi}{4}$	**8.** $\frac{3}{4}\log 3 - \log 2$
9. $\frac{\pi}{2} - 1$	**10.** $\frac{1}{2}\log 6$	**11.** $\frac{\pi}{12}$	**12.** $\frac{\pi}{4}$
13. $\left(\frac{\pi}{2} - 1\right)$	**14.** $\frac{2}{3}\tan^{-1}\frac{1}{3}$	**15.** $\frac{1}{2\sqrt{13}}\tan^{-1}\frac{2}{\sqrt{13}}$	**16.** 1

3

Properties of Definite Integrals

3.1. PROPERTIES OF DEFINITE INTEGRALS

Property 1. *Prove that* $\int_a^b f(x)\,dx = \int_a^b f(z)\,dz.$

Proof. Let $\int f(x)\,dx = F(x)$

$$\Rightarrow \qquad \int f(z)\,dz = F(z)$$

$$\therefore \quad \text{L.H.S.} = \int_a^b f(x)\,dx = \Big[F(x)\Big]_a^b = F(b) - F(a) \qquad \ldots(1)$$

$$\text{R.H.S.} = \int_a^b f(z)\,dz = \Big[F(z)\Big]_a^b = F(b) - F(a) \qquad \ldots(2)$$

$\therefore$ From equations (1) and (2), we have

$$\int_a^b f(x)\,dx = \int_a^b f(z)\,dz$$

The value of a definite integral remains unchanged if the variable is changed, provided the function and limits of integration remain the same.

Property 2. *Prove that* $\int_a^b f(x)\,dx = -\int_b^a f(x)\,dx$

Proof. Let $\int f(x)\,dx = F(x)$

$$\therefore \quad \text{L.H.S.} = \int_a^b f(x)\,dx = \Big[F(x)\Big]_a^b = F(b) - F(a) \qquad \ldots(1)$$

$$\text{R.H.S.} = -\int_b^a f(x)\,dx = -\Big[F(x)\Big]_b^a = -[F(a) - F(b)] = F(b) - F(a) \qquad \ldots(2)$$

$\therefore$ From equations (1) and (2), we have

$$\int_a^b f(x)\,dx = -\int_b^a f(x)\,dx$$

An interchange of the limits of integration only changes the sign of the integral.

Property 3. *Prove that* $\int_a^b f(x)\,dx = \int_a^c f(x)\,dx + \int_c^b f(x)\,dx$ *where* $a < c < b.$

Proof. Let $\int f(x)\,dx = F(x)$.

$\therefore$ L.H.S. $= \int_a^b f(x)\,dx = \Big[F(x)\Big]_a^b = F(b) - F(a)$ ...(1)

R.H.S. $= \int_a^c f(x)\,dx + \int_c^b f(x)\,dx = \Big[F(x)\Big]_a^c + \Big[F(x)\Big]_c^b = [F(c) - F(a)] + [F(b) - F(c)]$

$= F(b) - F(a)$ ...(2)

$\therefore$ From equations (1) and (2), we have

$$\int_a^b f(x)\,dx = \int_a^c f(x)\,dx + \int_c^b f(x)\,dx.$$

Generalisation of Property 3.

Prove that :

$$\int_a^b f(x)\,dx = \int_a^{a_1} f(x)\,dx + \int_{a_1}^{a_2} f(x)\,dx + \int_{a_2}^{a_3} f(x)\,dx + \ldots\ldots + \int_{a_{n-1}}^{a_n} f(x)\,dx + \int_{a_n}^b f(x)\,dx.$$

where $a < a_1 < a_2 < a_3 \ldots\ldots < a_{n-1} < a_n < b.$

Proof. Let $\int f(x)\,dx = F(x)$

L.H.S. $= \int_a^b f(x)\,dx = \Big[F(x)\Big]_a^b = F(b) - F(a)$ ...(1)

R.H.S. $= \int_a^{a_1} f(x)\,dx + \int_{a_1}^{a_2} f(x)\,dx + \int_{a_2}^{a_3} f(x)\,dx + \ldots\ldots + \int_{a_{n-1}}^{a_n} f(x)\,dx + \int_{a_n}^b f(x)\,dx$

$= \Big[F(x)\Big]_a^{a_1} + \Big[F(x)\Big]_{a_1}^{a_2} + \Big[F(x)\Big]_{a_2}^{a_3} + \ldots\ldots + \Big[F(x)\Big]_{a_{n-1}}^{a_n} + \Big[F(x)\Big]_{a_n}^{b}$

$= F(a_1) - F(a)$
$+ F(a_2) - F(a_1)$
$+ F(a_3) - F(a_2)$
....
....
$+ F(a_n) - F(a_{n-1})$ ...(2)
$+ F(b) - F(a_n) = F(b) - F(a)$

$\therefore$ From equations (1) and (2), we have

L.H.S. = R.H.S.

The range of integration can be split up into any finite number of sub-intervals.

Property 4. *Prove that* $\int_0^a f(x)\,dx = \int_0^a f(a-x)\,dx$.

Proof. Let $I = \int_0^a f(x)\,dx$

Put $x = a - z \Rightarrow dx = -dz$

When $x = 0,\ (a - z) = 0 \Rightarrow z = a$ and when $x = a,\ (a - z) = a \Rightarrow z = 0$

$$\therefore \qquad I = \int_0^a f(x)\,dx = \int_a^0 f(a-z)\,.\,(-dz)$$

$$= -\int_a^0 f(a-z)\,dz = \int_0^a f(a-z)\,dz \qquad \left[\because \int_a^b f(x)\,dx = -\int_b^a f(x)\,dx\right]$$

$$= \int_0^a f(a-x)\,dx \qquad \left[\because \int_a^b f(x)\,dx = \int_a^b f(z)\,dz\right]$$

If the lower limit in any definite integral is zero, the value of the integral remains unchanged on replacing x by (upper limit $- x$) in the integrand.

Property 5. *Prove that* $\int_0^{2a} f(x)\,dx = \int_0^a f(x)\,dx + \int_0^a f(2a-x)\,dx.$

Proof. Since, we know that

$$\int_a^b f(x)\,dx = \int_a^c f(x)\,dx + \int_c^b f(x)\,dx, \quad \text{where} \quad a < c < b.$$

$$\therefore \qquad \int_0^{2a} f(x)\,dx = \int_0^a f(x)\,dx + \int_a^{2a} f(x)\,dx \qquad ...(1)$$

Now, take the second integral on R.H.S. of equation (1), we have

$$= \int_a^{2a} f(x)\,dx$$

Put $\quad x = 2a - z \;\Rightarrow\; dx = -dz$

When $\quad x = a,\; 2a - z = a \;\Rightarrow\; z = a$ and when $x = 2a,\; 2a - z = 2a \;\Rightarrow\; z = 0$

$$\therefore \quad \int_a^{2a} f(x)\,dx = \int_a^0 f(2a-z)\,(-dz) = -\int_a^0 f(2a-z)\,dz$$

$$= \int_0^a f(2a-z)\,dz \qquad \left[\because \int_a^b f(x)\,dx = -\int_b^a f(x)\,dx\right]$$

$$= \int_0^a f(2a-x)\,dx \qquad \left[\because \int_a^b f(x)\,dx = \int_a^b f(z)\,dz\right]$$

$\therefore$ From equation (1), we have

$$\int_0^{2a} f(x)\,dx = \int_0^a f(x)\,dx + \int_0^a f(2a-x)\,dx.$$

Property 6. *Prove that :*

(*i*) $\int_0^{2a} f(x)\,dx = 2\int_0^a f(x)\,dx,$ *if* $f(2a-x) = f(x)$

(*ii*) $\int_0^{2a} f(x)\,dx = 0,$ *if* $f(2a-x) = -f(x).$

Proof. Since, we know that

$$\int_a^b f(x)\,dx = \int_a^c f(x)\,dx + \int_c^b f(x)\,dx, \qquad a < c < b$$

$$\therefore \qquad \int_0^{2a} f(x)\,dx = \int_0^a f(x)\,dx + \int_a^{2a} f(x)\,dx \qquad ...(1)$$

Now, take the second integral on the R.H.S. of equation (1), we have

Let $\qquad I = \int_a^{2a} f(x)\,dx$

Put $x = 2a - z \Rightarrow dx = -dz$

When $x = a$, $2a - z = a \Rightarrow z = a$ and when $x = 2a$, $2a - z = 2a \Rightarrow z = 0$

$$\therefore \quad I = \int_a^{2a} f(x)\,dx = \int_a^0 f(2a - z)(-dz) = -\int_a^0 f(2a - z)\,dz$$

$$= \int_0^a f(2a - z)\,dz \qquad \left[\because \int_a^b f(x)\,dx = -\int_b^a f(x)\,dx\right]$$

$$= \int_0^a f(2a - x\,dx \qquad \left[\because \int_a^b f(x)\,dx = \int_a^b f(z)\,dz\right]$$

$\therefore$ From equation (1), we have

$$\int_0^{2a} f(x)\,dx = \int_0^a f(x)\,dx + \int_0^a f(2a - x)\,dx \qquad ...(2)$$

Case I. When $f(2a - x) = f(x)$

$\therefore$ From equation (2),

$$\int_0^{2a} f(x)\,dx = \int_0^a f(x)\,dx + \int_0^a f(x)\,dx = 2\int_0^a f(x)\,dx$$

Case II. When $f(2a - x) = -f(x)$

$\therefore$ From equation (2), we have

$$\int_0^{2a} f(x)\,dx = \int_0^a f(x)\,dx - \int_0^a f(x)\,dx = 0.$$

Remark. If the lower limit of integration is zero, and

(*i*) The integrand remains unaltered when x is replaced by (upper limit $- x$), then we can multiply the integral by 2 and divide the upper limit of integration by 2.

This property is extremely important. It is generally used when the integral involves trigonometric functions and the limits of integration are 0 to π or 0 to 2π.

(*ii*) If the integrand changes in sign only when x is replaced by (upper limit $- x$), then the value of the integral is zero.

Property 7. *If $f(x) = f(a + x)$, then prove that*

$$\int_0^{na} f(x)\,dx = n\int_0^a f(x)\,dx.$$

Proof. By the generalised form of the property,

$$\int_a^b f(x)\,dx = \int_a^c f(x)\,dx + \int_c^b f(x)\,dx, \qquad a < c < b$$

We have

$$\int_0^{na} f(x)\,dx = \int_0^a f(x)\,dx + \int_a^{2a} f(x)\,dx + \int_{2a}^{3a} f(x)\,dx + ... + \int_{(n-1)a}^{na} f(x)\,dx \qquad ...(1)$$

Now, take the second integral on the R.H.S. of equation (1), we have

Let $I = \int_a^{2a} f(x)\,dx$

Put $x = a + z \Rightarrow dx = dz$

When $x = a$, $z + a = a \Rightarrow z = 0$ and when $x = 2a$, $z + a = 2a \Rightarrow z = +a$

$$\therefore \quad I = \int_a^{2a} f(x)\,dx = \int_0^a f(a + z)\,dz \qquad \left[\because \int_a^b f(x)\,dx = \int_a^b f(z)\,dz\right]$$

$$= \int_0^a f(a+x)\, dx$$

$$= \int_0^a f(x)\, dx \qquad [\because \ f(a+x) = f(x)]$$

= the first integral on R.H.S. of equation (1).

Now, take the third integral on the R.H.S. of equation (1), we have

Let $\quad I_1 = \int_{2a}^{3a} f(x)\, dx$

Put $\quad x = a + z \Rightarrow dx = dz$

When $x = 2a$, $z + a = 2a \Rightarrow z = a$ and when $x = 3a$, $z + a = 3a \Rightarrow z = 2a$

$$\therefore \quad I_1 = \int_{2a}^{3a} f(x)\, dx = \int_a^{2a} f(a+z)\, dz = \int_a^{2a} f(a+x)\, dx \qquad \left[\because \int_a^b f(x)\, dx = \int_a^b f(z)\, dz\right]$$

$$= \int_a^{2a} f(x)\, dx \qquad [\because \ f(a+x) = f(x)]$$

= the second integral on R.H.S. of equation (1)

= the first integral on R.H.S. of equation (1).

Similarly, every integral on R.H.S. of equation (1) is equal to the first integral on R.H.S. of equation (1).

$$\therefore \quad \int_0^{na} f(x)\, dx = \int_0^a f(x)\, dx + \int_0^a f(x)\, dx + \int_0^a f(x)\, dx \ \ldots\ldots \ n \text{ times.}$$

$$= n \int_0^a f(x)\, dx.$$

3.2. DEFINITION

Even Function

A function $f(x)$ is said to be an even function of x if $f(-x) = f(x)$

Odd Function

A function $f(x)$ is said to be odd function of x if $f(-x) = -f(x)$.

Property 8. *Prove that :*

(*i*) *If $f(x)$ is an even function of x, then* $\int_{-a}^{a} f(x)\, dx = 2\int_0^a f(x)\, dx$

(*ii*) *If $f(x)$ is an odd function of x, then* $\int_{-a}^{a} f(x)\, dx = 0.$

Proof. Since $\int_a^b f(x)\, dx = \int_a^c f(x)\, dx + \int_c^b f(x)\, dx, \quad a < c < b$

$$\therefore \quad \int_{-a}^{a} f(x)\, dx = \int_{-a}^{0} f(x)\, dx + \int_0^a f(x)\, dx \qquad \ldots(1)$$

Now, take the first integral in the R.H.S. of equation (1), we have

Let $\quad I = \int_{-a}^{0} f(x)\, dx$

Put $\quad x = -z \Rightarrow dx = -dz$

When $x = -a$, $-z = -a \Rightarrow z = a$ and when $x = 0$, $-z = 0 \Rightarrow z = 0$

$\therefore \qquad I = \int_{-a}^{0} f(x)\,dx = \int_{a}^{0} f(-z)\,(-dz).$

$$= -\int_{a}^{0} f(-z)\,dz = \int_{0}^{a} f(-z)\,dz \qquad \left[\because \int_a^b f(x)\,dx = -\int_b^a f(x)\,dx\right]$$

$$= \int_{0}^{a} f(-x)\,dx \qquad \left[\because \int_a^b f(x)\,dx = \int_a^b f(z)\,dz\right]$$

$\therefore$ From equation (1), we have

$$\int_{-a}^{a} f(x)\,dx = \int_{0}^{a} f(-x)\,dx + \int_{0}^{a} f(x)\,dx \qquad ...(2)$$

Case I. If $f(x)$ is an even function of x, then $f(-x) = f(x)$.

$\therefore$ From equation (2), we have

$$\int_{-a}^{a} f(x)\,dx = \int_{0}^{a} f(x)\,dx + \int_{0}^{a} f(x)\,dx = 2\int_{0}^{a} f(x)\,dx.$$

Case II. If $f(x)$ is an odd function of x, then $f(-x) = -f(x)$.

$\therefore$ From equation (2), we have

$$\int_{-a}^{a} f(x)\,dx = -\int_{0}^{a} f(x)\,dx + \int_{0}^{a} f(x)\,dx = 0.$$

Property 9. *Prove that* $\int_a^b f(x)\,dx = \int_a^b f(a+b-x)\,dx.$

Proof. Let $I = \int_a^b f(x)\,dx$

Put $\qquad x = a + b - z \Rightarrow dx = -dz$

When $\qquad x = a,\ a + b - z = a \Rightarrow z = b$ and when $x = b,\ a + b - z = b \Rightarrow z = a.$

$\therefore \qquad \int_a^b f(x)\,dx = \int_b^a f(a+b-z)\,(-dz)$

$$= -\int_b^a f(a+b-z)\,dz = \int_a^b f(a+b-z)\,dz \qquad \left[\because \int_a^b f(x)\,dx = -\int_b^a f(x)\,dx\right]$$

$$= \int_a^b f(a+b-x)\,dx \qquad \left[\because \int_a^b f(x)\,dx = \int_a^b f(z)\,dz\right]$$

Hence, $\int_a^b f(x)\,dx = \int_a^b f(a+b-x)\,dx.$

SOME SOLVED EXAMPLES

Example 1. *Evaluate the following integrals :*

(*i*) $\int_0^{\pi/2} \frac{\sin x}{\sin x + \cos x}\,dx$ $\qquad$ (*ii*) $\int_0^{\pi/2} \frac{\cos x}{\sin x + \cos x}\,dx$

(*iii*) $\int_0^{\pi/2} \frac{1}{1 + \tan x}\,dx$ $\qquad$ (*iv*) $\int_0^{\pi/2} \frac{\sqrt{\cos x}}{\sqrt{\sin x} + \sqrt{\cos x}}\,dx$

(*v*) $\int_0^{\pi/2} \frac{\sqrt{\sin x}}{\sqrt{\sin x} + \sqrt{\cos x}}\,dx$ $\qquad$ (*vi*) $\int_0^{\pi/2} \frac{1}{1 + \cot x}\,dx.$

Solution. (*i*) Let $\quad I = \int_0^{\pi/2} \frac{\sin x}{\sin x + \cos x}\, dx \qquad ...(1)$

$$\Rightarrow \qquad I = \int_0^{\pi/2} \frac{\sin\left(\frac{\pi}{2} - x\right)}{\sin\left(\frac{\pi}{2} - x\right) + \cos\left(\frac{\pi}{2} - x\right)}\, dx \qquad \left[\because \int_0^a f(x)\, dx = \int_0^a f(a - x)\, dx\right]$$

$$\therefore \qquad I = \int_0^{\pi/2} \frac{\cos x}{\cos x + \sin x}\, dx \qquad ...(2)$$

Adding equations (1) and (2), we get

$$2I = \int_0^{\pi/2} \frac{\sin x}{\sin x + \cos x}\, dx + \int_0^{\pi/2} \frac{\cos x}{\cos x + \sin x}\, dx$$

$$= \int_0^{\pi/2} \left[\frac{\sin x}{\sin x + \cos x} + \frac{\cos x}{\cos x + \sin x}\right] dx = \int_0^{\pi/2} \frac{\sin x + \cos x}{\sin x + \cos x}\, dx$$

$$= \int_0^{\pi/2} 1 .\, dx = \left[x\right]_0^{\pi/2} = \left[\frac{\pi}{2} - 0\right] = \frac{\pi}{2}$$

$$\therefore \qquad I = \frac{1}{2} . \frac{\pi}{2} = \frac{\pi}{4} .$$

(*ii*) Let $\quad I = \int_0^{\pi/2} \frac{\cos x}{\sin x + \cos x}\, dx \qquad ...(1)$

$$= \int_0^{\pi/2} \frac{\cos\left(\frac{\pi}{2} - x\right)}{\sin\left(\frac{\pi}{2} - x\right) + \cos\left(\frac{\pi}{2} - x\right)}\, dx \qquad \left[\because \int_0^a f(x)\, dx = \int_0^a f(a - x)\, dx\right]$$

$$\therefore \qquad I = \int_0^{\pi/2} \frac{\sin x}{\cos x + \sin x}\, dx \qquad ...(2)$$

Adding equations (1) and (2), we get

$$2I = \int_0^{\pi/2} \frac{\cos x}{\sin x + \cos x}\, dx + \int_0^{\pi/2} \frac{\sin x}{\cos x + \sin x}\, dx$$

$$= \int_0^{\pi/2} \left[\frac{\cos x}{\sin x + \cos x} + \frac{\sin x}{\cos x + \sin x}\right] dx = \int_0^{\pi/2} \frac{\cos x + \sin x}{\cos x + \sin x}\, dx$$

$$= \int_0^{\pi/2} 1 .\, dx = \left[x\right]_0^{\pi/2} = \left[\frac{\pi}{2} - 0\right] = \frac{\pi}{2}$$

$$\therefore \qquad I = \frac{1}{2} . \frac{\pi}{2} = \frac{\pi}{4} .$$

(*iii*) Let $\quad I = \int_0^{\pi/2} \frac{1}{1 + \tan x}\, dx = \int_0^{\pi/2} \frac{1}{1 + \frac{\sin x}{\cos x}}\, dx$

$$= \int_0^{\pi/2} \frac{\cos x}{\cos x + \sin x}\, dx \qquad ...(1)$$

$$= \int_0^{\pi/2} \frac{\cos\left(\frac{\pi}{2} - x\right)}{\cos\left(\frac{\pi}{2} - x\right) + \sin\left(\frac{\pi}{2} - x\right)}\, dx \qquad \left[\because \int_0^a f(x)\, dx = \int_0^a f(a-x)\, dx\right]$$

$$\therefore \quad I = \int_0^{\pi/2} \frac{\sin x}{\sin x + \cos x}\, dx \qquad ...(2)$$

Adding equation (1) and (2), we get

$$2I = \int_0^{\pi/2} \frac{\cos x}{\cos x + \sin x}\, dx + \int_0^{\pi/2} \frac{\sin x}{\sin x + \cos x}\, dx$$

$$= \int_0^{\pi/2} \left[\frac{\cos x}{\cos x + \sin x} + \frac{\sin x}{\sin x + \cos x}\right] dx = \int_0^{\pi/2} \left[\frac{\cos x + \sin x}{\cos x + \sin x}\right] dx$$

$$= \int_0^{\pi/2} 1 \,.\, dx = \left[x\right]_0^{\pi/2} = \left[\frac{\pi}{2} - 0\right] = \frac{\pi}{2}$$

$$\therefore \quad I = \frac{1}{2} \cdot \frac{\pi}{2} = \frac{\pi}{4}.$$

(*iv*) Let $I = \int_0^{\pi/2} \frac{\sqrt{\cos x}}{\sqrt{\sin x} + \sqrt{\cos x}}\, dx$ ...(1)

$$= \int_0^{\pi/2} \frac{\sqrt{\cos\left(\frac{\pi}{2} - x\right)}}{\sqrt{\sin\left(\frac{\pi}{2} - x\right)} + \sqrt{\cos\left(\frac{\pi}{2} - x\right)}}\, dx \qquad \left[\because \int_0^a f(x)\, dx = \int_0^a f(a-x)\, dx\right]$$

$$\therefore \quad I = \int_0^{\pi/2} \frac{\sqrt{\sin x}}{\sqrt{\cos x} + \sqrt{\sin x}}\, dx \qquad ...(2)$$

Adding equations (1) and (2), we get

$$2I = \int_0^{\pi/2} \frac{\sqrt{\cos x}}{\sqrt{\sin x} + \sqrt{\cos x}}\, dx + \int_0^{\pi/2} \frac{\sqrt{\sin x}}{\sqrt{\cos x} + \sqrt{\sin x}}\, dx$$

$$= \int_0^{\pi/2} \left[\frac{\sqrt{\cos x}}{\sqrt{\sin x} + \sqrt{\cos x}} + \frac{\sqrt{\sin x}}{\sqrt{\cos x} + \sqrt{\sin x}}\right] . \, dx = \int_0^{\pi/2} \left[\frac{\sqrt{\cos x} + \sqrt{\sin x}}{\sqrt{\cos x} + \sqrt{\sin x}}\right] dx$$

$$= \int_0^{\pi/2} 1 \,.\, dx = \left[x\right]_0^{\pi/2} = \left[\frac{\pi}{2} - 0\right] = \frac{\pi}{2}$$

$$\therefore \quad I = \frac{1}{2} \cdot \frac{\pi}{2} = \frac{\pi}{4}.$$

(*v*) Let $I = \int_0^{\pi/2} \frac{\sqrt{\sin x}}{\sqrt{\sin x} + \sqrt{\cos x}}\, dx$ ...(1)

$$= \int_0^{\pi/2} \frac{\sqrt{\sin\left(\frac{\pi}{2} - x\right)}}{\sqrt{\sin\left(\frac{\pi}{2} - x\right)} + \sqrt{\cos\left(\frac{\pi}{2} - x\right)}}\, dx \qquad \left[\because \int_0^a f(x)\, dx = \int_0^a f(a - x)\, dx\right]$$

$$\therefore \qquad I = \int_0^{\pi/2} \frac{\sqrt{\cos x}}{\sqrt{\cos x} + \sqrt{\sin x}}\, dx \qquad ...(2)$$

Adding equations (1) and (2), we get

$$2I = \int_0^{\pi/2} \left[\frac{\sqrt{\sin x}}{\sqrt{\sin x} + \sqrt{\cos x}}\right] dx + \int_0^{\pi/2} \left[\frac{\sqrt{\cos x}}{\sqrt{\cos x} + \sqrt{\sin x}}\right] dx$$

$$= \int_0^{\pi/2} \left[\frac{\sqrt{\sin x}}{\sqrt{\sin x} + \sqrt{\cos x}} + \frac{\sqrt{\cos x}}{\sqrt{\cos x} + \sqrt{\sin x}}\right] dx = \int_0^{\pi/2} \frac{\sqrt{\sin x} + \sqrt{\cos x}}{\sqrt{\sin x} + \sqrt{\cos x}}\, dx$$

$$= \int_0^{\pi/2} 1 . dx = [x]_0^{\pi/2} = \left[\frac{\pi}{2} - 0\right] = \frac{\pi}{2}$$

$$\therefore \qquad I = \frac{1}{2} . \frac{\pi}{2} = \frac{\pi}{4} .$$

(*vi*) Let $I = \int_0^{\pi/2} \frac{1}{1 + \cot x}\, dx = \int_0^{\pi/2} \frac{1}{1 + \frac{\cos x}{\sin x}}\, dx$

$$= \int_0^{\pi/2} \frac{\sin x}{\sin x + \cos x}\, dx \qquad ...(1)$$

$$= \int_0^{\pi/2} \frac{\sin\left(\frac{\pi}{2} - x\right)}{\sin\left(\frac{\pi}{2} - x\right) + \cos\left(\frac{\pi}{2} - x\right)}\, dx \qquad \left[\because \int_0^a f(x)\, dx - \int_0^a f(a - x)\, dx\right]$$

$$\therefore \qquad I = \int_0^{\pi/2} \frac{\cos x}{\cos x + \sin x}\, dx \qquad ...(2)$$

Adding equations (1) and (2), we get

$$2I = \int_0^{\pi/2} \frac{\sin x}{\sin x + \cos x}\, dx + \int_0^{\pi/2} \frac{\cos x}{\cos x + \sin x}\, dx$$

$$= \int_0^{\pi/2} \left[\frac{\sin x}{\sin x + \cos x} + \frac{\cos x}{\cos x + \sin x}\right] dx = \int_0^{\pi/2} \left[\frac{\sin x + \cos x}{\sin x + \cos x}\right] dx$$

$$= \int_0^{\pi/2} 1 . dx = \Big[x\Big]_0^{\pi/2} = \left[\frac{\pi}{2} - 0\right] = \frac{\pi}{2}$$

$$\therefore \qquad I = \frac{1}{2} . \frac{\pi}{2} = \frac{\pi}{4} .$$

Example 2. *Evaluate the following integrals :*

(*i*) $\int_0^{\pi/2} \frac{\sqrt{\cot x}}{\sqrt{\cot x} + \sqrt{\tan x}}\, dx$ (*ii*) $\int_a^b \frac{f(x)}{f(a+b-x) + f(x)}\, dx$

(*iii*) $\int_0^{\pi/2} \frac{\sqrt{\tan x}}{1 + \sqrt{\tan x}}\, dx$ (*iv*) $\int_0^{\pi/2} \frac{\sin^{3/2} x}{\sin^{3/2} x + \cos^{3/2} x}\, dx$

(*v*) $\int_0^{\pi/2} \frac{1}{1 + \sqrt{\tan x}}\, dx$ (*vi*) $\int_0^{\pi/2} \frac{\cos^5 x}{\sin^5 x + \cos^5 x}\, dx.$

Solution. (*i*) Let $I = \int_0^{\pi/2} \frac{\sqrt{\cot x}}{\sqrt{\cot x} + \sqrt{\tan x}}\, dx$ ...(1)

$$= \int_0^{\pi/2} \frac{\sqrt{\cot\left(\frac{\pi}{2} - x\right)}}{\sqrt{\cot\left(\frac{\pi}{2} - x\right)} + \sqrt{\tan\left(\frac{\pi}{2} - x\right)}}\, dx \qquad \left[\because \int_0^a f(x)\, dx = \int_0^a f(a-x)\, dx\right]$$

$$\therefore \qquad I = \int_0^{\pi/2} \frac{\sqrt{\tan x}}{\sqrt{\tan x} + \sqrt{\cot x}}\, dx \qquad ...(2)$$

Adding equations (1) and (2), we get

$$2I = \int_0^{\pi/2} \frac{\sqrt{\cot x}}{\sqrt{\cot x} + \sqrt{\tan x}}\, dx + \int_0^{\pi/2} \frac{\sqrt{\tan x}}{\sqrt{\tan x} + \sqrt{\cot x}}\, dx$$

$$= \int_0^{\pi/2} \left[\frac{\sqrt{\cot x}}{\sqrt{\cot x} + \sqrt{\tan x}} + \frac{\sqrt{\tan x}}{\sqrt{\tan x} + \sqrt{\cot x}}\right] dx = \int_0^{\pi/2} \left[\frac{\sqrt{\cot x} + \sqrt{\tan x}}{\sqrt{\cot x} + \sqrt{\tan x}}\right] dx$$

$$= \int_0^{\pi/2} 1 \,.\, dx = \left[x\right]_0^{\pi/2} = \left[\frac{\pi}{2} - 0\right] = \frac{\pi}{2}$$

$$\therefore \qquad I = \frac{1}{2} \cdot \frac{\pi}{2} = \frac{\pi}{4}\,.$$

(*ii*) Let $I = \int_a^b \frac{f(x)}{f(a+b-x) + f(x)}\, dx$ ...(1)

$$= \int_a^b \frac{f(a+b-x)}{f[a+b-(a+b-x)] + f(a+b-x)}\, dx$$

$$\left[\because \int_a^b f(x)\, dx = \int_a^b f(a+b-x)\, dx\right]$$

$$\therefore \qquad I = \int_a^b \frac{f(a+b-x)}{f(x) + f(a+b-x)}\, dx \qquad ...(2)$$

Adding equations (1) and (2), we get

$$2I = \int_a^b \frac{f(x)}{f(a+b-x) + f(x)}\, dx + \int_a^b \frac{f(a+b-x)}{f(x) + f(a+b-x)}\, dx$$

$$= \int_a^b \left[\frac{f(x)}{f(a+b-x) + f(x)} + \frac{f(a+b-x)}{f(x) + f(a+b-x)}\right] dx$$

$$= \int_a^b \frac{f(x) + f(a+b-x)}{f(x) + f(a+b-x)}\, dx = \int_a^b 1 \,.\, dx = \Big[x\Big]_a^b = [b-a]$$

$$\therefore \qquad I = \frac{1}{2}(b-a).$$

(*iii*) Let $\quad I = \int_0^{\pi/2} \frac{\sqrt{\tan x}}{1 + \sqrt{\tan x}}\, dx = \int_0^{\pi/2} \frac{\sqrt{\frac{\sin x}{\cos x}}}{1 + \sqrt{\frac{\sin x}{\cos x}}}\, dx$

$$= \int_0^{\pi/2} \frac{\sqrt{\sin x}}{\sqrt{\cos x} + \sqrt{\sin x}}\, dx \qquad \text{...(1)}$$

$$= \int_0^{\pi/2} \frac{\sqrt{\sin\left(\frac{\pi}{2} - x\right)}}{\sqrt{\cos\left(\frac{\pi}{2} - x\right)} + \sqrt{\sin\left(\frac{\pi}{2} - x\right)}}\, dx \qquad \left[\because \int_0^a f(x)\, dx = \int_0^a f(a-x)\, dx\right]$$

$$\therefore \qquad I = \int_0^{\pi/2} \frac{\sqrt{\cos x}}{\sqrt{\sin x} + \sqrt{\cos x}}\, dx \qquad \text{...(2)}$$

Adding equations (1) and (2), we get

$$2I = \int_0^{\pi/2} \frac{\sqrt{\sin x}}{\sqrt{\cos x} + \sqrt{\sin x}}\, dx + \int_0^{\pi/2} \frac{\sqrt{\cos x}}{\sqrt{\sin x} + \sqrt{\cos x}}\, dx$$

$$= \int_0^{\pi/2} \left[\frac{\sqrt{\sin x}}{\sqrt{\cos x} + \sqrt{\sin x}} + \frac{\sqrt{\cos x}}{\sqrt{\sin x} + \sqrt{\cos x}}\right] dx$$

$$= \int_0^{\pi/2} \frac{\sqrt{\sin x} + \sqrt{\cos x}}{\sqrt{\sin x} + \sqrt{\cos x}}\, dx$$

$$= \int_0^{\pi/2} 1 \,.\, dx = \Big[x\Big]_0^{\pi/2} = \left[\frac{\pi}{2} - 0\right] = \frac{\pi}{2}$$

$$\therefore \qquad I = \frac{1}{2} \cdot \frac{\pi}{2} = \frac{\pi}{4}.$$

(*iv*) Let $\quad I = \int_0^{\pi/2} \frac{\sin^{3/2} x}{\sin^{3/2} x + \cos^{3/2} x}\, dx \qquad \text{...(1)}$

$$= \int_0^{\pi/2} \frac{\sin^{3/2}\left(\frac{\pi}{2} - x\right)}{\sin^{3/2}\left(\frac{\pi}{2} - x\right) + \cos^{3/2}\left(\frac{\pi}{2} - x\right)}\, dx \qquad \left[\because \int_0^a f(x)\, dx = \int_0^a f(a-x)\, dx\right]$$

$$\therefore \qquad I = \int_0^{\pi/2} \frac{\cos^{3/2} x}{\cos^{3/2} x + \sin^{3/2} x}\, dx \qquad \text{...(2)}$$

Adding equations (1) and (2), we get

$$2I = \int_0^{\pi/2} \frac{\sin^{3/2} x}{\sin^{3/2} x + \cos^{3/2} x}\, dx + \int_0^{\pi/2} \frac{\cos^{3/2} x}{\cos^{3/2} x + \sin^{3/2} x}\, dx$$

$$= \int_0^{\pi/2} \left[\frac{\sin^{3/2} x}{\sin^{3/2} x + \cos^{3/2} x} + \frac{\cos^{3/2} x}{\cos^{3/2} x + \sin^{3/2} x} \right] dx$$

$$= \int_0^{\pi/2} \frac{\sin^{3/2} x + \cos^{3/2} x}{\sin^{3/2} x + \cos^{3/2} x}\, dx$$

$$= \int_0^{\pi/2} 1 . dx = [x]_0^{\pi/2} = \left[\frac{\pi}{2} - 0\right] = \frac{\pi}{2}$$

$\therefore$ $$I = \frac{1}{2} . \frac{\pi}{2} = \frac{\pi}{4} .$$

(*v*) Let $I = \int_0^{\pi/2} \frac{1}{1+\sqrt{\tan x}}\, dx = \int_0^{\pi/2} \frac{1}{1 + \frac{\sqrt{\sin x}}{\sqrt{\cos x}}}\, dx$

$\therefore$ $$I = \int_0^{\pi/2} \frac{\sqrt{\cos x}}{\sqrt{\cos x} + \sqrt{\sin x}}\, dx$$

Proceed as in example 1 (*iv*).

[**Hint.** Using $\int_0^a f(x)\, dx = \int_0^a f(a - x)\, dx$.] $\left[\textbf{Ans. } \frac{\pi}{4}.\right]$

(*vi*) Let $I = \int_0^{\pi/2} \frac{\cos^5 x}{\sin^5 x + \cos^5 x}\, dx$ ...(1)

$$= \int_0^{\pi/2} \frac{\cos^5\left(\frac{\pi}{2} - x\right)}{\sin^5\left(\frac{\pi}{2} - x\right) + \cos^5\left(\frac{\pi}{2} - x\right)}\, dx \qquad \left[\because \int_0^a f(x)\, dx = \int_0^a f(a - x)\, dx\right]$$

$\therefore$ $$I = \int_0^{\pi/2} \frac{\sin^5 x}{\cos^5 x + \sin^5 x}\, dx \qquad ...(2)$$

Adding equations (1) and (2), we get

$$2I = \int_0^{\pi/2} \frac{\cos^5 x}{\sin^5 x + \cos^5 x}\, dx + \int_0^{\pi/2} \frac{\sin^5 x}{\cos^5 x + \sin^5 x}\, dx$$

$$= \int_0^{\pi/2} \left[\frac{\cos^5 x}{\sin^5 x + \cos^5 x} + \frac{\sin^5 x}{\cos^5 x + \sin^5 x} \right] dx$$

$$= \int_0^{\pi/2} \left[\frac{\cos^5 x + \sin^5 x}{\sin^5 x + \cos^5 x} \right] dx$$

$$= \int_0^{\pi/2} 1 . dx = \Big[x \Big]_0^{\pi/2} = \left[\frac{\pi}{2} - 0\right] = \frac{\pi}{2}$$

$\therefore$ $$I = \frac{1}{2} . \frac{\pi}{2} = \frac{\pi}{4} .$$

Example 3. *Evaluate the following integrals :*

(*i*) $\int_0^{\pi/2} \frac{\sqrt{\cot x}}{1+\sqrt{\cot x}}\,dx$ (*ii*) $\int_{\pi/6}^{\pi/3} \frac{1}{1+\sqrt{\tan x}}\,dx$

(*iii*) $\int_0^{a} \frac{\sqrt{x}}{\sqrt{x}+\sqrt{a-x}}\,dx$ (*iv*) $\int_0^{\pi/2} \frac{1}{1+\sqrt{\cot x}}\,dx$

(*v*) $\int_0^{\pi/2} \frac{\sin^n x}{\sin^n x+\cos^n x}\,dx$ (*vi*) $\int_0^{\pi/2} \frac{1}{1+\tan^3 x}\,dx.$

Solution. (*i*) Let $\text{I} = \int_0^{\pi/2} \frac{\sqrt{\cot x}}{1+\sqrt{\cot x}}\,dx = \int_0^{\pi/2} \frac{\sqrt{\frac{\cos x}{\sin x}}}{1+\sqrt{\frac{\cos x}{\sin x}}}\,dx$

$$\therefore \qquad \text{I} = \int_0^{\pi/2} \frac{\sqrt{\cos x}}{\sqrt{\sin x}+\sqrt{\cos x}}\,dx \qquad \text{...(1)}$$

Proceed as in example 1(*iv*).

$\left[\textbf{Hint.} \text{ Using } \int_0^a f(x) = \int_0^a f(a-x)\,dx\right]$ $\left[\textbf{Ans. } \frac{\pi}{4}.\right]$

(*ii*) Let $\text{I} = \int_{\pi/6}^{\pi/3} \frac{1}{1+\sqrt{\tan x}}\,dx = \int_{\pi/6}^{\pi/3} \frac{1}{1+\frac{\sqrt{\sin x}}{\sqrt{\cos x}}}\,dx$

$$= \int_{\pi/6}^{\pi/3} \frac{\sqrt{\cos x}}{\sqrt{\cos x}+\sqrt{\sin x}}\,dx \qquad \text{...(1)}$$

$$= \int_{\pi/6}^{\pi/3} \frac{\sqrt{\cos\left(\frac{\pi}{3}+\frac{\pi}{6}-x\right)}}{\sqrt{\cos\left(\frac{\pi}{3}+\frac{\pi}{6}-x\right)}+\sqrt{\sin\left(\frac{\pi}{3}+\frac{\pi}{6}-x\right)}}\,dx$$

$$\left[\because \int_a^b f(x)\,dx = \int_a^b f(a+b-x)\,dx\right]$$

$$= \int_{\pi/6}^{\pi/3} \frac{\sqrt{\cos\left(\frac{\pi}{2}-x\right)}}{\sqrt{\cos\left(\frac{\pi}{2}-x\right)}+\sqrt{\sin\left(\frac{\pi}{2}-x\right)}}\,dx \qquad \left[\because \frac{\pi}{3}+\frac{\pi}{6} = \frac{2\pi+\pi}{6} = \frac{3\pi}{6} = \frac{\pi}{2}\right]$$

$$\therefore \qquad \text{I} = \int_{\pi/6}^{\pi/3} \frac{\sqrt{\sin x}}{\sqrt{\sin x}+\sqrt{\cos x}}\,dx \qquad \text{...(2)}$$

Adding equations (1) and (2), we get

$$2\text{I} = \int_{\pi/6}^{\pi/3} \frac{\sqrt{\cos x}}{\sqrt{\cos x}+\sqrt{\sin x}}\,dx + \int_{\pi/6}^{\pi/3} \frac{\sqrt{\sin x}}{\sqrt{\sin x}+\sqrt{\cos x}}\,dx$$

$$= \int_{\pi/6}^{\pi/3} \left[\frac{\sqrt{\cos x}}{\sqrt{\cos x} + \sqrt{\sin x}} + \frac{\sqrt{\sin x}}{\sqrt{\sin x} + \sqrt{\cos x}} \right] dx$$

$$= \int_{\pi/6}^{\pi/3} \left[\frac{\sqrt{\cos x} + \sqrt{\sin x}}{\sqrt{\cos x} + \sqrt{\sin x}} \right] dx$$

$$= \int_{\pi/6}^{\pi/3} 1 \,.\, dx = \Big[x \Big]_{\pi/6}^{\pi/3} = \left[\frac{\pi}{3} - \frac{\pi}{6} \right] = \frac{\pi}{6}$$

$\therefore \qquad I = \frac{1}{2} \cdot \frac{\pi}{6} = \frac{\pi}{12}$.

(*iii*) Let $\qquad I = \int_0^a \frac{\sqrt{x}}{\sqrt{x} + \sqrt{a - x}}\, dx \qquad ...(1)$

$$= \int_0^a \frac{\sqrt{a - x}}{\sqrt{a - x} + \sqrt{a - (a - x)}}\, dx \qquad \left[\because \int_0^a f(x)\, dx = \int_0^a f(a - x)\, dx \right]$$

$\therefore \qquad I = \int_0^a \frac{\sqrt{a - x}}{\sqrt{a - x} + \sqrt{x}}\, dx \qquad ...(2)$

Adding equations (1) and (2), we get

$$2I = \int_0^a \frac{\sqrt{x}}{\sqrt{x} + \sqrt{a - x}}\, dx + \int_0^a \frac{\sqrt{a - x}}{\sqrt{a - x} + \sqrt{x}}\, dx$$

$$= \int_0^a \left[\frac{\sqrt{x}}{\sqrt{x} + \sqrt{a - x}} + \frac{\sqrt{a - x}}{\sqrt{a - x} + \sqrt{x}} \right] dx = \int_0^a \left[\frac{\sqrt{x} + \sqrt{a - x}}{\sqrt{x} + \sqrt{a - x}} \right] dx$$

$$= \int_0^a 1 \,.\, dx = \Big[x \Big]_0^a = [a - 0] = a$$

$\therefore \qquad I = \frac{1}{2} \,.\, (a) = \frac{a}{2}$.

(*iv*) Let $\qquad I = \int_0^{\pi/2} \frac{1}{1 + \sqrt{\cot x}}\, dx = \int_0^{\pi/2} \frac{1}{1 + \dfrac{\sqrt{\cos x}}{\sqrt{\sin x}}}\, dx$

$\therefore \qquad I = \int_0^{\pi/2} \frac{\sqrt{\sin x}}{\sqrt{\sin x} + \sqrt{\cos x}}\, dx$

Proceed as in example 1(*v*).

$\left[\textbf{Hint.}\ \text{Using} \int_0^a f(x)\, dx = \int_0^a f(a - x)\, dx \right] \qquad \left[\textbf{Ans.}\ \frac{\pi}{4}. \right]$

(*v*) Let $\qquad I = \int_0^{\pi/2} \frac{\sin^n x}{\sin^n x + \cos^n x}\, dx \qquad ...(1)$

$$= \int_0^{\pi/2} \frac{\sin^n \left(\frac{\pi}{2} - x \right)}{\sin^n \left(\frac{\pi}{2} - x \right) + \cos^n \left(\frac{\pi}{2} - x \right)}\, dx \qquad \left[\because \int_0^a f(x)\, dx = \int_0^a f(a - x)\, dx \right]$$

$$\therefore \qquad I = \int_0^{\pi/2} \frac{\cos^n x}{\cos^n x + \sin^n x}\, dx \qquad ...(2)$$

Adding equations (1) and (2), we get

$$2I = \int_0^{\pi/2} \frac{\sin^n x}{\sin^n x + \cos^n x}\, dx + \int_0^{\pi/2} \frac{\cos^n x}{\cos^n x + \sin^n x}\, dx$$

$$= \int_0^{\pi/2} \left[\frac{\sin^n x}{\sin^n x + \cos^n x} + \frac{\cos^n x}{\cos^n x + \sin^n x} \right] dx$$

$$= \int_0^{\pi/2} \left[\frac{\sin^n x + \cos^n x}{\sin^n x + \cos^n x} \right] dx$$

$$= \int_0^{\pi/2} 1 \,.\, dx = \Big[x \Big]_0^{\pi/2} = \left[\frac{\pi}{2} - 0 \right] = \frac{\pi}{2}$$

$$\therefore \qquad I = \frac{1}{2} \cdot \frac{\pi}{2} = \frac{\pi}{4}\,.$$

(*vi*) Let $I = \int_0^{\pi/2} \frac{1}{1 + \tan^3 x}\, dx = \int_0^{\pi/2} \frac{1}{1 + \frac{\sin^3 x}{\cos^3 x}}\, dx$

$$= \int_0^{\pi/2} \frac{\cos^3 x}{\cos^3 x + \sin^3 x}\, dx \qquad ...(1)$$

$$= \int_0^{\pi/2} \frac{\cos^3 \left(\frac{\pi}{2} - x \right)}{\cos^3 \left(\frac{\pi}{2} - x \right) + \sin^3 \left(\frac{\pi}{2} - x \right)}\, dx \qquad \left[\because \int_0^a f(x)\, dx = \int_0^a f(a - x)\, dx \right]$$

$$\therefore \qquad I = \int_0^{\pi/2} \frac{\sin^3 x}{\sin^3 x + \cos^3 x}\, dx \qquad ...(2)$$

Adding equations (1) and (2), we get

$$2I = \int_0^{\pi/2} \frac{\cos^3 x}{\cos^3 x + \sin^3 x}\, dx + \int_0^{\pi/2} \frac{\sin^3 x}{\sin^3 x + \cos^3 x}\, dx$$

$$= \int_0^{\pi/2} \left[\frac{\cos^3 x}{\cos^3 x + \sin^3 x} + \frac{\sin^3 x}{\sin^3 x + \cos^3 x} \right] dx$$

$$= \int_0^{\pi/2} \left[\frac{\cos^3 x + \sin^3 x}{\cos^3 x + \sin^3 x} \right] dx$$

$$= \int_0^{\pi/2} 1 \,.\, dx = \Big[x \Big]_0^{\pi/2} = \left[\frac{\pi}{2} - 0 \right] = \frac{\pi}{2}$$

$$\therefore \qquad I = \frac{1}{2} \cdot \frac{\pi}{2} = \frac{\pi}{4}\,.$$

Example 4. *Evaluate the following integrals :*

(*i*) $\int_0^{\pi/2} \log(\tan x)\, dx$ (*ii*) $\int_0^{\pi/2} \log(\cot x)\, dx$

(*iii*) $\int_0^1 x(1-x)^n\, dx$ (*iv*) $\int_0^{\pi/4} \log(1+\tan x)\, dx$

(*v*) *Show that :* $\int_0^{\pi/2} \phi(\sin x)\, dx = \int_0^{\pi/2} \phi(\cos x)\, dx$

(*vi*) $\int_0^1 \frac{\log(1+x)}{1+x^2}\, dx.$

Solution. (*i*) Let $I = \int_0^{\pi/2} \log(\tan x)\, dx$ ...(1)

$$= \int_0^{\pi/2} \log\left[\tan\left(\frac{\pi}{2} - x\right)\right] dx \qquad \left[\because \int_0^a f(x)\, dx = \int_0^a f(a-x)\, dx\right]$$

$$\therefore \qquad I = \int_0^{\pi/2} \log(\cot x)\, dx \qquad ...(2)$$

Adding equations (1) and (2), we get

$$2I = \int_0^{\pi/2} \log(\tan x)\, dx + \int_0^{\pi/2} \log(\cot x)\, dx$$

$$= \int_0^{\pi/2} [\log(\tan x) + \log(\cot x)]\, dx$$

$$= \int_0^{\pi/2} \log(\tan x \,.\, \cot x)\, dx \qquad [\because \log m + \log n = \log(m.n)]$$

$$= \int_0^{\pi/2} \log 1 \,.\, dx = \int_0^{\pi/2} 0 \,.\, dx = 0 \qquad [\because \log 1 = 0]$$

$$\therefore \qquad I = 0.$$

(*ii*) Let $I = \int_0^{\pi/2} \log(\cot x)\, dx$ ...(1)

$$= \int_0^{\pi/2} \log\left[\cot\left(\frac{\pi}{2} - x\right)\right] dx \qquad \left[\because \int_0^a f(x)\, dx = \int_0^a f(a-x)\, dx\right]$$

$$\therefore \qquad I = \int_0^{\pi/2} \log(\tan x)\, dx \qquad ...(2)$$

Adding equations (1) and (2), we get

$$2I = \int_0^{\pi/2} \log(\cot x)\, dx + \int_0^{\pi/2} \log(\tan x)\, dx$$

$$= \int_0^{\pi/2} [\log(\cot x) + \log(\tan x)]\, dx$$

$$= \int_0^{\pi/2} \log(\cot x \,.\, \tan x)\, dx \qquad [\because \log m + \log n = \log(m.n)]$$

$$= \int_0^{\pi/2} \log 1 \,.\, dx \qquad \textbf{[Ans. } 0]$$

Proceed as in part (*i*).

(*iii*) Let $\mathrm{I} = \int_0^1 x(1-x)^n \, dx$ ...(1)

$$= \int_0^1 (1-x)\,[1-(1-x)]^n \, dx \qquad \left[\because \int_0^a f(x)\,dx = \int_0^a f(a-x)\,dx\right]$$

$$= \int_0^1 (1-x)\, x^n \, dx = \int_0^1 (x^n - x^{n+1})\, dx = \int_0^1 x^n \, dx - \int_0^1 x^{n+1} \, dx$$

$$= \left[\frac{x^{n+1}}{n+1}\right]_0^1 - \left[\frac{x^{n+2}}{n+2}\right]_0^1 = \left[\frac{(1)^{n+1}}{n+1} - 0\right] - \left[\frac{(1)^{n+2}}{n+2} - 0\right] = \frac{1}{n+1} - \frac{1}{n+2}$$

$$= \frac{(n+2)-(n+1)}{(n+1)\,(n+2)} = \frac{1}{(n+1)\,(n+2)}.$$

(*iv*) Let $\mathrm{I} = \int_0^{\pi/4} \log\,(1 + \tan x)\, dx$ ...(1)

$$\therefore \quad \mathrm{I} = \int_0^{\pi/4} \log\left[1 + \tan\left(\frac{\pi}{4} - x\right)\right] dx \qquad \left[\because \int_0^a f(x)\,dx = \int_0^a f(a-x)\,dx\right]$$

$$= \int_0^{\pi/4} \log\left[1 + \frac{1-\tan x}{1+\tan x}\right] dx \qquad \left[\because \tan\,(45° - \mathrm{A}) = \frac{1-\tan \mathrm{A}}{1+\tan \mathrm{A}}\right]$$

$$= \int_0^{\pi/4} \log\left[\frac{1+\tan x + 1 - \tan x}{1+\tan x}\right] dx = \int_0^{\pi/4} \log\left[\frac{2}{1+\tan x}\right] dx$$

$$\therefore \quad \mathrm{I} = \int_0^{\pi/4} [\log 2 - \log\,(1+\tan x)]\, dx \quad ...(2) \qquad \left[\because \log\left(\frac{m}{n}\right) = \log m - \log n\right]$$

Adding equations (1) and (2), we get

$$2\mathrm{I} = \int_0^{\pi/4} \log\,(1+\tan x)\, dx + \int_0^{\pi/4} [\log 2 - \log\,(1+\tan x)]\, dx$$

$$= \int_0^{\pi/4} [\log\,(1+\tan x) + \log 2 - \log\,(1+\tan x)]\, dx$$

$$= \int_0^{\pi/4} \log 2 \, dx = \log 2 \int_0^{\pi/4} 1 \,.\, dx$$

$$= \log 2 \left[x\right]_0^{\pi/4} = \log 2 \left[\frac{\pi}{4} - 0\right] = \frac{\pi}{4} \log 2$$

$$\therefore \quad \mathrm{I} = \frac{\pi}{8} \log 2.$$

(*v*) Let $\mathrm{I} = \int_0^{\pi/2} \phi(\sin x)\, dx$

$$= \int_0^{\pi/4} \phi\left[\sin\left(\frac{\pi}{2} - x\right)\right] dx \qquad \left[\because \int_0^a f(x)\,dx = \int_0^a f(a-x)\,dx\right]$$

$$= \int_0^{\pi/2} \phi(\cos x)\, dx.$$

(*vi*) Let $\quad I = \int_0^1 \frac{\log(1+x)}{1+x^2}\, dx$

Put $\quad x = \tan\theta \Rightarrow dx = \sec^2\theta\, d\theta$

When $\quad x = 0, \quad \tan\theta = 0 \Rightarrow \theta = 0$ and when $x = 1, \quad \tan\theta = 1 \Rightarrow \theta = \frac{\pi}{4}$

$$\therefore \quad I = \int_0^{\pi/4} \frac{\log(1+\tan\theta)}{(1+\tan^2\theta)} \cdot \sec^2\theta\, d\theta$$

$$= \int_0^{\pi/4} \frac{\log(1+\tan\theta)}{\sec^2\theta} \cdot \sec^2\theta\, d\theta \qquad [\because \sec^2 A - \tan^2 A = 1]$$

$$= \int_0^{\pi/4} \log(1+\tan\theta) \cdot d\theta$$

Proceed as in part (*iv*).

$\left[\textbf{Hint.} \text{ Using } \int_0^a f(x)\,dx = \int_0^a f(a-x)\,dx\right]$ $\qquad \left[\textbf{Ans. } \frac{\pi}{8}\log 2\right]$

Example 5. *Evaluate the following integrals :*

(*i*) $\int_0^a \frac{1}{x+\sqrt{a^2-x^2}}\, dx$ $\qquad$ (*ii*) $\int_{-1}^1 \log\left(\frac{4-x}{4+x}\right) dx$

(*iii*) $\int_0^1 \log\left(\frac{1}{x}-1\right) dx$ $\qquad$ (*iv*) $\int_0^\infty \frac{\log x}{1+x^2}\, dx$

(*v*) $\int_0^{\pi/2} \sin 2x \log(\tan x) \cdot dx$ $\qquad$ (*vi*) $\int_{\pi/6}^{\pi/3} \frac{\sqrt{\sin x}}{\sqrt{\sin x}+\sqrt{\cos x}}\, dx.$

Solution. (*i*) Let $I = \int_0^a \frac{1}{x+\sqrt{a^2-x^2}}\, dx$

Put $\quad x = a\sin\theta \Rightarrow dx = a\cos\theta\, d\theta$

When $\quad x = 0, \quad a\sin\theta = 0 \Rightarrow \sin\theta = 0 \Rightarrow \theta = 0$

and $\quad$ when $x = a, \quad a\sin\theta = a \Rightarrow \sin\theta = 1 \Rightarrow \theta = \frac{\pi}{2}$

$$\therefore \quad I = \int_0^{\pi/2} \frac{1}{a\sin\theta + \sqrt{a^2 - a^2\sin^2\theta}} \cdot a\cos\theta\, d\theta$$

$$= \int_0^{\pi/2} \frac{a\cos\theta}{a\sin\theta + a\sqrt{1-\sin^2\theta}}\, d\theta$$

$$= \int_0^{\pi/2} \frac{a\cos\theta}{a[\sin\theta + \cos\theta]}\, d\theta \qquad [\because \sin^2 A + \cos^2 A = 1]$$

$$= \int_0^{\pi/2} \frac{\cos\theta}{\sin\theta + \cos\theta}\, d\theta \qquad ...(1)$$

$$= \int_0^{\pi/2} \frac{\cos\left(\frac{\pi}{2} - \theta\right)}{\sin\left(\frac{\pi}{2} - \theta\right) + \cos\left(\frac{\pi}{2} - \theta\right)} \, d\theta \qquad \left[\because \int_0^a f(x)\,dx = \int_0^a f(a-x)\,dx\right]$$

$$\therefore \qquad I = \int_0^{\pi/2} \frac{\sin\theta}{\cos\theta + \sin\theta} \, d\theta \qquad \ldots(2)$$

Adding equations (1) and (2), we get

$$2I = \int_0^{\pi/2} \frac{\cos\theta}{\sin\theta + \cos\theta} \, d\theta + \int_0^{\pi/2} \frac{\sin\theta}{\cos\theta + \sin\theta} \, d\theta$$

$$= \int_0^{\pi/2} \left[\frac{\cos\theta}{\sin\theta + \cos\theta} + \frac{\sin\theta}{\cos\theta + \sin\theta}\right] d\theta = \int_0^{\pi/2} \left[\frac{\cos\theta + \sin\theta}{\sin\theta + \cos\theta}\right] d\theta$$

$$= \int_0^{\pi/2} 1 \,.\, d\theta = \Big[\theta\Big]_0^{\pi/2} = \left[\frac{\pi}{2} - 0\right] = \frac{\pi}{2}$$

$$\therefore \qquad I = \frac{1}{2} \cdot \frac{\pi}{2} = \frac{\pi}{4}\,.$$

(*ii*) Let $\qquad I = \int_{-1}^{1} \log\left(\frac{4-x}{4+x}\right) dx$

Let $\qquad f(x) = \log\left(\frac{4-x}{4+x}\right)$

$$\Rightarrow \qquad f(-x) = \log\left[\frac{4-(-x)}{4+(-x)}\right] = \log\left(\frac{4+x}{4-x}\right) = \log\left(\frac{4-x}{4+x}\right)^{-1}$$

$$= -\log\left(\frac{4-x}{4+x}\right) = -f(x). \qquad [\because \log m^n = n \log m]$$

$\Rightarrow$ $f(x)$ is an odd function. $\qquad \left[\because \int_{-a}^{a} f(x)\,dx = 0;\text{ where } f(x) \text{ is an odd function}\right]$

$$\therefore \quad \int_{-1}^{1} \log\left(\frac{4-x}{4+x}\right) dx = 0.$$

(*iii*) Let $\qquad I = \int_0^1 \log\left(\frac{1}{x} - 1\right) dx = \int_0^1 \log\left(\frac{1-x}{x}\right) dx \qquad \ldots(1)$

$$= \int_0^1 \log\left[\frac{1-(1-x)}{(1-x)}\right] . \, dx \qquad \left[\because \int_0^a f(x)\,dx = \int_0^a f(a-x)\,dx\right]$$

$$\therefore \qquad I = \int_0^1 \log\left(\frac{x}{1-x}\right) dx \qquad \ldots(2)$$

Adding equations (1) and (2), we get

$$2I = \int_0^1 \log\left(\frac{1-x}{x}\right) dx + \int_0^1 \log\left(\frac{x}{1-x}\right) dx$$

$$= \int_0^1 \left[\log\left(\frac{1-x}{x}\right) + \log\left(\frac{x}{1-x}\right)\right] dx$$

$$= \int_0^1 \log\left(\frac{1-x}{x} \cdot \frac{x}{1-x}\right) dx \qquad [\because \ \log m + \log n = \log m.n]$$

$$= \int_0^1 \log 1 \, . \, dx = \int_0^1 0 \, . \, dx = 0$$

$\therefore \qquad 2\text{I} = 0 \ \Rightarrow \ \text{I} = 0.$

(*iv*) Let $\quad \text{I} = \int_0^\infty \frac{\log x}{1+x^2}\, dx$

Put $\quad x = \tan\theta \ \Rightarrow \ dx = \sec^2\theta \, d\theta$

When $\quad x = 0, \ \tan\theta = 0 \ \Rightarrow \ \theta = 0$ and when $x = \infty, \ \tan\theta = \infty \ \Rightarrow \ \theta = \frac{\pi}{2}$

$$\therefore \qquad \text{I} = \int_0^{\pi/2} \frac{\log(\tan\theta)}{1+\tan^2\theta} \, . \, \sec^2\theta \, d\theta$$

$$= \int_0^{\pi/2} \frac{\log(\tan\theta)}{\sec^2\theta} \, . \, \sec^2\theta \, d\theta \qquad [\because \ \sec^2 \text{A} - \tan^2 \text{A} = 1]$$

$$\Rightarrow \qquad \text{I} = \int_0^{\pi/2} \log(\tan\theta)\, d\theta \qquad ...(1)$$

Proceed as in example 4(*i*). **[Ans.** 0.]

(*v*) Let $\quad \text{I} = \int_0^{\pi/2} \sin 2x \log(\tan x)\, dx \qquad ...(1)$

$$= \int_0^{\pi/2} \sin\left\{2\left(\frac{\pi}{2} - x\right)\right\} . \log\left\{\tan\left(\frac{\pi}{2} - x\right)\right\} . \, dx \qquad \left[\because \ \int_0^a f(x)\, dx = \int_0^a f(a-x)\, dx\right]$$

$$= \int_0^{\pi/2} \sin(\pi - 2x) \log(\cot x)\, dx \qquad [\because \ \sin(\pi - \text{A}) = \sin \text{A}]$$

$$\therefore \qquad \text{I} = \int_0^{\pi/2} \sin 2x \log(\cot x)\, dx \qquad ...(2)$$

Adding equation (1) and (2), we get

$$2\text{I} = \int_0^{\pi/2} \sin 2x \log(\tan x)\, dx + \int_0^{\pi/2} \sin 2x \log(\cot x)\, dx$$

$$= \int_0^{\pi/2} [\sin 2x \log(\tan x) + \sin 2x \log(\cot x)]\, dx$$

$$= \int_0^{\pi/2} \sin 2x\, [\log(\tan x) + \log(\cot x)]\, dx$$

$$= \int_0^{\pi/2} \sin 2x\, [\log(\tan x \, . \, \cot x)]\, dx \qquad [\because \ \log m + \log n = \log m.n]$$

$$= \int_0^{\pi/2} \sin 2x \log 1 \, . \, dx = \int_0^{\pi/2} 0 \, . \, dx = 0 \qquad [\because \ \log 1 = 0]$$

$\therefore \qquad \text{I} = 0.$

(*vi*) Let $I = \int_{\pi/6}^{\pi/3} \frac{\sqrt{\sin x}}{\sqrt{\sin x} + \sqrt{\cos x}}\, dx$ ...(1)

$$= \int_{\pi/6}^{\pi/3} \frac{\sqrt{\sin\left(\frac{\pi}{3} + \frac{\pi}{6} - x\right)}}{\sqrt{\sin\left(\frac{\pi}{3} + \frac{\pi}{6} - x\right)} + \sqrt{\cos\left(\frac{\pi}{3} + \frac{\pi}{6} - x\right)}}\, dx$$

$$\left[\because \int_a^b f(x)\, dx = \int_a^b f(a+b-x)\, dx\right]$$

$$= \int_{\pi/6}^{\pi/3} \frac{\sqrt{\sin\left(\frac{\pi}{2} - x\right)}}{\sqrt{\sin\left(\frac{\pi}{2} - x\right)} + \sqrt{\cos\left(\frac{\pi}{2} - x\right)}}\, dx \qquad \left[\because \frac{\pi}{3} + \frac{\pi}{6} = \frac{2\pi + \pi}{6} = \frac{3\pi}{6} = \frac{\pi}{2}\right]$$

$\therefore$ $I = \int_{\pi/6}^{\pi/3} \frac{\sqrt{\cos x}}{\sqrt{\cos x} + \sqrt{\sin x}}\, dx$ ...(2)

Adding equations (1) and (2), we get

$$2I = \int_{\pi/6}^{\pi/3} \frac{\sqrt{\sin x}}{\sqrt{\sin x} + \sqrt{\cos x}}\, dx + \int_{\pi/6}^{\pi/3} \frac{\sqrt{\cos x}}{\sqrt{\cos x} + \sqrt{\sin x}}\, dx$$

$$= \int_{\pi/6}^{\pi/3} \left[\frac{\sqrt{\sin x}}{\sqrt{\sin x} + \sqrt{\cos x}} + \frac{\sqrt{\cos x}}{\sqrt{\cos x} + \sqrt{\sin x}}\right] dx = \int_{\pi/6}^{\pi/3} \left[\frac{\sqrt{\sin x} + \sqrt{\cos x}}{\sqrt{\sin x} + \sqrt{\cos x}}\right] . dx$$

$$= \int_{\pi/6}^{\pi/3} 1 .\, dx = \Big[x\Big]_{\pi/6}^{\pi/3} = \left[\frac{\pi}{3} - \frac{\pi}{6}\right] = \frac{\pi}{6}.$$

$\therefore$ $I = \frac{1}{2}\left(\frac{\pi}{6}\right) = \frac{\pi}{12}$.

Example 6. *Evaluate the following integrals :*

(*i*) $\int_0^5 \frac{\sqrt[3]{x}}{\sqrt[3]{x} + \sqrt[3]{5-x}}\, dx$ (*ii*) $\int_0^{\pi} \frac{x \tan x}{\sec x + \cos x}\, dx$

(*iii*) $\int_0^{\pi/2} [2 \log \sin x - \log \sin 2x]\, dx$ (*iv*) $\int_{-\pi/4}^{\pi/4} x^3 \sin^4 x\, dx$

(*v*) $\int_0^{\pi/2} \log (\sin x)\, dx$ (*vi*) $\int_{-a}^{a} \sqrt{\frac{a-x}{a+x}}\, dx.$

Solution. (*i*) Let $I = \int_0^5 \frac{\sqrt[3]{x}}{\sqrt[3]{x} + \sqrt[3]{5-x}}\, dx$ ...(1)

$$= \int_0^5 \frac{\sqrt[3]{5-x}}{\sqrt[3]{5-x} + \sqrt[3]{5-(5-x)}}\, dx \qquad \left[\because \int_0^a f(x)\, dx = \int_0^a f(a-x)\, dx\right]$$

$\therefore$ $I = \int_0^5 \frac{\sqrt[3]{5-x}}{\sqrt[3]{5-x} + \sqrt[3]{x}}\, dx$ ...(2)

Adding equations (1) and (2), we get

$$2I = \int_0^5 \frac{\sqrt[3]{x}}{\sqrt[3]{x} + \sqrt[3]{5-x}}\, dx + \int_0^5 \frac{\sqrt[3]{5-x}}{\sqrt[3]{5-x} + \sqrt[3]{x}}\, dx$$

$$= \int_0^5 \left[\frac{\sqrt[3]{x}}{\sqrt[3]{x} + \sqrt[3]{5-x}} + \frac{\sqrt[3]{5-x}}{\sqrt[3]{5-x} + \sqrt[3]{x}} \right] dx = \int_0^5 \left[\frac{\sqrt[3]{x} + \sqrt[3]{5-x}}{\sqrt[3]{x} + \sqrt[3]{5-x}} \right] . dx$$

$$= \int_0^5 1 . dx = \left[x \right]_0^5 = [5 - 0] = 5$$

$$\therefore \qquad I = \frac{5}{2} .$$

(*ii*) Let $I = \int_0^\pi \frac{x \tan x}{\sec x + \cos x}\, dx$

$$= \int_0^\pi \frac{(\pi - x) \tan (\pi - x)}{\sec (\pi - x) + \cos (\pi - x)}\, dx \qquad \left[\because \int_0^a f(x)\, dx = \int_0^a f(a-x)\, dx \right]$$

$$= \int_0^\pi \frac{(\pi - x)(-\tan x)}{-\sec x - \cos x}\, dx = \int_0^\pi \frac{(\pi - x) \tan x}{\sec x + \cos x}\, dx$$

$$= \int_0^\pi \frac{\pi \tan x}{\sec x + \cos x}\, dx - \int_0^\pi \frac{x \tan x}{\sec x + \cos x}\, dx$$

$$= \pi \int_0^\pi \frac{\frac{\sin x}{\cos x}}{\frac{1}{\cos x} + \cos x} . dx - I$$

$$\therefore \qquad 2I = \pi \int_0^\pi \frac{\sin x}{1 + \cos^2 x}\, dx$$

Put $\cos x = z \Rightarrow -\sin x\, dx = dz \Rightarrow \sin x\, dx = -dz$

When $x = 0$, $z = \cos 0° = 1$ and when $x = \pi$, $z = \cos \pi = -1$

$$\therefore \qquad 2I = \pi \int_{+1}^{-1} \frac{1}{1+z^2} (-dz) = -\pi \int_{+1}^{-1} \frac{1}{1+z^2}\, dz$$

$$= \pi \int_{-1}^{+1} \frac{1}{1+z^2}\, dz \qquad \left[\because \int_a^b f(x)\, dx = -\int_b^a f(x)\, dx \right]$$

$$= \pi \left[\tan^{-1} z \right]_{-1}^{1} \qquad \left[\because \int \frac{1}{a^2 + x^2}\, dx = \frac{1}{a} \tan^{-1} \frac{x}{a} + c \right]$$

$$= \pi [\tan^{-1} (1) - \tan^{-1} (-1)] = \pi \left[\frac{\pi}{4} + \frac{\pi}{4} \right] = \pi \left[\frac{\pi}{2} \right]$$

$$\Rightarrow \qquad 2I = \frac{\pi^2}{2} \Rightarrow I = \frac{\pi^2}{4}.$$

(*iii*) Let $\quad I = \int_0^{\pi/2} [2 \log \sin x - \log \sin 2x]\, dx \qquad ...(1)$

$$= \int_0^{\pi/2} \left[2 \log \left\{\sin\left(\frac{\pi}{2} - x\right)\right\} - \log\left\{\sin 2\left(\frac{\pi}{2} - x\right)\right\}\right] dx$$

$$\left[\because \int_0^a f(x)\, dx = \int_0^a f(a-x)\, dx\right]$$

$$= \int_0^{\pi/2} [2 \log (\cos x) - \log \{\sin (\pi - 2x)\}]\, dx$$

$$= \int_0^{\pi/2} [2 \log (\cos x) - \log (\sin 2x)]\, dx \qquad ...(2) \quad [\because \sin(\pi - A) = \sin A]$$

Adding equations (1) and (2), we get

$$2I = \int_0^{\pi/2} [2 \log (\sin x) - \log (\sin 2x)]\, dx + \int_0^{\pi/2} [2 \log (\cos x) - \log (\sin 2x)]\, dx$$

$$= \int_0^{\pi/2} [2 \log (\sin x) - \log (\sin 2x) + 2 \log (\cos x) - \log (\sin 2x)]\, dx$$

$$= \int_0^{\pi/2} [2 (\log (\sin x) + \log (\cos x)) - 2 \log (\sin 2x)]\, dx$$

$$= 2 \int_0^{\pi/2} [\log (\sin x \,.\, \cos x) - \log (\sin 2x)]\, dx \qquad [\because \log m + \log n = \log (m.n)]$$

$$\Rightarrow \qquad I = \int_0^{\pi/2} [\log (\sin x \cos x) - \log (\sin 2x)\, dx$$

$$= \int_0^{\pi/2} \log\left[\frac{\sin x \cos x}{\sin 2x}\right] dx \qquad \left[\because \log m - \log n = \log\left(\frac{m}{n}\right)\right]$$

$$= \int_0^{\pi/2} \log\left[\frac{\sin x \cos x}{2 \sin x \cos x}\right] dx = \int_0^{\pi/2} \log\left(\frac{1}{2}\right) dx \qquad [\because \sin 2A = 2 \sin A \cos A]$$

$$= \log\left(\frac{1}{2}\right) \int_0^{\pi/2} 1 \,.\, dx$$

$$= \log (2)^{-1} \Big[x\Big]_0^{\pi/2} = -\log 2 \left[\frac{\pi}{2} - 0\right] \qquad [\because \log m^n = n \log m]$$

$$= -\frac{\pi}{2} \log 2.$$

(*iv*) Let $\quad I = \int_{-\pi/4}^{\pi/4} x^3 \sin^4 x\, dx$

Let $\quad f(x) = x^3 \sin^4 x$

$$\Rightarrow \quad f(-x) = (-x)^3 \sin^4 (-x) = -x^3 [\sin (-x)]^4$$

$$= -x^3 [-\sin x]^4 \qquad [\because \sin(-\theta) = -\sin\theta]$$

$$= -x^3 \sin^4 x$$

$$= -f(x)$$

$\Rightarrow$ $f(x)$ is an odd function. $\qquad \left[\because \int_{-a}^{a} f(x)\, dx = 0; \text{ where } f(x) \text{ is an odd function}\right]$

$\therefore \quad \int_{-\pi/4}^{\pi/4} x^3 \sin^4 x \, dx = 0.$

(*v*) Let $\quad I = \int_0^{\pi/2} \log (\sin x) \, dx \qquad ...(1)$

$= \int_0^{\pi/2} \log \left[\sin \left(\frac{\pi}{2} - x \right) \right] dx \qquad \left[\because \int_0^a f(x) \, dx = \int_0^a f(a-x) \, dx \right]$

$\therefore \quad I = \int_0^{\pi/2} \log (\cos x) \, dx \qquad ...(2)$

Adding equations (1) and (2), we get

$$2I = \int_0^{\pi/2} \log (\sin x) \, dx + \int_0^{\pi/2} \log (\cos x) \, dx$$

$$= \int_0^{\pi/2} [\log (\sin x) + \log (\cos x)] \, dx$$

$$= \int_0^{\pi/2} \log (\sin x \, . \cos x) \, . \, dx \qquad [\because \log m + \log n = \log (m.n)]$$

$$= \int_0^{\pi/2} \log \left(\frac{2 \sin x \cos x}{2} \right) dx \qquad \text{[Note this step]} \quad \text{[Multily and divded by 2]}$$

$$= \int_0^{\pi/2} \log \left(\frac{\sin 2x}{2} \right) dx \qquad [\because \sin 2A = 2 \sin A \cos A]$$

$$= \int_0^{\pi/2} [\log (\sin 2x) - \log 2] \, dx \qquad \left[\because \log \left(\frac{m}{n} \right) = \log m - \log n \right]$$

$$= \int_0^{\pi/2} \log (\sin 2x) \, dx - \int_0^{\pi/2} \log 2 \, . \, dx$$

Put $\quad 2x = z \Rightarrow 2dx = dz \Rightarrow dx = \frac{1}{2} dz$

When $x = 0$, $\; z = 2(0) = 0 \;$ and $\;$ when $x = \frac{\pi}{2}$, $\; z = 2 \left(\frac{\pi}{2} \right) = \pi$

$$\therefore \quad 2I = \int_0^{\pi} \log (\sin z) \left(\frac{1}{2} dz \right) - \log 2 \int_0^{\pi/2} 1 \, . \, dx = \frac{1}{2} \int_0^{\pi} \log (\sin z) \, dz - \log 2 \left[x \right]_0^{\pi/2}$$

$$= \frac{1}{2} \times 2 \int_0^{\pi/2} \log (\sin z) \, dz - \log 2 \left[\frac{\pi}{2} - 0 \right]$$

$$= \int_0^{\pi/2} \log (\sin x) \, dx - \frac{\pi}{2} \log 2 \qquad \left[\because \int_a^b f(x) \, dx = \int_a^b f(z) \, dz \right]$$

$$= I - \frac{\pi}{2} \log 2$$

$$\Rightarrow \quad I = -\frac{\pi}{2} \log 2.$$

(*vi*) Let $\quad I = \int_{-a}^{a} \sqrt{\frac{a-x}{a+x}} \, dx = \int_{-a}^{a} \sqrt{\frac{a-x}{a+x} \times \frac{a-x}{a-x}} \, dx \qquad ...(1) \quad$ [Rationalization]

$$= \int_{-a}^{a} \sqrt{\frac{(a-x)^2}{a^2 - x^2}} \, dx = \int_{-a}^{a} \frac{a-x}{\sqrt{a^2 - x^2}} \, dx$$

$$\therefore \quad I = a\int_{-a}^{a}\frac{1}{\sqrt{a^2-x^2}}\,dx - \int_{-a}^{a}\frac{x}{\sqrt{a^2-x^2}}\,dx = aI_1 - I_2 \quad \text{(say)} \qquad ...(2)$$

Let $f(x) = \dfrac{1}{\sqrt{a^2-x^2}}$

$$\Rightarrow \quad f(-x) = \frac{1}{\sqrt{a^2-(-x)^2}} = \frac{1}{\sqrt{a^2-x^2}} = f(x)$$

$\therefore$ $f(x)$ is an even function.

Let $g(x) = \dfrac{x}{\sqrt{a^2-x^2}}$

$$\Rightarrow \quad g(-x) = \frac{-x}{\sqrt{a^2-(-x)^2}} = -\frac{x}{\sqrt{a^2-x^2}} = -g(x)$$

$\therefore$ $g(x)$ is an odd function.

Now $I_1 = \displaystyle\int_{-a}^{a}\frac{1}{\sqrt{a^2-x^2}}\,dx$

$$= 2\int_{0}^{a}\frac{1}{\sqrt{a^2-x^2}}\,dx \qquad \left[\because \int_{-a}^{a} f(x)\,dx = 0; \text{ when } f(x) \text{ is an odd function.} \atop = 2\int_{0}^{a} f(x)\,dx; \text{ when } f(x) \text{ is an even function}\right]$$

$$= 2\left[\sin^{-1}\frac{x}{a}\right]_0^a = 2\,[\sin^{-1} 1 - \sin^{-1} 0] \qquad \left[\because \int \frac{1}{\sqrt{a^2-x^2}}\,dx = \sin^{-1}\frac{x}{a} + c\right]$$

$$\Rightarrow \quad I_1 = 2\left[\frac{\pi}{2} - 0\right] = \pi \qquad ...(3)$$

and $I_2 = \displaystyle\int_{-a}^{a}\frac{x}{\sqrt{a^2-x^2}}\,dx$

$\Rightarrow$ $I_2 = 0$ ...(4) [$\because$ $g(x)$ is an odd function]

$\therefore$ From equation (2), we have

$I = aI_1 - I_2$ [Using (3) and (4)]

$\Rightarrow$ $I = a\pi - 0 = a\pi.$

Example 7. *Evaluate the following integrals :*

(*i*) $\displaystyle\int_0^{\pi/2}\frac{\sin^2 x}{\sin x + \cos x}\,dx$ (*ii*) $\displaystyle\int_0^{\pi}\frac{x \sin x}{1+\cos^2 x}\,dx$

(*iii*) $\displaystyle\int_0^{1} x(1-x)^{3/2}\,dx$ (*iv*) $\displaystyle\int_0^{1}\frac{\sin^{-1} x}{x}\,dx$

(*v*) $\displaystyle\int_{-\pi/2}^{\pi/2}\sin^2 x\,dx$ (*vi*) $\displaystyle\int_{-\pi/2}^{\pi/2}\cos x\,dx.$

Solution. (*i*) Let $I = \int_0^{\pi/2} \frac{\sin^2 x}{\sin x + \cos x}\, dx$ ...(1)

$$= \int_0^{\pi/2} \frac{\sin^2\left(\frac{\pi}{2} - x\right)}{\sin\left(\frac{\pi}{2} - x\right) + \cos\left(\frac{\pi}{2} - x\right)}\, dx \qquad \left[\because \int_0^a f(x)\, dx = \int_0^a f(a-x)\, dx\right]$$

$$\therefore \qquad I = \int_0^{\pi/2} \frac{\cos^2 x}{\cos x + \sin x}\, dx \qquad ...(2)$$

Adding equations (1) and (2), we get

$$2I = \int_0^{\pi/2} \frac{\sin^2 x}{\sin x + \cos x}\, dx + \int_0^{\pi/2} \frac{\cos^2 x}{\cos x + \sin x}\, dx$$

$$= \int_0^{\pi/2} \left[\frac{\sin^2 x}{\sin x + \cos x} + \frac{\cos^2 x}{\cos x + \sin x}\right] dx$$

$$= \int_0^{\pi/2} \left[\frac{\sin^2 x + \cos^2 x}{\sin x + \cos x}\right] dx = \int_0^{\pi/2} \frac{1}{\sin x + \cos x}\, .\, dx \qquad [\because \sin^2 A + \cos^2 A = 1]$$

$$= \frac{1}{\sqrt{2}} \int_0^{\pi/2} \frac{1}{\frac{1}{\sqrt{2}} \sin x + \frac{1}{\sqrt{2}} \cos x}\, dx \qquad [\text{Multiply and divided by } \sqrt{2}]$$

$$= \frac{1}{\sqrt{2}} \int_0^{\pi/2} \frac{1}{\sin x \sin \frac{\pi}{4} + \cos x \cos \frac{\pi}{4}}\, dx$$

$$= \frac{1}{\sqrt{2}} \int_0^{\pi/2} \frac{1}{\cos\left(x - \frac{\pi}{4}\right)}\, dx \qquad [\because \cos(A - B) = \sin A \sin B + \cos A \cos B]$$

$$= \frac{1}{\sqrt{2}} \int_0^{\pi/2} \sec\left(x - \frac{\pi}{4}\right) dx$$

$$= \frac{1}{\sqrt{2}} \left[\log\left|\sec\left(x - \frac{\pi}{4}\right) + \tan\left(x - \frac{\pi}{4}\right)\right|\right]_0^{\pi/2} \qquad \left[\because \int \sec x\, dx = \log|\sec x + \tan x| + c\right]$$

$$= \frac{1}{\sqrt{2}} \left[\log\left|\sec\left(\frac{\pi}{2} - \frac{\pi}{4}\right) + \tan\left(\frac{\pi}{2} - \frac{\pi}{4}\right)\right| - \log\left|\sec\left(0 - \frac{\pi}{4}\right) + \tan\left(0 - \frac{\pi}{4}\right)\right|\right]$$

$$= \frac{1}{\sqrt{2}} \left[\log\left|\sec\frac{\pi}{4} + \tan\frac{\pi}{4}\right| - \log\left|\sec\left(-\frac{\pi}{4}\right) + \tan\left(-\frac{\pi}{4}\right)\right|\right]$$

$$= \frac{1}{\sqrt{2}} \left[\log\left|\sqrt{2} + 1\right| - \log\left|\sec\frac{\pi}{4} - \tan\frac{\pi}{4}\right|\right]$$

$$= \frac{1}{\sqrt{2}} [\log(\sqrt{2} + 1) - \log(\sqrt{2} - 1)]$$

$$= \frac{1}{\sqrt{2}}\left[\log\left(\frac{\sqrt{2}+1}{\sqrt{2}-1}\right)\right] \qquad \left[\because \log m - \log n = \log\left(\frac{m}{n}\right)\right]$$

$$= \frac{1}{\sqrt{2}}\left[\log\left(\frac{\sqrt{2}+1}{\sqrt{2}-1}\times\frac{\sqrt{2}+1}{\sqrt{2}+1}\right)\right] \qquad \text{[Rationalization]}$$

$$= \frac{1}{\sqrt{2}}\left[\log\frac{(\sqrt{2}+1)^2}{2-1}\right] = \frac{1}{\sqrt{2}}\log(\sqrt{2}+1)^2$$

$$= \frac{2}{\sqrt{2}}\log(\sqrt{2}+1) \qquad [\because \log m^n = n\log m)$$

$$\therefore \qquad I = \frac{1}{\sqrt{2}}\log(\sqrt{2}+1).$$

(*ii*) Let $\quad I = \int_0^{\pi} \frac{x\sin x}{1+\cos^2 x}\,dx \qquad ...(1)$

$$= \int_0^{\pi} \frac{(\pi - x)\sin(\pi - x)}{1+\cos^2(\pi - x)}\,dx \qquad \left[\because \int_0^a f(x)\,dx = \int_0^a f(a-x)\,dx\right]$$

$$\therefore \qquad I = \int_0^{\pi} \frac{(\pi - x)\sin x}{1+\cos^2 x}\,dx \qquad ...(2)$$

Adding equations (1) and (2), we get

$$2I = \int_0^{\pi} \frac{x\sin x}{1+\cos^2 x}\,dx + \int_0^{\pi} \frac{(\pi - x)\sin x}{1+\cos^2 x}\,dx$$

$$= \int_0^{\pi}\left[\frac{x\sin x}{1+\cos^2 x} + \frac{(\pi - x)\sin x}{1+\cos^2 x}\right]dx = \int_0^{\pi}\left[\frac{x\sin x + \pi\sin x - x\sin x}{1+\cos^2 x}\right]dx$$

$$= \int_0^{\pi} \frac{\pi\sin x}{1+\cos^2 x}\,dx = \pi\int_0^{\pi} \frac{\sin x}{1+\cos^2 x}\,dx$$

Put $\quad \cos x = z \;\Rightarrow\; -\sin x\,dx = dz \;\Rightarrow\; \sin x\,dx = -dz$

When $\quad x = 0,\quad z = \cos 0° = 1 \quad$ and $\quad$ when $x = \pi,\quad z = \cos\pi = -1$

$$\therefore \qquad 2I = \pi\int_1^{-1} \frac{1}{1+z^2}(-dz) = -\pi\int_1^{-1}\frac{1}{1+z^2}\,dz$$

$$= \pi\int_{-1}^{1}\frac{1}{1+z^2}\,dz \qquad \left[\because \int_a^b f(x)\,dx = -\int_b^a f(x)\,dx\right]$$

$$= \pi\left[\tan^{-1} z\right]_{-1}^{1} \qquad \left[\because \int\frac{1}{a^2+x^2}\,dx = \frac{1}{a}\tan^{-1}\frac{x}{a} + c\right]$$

$$= \pi[\tan^{-1}1 - \tan^{-1}(-1)] = \pi\left[\frac{\pi}{4} - \left(-\frac{\pi}{4}\right)\right] = \pi\left[\frac{\pi}{4}+\frac{\pi}{4}\right] = \pi\left[\frac{\pi}{2}\right]$$

$$\Rightarrow \qquad I = \frac{\pi^2}{4}.$$

(*iii*) Let $\quad I = \int_0^1 x(1-x)^{3/2}\,dx$ ...(1)

$$= \int_0^1 (1-x)\,[1-(1-x)]^{3/2}\,dx \qquad \left[\because \int_0^a f(x)\,dx = \int_0^a f(a-x)\,dx\right]$$

$$= \int_0^1 (1-x)\,(x^{3/2})\,dx = \int_0^1 (x^{3/2} - x^{5/2})\,dx = \int_0^1 x^{3/2}\,dx - \int_0^1 x^{5/2}\,dx$$

$$= \left[\frac{x^{\frac{3}{2}+1}}{\frac{3}{2}+1}\right]_0^1 - \left[\frac{x^{\frac{5}{2}+1}}{\frac{5}{2}+1}\right]_0^1 = \frac{2}{5}\left[x^{5/2}\right]_0^1 - \frac{2}{7}\left[x^{7/2}\right]_0^1 = \frac{2}{5}[(1)^{5/2} - 0] - \frac{2}{7}[(1)^{7/2} - 0]$$

$$= \frac{2}{5} - \frac{2}{7} = \frac{14-10}{35} = \frac{4}{35}.$$

(*iv*) Let $\quad I = \int_0^1 \frac{\sin^{-1} x}{x}\,dx$

Put $\quad x = \sin z \;\Rightarrow\; \sin^{-1} x = z \;\Rightarrow\; dx = \cos z\,dz$

When $\quad x = 0,\ \sin z = 0 \;\Rightarrow\; z = 0$ and when $x = 1,\ \sin z = 1 \;\Rightarrow\; z = \frac{\pi}{2}$

$$\therefore \quad I = \int_0^{\pi/2} \frac{z}{\sin z}\cos z\,dz$$

$$= \int_0^{\pi/2} \underset{\text{I}}{z}\ \underset{\text{II}}{\cot z}\,dz$$

Integrating by parts, we get

$$I = \left[z \,.\, \log(\sin z)\right]_0^{\pi/2} - \int_0^{\pi/2} \frac{d}{dz}(z)\,.\,\log(\sin z)\,dz$$

$$= \left[\frac{\pi}{2}\log\sin\frac{\pi}{2} - 0\right] - \int_0^{\pi/2} 1\,.\,\log(\sin z)\,dz$$

$$= \left(\frac{\pi}{2}\log 1\right) - \int_0^{\pi/2} \log(\sin z)\,dz$$

$$= \frac{\pi}{2}(0) - \left(-\frac{\pi}{2}\log 2\right) = 0 + \frac{\pi}{2}\log 2 \qquad [\because \text{ See example 6 } (v).]$$

$$= \frac{\pi}{2}\log 2.$$

(*v*) Let $\quad I = \int_{-\pi/2}^{\pi/2} \sin^2 x\,dx$

Let $\quad f(x) = \sin^2 x$

$\Rightarrow \quad f(-x) = \sin^2(-x) = [\sin(-x)]^2 = [-\sin x]^2 = \sin^2 x = f(x)$

$\therefore\ f(x)$ is an even function.

$$\therefore \quad \int_{-\pi/2}^{\pi/2} \sin^2 x\,dx = 2\int_0^{\pi/2} \sin^2 x\,dx \qquad \left[\because \int_{-a}^{a} f(x)\,dx = 2\int_0^a f(x)\,dx;\ \text{if } f(x) \text{ is an even function}\right]$$

$$= 2\int_0^{\pi/2}\left(\frac{1-\cos 2x}{2}\right)dx \qquad \left[\begin{array}{l}\because\ 1-\cos 2A = 2\sin^2 A \\ \Rightarrow\ \dfrac{1-\cos 2A}{2} = \sin^2 A\end{array}\right]$$

$$= \int_0^{\pi/2}(1-\cos 2x)\,dx = \left[x - \frac{\sin 2x}{2}\right]_0^{\pi/2}$$

$$= \left[\left(\frac{\pi}{2} - \frac{\sin^2\left(\frac{\pi}{2}\right)}{2}\right) - \left(0 - \frac{\sin 0^\circ}{2}\right)\right] = \left(\frac{\pi}{2} - 0\right) - (0-0)$$

$$= \frac{\pi}{2}.$$

(*vi*) Let $\quad I = \int_{-\pi/2}^{\pi/2} \cos x\,dx$

Let $\quad f(x) = \cos x$

$\Rightarrow \quad f(-x) = \cos(-x) = \cos x = f(x)$

$\therefore$ $f(x)$ is an even function.

$$\therefore \quad \int_{-\pi/2}^{\pi/2}\cos x\,dx = 2\int_0^{\pi/2}\cos x\,dx \qquad \left[\begin{array}{l}\because\ \displaystyle\int_{-a}^{a} f(x)\,dx = 2\int_0^a f(x)\,dx;\\ \text{if } f(x) \text{ is an even function}\end{array}\right]$$

$$= 2\left[\sin x\right]_0^{\pi/2} = 2\left[\sin\frac{\pi}{2} - \sin 0^\circ\right] = 2\,[1-0]$$

$$= 2.$$

Example 8. *Evaluate the following integrals :*

(*i*) $\displaystyle\int_0^{\pi} \frac{x}{1+\sin x}\,dx$ (*ii*) $\displaystyle\int_0^{\pi} \frac{x\tan x}{\sec x + \tan x}\,dx$

(*iii*) $\displaystyle\int_0^{\infty} \frac{x}{(1+x)(1+x^2)}\,dx$ (*iv*) $\displaystyle\int_{-\pi/2}^{\pi/2} \sin^7 x\,dx$

(*v*) $\displaystyle\int_0^{2\pi} \cos^5 x\,dx$ (*vi*) $\displaystyle\int_0^{\pi} \frac{1}{1+\sin x}\,dx.$

Solution. (*i*) Let $I = \displaystyle\int_0^{\pi} \frac{x}{1+\sin x}\,dx$ ...(1)

$$\Rightarrow \quad I = \int_0^{\pi}\frac{\pi - x}{1+\sin(\pi - x)}\,dx \qquad \left[\because\ \int_0^a f(x)\,dx = \int_0^a f(a-x)\,dx\right]$$

$$\therefore \quad I = \int_0^{\pi}\frac{(\pi - x)}{1+\sin x}\,dx \qquad ...(2)$$

Adding equations (1) and (2), we get

$$2I = \int_0^{\pi}\frac{x}{1+\sin x}\,dx + \int_0^{\pi}\frac{(\pi-x)}{1+\sin x}\,dx$$

$$= \int_0^{\pi}\left[\frac{x}{1+\sin x} + \frac{\pi - x}{1+\sin x}\right]dx = \int_0^{\pi}\frac{x+\pi-x}{1+\sin x}\,dx = \int_0^{\pi}\frac{\pi}{1+\sin x}\,dx$$

$$= \pi \int_0^{\pi} \left[\frac{1}{1+\sin x} \times \frac{1-\sin x}{1-\sin x}\right] dx \qquad \text{[Rationalization]}$$

$$= \pi \int_0^{\pi} \frac{1-\sin x}{1-\sin^2 x}\, dx = \pi \int_0^{\pi} \frac{1-\sin x}{\cos^2 x}\, dx = \pi \int_0^{\pi} \left[\frac{1}{\cos^2 x} - \frac{\sin x}{\cos^2 x}\right] dx$$

$$= \pi \int_0^{\pi} (\sec^2 x - \sec x \tan x)\, dx = \pi \left[\tan x - \sec x\right]_0^{\pi}$$

$$= \pi\, [(\tan \pi - \sec \pi) - (\tan 0° - \sec 0°)]$$

$$= \pi\, [0 - (-1) - (0 - 1)] = \pi\, [1 + 1] = 2\pi$$

$\therefore \qquad I = \pi.$

(*ii*) Let $\quad I = \int_0^{\pi} \frac{x \tan x}{\sec x + \tan x}\, dx = \int_0^{\pi} \frac{x \,.\, \frac{\sin x}{\cos x}}{\frac{1}{\cos x} + \frac{\sin x}{\cos x}}\, dx$

$$= \int_0^{\pi} \frac{x \sin x}{1+\sin x}\, dx \qquad ...(1)$$

$$= \int_0^{\pi} \frac{(\pi - x)\sin(\pi - x)}{1+\sin(\pi - x)}\, dx \qquad \left[\because \int_0^{a} f(x)\, dx = \int_0^{a} f(a-x)\, dx\right]$$

$$\therefore \qquad I = \int_0^{\pi} \frac{(\pi - x)\sin x}{1+\sin x}\, dx \qquad ...(2)$$

Adding equations (1) and (2), we get

$$2I = \int_0^{\pi} \frac{x \sin x}{1+\sin x}\, dx + \int_0^{\pi} \frac{(\pi - x)\sin x}{1+\sin x}\, dx = \int_0^{\pi} \left[\frac{x\sin x}{1+\sin x} + \frac{(\pi - x)\sin x}{1+\sin x}\right] dx$$

$$= \int_0^{\pi} \left[\frac{x\sin x + \pi \sin x - x \sin x}{1+\sin x}\right] dx = \int_0^{\pi} \frac{\pi \sin x}{1+\sin x}\, dx$$

$$= \pi \int_0^{\pi} \frac{1+\sin x - 1}{1+\sin x}\, dx \qquad \text{[Add and subtract 1 to the numerator]}$$

$$= \pi \int_0^{\pi} \left(\frac{1+\sin x}{1+\sin x} - \frac{1}{1+\sin x}\right) dx = \pi \int_0^{\pi} 1 \,.\, dx - \pi \int_0^{\pi} \frac{1}{1+\sin x}\, dx$$

$$= \pi \left[x\right]_0^{\pi} - \pi \int_0^{\pi} \left[\frac{1}{1+\sin x} \times \frac{1-\sin x}{1-\sin x}\right] dx \qquad \text{[Rationalization]}$$

$$= \pi\, [\pi - 0] - \pi \int_0^{\pi} \frac{1-\sin x}{1-\sin^2 x}\, dx$$

$$= \pi^2 - \pi \int_0^{\pi} \frac{1-\sin x}{\cos^2 x}\, dx \qquad [\because \sin^2 A + \cos^2 A = 1]$$

$$= \pi^2 - \pi \int_0^{\pi} \left[\frac{1}{\cos^2 x} - \frac{\sin x}{\cos^2 x}\right] dx = \pi^2 - \pi \int_0^{\pi} (\sec^2 x - \sec x \tan x)\, dx$$

$$= \pi^2 - \pi \left[\tan x - \sec x\right]_0^\pi = \pi^2 - \pi\,[(\tan \pi - \sec \pi) - (\tan 0° - \sec 0°)]$$

$$= \pi^2 - \pi\,[(0 - (-1)) - (0 - 1)] = \pi^2 - \pi\,[1 + 1] = \pi^2 - 2\pi$$

$$\Rightarrow \qquad I = \frac{\pi}{2}(\pi - 2).$$

(*iii*) Let $I = \int_0^\infty \frac{x}{(1+x)(1+x^2)}\,dx$

Put $x = \tan \theta \Rightarrow dx = \sec^2 \theta\, d\theta$

When $x = 0$, $\tan \theta = 0 \Rightarrow \theta = 0$ and when $x = \infty$, $\tan \theta = \infty \Rightarrow \theta = \frac{\pi}{2}$.

$$\therefore \qquad I = \int_0^{\pi/2} \frac{\tan \theta}{(1+\tan \theta)(1+\tan^2 \theta)} . \sec^2 \theta\, d\theta$$

$$= \int_0^{\pi/2} \frac{\tan \theta}{(1+\tan \theta)\sec^2 \theta} . \sec^2 \theta\, d\theta \qquad [\because \sec^2 A - \tan^2 A = 1]$$

$$= \int_0^{\pi/2} \frac{\tan \theta}{1+\tan \theta}\, d\theta \qquad ...(1)$$

$$= \int_0^{\pi/2} \frac{\tan\left(\frac{\pi}{2} - \theta\right)}{1 + \tan\left(\frac{\pi}{2} - \theta\right)}\, d\theta \qquad \left[\because \int_0^a f(x)\,dx = \int_0^a f(a-x)\,dx\right]$$

$$= \int_0^{\pi/2} \frac{\cot \theta}{1+\cot \theta}\, d\theta = \int_0^{\pi/2} \frac{\frac{1}{\tan \theta}}{1 + \frac{1}{\tan \theta}}\, d\theta$$

$$\therefore \qquad I = \int_0^{\pi/2} \frac{1}{1+\tan \theta}\, d\theta \qquad ...(2)$$

Adding equations (1) and (2), we get

$$2I = \int_0^{\pi/2} \frac{\tan \theta}{1+\tan \theta}\, d\theta + \int_0^{\pi/2} \frac{1}{1+\tan \theta}\, d\theta$$

$$= \int_0^{\pi/2} \left[\frac{\tan \theta}{1+\tan \theta} + \frac{1}{1+\tan \theta}\right] d\theta = \int_0^{\pi/2} \left(\frac{1+\tan \theta}{1+\tan \theta}\right) d\theta$$

$$= \int_0^{\pi/2} 1 . d\theta = \left[\theta\right]_0^{\pi/2} = \left[\frac{\pi}{2} - 0\right] = \frac{\pi}{2}$$

$$\therefore \qquad I = \frac{1}{2} . \frac{\pi}{2} = \frac{\pi}{4}.$$

(*iv*) Let $\quad I = \int_{-\pi/2}^{\pi/2} \sin^7 x \, dx$

Let $\quad f(x) = \sin^7 x$

$\Rightarrow \quad f(x) = \sin^7(-x) = [\sin(-x)]^7 = [-\sin x]^7 = -\sin^7 x = -f(x)$

$\therefore \quad f(x)$ is an odd function.

Hence $\quad \int_{-\pi/2}^{\pi/2} \sin^7 x \, dx = 0.$ $\quad \left[\because \int_{-a}^{a} f(x)\, dx = 0;\text{ where } f(x) \text{ is an odd function}\right]$

(*v*) Let $\quad I = \int_0^{2\pi} \cos^5 x \, dx$

Let $\quad f(x) = \cos^5 x$

$\Rightarrow f(2\pi - x) = \cos^5(2\pi - x) = \cos^5 x = f(x).$

$$\therefore \quad I = \int_0^{2\pi} \cos^5 x \, dx = 2\int_0^{\pi} \cos^5 x \, dx$$

Now, let $g(x) = \cos^5 x$

$\Rightarrow \quad g(\pi - x) = \cos^5(\pi - x) = [\cos(\pi - x)]^5 = [-\cos x]^5 = -\cos^5 x = -f(x).$

$\therefore \quad I = 2\int_0^{\pi} \cos^5 x \, dx = 0.$ $\quad \left[\because \int_0^{2a} f(x)\, dx = 0 \text{ if } f(2a - x) = -f(x)\right]$

$\left[\because \int_0^{2a} f(x)\, dx = 2\cdot\int_0^{a} f(x)\, dx;\ \text{if } f(2a - x) = f(x)\right]$

(*vi*) Let $\quad I = \int_0^{\pi} \frac{1}{1 + \sin x}\, dx$

Let $\quad f(x) = \frac{1}{1 + \sin x}$

$$\Rightarrow \quad f(\pi - x) = \frac{1}{1 + \sin(\pi - x)} = \frac{1}{1 + \sin x} = f(x)$$

$$\therefore \quad I = 2\int_0^{\pi/2} \frac{1}{1 + \sin x}\, dx$$

$$= 2\int_0^{\pi/2} \frac{1}{1 + \sin\left(\frac{\pi}{2} - x\right)}\, dx \quad \left[\because \int_0^{a} f(x)\, dx = \int_0^{a} f(a - x)\, dx\right]$$

$$= 2\int_0^{\pi/2} \frac{1}{1 + \cos x}\, dx = 2\int_0^{\pi/2} \frac{1}{2\cos^2 \frac{x}{2}}\, dx \quad \left[\begin{array}{l}\because\ 1 + \cos 2A = 2\cos^2 A \\ \Rightarrow\ 1 + \cos A = 2\cos^2 \frac{A}{2}\end{array}\right]$$

$$= \int_0^{\pi/2} \sec^2 \frac{x}{2}\, dx = \left[\frac{\tan \frac{x}{2}}{\frac{1}{2}}\right]_0^{\pi/2} = 2\left[\tan \frac{x}{2}\right]_0^{\pi/2}$$

$$\Rightarrow \quad I = 2\left[\tan \frac{\pi}{4} - \tan 0°\right] = 2(1 - 0) = 2.$$

Example 9. *Evaluate the following integrals :*

(*i*) $\int_0^{\pi/2} \frac{\sin x - \cos x}{1 + \sin x \cos x}\, dx$ (*ii*) $\int_0^{\pi/2} (\sin x - \cos x) \log (\sin x + \cos x)\, dx$

(*iii*) $\int_{-\pi/2}^{\pi/2} \sin^{199} x\, dx$ (*iv*) $\int_0^{\pi} \sin^2 x \cos^3 x\, dx$

(*v*) $\int_0^{\sqrt{3}} \tan^{-1}\left(\frac{2x}{1-x^2}\right) dx.$

Solution. (*i*) Let $I = \int_0^{\pi/2} \frac{\sin x - \cos x}{1 + \sin x \cos x}\, dx$ ...(1)

$$= \int_0^{\pi/2} \frac{\sin\left(\frac{\pi}{2} - x\right) - \cos\left(\frac{\pi}{2} - x\right)}{1 + \sin\left(\frac{\pi}{2} - x\right)\cos\left(\frac{\pi}{2} - x\right)}\, dx \qquad \left[\because \int_0^a f(x)\, dx = \int_0^a f(a-x)\, dx\right]$$

$\therefore$ $$I = \int_0^{\pi/2} \frac{\cos x - \sin x}{1 + \cos x \sin x}\, dx \qquad ...(2)$$

Adding equations (1) and (2), we get

$$2I = \int_0^{\pi/2} \frac{\sin x - \cos x}{1 + \sin x \cos x}\, dx + \int_0^{\pi/2} \frac{\cos x - \sin x}{1 + \cos x \sin x}\, dx$$

$$= \int_0^{\pi/2} \left[\frac{\sin x - \cos x}{1 + \sin x \cos x} + \frac{\cos x - \sin x}{1 + \cos x \sin x}\right] dx$$

$$= \int_0^{\pi/2} \frac{\sin x - \cos x + \cos x - \sin x}{1 + \sin x \cos x}\, dx = \int_0^{\pi/2} 0\, dx = 0$$

$\Rightarrow$ $I = 0.$

(*ii*) Let $$I = \int_0^{\pi/2} (\sin x - \cos x) \log (\sin x + \cos x)\, dx \qquad ...(1)$$

$$= \int_0^{\pi/2} \left[\left\{\sin\left(\frac{\pi}{2} - x\right) - \cos\left(\frac{\pi}{2} - x\right)\right\} \log \left\{\sin\left(\frac{\pi}{2} - x\right) + \cos\left(\frac{\pi}{2} - x\right)\right\}\right] dx$$

$$\left[\because \int_0^a f(x)\, dx = \int_0^a f(a-x)\, dx\right]$$

$\therefore$ $$I = \int_0^{\pi/2} (\cos x - \sin x) \log (\cos x + \sin x)\, .\, dx \qquad ...(2)$$

Adding equations (1) and (2), we get

$$2I = \int_0^{\pi/2} (\sin x - \cos x) \log (\sin x + \cos x)\, dx + \int_0^{\pi/2} (\cos x - \sin x) \log (\cos x + \sin x)\, dx$$

$$= \int_0^{\pi/2} [(\sin x - \cos x) \log (\sin x + \cos x) + (\cos x - \sin x) \log (\cos x + \sin x)]\, dx$$

$$= \int_0^{\pi/2} [(\sin x - \cos x) \log (\sin x + \cos x) - (\sin x - \cos x) \log (\cos x + \sin x)]\, dx$$

$$= \int_0^{\pi/2} 0 \,.\, dx = 0$$

$\Rightarrow \quad I = 0.$

(*iii*) Let $\quad I = \int_{-\pi/2}^{\pi/2} \sin^{199} x \, dx$

Let $\quad f(x) = \sin^{199} x$

$\Rightarrow \quad f(-x) = \sin^{199}(-x) = [\sin(-x)]^{199} = [-\sin x]^{199} = -\sin^{199} x = -f(x)$

$\therefore \quad f(x)$ is an odd function.

Hence $\int_{-\pi/2}^{\pi/2} \sin^{199} x \, dx = 0. \qquad \left[\because \int_{-a}^{a} f(x)\, dx = 0;\text{ where } f(x) \text{ is an odd function}\right]$

(*iv*) Let $\quad I = \int_0^{\pi} \sin^2 x \cos^3 x \, dx$

Let $\quad f(x) = \sin^2 x \cos^3 x \, dx$

$\Rightarrow \quad f(\pi - x) = \sin^2(\pi - x)\cos^3(\pi - x) = \sin^2 x(-\cos x)^3 = -\sin^2 x \cos^3 x = -f(x)$

$\therefore \quad \int_0^{\pi} \sin^2 x \cos^3 x \, dx = 0. \qquad \left[\because \int_0^{2a} f(x)\, dx = 0, \text{ if } f(2a - x) = -f(x) \quad (\text{Here } 2a = \pi)\right]$

(*v*) Let $I = \int_0^{\sqrt{3}} \tan^{-1}\left(\frac{2x}{1-x^2}\right) dx$

Since, we have $\tan^{-1}\left(\frac{2x}{1-x^2}\right) = \begin{cases} 2\tan^{-1} x, \text{ if } -1 < x < 1 \\ -\pi + 2\tan^{-1} x, \text{ if } \quad x > 1 \\ \pi + 2\tan^{-1} x, \text{ if } \quad x < -1 \end{cases}$

$$\therefore \quad \int_0^{\sqrt{3}} \tan^{-1}\left(\frac{2x}{1-x^2}\right) dx = \int_0^1 \tan^{-1}\left(\frac{2x}{1-x^2}\right) dx + \int_1^{\sqrt{3}} \tan^{-1}\left(\frac{2x}{1-x^2}\right) dx$$

$$= \int_0^1 2\tan^{-1} x \, dx + \int_1^{\sqrt{3}} (-\pi + 2\tan^{-1} x)\, dx$$

$$= \int_0^1 2\tan^{-1} x \, dx + \int_1^{\sqrt{3}} -\pi \, dx + \int_1^{\sqrt{3}} 2\tan^{-1} x \, dx$$

$$= \int_0^1 2\tan^{-1} x \, dx + \int_1^{\sqrt{3}} 2\tan^{-1} x \, dx - \pi \int_1^{\sqrt{3}} 1 \,.\, dx$$

$$= \int_0^{\sqrt{3}} 2\tan^{-1} x \, dx - \pi \int_1^{\sqrt{3}} 1 \,.\, dx$$

$$= 2\int_0^{\sqrt{3}} \underset{\text{II}}{1} \,.\, \underset{\text{I}}{\tan^{-1} x} \, dx - \pi \Big[x\Big]_1^{\sqrt{3}}$$

Integrating the first integral by parts, we get

$$I = 2\Big[\tan^{-1} x\Big]_0^{\sqrt{3}} - 2\int_0^{\sqrt{3}} \frac{d}{dx}(\tan^{-1} x) \,.\, x \, dx - \pi[\sqrt{3} - 1]$$

$$= 2[\sqrt{3}\tan^{-1}\sqrt{3} - 0] - 2\int_0^{\sqrt{3}} \frac{x}{1+x^2} dx - \pi\left(\sqrt{3} - 1\right)$$

$$= 2\left[\sqrt{3}\,.\,\frac{\pi}{3}\right] - \int_0^{\sqrt{3}} \frac{2x}{1+x^2}\,dx - \pi\left(\sqrt{3}-1\right)$$

$$= \frac{2\pi}{\sqrt{3}} - \left[\log|1+x^2|\right]_0^{\sqrt{3}} - \pi\left(\sqrt{3}-1\right) \qquad \left[\because \int \frac{f'(x)}{f(x)}\,dx = \log|f(x)| + c\right]$$

$$= \frac{2\pi}{\sqrt{3}} - \left[\log\left|1+\left(\sqrt{3}\right)^2\right| - \log|1+0|\right] - \pi\left(\sqrt{3}-1\right)$$

$$= \frac{2\pi}{\sqrt{3}} - (\log 4 - \log 1) - \pi\left(\sqrt{3}-1\right) = \frac{2\pi}{\sqrt{3}} - \log 4 - \pi\left(\sqrt{3}-1\right)$$

$$= \pi\left(1 - \frac{1}{\sqrt{3}}\right) - \log 4.$$

Example 10. *Evaluate the following integrals :*

(*i*) $\int_0^{\pi} \frac{x \tan x}{\sec x \operatorname{cosec} x}\,dx$ (*ii*) $\int_0^{\pi} \sec^{2m} x \cos^{2n+1} x\,dx$

(*iii*) $\int_{-\pi/2}^{\pi/2} \log\left(\frac{2-\sin x}{2+\sin x}\right) dx$ (*iv*) $\int_0^{a} \frac{a}{\left(x+\sqrt{a^2-x^2}\right)^2}\,dx$

(*v*) $\int_{-a}^{a} x^3\sqrt{a^2-x^2}\,dx$ (*vi*) $\int_0^{1} \cot^{-1}(1-x+x^2)\,dx$

(*vii*) $\int_0^{\pi} \frac{x}{1+\sin^2 x}\,dx$ (*viii*) $\int_0^{\pi/2} \frac{x}{\sin x + \cos x}\,dx.$

Solution. (*i*) Let $I = \int_0^{\pi} \frac{x \tan x}{\sec x \operatorname{cosec} x}\,dx$ ...(1)

$$= \int_0^{\pi} \frac{(\pi-x)\tan(\pi-x)}{\sec(\pi-x)\operatorname{cosec}(\pi-x)}\,dx \qquad \left[\because \int_0^a f(x)\,dx = \int_0^a f(a-x)\,dx\right]$$

$$= \int_0^{\pi} \frac{(\pi-x)(-\tan x)}{(-\sec x)\operatorname{cosec} x}\,dx$$

$\therefore$ $I = \int_0^{\pi} \frac{(\pi-x)\tan x}{\sec x \operatorname{cosec} x}\,dx$ ...(2)

Adding equations (1) and (2), we get

$$2I = \int_0^{\pi} \frac{x\tan x}{\sec x \operatorname{cosec} x}\,dx + \int_0^{\pi} \frac{(\pi-x)\tan x}{\sec x \operatorname{cosec} x}\,dx = \int_0^{\pi}\left[\frac{x\tan x}{\sec x \operatorname{cosec} x} + \frac{(\pi-x)\tan x}{\sec x \operatorname{cosec} x}\right] dx$$

$$= \int_0^{\pi}\left[\frac{x\tan x + \pi\tan x - x\tan x}{\sec x \operatorname{cosec} x}\right] dx$$

$$= \int_0^{\pi} \frac{\pi\tan x}{\sec x \operatorname{cosec} x}\,dx = \int_0^{\pi} \frac{\pi\,\frac{\sin x}{\cos x}}{\frac{1}{\cos x}\cdot\frac{1}{\sin x}}\,dx = \pi\int_0^{\pi} \sin^2 x\,dx$$

$$= \pi\int_0^{\pi}\left(\frac{1-\cos 2x}{2}\right) dx \qquad \left[\begin{array}{l}\because\ 1-\cos 2A = 2\sin^2 A \\ \Rightarrow\ \frac{1-\cos 2A}{2} = \sin^2 A\end{array}\right]$$

$$= \frac{\pi}{2}\int_0^{\pi}(1-\cos 2x)\,dx = \frac{\pi}{2}\left[x - \frac{\sin 2x}{2}\right]_0^{\pi}$$

$$= \frac{\pi}{2}\left[\left(\pi - \frac{\sin 2\pi}{2}\right) - \left(0 - \frac{\sin 0^\circ}{2}\right)\right] = \frac{\pi}{2}[(\pi - 0) - (0 - 0)] = \frac{\pi^2}{2}$$

$\therefore$ $$I = \frac{\pi^2}{4}.$$

(*ii*) Let $$I = \int_0^{\pi} \sin^{2m} x \cos^{2n+1} x\, dx$$

$$= \int_0^{\pi} \sin^{2m}(\pi - x)\cos^{2n+1}(\pi - x)\,.\,dx \qquad \left[\because \int_0^a f(x)\,dx = \int_0^a f(a-x)\,dx\right]$$

$$= \int_0^{\pi} \sin^{2m} x\,(-\cos x)^{2n+1}\,.\,dx$$

$$= \int_0^{\pi} \sin^{2m} x\,(-\cos^{2n+1} x)\,dx \qquad [\because (-1)^{2n+1} = (-1)]$$

$$= -\int_0^{\pi} \sin^{2m} x \cos^{2n+1} x\,dx$$

$\Rightarrow$ $$I = -\int_0^{\pi} \sin^{2m} x \cos^{2n+1} x\,dx = -I$$

$\Rightarrow$ $$I + I = 0 \Rightarrow 2I = 0 \Rightarrow I = 0.$$

(*iii*) Let $$I = \int_{-\pi/2}^{\pi/2} \log\left(\frac{2-\sin x}{2+\sin x}\right) dx$$

Let $$f(x) = \log\left(\frac{2-\sin x}{2+\sin x}\right)$$

$\Rightarrow$ $$f(-x) = \log\left[\frac{2-\sin(-x)}{2+\sin(-x)}\right] = \log\left[\frac{2+\sin x}{2-\sin x}\right] \qquad [\because \sin(-\theta) = -\sin\theta]$$

$$= \log\left(\frac{2-\sin x}{2+\sin x}\right)^{-1} = -\log\left(\frac{2-\sin x}{2+\sin x}\right) \qquad [\because m\log n = \log n^m]$$

$$= -f(x).$$

$\therefore$ $f(x)$ is an odd function.

Hence $$\int_{-\pi/2}^{\pi/2} \log\left(\frac{2-\sin x}{2+\sin x}\right) dx = 0. \qquad \left[\because \int_{-a}^{a} f(x)\,dx = 0; \text{ if } f(x) \text{ is an odd function}\right]$$

(*iv*) Let $$I = \int_0^a \frac{a}{\left(x + \sqrt{a^2 - x^2}\right)^2}\,dx$$

Put $x = a\sin\theta \Rightarrow dx = a\cos\theta\,d\theta$

When $x = 0,\ a\sin\theta = 0 \Rightarrow \sin\theta = 0 \Rightarrow \theta = 0$

and when $x = a,\ a\sin\theta = a \Rightarrow \sin\theta = 1 \Rightarrow \theta = \dfrac{\pi}{2}$

$$\therefore \qquad I = \int_0^{\pi/2} \frac{a}{\left[a \sin\theta + \sqrt{a^2 - a^2 \sin^2\theta}\right]^2}\, a \cos\theta\, d\theta$$

$$= \int_0^{\pi/2} \frac{a^2 \cos\theta}{\left[a \sin\theta + a\sqrt{1 - \sin^2\theta}\right]^2}\, d\theta = \int_0^{\pi/2} \frac{a^2 \cos\theta}{\left[a \sin\theta + a\sqrt{\cos^2\theta}\right]^2}\, d\theta$$

$$= \int_0^{\pi/2} \frac{a^2 \cos\theta}{a^2 [\sin\theta + \cos\theta]^2}\, d\theta = \int_0^{\pi/2} \frac{\cos\theta}{(\sin\theta + \cos\theta)^2}\, d\theta \qquad ...(1)$$

$$= \int_0^{\pi/2} \frac{\cos\left(\frac{\pi}{2} - \theta\right)}{\left[\sin\left(\frac{\pi}{2} - \theta\right) + \cos\left(\frac{\pi}{2} - \theta\right)\right]^2}\, d\theta \qquad \left[\because \int_0^a f(x)\, dx = \int_0^a f(a - x)\, dx\right]$$

$$\Rightarrow \qquad I = \int_0^{\pi/2} \frac{\sin\theta}{(\cos\theta + \sin\theta)^2}\, d\theta \qquad ...(2)$$

Adding equations (1) and (2), we get

$$2I = \int_0^{\pi/2} \frac{\cos\theta}{(\sin\theta + \cos\theta)^2}\, d\theta + \int_0^{\pi/2} \frac{\sin\theta}{(\cos\theta + \sin\theta)^2}\, d\theta$$

$$= \int_0^{\pi/2} \left[\frac{\cos\theta}{(\sin\theta + \cos\theta)^2} + \frac{\sin\theta}{(\cos\theta + \sin\theta)^2}\right] d\theta$$

$$= \int_0^{\pi/2} \frac{\cos\theta + \sin\theta}{(\sin\theta + \cos\theta)^2} \,.\, d\theta = \int_0^{\pi/2} \frac{1}{\sin\theta + \cos\theta}\, d\theta$$

$$= \frac{1}{\sqrt{2}} \int_0^{\pi/2} \left[\frac{1}{\frac{1}{\sqrt{2}} \sin\theta + \frac{1}{\sqrt{2}} \cos\theta}\right] d\theta$$

[Multiply and dividing the denominator by $\sqrt{2}$]

$$= \frac{1}{\sqrt{2}} \int_0^{\pi/2} \frac{1}{\left(\sin\theta \sin\frac{\pi}{4} + \cos\theta \cos\frac{\pi}{4}\right)}\, d\theta$$

$$= \frac{1}{\sqrt{2}} \int_0^{\pi/2} \frac{1}{\cos\left(\theta - \frac{\pi}{4}\right)}\, d\theta$$

[$\because$ cos (A – B) = sin A sin B + cos A cos B]

$$= \frac{1}{\sqrt{2}} \int_0^{\pi/2} \sec\left(\theta - \frac{\pi}{4}\right) d\theta$$

$$= \frac{1}{\sqrt{2}} \left[\log\left|\sec\left(\theta - \frac{\pi}{4}\right) + \tan\left(\theta - \frac{\pi}{4}\right)\right|\right]_0^{\pi/2}$$

$$\left[\because \int \sec x\, dx = \log|\sec x + \tan x| + c\right]$$

$$= \frac{1}{\sqrt{2}}\left[\log\left|\sec\left(\frac{\pi}{2}-\frac{\pi}{4}\right)+\tan\left(\frac{\pi}{2}-\frac{\pi}{4}\right)\right| - \log\left|\sec\left(0-\frac{\pi}{4}\right)+\tan\left(0-\frac{\pi}{4}\right)\right|\right]$$

$$= \frac{1}{\sqrt{2}}\left[\log\left|\sec\frac{\pi}{4}+\tan\frac{\pi}{4}\right| - \log\left|\sec\left(-\frac{\pi}{4}\right)+\tan\left(-\frac{\pi}{4}\right)\right|\right]$$

$$= \frac{1}{\sqrt{2}}\left[\log\left|\sqrt{2}+1\right| - \log\left|\sec\frac{\pi}{4}-\tan\frac{\pi}{4}\right|\right]$$

$$= \frac{1}{\sqrt{2}}[\log(\sqrt{2}+1) - \log(\sqrt{2}-1)]$$

$$= \frac{1}{\sqrt{2}}\log\left(\frac{\sqrt{2}+1}{\sqrt{2}-1}\right) \qquad \left[\because\ \log m - \log n = \log\frac{m}{n}\right]$$

$$= \frac{1}{\sqrt{2}}\log\left(\frac{\sqrt{2}+1}{\sqrt{2}-1}\times\frac{\sqrt{2}+1}{\sqrt{2}+1}\right) \qquad \text{[Rationalization]}$$

$$= \frac{1}{\sqrt{2}}\log\left(\frac{\sqrt{2}+1}{2-1}\right)^2 = \frac{1}{\sqrt{2}}\log\left(\sqrt{2}+1\right)^2$$

$$= \frac{2}{\sqrt{2}}\log\left(\sqrt{2}+1\right) \qquad [\because\ m\log n = \log n^m]$$

$$\therefore \qquad I = \frac{1}{\sqrt{2}}\log\left(\sqrt{2}+1\right).$$

(*v*) Let $\quad I = \int_{-a}^{a} x^3\sqrt{a^2-x^2}\, dx$

Let $\quad f(x) = x^3\sqrt{a^2-x^2}$

$\Rightarrow \quad f(-x) = (-x)^3\sqrt{a^2-(-x)^2} = -x^3\sqrt{a^2-x^2} = -f(x)$

$\therefore\ f(x)$ is an odd function.

Hence $\int_{-a}^{a} x^3\sqrt{a^2-x^2}\, dx = 0. \qquad \left[\because\ \int_{-a}^{a} f(x)\,dx = 0;\text{ if } f(x) \text{ is an odd function}\right]$

(*vi*) Let $\quad I = \int_0^1 \cot^{-1}(1-x+x^2)\, dx$

$$\therefore \qquad I = \int_0^1 \tan^{-1}\left(\frac{1}{1-x+x^2}\right)dx \qquad \left[\because\ \cot^{-1}x = \tan^{-1}\frac{1}{x}\right]$$

$$= \int_0^1 \tan^{-1}\left(\frac{x+(1-x)}{1-x+x^2}\right)dx \qquad \text{[Note this step]}$$

$$= \int_0^1 \tan^{-1}\left[\frac{x+(1-x)}{1-x(1-x)}\right]dx$$

$$= \int_0^1 [\tan^{-1}x + \tan^{-1}(1-x)]\, dx \qquad \left[\because\ \tan^{-1}\left(\frac{a+b}{1-ab}\right) = \tan^{-1}a + \tan^{-1}b\right]$$

$$= \int_0^1 \tan^{-1} x\, dx + \int_0^1 \tan^{-1} (1 - x)\, dx$$

$$= \int_0^1 \tan^{-1} x\, dx + \int_0^1 \tan^{-1} [1 - (1 - x)]\, dx \qquad \left[\because \int_0^a f(x)\, dx = \int_0^a f(a - x)\, dx\right]$$

$$= \int_0^1 \tan^{-1} x\, dx + \int_0^1 \tan^{-1} x\, dx$$

$$I = 2 \int_0^1 \underset{\text{II}}{1} . \underset{\text{I}}{\tan^{-1} x} . dx \qquad \text{[Taking unity as the second function]}$$

Integrating by parts, we get

$$I = 2 \Big[\tan^{-1} x . x\Big]_0^1 - 2 \int_0^1 \frac{d}{dx} (\tan^{-1} x) . x\, dx$$

$$= 2 \Big[x \tan^{-1} x\Big]_0^1 - \int_0^1 \frac{2x}{1 + x^2}\, dx$$

$$= 2\, [1 \tan^{-1} 1 - 0] - \Big[\log |1 + x^2|\Big]_0^1 \qquad \left[\because \int \frac{f'(x)}{f(x)}\, dx = \log |f(x)| + c\right]$$

$$= 2 \left[\frac{\pi}{4} - 0\right] - [\log |1 + 1| - \log |1 + 0|] = \frac{\pi}{2} - \log 2 + \log 1$$

$$= \frac{\pi}{2} - \log 2. \qquad [\because \log 1 = 0]$$

(*vii*) Let $\quad I = \int_0^\pi \frac{x}{1 + \sin^2 x}\, dx \qquad ...(1)$

$$= \int_0^\pi \frac{(\pi - x)}{1 + \sin^2 (\pi - x)}\, dx \qquad \left[\because \int_0^a f(x)\, dx = \int_0^a f(a - x)\, dx\right]$$

$\therefore \quad I = \int_0^\pi \frac{(\pi - x)}{1 + \sin^2 x}\, dx \qquad ...(2)$

Adding equations (1) and (2), we get

$$2I = \int_0^\pi \frac{x}{1 + \sin^2 x}\, dx + \int_0^\pi \frac{\pi - x}{1 + \sin^2 x}\, dx$$

$$= \int_0^\pi \left[\frac{x}{1 + \sin^2 x} + \frac{\pi - x}{1 + \sin^2 x}\right] dx$$

$$= \int_0^\pi \frac{x + \pi - x}{1 + \sin^2 x} . dx = \pi \int_0^\pi \frac{1}{1 + \sin^2 x}\, dx$$

$$= \pi \int_0^\pi \frac{\frac{1}{\cos^2 x}}{\frac{1}{\cos^2 x} + \frac{\sin^2 x}{\cos^2 x}}\, dx \qquad \left[\begin{array}{l}\text{Dividing the numerator and} \\ \text{the denominator by } \cos^2 x\end{array}\right]$$

$$= \pi \int_0^{\pi} \frac{\sec^2 x}{\sec^2 x + \tan^2 x} \, dx$$

$$= \pi \int_0^{\pi} \frac{\sec^2 x}{(1+\tan^2 x) + \tan^2 x} \, dx \qquad [\because \ \sec^2 A - \tan^2 A = 1]$$

$$= \pi \int_0^{\pi} \frac{\sec^2 x}{1 + 2\tan^2 x} \, dx = 2\pi \int_0^{\pi/2} \frac{\sec^2 x}{1+2\tan^2 x} \, dx$$

Put $\tan x = z \ \Rightarrow \ \sec^2 x \, dx = dz$

When $x = 0$, $z = \tan 0° = 0$ and when $x = \frac{\pi}{2}$, $z = \tan \frac{\pi}{2} = \infty$

$$\therefore \qquad 2I = 2\pi \int_0^{\infty} \frac{1}{1+2z^2} \, dz = 2\pi \cdot \frac{1}{2} \int_0^{\infty} \frac{1}{\frac{1}{2} + z^2} \, dz$$

$$= \pi \int_0^{\infty} \frac{1}{\left(\frac{1}{\sqrt{2}}\right)^2 + z^2} \, dz$$

$$= \pi \left[\frac{1}{\frac{1}{\sqrt{2}}} \cdot \tan^{-1} \frac{z}{\frac{1}{\sqrt{2}}} \right]_0^{\infty} \qquad \left[\because \ \int \frac{1}{a^2 + x^2} \, dx = \frac{1}{a} \tan^{-1} \frac{x}{a} + c\right]$$

$$= \sqrt{2}\pi \left[\tan^{-1} \sqrt{2} z \right]_0^{\infty} = \sqrt{2}\pi \, [\tan^{-1} \infty - \tan^{-1} 0]$$

$$= \sqrt{2}\pi \left[\frac{\pi}{2} - 0\right] = \frac{\pi^2}{\sqrt{2}}$$

$$\therefore \qquad I = \frac{\pi^2}{2\sqrt{2}}.$$

(*viii*) Let $I = \int_0^{\pi/2} \frac{x}{\sin x + \cos x} \, dx$ ...(1)

$$= \int_0^{\pi/2} \frac{\left(\frac{\pi}{2} - x\right)}{\sin\left(\frac{\pi}{2} - x\right) + \cos\left(\frac{\pi}{2} - x\right)} \cdot dx \qquad \left[\because \ \int_0^a f(x) \, dx = \int_0^a f(a-x) \, dx\right]$$

$$\therefore \qquad I = \int_0^{\pi/2} \frac{\left(\frac{\pi}{2} - x\right)}{\cos x + \sin x} \, dx \qquad ...(2)$$

Adding equations (1) and (2), we get

$$2I = \int_0^{\pi/2} \frac{x}{\sin x + \cos x}\, dx + \int_0^{\pi/2} \frac{\left(\frac{\pi}{2} - x\right)}{\cos x + \sin x} \,.\, dx$$

$$= \int_0^{\pi/2} \left[\frac{x}{\sin x + \cos x} + \frac{\left(\frac{\pi}{2} - x\right)}{\cos x + \sin x} \right] dx = \int_0^{\pi/2} \frac{x + \frac{\pi}{2} - x}{\sin x + \cos x}\, dx$$

$$\Rightarrow \quad 2I = \frac{\pi}{2} \int_0^{\pi/2} \frac{1}{\sin x + \cos x}\, dx$$

Proceed as in part (*iv*).

[**Hint.** Multiply and dividing the denominator by $\sqrt{2}$] $\left[\textbf{Ans. } \frac{\pi}{2\sqrt{2}} \log\left(\sqrt{2} + 1\right).\right]$

Example 11. *Evaluate the following integrals :*

(*i*) $\int_0^{\pi} \frac{x}{1 + e \sin x}\, dx$ (*ii*) $\int_0^{1} \tan^{-1}\left(\frac{3x - x^3}{1 - 3x^2}\right) dx$

(*iii*) $\int_0^{\pi/2} \frac{\sin^2 x}{1 + \sin x \cos x}\, dx$ (*iv*) $\int_0^{\pi} \frac{x}{1 + \cos \alpha \sin x}\, dx$

(*v*) $\int_0^{\pi/2} \frac{1}{(a^2 \cos^2 x + b^2 \sin^2 x)^2}\, dx$ (*vi*) $\int_0^{\pi} x \cos^2 x\, dx.$

Solution. (*i*) Let $I = \int_0^{\pi} \frac{x}{1 + e \sin x}\, dx$ ...(1)

$$= \int_0^{\pi} \frac{(\pi - x)}{1 + e \sin (\pi - x)}\, dx \qquad \left[\because \int_0^a f(x)\, dx = \int_0^a f(a - x)\, dx\right]$$

$$\therefore \quad I = \int_0^{\pi} \frac{\pi - x}{1 + e \sin x}\, dx \qquad ...(2)$$

Adding equations (1) and (2), we get

$$2I = \int_0^{\pi} \frac{x}{1 + e \sin x}\, dx + \int_0^{\pi} \frac{\pi - x}{1 + e \sin x}\, dx = \int_0^{\pi} \left[\frac{x}{1 + e \sin x} + \frac{\pi - x}{1 + e \sin x}\right] dx$$

$$= \int_0^{\pi} \frac{x + \pi - x}{1 + e \sin x}\, dx = \pi \int_0^{\pi} \frac{1}{1 + e \sin x}\, dx \qquad [\because f(\pi - x) = f(x)]$$

$$\therefore \quad 2I = 2\pi \int_0^{\pi/2} \frac{1}{1 + e \sin x}\, dx \qquad \left[\because \int_0^{2a} f(x)\, dx = 2\int_0^a f(x)\, dx \,;\text{ if } f(2a - x) = f(x)\right]$$

$$\Rightarrow \quad I = \pi \int_0^{\pi/2} \frac{1}{1 + e \sin x}\, dx \qquad [\text{Here, } 2a = \pi]$$

$$= \pi \int_0^{\pi/2} \frac{1}{1 + e \sin\left(\frac{\pi}{2} - x\right)}\, dx \qquad \left[\because \int_0^a f(x)\, dx = \int_0^a f(a - x)\, dx\right]$$

$$= \pi \int_0^{\pi/2} \frac{1}{1 + e\cos x}\, dx$$

$$\therefore \qquad I = \pi \left[\frac{1}{\sqrt{1-e^2}} \cos^{-1}\left(\frac{e + 1 . \cos x}{1 + e\cos x}\right) \right]_0^{\pi/2}$$

$$\left[\because \text{ if } a^2 > b^2, \int \frac{1}{a + b\cos x}\, dx = \frac{1}{\sqrt{a^2 - b^2}} \cos^{-1}\left(\frac{b + a\cos x}{a + b\cos x}\right)\right]$$

$$= \frac{\pi}{\sqrt{1-e^2}} \left[\cos^{-1}\left(\frac{e + \cos\frac{\pi}{2}}{1 + e\cos\frac{\pi}{2}}\right) - \cos^{-1}\left(\frac{e + \cos 0^\circ}{1 + e\cos 0^\circ}\right) \right]$$

$$= \frac{\pi}{\sqrt{1-e^2}} \left[\cos^{-1}\left(\frac{e+0}{1+0}\right) - \cos^{-1}\left(\frac{e+1}{1+e}\right) \right]$$

$$= \frac{\pi}{\sqrt{1-e^2}} [\cos^{-1} e - \cos^{-1} 1] = \frac{\pi}{\sqrt{1-e^2}} [\cos^{-1} e - 0]$$

$$= \frac{\pi \cos^{-1} e}{\sqrt{1-e^2}}.$$

(ii) Let $\quad I = \int_0^1 \tan^{-1}\left(\frac{3x - x^3}{1 - 3x^2}\right) dx$

Since, we have

$$\tan^{-1}\left(\frac{3x - x^3}{1 - 3x^2}\right) = \begin{cases} 3\tan^{-1} x, \text{ if } -\frac{1}{\sqrt{3}} < x < \frac{1}{\sqrt{3}} \\ -\pi + 3\tan^{-1} x, \text{ if } \quad x > \frac{1}{\sqrt{3}} \\ \pi + 3\tan^{-1} x, \text{ if } \quad x < -\frac{1}{\sqrt{3}} \end{cases}$$

$$\therefore \quad \int_0^1 \tan^{-1}\left(\frac{3x - x^3}{1 - 3x^2}\right) dx = \int_0^{1/\sqrt{3}} \tan^{-1}\left(\frac{3x - x^3}{1 - 3x^2}\right) dx + \int_{1/\sqrt{3}}^1 \tan^{-1}\left(\frac{3x - x^3}{1 - 3x^2}\right) dx$$

$$= \int_0^{1/\sqrt{3}} 3\tan^{-1} x\, dx + \int_{1/\sqrt{3}}^1 (-\pi + 3\tan^{-1} x)\, dx$$

$$= \int_0^{1/\sqrt{3}} 3\tan^{-1} x\, dx - \pi \int_{1/\sqrt{3}}^1 1 . dx + \int_{1/\sqrt{3}}^1 3\tan^{-1} x\, dx$$

$$= \int_0^1 3\tan^{-1} x - \pi \int_{1/\sqrt{3}}^1 1 . dx \qquad \left[\because \int_a^b f(x)\, dx = \int_a^c f(x)\, dx + \int_c^b f(x)\, dx, \text{ where } a < c < b.\right]$$

$$= 3\int_0^1 \underset{\text{II}}{1} . \underset{\text{I}}{\tan^{-1} x}\, dx - \pi \int_{1/\sqrt{3}}^1 1 . dx \qquad \text{[Taking unity as second function]}$$

Integrating the first integral by parts, we have

$$I = 3\left[\tan^{-1} x \,.\, x\right]_0^1 - 3\int_0^1 \frac{d}{dx}(\tan^{-1} x)\,.\, x\,dx - \pi\left[x\right]_{1/\sqrt{3}}^1$$

$$= 3\left[x\tan^{-1} x\right]_0^1 - 3\int_0^1 \frac{x}{1+x^2}\,dx - \pi\left[1-\frac{1}{\sqrt{3}}\right]$$

$$= 3\,[1\tan^{-1} 1 - 0] - \frac{3}{2}\int_0^1 \frac{2x}{1+x^2}\,dx - \pi\left[1-\frac{1}{\sqrt{3}}\right] \qquad \text{[Multiply and divided by 2]}$$

$$= 3\left[\frac{\pi}{4}\right] - \frac{3}{2}\left[\log|1+x^2|\right]_0^1 - \pi\left[1-\frac{1}{\sqrt{3}}\right] \qquad \left[\because \int \frac{f'(x)}{f(x)}\,dx = \log|f(x)| + c\right]$$

$$= \frac{3\pi}{4} - \pi + \frac{\pi}{\sqrt{3}} - \frac{3}{2}\,[\log|1+1| - \log|1+0|]$$

$$= \frac{\pi}{\sqrt{3}} - \frac{\pi}{4} - \frac{3}{2}\,[\log 2 - \log 1]$$

$$= \pi\left(\frac{1}{\sqrt{3}} - \frac{1}{4}\right) - \frac{3}{2}\log 2.$$

(*iii*) Let $\quad I = \int_0^{\pi/2} \frac{\sin^2 x}{1+\sin x\cos x}\,dx \qquad ...(1)$

$$= \int_0^{\pi/2} \frac{\sin^2\left(\frac{\pi}{2}-x\right)}{1+\sin\left(\frac{\pi}{2}-x\right)\cos\left(\frac{\pi}{2}-x\right)}\,dx \qquad \left[\because \int_0^a f(x)\,dx = \int_0^a f(a-x)\,dx\right]$$

$\therefore \quad I = \int_0^{\pi/2} \frac{\cos^2 x}{1+\cos x\sin x}\,dx \qquad ...(2)$

Adding equations (1) and (2), we get

$$2I = \int_0^{\pi/2} \frac{\sin^2 x}{1+\sin x\cos x}\,dx + \int_0^{\pi/2} \frac{\cos^2 x}{1+\cos x\sin x}\,dx$$

$$= \int_0^{\pi/2}\left[\frac{\sin^2 x}{1+\sin x\cos x} + \frac{\cos^2 x}{1+\cos x\sin x}\right]dx = \int_0^{\pi/2}\left[\frac{\sin^2 x+\cos^2 x}{1+\sin x\cos x}\right]dx$$

$$= \int_0^{\pi/2} \frac{1}{1+\sin x\cos x}\,dx \qquad [\because \sin^2 A + \cos^2 A = 1]$$

$$= \int_0^{\pi/2} \frac{\frac{1}{\cos^2 x}}{\frac{1}{\cos^2 x} + \frac{\sin x\cos x}{\cos^2 x}}\,dx \qquad \left[\begin{array}{l}\text{Dividing the numerator and}\\ \text{the denominator by } \cos^2 x\end{array}\right]$$

$$= \int_0^{\pi/2} \frac{\sec^2 x}{\sec^2 x + \tan x}\,dx$$

$$= \int_0^{\pi/2} \frac{\sec^2 x}{1+\tan^2 x + \tan x}\, dx \qquad [\because \sec^2 A - \tan^2 A = 1]$$

Put $\tan x = z \Rightarrow \sec^2 x\, dx = dz$

When $x = 0$, $z = \tan 0° = 0$ and when $x = \frac{\pi}{2}$, $z = \tan\frac{\pi}{2} = \infty$

$$\therefore \quad 2I = \int_0^{\infty} \frac{1}{z^2 + z + 1}\, dz$$

$$= \int_0^{\infty} \frac{1}{\left(z^2 + z + \frac{1}{4}\right) + \left(1 - \frac{1}{4}\right)}\, dz \qquad \left[\begin{array}{l}\text{Add and subtract } \frac{1}{4} \text{ to the denom.}\\ \because \left(\frac{1}{2}\text{ co-eff. of } z\right)^2 = \frac{1}{4}\end{array}\right]$$

$$= \int_0^{\infty} \frac{1}{\left(z+\frac{1}{2}\right)^2 + \frac{3}{4}}\, dz = \int_0^{\infty} \frac{1}{\left(z+\frac{1}{2}\right)^2 + \left(\frac{\sqrt{3}}{2}\right)^2}\, dz$$

$$= \left[\frac{1}{\frac{\sqrt{3}}{2}} \tan^{-1} \frac{\left(z+\frac{1}{2}\right)}{\frac{\sqrt{3}}{2}}\right]_0^{\infty} \qquad \left[\because \int \frac{1}{a^2+x^2}\, dx = \frac{1}{a}\tan^{-1}\frac{x}{a} + c\right]$$

$$= \frac{2}{\sqrt{3}}\left[\tan^{-1}\left(\frac{2z+1}{\sqrt{3}}\right)\right]_0^{\infty} = \frac{2}{\sqrt{3}}\left[\tan^{-1}\infty - \tan^{-1}\left(\frac{0+1}{\sqrt{3}}\right)\right]$$

$$= \frac{2}{\sqrt{3}}\left[\frac{\pi}{2} - \frac{\pi}{6}\right] = \frac{2}{\sqrt{3}}\left[\frac{3\pi - \pi}{6}\right] = \frac{2}{\sqrt{3}}\left[\frac{2\pi}{6}\right] = \frac{2\pi}{3\sqrt{3}}$$

$$\therefore \quad I = \frac{\pi}{3\sqrt{3}}.$$

(*iv*) Let $$I = \int_0^{\pi} \frac{x}{1+\cos\alpha \sin x}\, dx \qquad ...(1)$$

$$\Rightarrow \quad I = \int_0^{\pi} \frac{(\pi - x)}{1+\cos\alpha \sin(\pi - x)}\, dx \qquad \left[\because \int_0^a f(x)\, dx = \int_0^a f(a-x)\, dx\right]$$

$$\Rightarrow \quad I = \int_0^{\pi} \frac{\pi - x}{1+\cos\alpha \sin x}\, dx \qquad ...(2)$$

Adding equations (1) and (2), we get

$$2I = \int_0^{\pi} \frac{x}{1+\cos\alpha \sin x}\, dx + \int_0^{\pi} \frac{\pi - x}{1+\cos\alpha \sin x}\, dx$$

$$= \int_0^{\pi} \left[\frac{x}{1+\cos\alpha \sin x} + \frac{\pi - x}{1+\cos\alpha \sin x}\right] dx$$

$$= \int_0^{\pi} \left[\frac{x + \pi - x}{1+\cos\alpha \sin x}\right] dx = \int_0^{\pi} \frac{\pi}{1+\cos\alpha \sin x}\, dx$$

$$= \pi \int_0^{\pi} \frac{1}{1 + \cos\alpha\left(2\sin\frac{x}{2}\cos\frac{x}{2}\right)}\, dx \qquad \left[\begin{array}{l} \because \ \sin 2A = 2\sin A\cos A \\ \Rightarrow \ \sin A = 2\sin\frac{A}{2}\cos\frac{A}{2} \end{array}\right]$$

$$= \pi \int_0^{\pi} \frac{\dfrac{1}{\cos^2\frac{x}{2}}}{\dfrac{1}{\cos^2\frac{x}{2}} + \dfrac{\cos\alpha\left(2\sin\frac{x}{2}\cos\frac{x}{2}\right)}{\cos^2\frac{x}{2}}}\, dx \qquad \left[\begin{array}{l}\text{Dividing the numerator and} \\ \text{the denominator by } \cos^2\frac{x}{2}\end{array}\right]$$

$$= \pi \int_0^{\pi} \frac{\sec^2\frac{x}{2}}{\sec^2\frac{x}{2} + 2\cos\alpha\tan\frac{x}{2}}\, dx$$

$$= \pi \int_0^{\pi} \frac{\sec^2\frac{x}{2}}{1 + \tan^2\frac{x}{2} + 2\cos\alpha \,.\, \tan\frac{x}{2}}\, dx \qquad [\because \ \sec^2 A - \tan^2 A = 1]$$

Put $\tan\frac{x}{2} = z \ \Rightarrow \ \frac{1}{2}\sec^2\frac{x}{2}\, dx = dz \ \Rightarrow \ \sec^2\frac{x}{2}\, dx = 2dz$

When $x = 0$, $z = \tan 0° = 0$ and when $x = \pi$, $z = \tan\frac{\pi}{2} = \infty$

$$\therefore \qquad 2I = \pi \int_0^{\infty} \frac{1}{1 + z^2 + 2z\cos\alpha} \,.\, (2dz) = 2\pi \int_0^{\infty} \frac{1}{z^2 + 2z\cos\alpha + 1}\, dz$$

$$\Rightarrow \qquad I = \pi \int_0^{\infty} \frac{1}{(z^2 + 2z\cos\alpha + \cos^2\alpha) + (1 - \cos^2\alpha)}\, dz$$

$$\left[\begin{array}{l}\text{Add and subtract } \cos^2\alpha \text{ to the denom.} \\ \because \ \left(\frac{1}{2}\text{ co-eff. of } z\right)^2 = \cos^2\alpha\end{array}\right]$$

$$= \pi \int_0^{\infty} \frac{1}{(z + \cos\alpha)^2 + \sin^2\alpha}\, dz \qquad [\because \ \sin^2 A + \cos^2 A = 1]$$

$$= \pi \left[\frac{1}{\sin\alpha} \,.\, \tan^{-1}\left(\frac{z + \cos\alpha}{\sin\alpha}\right)\right]_0^{\infty} \qquad \left[\because \int \frac{1}{a^2 + x^2}\, dx = \frac{1}{a}\tan^{-1}\frac{x}{a} + c\right]$$

$$= \frac{\pi}{\sin\alpha}\left[\tan^{-1}\infty - \tan^{-1}\left(\frac{0 + \cos\alpha}{\sin\alpha}\right)\right] = \frac{\pi}{\sin\alpha}\left[\frac{\pi}{2} - \tan^{-1}(\cot\alpha)\right]$$

$$= \frac{\pi}{\sin\alpha}\left[\frac{\pi}{2} - \tan^{-1}\left\{\tan\left(\frac{\pi}{2} - \alpha\right)\right\}\right] = \frac{\pi}{\sin\alpha}\left[\frac{\pi}{2} - \left(\frac{\pi}{2} - \alpha\right)\right] = \frac{\pi\alpha}{\sin\alpha}$$

$= \pi\alpha \operatorname{cosec}\alpha.$

(*v*) Let $\quad I = \int_0^{\pi/2} \frac{1}{(a^2 \cos^2 x + b^2 \sin^2 x)^2}\, dx$

$$= \int_0^{\pi/2} \frac{\sec^4 x}{(a^2 + b^2 \tan^2 x)^2}\, dx \qquad \left[\begin{array}{l}\text{Dividing the numerator and} \\ \text{the denominator by } \cos^4 x\end{array}\right]$$

$$= \int_0^{\pi/2} \frac{\sec^2 x \cdot \sec^2 x}{(a^2 + b^2 \tan^2 x)^2}\, dx$$

Put $\quad b \tan x = a \tan \theta \qquad \Rightarrow \quad b \sec^2 x\, dx = a \sec^2 \theta\, d\theta$

$\Rightarrow \quad \sec^2 x\, dx = \frac{a}{b} \sec^2 \theta\, d\theta \quad \Rightarrow \quad \sec^2 x\, dx = \frac{a}{b}(1 + \tan^2 \theta)\, d\theta$

When $\quad x = 0, \quad a \tan \theta = b \tan 0° \quad \Rightarrow \quad a \tan \theta = 0 \quad \Rightarrow \quad \theta = 0$

and when $x = \frac{\pi}{2}$, $a \tan \theta = b \tan \frac{\pi}{2} \quad \Rightarrow \quad a \tan \theta = \infty \Rightarrow \quad \tan \theta = \infty \quad \Rightarrow \quad \theta = \tan^{-1}(\infty) = \frac{\pi}{2}$.

$$\therefore \qquad I = \int_0^{\pi/2} \frac{(1 + \tan^2 x) \cdot \frac{a}{b}(1 + \tan^2 \theta)}{(a^2 + a^2 \tan^2 \theta)^2} \cdot d\theta = \int_0^{\pi/2} \frac{\left(1 + \frac{a^2}{b^2} \tan^2 \theta\right) \cdot \frac{a}{b}(1 + \tan^2 \theta)}{a^4 (1 + \tan^2 \theta)^2}\, d\theta$$

$$= \int_0^{\pi/2} \frac{(b^2 + a^2 \tan^2 \theta)\, a \sec^2 \theta}{b^3 \cdot a^4 (\sec^2 \theta)^2}\, d\theta = \int_0^{\pi/2} \frac{1}{a^3 b^3} (b^2 + a^2 \tan^2 \theta) \cdot \cos^2 \theta\, d\theta$$

$$= \frac{1}{a^3 b^3} \int_0^{\pi/2} \left(b^2 + \frac{a^2 \sin^2 \theta}{\cos^2 \theta}\right) \cos^2 \theta\, d\theta$$

$$= \frac{1}{a^3 b^3} \int_0^{\pi/2} (b^2 \cos^2 \theta + a^2 \sin^2 \theta)\, d\theta$$

$$= \frac{1}{a^3 b^3} \int_0^{\pi/2} \left[b^2 \left(\frac{1 + \cos 2\theta}{2}\right) + a^2 \left(\frac{1 - \cos 2\theta}{2}\right)\right] d\theta \qquad \left[\begin{array}{l}\because \; 1 + \cos 2A = 2 \cos^2 A \\ 1 - \cos 2A = 2 \sin^2 A\end{array}\right]$$

$$= \frac{1}{2a^3 b^3} \int_0^{\pi/2} [(b^2 + a^2) + (b^2 - a^2) \cos 2\theta]\, d\theta$$

$$= \frac{1}{2a^3 b^3} \left[(b^2 + a^2) \cdot \theta + (b^2 - a^2) \cdot \frac{\sin 2\theta}{2}\right]_0^{\pi/2}$$

$$= \frac{1}{2a^3 b^3} \left[\left((b^2 + a^2) \frac{\pi}{2} + (b^2 - a^2) \frac{\sin \pi}{2}\right) - 0 + \frac{(b^2 - a^2) \sin 0°}{2}\right]$$

$$= \frac{1}{2a^3 b^3} \left[(b^2 + a^2) \frac{\pi}{2} + 0 - 0 + 0\right]$$

$$= \frac{\pi (a^2 + b^2)}{4a^3 b^3}.$$

(*vi*) Let $I = \int_0^{\pi} x \cos^2 x\, dx \qquad$ ...(1)

$$= \int_0^{\pi} (\pi - x) \cos^2 (\pi - x)\, dx \qquad \left[\because \; \int_0^a f(x)\, dx = \int_0^a f(a - x)\, dx\right]$$

$$= \int_0^{\pi} (\pi - x)(-\cos x)^2\, dx$$

$$\therefore \quad I = \int_0^{\pi} (\pi - x) \cos^2 x \, dx \quad ...(2)$$

Adding equations (1) and (2), we get

$$2I = \int_0^{\pi} x \cos^2 x \, dx + \int_0^{\pi} (\pi - x) \cos^2 x \, dx$$

$$= \int_0^{\pi} [x \cos^2 x + (\pi - x) \cos^2 x] \, dx$$

$$= \int_0^{\pi} \pi \cos^2 x \, dx$$

Let $\quad f(x) = \cos^2 x$

$\Rightarrow \quad f(\pi - x) = \cos^2 (\pi - x) = (-\cos x)^2 = \cos^2 x = f(x).$

$$\left[\because \int_0^{2a} f(x) \, dx = 2\int_0^{a} f(x) \, dx, \text{ if } f(2a - x) = f(x)\right]$$

$$\therefore \quad 2I = 2\pi \int_0^{\pi/2} \cos^2 x \, dx \quad [\text{Here } 2a = \pi]$$

$$\Rightarrow \quad I = \pi \int_0^{\pi/2} \cos^2 x \, dx$$

$$= \pi \int_0^{\pi/2} \frac{1 + \cos 2x}{2} \, . \, dx \quad \left[\begin{array}{l} \because \ 1 + \cos 2A = 2\cos^2 A \\ \Rightarrow \ \dfrac{1 + \cos 2A}{2} = \cos^2 A \end{array}\right]$$

$$= \frac{\pi}{2} \int_0^{\pi/2} (1 + \cos 2x) \, dx$$

$$= \frac{\pi}{2}\left[x + \frac{\sin 2x}{2}\right]_0^{\pi/2} = \frac{\pi}{2}\left[\left(\frac{\pi}{2} + \frac{\sin 2\pi}{2}\right) - \left(0 + \frac{\sin 0°}{2}\right)\right] = \frac{\pi}{2}\left[\left(\frac{\pi}{2} + 0\right) - (0 + 0)\right]$$

$$= \frac{\pi^2}{4} .$$

Example 12. *Evaluate the following integrals :*

(*i*) $\int_0^{\pi} x \log (\sin x) \, dx$ (*ii*) $\int_0^{\pi/2} x \cot x \, dx$

(*iii*) $\int_0^{\pi} \frac{x}{a^2 \cos^2 x + b^2 \sin^2 x} \, dx$ (*iv*) $\int_0^{\pi/2} \frac{x \sin x \cos x}{\cos^4 x + \sin^4 x} \, dx$

(*v*) $\int_0^{\pi/2} \log (\tan \theta + \cot \theta) \, d\theta$ (*vi*) $\int_0^{\infty} \log\left(x + \frac{1}{x}\right) . \frac{1}{1 + x^2} \, dx$

(*vii*) $\int_0^{\pi} \log (1 + \cos x) \, dx$ (*viii*) $\int_0^{1} \frac{\log x}{\sqrt{1 - x^2}} \, dx.$

Solution. (*i*) Let $I = \int_0^{\pi} x \log (\sin x) \, dx \quad ...(1)$

$$= \int_0^{\pi} (\pi - x) \log [\sin (\pi - x)] \, dx \quad \left[\because \int_0^{a} f(x) \, dx = \int_0^{a} f(a - x) \, dx\right]$$

$$\therefore \quad I = \int_0^{\pi} (\pi - x) \log (\sin x) \, dx \quad ...(2)$$

Adding equations (1) and (2), we get

$$2I = \int_0^{\pi} x \log (\sin x)\, dx + \int_0^{\pi} (\pi - x) \log (\sin x)\, dx$$

$$= \int_0^{\pi} [x \log (\sin x) + (\pi - x) \log (\sin x)]\, dx$$

$$= \int_0^{\pi} [x \log (\sin x) + \pi \log (\sin x) - x \log (\sin x)]\, dx$$

$$= \int_0^{\pi} \pi \log (\sin x)\, dx$$

$\therefore \quad 2I = 2\pi \int_0^{\pi/2} \log (\sin x)\, dx \qquad \left[\because \int_0^{2a} f(x)\, dx = 2\int_0^{a} f(x)\, dx, \text{ if } f(2a - x) = f(x)\right]$

$\Rightarrow \quad I = \pi \left[-\frac{\pi}{2} \log 2\right] \qquad \left[\because \int_0^{\pi/2} \log (\sin x)\, dx = -\frac{\pi}{2} \log 2\right.$ See Example 6(*v*).]

$$= \frac{\pi^2}{2} \log (2)^{-1} \qquad [\because \; m \log n = \log n^m]$$

$$= \frac{\pi^2}{2} \log \frac{1}{2}.$$

(*ii*) Let $\quad I = \int_0^{\pi/2} \underset{\text{I}}{x^2} \; \underset{\text{II}}{\cot x}\, dx$

Integrating by parts, we have

$$= \Big[x \,.\, \log \sin x\Big]_0^{\pi/2} - \int_0^{\pi/2} \frac{d}{dx}(x).\, \log (\sin x)\, dx$$

$$= \left[\frac{\pi}{2} \log \left(\sin \frac{\pi}{2}\right) - 0\right] - \int_0^{\pi/2} 1 \,.\, \log (\sin x)\, dx$$

$$= \left(\frac{\pi}{2} \log 1\right) - \int_0^{\pi/2} \log (\sin x)\, dx$$

$= 0 - \left(-\frac{\pi}{2} \log 2\right) \qquad \left[\because \int_0^{\pi/2} \log (\sin x)\, dx = -\frac{\pi}{2} \log 2\right.$ See Example 6(*v*).]

$$= \frac{\pi}{2} \log 2.$$

(*iii*) Let $\quad I = \int_0^{\pi} \frac{x}{a^2 \cos^2 x + b^2 \sin^2 x}\, dx \qquad ...(1)$

$$= \int_0^{\pi} \frac{(\pi - x)}{a^2 \cos^2 (\pi - x) + b^2 \sin^2 (\pi - x)}\, dx \qquad \left[\because \int_0^{a} f(x)\, dx = \int_0^{a} f(a - x)\, dx\right]$$

$$= \int_0^{\pi} \frac{(\pi - x)}{a^2 (-\cos x)^2 + b^2 \sin^2 x}\, dx$$

$\therefore \quad I = \int_0^{\pi} \frac{(\pi - x)}{a^2 \cos^2 x + b^2 \sin^2 x}\, dx \qquad ...(2)$

Adding equations (1) and (2), we get

$$2I = \int_0^{\pi} \frac{x}{a^2\cos^2 x + b^2 \sin^2 x}\, dx + \int_0^{\pi} \frac{(\pi - x)}{a^2\cos^2 x + b^2\sin^2 x}\, dx$$

$$= \int_0^{\pi} \left[\frac{x}{a^2\cos^2 x + b^2\sin^2 x} + \frac{\pi - x}{a^2\cos^2 x + b^2\sin^2 x}\right] dx$$

$$= \int_0^{\pi} \frac{(x + \pi - x)}{a^2\cos^2 x + b^2\sin^2 x}\, dx$$

$$= \int_0^{\pi} \frac{\pi}{a^2\cos^2 x + b^2\sin^2 x}\, dx$$

$$\therefore \quad 2I = 2\pi \int_0^{\pi/2} \frac{1}{a^2\cos^2 x + b^2\sin^2 x}\, dx$$

$$\left[\because \int_0^{2a} f(x)\, dx = 2\int_0^a f(x)\, dx, \text{ if } f(2a - x) = f(x) \quad \text{Here } (2a = \pi).\right]$$

$$\Rightarrow \quad I = \pi \int_0^{\pi/2} \frac{\frac{1}{\cos^2 x}}{\left[\frac{a^2\cos^2 x}{\cos^2 x} + \frac{b^2\sin^2 x}{\cos^2 x}\right]}\, dx$$

$\left[\text{Dividing the numerator and the denominator by } \cos^2 x\right]$

$$= \pi \int_0^{\pi/2} \frac{\sec^2 x}{a^2 + b^2\tan^2 x}\, dx$$

Put $\tan x = z \Rightarrow \sec^2 x\, dx = dz$

When $x = 0$, $z = \tan 0° \Rightarrow z = 0$ and when $x = \frac{\pi}{2}$, $z = \tan\frac{\pi}{2} \Rightarrow z = \infty$

$$\therefore \quad I = \pi \int_0^{\infty} \frac{1}{a^2 + b^2 z^2}\, dz = \frac{\pi}{b^2}\int_0^{\infty} \frac{1}{\frac{a^2}{b^2} + z^2}\, dz$$

$$= \frac{\pi}{b^2}\left[\frac{1}{\left(\frac{a}{b}\right)}\tan^{-1}\frac{z}{\left(\frac{a}{b}\right)}\right]_0^{\infty} \qquad \left[\because \int \frac{1}{a^2 + x^2}\, dx = \frac{1}{a}\tan^{-1}\frac{x}{a} + c\right]$$

$$= \frac{\pi}{b^2}\cdot\frac{b}{a}\left[\tan^{-1}\frac{bz}{a}\right]_0^{\infty} = \frac{\pi}{ab}\,[\tan^{-1}\infty - \tan^{-1} 0] = \frac{\pi}{ab}\left[\frac{\pi}{2} - 0\right]$$

$$= \frac{\pi^2}{2ab}.$$

(*iv*) Let $I = \int_0^{\pi/2} \frac{x\sin x\cos x}{\cos^4 x + \sin^4 x}\, dx \qquad ...(1)$

$$= \int_0^{\pi/2} \frac{\left(\frac{\pi}{2} - x\right)\sin\left(\frac{\pi}{2} - x\right)\cos\left(\frac{\pi}{2} - x\right)}{\cos^4\left(\frac{\pi}{2} - x\right) + \sin^4\left(\frac{\pi}{2} - x\right)}\, dx \qquad \left[\because \int_0^a f(x)\, dx = \int_0^a f(a - x)\, dx\right]$$

$$\therefore \qquad I = \int_0^{\pi/2} \frac{\left(\frac{\pi}{2} - x\right)\cos x \sin x}{\sin^4 x + \cos^4 x}\,dx \qquad ...(2)$$

Adding equations (1) and (2), we get

$$2I = \int_0^{\pi/2} \frac{x \sin x \cos x}{\cos^4 x + \sin^4 x}\,dx + \int_0^{\pi/2} \frac{\left(\frac{\pi}{2} - x\right)\cos x \sin x}{\sin^4 x + \cos^4 x}\,dx$$

$$= \int_0^{\pi/2} \left[\frac{x \sin x \cos x}{\cos^4 x + \sin^4 x} + \frac{\left(\frac{\pi}{2} - x\right)\sin x \cos x}{\cos^4 x + \sin^4 x}\right] dx$$

$$= \int_0^{\pi/2} \frac{\left(x + \frac{\pi}{2} - x\right)\sin x \cos x}{\cos^4 x + \sin^4 x}\,dx = \frac{\pi}{2}\int_0^{\pi/2} \frac{\sin x \cos x}{\cos^4 x + \sin^4 x}\,dx$$

$$= \frac{\pi}{2}\int_0^{\pi/2} \frac{\frac{\sin x \cos x}{\cos^4 x}}{\frac{\cos^4 x}{\cos^4 x} + \frac{\sin^4 x}{\cos^4 x}}\,dx \qquad \left[\text{Dividing the numerator and the denominator by } \cos^4 x\right]$$

$$= \frac{\pi}{2}\int_0^{\pi/2} \frac{\tan x \sec^2 x}{1 + \tan^4 x}\,dx$$

Put $\tan^2 x = z \Rightarrow 2 \tan x \sec^2 x\,dx = dz \Rightarrow \tan x \sec^2 x\,dx = \frac{1}{2}\,dz$

When $x = 0,\ z = \tan^2 0 \Rightarrow z = 0$ and when $x = \frac{\pi}{2},\ z = \tan^2 \frac{\pi}{2} \Rightarrow z = \infty$

$$\therefore \qquad 2I = \frac{\pi}{2}\int_0^{\infty} \frac{1}{1+z^2}\left(\frac{1}{2}\,dz\right) = \frac{\pi}{4}\int_0^{\infty} \frac{1}{1+z^2}\,dz$$

$$= \frac{\pi}{4}\Big[\tan^{-1} z\Big]_0^{\infty} \qquad \left[\because \int \frac{1}{a^2 + x^2}\,dx = \frac{1}{a}\tan^{-1}\frac{x}{a} + c\right]$$

$$= \frac{\pi}{4}[\tan^{-1}\infty - \tan^{-1} 0] = \frac{\pi}{4}\left[\frac{\pi}{2} - 0\right] = \frac{\pi^2}{8}$$

$$\Rightarrow \qquad I = \frac{\pi^2}{16}.$$

(*v*) Let $I = \int_0^{\pi/2} \log(\tan\theta + \cot\theta)\,d\theta = \int_0^{\pi/2} \log\left[\frac{\sin\theta}{\cos\theta} + \frac{\cos\theta}{\sin\theta}\right] d\theta$

$$= \int_0^{\pi/2} \log\left[\frac{\sin^2\theta + \cos^2\theta}{\sin\theta\cos\theta}\right] d\theta$$

$$= \int_0^{\pi/2} \log\left(\frac{1}{\sin\theta\cos\theta}\right) d\theta \qquad [\because \sin^2 A + \cos^2 A = 1]$$

$$= -\int_0^{\pi/2} \log(\sin\theta\cos\theta)\, d\theta.$$

$$= -\int_0^{\pi/2} [\log(\sin\theta) + \log(\cos\theta)]\, d\theta \qquad [\because \log(m.n) = \log m + \log n]$$

$$= -\int_0^{\pi/2} \log(\sin\theta)\, d\theta - \int_0^{\pi/2} \log(\cos\theta)\, d\theta$$

$$= -\int_0^{\pi/2} \log(\sin\theta)\, d\theta - \int_0^{\pi/2} \log\left[\cos\left(\frac{\pi}{2} - \theta\right)\right] d\theta$$

$$\left[\because \int_0^a f(x)\, dx = \int_0^a f(a-x)\, dx\right]$$

$$= -\int_0^{\pi/2} \log(\sin\theta)\, d\theta - \int_0^{\pi/2} \log(\sin\theta)\, d\theta$$

$$= -2\int_0^{\pi/2} \log(\sin\theta)\, d\theta$$

$$= -2\left[-\frac{\pi}{2}\log 2\right] \qquad \left[\because \int_0^{\pi/2} \log(\sin x)\, dx = -\frac{\pi}{2}\log 2 \text{ See Example 6}(v).\right]$$

$\therefore$ $\quad I = \pi \log 2.$

(*vi*) Let $\quad I = \int_0^{\infty} \log\left(x + \frac{1}{x}\right) . \frac{1}{1+x^2}\, dx$

Put $\quad x = \tan\theta \Rightarrow dx = \sec^2\theta\, d\theta$

When $\quad x = 0, \tan\theta = 0 \Rightarrow \theta = 0$ and when $x = \infty, \tan\theta = \infty \Rightarrow \theta = \frac{\pi}{2}$

$$\therefore \quad I = \int_0^{\pi/2} \log\left[\tan\theta + \frac{1}{\tan\theta}\right] . \left(\frac{1}{1+\tan^2\theta}\right) . \sec^2\theta\, d\theta \qquad [\because \sec^2 A - \tan^2 A = 1]$$

$$= \int_0^{\pi/2} \log(\tan\theta + \cot\theta) . \frac{\sec^2\theta}{\sec^2\theta}\, d\theta$$

$$\Rightarrow \quad I = \int_0^{\pi/2} \log(\tan\theta + \cot\theta)\, d\theta$$

Proceed as in part (*v*). **[Ans.** $\pi \log 2$.**]**

(*vii*) Let $\quad I = \int_0^{\pi} \log(1 + \cos x)\, dx \qquad \ldots(1)$

$$= \int_0^{\pi} \log[1 + \cos(\pi - x)]\, dx \qquad \left[\because \int_0^a f(x)\, dx = \int_0^a f(a-x)\, dx\right]$$

$$\therefore \quad I = \int_0^{\pi} \log[1 - \cos x]\, dx \qquad \ldots(2)$$

Adding equations (1) and (2), we get

$$2I = \int_0^{\pi} \log(1 + \cos x)\, dx + \int_0^{\pi} \log(1 - \cos x)\, dx$$

$$= \int_0^{\pi} [\log(1 + \cos x) + \log(1 - \cos x)]\, dx$$

$$= \int_0^{\pi} \log[(1 + \cos x)(1 - \cos x)]\, dx \qquad [\because \log m + \log n = \log(m.n)]$$

$$= \int_0^{\pi} \log(1 - \cos^2 x)\, dx$$

$$= \int_0^{\pi} \log(\sin^2 x)\, dx \qquad [\because \sin^2 A + \cos^2 A = 1]$$

$$= 2\int_0^{\pi} \log(\sin x)\, dx$$

$\therefore$ $$I = 2\left(-\frac{\pi}{2}\log 2\right) \qquad \left[\because \int_0^{\pi/2} \log(\sin x)\, dx = -\frac{\pi}{2}\log 2 \text{ See Example 6}(v).\right]$$

$\Rightarrow$ $$I = -\pi \log 2.$$

(*viii*) Let $I = \int_0^1 \frac{\log x}{\sqrt{1-x^2}}\, dx$

Put $x = \sin\theta \Rightarrow dx = \cos\theta\, d\theta$

When $x = 0$, $\sin\theta = 0 \Rightarrow \theta = 0$ and when $x = 1$, $\sin\theta = 1 \Rightarrow \theta = \frac{\pi}{2}$

$\therefore$ $$I = \int_0^{\pi/2} \frac{\log(\sin\theta)}{\sqrt{1-\sin^2\theta}} . \cos\theta\, d\theta = \int_0^{\pi/2} \frac{\log(\sin\theta)}{\sqrt{\cos^2\theta}} . \cos\theta\, d\theta$$

$$= \int_0^{\pi/2} \frac{\log(\sin\theta)}{\cos\theta} . \cos\theta\, d\theta = \int_0^{\pi/2} \log(\sin\theta)\, d\theta$$

$$= -\frac{\pi}{2}\log 2. \qquad \left[\because \int_0^{\pi/2} \log(\sin x)\, dx = -\frac{\pi}{2}\log 2 \text{ See Example 6}(v).\right]$$

Example 13. *Evaluate the following integrals :*

(*i*) $\int_0^{2\pi} \frac{\sin 2\theta}{a - b\cos\theta}\, d\theta,\ a > b > 0$ (*ii*) $\int_{-\pi}^{\pi} x^{10} \sin^7 x\, dx$

(*iii*) $\int_{-\pi}^{\pi} \sin^5 x \cos x\, dx$ (*iv*) $\int_{-1}^{1} \log\left(\frac{2-x}{2+x}\right) dx$

(*v*) $\int_{-\pi/4}^{\pi/4} \log(\sin x + \cos x)\, dx$ (*vi*) $\int_{-1}^{1} e^{|x|}\, dx$

(*vii*) $\int_{\pi/6}^{\pi/3} \frac{1}{1+\sqrt{\cot x}}\, dx.$

Solution. (*i*) Let $I = \int_0^{2\pi} \frac{\sin 2\theta}{a - b\cos\theta}\, d\theta$ ...(1)

$$= \int_0^{2\pi} \frac{\sin 2(2\pi - \theta)}{a - b\cos(2\pi - \theta)}\, d\theta \qquad \left[\because \int_0^a f(x)\, dx = \int_0^a f(a-x)\, dx\right]$$

$$= \int_0^{2\pi} \frac{\sin(4\pi - 2\theta)}{a - b\cos\theta}\, d\theta$$

$$\therefore \qquad I = \int_0^{2\pi} \frac{-\sin 2\theta}{a - b\cos 2\theta}\, d\theta \qquad \text{...(2)}$$

Adding equations (1) and (2), we get

$$2I = \int_0^{2\pi} \frac{\sin 2\theta}{a - b\cos\theta}\, d\theta - \int_0^{2\pi} \frac{\sin 2\theta}{a - b\cos\theta}\, d\theta$$

$$\Rightarrow \qquad 2I = 0$$

$$\Rightarrow \qquad I = 0.$$

(ii) Let $\quad I = \int_{-\pi}^{\pi} x^{10} \sin^7 x\, dx$

Let $\quad f(x) = x^{10} \sin^7 x$

$\Rightarrow \quad f(-x) = (-x)^{10} [\sin(-x)]^7 = x^{10} (-\sin x)^7 = -x^{10} \sin^7 x = -f(x).$

$\therefore$ $f(x)$ is an odd function.

$\therefore \quad \int_{-\pi}^{\pi} x^{10} \sin^7 x\, dx = 0 \qquad \left[\because \int_{-a}^{a} f(x)\, dx = 0; \text{ if } f(x) \text{ is an odd function}\right]$

(iii) Let $\quad I = \int_{-\pi}^{\pi} \sin^5 x \cos x\, dx$

Let $\quad f(x) = \sin^5 x \cos x$

$\Rightarrow \quad f(-x) = \sin^5(-x)\cos(-x) = [\sin(-x)]^5 \cos x = (-\sin x)^5 \cos x = -\sin^5 x \cos x = -f(x)$

$\Rightarrow$ $f(x)$ is an odd function.

$\therefore \quad \int_{-\pi}^{\pi} \sin^5 x \cos x\, dx = 0. \qquad \left[\because \int_{-a}^{a} f(x)\, dx = 0; \text{ if } f(x) \text{ is an odd function}\right]$

(iv) Let $\quad I = \int_{-1}^{1} \log\left(\frac{2-x}{2+x}\right) dx$

Let $\quad f(x) = \log\left(\frac{2-x}{2+x}\right)$

$$\Rightarrow \qquad f(-x) = \log\left[\frac{2-(-x)}{2+(-x)}\right] = \log\left(\frac{2+x}{2-x}\right)$$

$$= \log\left(\frac{2-x}{2+x}\right)^{-1} = -\log\left(\frac{2-x}{2+x}\right) = -f(x) \qquad [\because \; m\log n = \log n^m]$$

$\therefore$ $f(x)$ is an odd function. $\qquad \left[\because \int_{-a}^{a} f(x)\, dx = 0; \text{ if } f(x) \text{ is an odd function}\right]$

$\therefore \quad \int_{-1}^{1} \log\left(\frac{2-x}{2+x}\right) dx = 0.$

(v) Let $\quad I = \int_{-\pi/4}^{\pi/4} \log(\sin x + \cos x)\, dx$

$$= \int_{-\pi/4}^{\pi/4} \log\left[\sqrt{2}\left(\frac{1}{\sqrt{2}}\sin x + \frac{1}{\sqrt{2}}\cos x\right)\right] dx \qquad [\text{Multiply and divided by } \sqrt{2}]$$

$$= \int_{-\pi/4}^{\pi/4} \log\left[\sqrt{2}\left(\sin x \cos\frac{\pi}{4} + \cos x \sin\frac{\pi}{4}\right)\right] dx$$

$$I = \int_{-\pi/4}^{\pi/4} \log\left[\sqrt{2}\sin\left(x+\frac{\pi}{4}\right)\right] dx \quad [\because \sin(A+B) = \sin A \cos B + \cos A \sin B]$$

Put $x + \frac{\pi}{4} = \theta \Rightarrow dx = d\theta$

When $x = -\frac{\pi}{4}$, $\theta = -\frac{\pi}{4} + \frac{\pi}{4} = 0$ and when $x = \frac{\pi}{4}$, $\theta = \frac{\pi}{4} + \frac{\pi}{4} = \frac{2\pi}{4} = \frac{\pi}{2}$.

$$\therefore \quad I = \int_0^{\pi/2} \log\left(\sqrt{2}\sin\theta\right) d\theta$$

$$= \int_0^{\pi/2} [\log\sqrt{2} + \log(\sin\theta)]\, d\theta \qquad [\because \log(m.n) = \log m + \log n]$$

$$= \log\sqrt{2}\int_0^{\pi/2} 1.\, d\theta + \int_0^{\pi/2} \log(\sin\theta)\, d\theta$$

$$= \frac{1}{2}\log 2\,[\theta]_0^{\pi/2} + \left(-\frac{\pi}{2}\log 2\right) \qquad \left[\because \int_0^{\pi/2} \log(\sin x)\, dx = -\frac{\pi}{2}\log 2 \text{ See Example 6}(v).\right]$$

$$= \frac{1}{2}\log 2\left[\frac{\pi}{2} - 0\right] - \frac{\pi}{2}\log 2$$

$$= \frac{\pi}{4}\log 2 - \frac{\pi}{2}\log 2$$

$$= -\frac{\pi}{4}\log 2.$$

(*vi*) Let $I = \int_{-1}^{1} e^{|x|}\, .\, dx$

Let $f(x) = e^{|x|}$

$\Rightarrow \quad f(-x) = e^{|-x|} = e^{|x|} = f(x).$

$\Rightarrow$ $f(x)$ is an even function.

$$\therefore \quad I = \int_{-1}^{1} e^{|x|}\, dx = 2\int_0^1 e^{|x|}\, .\, dx = 2\int_0^1 e^x\, dx \qquad [\because |x| = x, \text{ for } 0 \le x \le 1]$$

$$= 2\left[e^x\right]_0^1 = 2[e^1 - e^0]$$

$$= 2(e-1).$$

(*vii*) Let $I = \int_{\pi/6}^{\pi/3} \frac{1}{1+\sqrt{\cot x}}\, dx$

$$\Rightarrow \quad I = \int_{\pi/6}^{\pi/3} \frac{1}{1+\sqrt{\frac{\cos x}{\sin x}}}\, dx = \int_{\pi/6}^{\pi/3} \frac{\sqrt{\sin x}}{\sqrt{\sin x}+\sqrt{\cos x}}\, dx \qquad ...(1)$$

Put $x = \left(\frac{\pi}{2} - z\right) \Rightarrow dx = -dz$

When $x = \frac{\pi}{6}$, $\frac{\pi}{2} - z = \frac{\pi}{6} \Rightarrow z = \left(\frac{\pi}{2} - \frac{\pi}{6}\right) = \frac{2\pi}{6} = \frac{\pi}{3}$.

and when $x = \frac{\pi}{3}$, $\frac{\pi}{2} - z = \frac{\pi}{3} \Rightarrow z = \left(\frac{\pi}{2} - \frac{\pi}{3}\right) = \frac{\pi}{6}$.

$$\therefore \quad I = \int_{\pi/3}^{\pi/6} \frac{\sqrt{\sin\left(\frac{\pi}{2} - z\right)}}{\sqrt{\sin\left(\frac{\pi}{2} - z\right)} + \sqrt{\cos\left(\frac{\pi}{2} - z\right)}} \cdot (-dz)$$

$$= -\int_{\pi/3}^{\pi/6} \frac{\sqrt{\cos z}}{\sqrt{\cos z} + \sqrt{\sin z}}\, dz$$

$$= \int_{\pi/6}^{\pi/3} \frac{\sqrt{\cos z}}{\sqrt{\cos z} + \sqrt{\sin z}}\, dz \qquad \left[\because \int_a^b f(x)\,dx = -\int_b^a f(x)\,dx\right]$$

$$\Rightarrow \quad I = \int_{\pi/6}^{\pi/3} \frac{\sqrt{\cos x}}{\sqrt{\cos x} + \sqrt{\sin x}} \cdot dx \qquad \ldots(2) \qquad \left[\because \int_a^b f(x)\,dx = \int_a^b f(z)\,dz\right]$$

Adding equations (1) and (2), we get

$$2I = \int_{\pi/6}^{\pi/3} \frac{\sqrt{\sin x}}{\sqrt{\sin x} + \sqrt{\cos x}}\, dx + \int_{\pi/6}^{\pi/3} \frac{\sqrt{\cos x}}{\sqrt{\cos x} + \sqrt{\sin x}}\, dx$$

$$= \int_{\pi/6}^{\pi/3} \left[\frac{\sqrt{\sin x}}{\sqrt{\sin x} + \sqrt{\cos x}} + \frac{\sqrt{\cos x}}{\sqrt{\cos x} + \sqrt{\sin x}}\right] dx$$

$$= \int_{\pi/6}^{\pi/3} \left[\frac{\sqrt{\sin x} + \sqrt{\cos x}}{\sqrt{\sin x} + \sqrt{\cos x}}\right] \cdot dx$$

$$= \int_{\pi/6}^{\pi/3} 1 \cdot dx = [x]_{\pi/6}^{\pi/3} = \left[\frac{\pi}{3} - \frac{\pi}{6}\right] = \frac{\pi}{6}$$

$$\therefore \quad I = \frac{1}{2}\left(\frac{\pi}{6}\right) = \frac{\pi}{12}.$$

Example 14. *Evaluate the following integrals :*

(*i*) $\int_{-2}^{2} |x + 1| \cdot dx$ (*ii*) $\int_0^3 |3x - 1|\, dx$

(*iii*) $\int_2^8 |x - 5|\, dx$ (*iv*) $\int_{-5}^{5} |x + 2|\, dx$

(*v*) $\int_{1/4}^{1} |2x - 1|\, dx$ (*vi*) $\int_0^1 |5x - 3|\, dx$.

[**Note.** $|x| = x$. if $x \geq 0$ and $|x| = -x$, if $x < 0$]

Solution. (*i*) Let $I = \int_{-2}^{2} |x + 1|\, dx$

Since, we have

$$|x + 1| = \begin{cases} -(x+1); \text{ if } -2 < x < -1 \\ (x+1); \text{ if } \quad -1 < x < 2 \end{cases}$$

$$\therefore \quad \int_{-2}^{2}|x+1|\,dx=\int_{-2}^{-1}|x+1|\,dx+\int_{-1}^{2}|x+1|\,dx=\int_{-2}^{-1}-(x+1)\,dx+\int_{-1}^{2}(x+1)\,dx$$

$$=-\left[\frac{x^2}{2}+x\right]_{-2}^{-1}+\left[\frac{x^2}{2}+x\right]_{-1}^{2}=-\left[\left(\frac{1}{2}-1\right)-\left(\frac{4}{2}-2\right)\right]+\left[\left(\frac{4}{2}+2\right)-\left(\frac{1}{2}-1\right)\right]$$

$$=-\left[-\frac{1}{2}-0\right]+\left[4-\left(-\frac{1}{2}\right)\right]=\frac{1}{2}+4+\frac{1}{2}=4+1$$

$$=5.$$

(*ii*) Let $\quad I=\int_{0}^{3}|3x-1|\,dx$

Since, we have $|3x-1|=\begin{cases}-(3x-1)\,;\text{ if } 0<x<\frac{1}{3}\\(3x-1)\,;\text{ if } \frac{1}{3}<x<3\end{cases}$

$$\therefore \quad \int_{0}^{3}|3x-1|.\,dx=\int_{0}^{1/3}|3x-1|\,dx+\int_{1/3}^{3}|3x-1|\,dx=\int_{0}^{1/3}-(3x-1)\,dx+\int_{1/3}^{3}(3x-1)\,dx$$

$$=-\left[\frac{3x^2}{2}-x\right]_{0}^{1/3}+\left[\frac{3x^2}{2}-x\right]_{1/3}^{3}$$

$$=-\left[\left(\frac{3}{2}\left(\frac{1}{3}\right)^2-\frac{1}{3}\right)-(0-0)\right]+\left[\left(\frac{3(3)^2}{2}-3\right)-\left(\frac{3}{2}\left(\frac{1}{3}\right)^2-\frac{1}{3}\right)\right]$$

$$=-\left[\frac{3}{2}.\frac{1}{9}-\frac{1}{3}\right]+\left[\left(\frac{3(9)}{2}-3\right)-\left(\frac{3}{2}.\frac{1}{9}-\frac{1}{3}\right)\right]$$

$$=-\left[\frac{1}{6}-\frac{1}{3}\right]+\left[\frac{27}{2}-3-\frac{1}{6}+\frac{1}{3}\right]=-\left(-\frac{1}{6}\right)+\left[\frac{81-18-1+2}{6}\right]$$

$$=\frac{1}{6}+\frac{64}{6}=\frac{65}{6}.$$

(*iii*) Let $\quad I=\int_{2}^{8}|x-5|\,dx$

Since, we have $|x-5|=\begin{cases}-(x-5)\,;\text{ if } 2\le x<5\\(x-5)\,;\text{ if } 5<x\le 8\end{cases}$

$$\therefore \quad \int_{2}^{8}|x-5|.\,dx=\int_{2}^{5}|x-5|\,dx+\int_{5}^{8}|x-5|\,dx$$

$$=\int_{2}^{5}-(x-5)\,dx+\int_{5}^{8}(x-5)\,dx=\int_{2}^{5}(5-x)\,dx+\int_{5}^{8}(x-5)\,dx$$

$$=\left[5x-\frac{x^2}{2}\right]_{2}^{5}+\left[\frac{x^2}{2}-5x\right]_{5}^{8}$$

$$=\left[\left(5(5)-\frac{25}{2}\right)-\left(5(2)-\frac{4}{2}\right)\right]+\left[\left(\frac{64}{2}-5(8)\right)-\left(\frac{25}{2}-5(5)\right)\right]$$

$$= \left[\left(25 - \frac{25}{2}\right) - (10 - 2)\right] + \left[(32 - 40) - \left(\frac{25}{2} - 25\right)\right]$$

$$= \frac{25}{2} - 8 + (-8) - \left(-\frac{25}{2}\right) = \frac{25}{2} - 8 - 8 + \frac{25}{2}$$

$$= 25 - 16 = 9.$$

(*iv*) Let $\quad I = \int_{-5}^{5} |x + 2|\, dx$

Since, we have

$$|x + 2| = \begin{cases} -(x + 2)\,; \text{ if } & x + 2 < 0 \\ (x + 2)\,; \text{ if } & x + 2 \geq 0 \end{cases}$$

$$\therefore \quad \int_{-5}^{5} |x + 2|\,.\, dx = \int_{-5}^{-2} |x + 2|\, dx + \int_{-2}^{5} |x + 2|\, dx$$

$$= \int_{-5}^{-2} -(x + 2)\, dx + \int_{-2}^{5} (x + 2)\, dx = -\int_{-5}^{-2} (x + 2)\, dx + \int_{-2}^{5} (x + 2)\, dx$$

$$= -\left[\frac{x^2}{2} + 2x\right]_{-5}^{-2} + \left[\frac{x^2}{2} + 2x\right]_{-2}^{5}$$

$$= -\left[\left(\frac{(-2)^2}{2} + 2(-2)\right) - \left(\frac{(-5)^2}{2} + 2(-5)\right)\right] + \left[\left(\frac{25}{2} + 10\right) - \left(\frac{(-2)^2}{2} + 2(-2)\right)\right]$$

$$= -\left[\left(\frac{4}{2} - 4\right) - \left(\frac{25}{2} - 10\right)\right] + \left[\left(\frac{25 + 20}{2}\right) - \left(\frac{4}{2} - 4\right)\right]$$

$$= -\left[-2 - \frac{5}{2}\right] + \left[\frac{45}{2} - (-2)\right] = -\left(-\frac{9}{2}\right) + \left(\frac{45}{2} + 2\right) = \frac{9}{2} + \frac{49}{2}$$

$$= \frac{58}{2} = 29.$$

(*v*) Let $\quad I = \int_{1/4}^{1} |2x - 1|\,.\, dx$

Since, we have

$$|2x - 1| = \begin{cases} -(2x - 1)\,; \text{ if } & \frac{1}{4} \leq x < \frac{1}{2} \\ (2x - 1)\,; \text{ if } & \frac{1}{2} < x \leq 1 \end{cases}$$

$$\therefore \quad \int_{1/4}^{1} |2x - 1|\,.\, dx = \int_{1/4}^{1/2} |2x - 1|\, dx + \int_{1/2}^{1} |2x - 1|\, dx$$

$$= \int_{1/4}^{1/2} -(2x - 1)\, dx + \int_{1/2}^{1} (2x - 1)\, dx$$

$$= \int_{1/4}^{1/2} (1 - 2x)\, dx + \int_{1/2}^{1} (2x - 1)\, dx$$

$$= \left[x - x^2\right]_{1/4}^{1/2} + \left[x^2 - x\right]_{1/2}^{1} = \left[\left(\frac{1}{2} - \frac{1}{4}\right) - \left(\frac{1}{4} - \frac{1}{16}\right)\right] + \left[(1-1) - \left(\frac{1}{4} - \frac{1}{2}\right)\right]$$

$$= \left[\left(\frac{1}{4} - \frac{3}{16}\right)\right] + \left[0 - \left(-\frac{1}{4}\right)\right] = \frac{1}{4} - \frac{3}{16} + \frac{1}{4} = \frac{4-3+4}{16}$$

$$= \frac{5}{16}.$$

(*vi*) Let $\quad I = \int_0^1 |5x - 3|\, dx$

Since, we have

$$|5x - 3| = \begin{cases} -(5x-3)\,; \text{if} & 0 < x < \dfrac{3}{5} \\ (5x-3)\,; \text{if} & \dfrac{3}{5} \geq x > 1 \end{cases}$$

$$\therefore \quad \int_0^1 |5x-3|\, dx = \int_0^{3/5} |5x-3|\, dx + \int_{3/5}^1 |5x-3|\, dx$$

$$= \int_0^{3/5} -(5x-3)\, dx + \int_{3/5}^1 (5x-3)\, dx = \int_0^{3/5} (3-5x)\, dx + \int_{3/5}^1 (5x-3)\, dx$$

$$= \left[3x - \frac{5x^2}{2}\right]_0^{3/5} + \left[\frac{5x^2}{2} - 3x\right]_{3/5}^1$$

$$= \left[\left\{3\left(\frac{3}{5}\right) - \frac{5}{2}\left(\frac{3}{5}\right)^2\right\} - \{0-0\}\right] + \left[\left(\frac{5}{2} - 3\right) - \left\{\frac{5}{2}\left(\frac{3}{5}\right)^2 - 3\left(\frac{3}{5}\right)\right\}\right]$$

$$= \left(\frac{9}{5} - \frac{9}{10}\right) + \left[-\frac{1}{2} - \left(\frac{9}{10} - \frac{9}{5}\right)\right] = \frac{9}{10} - \frac{1}{2} + \frac{9}{10}$$

$$= \frac{9-5+9}{10} = \frac{13}{10}.$$

Example 15. *Evaluate the following integrals :*

(*i*) $\int_{-4}^{4} |x+1|\, dx$ (*ii*) $\int_{-1}^{1} |2x+1|\, dx$

(*iii*) $\int_0^2 |x^2 + 2x - 3|\, dx$ (*iv*) $\int_1^4 (|x-1| + |x-2| + |x-3|)\, dx$

(*v*) $\int_0^3 |x-2|\, dx$ (*vi*) $\int_{-2}^{2} |2x+3|\, dx$

(*vii*) $\int_1^3 |x-4|\, dx.$

Solution. (*i*) Let $I = \int_{-4}^{4} |x+1|\, dx$

Since, we have

$$|x+1| = \begin{cases} -(x+1)\,; \text{if} & -4 \leq x \leq -1 \\ (x+1)\,; \text{if} & -1 \geq x \geq 4 \end{cases}$$

$\therefore \quad \int_{-4}^{4} |x+1|\, dx = \int_{-4}^{-1} |x+1|\, dx + \int_{-1}^{4} |x+1|\, dx = \int_{-4}^{-1} -(x+1)\, dx + \int_{-1}^{4} (x+1)\, dx$

$$= -\left[\frac{x^2}{2} + x\right]_{-4}^{-1} + \left[\frac{x^2}{2} + x\right]_{-1}^{4}$$

$$= -\left[\left(\frac{(-1)^2}{2} + (-1)\right) - \left(\frac{(-4)^2}{2} + (-4)\right)\right] + \left[\left(\frac{(4)^2}{2} + 4\right) - \left(\frac{(-1)^2}{2} + (-1)\right)\right]$$

$$= -\left[\left(\frac{1}{2} - 1\right) - (8-4)\right] + \left[12 - \left(\frac{1}{2} - 1\right)\right] = -\left(-\frac{1}{2}\right) + 4 + 12 - \left(-\frac{1}{2}\right) = \frac{1}{2} + 16 + \frac{1}{2}$$

$$= 16 + 1 = 17.$$

(*ii*) Let $\quad I = \int_{-1}^{1} |2x+1|\, dx$

Since, we have

$$|2x+1| = \begin{cases} -(2x+1)\,; \text{ if } & -1 < x < -\frac{1}{2} \\ (2x+1)\,; \text{ if } & -\frac{1}{2} \geq x \geq 1 \end{cases}$$

$\therefore \quad \int_{-1}^{1} |2x+1|\, dx = \int_{-1}^{-1/2} |2x+1|\, dx + \int_{-1/2}^{1} |2x+1|\, dx$

$$= \int_{-1}^{-1/2} -(2x+1)\, dx + \int_{-1/2}^{1} (2x+1)\, dx$$

$$= -\left[x^2 + x\right]_{-1}^{-1/2} + \left[x^2 + x\right]_{-1/2}^{1}$$

$$= -\left[\left\{\left(-\frac{1}{2}\right)^2 + \left(-\frac{1}{2}\right)\right\} - \left\{(-1)^2 + (-1)\right\}\right] + \left[(1+1) - \left\{\left(-\frac{1}{2}\right)^2 + \left(-\frac{1}{2}\right)\right\}\right]$$

$$= -\left[\frac{1}{4} - \frac{1}{2} - 0\right] + \left[2 - \left(\frac{1}{4} - \frac{1}{2}\right)\right] = \frac{1}{4} + 2 + \frac{1}{4} = \frac{1+8+1}{4}$$

$$= \frac{10}{4} = \frac{5}{2}.$$

(*iii*) Let $\quad I = \int_{0}^{2} |x^2 + 2x - 3|\, dx$

Since, we have

$$|x^2 + 2x - 3| = \begin{cases} -(x^2 + 2x - 3)\,; \text{ if } 0 < x < 1 \\ (x^2 + 2x - 3)\,; \text{ if } 1 \leq x \leq 2 \end{cases}$$

$\therefore \quad \int_{0}^{2} |x^2 + 2x - 3|.\, dx = \int_{0}^{1} |x^2 + 2x - 3|\, dx + \int_{1}^{2} |x^2 + 2x - 3|\, dx$

$$= \int_{0}^{1} -(x^2 + 2x - 3)\, dx + \int_{1}^{2} (x^2 + 2x - 3)\, dx$$

$$= -\left[\frac{x^3}{3} + x^2 - 3x\right]_{0}^{1} + \left[\frac{x^3}{3} + x^2 - 3x\right]_{1}^{2}$$

$$= -\left[\left(\frac{1}{3}+1-3\right)-(0)\right]+\left[\left\{\frac{(2)^3}{3}+(2)^2-3(2)\right\}-\left\{\frac{(1)^3}{3}+(1)^2-3(1)\right\}\right]$$

$$= -\left[\frac{1}{3}-2\right]+\left[\left(\frac{8}{3}+4-6\right)-\left(\frac{1}{3}+1-2\right)\right] = -\left(-\frac{5}{3}\right)+\frac{2}{3}+\frac{5}{3} = \frac{5}{3}+\frac{2}{3}+\frac{5}{3}$$

$$= \frac{12}{3} = 4.$$

(*iv*) Let $\quad I = \int_1^4 (|x-1|+|x-2|+|x-3|)\, dx$

Let $\quad f(x) = |x-1| + |x-2| + |x-3|$

$$\therefore \quad f(x) = \begin{cases} -(x-1)-(x-2)-(x-3); \text{ if } x<1 \\ (x-1)-(x-2)-(x-3); \text{ if } 1 \le x < 2 \\ (x-1)+(x-2)-(x-3); \text{ if } 2 \le x < 3 \\ (x-1)+(x+2)+(x-3); \text{ if } x \ge 3 \end{cases}$$

$$\Rightarrow \quad f(x) = \begin{cases} -3x+6; \text{ if } x<1 \\ -x+4; \text{ if } 1 \le x < 2 \\ x; \text{ if } 2 \le x < 3 \\ 3x-6; \text{ if } x \ge 3 \end{cases}$$

$$\therefore \quad I = \int_1^4 f(x)\, dx = \int_1^2 f(x)\, dx + \int_2^3 f(x)\, dx + \int_3^4 f(x)\, dx$$

$$= \int_1^2 (-x+4)\, dx + \int_2^3 x\, dx + \int_3^4 (3x-6)\, dx$$

$$= \left[-\frac{x^2}{2}+4x\right]_1^2 + \left[\frac{x^2}{2}\right]_2^3 + \left[\frac{3x^2}{2}-6x\right]_3^4$$

$$= \left[\left\{\frac{-(2)^2}{2}+4(2)\right\}-\left\{\frac{-1}{2}+4\right\}\right]+\left[\frac{9}{2}-\frac{4}{2}\right]+\left[\left\{\frac{3(4)^2}{2}-6(4)\right\}-\left\{\frac{3(3)^2}{2}-6(3)\right\}\right]$$

$$= \left[(-2+8)-\left(\frac{7}{2}\right)\right]+\left[\frac{5}{2}\right]+\left[(24-24)-\left(\frac{27}{2}-18\right)\right]$$

$$= \left(6-\frac{7}{2}\right)+\frac{5}{2}+\left(0-\left(-\frac{9}{2}\right)\right) = 6-\frac{7}{2}+\frac{5}{2}+\frac{9}{2}$$

$$= \frac{12-7+5+9}{2} = \frac{19}{2}.$$

(*v*) Let $\quad I = \int_0^3 |x-2|\, dx$

Since, we have

$$|x-2| = \begin{cases} -(x-2); \text{ if } & x-2 \le 0 \\ (x-2); \text{ if } & x-2 \ge 0 \end{cases}$$

$$= \int_0^3 |x-2|\, dx = \int_0^2 |x-2|\, dx + \int_2^3 |x-2|\, dx$$

$$= \int_0^2 -(x-2)\,dx + \int_2^3 (x-2)\,dx = \int_0^2 (2-x)\,dx + \int_2^3 (x-2)\,dx$$

$$= \left[2x - \frac{x^2}{2}\right]_0^2 + \left[\frac{x^2}{2} - 2x\right]_2^3$$

$$= \left[\left(2(2) - \frac{(2)^2}{2}\right) - 0\right] + \left[\left(\frac{(3)^2}{2} - 2(3)\right) - \left(\frac{(2)^2}{2} - 2(2)\right)\right]$$

$$= \left[(4-2) + \left(\frac{9}{2} - 6\right) - (-2)\right] = 2 + \left(-\frac{3}{2}\right) + 2$$

$$= 4 - \frac{3}{2} = \frac{5}{2}.$$

(*vi*) Let $\quad I = \int_{-2}^{+2} |2x+3|\,dx$

Since, we have

$$|2x+3| = \begin{cases} -(2x+3)\,; \text{ if } & -2 \le x \le -\frac{3}{2} \\ (2x+3)\,; \text{ if } & -\frac{3}{2} \le x \le 2 \end{cases}$$

$$\therefore \quad \int_{-2}^{2} |2x+3|\,dx = \int_{-2}^{-3/2} |2x+3|\,dx + \int_{-3/2}^{2} |2x+3|\,dx$$

$$= \int_{-2}^{-3/2} -(2x+3)\,dx + \int_{-3/2}^{2} (2x+3)\,dx$$

$$= -\left[x^2 + 3x\right]_{-2}^{-3/2} + \left[x^2 + 3x\right]_{-3/2}^{2}$$

$$= -\left[\left\{\left(-\frac{3}{2}\right)^2 + 3\left(-\frac{3}{2}\right)\right\} - \left\{(-2)^2 + 3(-2)\right\}\right] + \left[(4+6) - \left\{\left(-\frac{3}{2}\right)^2 + 3\left(-\frac{3}{2}\right)\right\}\right]$$

$$= -\left[\left(\frac{9}{4} - \frac{9}{2}\right) - (4-6)\right] + \left[10 - \left(\frac{9}{4} - \frac{9}{2}\right)\right] = +\frac{9}{4} - 2 + 10 + \frac{9}{4} = \frac{9}{2} + 8 = \frac{9+16}{2}$$

$$= \frac{25}{2}.$$

(*vii*) Let $\quad I = \int_1^3 |x-4|\,dx$

Since, we have

$$|x-4| = \begin{cases} -(x-4)\,; \text{ if } & -x < 4 \\ (x-4)\,; \text{ if } & x \ge 4 \end{cases}$$

$$\therefore \quad \int_1^3 |x-4|\,dx = \int_1^3 -(x-4)\,.\,dx = \int_1^3 (4-x)\,dx$$

$$= \left[4x - \frac{x^2}{2}\right]_1^3 = \left[4(3) - \frac{(3)^2}{2}\right] - \left[4(1) - \frac{(1)^2}{2}\right]$$

$$= \left(12 - \frac{9}{2}\right) - \left(4 - \frac{1}{2}\right) = 12 - \frac{9}{2} - 4 + \frac{1}{2} = 8 - \frac{8}{2}$$

$$= 8 - 4 = 4.$$

Example 16. *Evaluate the following integrals :*

(i) $\displaystyle\int_0^{\pi} \frac{x \sin(2x) \sin\left(\frac{\pi}{2}\cos x\right)}{(2x - \pi)}\, dx$ (ii) $\displaystyle\int_0^{2\pi} \frac{x \sin^{2n} x}{\sin^{2n} x + \cos^{2n} x}\, dx$

(iii) $\displaystyle\int_0^{\pi} \cos 2x \log(\sin x)\, dx$ (iv) $\displaystyle\int_{-1}^{1} \frac{x \sin^{-1} x}{\sqrt{1 - x^2}}\, dx$

(v) $\displaystyle\int_{-\pi/4}^{\pi/4} \frac{\left(x + \frac{\pi}{4}\right)}{2 - \cos 2x}\, dx$ (vi) $\displaystyle\int_{-\pi}^{\pi} \frac{2x(1 + \sin x)}{1 + \cos^2 x}\, dx.$

Solution. (*i*) Let $\displaystyle I = \int_0^{\pi} \frac{x \sin(2x) \sin\left(\frac{\pi}{2}\cos x\right)}{(2x - \pi)}\, dx$ ...(1)

$$= \int_0^{\pi} \frac{(\pi - x) \sin 2(\pi - x) \sin\left(\frac{\pi}{2}\cos(\pi - x)\right)}{2(\pi - x) - \pi}\, dx \quad \left[\because \int_0^a f(x)\, dx = \int_0^a f(a - x)\, dx.\right]$$

$$= \int_0^{\pi} \frac{(\pi - x) \sin(2\pi - 2x) \sin\left(\frac{\pi}{2}(-\cos x)\right)}{2\pi - 2x - \pi}\, dx$$

$$= \int_0^{\pi} \frac{(\pi - x)(-\sin 2x)\left(-\sin\left(\frac{\pi}{2}\cos x\right)\right)}{-(2x - \pi)}\, dx$$

$$= -\int_0^{\pi} \frac{(\pi - x)\{-(\sin 2x)\} \sin\left(\frac{\pi}{2}\cos x\right)}{(2x - \pi)}\, dx$$

$\therefore$ $\displaystyle I = \int_0^{\pi} \frac{(x - \pi) \sin(2x) \sin\left(\frac{\pi}{2}\cos x\right)}{(2x - \pi)}\, dx$ ...(2)

Adding equations (1) and (2), we get

$$2I = \int_0^{\pi} \frac{x \sin(2x) \sin\left(\frac{\pi}{2}\cos x\right)}{2x - \pi}\, dx + \int_0^{\pi} \frac{(x - \pi) \sin(2x) \sin\left(\frac{\pi}{2}\cos x\right)}{2x - \pi}\, dx$$

$$= \int_0^{\pi} \left[\frac{x \sin(2x) \sin\left(\frac{\pi}{2}\cos x\right)}{2x - \pi} + \frac{(x - \pi) \sin(2x) \sin\left(\frac{\pi}{2}\cos x\right)}{2x - \pi}\right] dx$$

$$= \int_0^{\pi} \frac{(2x - \pi) \sin (2x) \sin\left(\frac{\pi}{2} \cos x\right)}{2x - \pi} . dx = \int_0^{\pi} \sin 2x \sin\left(\frac{\pi}{2} \cos x\right) . dx$$

$$\Rightarrow \quad I = \frac{1}{2} \int_0^{\pi} 2 \sin x \cos x \sin\left(\frac{\pi}{2} \cos x\right) dx \qquad [\because \ \sin 2A = 2 \sin A \cos A]$$

$$\Rightarrow \quad I = \int_0^{\pi} \sin x \cos x \sin\left(\frac{\pi}{2} \cos x\right) dx$$

Put $\cos x = z \Rightarrow - \sin x \, dx = dz \Rightarrow \sin x \, dx = - dz$

When $x = 0$; $z = \cos 0° = 1$ and when $x = \pi$; $z = \cos \pi = -1$

$$\therefore \quad I = \int_{+1}^{-1} z \sin\left(\frac{\pi z}{2}\right) (- dz) = - \int_{+1}^{-1} z \sin\left(\frac{\pi z}{2}\right) dz$$

$$= \int_{-1}^{1} z \sin\left(\frac{\pi z}{2}\right) dz \qquad \left[\because \ \int_a^b f(x) \, dx = - \int_b^a f(x) \, dx\right]$$

$$\Rightarrow \quad I = 2 \int_0^1 \underset{\text{I}}{z} \ \underset{\text{II}}{\sin\left(\frac{\pi z}{2}\right)} dz \qquad \left[\because \ \int_0^{2a} f(x) \, dx = 2 \int_0^a f(x) \, dx ; \text{ if } f(2a - x) = f(x)\right]$$

Integrating by parts, we get

$$= 2 \left[z . \left(\frac{- \cos\left(\frac{\pi z}{2}\right)}{\frac{\pi}{2}} \right) \right]_0^1 - 2 \int_0^1 \frac{d}{dz} (z) . \frac{- \cos\left(\frac{\pi z}{2}\right)}{\frac{\pi}{2}} dz$$

$$= 2 \left[- z \cos\left(\frac{\pi z}{2}\right) . \frac{2}{\pi} \right]_0^1 - 2 \int_0^1 - \frac{2}{\pi} \cos\left(\frac{\pi z}{2}\right) dz$$

$$= - \frac{4}{\pi} \left[\cos\left(\frac{\pi}{2}\right) - 0 \right] + \frac{4}{\pi} \left[\sin\left(\frac{\pi z}{2}\right) . \frac{2}{\pi} \right]_0^1 = - \frac{4}{\pi} [0 - 0] + \frac{8}{\pi^2} \left[\sin\left(\frac{\pi}{2}\right) - \sin 0° \right]$$

$$= 0 + \frac{8}{\pi^2} [1 - 0] = \frac{8}{\pi^2} .$$

(*ii*) Let
$$I = \int_0^{2\pi} \frac{x \sin^{2n} x}{\sin^{2n} x + \cos^{2n} x} dx \qquad \text{...(1)}$$

$$= \int_0^{2\pi} \frac{(2\pi - x) \sin^{2n} (2\pi - x)}{\sin^{2n} (2\pi - x) + \cos^{2n} (2\pi - x)} dx \qquad \left[\because \ \int_0^a f(x) \, dx = \int_0^a f(a - x) \, dx\right]$$

$$\therefore \quad I = \int_0^{2\pi} \frac{(2\pi - x) \sin^{2n} x}{\sin^{2n} x + \cos^{2n} x} dx \qquad \text{...(2)}$$

Adding equations (1) and (2), we get

$$2I = \int_0^{2\pi} \frac{x \sin^{2n} x}{\sin^{2n} x + \cos^{2n} x} dx + \int_0^{2\pi} \frac{(2\pi - x) \sin^{2n} x}{\sin^{2n} x + \cos^{2n} x} dx$$

$$= \int_0^{2\pi} \left[\frac{x \sin^{2n} x}{\sin^{2n} x + \cos^{2n} x} + \frac{(2\pi - x) \sin^{2n} x}{\sin^{2n} x + \cos^{2n} x} \right] dx$$

$$= \int_0^{2\pi} \frac{2\pi \sin^{2n} x}{\sin^{2n} x + \cos^{2n} x}\, dx$$

$$\therefore \quad I = \pi \int_0^{2\pi} \frac{\sin^{2n} x}{\sin^{2n} x + \cos^{2n} x}\, dx$$

$$= 2\pi \int_0^{\pi} \frac{\sin^{2n} x}{\sin^{2n} x + \cos^{2n} x}\, dx \qquad \left[\because \int_0^{2a} f(x)\, dx = 2 \int_0^a f(x)\, dx; \text{ if } f(2a - x) = f(x) \right]$$

[Again using the same property as above]

$$= 4\pi \int_0^{\pi/2} \frac{\sin^{2n} x}{\sin^{2n} x + \cos^{2n} x}\, dx \qquad ...(3)$$

$$= 4\pi \int_0^{\pi/2} \frac{\sin^{2n}\left(\frac{\pi}{2} - x\right)}{\sin^{2n}\left(\frac{\pi}{2} - x\right) + \cos^{2n}\left(\frac{\pi}{2} - x\right)}\, dx \qquad \left[\because \int_0^a f(x)\, dx = \int_0^a f(a - x)\, dx \right]$$

$$= 4\pi \int_0^{\pi/2} \frac{\cos^{2n} x}{\cos^{2n} x + \sin^{2n} x}\, dx \qquad ...(4)$$

Adding equations (3) and (4), we get

$$2I = 4\pi \int_0^{\pi/2} \frac{\sin^{2n} x}{\sin^{2n} x + \cos^{2n} x}\, dx + 4\pi \int_0^{\pi/2} \frac{\cos^{2n} x}{\cos^{2n} x + \sin^{2n} x}\, dx$$

$$= 4\pi \int_0^{\pi/2} \left[\frac{\sin^{2n} x}{\sin^{2n} x + \cos^{2n} x} + \frac{\cos^{2n} x}{\cos^{2n} x + \sin^{2n} x} \right] dx$$

$$= 4\pi \int_0^{\pi/2} \left[\frac{\sin^{2n} x + \cos^{2n} x}{\sin^{2n} x + \cos^{2n} x} \right] dx$$

$$\Rightarrow \quad I = 2\pi \int_0^{\pi/2} 1 \,.\, dx = 2\pi \left[x\right]_0^{\pi/2} = 2\pi \left[\frac{\pi}{2} - 0 \right] = \pi^2$$

$$\Rightarrow \quad I = \pi^2.$$

(*iii*) Let $\quad I = \int_0^{\pi} \cos 2x \log (\sin x)\, dx$

$$\therefore \quad I = 2 \int_0^{\pi/2} \underset{\text{II}}{\cos 2x} \log \underset{\text{I}}{(\sin x)}\, dx \qquad \left[\because \int_0^{2a} f(x)\, dx = 2 \int_0^a f(x)\, dx; \text{ if } f(2a - x) = f(x) \right]$$

Integrating by parts, we get

$$I = 2 \left[\log (\sin x) \,.\, \frac{\sin 2x}{2} \right]_0^{\pi/2} - 2 \int_0^{\pi/2} \frac{d}{dx} \{\log (\sin x)\} \,.\, \frac{\sin 2x}{2}\, dx$$

$$= \Big[\sin 2x \log(\sin x)\Big]_0^{\pi/2} - \int_0^{\pi/2} \frac{1}{\sin x} \cdot \cos x \cdot \sin 2x \, dx$$

[∵ $\sin 2A = 2 \sin A \cos A$]

$$= \left[\sin \pi \log\left\{\sin\left(\frac{\pi}{2}\right)\right\} - \sin 0° \log(\sin 0°)\right] - \int_0^{\pi/2} \frac{\cos x}{\sin x} \cdot 2 \sin x \cos x \, dx$$

$$= (0 - 0) - 2\int_0^{\pi/2} \cos^2 x \, dx$$

$$= -2\int_0^{\pi/2} \left(\frac{1 + \cos 2x}{2}\right) dx \qquad \left[\begin{array}{l} \because \ 1 + \cos 2A = 2\cos^2 A \\ \Rightarrow \ \dfrac{1 + \cos 2A}{2} = \cos^2 A \end{array}\right]$$

$$= -\int_0^{\pi/2} (1 + \cos 2x) \, dx$$

$$= -\left[x + \frac{\sin 2x}{2}\right]_0^{\pi/2} = -\left[\left(\frac{\pi}{2} + \frac{\sin \pi}{2}\right) - \left(0 + \frac{\sin 0°}{2}\right)\right]$$

$$= -\left[\left(\frac{\pi}{2} + 0\right) - (0 + 0)\right]$$

∴ $$I = -\frac{\pi}{2}.$$

(*iv*) Let $$I = \int_{-1}^{1} \frac{x \sin^{-1} x}{\sqrt{1 - x^2}} \, dx$$

Let $$f(x) = \frac{x \sin^{-1} x}{\sqrt{1 - x^2}}$$

⇒ $$f(-x) = \frac{(-x)\sin^{-1}(-x)}{\sqrt{1 - (-x)^2}} = \frac{(-x)(-\sin^{-1} x)}{\sqrt{1 - x^2}} = \frac{x \sin^{-1} x}{\sqrt{1 - x^2}} = f(x)$$

∴ $f(x)$ is an even function.

∴ $$I = \int_{-1}^{1} \frac{x \sin^{-1} x}{\sqrt{1 - x^2}} \, dx = 2\int_0^{1} \frac{x \sin^{-1} x}{\sqrt{1 - x^2}} \, dx \qquad \left[\begin{array}{l} \because \ \int_{-a}^{a} f(x)\, dx = 2\int_0^{a} f(x)\, dx; \\ \text{if } f(x) \text{ is an even function} \end{array}\right]$$

Put $x = \sin \theta \ \Rightarrow \ dx = \cos \theta \, d\theta$

When $x = 0$; $\sin \theta = 0 \ \Rightarrow \ \theta = 0$ and when $x = 1$; $\sin \theta = 1 \ \Rightarrow \ \theta = \frac{\pi}{2}$

∴ $$I = 2\int_0^{\pi/2} \frac{\theta \sin \theta}{\sqrt{1 - \sin^2 \theta}} \cdot \cos \theta \, d\theta$$

$$= 2\int_0^{\pi/2} \frac{\theta \sin \theta \cos \theta}{\sqrt{\cos^2 \theta}} \, d\theta \qquad [\because \ \sin^2 A + \cos^2 A = 1]$$

$$= 2\int_0^{\pi/2} \underset{\text{I}}{\theta} \ \underset{\text{II}}{(\sin \theta)} \, d\theta$$

Integrating by parts, we get

$$= 2\left[\theta.(-\cos\theta)\right]_0^{\pi/2} - 2\int_0^{\pi/2}\frac{d}{d\theta}(\theta).(-\cos\theta)\,d\theta$$

$$= -2\left[\theta\cos\theta\right]_0^{\pi/2} + 2\int_0^{\pi/2}\cos\theta\,d\theta = -2\left[\frac{\pi}{2}\cos\frac{\pi}{2} - 0\right] + 2\left[\sin\theta\right]_0^{\pi/2}$$

$$= -2\,[0-0] + 2\left[\sin\frac{\pi}{2} - \sin 0^\circ\right] = 0 + 2\,[1-0]$$

$$= 2.$$

(*v*) Let $$\mathrm{I} = \int_{-\pi/4}^{\pi/4}\frac{\left(x+\frac{\pi}{4}\right)}{(2-\cos 2x)}\,dx$$

$$= \int_{-\pi/4}^{\pi/4}\frac{x}{(2-\cos 2x)}\,dx + \int_{-\pi/4}^{\pi/4}\frac{\frac{\pi}{4}}{(2-\cos 2x)}\,dx$$ [Note this step]

$$= 0 + \frac{\pi}{4}\int_{-\pi/4}^{\pi/4}\frac{1}{2-\cos 2x}\,dx$$

$\left[\because\ \int_{-a}^{a} f(x)\,dx = 0\text{; if } f(x) \text{ is an odd function}\right.$

$\int_{-a}^{a} f(x)\,dx = 2\int_0^a f(x)\,dx\text{; if } f(x) \text{ is an even function}$

Here $\frac{x}{2-\cos 2x}$ is an odd function

$\left.\text{and } \frac{1}{2-\cos 2x} \text{ is an even function.}\right]$

$$= 2\,.\,\frac{\pi}{4}\int_0^{\pi/4}\frac{1}{2-\cos 2x}\,dx$$

$$= \frac{\pi}{2}\int_0^{\pi/4}\frac{1}{2-\frac{1-\tan^2 x}{1+\tan^2 x}}\,dx$$ $\left[\because\ \cos 2\mathrm{A} = \frac{1-\tan^2\mathrm{A}}{1+\tan^2\mathrm{A}}\right]$

$$= \frac{\pi}{2}\int_0^{\pi/4}\frac{1+\tan^2 x}{2(1+\tan^2 x)-(1-\tan^2 x)}\,dx = \frac{\pi}{2}\int_0^{\pi/4}\frac{1+\tan^2 x}{1+3\tan^2 x}\,dx$$

$\Rightarrow$ $$\mathrm{I} = \frac{\pi}{2}\int_0^{\pi/2}\frac{\sec^2 x}{1+3\tan^2 x}\,dx$$ $\left[\because\ \sec^2\mathrm{A} - \tan^2\mathrm{A} = 1 \Rightarrow \sec^2\mathrm{A} = 1+\tan^2\mathrm{A}\right]$

Put $\tan x = z \Rightarrow \sec^2 x\,dx = dz$

When $x = 0$; $z = \tan 0^\circ = 0$ and when $x = \frac{\pi}{2}$; $z = \tan\frac{\pi}{2} = \infty$

$$\mathrm{I} = \frac{\pi}{2}\int_0^{\infty}\frac{1}{1+3z^2}\,dz$$

$$= \frac{\pi}{6}\int_0^{\infty}\frac{1}{\frac{1}{3}+z^2}\,dz = \frac{\pi}{6}\int_0^{\infty}\frac{1}{\left(\frac{1}{\sqrt{3}}\right)^2+z^2}\,dz$$

$$= \frac{\pi}{6} \cdot \left[\frac{1}{\frac{1}{\sqrt{3}}} \tan^{-1} \left(\frac{z}{\frac{1}{\sqrt{3}}} \right) \right]_0^{\infty} \qquad \left[\because \int \frac{1}{a^2 + x^2}\, dx = \frac{1}{a} \tan^{-1} \frac{x}{a} + c \right]$$

$$= \frac{\sqrt{3}\pi}{6} \cdot \left[\tan^{-1} \sqrt{3} z \right]_0^{\infty} = \frac{\sqrt{3}\pi}{2 \times \sqrt{3} \times \sqrt{3}} [\tan^{-1} \infty - \tan^{-1} 0] = \frac{\pi}{2\sqrt{3}} \left[\frac{\pi}{2} - 0 \right]$$

$$= \frac{\pi^2}{4\sqrt{3}} .$$

(*vi*) Let $\quad I = \int_{-\pi}^{\pi} \frac{2x(1 + \sin x)}{1 + \cos^2 x}\, dx = \int_{-\pi}^{\pi} \frac{(2x + 2x \sin x)}{1 + \cos^2 x} \cdot dx$

$$= \int_{-\pi}^{\pi} \frac{2x}{1 + \cos^2 x}\, dx + \int_{-\pi}^{\pi} \frac{2x \sin x}{1 + \cos^2 x}\, dx$$

$$= 0 + 2 \int_0^{\pi} \frac{2x \sin x}{1 + \cos^2 x}\, dx$$

$\Bigg[\because \int_{-a}^{a} f(x)\, dx = 0$; if $f(x)$ is an odd function

$\int_{-a}^{a} f(x)\, dx = 2 \int_0^{a} f(x)\, dx$; if $f(x)$ is an even function

Here $\frac{2x}{1 + \cos^2 x}$ is an odd function

and $\frac{2x \sin x}{1 + \cos^2 x}$ is an even function. $\Bigg]$

$$= 2 \int_0^{\pi} \frac{2x \sin x}{1 + \cos^2 x}\, dx$$

$$= 4 \int_0^{\pi} \frac{x \sin x}{1 + \cos^2 x}\, dx \qquad \text{...(1)}$$

$$= 4 \int_0^{\pi} \frac{(\pi - x) \sin (\pi - x)}{1 + \cos^2 (\pi - x)}\, dx \qquad \left[\because \int_0^{a} f(x)\, dx = \int_0^{a} f(a - x)\, dx \right]$$

$\therefore \quad I = 4 \int_0^{\pi} \frac{(\pi - x) \sin x}{1 + \cos^2 x}\, dx \qquad$ ...(2)

Adding equations (1) and (2), we get

$$2I = 4 \int_0^{\pi} \frac{x \sin x}{1 + \cos^2 x}\, dx + 4 \int_0^{\pi} \frac{(\pi - x) \sin x}{1 + \cos^2 x}\, dx$$

$$= 4 \int_0^{\pi} \left[\frac{x \sin x}{1 + \cos^2 x} + \frac{(\pi - x) \sin x}{1 + \cos^2 x} \right] dx$$

$\Rightarrow \quad I = 2 \int_0^{\pi} \frac{\pi \sin x}{1 + \cos^2 x}\, dx = 2\pi \int_0^{\pi} \frac{\sin x}{1 + \cos^2 x}\, dx$

Put $\quad \cos x = z \;\Rightarrow\; - \sin x\, dx = dz \;\Rightarrow\; \sin x\, dx = - dz$

When $\quad x = 0$; $z = \cos 0° = 1$ and when $x = \pi$; $z = \cos \pi = - 1$

$$\therefore \qquad I = 2\pi \int_{+1}^{-1} \frac{1}{1+z^2} (-dz) = -2\pi \int_{1}^{-1} \frac{1}{1+z^2}\, dz$$

$$= 2\pi \int_{-1}^{1} \frac{1}{1+z^2}\, dz \qquad \left[\because \int_a^b f(x)\,dx = -\int_b^a f(x)\,dx\right]$$

$$= 2\pi \left[\tan^{-1} z\right]_{-1}^{1} \qquad \left[\because \int \frac{1}{a^2+x^2}\,dx = \frac{1}{a}\tan^{-1}\frac{x}{a} + c\right]$$

$$= 2\pi\,[\tan^{-1} 1 - \tan^{-1}(-1)] = 2\pi\left[\frac{\pi}{4} - \left(-\frac{\pi}{4}\right)\right] = 2\pi\left[\frac{\pi}{4} + \frac{\pi}{4}\right]$$

$$= 2\pi\left(\frac{\pi}{2}\right) = \pi^2.$$

Example 17. *Evaluate the following integrals :*

(*i*) $\int_{-\pi/2}^{\pi/2} |\sin x|\, dx$ (*ii*) $\int_0^{2\pi} |\sin x|\, dx$

(*iii*) $\int_{-1}^{3/2} |x \sin \pi x|\, dx$ (*iv*) $\int_0^{\pi} |\cos x|\, dx$

(*v*) $\int_0^{\pi/2} |\cos 2x|\, dx$ (*vi*) $\int_{-\pi/2}^{\pi/2} (\sin |x| - \cos |x|)\, dx$

(*vii*) $\int_0^{3/2} |x \cos \pi x|\, dx$ (*viii*) $\int_{-\pi/4}^{\pi/4} |\sin x|\, dx$

(*ix*) $\int_{1/e}^{e} |\log_e x|\, dx$ (*x*) $\int_0^{2\pi} |\cos x|\, dx$

(*xi*) $\int_{-1/2}^{1/2} \left| x \cos \frac{\pi}{2} x \right| dx.$

Solution. (*i*) Let $I = \int_{-\pi/2}^{\pi/2} |\sin x|\, dx$

Since, we have

$$|\sin x| = \begin{cases} -\sin x; \text{ if } & -\frac{\pi}{2} \le x < 0 \\ \sin x; \text{ if } & 0 < x \le \frac{\pi}{2} \end{cases}$$

$$\therefore \qquad I = \int_{-\pi/2}^{\pi/2} |\sin x|\, dx = \int_{-\pi/2}^{0} |\sin x|\, dx + \int_0^{\pi/2} |\sin x|\, dx$$

$$= \int_{-\pi/2}^{0} -\sin x\, dx + \int_0^{\pi/2} \sin x\, dx$$

$$= -\left[-\cos x\right]_{-\pi/2}^{0} + \left[-\cos x\right]_0^{\pi/2} = \left[\cos x\right]_{-\pi/2}^{0} - \left[\cos x\right]_0^{\pi/2}$$

$$= \left[\cos 0° - \cos\left(-\frac{\pi}{2}\right)\right] - \left[\cos\frac{\pi}{2} - \cos 0°\right] = [1-0] - [0-1]$$

$$= 1 + 1 = 2.$$

(*ii*) Let $\quad I = \int_0^{2\pi} |\sin x|\, dx$

Since, we have

$$|\sin x| = \begin{cases} -\sin x ; \text{ if } & \pi \le x \le 2\pi \\ \sin x ; \text{ if } & 0 \le x \le \pi \end{cases}$$

$$\therefore \quad I = \int_0^{2\pi} |\sin x|\, dx = \int_0^{\pi} |\sin x|\, dx + \int_{\pi}^{2\pi} |\sin x|\, dx$$

$$= \int_0^{\pi} \sin x\, dx + \int_{\pi}^{2\pi} -\sin x\, dx$$

$$= \Big[-\cos x\Big]_0^{\pi} + \Big[\cos x\Big]_{\pi}^{2\pi} = -[(\cos \pi - \cos 0°)] + [\cos 2\pi - \cos \pi]$$

$$= -[-1-1] + [1-(-1)] = -[-2] + [1+1]$$

$$= 2 + 2 = 4.$$

(*iii*) Let $\quad I = \int_{-1}^{3/2} |x \sin \pi x|\, dx$

Since, we have if $-1 < x < \frac{3}{2} \Rightarrow -\pi < \pi x < \frac{3\pi}{2}$

$\therefore$ when, $-1 < x < 0 \Rightarrow -\pi < \pi x < 0$

$\Rightarrow \sin \pi x < 0$

$\Rightarrow x \sin \pi x > 0$

$$\therefore \quad |x \sin \pi x| = x \sin \pi x \qquad ...(1)$$

when, $0 < x < 1 \Rightarrow 0 < \pi x < \pi$

$\Rightarrow \sin \pi x > 0$

$\Rightarrow x \sin \pi x > 0$

$$\therefore \quad |x \sin \pi x| = x \sin \pi x \qquad ...(2)$$

when $1 < x < \frac{3}{2} \Rightarrow \pi < \pi x < \frac{3\pi}{2}$

$\Rightarrow \sin \pi x < 0$

$\Rightarrow x \sin \pi x < 0$

$$\therefore \quad |x \sin \pi x| = -x \sin \pi x \qquad ...(3)$$

$\therefore$ From equations (1) to (3), we have

$$|x \sin \pi x| = \begin{cases} x \sin \pi x ; \text{ if } & -1 < x < 1 \\ -x \sin \pi x ; \text{ if } & 1 < x < \frac{3}{2} \end{cases}$$

$$\therefore \quad I = \int_{-1}^{3/2} |x \sin \pi x|\, dx = \int_{-1}^{0} |x \sin \pi x|\, dx + \int_0^1 |x \sin \pi x|\, dx + \int_1^{3/2} |x \sin \pi x|\, dx$$

$$= \int_{-1}^{0} x \sin \pi x\, dx + \int_0^1 x \sin \pi x\, dx + \int_1^{3/2} -x \sin \pi x\, dx$$

$$\left[\because \int_a^b f(x)\, dx = \int_a^c f(x)\, dx + \int_c^b f(x)\, dx ; a < c < b\right]$$

$$I = \int_{-1}^{1} \underset{\text{I}}{x} \underset{\text{II}}{\sin \pi x}\, dx - \int_{1}^{3/2} \underset{\text{I}}{x} \underset{\text{II}}{\sin \pi x}\, dx$$

Integrating by parts, we get

$$= \left[x.\left(-\frac{\cos \pi x}{\pi}\right)\right]_{-1}^{1} - \int_{-1}^{1} \frac{d}{dx}(x).\left(-\frac{\cos \pi x}{\pi}\right) dx - \left[x.\left(-\frac{\cos \pi x}{\pi}\right)\right]_{1}^{3/2} + \int_{1}^{3/2} \frac{d}{dx}(x).\left(-\frac{\cos \pi x}{\pi}\right) dx$$

$$= -\frac{1}{\pi}\left[x \cos \pi x\right]_{-1}^{1} + \frac{1}{\pi}\int_{-1}^{1} \cos \pi x\, dx + \frac{1}{\pi}\left[x \cos \pi x\right]_{1}^{3/2} - \frac{1}{\pi}\int_{1}^{3/2} \cos \pi x\, dx$$

$$= -\frac{1}{\pi}[\cos \pi - \{-\cos(-\pi)\}] + \frac{1}{\pi}\left[\frac{\sin \pi x}{\pi}\right]_{-1}^{1} + \frac{1}{\pi}\left[\frac{3}{2}\cos\frac{3\pi}{2} - \cos \pi\right] - \frac{1}{\pi}\left[\frac{\sin \pi x}{\pi}\right]_{1}^{3/2}$$

$$= -\frac{1}{\pi}[(-1) + (-1)] + \frac{1}{\pi^2}[\sin \pi - \sin(-\pi)] + \frac{1}{\pi}\left[\frac{3}{2}(0) - (-1)\right] - \frac{1}{\pi^2}\left[\sin\frac{3\pi}{2} - \sin \pi\right]$$

$$= \frac{2}{\pi} + \frac{1}{\pi^2}[0 + 0] + \frac{1}{\pi}[0 + 1] - \frac{1}{\pi^2}[-1 - 0] = \frac{2}{\pi} + 0 + \frac{1}{\pi} + \frac{1}{\pi^2} = \frac{3}{\pi} + \frac{1}{\pi^2}$$

$$= \frac{3\pi + 1}{\pi^2}.$$

(*iv*) Let $I = \int_0^{\pi} |\cos x|\, dx$

Since, we have $|\cos x| = \begin{cases} \cos x; \text{ if } & 0 \le x \le \frac{\pi}{2} \\ -\cos x; \text{ if } & \frac{\pi}{2} \le x \le \pi \end{cases}$

$$\therefore \quad I = \int_0^{\pi} |\cos x|\, dx = \int_0^{\pi/2} |\cos x|\, dx + \int_{\pi/2}^{\pi} |\cos x|\, dx$$

$$= \int_0^{\pi/2} \cos x\, dx + \int_{\pi/2}^{\pi} -\cos x\, dx$$

$$= \left[\sin x\right]_0^{\pi/2} - \left[\sin x\right]_{\pi/2}^{\pi} = \left[\sin\frac{\pi}{2} - \sin 0°\right] - \left[\sin \pi - \sin\frac{\pi}{2}\right]$$

$$= [1 - 0] - [0 - 1] = 1 + 1 = 2.$$

(*v*) Let $I = \int_0^{\pi/2} |\cos 2x|\, dx$

Since, we have $|\cos 2x| = \begin{cases} \cos 2x; \text{ if } 0 \le x < \frac{\pi}{4} \Rightarrow 0 \le 2x < \frac{\pi}{2} \\ -\cos 2x; \text{ if } \frac{\pi}{4} < x \le \frac{\pi}{2} \Rightarrow \frac{\pi}{2} < 2x \le \pi \end{cases}$

$$\therefore \quad I = \int_0^{\pi/2} |\cos 2x|\, dx = \int_0^{\pi/4} |\cos 2x|\, dx + \int_{\pi/4}^{\pi/2} |\cos 2x|\, dx$$

$$= \int_0^{\pi/4} \cos 2x \, dx + \int_{\pi/4}^{\pi/2} - \cos 2x \, dx$$

$$= \left[\frac{\sin 2x}{2}\right]_0^{\pi/4} - \left[\frac{\sin 2x}{2}\right]_{\pi/4}^{\pi/2} = \frac{1}{2}\left[\sin 2\left(\frac{\pi}{4}\right) - \sin 0°\right] - \frac{1}{2}\left[\sin 2\left(\frac{\pi}{2}\right) - \sin 2\left(\frac{\pi}{4}\right)\right]$$

$$= \frac{1}{2}\left[\sin\frac{\pi}{2} - 0\right] - \frac{1}{2}\left[\sin \pi - \sin\frac{\pi}{2}\right] = \frac{1}{2}[1 - 0] - \frac{1}{2}[0 - 1]$$

$$= \frac{1}{2} + \frac{1}{2} = 1.$$

(*vi*) Let $\quad I = \int_{-\pi/2}^{\pi/2} (\sin |x| - \cos |x|) \, dx$

$$= \int_{-\pi/2}^{\pi/2} \sin |x| \, dx - \int_{-\pi/2}^{\pi/2} \cos |x| \, dx$$

Since $\sin |x|$ and $\cos |x|$ are even functions.

$\therefore \quad I = 2\int_0^{\pi/2} \sin |x| \, dx - 2\int_0^{\pi/2} \cos |x| \, dx \quad \left[\because \int_{-a}^{a} f(x) \, dx = 2\int_0^a f(x) \, dx; \text{ if } f(x) \text{ is an even function}\right]$

$$= 2\int_0^{\pi/2} \sin x \, dx - 2\int_0^{\pi/2} \cos x \, dx \quad \left[\because |x| = x \text{ on } \left[0, \frac{\pi}{2}\right]\right]$$

$$= 2\left[-\cos x\right]_0^{\pi/2} - 2\left[\sin x\right]_0^{\pi/2}$$

$$= -2\left[\cos\frac{\pi}{2} - \cos 0°\right] - 2\left[\sin\frac{\pi}{2} - \sin 0°\right] = -2[0 - 1] - 2[1 - 0]$$

$$= 2 - 2 = 0.$$

(*vii*) Let $\quad I = \int_0^{3/2} |x \cos \pi x| \, dx$

Put $\quad \pi x = z \Rightarrow \pi dx = dz \Rightarrow dx = \frac{1}{\pi} dz$

When $\quad x = 0; \; z = 0$ and when $x = \frac{3}{2}; \; z = \frac{3\pi}{2}$

$\therefore \quad I = \int_0^{3\pi/2} \left|\frac{z}{\pi} \cos z\right| . \frac{1}{\pi} \, dz$

$$= \frac{1}{\pi^2}\int_0^{3\pi/2} |z \cos z| \, dz$$

$$= \frac{1}{\pi^2}\int_0^{3\pi/2} |z| \, . \, |\cos z| \, dz$$

$$= \frac{1}{\pi^2}\int_0^{3\pi/2} z \, |\cos z| \, dz$$

Since, we have

$$|\cos z| = \begin{cases} \cos z\,; \text{ if } \cos z \geq 0 \text{ on } \left[0, \dfrac{\pi}{2}\right] \\ -\cos z\,; \text{ if } \cos z \leq 0 \text{ on } \left[\dfrac{\pi}{2}, \dfrac{3\pi}{2}\right] \end{cases}$$

$$\therefore \quad I = \frac{1}{\pi^2}\left[\int_0^{\pi/2} z\,|\cos z|\,dz + \int_{\pi/2}^{3\pi/2} z\,|\cos z|\,dz\right] = \frac{1}{\pi^2}\left[\int_0^{\pi/2} z\cos z\,dz + \int_{\pi/2}^{3\pi/2} z\,(-\cos z)\,dz\right]$$

$$= \frac{1}{\pi^2}\left[\int_0^{\pi/2} \underset{\text{I}}{z}\ \underset{\text{II}}{\cos z}\,dz - \int_{\pi/2}^{3\pi/2} \underset{\text{I}}{z}\ \underset{\text{II}}{\cos z}\,dz\right]$$

Integrating by parts, we get

$$= \frac{1}{\pi^2}\Big[z.\sin z\Big]_0^{\pi/2} - \frac{1}{\pi^2}\int_0^{\pi/2} \sin z\,dz - \frac{1}{\pi^2}\Big[z\sin z\Big]_{\pi/2}^{3\pi/2} - \frac{1}{\pi^2}\int_{\pi/2}^{3\pi/2} \sin z\,dz$$

$$= \frac{1}{\pi^2}\left[\frac{\pi}{2}\sin\frac{\pi}{2} - 0\right] - \frac{1}{\pi^2}\Big[\cos z\Big]_0^{\pi/2} - \frac{1}{\pi^2}\left[\frac{3\pi}{2}\sin\frac{3\pi}{2} - \frac{\pi}{2}\sin\frac{\pi}{2}\right] - \frac{1}{\pi^2}\Big[\cos z\Big]_{\pi/2}^{3\pi/2}$$

$$= \frac{1}{\pi^2}\left[\frac{\pi}{2} - 0\right] - \frac{1}{\pi^2}\left[\cos\frac{\pi}{2} - \cos 0°\right] - \frac{1}{\pi^2}\left[-\frac{3\pi}{2} - \frac{\pi}{2}\right] - \frac{1}{\pi^2}\left[\cos\frac{3\pi}{2} - \cos\frac{\pi}{2}\right]$$

$$= \frac{1}{2\pi} - \frac{1}{\pi^2}[0 - 1] + \frac{3}{2\pi} + \frac{1}{2\pi} - \frac{1}{\pi^2}[0 - 0] = \frac{1}{2\pi} - \frac{1}{\pi^2} + \frac{3}{2\pi} + \frac{1}{2\pi}$$

$$= \frac{5}{2\pi} - \frac{1}{\pi^2} = \left(\frac{5\pi - 2}{2\pi^2}\right).$$

(*viii*) Let $\quad I = \displaystyle\int_{-\pi/4}^{\pi/4} |\sin x|\ dx$

Since, we have

$$|\sin x| = \begin{cases} \sin x\,; \text{ if } 0 \leq x \leq \dfrac{\pi}{4} \\ -\sin x\,; \text{ if } -\dfrac{\pi}{4} \leq x \leq 0 \end{cases}$$

$$\therefore \quad I = \int_{-\pi/4}^{\pi/4} |\sin x|\,dx = \int_{-\pi/4}^{0} |\sin x|\,dx + \int_0^{\pi/4} |\sin x|\ dx$$

$$= \int_{-\pi/4}^{0} -\sin x\,dx + \int_0^{\pi/4} \sin x\,dx$$

$$= -\Big[-\cos x\Big]_{-\pi/4}^{0} + \Big[-\cos x\Big]_0^{\pi/4} = \Big[\cos x\Big]_{-\pi/4}^{0} - \Big[\cos x\Big]_0^{\pi/4}$$

$$= \left[\cos 0° - \cos\left(-\frac{\pi}{4}\right)\right] - \left[\cos\frac{\pi}{4} - \cos 0°\right] = \left[1 - \frac{1}{\sqrt{2}}\right] - \left[\frac{1}{\sqrt{2}} - 1\right]$$

$$= 2 - \sqrt{2}\ .$$

(*ix*) Let $I = \int_{1/e}^{e} |\log_e x| \, . \, dx$

Since, we have

$$|\log_e x| = \begin{cases} -\log_e x ; \text{ if } & \frac{1}{e} < x < 1 \\ \log_e x ; \text{ if } & -1 < x < e \end{cases}$$

$\therefore$ $$I = \int_{1/e}^{e} |\log_e x| . dx = \int_{1/e}^{1} -\log_e x \, dx + \int_{1}^{e} \log_e x \, dx$$

$$= -\int_{1/e}^{1} \log_e x \, dx + \int_{1}^{e} \log_e x \, dx$$

$$= -\Big[x(\log_e x - 1)\Big]_{1/e}^{1} + \Big[x(\log_e x - 1)\Big]_{1}^{e}$$

$$= -\left[1.(\log_e 1 - 1) - \left\{\frac{1}{e}\log_e\left(\frac{1}{e}\right) - 1\right\}\right] + [e(\log_e e - 1) - 1(\log_e 1 - 1)]$$

$$= -\left[(0-1) - \frac{1}{e}(-1-1)\right] + [e(1-1) - 1(0-1)]$$

$$= -\left[-1 + \frac{2}{e}\right] + [0 + 1] = 1 - \frac{2}{e} + 1$$

$$= 2 - \frac{2}{e} .$$

(*x*) Let $I = \int_{0}^{2\pi} |\cos x| \, dx$

Since, we have

$$|\cos x| = \begin{cases} \cos x ; \text{ when } 0 \le x \le \frac{\pi}{2} \\ -\cos x ; \text{ when } \frac{\pi}{2} \le x \le \frac{3\pi}{2} \\ \cos x ; \text{ when } \frac{3\pi}{2} \le x \le 2\pi \end{cases}$$

$\therefore$ $$I = \int_{0}^{\pi/2} |\cos x| \, dx + \int_{\pi/2}^{3\pi/2} |\cos x| \, dx + \int_{3\pi/2}^{2\pi} |\cos x| \, dx$$

$$= \int_{0}^{\pi/2} \cos x \, dx + \int_{\pi/2}^{3\pi/2} -\cos x \, dx + \int_{3\pi/2}^{2\pi} \cos x \, dx$$

Note. $\because \int \underset{\text{II}}{1} . \underset{\text{I}}{\log_e x} \, dx$ [Taking unity as second function]

Integrating by parts, we have

$$= \log_e x . x - \int \frac{d}{dx}(\log_e x) . x \, dx = x \log_e x - \int \frac{1}{x} . x \, dx$$

$$= x \log_e x - \int 1 . dx = x \log_e x - x = x(\log_e x - 1)$$

$$= \Big[\sin x\Big]_0^{\pi/2} - \Big[\sin x\Big]_{\pi/2}^{3\pi/2} + \Big[\sin x\Big]_{3\pi/2}^{2\pi}$$

$$= \left[\sin\frac{\pi}{2} - \sin 0°\right] - \left[\sin\frac{3\pi}{2} - \sin\frac{\pi}{2}\right] + \left[\sin 2\pi - \sin\frac{3\pi}{2}\right]$$

$$= [1 - 0] - [-1 - 1] + [0 - (-1)] = 1 - (-2) + 1$$

$$= 4.$$

(*xi*) Let $\quad I = \int_{-1/2}^{1/2} \left| x \cos\frac{\pi}{2} x \right| dx$

Put $\quad \frac{\pi}{2} x = z \;\Rightarrow\; x = \frac{2z}{\pi} \;\Rightarrow\; dx = \frac{2}{\pi} dz$

When $x = -\frac{1}{2}$; $z = -\frac{1}{2}\left(\frac{\pi}{2}\right) = -\frac{\pi}{4}$ and when $x = \frac{1}{2}$; $z = \frac{1}{2}\left(\frac{\pi}{2}\right) = \frac{\pi}{4}$.

$\therefore \quad I = \int_{-1/2}^{1/2} \left| x \cos\frac{\pi}{2} x \right| dx = \int_{-\pi/4}^{\pi/4} \left| \frac{2}{\pi} z \cos z \right| \frac{2}{\pi} dz$

$$= \frac{4}{\pi^2} \int_{-\pi/4}^{\pi/4} |z \cos z|\, dz$$

Since $|z \cos z|$ is an even function.

$\therefore \quad \frac{4}{\pi^2} \int_{-\pi/4}^{\pi/4} |z \cos z|\, dz = \frac{4}{\pi^2} . 2 \int_0^{\pi/4} |z \cos z|\, dz.$ $\quad \left[\because \int_{-a}^{a} f(x)\, dx = 2\int_0^a f(x)\, dx ; \text{ if } f(x) \text{ is an even function}\right]$

$\therefore \quad I = \frac{8}{\pi^2} \int_0^{\pi/4} \underset{\text{I}}{z}\, \underset{\text{II}}{\cos z}\, dz$ $\quad \left[\because |z \cos z| = z \cos z \text{ on } \left[0, \frac{\pi}{4}\right]\right]$

Integrating by parts, we get

$$= \frac{8}{\pi^2}\Big[z . \sin z\Big]_0^{\pi/4} - \frac{8}{\pi^2}\int_0^{\pi/4} 1 . \sin z\, dz$$

$$= \frac{8}{\pi^2}\left[\frac{\pi}{4} \sin\frac{\pi}{4} - 0\right] - \frac{8}{\pi^2}\Big[-\cos z\Big]_0^{\pi/4}$$

$$= \frac{8}{\pi^2}\left[\frac{\pi}{4} . \left(\frac{1}{\sqrt{2}}\right) - 0\right] + \frac{8}{\pi^2}\left[\cos\frac{\pi}{4} - \cos 0°\right] = \frac{8}{\pi^2}\left[\frac{\pi}{4\sqrt{2}}\right] + \frac{8}{\pi^2}\left[\frac{1}{\sqrt{2}} - 1\right]$$

$$= \frac{8}{\pi^2}\left[\frac{\pi}{4\sqrt{2}} + \frac{1}{\sqrt{2}} - 1\right].$$

Example 18. *Evaluate the following integrals :*

(*i*) $\int_{-1}^{1} f(x)\, dx$; *where* $f(x) = \begin{cases} 1 - 2x ; \text{ if } x \le 0 \\ 1 + 2x ; \text{ if } x \ge 0 \end{cases}$

(*ii*) $\int_{1}^{4} f(x)\, dx$; *where* $f(x) = \begin{cases} 7x + 3 ; \text{ if } 1 \le x \le 3 \\ 8x ; \text{ if } 3 \le x \le 4 \end{cases}$

(*iii*) $\int_{1}^{3} f(x)\, dx$; *where* $f(x) = \begin{cases} x^2 - 8 ; \text{ if } 1 \le x \le 2 \\ -2x ; \text{ if } 2 \le x \le 3 \end{cases}$

(iv) $\int_1^4 f(x)\,dx$; where $f(x) = \begin{cases} 2x + 8 \text{ ; if } 1 \le x \le 2 \\ 6x \text{ ; if } 2 \le x \le 4 \end{cases}$

(v) $\int_1^4 f(x)\,dx$; where $f(x) = \begin{cases} 4x + 3 \text{ ; if } 1 \le x \le 2 \\ 3x + 5 \text{ ; if } 2 \le x \le 4 \end{cases}$

(vi) $\int_1^3 f(x)\,dx$; where $f(x) = \begin{cases} x^2 + x + 1 \text{ ; if } 1 \le x \le 2 \\ 2x + 3 \text{ ; if } 2 \le x \le 3 \end{cases}$

(vii) $\int_{-2}^3 f(x)\,dx$; where $f(x) = \begin{cases} 1 - 3x \text{ ; if } x \le 0 \\ 1 + 3x \text{ ; if } x \ge 0 \end{cases}$

(viii) $\int_0^9 f(x)\,dx$; where $f(x) = \begin{cases} \sin x \text{ ; if } 0 \le x \le \frac{\pi}{2} \\ 1 \text{ ; if } \frac{\pi}{2} \le x \le 3 \\ e^{x-3} \text{ ; if } 3 \le x \le 9 \end{cases}$.

Solution. (i) $\int_{-1}^1 f(x)\,dx = \int_{-1}^0 f(x)\,dx + \int_0^1 f(x)\,dx$

$$= \int_{-1}^0 (1 - 2x)\,dx + \int_0^1 (1 + 2x)\,dx$$

$$= \Big[x - x^2\Big]_{-1}^0 + \Big[x + x^2\Big]_0^1$$

$$= (0 - 0) - [(-1) - (-1)^2] + [1 + 1] - [0 + 0]$$

$$= 0 - (-1 - 1) + 2 - 0 = 2 + 2$$

$$= 4.$$

(ii) $\int_1^4 f(x)\,dx = \int_1^3 f(x)\,dx + \int_3^4 f(x)\,dx = \int_1^3 (7x + 3)\,dx + \int_3^4 8x\,dx$

$$= \left[\frac{7x^2}{2} + 3x\right]_1^3 + \left[\frac{8x^2}{2}\right]_3^4$$

$$= \left[\left\{\frac{7}{2}(3)^2 + 3(3)\right\} - \left\{\frac{7}{2}(1)^2 + 3(1)\right\}\right] + [4(4)^2 - 4(3)^2]$$

$$= \left[\left(\frac{63}{2} + 9\right) - \left(\frac{7}{2} + 3\right)\right] + [64 - 36] = \frac{63}{2} + 9 - \frac{7}{2} - 3 + 28$$

$$= \frac{56}{2} + 34 = 28 + 34 = 62.$$

(iii) $\int_1^3 f(x)\,dx = \int_1^2 f(x)\,dx + \int_2^3 f(x)\,dx = \int_1^2 (x^2 - 8)\,dx + \int_2^3 -2x\,dx$

$$= \left[\frac{x^3}{3} - 8x\right]_1^2 - \left[\frac{2x^2}{2}\right]_2^3 = \left[\left\{\frac{(2)^3}{3} - 8(2)\right\} - \left\{\frac{(1)^3}{3} - 8(1)\right\}\right] - [(3)^2 - (2)^2]$$

$$= \left[\left(\frac{8}{3} - 16\right) - \frac{1}{3} + 8\right] - [9 - 4] = \frac{7}{3} - 8 - 5 = \frac{7}{3} - 13 = \frac{7 - 39}{3}$$

$$= -\frac{32}{3}.$$

(iv) $$\int_1^4 f(x)\,dx = \int_1^2 f(x)\,dx + \int_2^4 f(x)\,dx = \int_1^2 (2x + 8)\,dx + \int_2^4 6x\,dx$$

$$= \left[\frac{2x^2}{2} + 8x\right]_1^2 + \left[\frac{6x^2}{2}\right]_2^4 = \left[x^2 + 8x\right]_1^2 + \left[3x^2\right]_2^4$$

$$= [\{(2)^2 + 8(2)\} - \{(1)^2 + 8(1)\}] + [3(4)^2 - 3(2)^2]$$

$$= [4 + 16 - 1 - 8] + [48 - 12] = 11 + 36$$

$$= 47.$$

(v) $$\int_1^4 f(x)\,dx = \int_1^2 f(x)\,dx + \int_2^4 f(x)\,dx = \int_1^2 (4x + 3)\,dx + \int_2^4 (3x + 5)\,dx$$

$$= \left[\frac{4x^2}{2} + 3x\right]_1^2 + \left[\frac{3x^2}{2} + 5x\right]_2^4 = \left[2x^2 + 3x\right]_1^2 + \left[\frac{3x^2}{2} + 5x\right]_2^4$$

$$= [\{2(2)^2 + 3(2)\} - \{2(1)^2 + 3(1)\}] + \left[\left\{\frac{3}{2}(4)^2 + 5(4)\right\} - \left\{\frac{3}{2}(2)^2 + 5(2)\right\}\right]$$

$$= [(8 + 6) - (2 + 3)] + [(24 + 20) - (6 + 10)] = 14 - 5 + 44 - 16$$

$$= 37.$$

(vi) $$\int_1^3 f(x)\,dx = \int_1^2 f(x)\,dx + \int_2^3 f(x)\,dx = \int_1^2 (x^2 + x + 1)\,dx + \int_2^3 (2x + 3)\,dx$$

$$= \left[\frac{x^3}{3} + \frac{x^2}{2} + x\right]_1^2 + \left[\frac{2x^2}{2} + 3x\right]_2^3$$

$$= \left[\left\{\frac{(2)^3}{3} + \frac{(2)^2}{2} + 2\right\} - \left\{\frac{1}{3} + \frac{1}{2} + 1\right\}\right] + [\{(3)^2 + 3(3)\} - \{(2)^2 + 3(2)\}]$$

$$= \left(\frac{8}{3} + \frac{4}{2} + 2\right) - \left(\frac{2 + 3 + 6}{2}\right) + [(9 + 9) - (4 + 6)]$$

$$= \left(\frac{16 + 12 + 12}{6}\right) - \left(\frac{11}{6}\right) + [18 - 10] = \frac{40}{6} - \frac{11}{6} + 8 = \left(\frac{40 - 11 + 48}{6}\right)$$

$$= \frac{77}{6}.$$

(vii) $$\int_{-2}^3 f(x)\,dx = \int_{-2}^0 f(x)\,dx + \int_0^3 f(x)\,dx = \int_{-2}^0 (1 - 3x)\,dx + \int_0^3 (1 + 3x)\,dx$$

$$= \left[x - \frac{3x^2}{2}\right]_{-2}^0 + \left[x + \frac{3x^2}{2}\right]_0^3$$

$$= \left[(0 - 0) - \left\{(-2) - \frac{3(-2)^2}{2}\right\}\right] + \left[\left\{3 + \frac{3(3)^2}{2}\right\} - (0 + 0)\right]$$

$$= -[-2-6] + \left[3 + \frac{27}{2}\right] = 8 + 3 + \frac{27}{2} = 11 + \frac{27}{2} = \frac{22+27}{2}$$

$$= \frac{49}{2}.$$

(*viii*) $$\int_0^9 f(x)\,dx = \int_0^{\pi/2} f(x)\,dx + \int_{\pi/2}^9 f(x)\,dx$$

$$= \int_0^{\pi/2} f(x)\,dx + \int_{\pi/2}^3 f(x)\,dx + \int_3^9 f(x)\,dx$$

$$= \int_0^{\pi/2} \sin x\,dx + \int_{\pi/2}^3 1.\,dx + \int_3^9 e^{x-3}\,dx$$

$$= \Big[-\cos x\Big]_0^{\pi/2} + \Big[x\Big]_{\pi/2}^3 + \Big[e^{x-3}\Big]_3^9$$

$$= -\left[\cos\frac{\pi}{2} - \cos 0^\circ\right] + \left[3 - \frac{\pi}{2}\right] + [e^{9-3} - e^{3-3}]$$

$$= -[0-1] + \left[3 - \frac{\pi}{2}\right] + [e^6 - e^0] = 1 + 3 - \frac{\pi}{2} + e^6 - 1$$

$$= 3 - \frac{\pi}{2} + e^6.$$

Example 19. *Prove that :* $\int_0^{2a} f(x)\,dx = \int_0^{2a} f(2a-x)\,dx.$

Solution. Let $\quad I = \int_0^{2a} f(x)\,dx$

Put $\quad (2a - x) = z \;\Rightarrow\; -dx = dz \;\Rightarrow\; dx = -dz$

When $\quad x = 0 \;\Rightarrow\; z = 2a - 0 \;\Rightarrow\; z = 2a$

and $\quad$ when $x = 2a \;\Rightarrow\; z = 2a - 2a \;\Rightarrow\; z = 0$

$\therefore \quad I = \int_{2a}^0 f(2a - z)(-dz) = -\int_{2a}^0 f(2a-z)\,dz$

$\quad = \int_0^{2a} f(2a-z)\,dz \qquad \left[\because \int_a^b f(x)\,dx = -\int_b^a f(x)\,dx\right]$

$\Rightarrow \quad I = \int_0^{2a} f(2a - x)\,dx. \qquad \left[\because \int_a^b f(x)\,dx = \int_a^b f(z)\,dz\right]$

Example 20. *Evaluate :* $\int_0^{\pi} \frac{e^{\cos x}}{e^{\cos x} + e^{-\cos x}}\,dx.$

Solution. Let $\quad I = \int_0^{\pi} \frac{e^{\cos x}}{e^{\cos x} + e^{-\cos x}}\,dx \qquad \text{...(1)}$

$\quad = \int_0^{\pi} \frac{e^{\cos(\pi - x)}}{e^{\cos(\pi-x)} + e^{-\cos(\pi-x)}}\,dx \qquad \left[\because \int_0^a f(x)\,dx = \int_0^a f(a-x)\,dx\right]$

$\therefore \quad I = \int_0^{\pi} \frac{e^{-\cos x}}{e^{-\cos x} + e^{\cos x}}\,dx \qquad \text{...(2)} \quad [\because \cos(\pi - x) = -\cos x]$

Adding equations (1) and (2), we get

$$2I = \int_0^{\pi} \frac{e^{\cos x}}{e^{\cos x} + e^{-\cos x}}\, dx + \int_0^{\pi} \frac{e^{-\cos x}}{e^{-\cos x} + e^{\cos x}}\, dx$$

$$= \int_0^{\pi} \frac{e^{\cos x} + e^{-\cos x}}{e^{\cos x} + e^{-\cos x}} \,.\, dx$$

$$= \int_0^{\pi} 1 \,.\, dx = \Big[x \Big]_0^{\pi} = [\pi - 0] = \pi$$

$\therefore$ $$I = \frac{\pi}{2}\,.$$

EXERCISE FOR PRACTICE

Evaluate the following definite integrals by using the Properties : (1—14).

1. $\displaystyle\int_0^{\pi/2} \frac{\sqrt{\cos x}}{\sqrt{\sin x} + \sqrt{\cos x}}\, dx$

2. $\displaystyle\int_0^{1} \log\left(\frac{1}{x} - 1\right) dx$

3. $\displaystyle\int_{-\pi}^{\pi} (\sin^{93} x + x^{295})\, dx$

4. $\displaystyle\int_{-3}^{2} f(x)\, dx$; where $f(x) = \begin{cases} x^2 \,;\, \text{if } -3 \le x \le 1 \\ 2 - x \,;\, \text{if } 1 \le x \le 2 \end{cases}$

5. $\displaystyle\int_{-6}^{6} |x + 2|\, dx$

6. $\displaystyle\int_0^{2\pi} |\cos x|\, dx$

7. $\displaystyle\int_{-1/2}^{1/2} \left| x \cos \frac{\pi x}{2} \right| dx$

8. $\displaystyle\int_0^{\pi/2} \frac{x \sin x \cos x}{\cos^4 x + \sin^4 x}\, dx$

9. $\displaystyle\int_0^{a} x^2 (a - x)^{3/2}\, dx$

10. $\displaystyle\int_0^{\pi/2} \log (\sin 2x)\, dx$

11. $\displaystyle\int_0^{\pi/2} \frac{\cos^2 x}{\sin x + \cos x}\, dx$

12. $\displaystyle\int_{-\pi/4}^{\pi/4} x^3 \sin^4 x\, dx$

13. $\displaystyle\int_0^{1} |3x - 1|\, dx$

14. $\displaystyle\int_0^{\pi/2} \log (\tan \theta + \cot \theta)\, d\theta.$

Answers

1. $\frac{\pi}{4}$ **2.** 0 **3.** 0 **4.** $\frac{59}{6}$

5. 40 **6.** 4 **7.** $\frac{\sqrt{2}}{\pi^2}\left(\pi + 4 - 4\sqrt{2}\right)$ **8.** $\frac{\pi^2}{16}$

9. $\frac{16}{315} a^{9/2}$ **10.** $-\frac{\pi}{2} \log 2$ **11.** $\frac{\log \sqrt{2} + 1}{\sqrt{2}}$ **12.** 0

13. $\frac{5}{6}$ **14.** $\pi \log 2.$

4

Area of Bounded Regions Using Definite Integrals

4.1. INTRODUCTION

We know the methods of evaluating definite integrals. In this chapter we shall develop the idea of calculating area of the bounded regions by using definite integral. The first step in finding the areas of bounded regions is to identify the region whose area is to be determined. For this, we shall require to draw rough sketch of the given function. The process of drawing rough sketch of a given function is called curve sketching. So, we first discuss drawing of rough sketch and symmetry of common curves.

4.2. STEPS TO DRAW A ROUGH SKETCH

Step I. Symmetry : The following rules are applied to ascertain whether the given curve is symmetrical about any line or not.

(*i*) **Symmetry about x-axis :** If the equation of the curve remains unaltered when y is changed to $-y$, then the curve is symmetrical about x-axis.
e.g., The curve of $y^2 = 4ax$ is symmetrical about x-axis.

(*ii*) **Symmetry about y-axis :** If the equation of the curve remains unaltered when x is changed to $-x$, then the curve is symmetrical about y-axis.
e.g., $x^4 - x^2 + 5y = 0$ is symmetrical about y-axis.

(*iii*) **Symmetry in opposite quadrants :** If the equation of the curve remains unaltered when both x and y are changed to $-x$ and $-y$ respectively, then the curve is symmetrical in opposite quadrants.
e.g., the curve $y = \sin x$ is symmetrical in opposite quadrants.

(*iv*) **Symmetry about the line y = x :** If the equation of the curve remains unaltered when both x and y are changed to y and x respectively, then the curve is symmetrical about the line $y = x$.
e.g., the curve $x^2 + y^2 = 8axy$ is symmetrical about $y = x$.

Step II. Origin : Find whether the curve is passing through the origin or not.

Step III. Axes intersection : The points where the curve intersects the axes are found out by putting $x = 0$ and $y = 0$ in turn, in the equation of the curve.

Step IV. Monotonicity : The regions where the function is strictly increasing or strictly decreasing are found out by studying the sign of the derivative.

Step V. Extreme values : The extreme points are the turning points of the curve. These points are very useful for curve sketching.

Step VI. Some additional points : Finally, some points are also found out on the curve. The graph of the function is drawn by free hand.

4.3. ROUGH SKETCHES OF SOME COMMON CURVES : i.e., (STRAIGHT LINE, CIRCLE, PARABOLA, ELLIPSE)

In the problems of finding areas of bounded regions, a circle or a parabola, or an ellipse or a straight line may be involved.

Let us discuss some of the basic characteristics of these curves.

4.3.1 I. Straight Line : Every first degree equation $ax + by + c = 0$ in x and y represents a straight line and every inequation of the form $ax + by \geq c$ or $ax + by \leq c$ represents the region on either side of the line $ax + by = c$.

The rough sketch of a straight line is as shown in the Fig. (4.1).

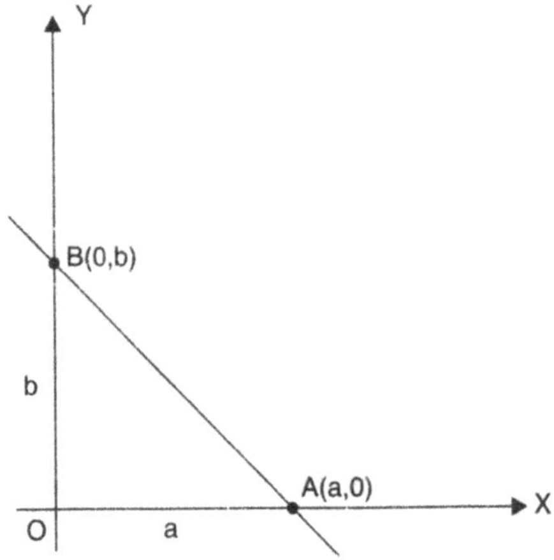

Fig. 4.1

4.3.2 II. Circle : (*i*) The equation of an circle having centre at the origin *i.e.*, (0, 0) and radius r is given by :

$$x^2 + y^2 = r^2$$

(*ii*) The equation of a circle having centre at (h, k) and radius r is given by :

$$(x - h)^2 + (y - k)^2 = r^2$$

(*iii*) The general equation of a circle is given by :

$$x^2 + y^2 + 2gx + 2fy + c = 0.$$

This represents the circle whose centre is at $(-g, -f)$ and radius equal to $\sqrt{g^2 + f^2 - c}$.

Note. (*i*) The circle represented by $x^2 + y^2 = r^2$ is symmetrical about both x-axis as well as y-axis as it contains only even powers of x and y.

(*ii*) The circle represented by $(x - h)^2 + (y - k)^2 = r^2$ is symmetrical about the lines

$$x = h \quad \text{and} \quad y = k.$$

The rough sketch of the circle in the standard form and in the central form refer to Fig. 4.2 (*a*) and 4.2 (*b*) respectively.

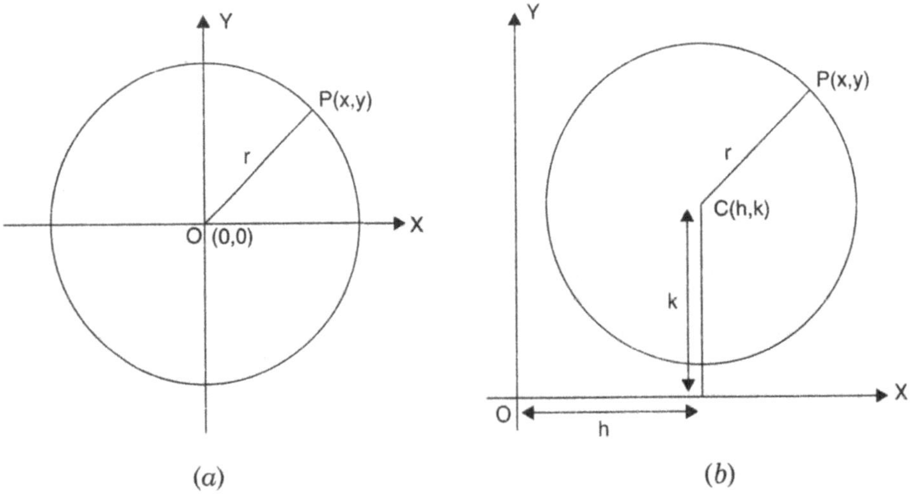

Fig. 4.2

4.3.3 III. Parabola : There are four standard forms of parabola with vertex at the origin and axis along either of co-ordinate axis.

1. Right Handed Parabola : The equation of this type of parabola is of the form

$$y^2 = 4ax \; ; \; a > 0.$$

For this Parabola :

(*i*) Vertex : (0, 0)
(*ii*) Focus : $(a, 0)$
(*iii*) Directrix : $x + a = 0$
(*iv*) Latus rectum : $4a$
(*v*) Axis : $y = 0$
(*vi*) Symmetry : It is symmetric about x-axis.

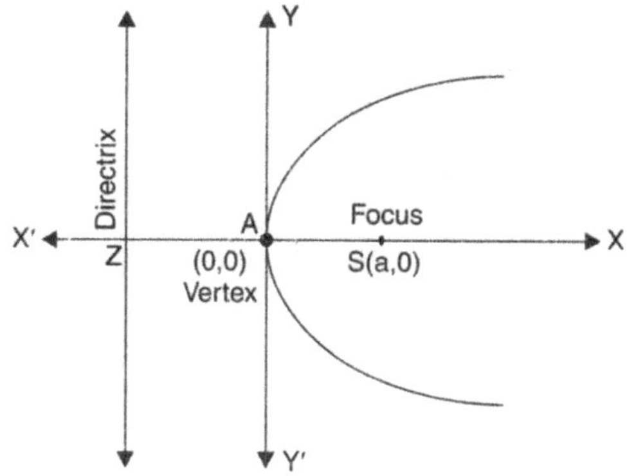

Fig. 4.3 (*a*)

2. Left Handed Parabola : The equation of this type of parabola is of the form

$$y^2 = -4ax \; ; \; a > 0.$$

For this Parabola :

(*i*) Vertex : (0, 0)
(*ii*) Focus : $(-a, 0)$
(*iii*) Directrix : $x - a = 0$
(*iv*) Latus rectum : $4a$
(*v*) Axis : $y = 0$
(*vi*) Symmetry : It is symmetric about x-axis.

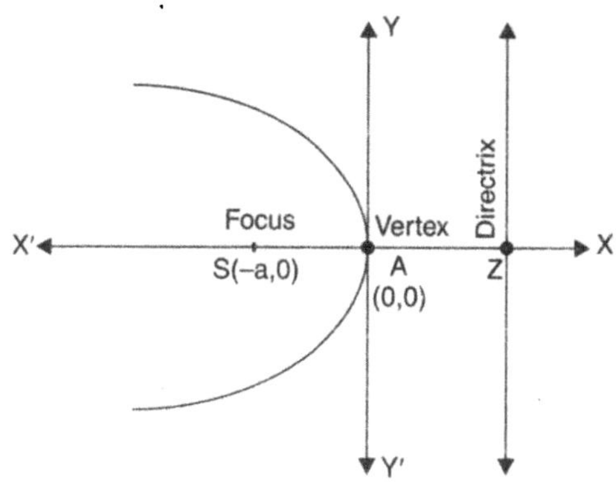

Fig. 4.3 (*b*)

3. Upward Parabola : The equation of this type of parabola is of the form

$$x^2 = 4ay, \; a > 0.$$

For this parabola :

(*i*) Vertex : (0, 0)
(*ii*) Focus : $(0, a)$
(*iii*) Directrix : $y + a = 0$
(*iv*) Latus rectum : $4a$
(*v*) Axis : $x = 0$
(*vi*) Symmetry : It is symmetric about y-axis.

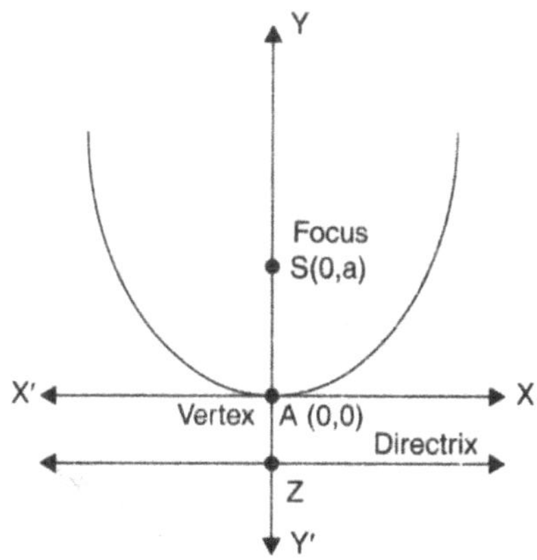

Fig. 4.3 (*c*)

4. Downward Parabola : The equation of this type of parabola is of the from $x^2 = -4ay$; $a > 0$.

For this parabola :

(*i*) Vertex : (0, 0)
(*ii*) Focus : (0, $-a$)
(*iii*) Directrix : $y - a = 0$
(*iv*) Latus rectum : $4a$
(*v*) Axis : $x = 0$
(*vi*) Symmetry : It is symmetric about y-axis.

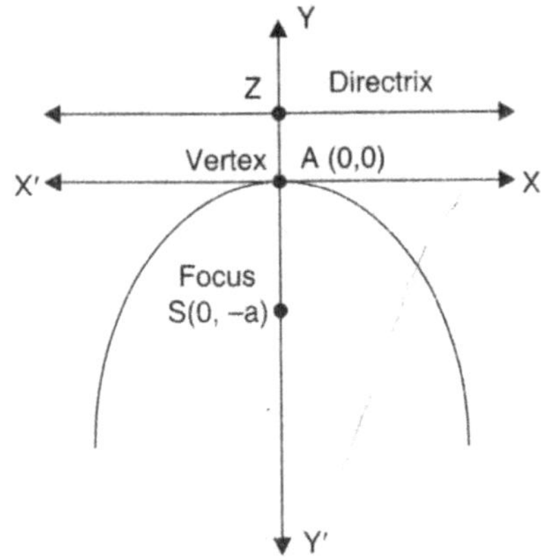

Fig. 4.3 (*d*)

4.3.4 IV. Ellipse : There are two standard forms of ellipse :

1. Foci on x-axis : The equation of this type of ellipse is of the form

$$\frac{x^2}{a^2} + \frac{y^2}{b^2} = 1,\ a > b > 0.$$

For this ellipse :

(*i*) Centre : (0, 0)
(*ii*) Vertices : ($\pm a$, 0)
(*iii*) Foci : ($\pm ae$, 0)
(*iv*) Directrices : $x = \pm \dfrac{a}{e}$
(*v*) Major axis : $2a$
(*vi*) Minor axis : $2b$
(*vii*) Equation of major axis : $y = 0$
(*viii*) Equation of minor axis : $x = 0$
(*ix*) Latus rectum $= \dfrac{2b^2}{a}$
(*x*) Symmetry : It is symmetric about both axis.

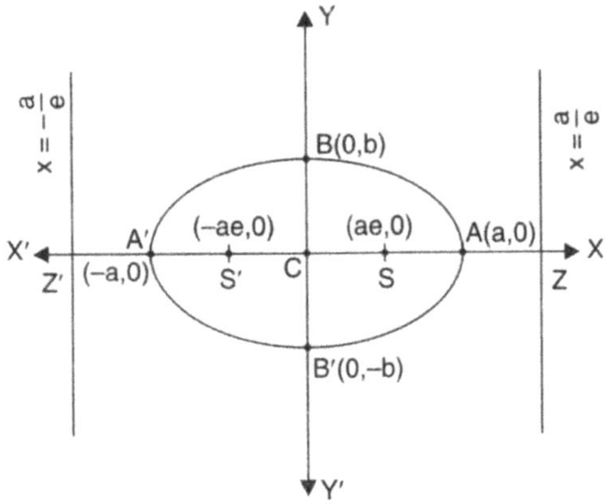

Fig. 4.4 (*a*)

2. Foci on y-axis : The equation of this type of ellipse is of the form :

$$\frac{x^2}{a^2} + \frac{y^2}{b^2} = 1,\ a > b > 0.$$

For this ellipse :

(*i*) Centre : (0, 0)
(*ii*) Vertices : (0, $\pm a$)
(*iii*) Foci : (0, $\pm ae$)
(*iv*) Directrices : $y = \pm \dfrac{a}{e}$
(*v*) Major axis = $2a$
(*vi*) Minor axis = $2b$
(*vii*) Equation of major axis : $x = 0$
(*viii*) Equation of minor axis : $y = 0$
(*ix*) Latus rectum $= \dfrac{2b^2}{a}$
(*x*) Symmetry : It is symmetric about both the axis.

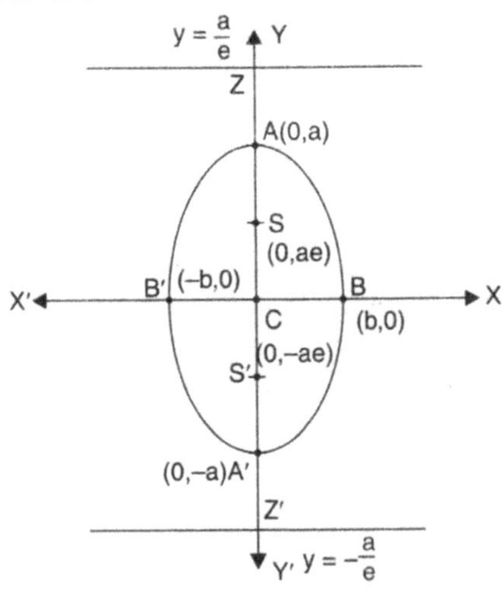

Fig. 4.4 (*b*)

Remark. The equation $\frac{x^2}{l} + \frac{y^2}{m} = 1$ always represents an ellipse whenever $l \neq m$.

(*a*) If $l > m$, then the foci of the ellipse are along x-axis.

(*b*) If $l < m$, then the foci of the ellipse are along y-axis.

4.4. AREA OF BOUNDED REGIONS

Theorem 1. *Let $y = f(x)$ be a continuous and non-negative function defined in [a, b]. Then, prove that the area bounded by the curve $y = f(x)$, the x-axis and between the ordinates is $x = a$, $x = b$ is given by :*

$$\int_a^b y \, . \, dx.$$

Proof. Let AB be the curve $y = f(x)$ and CA, DB be the two ordinates at $x = a$ and $x = b$ respectively.

Let y be an increasing function of x in $[a, b]$.

Let P(x, y) be any point on the curve and Q$(x + \delta x, y + \delta y)$ be a neighbouring point.

Draw PM and QN.

As A changes, the area ACMP also changes and hence is clearly a function of x.

Let A be area ACMP, then the area

$$\text{ACNQ} = \text{A} + \delta\text{A}.$$

$\therefore$ Area PMNQ = δA.

Complete the rectangle PRQS. Then the area PMNQ lies between the area of rectangles PMNR and SMNQ.

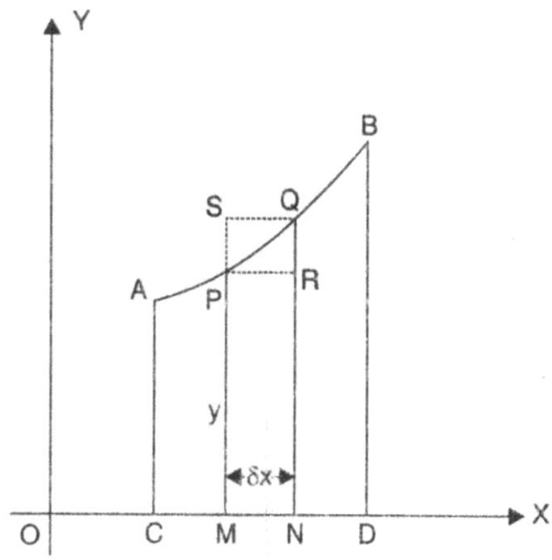

Fig. 4.5

i.e., δA lies between $y\delta x$ and $(y + \delta y)\,\delta x$.

Dividing throughout by δx ; $\frac{\delta \text{A}}{\delta x}$ lies between y and $y + \delta y$.

In the limiting position when Q $\to$ P, $\delta x \to 0$ and $\delta y \to 0$

$\therefore$ $\lim\limits_{\delta x \to 0} \frac{\delta \text{A}}{\delta x}$ lies between y and $\lim\limits_{\delta y \to 0} (y + \delta y)$.

$\Rightarrow$ $\frac{d\text{A}}{dx}$ lies between y and a quantity which tends to y.

$$\Rightarrow \quad \frac{d\text{A}}{dx} = y$$

Integrating both sides w.r.t. x ; from $x = a$ to $x = b$, we have

$$\int_a^b y \, dx = \int_a^b \frac{d\text{A}}{dx} \, . \, dx = \Big[\text{A}\Big]_a^b = (\text{value of A when } x = b) - (\text{value of A when } x = a)$$

$$= \text{Area ACDB} - 0$$

$$\therefore \quad \int_a^b y \, . \, dx = \text{Area ACDB}.$$

The area bounded by the curve AB, the ordinates A and B, and the x-axis is often called 'the area under the curve AB'.

4.5. IMPORTANT REMARKS

1. The area under the curve $y = f(x)$, above x-axis between the ordinates $x = a$ and $x = b$ is given by

$$\int_a^b y \, . \, dx = \int_a^b f(x) \, dx .$$

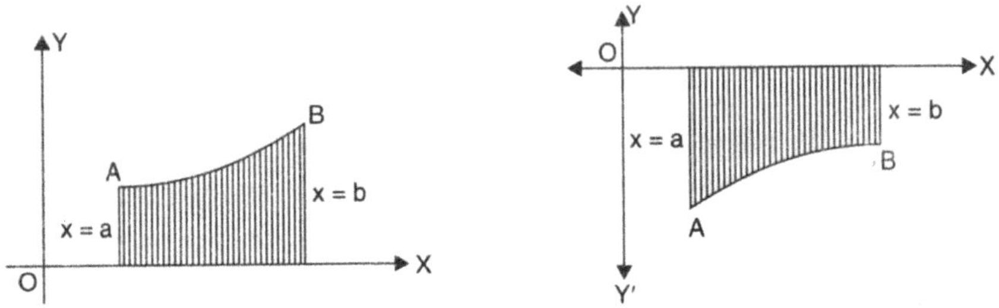

(*a*) (*b*)

Fig. 4.6

2. The area bounded by the curve $y = f(x)$, below x-axis between the ordinates $x = a$ and $x = b$ is given by

$$\int_a^b - y \, dx = - \int_a^b y \, dx = - \int_a^b f(x) \, dx .$$

3. The area bounded by the curve $x = f(y)$, y-axis between the abscissae $y = a$ and $y = b$ is given by

$$\int_a^b x \, dy = \int_a^b f(y) \, dy .$$

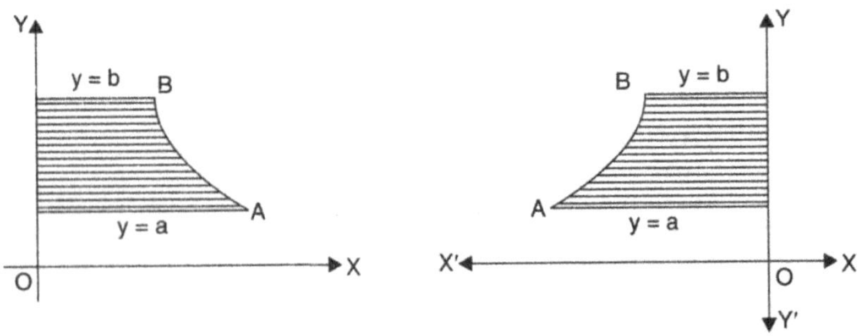

(*c*) (*d*)

Fig. 4.6

4. The area bounded by the curve $x = f(y)$, y-axis between abscissae $y = a$ and $y = b$ is given by

$$\int_a^b - x \, dy = - \int_a^b f(y) \, dy .$$

5. If $f(x) \geq 0$ for $a \leq x \leq c$ and $f(x) \leq 0$ for $c \leq x \leq b$, then the area bounded by the curve $y = f(x)$, x-axis and the ordinates $x = a$ and $x = b$ is given by

$$\int_a^c f(x)\,dx + \int_c^b -f(x)\,dx$$

$$= \int_a^c f(x)\,dx - \int_c^b f(x)\,dx.$$

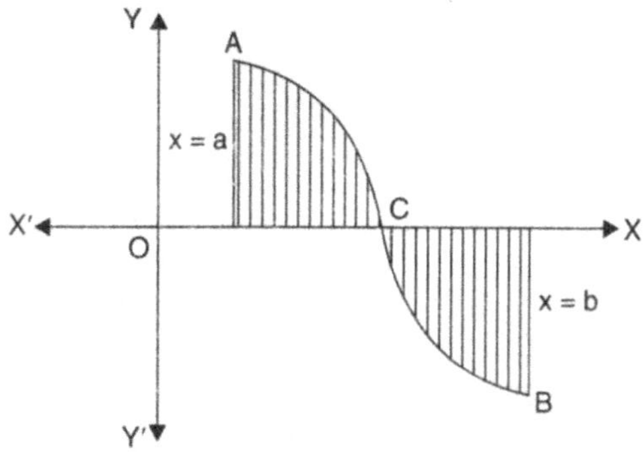

Fig. 4.6 (*e*)

Note. The rough sketch is drawn to see whether the region is above x-axis or is below x-axis or is partly above and partly below the x-axis. If it is difficult to draw the sketch of a function or is not specifically asked in the question, then we can avoid drawing the sketch and observe the sign of the function on the interval under consideration.

SOME SOLVED EXAMPLES

Example 1. *Find the area bounded by the curve $y = \cos x$, x-axis and the ordinates $x = 0$ and $x = 2\pi$.*

Solution. The equation of the given curve is $y = \cos x$.

Now $\cos x > 0$ when $x \in \left(0, \frac{\pi}{2}\right)$

$\cos x < 0$ when $x \in \left(\frac{\pi}{2}, \frac{3\pi}{2}\right)$

$\cos x > 0$ when $x \in \left(\frac{3\pi}{2}, 2\pi\right)$

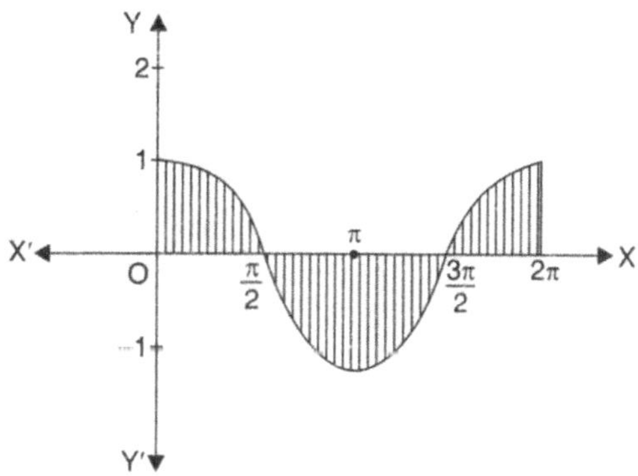

$\Rightarrow$ $f(x)$ changes sign in the given interval.

Table of values of y corresponding to values of x from 0 to 2π is given by

x	0	$\pi/2$	π	$3\pi/2$	2π
y	1	0	-1	0	1

By joining these points with a free hand we obtain a rough sketch of the curve as shown in the figure.

$\therefore$ Required area = Area of shaded region.

$$= \int_0^{2\pi} |y|.\,dx = \int_0^{\pi/2} |y|.\,dx + \int_{\pi/2}^{3\pi/2} |y|.\,dx + \int_{3\pi/2}^{2\pi} |y|.\,dx$$

$$= \int_0^{\pi/2} \cos x\,dx + \int_{\pi/2}^{3\pi/2} -\cos x\,dx + \int_{3\pi/2}^{2\pi} \cos x\,dx$$

$$= \left[\sin x\right]_0^{\pi/2} - \left[\sin x\right]_{\pi/2}^{3\pi/2} + \left[\sin x\right]_{3\pi/2}^{2\pi}$$

$$= \left[\sin\frac{\pi}{2} - \sin 0\right] - \left[\sin\frac{3\pi}{2} - \sin\frac{\pi}{2}\right] + \left[\sin 2\pi - \sin\frac{3\pi}{2}\right]$$

$$= (1 - 0) - (-1 - 1) + [0 - (-1)] = 1 + 2 + 1$$

$$= 4 \text{ sq. units.}$$

Example 2. *Find the area bounded by the curve $y = \sin x$ between $x = 0$ and $x = 2\pi$.*

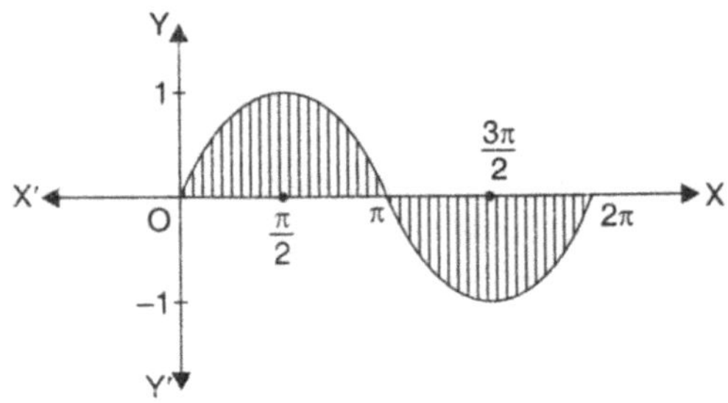

Solution. The equation of the given curve is

$$y = \sin x.$$

Now $\qquad \sin x \geq 0$, for $0 \leq x \leq \pi$

and $\qquad \sin x \leq 0$, for $\pi \leq x \leq 2\pi$.

Thus the curve will lie above the x-axis for $0 \leq x \leq \pi$ and below the x-axis for $\pi \leq x \leq 2\pi$.

Table of values of y corresponding to values of x from 0 to 2π is given by

x	0	$\pi/2$	π	$3\pi/2$	2π
y	0	1	0	-1	0

By joining these points with a free hand we obtain a rough sketch of the curve as shown in the figure.

$\therefore$ Required area = Area of shaded region.

$$= \int_0^{2\pi} |y|.\,dx = \int_0^{\pi} |y|.\,dx + \int_{\pi}^{2\pi} |y|.\,dx$$

$$= \int_0^{\pi} y.\,dx + \int_{\pi}^{2\pi} -y.\,dx = \int_0^{\pi} \sin x\,dx - \int_{\pi}^{2\pi} \sin x\,dx$$

$$= \left[-\cos x\right]_0^{\pi} - \left[-\cos x\right]_{\pi}^{2\pi} = -[\cos\pi - \cos 0°] + [\cos 2\pi - \cos\pi]$$

$$= -[-1 - 1] + [1 - (-1)] = -(-2) + 2 = 2 + 2$$

$$= 4 \text{ sq. units.}$$

Example 3. *Draw a rough sketch of $y = \sin 2x$ and determine the area enclosed by the curve, x-axis and the lines $x = \frac{\pi}{4}$ and $x = \frac{3\pi}{4}$.*

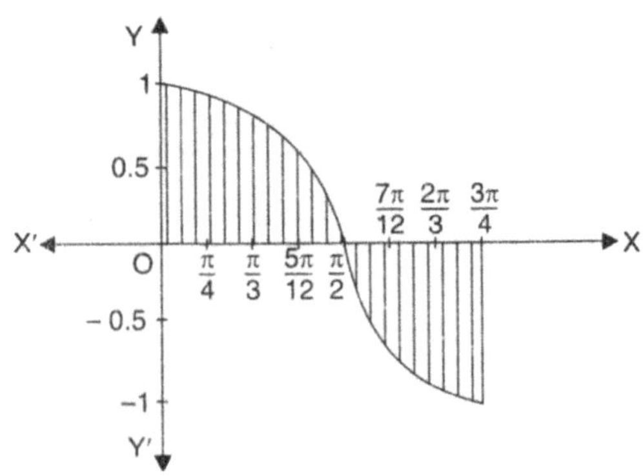

Solution. The equation of the given curve is

$$y = \sin 2x.$$

Since $\qquad \frac{\pi}{4} \leq x \leq \frac{3\pi}{4} \;\Rightarrow\; \frac{\pi}{2} \leq 2x \leq \frac{3\pi}{2}$

Table of values of y corresponding to values of x from $\frac{\pi}{4}$ to $\frac{3\pi}{4}$ is given by

x	$\pi/4$	$\pi/3$	$5\pi/12$	$\pi/2$	$7\pi/12$	$2\pi/3$	$3\pi/4$
y	1	0.87	0.5	0	– 0.5	– 0.87	– 1

By joining these points with a free hand we obtain a rough sketch of the curve as shown in the figure.

$\therefore$ Required area = Area of shaded region.

$$= \int_{\pi/4}^{3\pi/4} |y|.\, dx = \int_{\pi/4}^{\pi/2} |y|.\, dx + \int_{\pi/2}^{3\pi/4} |y|.\, dx$$

$$= \int_{\pi/4}^{\pi/2} y.\, dx + \int_{\pi/2}^{3\pi/4} -y.\, dx = \int_{\pi/4}^{\pi/2} \sin 2x\, dx - \int_{\pi/2}^{3\pi/4} \sin 2x\, dx$$

$$= \left[\frac{-\cos 2x}{2}\right]_{\pi/4}^{\pi/2} - \left[\frac{-\cos 2x}{2}\right]_{\pi/2}^{3\pi/4} = \frac{-1}{2}\Big[\cos 2x\Big]_{\pi/4}^{\pi/2} + \frac{1}{2}\Big[\cos 2x\Big]_{\pi/2}^{3\pi/4}$$

$$= \frac{-1}{2}\left[\cos \pi - \cos \frac{\pi}{2}\right] + \frac{1}{2}\left[\cos \frac{3\pi}{2} - \cos \pi\right] = \frac{-1}{2}\,[-1-0] + \frac{1}{2}\,[0-(-1)]$$

$$= \frac{1}{2} + \frac{1}{2} = 1 \text{ sq. unit.}$$

Example 4. *Make a rough sketch of the function $y = \cos 3x$, $0 \le x \le \frac{\pi}{6}$ and determine the area enclosed between the curve and the co-ordinate axis.*

Solution. The equation of the given curve is

$$y = \cos 3x.$$

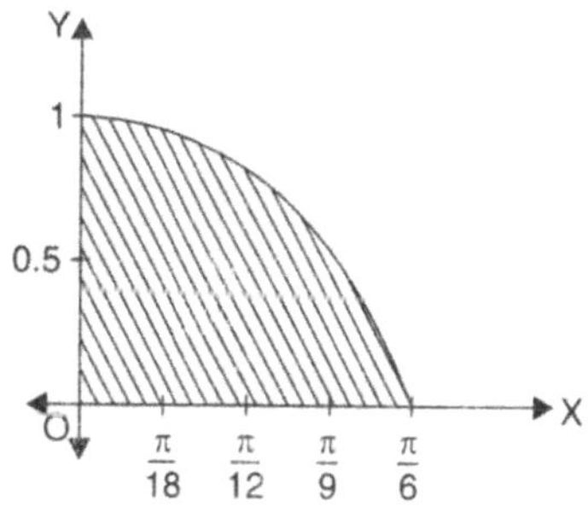

Since $\quad 0 \le x \le \frac{\pi}{6}$

$\Rightarrow \quad 0 \le 3x \le \frac{\pi}{2}$

$\Rightarrow \quad \cos 3x \ge 0.$

$\therefore$ The curve $y = \cos 3x$, $0 \le x \le \frac{\pi}{6}$ lies above the x-axis.

Table of values of y corresponding to values of x from 0 to $\frac{\pi}{6}$ is given by

x	$\pi/18$	$\pi/12$	$\pi/9$	$\pi/6$
y	0.87	0.71	0.5	0

By joining these points with a free hand we obtain a rough sketch of the curve as shown in the figure.

$\therefore$ Required area = Area between curve and the co-ordinate axis. *i.e.*, the area of the shaded region.

$$= \int_0^{\pi/6} y.\, dx = \int_0^{\pi/6} \cos 3x\, dx$$

$$= \left[\frac{\sin 3x}{3}\right]_0^{\pi/6} = \frac{1}{3}\left[\sin 3x\right]_0^{\pi/6} = \frac{1}{3}\left[\sin\frac{\pi}{2} - \sin 0°\right] = \frac{1}{3}[1-0]$$

$$= \frac{1}{3} \text{ sq. units.}$$

Example 5. *Find the area bounded by $y = 1 + 2\sin^2 x$, x-axis, $x = 0$ and $x = \pi$.*

Solution. We know that

$$y = 1 + 2\sin^2 x > 0 \text{ for } 0 \le x \le \pi$$

$\therefore$ The area bounded by $y = 1 + 2\sin^2 x$, x-axis, $x = 0$ and $x = \pi$ is above the x-axis.

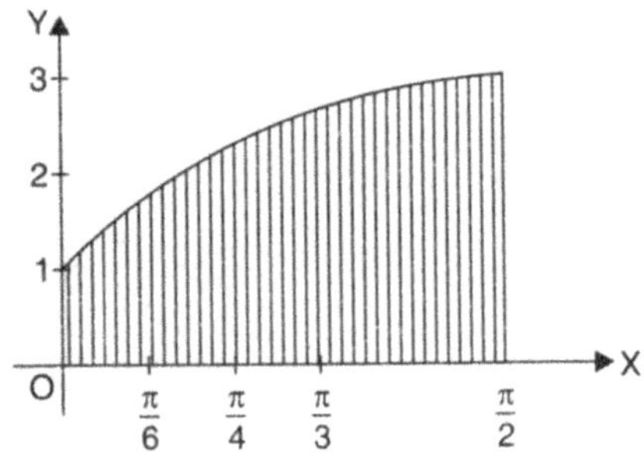

Table of values of y corresponding to values of x from 0 to π is given by

x	0	$\pi/6$	$\pi/4$	$\pi/3$	$\pi/2$
y	1	1.5	2	2.5	3

$$\therefore \quad \text{Required area} = \int_0^{\pi} y\,.\,dx = \int_0^{\pi} (1 + 2\sin^2 x)\,dx$$

$$= \int_0^{\pi} [1 + (1 - \cos 2x)]\,dx$$

$[\because \quad 1 - \cos 2A = 2\sin^2 A]$

$$= \int_0^{\pi} (2 - \cos 2x)\,dx$$

$$= \left[2x - \frac{\sin 2x}{2}\right]_0^{\pi} = \left[2\pi - \frac{\sin 2\pi}{2}\right] - \left[2(0) - \frac{\sin 0°}{2}\right]$$

$$= (2\pi - 0) - (0 - 0) = 2\pi \text{ sq. units.}$$

Example 6. *Draw a rough sketch of the curve $y = \frac{x}{\pi} + 2\sin^2 x$ and find the area between the x-axis, the curve and $x = 0$ and $x = \pi$.*

Solution. The equation of the given curve is

$$y = \frac{x}{\pi} + 2\sin^2 x.$$

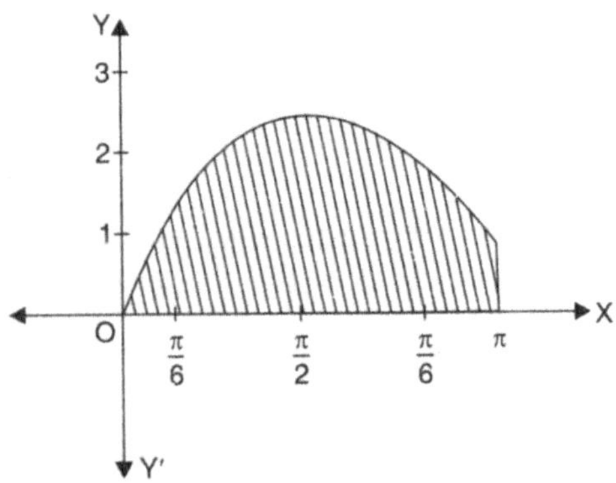

Table of values of y corresponding to values of x from 0 to π is given by

x	0	$\pi/6$	$\pi/2$	$5\pi/6$	π
y	0	0.66	2.5	1.33	1

By joining these points with a free hand, we obtain a rough sketch of the curve as shown in the figure.

$\therefore$ Required area = Area of the shaded region

$$= \int_0^\pi y.dx = \int_0^\pi \left(\frac{x}{\pi} + 2\sin^2 x\right) dx \qquad [\because \; 1 - \cos 2A = 2\sin^2 A]$$

$$= \int_0^\pi \left[\frac{x}{\pi} + 1 - \cos 2x\right] dx = \left[\frac{x^2}{2\pi} + x - \frac{\sin 2x}{2}\right]_0^\pi$$

$$= \left[\frac{\pi^2}{2\pi} + \pi - \frac{\sin 2\pi}{2}\right] - \left[\frac{0}{2\pi} + 0 - \frac{\sin 0°}{2}\right] = \left(\frac{\pi}{2} + \pi - 0\right) - (0 + 0 - 0)$$

$$= \frac{3\pi}{2} \text{ sq. units.}$$

Example 7. *Make a rough sketch of the function $y = 2\sin x$, $0 \le x \le \pi$ and determine the area enclosed between the curve and the co-ordinate axis.*

Solution. The equation of the given curve is

$$y = 2 \sin x$$

As increases from 0 to π ; $y = 2\sin x \ge 0$.

Table of values of y corresponding to values of x from 0 to π is given by

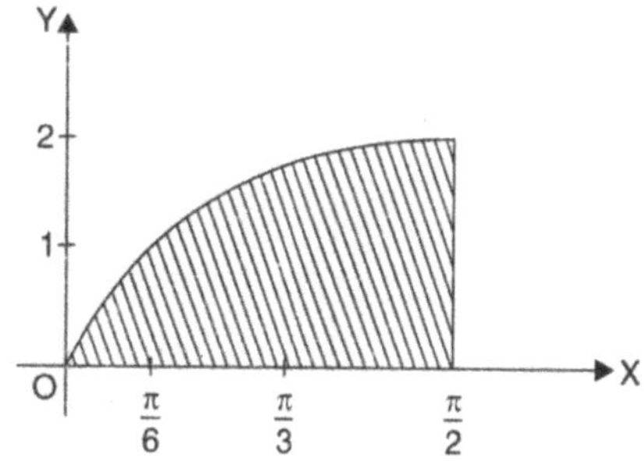

x	0	$\frac{\pi}{6}$	$\frac{\pi}{3}$	$\frac{\pi}{2}$
y	0	1	1.7	2

By joining these points with a free hand, we obtain a rough sketch of the curve as shown in the figure.

$$\therefore \text{ Required area} = \int_0^\pi y.dx$$

$$= \int_0^\pi 2\sin x\, dx = 2\left[-\cos x\right]_0^\pi$$

$$= -2\,[\cos \pi - \cos 0°] = -2\,[-1 - 1] = 4$$

$$= 4 \text{ sq. units.}$$

Example 8. *Find the area of the curve $y = x^2 - 2x$ lying below the x-axis.*

Solution. The equation of the given curve is :

$$y = x^2 - 2x \qquad ...(1)$$

Now, to find the limits of integration, we have to find out the intersection of the given curve with x-axis.

Put $y = 0$ in equation (1), we get

$$0 = x^2 - 2x$$

$$\Rightarrow \quad x(x - 2) = 0$$

$$\Rightarrow \quad x = 0,\ x = 2.$$

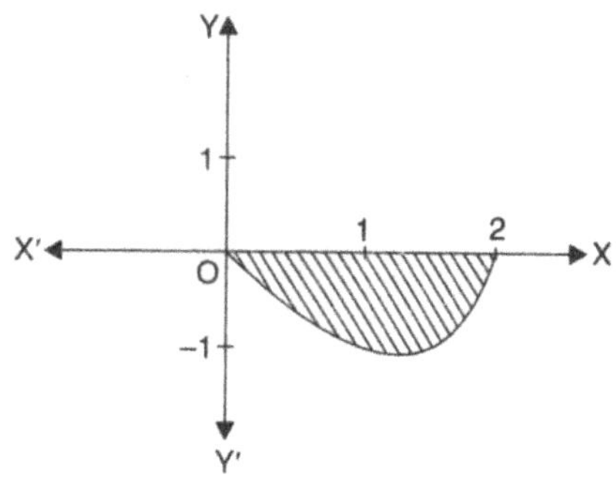

Table of values of y corresponding to the values of x is given by

x	0	1	2
y	0	-1	0

By joining these points with a free hand, we obtain a rough sketch of the curve as shown in the figure.

$\therefore$ Required Area = Area of shaded region

$$= \int_0^2 -y.dx = -\int_0^2 y.dx$$

$\left[\because \text{ If the curve between the lines } x = a \text{ and } x = b \text{ lies below } x\text{-axis, then area is given by } \int_a^b -y.dx = -\int_a^b y.dx\right]$

$$= -\int_0^2 (x^2 - 2x)\,dx = -\left[\frac{x^3}{3} - x^2\right]_0^2$$

$$= -\left[\frac{(2)^3}{3} - (2)^2\right] - 0 = -\left[\frac{8}{3} - 4\right] = -\left[\frac{8-12}{3}\right] = \frac{4}{3}$$

$$= \frac{4}{3} \text{ sq. units.}$$

Example 9. *Find the area bounded by the curve $2y = 8 - x$, x-axis and the lines $x = 2$ and $x = 4$.*

Solution. The equation of the given curve is

$$2y = -x + 8.$$

$$\Rightarrow \qquad y = -\frac{x}{2} + 4. \qquad ...(1)$$

Clearly, equation (1) represents a straight line.

Table of values of y corresponding to values of x from 2 to 4 is given by

x	2	3	4
y	3	2.5	2

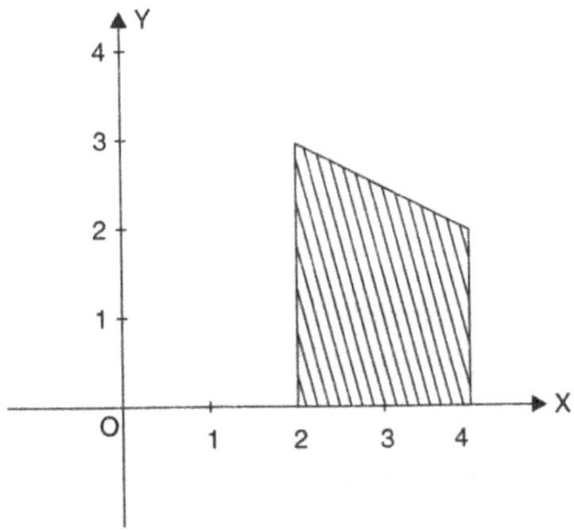

By joining these points with a free hand we obtain a rough sketch of the curve as shown in the figure.

$\therefore$ Required area = Area of shaded region.

$$= \int_2^4 y.dx = \int_2^4 \left(-\frac{x}{2} + 4\right) dx$$

$$= \left[-\frac{x^2}{4} + 4x \right]_2^4 = \left[\left\{ -\frac{(4)^2}{4} + 4\,(4) \right\} - \left\{ -\frac{(2)^2}{4} + 4\,(2) \right\} \right]$$

$$= \left[\left(-\frac{16}{4} + 16 \right) - \left(-\frac{4}{4} + 8 \right) \right]$$

$= + 12 - 7 = 5$ sq. units.

Example 10. *Sketch the region bounded by the curve $y = 2x - x^2$ and the x-axis and find its area.*

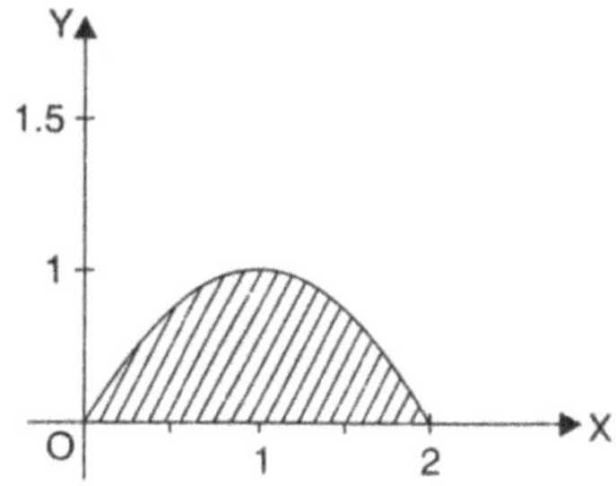

Solution. The equation of the given curve is

$$y = 2x - x^2$$

Table of values of y corresponding to values of x is given by

x	0	0.5	1	1.5	2
y	0	0.75	1	0.75	0

By joining these points with a free hand we obtain a rough sketch of the curve as shown in the figure.

To find the limits of integration, let us find the point of intersection of the curve with the x-axis.

Put $y = 0 \Rightarrow 2x - x^2 = 0 \Rightarrow x(2 - x) = 0$

$\Rightarrow \quad x = 0, \quad x = 2$

$\therefore$ Required area = Area of the shaded region

$$= \int_0^2 y.dx = \int_0^2 (2x - x^2)\,dx$$

$$= \left[\frac{2x^2}{2} - \frac{x^3}{3} \right]_0^2 = \left[\left((2)^2 - \frac{(2)^3}{3} \right) - 0 \right] = \left(4 - \frac{8}{3} \right) = \frac{12 - 8}{3}$$

$= \dfrac{4}{3}$ sq. units.

Example 11. *Find the area of the region bounded by the curve $y^2 = 2y - x$ and the y-axis.*

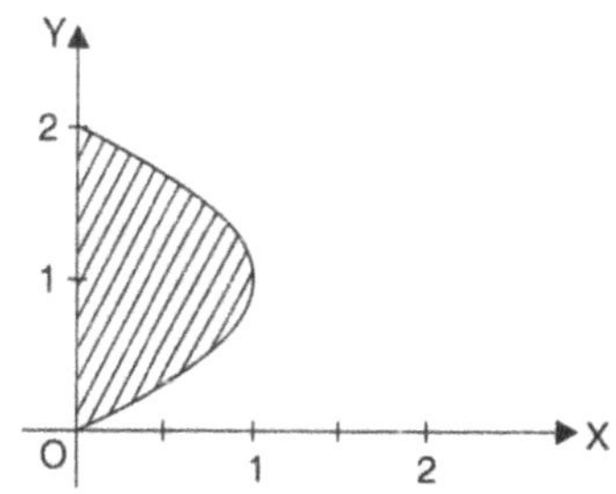

Solution. The equation of the given curve is

$$y^2 = 2y - x \qquad \ldots(1)$$

$\Rightarrow \quad x = 2y - y^2$

To find the limit of integration, let us find the point of intersection of the curve with the y-axis.

Put $x = 0$ in equation (1), we get

$$y^2 = 2y - 0$$

$\Rightarrow \quad y^2 - 2y = 0 \Rightarrow y(y - 2) = 0 \Rightarrow y = 0, y = 2$

Table of the values of x corresponding to the values of y is given by

y	0	1	2
x	0	1	0

By joining these points with a free hand, we obtain a rough sketch of the curve as shown in the figure.

$\therefore$ Required area = Area of the shaded region

$$= \int_0^2 x.dy = \int_0^2 (2y - y^2)\, dy$$

$$= \left[\frac{2y^2}{2} - \frac{y^3}{3}\right]_0^2 = \left[y^2 - \frac{y^3}{3}\right]_0^2 = \left[\left((2)^2 - \frac{(2)^3}{3}\right) - (0-0)\right] = \left[\left(4 - \frac{8}{3}\right)\right] = \frac{12-8}{3}$$

$$= \frac{4}{3} \text{ sq. units.}$$

Example 12. *Find the area bounded by the curve $y = x$, x-axis and the ordinates $x = -1$, $x = 2$.*

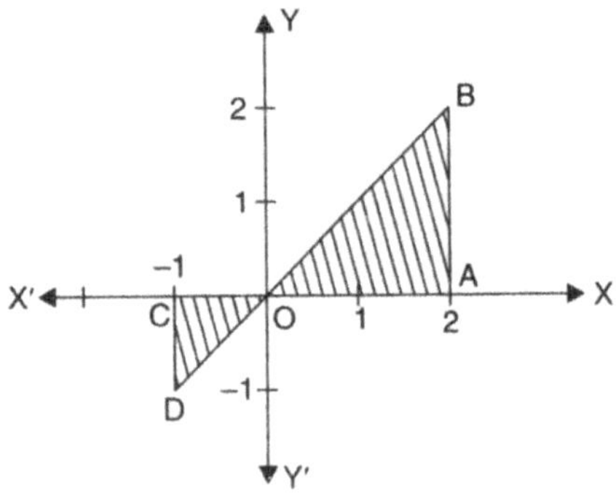

Solution. The equation of the given curve is

$$y = x.$$

The graph of $y = x$ is a straight line passing through the origin and making an angle 45° with the x-axis.

The function $y = x$ is negative on $[-1, 0]$ and positive on $[0, 2]$.

$\therefore$ Required area = Area of the shaded region

= Area OCD + Area OAB

$$= \int_{-1}^0 |y|.dx + \int_0^2 |y|.dx$$

$$= \int_{-1}^0 -y.dx + \int_0^2 y.dx = -\int_{-1}^0 x\,dx + \int_0^2 x\,dx$$

$$= -\left[\frac{x^2}{2}\right]_{-1}^0 + \left[\frac{x^2}{2}\right]_0^2 = -\left[0 - \frac{(-1)^2}{2}\right] + \left[\frac{(2)^2}{2} - 0\right]$$

$$= \frac{1}{2} + 2 = \frac{5}{2} \text{ sq. units.}$$

Example 13. *Make a rough sketch of the function $y = \sin 2x$, $0 \le x \le \frac{\pi}{4}$ and determine the area enclosed between the curve and the co-ordinate axis.*

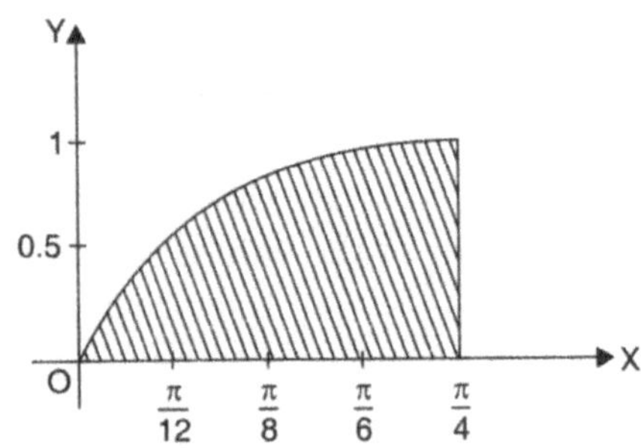

Solution. The equation of the given curve is

$$y = \sin 2x$$

Since $0 \le x \le \frac{\pi}{4} \Rightarrow 0 \le 2x \le \frac{\pi}{2}$

$\therefore$ $y = \sin 2x \ge 0$ when x increases from 0 to $\frac{\pi}{4}$.

Table of values of y corresponding to the values of x from 0 to $\frac{\pi}{4}$ is given by

x	0	$\frac{\pi}{12}$	$\frac{\pi}{8}$	$\frac{\pi}{6}$	$\frac{\pi}{4}$
y	0	0.5	0.7	0.8	1

By joining these points with a free hand, we obtain a rough sketch of the curve as shown in the figure.

$\therefore$ Required area = Area of shaded region

$$= \int_0^{\pi/4} y.dx$$

$$= \int_0^{\pi/4} \sin 2x\, dx = \left[-\frac{\cos 2x}{2}\right]_0^{\pi/4} = \frac{-1}{2}\left[\cos 2x\right]_0^{\pi/4}$$

$$= -\frac{1}{2}\left[\cos\frac{\pi}{2} - \cos 0^\circ\right] = -\frac{1}{2}[0-1]$$

$$= \frac{1}{2} \text{ sq. units.}$$

Example 14. *Find the area bounded by the parabola $y^2 = 4ax$ and its latus-rectum.*

Solution. The equation of the given curve is

$$y^2 = 4ax.$$

$$\Rightarrow \quad y = \sqrt{4ax}$$

Since the parabola contains only even powers to y, therefore, it is symmetrical about x-axis.

A rough sketch of the parabola is as shown in the figure.

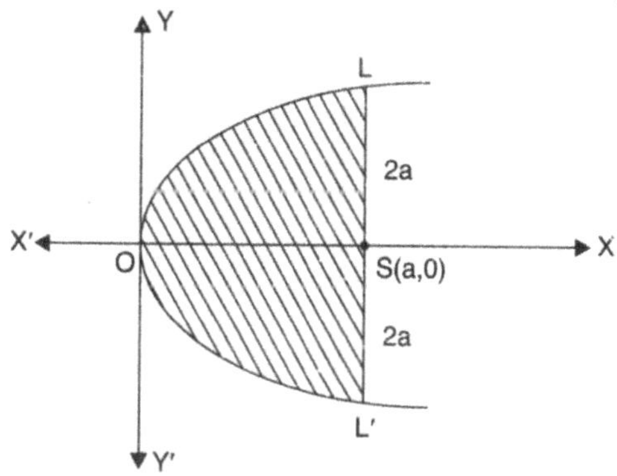

Let S(a, 0) be the focus and LSL′ be the directrix of the parabola $y^2 = 4ax$.

Here OS = a, LSL′ = $4a$

$\therefore$ Required Area = Area LOL′

$$= 2(\text{area of LOS})$$

$$= 2\int_0^a y.dx = 2\int_0^a \sqrt{4ax}\,dx$$

$$= 2\int_0^a 2\sqrt{a}\sqrt{x}.dx = 4\sqrt{a}\int_0^a x^{1/2}\,dx$$

$$= 4\sqrt{a}\left[\frac{x^{\frac{1}{2}+1}}{\frac{1}{2}+1}\right]_0^a = 4\sqrt{a}.\frac{2}{3}\left[x^{3/2}\right]_0^a = \frac{8\sqrt{a}}{3}[a^{3/2}-0] = \frac{8\sqrt{a}}{3}[a\sqrt{a}]$$

$$= \frac{8a^2}{3} \text{ sq. units.}$$

Example 15. *Find the area bounded by the curve $y = 2 \cos x$ and the x-axis from $x = 0$ to $x = 2\pi$.*

Solution. The equation of the given curve is

$$y = 2 \cos x$$

Table of values of y corresponding to the values of x from 0 to 2π is given by

x	0	$\frac{\pi}{2}$	π	$\frac{3\pi}{2}$	2π
y	2	0	-2	0	2

By joining these points with a free hand, we obtain a rough sketch of the curve as shown in the figure.

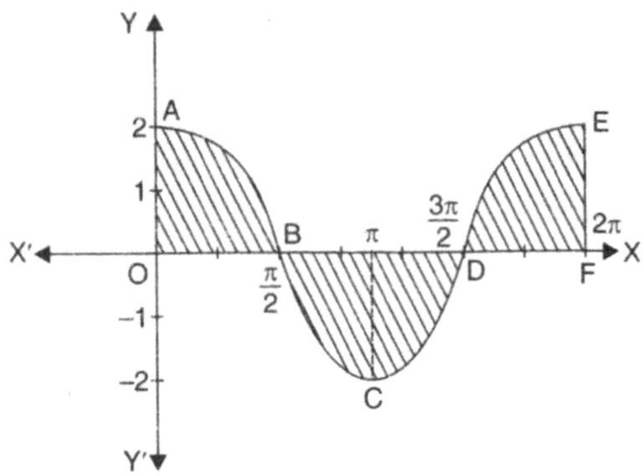

Our required area is the shaded area. If we integrate directly from $x = 0$ to $x = 2\pi$, the net result will be zero as half the area is above the x-axis and half is below the x-axis and therefore negative. Thus, to avoid this, we will find the area from $x = 0$ to $x = \frac{\pi}{2}$ (*i.e.*, area OAB) and multiply it by 4.

$\therefore$ Required area = 4 × area OAB

$$= 4\int_0^{\pi/2} y.dx = 4\int_0^{\pi/2} 2 \cos x \, dx$$

$$= 8\Big[\sin x\Big]_0^{\pi/2} = 8\left[\sin \frac{\pi}{2} - \sin 0°\right] = 8[1 - 0]$$

$$= 8 \text{ sq. units.}$$

Example 16. *Find the area of the region bounded by the parabola $y^2 = 16x$ and the line $x = 4$.*

Solution. The given equation of the curve is

$$y^2 = 16x$$

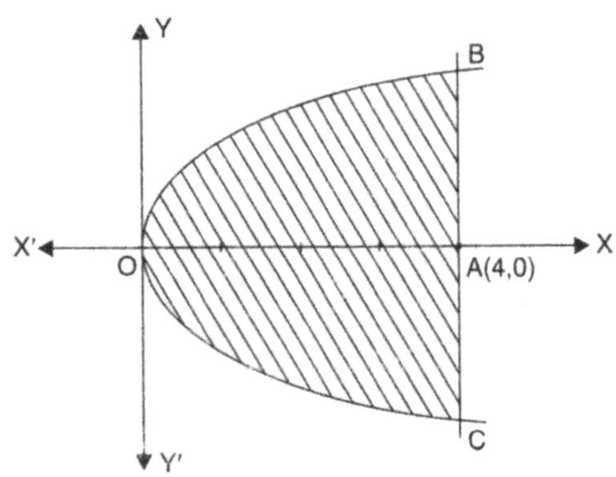

The equation $y^2 = 16x$ represents a parabola with vertex at the origin and the axis of symmetry along the positive direction of x-axis as shown in the figure.

$x = 4$ is a straight line parallel to y-axis at a distance of 4 units from it.

$\therefore$ Required area = Area of shaded region

= Area OBC

= 2 × (area OAB)

$$= 2\int_0^4 y.dx$$

$$= 2\int_0^4 4\sqrt{x} \, dx \qquad \left[\because y^2 = 16x \Rightarrow y = \sqrt{16x} = 4\sqrt{x}\right]$$

$$= 8\int_0^4 x^{1/2}\,dx = 8\left[\frac{x^{\frac{1}{2}+1}}{\frac{1}{2}+1}\right]_0^4 = 8.\frac{2}{3}\left[x^{3/2}\right]_0^4$$

$$= \frac{16}{3}\,[(4)^{3/2} - 0] = \frac{16}{3}[8 - 0] = \frac{128}{3} \text{ sq. units.}$$

Example 17. *Draw a rough sketch of the curve $y = \cos^2 x$ in $\left[0, \frac{\pi}{2}\right]$ and find the area enclosed by the curve, x-axis and the ordinates $x = 0$, $x = \frac{\pi}{2}$.*

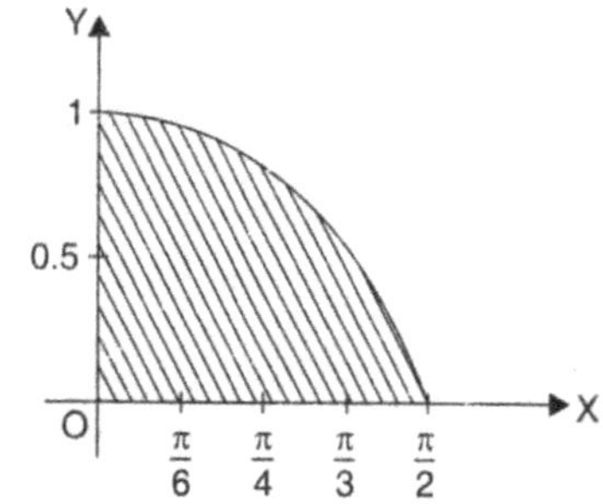

Solution. The equation of the given curve is

$$y = \cos^2 x\ .\ \ \forall\, x \in \left[0, \frac{\pi}{2}\right]$$

Now $y = \cos^2 x \geq 0$, when x increases from 0 to $\frac{\pi}{2}$.

Table of values of y corresponding to the values of x from 0 to $\frac{\pi}{2}$ is given by

x	0	$\frac{\pi}{6}$	$\frac{\pi}{4}$	$\frac{\pi}{3}$	$\frac{\pi}{2}$
y	1	0.7	0.5	0.25	0

By joining these points with a free hand, we obtain a rough sketch of the curve as shown in the figure.

$\therefore$ Required area = Area of shaded region

$$= \int_0^{\pi/2} y.dx = \int_0^{\pi/2} \cos^2 x\,dx$$

$$= \frac{1}{2}\int_0^{\pi/2} (1 + \cos^2 2x)\,dx \qquad \left[\begin{array}{l}\because\ 1 + \cos 2A = 2\cos^2 A \\ \Rightarrow\ \dfrac{1+\cos 2A}{2} = \cos^2 A\end{array}\right]$$

$$= \frac{1}{2}\left[x + \frac{\sin 2x}{2}\right]_0^{\pi/2} = \frac{1}{2}\left[\left\{\frac{\pi}{2} + \frac{\sin \pi}{2}\right\} - \left\{0 + \frac{\sin 0^\circ}{2}\right\}\right] = \frac{1}{2}\left[\left(\frac{\pi}{2} + 0\right) - (0 + 0)\right]$$

$$= \frac{\pi}{4} \text{ sq. units.}$$

Example 18. *Make a rough sketch of the function $y = 4 - x^2$, $0 \leq x \leq 2$ and determine the area enclosed between the curve and $x = 0$, $x = 2$ and x-axis.*

Solution. The given equation of the curve is

$$y = 4 - x^2.$$

The value of y is positive, as x increases from 0 to 2.

Table of values of y corresponding to the values of x from 0 to 2 is given by

x	0	1	2
y	4	3	0

By joining these points with a free hand, we obtain a rough sketch of the curve as shown in the figure.

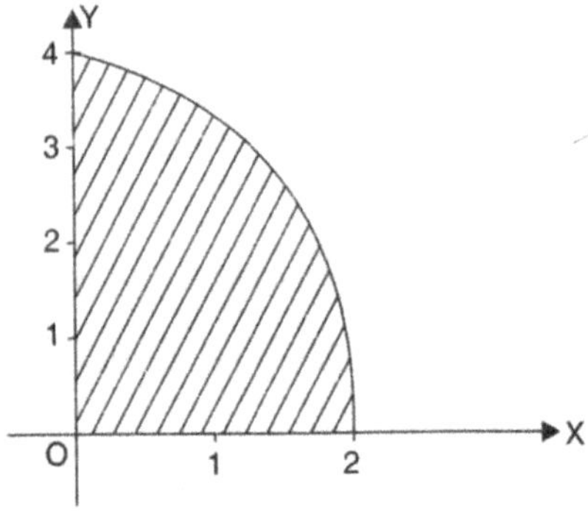

$\therefore$ Required area = Area of the shaded region

$$= \int_0^2 y.dx = \int_0^2 (4 - x^2)\, dx$$

$$= \left[4x - \frac{x^3}{3}\right]_0^2 = \left[\left\{4(2) - \frac{(2)^3}{3}\right\} - \{0 - 0\}\right]$$

$$= \left[\left(8 - \frac{8}{3}\right) - 0\right] = \frac{24 - 8}{3}$$

$$= \frac{16}{3} \text{ sq. units.}$$

Example 19. *Trace the curve $y = x^3$ and find the area bounded by the curve, x-axis and the lines $x = 0$ and $x = 4$.*

Solution. The given equation of the curve is

$$y = x^3.$$

The value of y is positive, as x increases from 0 to 4.

Table of values of y corresponding to the values of x from 0 to 4 is given by

x	0	1	2	3	4
y	0	1	8	27	64

By joining these points with a free hand, we obtain a rough sketch of the given curve as shown in the figure.

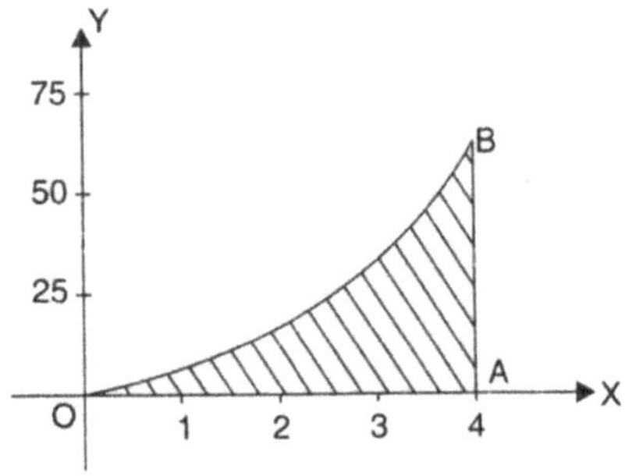

$\therefore$ Required area = Area of shaded region

= Area OAB

$$= \int_0^4 y.dx = \int_0^4 x^3 .dx = \left[\frac{x^4}{4}\right]_0^4$$

$$= \frac{1}{4}[(4)^4 - (0)^4] = \frac{1}{4} \times 256$$

$$= 64 \text{ sq. units.}$$

Example 20. *Find the area of the region bounded by the curve $xy - 3x - 2y - 10 = 0$, x-axis and the lines $x = 3$, $x = 4$.*

Solution. The given equation of the curve is

$$xy - 3x - 2y - 10 = 0$$

$$\Rightarrow \quad y(x - 2) = 3x + 10$$

$$\Rightarrow \qquad y = \frac{3x+10}{x-2}$$

$\therefore$ Required area $= \int_3^4 y.dx$

$$= \int_3^4 \frac{3x+10}{x-2}\,dx = 3\int_3^4 \frac{x+\frac{10}{3}}{x-2}\,dx$$

$$= 3\int_3^4 \left[\frac{x-2+\frac{10}{3}+2}{x-2}\right]dx \qquad \text{[Add and subtract 2 to the numerator]}$$

$$= 3\int_3^4 \left[\frac{x-2}{x-2}+\frac{\frac{16}{3}}{x-2}\right]dx = 3\int_3^4 1.dx + 3.\frac{16}{3}\int_3^4 \frac{1}{x-2}\,dx$$

$$= 3\Big[x\Big]_3^4 + 16\Big[\log|x-2|\Big]_3^4$$

$= 3[4-3] + 16[\log|4-2| - \log|3-2|] = 3 + 16\log[\log 2 - \log 1]$

$= (3 + 16\log 2)$ sq. units.

Example 21. *Find the area bounded by the curve $y^2 = 4ax$ and the line $y = 2a$ and y-axis.*

Solution. The given equation of the curve is

$$y^2 = 4ax.$$

The given equation represents a parabola with vertex at the origin and axis as x-axis, as shown in the figure. It is symmetrical about x-axis as it contains only even powers of y.

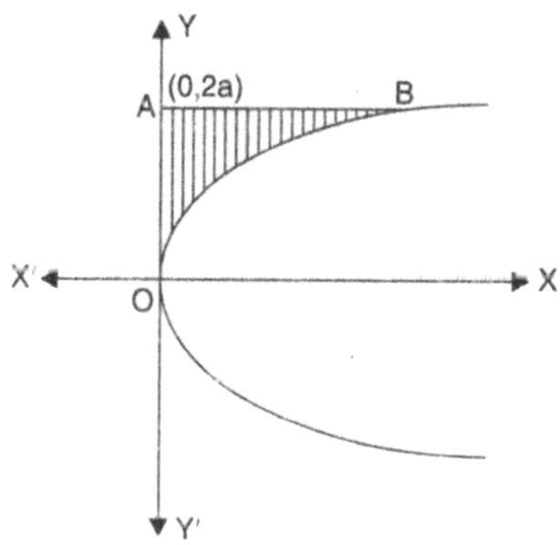

$y = 2a$ is a straight line parallel to x-axis at a distance of $2a$ from it.

$\therefore$ Required area = Area of shaded region.

= Area OAB

$$= \int_0^{2a} x.dy \qquad \left[\because y^2 = 4ax \Rightarrow x = \frac{y^2}{4a}\right]$$

$$= \int_0^{2a} \frac{y^2}{4a}\,dy = \frac{1}{4a}\int_0^{2a} y^2\,dy = \frac{1}{4a}\left[\frac{y^3}{3}\right]_0^{2a} = \frac{1}{12a}[(2a)^3 - (0)^3]$$

$$= \frac{8a^3}{12a} = \frac{2}{3}a^2 \text{ sq. units.}$$

Example 22. *Make a rough sketch of the graph of the curve $y = 9 - x^2$, $0 \le x \le 3$ and determine the area enclosed between the curve and the x-axis.*

Solution. The given equation of the curve is

$$y = 9 - x^2$$

The value of $y = 9 - x^2$ is positive, when x increases from 0 to 3.

Table of values of y corresponding to the values of x from 0 to 3 is given by

x	0	1	2	3
y	9	8	5	0

By joining these points with a free hand, we obtain a rough sketch of the curve as shown in the figure.

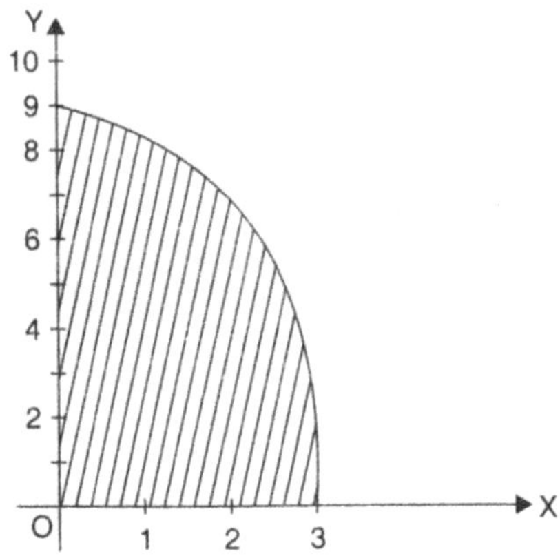

$\therefore$ Required area = Area of shaded region

$$= \int_0^3 y.dx = \int_0^3 (9 - x^2)\, dx$$

$$= \left[9x - \frac{x^3}{3}\right]_0^3$$

$$= \left[\left\{9\,(3) - \frac{(3)^3}{3}\right\} - (0 - 0)\right]$$

$$= \left[\left(27 - \frac{27}{3}\right) - 0\right] = 27 - 9$$

$$= 18 \text{ sq. units.}$$

Example 23. *Find the area bounded by the curve $xy = 4$, the x-axis and the lines $x = 1$, $x = 3$.*

Solution. The given equation of the curve is

$$xy = 4 \quad \Rightarrow \quad y = \frac{4}{x}$$

As x increases from 1 to 3, the value of $y = \frac{4}{x}$ decreases from $\frac{4}{1} = 4$ to $\frac{4}{3}$.

$\therefore$ The function $y = \frac{4}{x}$ is positive on [1, 3].

$\therefore$ Required area $= \int_1^3 y.dx = \int_1^3 \frac{4}{x}.dx = 4\int_1^3 \frac{1}{x}\, dx$

$$= 4\Big[\log|x|\Big]_1^3 = 4\,[\log|3| - \log|1|]$$

$$= 4 \log 3 \text{ sq. units.}$$

Example 24. *Draw a rough sketch of the curve $y = \cos^2 x$ in $[0, \pi]$ and find the area enclosed by the curve, the lines $x = 0$, $x = \pi$ and the x-axis.*

Solution. The given equation of the curve is

$$y = \cos^2 x.$$

Table of values of y corresponding to the values of x from 0 to π is given by

x	0	$\frac{\pi}{4}$	$\frac{\pi}{2}$	$\frac{3\pi}{4}$	π
y	1	0.5	0	0.5	1

By joining these points with a free hand, we obtain a rough sketch of the curve as shown in the figure.

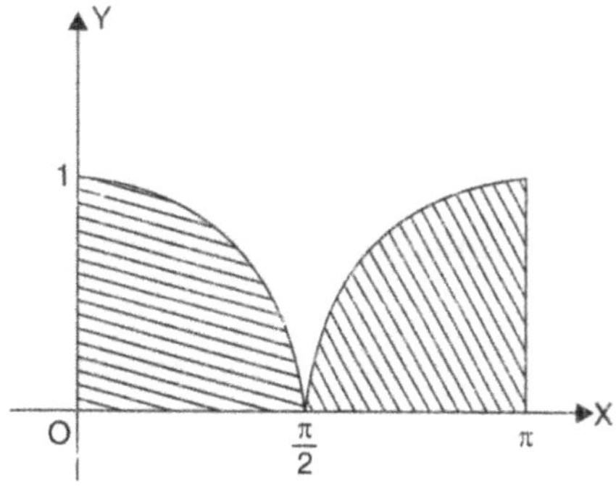

$\therefore$ Required area = Area of shaded region

$$= \int_0^{\pi} y.dx = \int_0^{\pi} \cos^2 x\, dx$$

$$\left[\because 1+\cos 2A = 2\cos^2 A \;\Rightarrow\; \frac{1+\cos 2A}{2} = \cos^2 A\right]$$

$$= \int_0^{\pi} \left(\frac{1+\cos 2x}{2}\right) dx$$

$$= \frac{1}{2}\int_0^{\pi} (1+\cos 2x)\, dx$$

$$= \frac{1}{2}\left[x + \frac{\sin 2x}{2}\right]_0^{\pi} = \frac{1}{2}\left[\left(\pi + \frac{\sin 2\pi}{2}\right) - \left(0 + \frac{\sin 0^\circ}{2}\right)\right] = \frac{1}{2}\,[(\pi + 0) - 0]$$

$$= \frac{\pi}{2} \text{ sq. units.}$$

Example 25. *Find the area in first quadrant bounded by the parabola $y = 4x^2$ and the lines $x = 0$, $y = 1$ and $y = 4$.*

Solution. The equation of the given curve is :

$$y = 4x^2 \;\Rightarrow\; x^2 = \frac{1}{4}y$$

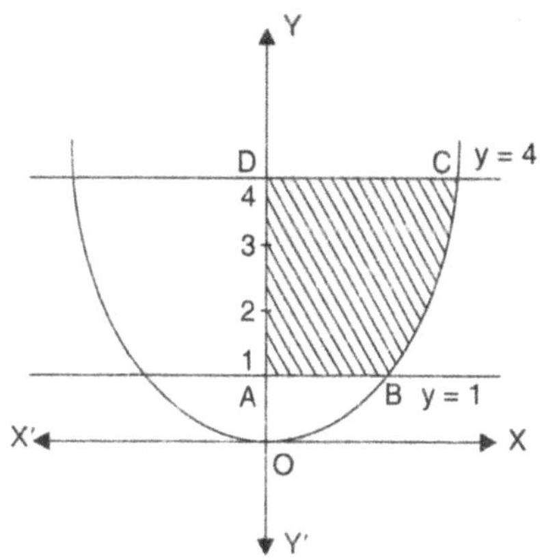

The given equation of the curve represents a parabola with vertex at the origin and axis is y-axis.

It is symmetrical about y-axis as it contains only the even powers of x.

$y = 1$ and $y = 4$ are the horizontal lines parallel to x-axis at a distance of 1 unit and 4 units from it respectively.

$\therefore$ Required area = Area of shaded region

= Area ABCD

$$= \int_1^4 |x|\, dy \qquad \left[\because x^2 = \frac{1}{4}y \;\Rightarrow\; x = \frac{1}{2}\sqrt{y}\right]$$

$$= \int_1^4 \frac{1}{2}\sqrt{y}\, dy = \frac{1}{2}\int_1^4 y^{1/2}\, dy$$

$$= \frac{1}{2}\left[\frac{y^{\frac{1}{2}+1}}{\frac{1}{2}+1}\right]_1^4 = \frac{1}{2}\cdot\frac{2}{3}\left[y^{3/2}\right]_1^4 = \frac{1}{3}\,[(4)^{3/2} - (1)^{3/2}]$$

$$= \frac{1}{3}\,[8-1] = \frac{7}{3} \text{ sq. units.}$$

Example 26. *Compare the areas under the curve $y = \sin x$ and $y = \sin 2x$ above x-axis and between $x = 0$ and $x = \frac{\pi}{3}$.*

Solution. Let the equations of the given curves are

$$y = \sin x \quad ...(1)$$

$$y = \sin 2x \quad ...(2)$$

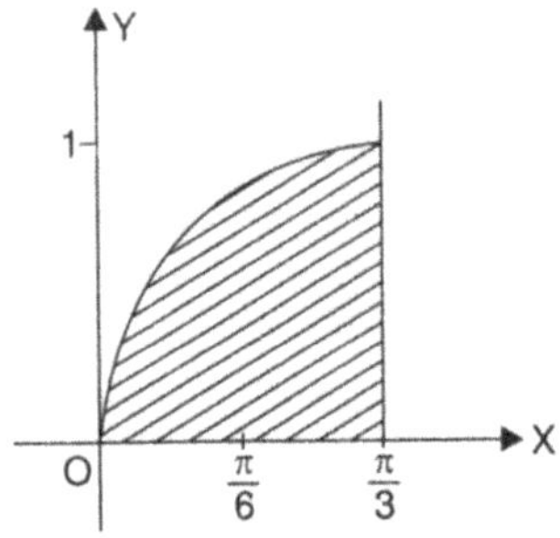

Case I. *Equation of the first curve is $y = \sin x$*

Since $\sin x$ is positive for $0 \le x \le \frac{\pi}{3}$, thus the curve lies above the x-axis.

Table of values of y corresponding to values of x from 0 to $\frac{\pi}{3}$ is given by

x	0	$\frac{\pi}{6}$	$\frac{\pi}{3}$
y	0	0.5	0.8

By joining these points with a free hand, we obtain a rough sketch of the given curve, as shown in the figure.

$\therefore$ Required area $= A_1$ = Area of shaded region

$$= \int_0^{\pi/3} y\,.\,dx = \int_0^{\pi/3} \sin x\, dx = \left[-\cos x\right]_0^{\pi/3}$$

$$= -\left[\cos x\right]_0^{\pi/3} = -\left[\cos\frac{\pi}{3} - \cos 0°\right] = -\left[\frac{1}{2} - 1\right] = -\left(\frac{-1}{2}\right)$$

$$= \frac{1}{2} \text{ sq. units.}$$

Case II. *Equation of the second curve is $y = \sin 2x$.*

Since $\sin x$ is positive for $0 \le x \le \frac{2\pi}{3}$, thus, $\sin 2x$ is positive for $0 \le x \le \frac{\pi}{3}$.

Therefore, the curve lies above x-axis.

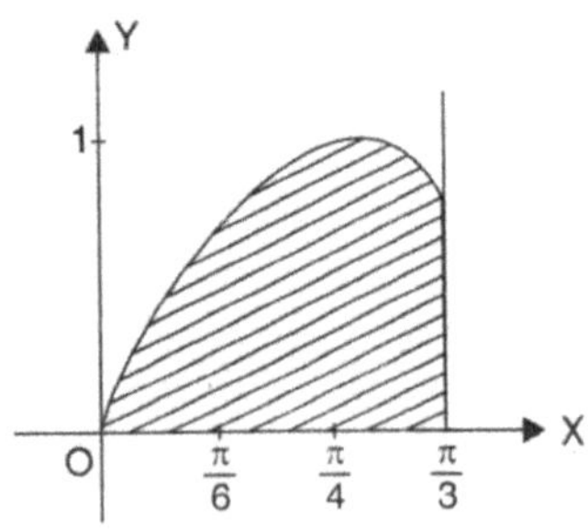

Table of values of y corresponding to the values of x from 0 to $\frac{\pi}{3}$ is given by

x	0	$\frac{\pi}{6}$	$\frac{\pi}{4}$	$\frac{\pi}{3}$
y	0	0.8	1	0.8

By joining these points with a free hand, we obtain a rough sketch of the given curve as shown in the figure.

$\therefore$ Required area = A_2 = Area of shaded region

$$= \int_0^{\pi/3} y \,.\, dx = \int_0^{\pi/3} \sin 2x \, dx$$

$$= \left[\frac{-\cos 2x}{2}\right]_0^{\pi/3} = \frac{-1}{2}\left[\cos 2x\right]_0^{\pi/3}$$

$$= \frac{-1}{2}\left[\cos\frac{2\pi}{3} - \cos 0\right] = \frac{-1}{2}\left[\frac{-1}{2} - 1\right] = \frac{-1}{2}\left(\frac{-3}{2}\right)$$

$$= \frac{3}{4} \text{ sq. units.}$$

$\therefore$ The ratio of the areas under the curves $y = \sin x$ and $y = \sin 2x$ between $x = 0$ and $x = \frac{\pi}{3}$ is given by

$$A_1 : A_2 = \frac{1}{2} : \frac{3}{4} = 2 : 3.$$

Example 27. *Sketch a rough graph of* $y = 4\sqrt{x-1}$, $1 \le x \le 3$ *and evaluate the area between the curve, x-axis and the line* $x = 3$.

Solution. The equation of the given curve is

$$y = 4\sqrt{x-1}.$$

Table of values of y corresponding to the values of x from 1 to 3 is given by

x	1	2	3
y	0	4	$4\sqrt{2} = 5.6$

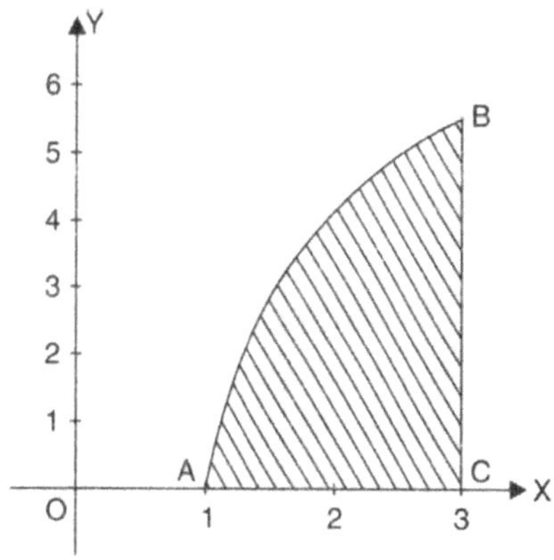

By joining these points with a free hand, we obtain a rough sketch of the given curve as shown in the figure.

$\therefore$ Required area = Area of the shaded region

$$= \int_1^3 y \,.\, dx = \int_1^3 4\sqrt{x-1}\, dx$$

$$= 4\int_1^3 (x-1)^{1/2} \,.\, dx$$

$$= 4\left[\frac{(x-1)^{\frac{1}{2}+1}}{\frac{1}{2}+1}\right]_1^3 = 4\cdot\frac{2}{3}\left[(x-1)^{3/2}\right]_1^3 = \frac{8}{3}\left[(3-1)^{3/2}-(1-1)^{3/2}\right]$$

$$= \frac{8}{3}\left[2^{3/2}-0\right] = \frac{16\sqrt{2}}{3} \text{ sq. units.}$$

Example 28. *Draw a rough sketch of the function $y = \sqrt{x} + 1$ in [0, 4] and determine the area of the region enclosed by the curve, x-axis and the lines $x = 0$, $x = 4$.*

Solution. The given equation of the curve is

$$y = \sqrt{x} + 1.$$

Table of the values of y corresponding to the values of x from 0 to 4 is given by

x	0	1	2	3	4
y	1	2	2.4	2.7	3

By joining these points with a free hand, we obtain a rough sketch of the given curve as shown in the figure.

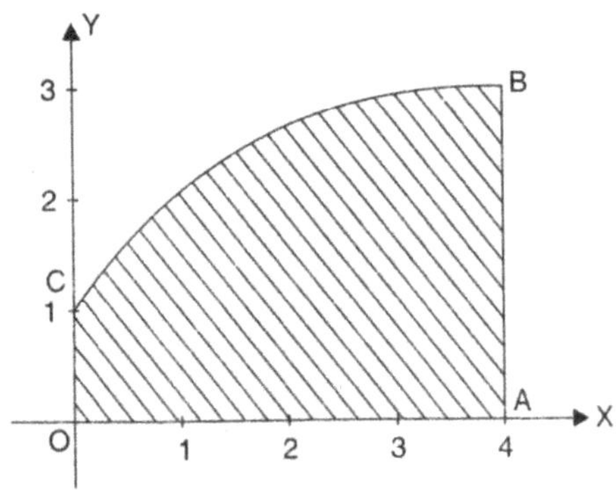

$\therefore$ Required area = Area of the shaded region

$$= \text{Area OABC}$$

$$= \int_0^4 y\,.\,dx = \int_0^4 (\sqrt{x}+1)\,dx$$

$$= \left[\frac{x^{\frac{1}{2}+1}}{\frac{1}{2}+1} + x\right]_0^4 = \left[\frac{2}{3}x^{3/2} + x\right]_0^4$$

$$= \left[\left\{\frac{2}{3}(4)^{3/2}+4\right\} - \{0+0\}\right] = \left[\frac{2}{3}(8)+4-0\right] = \left[\frac{16}{3}+4\right]$$

$$= \frac{16+12}{3} = \frac{28}{3} \text{ sq. units.}$$

Example 29. *Draw a rough sketch of the curve $y = \sqrt{3x+4}$ and find the area under the curve, above x-axis and between the lines $x = 0$ and $x = 4$.*

Solution. The equation of the given curve is

$$y = \sqrt{3x+4}\,.$$

The value of y is positive, when x increases from 0 to 4.

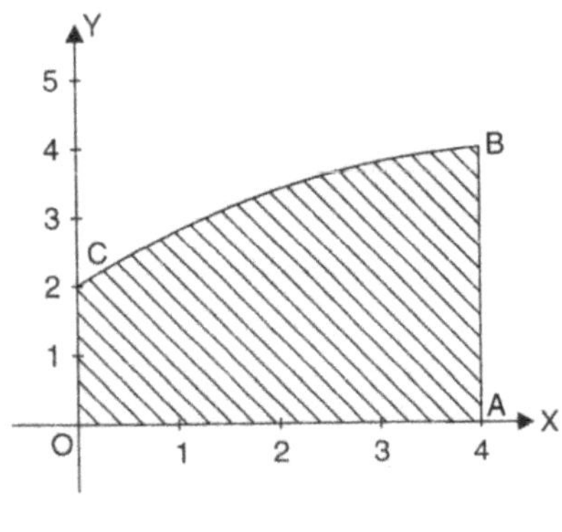

Table of values of y corresponding to the values of x from 0 to 4 is given by

x	0	1	2	3	4
y	2	2.6	3.1	3.6	4

By joining these points with a free hand, we obtain a rough sketch of the given curve as shown in the figure.

$\therefore$ Required area = Area of the shaded region

= Area OABC

$$= \int_0^4 y \, . \, dx = \int_0^4 \sqrt{3x+4} \, dx = \int_0^4 (3x+4)^{1/2} \, . \, dx$$

$$= \left[\frac{(3x+4)^{\frac{1}{2}+1}}{3.\left(\frac{1}{2}+1\right)} \right]_0^4 = \frac{2}{3} . \frac{1}{3} \left[(3x+4)^{3/2} \right]_0^4$$

$$= \frac{2}{9} [(3(4)+4)^{3/2} - (3(0)+4)^{3/2}]$$

$$= \frac{2}{9} [(16)^{3/2} - (4)^{3/2}] = \frac{2}{9} [64-8] = \frac{2}{9} \times 56$$

$$= \frac{112}{9} \text{ sq. units.}$$

Example 30. *Draw a rough sketch of the function* $y = 2\sqrt{1-x^2}$, $x \in [0, 1]$ *and evaluate the area enclosed between the curve and the x-axis.*

Solution. The given equation of the curve is :

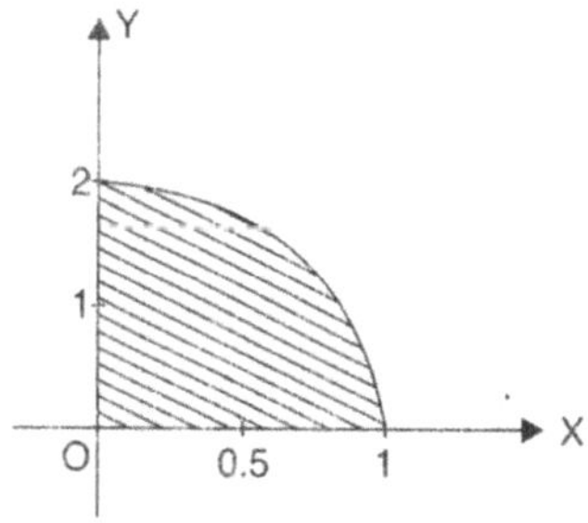

$$y = 2\sqrt{1-x^2} \, .$$

As $x \in [0, 1]$, therefore, value of y is non-negative.

Table of values of y corresponding to the values of x from 0 to 1 is given by

x	0	0.25	0.5	0.75	1
y	2	1.9	1.7	1.3	0

By joining these points with a free hand, we obtain a rough sketch of the given curve as shown in the figure.

$\therefore$ Required area = Area of shaded region

$$= \int_0^1 y \, . \, dx = \int_0^1 2\sqrt{1-x^2} \, dx$$

$$= 2 \int_0^1 \sqrt{(1)^2 - x^2} \, . \, dx$$

$$= 2\left[\frac{x\sqrt{1-x^2}}{2} + \frac{1}{2}\sin^{-1} x\right]_0^1 \qquad \left[\because \int \sqrt{a^2 - x^2}\, dx = \frac{x\sqrt{a^2 - x^2}}{2} + \frac{a^2}{2}\sin^{-1}\frac{x}{a} + c\right]$$

$$= 2\left[\frac{1\sqrt{1-1}}{2} + \frac{1}{2}\sin^{-1} 1 - 0 - \frac{1}{2}\sin^{-1} 0\right] = 2\left[0 + \frac{1}{2}.\frac{\pi}{2} - 0 - 0\right]$$

$$= \frac{\pi}{2} \text{ sq. units.}$$

Example 31. *Find the area bounded by the curve $y^2 = 4a^2(x - 3)$ and the lines $x = 3$, $y = 4a$.*

Solution. The given equation of the curve is

$$y^2 = 4a^2(x - 3).$$

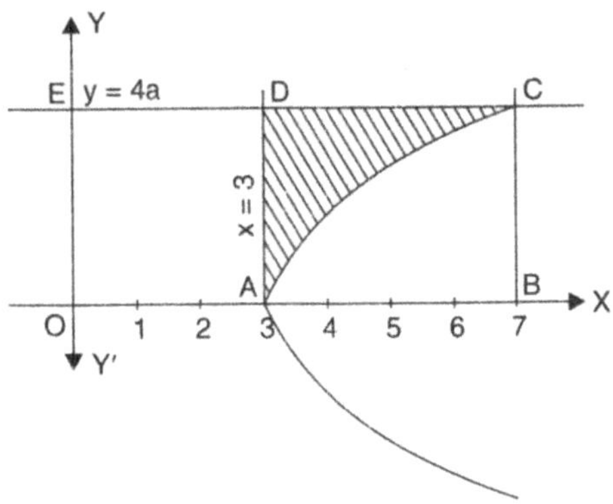

This equation represents a parabola with vertex at (3, 0) and axis along x-axis.

Its latus rectum is $4a^2$.

$x = 3$ is a line parallel to y-axis at a distance of 3 units from it and $y = 4a$ is a line parallel to x-axis at a distance of $4a$ units from it.

$\therefore$ Required area = Area of shaded region.

= Area ACD

= (Area OACDE) – (Area OADE)

$$= \int_0^{4a} (x \text{ of parabola}) \,.\, dy - \int_0^{4a} (x \text{ of line AD})\, dy$$

$$= \int_0^{4a} \left(\frac{y^2}{4a^2} + 3\right) dy - \int_0^{4a} 3 \,.\, dy$$

$$= \left[\frac{y^3}{12a^2} + 3y\right]_0^{4a} - 3\left[y\right]_0^{4a} = \left[\left\{\frac{(4a)^3}{12a^2} + 3(4a)\right\} - \{0 + 0\}\right] - 3[4a - 0]$$

$$= \frac{64a^3}{12a^2} + 12a - 0 - 12a$$

$$= \frac{16a}{3} \text{ sq. units.}$$

Example 32. *Find the area lying above the x-axis and under the parabola $y = 4x - x^2$.*

Solution. The equation of the given curve is

$$y = 4x - x^2. \qquad \text{...(1)}$$

Now, to find the limits of integration, we have to find the points of intersection of the curve with the x-axis.

Put $y = 0$ in equation (1), we get

$$0 = 4x - x^2 \quad \Rightarrow \quad x^2 - 4x = 0$$

$$\Rightarrow \quad x(x - 4) = 0 \quad \Rightarrow \quad x = 0, x = 4.$$

$\therefore$ The curve meets the x-axis in O(0, 0) and A(4, 0).

Table of the values of y corresponding to the values of x from 0 to 4 is given by

x	0	1	2	3	4
y	0	3	4	3	0

By joining these points, with a free hand, we obtain a rough sketch of the given curve as shown in the figure.

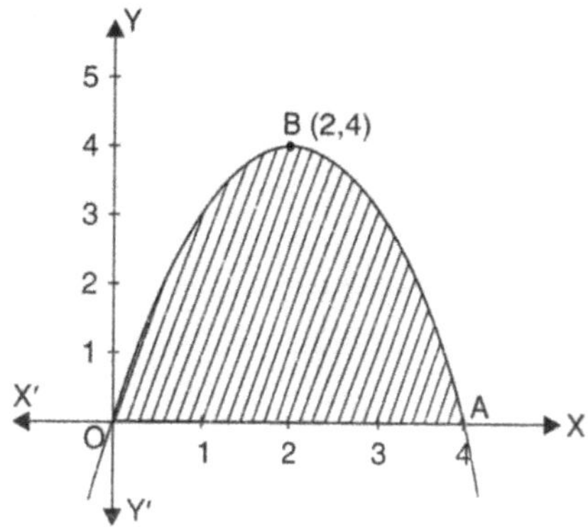

$\therefore$ Required area = Area of shaded region

$$= \text{Area } OAB$$

$$= \int_0^4 y \, . \, dx = \int_0^4 (4x - x^2) \, dx$$

$$= \left[\frac{4x^2}{2} - \frac{x^3}{3} \right]_0^4$$

$$= \left[\left\{ 2(4)^2 - \frac{(4)^3}{3} \right\} - \{0 - 0\} \right]$$

$$= \left[32 - \frac{64}{3} - 0 \right] = \frac{96 - 64}{3} = \frac{32}{3} \text{ sq. units.}$$

Example 33. *Sketch the graph of $y = | x + 1 |$. Evaluate $\int_{-4}^{2} |x + 1| dx$. What does the value of this integral represent on the graph.*

Solution. We have $y = | x + 1 |$

$$| x + 1 | = \begin{cases} -(x + 1); & \text{if } x + 1 < 0 \\ (x + 1); & \text{if } x + 1 \geq 0 \end{cases}$$

$$\Rightarrow \qquad y = \begin{bmatrix} (-x - 1); & \text{if } x < -1 \\ (x + 1); & \text{if } x \geq -1 \end{bmatrix}$$

Table of values of y corresponding to the values of x is given by :

For $x \leq -1$

x	-4	-2	-1
y	3	1	0

For $x \geq -1$

x	-1	0	2
y	0	1	3

By joining these points with a free hand, we obtain a rough sketch of the given curve as shown in the figure.

$\therefore$ Required area = Area of shaded region

= (Area ABD) + (Area ACE)

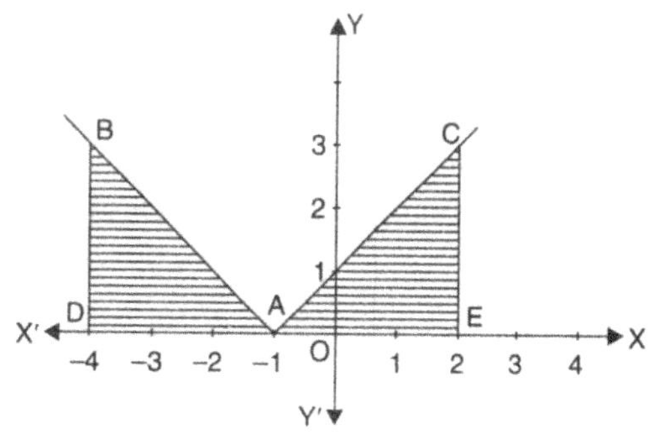

$$\int_{-4}^{2} y \, . \, dx = \int_{-4}^{2} |x+1| \, dx = \int_{-4}^{-1} |x+1| \, dx + \int_{-1}^{2} |x+1| \, dx$$

$$= \int_{-4}^{-1} (-x-1) \, dx + \int_{-1}^{2} (x+1) \, dx$$

$$= \left[\frac{-x^2}{2} - x\right]_{-4}^{-1} + \left[\frac{x^2}{2} + x\right]_{-1}^{2}$$

$$= \left[\left\{\frac{-(-1)^2}{2} - (-1)\right\} - \left\{\frac{-(-4)^2}{2} - (-4)\right\}\right] + \left[\left\{\frac{(2)^2}{2} + 2\right\} - \left\{\frac{(-1)^2}{2} + (-1)\right\}\right]$$

$$= \left[\left(\frac{-1}{2} + 1\right) - (-8+4)\right] + \left[4 - \left(\frac{1}{2} - 1\right)\right] = \left(\frac{1}{2} + 4\right) + \left(4 + \frac{1}{2}\right)$$

$= 9$ sq. units.

$\int_{-4}^{2} |x+1| \, . \, dx$ represents the area bounded by the curve $y = |x+1|$, x-axis and the lines $x = -4$ and $x = 2$. Thus, it is equal to the sum of areas of triangles ABD and ACE.

Example 34. *Using integration, find the area of the region bounded by the following curves after making a rough sketch*

$$y = 1 + |x + 1|, x = -3, x = 3, y = 0.$$

Solution. We have $y = 1 + |x + 1|$.

$$\Rightarrow \qquad y = 1 + |x+1| = \begin{cases} 1 + x + 1; & \text{if } x + 1 \ge 0 \\ 1 - (x+1); & \text{if } x < 0 \end{cases}$$

$$\Rightarrow \qquad y = \begin{cases} x + 2; & \text{if } x \ge -2 \\ -x; & \text{if } x < -1 \end{cases}$$

Clearly, $y = x + 2$ is a straight line intersecting x-axis and y-axis at $(-2, 0)$ and $(0, 2)$ respectively.

The equation $y = -x$ for $x < -1$ represents the part of the line which passes through the origin at angle of 135° with x-axis.

$x = -3$ and $x = 3$ are the lines parallel to y-axis.

$\therefore$ Table of the values of y corresponding to the values of x is given by :

For $y = x + 2$

x	-1	0	1	2	3
y	1	2	3	4	5

For $y = -x$

x	-1	-2	-3
y	1	2	3

By joining these points with a free hand, we obtain a rough sketch of the given curve as shown in the figure.

$\therefore$ Required area = Area of shaded region.

= (Area ABCD) + (Area ABEF)

$$= \int_{-1}^{3} y \, . \, dx + \int_{-3}^{-1} y \, . \, dx$$

$$= \int_{-1}^{3}(x+2)\,dx + \int_{-3}^{-1} -x\,dx$$

$$= \left[\frac{x^2}{2}+2x\right]_{-1}^{3} - \left[\frac{x^2}{2}\right]_{-3}^{-1}$$

$$= \left[\left\{\frac{(3)^2}{2}+2(3)\right\} - \left\{\frac{(-1)^2}{2}+2(-1)\right\}\right] - \left[\frac{(-1)^2}{2} - \frac{(-3)^2}{2}\right]$$

$$= \left[\left(\frac{9}{2}+6\right) - \left(\frac{1}{2}-2\right)\right] - \left[\frac{1}{2}-\frac{9}{2}\right]$$

$$= \frac{9}{2}+6-\frac{1}{2}+2-\left(\frac{-8}{2}\right)$$

$$= \frac{8}{2}+8+\frac{8}{2} = 8+8$$

$= 16$ sq. units.

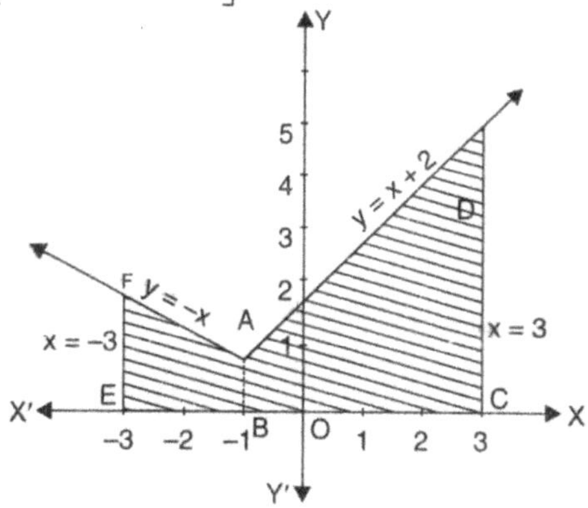

Example 35. *Find the area bounded by the parabola $x = 4 - y^2$ and the y-axis.*

Solution. The equation of the given curve is

$$x = 4 - y^2 \qquad ...(1)$$

$\Rightarrow \quad y^2 = 4 - x \quad \Rightarrow \quad y^2 = -(x-4)$

This equation represents a left handed parabola with vertex at (4, 0).

Now, to find the limit of integration, let us find out the intersection of the curve with the y-axis.

$\therefore$ Put $x = 0$ in equation (1), we get

$$0 = 4 - y^2 \Rightarrow y^2 = 4 \Rightarrow y = \pm 2$$

$\therefore$ The curve meets the y-axis at (0, – 2) and (0, 2).

The rough sketch of the curve is as shown in the figure.

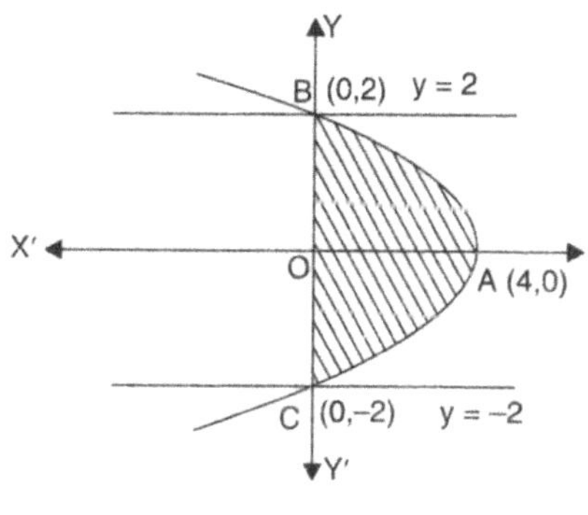

$y = +2$ and $y = -2$ are the lines parallel to x-axis.

$\therefore$ Required area = Area of shaded region

$$= \int_{-2}^{2} |x|\,.\,dy = \int_{-2}^{2}(4-y^2)\,dy$$

$$= \left[4y - \frac{y^3}{3}\right]_{-2}^{2} = \left[\left\{4(2) - \frac{(2)^3}{3}\right\} - \left\{4(-2) - \frac{(-2)^3}{3}\right\}\right]$$

$$= \left[\left(8-\frac{8}{3}\right) - \left(-8+\frac{8}{3}\right)\right]$$

$$= 8 - \frac{8}{3} + 8 - \frac{8}{3} = 16 - \frac{16}{3} = \frac{48 - 16}{3}$$

$$= \frac{32}{3} \text{ sq. units.}$$

Example 36. *Compare the areas under the curves $y = \cos^2 x$ and $y = \sin^2 x$ between $x = 0$ and $x = \pi$.*

Solution. As x increases from 0 to π, the functions $\cos^2 x$ and $\sin^2 x$ are both positive.

$\Rightarrow$ The curves $y = \sin^2 x$ and $y = \cos^2 x$ lie above the x-axis.

$\therefore$ Area under the curve $y = \sin^2 x$

$$\Rightarrow \quad A_1 = \int_0^\pi y \,.\, dx = \int_0^\pi \sin^2 x \, dx \qquad \left[\begin{array}{l} \because \; 1 - \cos 2A = 2\sin^2 A \\ \Rightarrow \; \dfrac{1 - \cos 2A}{2} = \sin^2 A \end{array}\right]$$

$$= \int_0^\pi \left(\frac{1 - \cos 2x}{2}\right) dx = \frac{1}{2}\int_0^\pi (1 - \cos 2x)\, dx$$

$$= \frac{1}{2}\left[x - \frac{\sin 2x}{2}\right]_0^\pi = \frac{1}{2}\left[\left\{\pi - \frac{\sin 2\pi}{2}\right\} - \left\{0 - \frac{\sin 0}{2}\right\}\right]$$

$$= \frac{1}{2}\,[(\pi - 0) - (0 - 0)] = \frac{\pi}{2} \text{ sq. units.}$$

The area under the curve $y = \cos^2 x$

$$A_2 = \int_0^\pi y \,.\, dx = \int_0^\pi \cos^2 x \, dx \qquad \left[\begin{array}{l} \because \; 1 + \cos 2A = 2\cos^2 A \\ \Rightarrow \; \dfrac{1 + \cos 2A}{2} = \cos^2 A \end{array}\right]$$

$$= \int_0^\pi \left(\frac{1 + \cos 2x}{2}\right) dx = \frac{1}{2}\left[x + \frac{\sin 2x}{2}\right]_0^\pi$$

$$= \frac{1}{2}\left[\left\{\pi + \frac{\sin 2\pi}{2}\right\} - \left\{0 + \frac{\sin 0}{2}\right\}\right]$$

$$= \frac{1}{2}\,[(\pi + 0) - (0 + 0)] = \frac{\pi}{2} \text{ sq. units.}$$

$$\therefore \quad A_1 : A_2 = \frac{\pi}{2} : \frac{\pi}{2} \quad \Rightarrow \quad 1 : 1$$

$$\Rightarrow \quad A_1 = A_2$$

$\therefore$ Area under the two given curves are equal.

Example 37. *Find the area bounded by the curve $x = at^2$, $y = 2at$ between the ordinates corresponding to $t = 1$ and $t = 2$.*

Solution. The given equations of the curve are

$$x = at^2$$

$$y = 2at \quad \Rightarrow \quad t = \frac{y}{2a}$$

$$\therefore \quad x = at^2$$

$$\Rightarrow \quad x = a \,.\, \left(\frac{y}{2a}\right)^2$$

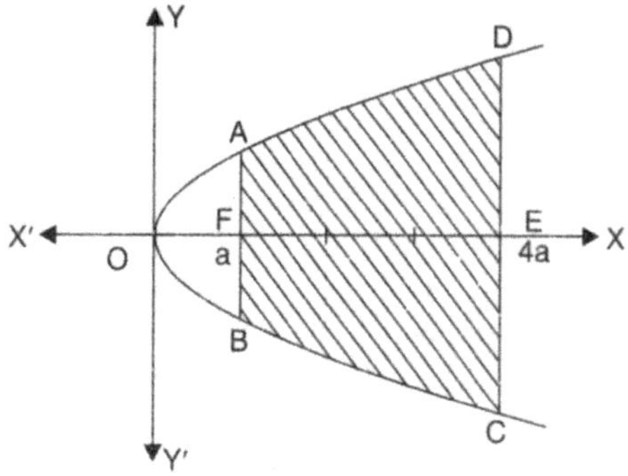

$\Rightarrow \qquad x = \frac{ay^2}{4a^2} \Rightarrow x = \frac{y^2}{4a}$

$\Rightarrow \qquad y^2 = 4ax \qquad ...(1)$

when $t = 1, \quad x = at^2 \Rightarrow x = a(1)^2 \Rightarrow x = a$

when $t = 2, \quad x = at^2 = a(2)^2 = 4a$

Clearly, the parabola is symmetrical about the x-axis.

The rough sketch of this parabola is as shown in the figure.

$\therefore$ Required area = Area of the shaded region.

= Area ABCD = 2 × Area ADEF

$$= 2\int_a^{4a} y \, . \, dx \qquad \left[\begin{array}{l} \because \ y^2 = 4ax \\ \Rightarrow \ y = \sqrt{4ax} = 2\sqrt{ax} \end{array}\right]$$

$$= 2\int_a^{4a} 2\sqrt{ax}\, dx$$

$$= 4\sqrt{a}\int_a^{4a} x^{1/2}\, dx = 4\sqrt{a}\left[\frac{x^{\frac{1}{2}+1}}{\frac{1}{2}+1}\right]_a^{4a}$$

$$= 4\sqrt{a}\,.\frac{2}{3}\left[x^{3/2}\right]_a^{4a} = \frac{8\sqrt{a}}{3}\,[(4a)^{3/2} - a^{3/2}] = \frac{8\sqrt{a}}{3}\,[8a^{3/2} - a^{3/2}]$$

$$= \frac{8\sqrt{a}}{3}\,.\,7a\sqrt{a} = \frac{56a^2}{3} \text{ sq. units.}$$

Example 38. *Find the area under the ellipse :*

$$\frac{x^2}{a^2} + \frac{y^2}{b^2} = 1.$$

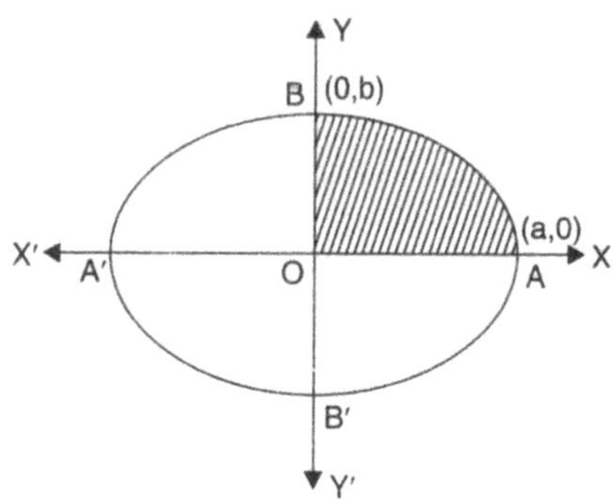

Solution. The equation of the given curve is

$$\frac{x^2}{a^2} + \frac{y^2}{b^2} = 1$$

The ellipse is symmetrical about both the axes as all powers of x and y both are even in the given equation of curve.

The rough sketch of the curve is as shown in the figure.

$\therefore$ Required area = 4 × area of the ellipse in first quadrant.

= 4 × area OAB

$$= 4\int_0^a y \, . \, dx \qquad ...(1)$$

The given equation of the curve is

$$\frac{x^2}{a^2}+\frac{y^2}{b^2}=1$$

$$\Rightarrow \qquad \frac{y^2}{b^2}=1-\frac{x^2}{a^2}$$

$$\Rightarrow \qquad y^2=\frac{b^2}{a^2}(a^2-x^2) \qquad [\because \ y \geq 0 \text{ in region OAB}]$$

$$\Rightarrow \qquad y=\frac{b}{a}\sqrt{a^2-x^2}$$

Putting this value of y in equation (1), we have

$$\therefore \text{ Required area} = \frac{4b}{a}\int_0^a \sqrt{a^2-x^2}\,dx$$

Put $x = a\sin\theta \Rightarrow dx = a\cos\theta\, d\theta$

When $x = 0 \Rightarrow 0 = a\sin\theta \Rightarrow \sin\theta = 0 \Rightarrow \theta = 0.$

and when $x = a \Rightarrow a = a\sin\theta \Rightarrow \sin\theta = 1 \Rightarrow \theta = \frac{\pi}{2}.$

$$\therefore \text{ Required area} = \frac{4b}{a}\int_0^{\pi/2}\sqrt{a^2-a^2\sin^2\theta}\ .\ a\cos\theta\, d\theta$$

$$= \frac{4b}{a}.\int_0^{\pi/2} a^2\sqrt{1-\sin^2\theta}\ .\cos\theta\, d\theta$$

$$= 4ab\int_0^{\pi/2}\cos\theta\,.\cos\theta\, d\theta \qquad [\because \ \sin^2 A+\cos^2 A=1]$$

$$= 4ab\int_0^{\pi/2}\cos^2\theta\, d\theta$$

$$= 4ab\int_0^{\pi/2}\left(\frac{1+\cos 2\theta}{2}\right)d\theta \qquad \left[\begin{array}{l}\because \ 1+\cos 2\theta = 2\cos^2\theta \\ \Rightarrow \left(\dfrac{1+\cos 2\theta}{2}\right)=\cos^2\theta\end{array}\right]$$

$$= 2ab\int_0^{\pi/2}(1+\cos 2\theta)\, d\theta$$

$$= 2ab\left[\theta+\frac{\sin 2\theta}{2}\right]_0^{\pi/2} = 2ab\left[\left\{\frac{\pi}{2}+\frac{\sin\pi}{2}\right\}-\left\{0+\frac{\sin 0^\circ}{2}\right\}\right]$$

$$= 2ab\left[\left(\frac{\pi}{2}+0\right)-(0+0)\right]$$

$$= \pi\, ab \text{ sq. units.}$$

Example 39. *Find the area of the region bounded by $x^2 = y - 3$, $y = 4$, $y = 6$ and the y-axis in the first quadrant.*

Solution. The given equation of the curve is $x^2 = y - 3$.

Clearly, the given equation of the curve is a parabola with vertex (0, 3) and axis along y-axis.

$y = 4$ and $y = 6$ are straight lines parallel to x-axis at a distance of 4 and 6 units from it respectively.

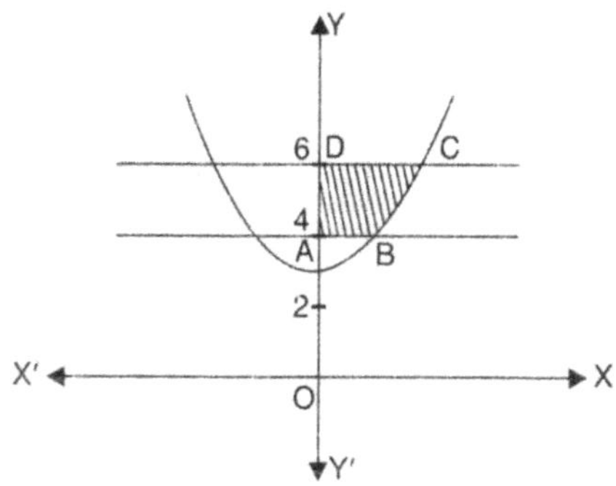

$\therefore$ Required area = Area of the shaded region

$$= \text{Area ABCD}$$

$$= \int_4^6 x\,.\,dy \qquad [\because \; x \geq 0]$$

$$= \int_4^6 \sqrt{y-3}\,.\,dy = \int_4^6 (y-3)^{1/2}\,.\,dy$$

$$= \left[\frac{(y-3)^{\frac{1}{2}+1}}{\frac{1}{2}+1}\right]_4^6 = \frac{2}{3}\left[(y-3)^{3/2}\right]_4^6$$

$$= \frac{2}{3}\,[(6-3)^{3/2} - (4-3)^{3/2}] = \frac{2}{3}\,[(3)^{3/2} - (1)^{3/2}]$$

$$= \frac{2}{3}\,.\,[3\sqrt{3} - 1] \text{ sq. units.}$$

Example 40. *Find the area of the circle $x^2 + y^2 = a^2$.*

Solution. The given equation of the curve is

$$x^2 + y^2 = a^2 \qquad \ldots(1)$$

The equation (1) represents the standard form of a circle with centre at the origin and radius is a.

It is symmetrical about both x-axis and y-axis as it contains only even powers of y and x.

The rough sketch of the curve is as shown in the figure.

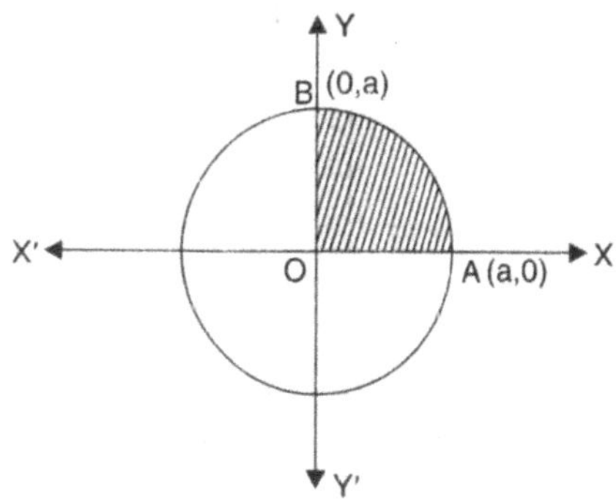

$\therefore$ Required area

$$= 4 \times (\text{Area of the circle in first quadrant})$$

$$= 4 \times \text{Area OAB} = 4\int_0^a y\,.\,dx$$

$$= 4\int_0^a \sqrt{a^2 - x^2}\; dx \qquad \left[\begin{aligned} &\because\; x^2 + y^2 = a^2 \\ &\Rightarrow\; y^2 = a^2 - x^2 \\ &\Rightarrow\; y = \sqrt{a^2 - x^2} \text{ for } y \geq 0 \end{aligned}\right]$$

Put $\quad x = a\sin\theta \;\Rightarrow\; dx = a\cos\theta\, d\theta$

When $x = 0$; $a\sin\theta = 0 \;\Rightarrow\; \sin\theta = 0 \;\Rightarrow\; \theta = 0$

and when $\quad x = a$; $a\sin\theta = a \;\Rightarrow\; \sin\theta = 1 \;\Rightarrow\; \theta = \dfrac{\pi}{2}$

$$= 4\int_0^{\pi/2} \sqrt{a^2 - a^2 \sin^2\theta} \;.\; a\cos\theta \, d\theta$$

$$= 4a^2 \int_0^{\pi/2} \sqrt{1 - \sin^2\theta} \;.\; \cos\theta \, d\theta \qquad [\because \sin^2 A + \cos^2 A = 1]$$

$$= 4a^2 \int_0^{\pi/2} \cos\theta \,.\, \cos\theta \, d\theta$$

$$= 4a^2 \int_0^{\pi/2} \cos^2\theta \, d\theta \qquad \left[\begin{array}{l} \because\ 1 + \cos 2A = 2\cos^2 A \\ \Rightarrow\ \dfrac{1+\cos 2A}{2} = \cos^2 A \end{array}\right]$$

$$= 4a^2 \int_0^{\pi/2} \left(\frac{1 + \cos 2\theta}{2}\right) d\theta$$

$$= 2a^2 \int_0^{\pi/2} (1 + \cos 2\theta)\, d\theta = 2a^2 \left[\theta + \frac{\sin 2\theta}{2}\right]_0^{\pi/2}$$

$$= 2a^2 \left[\left\{\frac{\pi}{2} + \frac{\sin \pi}{2}\right\} - \left\{0 + \frac{\sin 0^\circ}{2}\right\}\right] = 2a^2 \left[\left(\frac{\pi}{2} + 0\right) - (0 + 0)\right]$$

$= a^2\pi$ sq. units.

Example 41. *Find the area of the region bounded by $y^2 = x - 2$, $x = 4$, $x = 6$ and the x-axis in the first quadrant.*

Solution. The given equation of the curve is

$$y^2 = x - 2. \qquad ...(1)$$

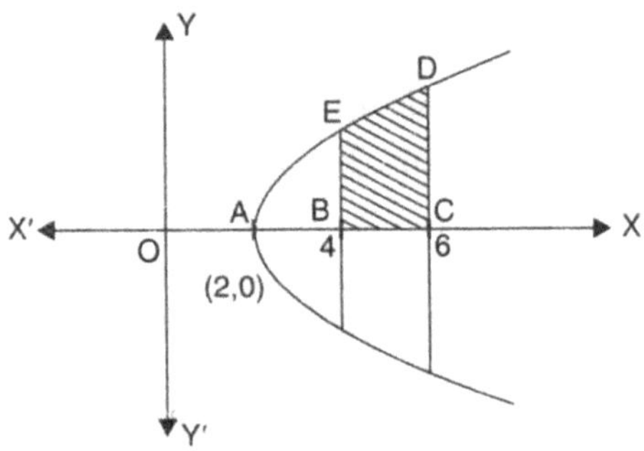

The equation (1) represents a parabola with vertex at (2, 0) and axis along x-axis.

$x = 4$ and $x = 6$ are straight lines parallel to y-axis at a distance of 4 and 6 units from it respectively.

$\therefore$ Required area = Area of the shaded region

= Area BCDE

$$= \int_4^6 y \,.\, dx = \int_4^6 \sqrt{x - 2} \,.\, dx \qquad \left[\begin{array}{l} \because\ y^2 = x - 2 \ \Rightarrow\ y = \pm\sqrt{x-2} \\ \qquad\qquad\quad\ \Rightarrow\ y = \sqrt{x-2} \text{ for } y \ge 0 \end{array}\right]$$

$$= \left[\frac{(x-2)^{\frac{1}{2}+1}}{\frac{1}{2}+1}\right]_4^6 = \frac{2}{3}\left[(x-2)^{3/2}\right]_4^6$$

$$= \frac{2}{3}\,[(6-2)^{3/2} - (4-2)^{3/2}] = \frac{2}{3}\,[4^{3/2} - 2^{3/2}] = \frac{2}{3}\,[8 - 2\sqrt{2}\,]$$

$$= \frac{16 - 4\sqrt{2}}{3} \text{ sq. units.}$$

Example 42. *Find the area of the region bounded by the ellipse* $\frac{x^2}{16}+\frac{y^2}{9}=1$.

Solution. The equation of the given curve is

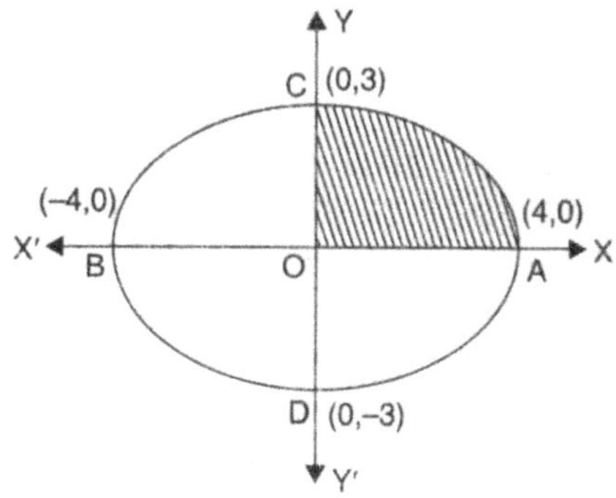

$$\frac{x^2}{16}+\frac{y^2}{9}=1$$

As the equation contains only even powers of x and y, so the ellipse is symmetrical about both the axis.

Also the curve is symmetrical in opposite quadrants as the given equation remains unchanged when x and y are replaced by $-x$ and $-y$.

$$\therefore \quad \frac{x^2}{16}+\frac{y^2}{9}=1 \Rightarrow \frac{y^2}{9}=1-\frac{x^2}{16}$$

$$\Rightarrow \quad y^2=\frac{9}{16}(16-x^2) \Rightarrow y=\pm\frac{3}{4}\sqrt{16-x^2}$$

Table of values of y corresponding to the values of x satisfying the given equation is given by

x	0	4	− 4	2	− 2
y	± 3	0	0	± 2.5	± 2.5

By joining these points with a free hand, we obtain a rough sketch of the curve as shown in the figure.

$\therefore$ Required area = 4 × (Area of ellipse in the first quadrant.)

$$= 4 \times (\text{Area OAC}) = 4\int_0^4 y\,.\,dx \qquad [\because \ y \geq 0]$$

$$= 4\int_0^4 \frac{3}{4}\sqrt{16-x^2}\,.\,dx = 3\int_0^4 \sqrt{16-x^2}\,dx = 3\int_0^4 \sqrt{(4)^2-x^2}\,.\,dx$$

$$= 3\,.\left[\frac{x\sqrt{16-x^2}}{2}+\frac{16}{2}\sin^{-1}\frac{x}{4}\right]_0^4$$

$$\left[\because \int \sqrt{a^2-x^2}\,dx = \frac{x\sqrt{a^2-x^2}}{2}+\frac{a^2}{2}\sin^{-1}\frac{x}{a}+c\right]$$

$$= 3\left[\left(\frac{4\sqrt{16-16}}{2}+8\sin^{-1}\frac{4}{4}\right)-\left(\frac{0}{2}+8\sin^{-1}0\right)\right]$$

$$= 3\left[\left(0+8\,.\,\frac{\pi}{2}\right)-(0+0)\right]$$

$$= 3(4\pi) = 12\pi \text{ sq. units.}$$

Example 43. *Find the area of the region bounded by the ellipse* $\frac{x^2}{4}+\frac{y^2}{9}=1$.

Solution. The given equation of the curve is

$$\frac{x^2}{4}+\frac{y^2}{9}=1$$

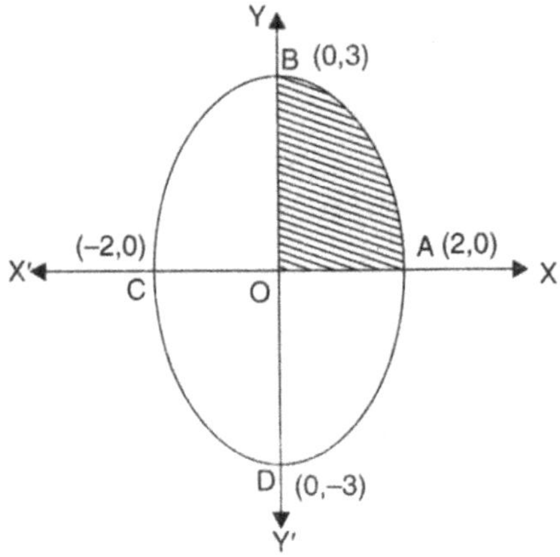

The given equation of curve represents an ellipse with centre at the origin, the length of major axis is 4 and that of minor axis is 9.

The given curve is symmetrical about both x-axis and y-axis as it contains only even powers of y and x.

The curve is symmetrical in the opposite quadrants also as the given equation remains unchanged when x and y are replaced by $-x$ and $-y$.

Now $\quad \frac{x^2}{4}+\frac{y^2}{9}=1$

$$\Rightarrow \quad \frac{y^2}{9}=1-\frac{x^2}{4} \quad \Rightarrow \quad y^2=\frac{9}{4}(4-x^2)$$

$$\Rightarrow \quad y=\pm\frac{3}{2}\sqrt{4-x^2}.$$

Table of values of y corresponding to the values of x satisfying the given equation is given by

x	0	1	-1	2	-2
y	± 3	± 2.5	± 2.5	0	0

By joining these points, with a free hand, we obtained a rough sketch of the given curve as shown in the figure.

$\therefore$ Required area

= 4 times the area of the region in the first quadrant.

= 4 × (Area OAB)

$$=4\int_0^2 y\,.\,dx \qquad [\because \; y \ge 0]$$

$$=4\int_0^2 \frac{3}{2}\sqrt{4-x^2}\,.\,dx=6\int_0^2\sqrt{4-x^2}\,dx=6\int_0^2\sqrt{(2)^2-x^2}\,dx$$

$$=6\left[\frac{x\sqrt{4-x^2}}{2}+\frac{4}{2}\sin^{-1}\frac{x}{2}\right]_0^2$$

$$\left[\because \int\sqrt{a^2-x^2}\,dx=\frac{x\sqrt{a^2-x^2}}{2}+\frac{a^2}{2}\sin^{-1}\frac{x}{a}+c\right]$$

$$=6\left[\left(\frac{2\sqrt{4-4}}{2}+2\sin^{-1}\frac{2}{2}\right)-(0+2\sin^{-1}0)\right]=6\left[\left(0+2\,.\,\frac{\pi}{2}\right)-(0+0)\right]$$

$= 6(\pi) = 6\pi$ sq. units.

Example 44. *Find the area of the region in the first quadrant enclosed by the x-axis, the line $x = \sqrt{3}y$ and the circle $x^2 + y^2 = 4$.*

Solution. The given equation of the curve is

$$x^2 + y^2 = 4.$$

It represents a circle with centre at the origin and radius 2.

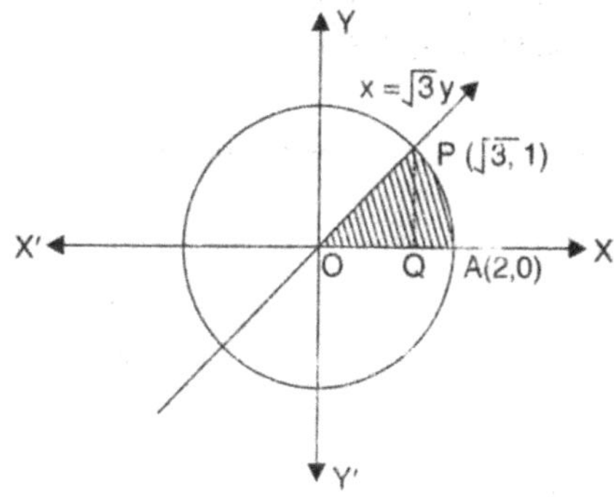

$x = \sqrt{3}y \Rightarrow y = \frac{1}{\sqrt{3}}x$ represents a straight line passing through the origin (0, 0) and intersecting the circle at point P($\sqrt{3}$, 1).

The rough sketch of the curve is as shown in the figure.

Now $\quad x^2 + y^2 = 4 \Rightarrow y^2 = 4 - x^2 \Rightarrow y = \pm\sqrt{4 - x^2}$

For the First Quadrant

$$y = \sqrt{4 - x^2}.$$

$\therefore$ Required area = Area of the shaded region

= Area OPQ + Area PAQ

$$= \int_0^{\sqrt{3}} y\,.\,dx + \int_{\sqrt{3}}^2 y\,.\,dx = \int_0^{\sqrt{3}} \frac{1}{\sqrt{3}} x\,dx + \int_{\sqrt{3}}^2 \sqrt{4 - x^2}\,dx$$

$$= \frac{1}{\sqrt{3}}\left[\frac{x^2}{2}\right]_0^{\sqrt{3}} + \int_{\sqrt{3}}^2 \sqrt{(2)^2 - x^2}\,.\,dx$$

$$= \frac{1}{2\sqrt{3}}[(\sqrt{3})^2 - 0] + \left[\frac{x\sqrt{4 - x^2}}{2} + \frac{2}{2}\sin^{-1}\frac{x}{2}\right]_{\sqrt{3}}^2$$

$$\left[\because \int \sqrt{a^2 - x^2} = \frac{x\sqrt{a^2 - x^2}}{2} + \frac{a}{2}\sin^{-1}\frac{x}{a} + c\right]$$

$$= \frac{1}{2\sqrt{3}}(3) + \left[\left(\frac{2\sqrt{4-4}}{2} + \frac{4}{2}\sin^{-1}\frac{2}{2}\right) - \left(\frac{\sqrt{3}\sqrt{4-3}}{2} + \frac{4}{2}\sin^{-1}\frac{\sqrt{3}}{2}\right)\right]$$

$$= \frac{\sqrt{3}}{2} + \left[(0 + 2\sin^{-1} 1) - \left(\frac{\sqrt{3}}{2} + 2\sin^{-1}\frac{\sqrt{3}}{2}\right)\right]$$

$$= \frac{\sqrt{3}}{2} + 2\,.\,\frac{\pi}{2} - \frac{\sqrt{3}}{2} - 2\,.\,\frac{\pi}{3} = 2\left(\frac{\pi}{2} - \frac{\pi}{3}\right) = 2\left(\frac{3\pi - 2\pi}{6}\right)$$

$$= \frac{\pi}{3} \text{ sq. units.}$$

4.6. AREA BETWEEN TWO CURVES

Let $y = f(x)$ and $y = g(x)$ be two functions such that $0 \le g(x) \le f(x)$ for $a \le x \le b$. *i.e.*, both the curves lie above the x-axis and the curve $y = f(x)$ lies above the curve $y = g(x)$.

The area between $y = f(x)$ and $y = g(x)$ for $a \le x \le b$ is as shown in the figure.

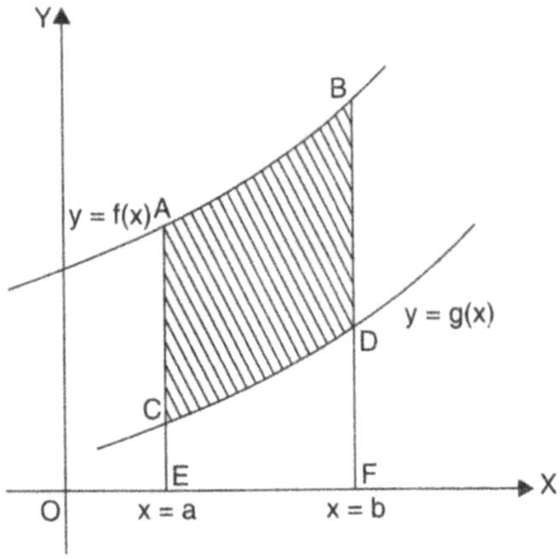

$\therefore$ Required area = Area of shaded region

= Area ABDC = Area ABFE – Area CDFE

= (Area bounded by $y = f(x)$, x-axis and the ordinates $x = a$, $x = b$)

– (Area bounded by $y = g(x)$, x-axis and the ordinates $x = a$, $x = b$).

$$= \int_a^b f(x)\,dx - \int_a^b g(x)\,dx = \int_a^b [f(x) - g(x)]\,.\,dx$$

$\therefore$ Area between $y = f(x)$ and $y = g(x)$, $a \le x \le b$.

$$= \int_a^b (y_{\text{upper curve}} - y_{\text{lower curve}})\,.\,dx$$

Remark. Sometimes, area bounded between two curves is asked. In such cases, to find the ordinates a and b *i.e.*, the limits of integration, we find the points of intersection of the two curves.

SOME SOLVED EXAMPLES

Example 1. *Using integration, find the area of the region bounded by the curves $y = x^2 + 2$, $y = x$, $x = 0$ and $x = 3$.*

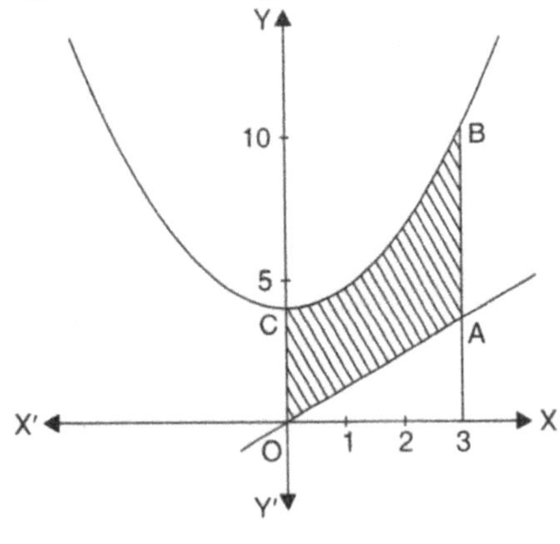

Solution. The given equation of curves are

$$y = x^2 + 2 \quad ...(1)$$

$$y = x \quad ...(2)$$

$$x = 0 \quad ...(3)$$

$$x = 3 \quad ...(4)$$

Now, from equation (1), we have

$$x^2 = y - 2$$

$\Rightarrow \qquad (x-0)^2 = 4\left(\frac{1}{4}\right)(y-2)$ (Note this step)

Clearly, this equation represents a parabola with vertex at (0, 2). The curve is symmetrical about y-axis, as it contains only the even powers of x.

Table of values of y corresponding to the values of x is given by

x	– 3	– 2	– 1	0	1	2	3
y	11	6	3	2	3	6	11

By joining these points with a free hand we obtain a rough sketch of the equation (1) as shown in the figure.

The equation $y = x$ represents a straight line.

Table of values of y corresponding to the values of x is given by

x	0	1	2	3
y	0	1	2	3

$\therefore$ Required area = Area of the shaded region

= Area OABC

$$= \int_0^3 (y_{\text{upper curve}} - y_{\text{lower curve}})\,.\,dx$$

$$= \int_0^3 [x^2 + 2 - x]\,dx$$

$$= \left[\frac{x^3}{3} + 2x - \frac{x^2}{2}\right]_0^3 = \left[\left\{\frac{(3)^3}{3} + 2(3) - \frac{(3)^2}{2}\right\} - \{0+0-0\}\right]$$

$$= \left[\left(\frac{27}{3} + 6 - \frac{9}{2}\right) - 0\right] = 9 + 6 - \frac{9}{2} = 15 - \frac{9}{2} = \frac{30-9}{2}$$

$$= \frac{21}{2} \text{ sq. units.}$$

Example 2. *Find the area bouded by the curve $x^2 = 4y$ and the straight line $x = 4y - 2$.*

Solution. The given equations of curves are :

$$x^2 = 4y \qquad ...(1)$$

$$x = 4y - 2 \qquad ...(2)$$

Clearly, equation (1) represents a parabola with vertex at the origin and axis along the positive direction of y-axis.

The curve is symmetrical about y-axis as it contains only the even powers of x.

Table of values of y corresponding to the values of x satisfying the equation (1) is given by

x	– 1	1	– 2	2	0
y	0.25	0.25	1	1	0

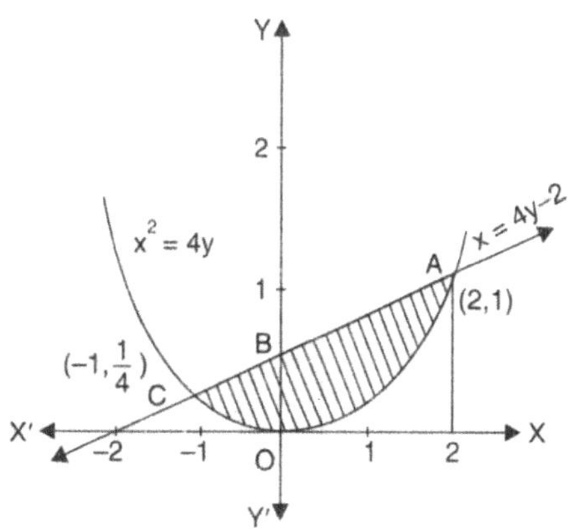

Equation (2) $x = 4y - 2$ represents a straight line.

Table of values of y corresponding to the values of x is given by

x	-2	2	0
y	0	1	0.5

By joining these points, we obtain a rough sketch of the given curves as shown in the figure.

From equation (1),

$$y = \frac{x^2}{4}$$

$$\Rightarrow \quad x = 4y - 2$$

$$\Rightarrow \quad x = 4\left(\frac{x^2}{4}\right) - 2 \quad \Rightarrow \quad x^2 - x - 2 = 0$$

$$\Rightarrow \quad (x-2)(x+1) = 0 \quad \Rightarrow \quad x = 2,\ x = -1$$

For $x = 2$, $\qquad y = \frac{(2)^2}{4} = \frac{4}{4} = 1$

For $x = -1$, $\qquad y = \frac{(-1)^2}{4} = \frac{1}{4}$

$\therefore$ The points of intersection of the given parabola and the line are (2, 1) and $\left(-1, \frac{1}{4}\right)$.

$\therefore$ Required area = Area of shaded region

= Area OABC

$$= \int_{-1}^{2} (y_{\text{upper curve}} - y_{\text{lower curve}})\,.\,dx$$

$$= \int_{-1}^{2} \left(\frac{x+2}{4} - \frac{x^2}{4}\right) dx = \frac{1}{4}\int_{-1}^{2} (x + 2 - x^2)\, dx \qquad \left[\because \begin{array}{l} \text{From (1)} : y = \frac{x^2}{4} \\ \text{From (2)} : y = \frac{x+2}{4} \end{array}\right]$$

$$= \frac{1}{4}\left[\frac{x^2}{2} + 2x - \frac{x^3}{3}\right]_{-1}^{2} = \frac{1}{4}\left[\left\{\frac{(2)^2}{2} + 2(2) - \frac{(2)^3}{3}\right\} - \left\{\frac{(-1)^2}{2} + 2(-1) - \frac{(-1)^3}{3}\right\}\right]$$

$$= \frac{1}{4}\left[\left(\frac{4}{2} + 4 - \frac{8}{3}\right) - \left(\frac{1}{2} - 2 + \frac{1}{3}\right)\right] = \frac{1}{4}\left[6 - \frac{8}{3} - \frac{1}{2} + 2 - \frac{1}{3}\right]$$

$$= \frac{1}{4}\left[8 - \frac{9}{3} - \frac{1}{2}\right] = \frac{1}{4}\left[8 - 3 - \frac{1}{2}\right] = \frac{1}{4}\left[5 - \frac{1}{2}\right]$$

$$= \frac{1}{4}\cdot\frac{9}{2} = \frac{9}{8} \text{ sq. units.}$$

Example 3. *Find the area bounded by the curve $y^2 = 4ax$ and the lines $y = 2a$ and y-axis.*

Solution. The given equations of the curve are

$$y^2 = 4ax \quad ...(1)$$

$$y = 2a \quad ...(2)$$

y-axis $\Rightarrow$ $x = 0$ ...(3)

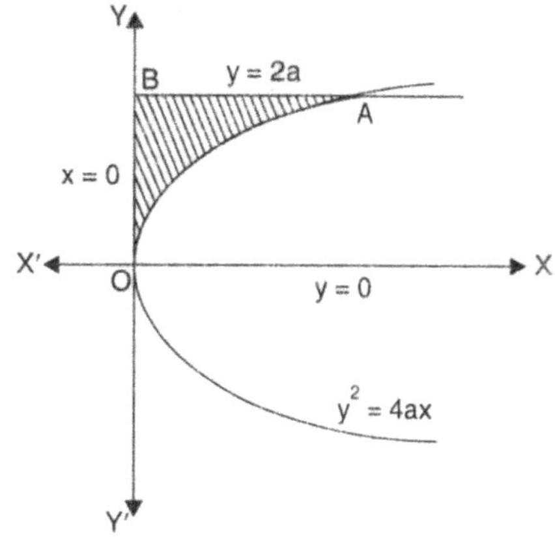

Clearly, equation (1) represents a right handed parabola with vertex at the origin *i.e.*, (0, 0)

Equation (2) represents a straight line parallel to x-axis at a distance of $2a$ units from it.

The rough sketches of these equations are as shown in the figure.

$\therefore$ Required area = Area of the shaded region

$$= \text{Area OAB} = \int_0^{2a} x \, . \, dy \qquad \left[\because \text{ Required area is the area bounded by the curve } x = \frac{y^2}{4a}, y\text{-axis and between the horizontal lines } y = 0, y = 2a\right]$$

$$= \int_0^{2a} \frac{y^2}{4a} \, . \, dy = \frac{1}{4a} \int_0^{2a} y^2 \, . \, dy$$

$$= \frac{1}{4a}\left[\frac{y^3}{3}\right]_0^{2a} = \frac{1}{12a} \, [(2a)^3 - 0] = \frac{1}{12a} \, . \, 8a^3$$

$$= \frac{2a^2}{3} \text{ sq. units.}$$

Example 4. *Find the area of the region bounded by the parabola $y^2 = 4x$ and the line $y = 2x$.*

Solution. The given equations of the curves are

$$y^2 = 4x \quad ...(1)$$

$$y = 2x \quad ...(2)$$

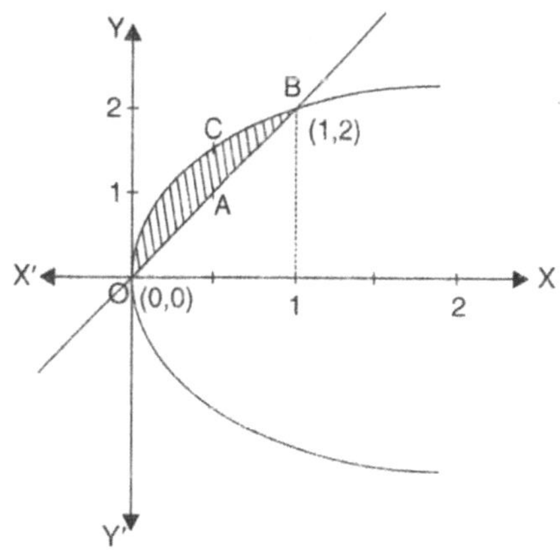

Clearly, the equation (1) represents a parabola with vertex at the origin (0, 0) and symmetrical about x-axis as it contains only even powers of y.

The equation (2) represents a straight line.

Table of values of y corresponding to the values of x satisfying equation (1) is given by

x	0	0.5	1
y	0	$\sqrt{2}$	2

Table of values of y corresponding to the values of x satisfying equation (2) is given by

x	0	0.5	1
y	0	1	2

By joining these points, we obtain a rough sketch of the given equation of curves as shown in the figure.

To find the points of intersection of equations (1) and (2).

Let us solve the given equations :

$$y^2 = 4x \quad \text{and} \quad y = 2x$$

$$\Rightarrow \quad y^2 = 4x \Rightarrow 4x^2 = 4x$$

$$\Rightarrow \quad 4x(x-1) = 0 \Rightarrow x = 0, x = 1$$

For $x = 0$, $\quad y^2 = 4(0) \quad \Rightarrow y = 0$

For $x = 1$, $\quad y^2 = 4 \quad \Rightarrow y = 2$

$\therefore$ The point of intersection of two given curves are (0, 0) and (1, 2).

$\therefore$ Required area = Area of the shaded region = Area OABC

$$= \int_0^1 (y_{\text{upper curve}} - y_{\text{lower curve}}) \, . \, dx$$

$$= \int_0^1 (2\sqrt{x} - 2x)\, dx = 2\int_0^1 (x^{1/2} - x)\, dx$$

$$= 2\left[\frac{x^{3/2}}{\frac{3}{2}} - \frac{x^2}{2}\right]_0^1 = 2\left[\frac{2}{3}x^{3/2} - \frac{x^2}{2}\right]_0^1$$

$$= 2\left[\left\{\frac{2}{3}(1)^{3/2} - \frac{(1)^2}{2}\right\} - \{0 - 0\}\right] = 2\left[\frac{2}{3} - \frac{1}{2} - 0\right] = 2\left[\frac{4-3}{6}\right]$$

$$= \frac{1}{3} \text{ sq. units.}$$

Example 5. *Find the area of the region included between the parabola* $y = \frac{3x^2}{4}$ *and the line* $3x - 2y + 12 = 0$.

Solution. The given equations of the curve are

$$y = \frac{3x^2}{4} \qquad \text{...(1)}$$

$$3x - 2y + 12 = 0 \qquad \text{...(2)}$$

Equation (1) $y = \frac{3x^2}{4} \Rightarrow x^2 = \frac{4y}{3}$ represents an upward parabola with vertex at the origin (0, 0) and symmetrical about y-axis as it contains only even powers of x.

Equation (2) represents a straight line passing through the points (– 2, 3) and (4, 12).

Table of the values of x and y satisfying equation (1) is given by :

x	-2	0	2	4
y	3	0	3	12

Table of the values of x and y satisfying equation (2) is given by :

x	-2	0	2	4
y	3	6	9	12

By joining these points, we obtain a rough sketch of the curves as shown in the figure.

Now, to find the points of intersection of the given parabola and the line, we solve equations (1) and (2) simultaneously.

From equation (2), we have

$$y = \frac{3x+12}{2}$$

Putting this value of y in equation (1), we have

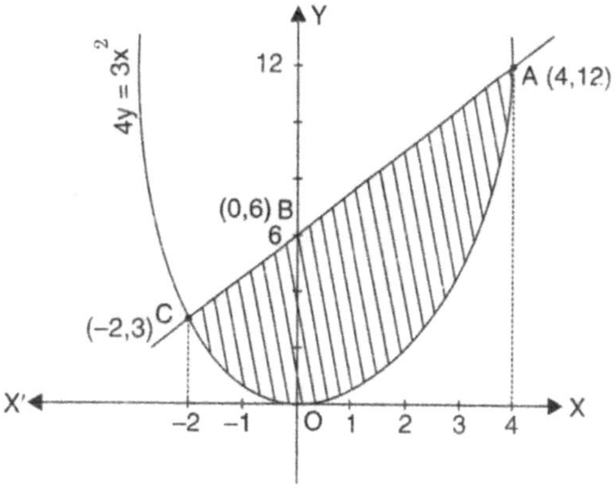

$$\Rightarrow \quad x^2 = \frac{4}{3}\left(\frac{3x+12}{2}\right)$$

$$\Rightarrow \quad x^2 = 2x + 8$$

$$\Rightarrow \quad x^2 - 2x - 8 = 0$$

$$\Rightarrow \quad x^2 - 4x + 2x - 8 = 0$$

$$\Rightarrow \quad x(x-4) + 2(x-4) = 0$$

$$\Rightarrow \quad (x-4)(x+2) = 0$$

$$\Rightarrow \quad x = 4, \quad x = -2$$

For $x = 4$, $\quad y = \dfrac{3(4)+12}{2} = \dfrac{24}{2} = 12$

For $x = -2$, $\quad y = \dfrac{3(-2)+12}{2} = \dfrac{6}{2} = 3$

$\therefore$ The points of intersection of the given parabola and the line are $(-2, 3)$ and $(4, 12)$.

$\therefore$ Required area = Area of shaded region = Area OABC

$$= \int_{-2}^{4} (y_{\text{upper curve}} - y_{\text{lower curve}}) \,.\, dx = \int_{-2}^{4} \left[\frac{3x+12}{2} - \frac{3x^2}{4}\right] dx$$

$$= \frac{3}{4}\int_{-2}^{4} (2x + 8 - x^2)\, dx = \frac{3}{4}\left[\frac{2x^2}{2} + 8x - \frac{x^3}{3}\right]_{-2}^{4}$$

$$= \frac{3}{4}\left[\left\{(4)^2 + 8(4) - \frac{(4)^3}{3}\right\} - \left\{(-2)^2 + 8(-2) - \frac{(-2)^3}{3}\right\}\right]$$

$$= \frac{3}{4}\left[\left(16 + 32 - \frac{64}{3}\right) - \left(4 - 16 + \frac{8}{3}\right)\right]$$

$$= \frac{3}{4}\left[48 - \frac{64}{3} + 12 - \frac{8}{3}\right] = \frac{3}{4}\left[60 - \frac{72}{3}\right] = \frac{3}{4}\left[\frac{180-72}{3}\right]$$

$$= \frac{108}{4} = 27 \text{ sq. units.}$$

Example 6. *Find the area of the region enclosed by the parabola $y^2 = 4ax$ and the chord $y = mx$.*

Solution. The given equations of the curves are

$$y^2 = 4ax \quad ...(1)$$

$$y = mx \quad ...(2)$$

Equation (1) represents the right handed parabola with vertex at the origin (0, 0) and symmetric to x-axis as it contains only even powers of y.

$y = mx$ represents a straight line passing through the origin and having the slope m.

To find the points of intersection of the parabola and the straight line we solve the equations (1) and (2).

From (1) and (2), we have

$$y^2 = 4ax$$

$$\Rightarrow \quad (mx)^2 = 4ax \Rightarrow m^2x^2 = 4ax$$

$$\Rightarrow \quad m^2x^2 - 4ax = 0 \Rightarrow x(m^2x - 4a) = 0$$

$$\Rightarrow \quad x = 0, \quad x = \frac{4a}{m^2}$$

For $x = 0$, $\quad y = m(0) = 0$

For $x = \frac{4a}{m^2}$, $\quad y = m\left(\frac{4a}{m^2}\right) = \frac{4a}{m}$

$\therefore$ The points of intersection of the given curve and chord are (0, 0) and $\left(\frac{4a}{m^2}, \frac{4a}{m}\right)$.

The rough sketch of the curves are as shown in the figure.

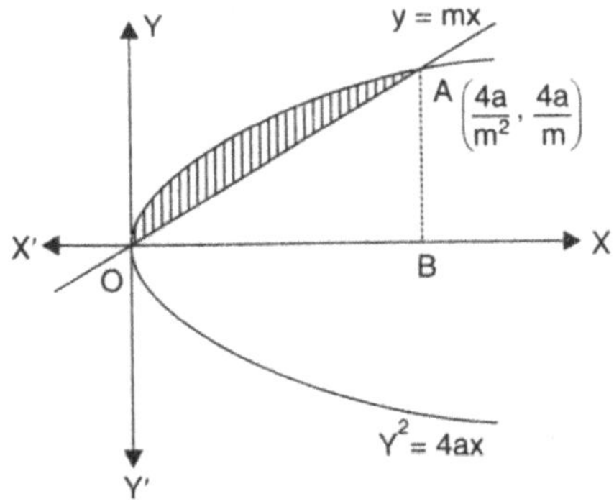

$\therefore$ Required area = Area of the shaded region

$$= \int_0^{4a/m^2} (y_{\text{upper curve}} - y_{\text{lower curve}}) \,.\, dx$$

$$= \int_0^{4a/m^2} (\sqrt{4ax} - mx)\, dx$$

$$= \left[\sqrt{4a}\,.\,\frac{x^{3/2}}{3/2} - \frac{mx^2}{2}\right]_0^{4a/m^2}$$

$$= \left[\frac{4\sqrt{a}}{3}\,x^{3/2} - \frac{mx^2}{2}\right]_0^{4a/m^2}$$

$$= \left[\left\{\frac{4\sqrt{a}}{3}\left(\frac{4a}{m^2}\right)^{3/2} - \frac{m}{2}\left(\frac{4a}{m^2}\right)^2\right\} - \{0 - 0\}\right]$$

$$= \left[\frac{4\sqrt{a}}{3}\cdot\frac{4a\sqrt{4a}}{m^3} - \frac{m}{2}\cdot\frac{16a^2}{m^4}\right] = \frac{32a^2}{3m^3} - \frac{8a^2}{m^3} = \frac{32a^2 - 24a^2}{3m^3}$$

$$= \frac{8a^2}{3m^3} \text{ sq. units.}$$

Example 7. *Find the area bounded by the curve $y = 2x - x^2$ and the straight line $y = -x$.*

Solution. The given equations of the curve are

$$y = 2x - x^2 \quad ...(1)$$

$$y = -x \quad ...(2)$$

Equation (1) represents a parabola passes through the origin.

Table of values of x and y, satisfying equation (1) is given by

x	0	1	− 1	2	− 2	3
y	0	1	− 3	0	− 8	− 3

Equation (2) represents a straight line passing through the origin.

Table of values of x and y, satisfying equation (2) is given by

x	0	1	− 1	2	3
y	0	− 1	1	− 2	− 3

By joining these points, we obtain a rough sketch of these curves as shown in the figure.

Now, to find the points of intersection of the two curves, let us solve equations (1) and (2) simultaneously.

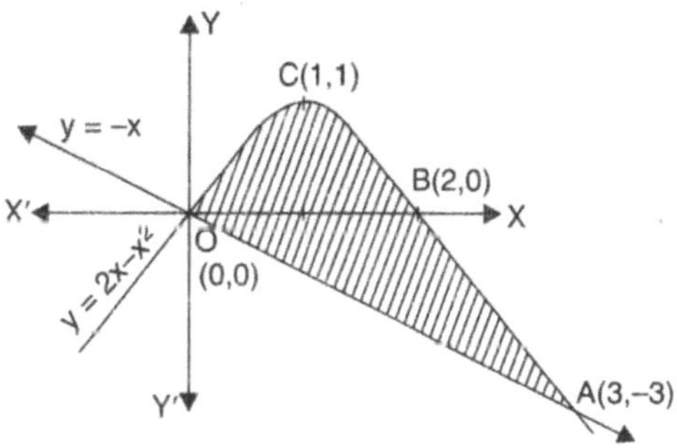

$\therefore$ From (1), $\quad y = 2x - x^2$

$$-x = 2x - x^2$$

$$\Rightarrow \quad x^2 - 3x = 0$$

$$\Rightarrow \quad x(x - 3) = 0$$

$$\Rightarrow \quad x = 0,\ x = 3$$

When $x = 0, y = 0$ and when $x = 3, y = -3$

$\therefore$ The points of intersection of two curves are (0, 0) and (3, − 3).

$\therefore$ Required area = Area of the shaded region.

= Area OABCO

$$= \int_0^3 (y_{\text{upper curve}} - y_{\text{lower curve}})\,.\,dx$$

$$= \int_0^3 [(2x - x^2) - (-x)]\,dx = \int_0^3 (2x - x^2 + x)\,dx$$

$$= \int_0^3 (3x - x^2)\,dx = \left[\frac{3x^2}{2} - \frac{x^3}{3}\right]_0^3$$

$$= \left[\left\{\frac{3(3)^2}{2} - \frac{(3)^3}{3}\right\} - 0\right] = \left[\frac{27}{2} - 9\right] = \frac{27 - 18}{2}$$

$$= \frac{9}{2} \text{ sq. units.}$$

Example 8. *Find the area of the region included between the parabola $y^2 = x$ and the straight line $x + y = 2$.*

Solution. The given equations of the curve are

$$y^2 = x \quad \text{...(1)}$$

$$x + y = 2 \quad \text{...(2)}$$

Equation (1) $y^2 = x$, represents a parabola with vertex at the origin (0, 0) and is symmetric about x-axis as it contains only even powers of y.

Equation (2) represents a straight line passing through the points (1, 1) and (4, – 2).

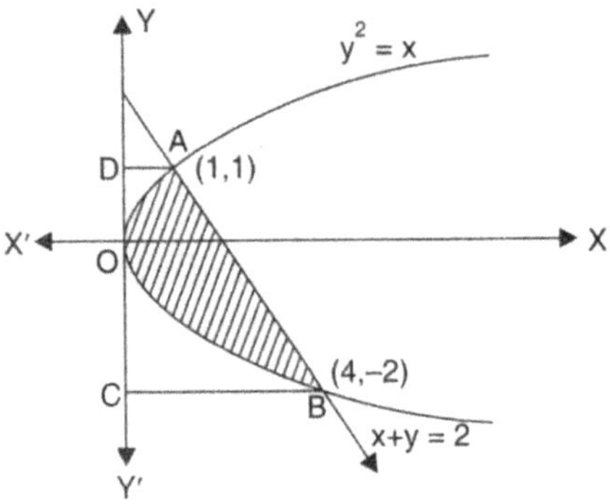

Now, to find the points of intersection of the two curves, let us solve equations (1) and (2) simultaneously.

$\therefore$ From (1), $y^2 = x$

$\Rightarrow \quad y^2 + y = 2 \quad \Rightarrow \quad y^2 + y - 2 = 0$

$\Rightarrow \quad y^2 + 2y - y - 2 = 0 \quad \Rightarrow \quad (y + 2)(y - 1) = 0$

$y = -2, \quad y = 1$

For $y = -2, \quad x = (-2)^2 = 4$

For $y = 1, \quad x = (1)^2 = 1$

$\therefore$ The points of intersection of the two curves are (1, 1) and (4, – 2).

The rough sketch of these curves is as shown in the figure.

$\therefore$ Required area = Area of the shaded region

= Area OABO

$$= \int_{-2}^{1} (x_{\text{upper curve}} - x_{\text{lower curve}})\, dy$$

$$= \int_{-2}^{1} [(2 - y) - y^2]\, dy = \int_{-2}^{1} (2 - y - y^2)\, dy$$

$$= \left[2y - \frac{y^2}{2} - \frac{y^3}{3}\right]_{-2}^{1} = \left[\left\{2(1) - \frac{(1)^2}{2} - \frac{(1)^3}{3}\right\} - \left\{2(-2) - \frac{(-2)^2}{2} - \frac{(-2)^3}{3}\right\}\right]$$

$$= \left[\left(2 - \frac{1}{2} - \frac{1}{3}\right) - \left(-4 - 2 + \frac{8}{3}\right)\right] = \left(\frac{12 - 3 - 2}{6}\right) - \left(-6 + \frac{8}{3}\right) = \frac{7}{6} + 6 - \frac{8}{3}$$

$$= \frac{7 + 36 - 16}{6} = \frac{27}{6} = \frac{9}{2} \text{ sq. units.}$$

Example 9. *Find the area of the region bounded by the parabola $y^2 = 16x$ and the line $x = 4$.*

Solution. The given equations of the curves are

$$y^2 = 16x \quad \text{...(1)}$$

$$x = 4 \quad \text{...(2)}$$

Equation (1) represents a parabola with vertex at the origin (0, 0). This is symmetrical to x-axis, as it contains only even powers of y.

Equation (2) is a straight line parallel to y-axis at a distance 4 units from it.

Table of values of x and y satisfying equation (1) is given by

x	0	1	2	4
y	0	± 4	$\pm 4\sqrt{2}$	± 8

By joihing these points, we obtain a rough sketch of the given curve as shown in the Fig.

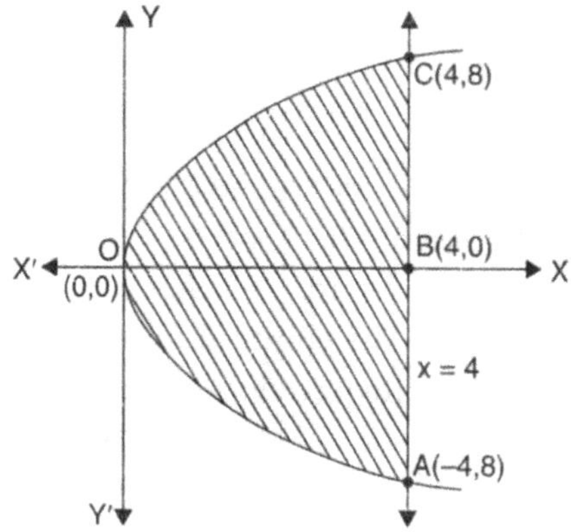

Now, to find the points of intersection of the two curves.

Put $x = 4$ in equation (1), we have

$$y^2 = 16(4) \Rightarrow y^2 = 64 \Rightarrow y \pm 8$$

For $x = 4$, $y = +8$

For $x = 4$, $y = -8$

$\therefore$ The points of intersection of these curves are $(4, -8)$ and $(4, 8)$.

$\therefore$ Required area = Area of the shaded region

= Area OABC

= Area OAB + Area OBC

= 2 area OBC

$$= 2\int_0^4 (y \text{ of parabola}).\, dy \qquad [\because \text{ For } y \geq 0]$$

$$= 2\int_0^4 4\sqrt{x}\, dx = 8\left[\frac{x^{\frac{1}{2}+1}}{\frac{1}{2}+1}\right]_0^4$$

$$= 8 \cdot \frac{2}{3}\left[x^{3/2}\right]_0^4 = \frac{16}{3}\left[(4)^{3/2} - (0)^{3/2}\right] = \frac{16}{3}(8 - 0)$$

$$= \frac{128}{3} \text{ sq. units.}$$

Example 10. *Find the area of the region included between the parabolas $y^2 = 4ax$ and $x^2 = 4ay$; $a > 0$.*

Solution. The given equations of the curve are

$$y^2 = 4ax \qquad \text{...(1)}$$

$$x^2 = 4ay \qquad \text{...(2)}$$

Equation (1) $y^2 = 4ax$ represents a parabola with vertex at the origin (0, 0) and symmetric to x-axis as it contains only even powers of y.

Equation (2) $x^2 = 4ay$ also represents a parabola with vertex at the origin (0, 0) and symmetric to y-axis as it contains only even power of x.

The rough sketch of the two curves are as shown in the figure.

Now, to find the point of intersection of the two curves, let us solve the equations (1) and (2) simultaneously.

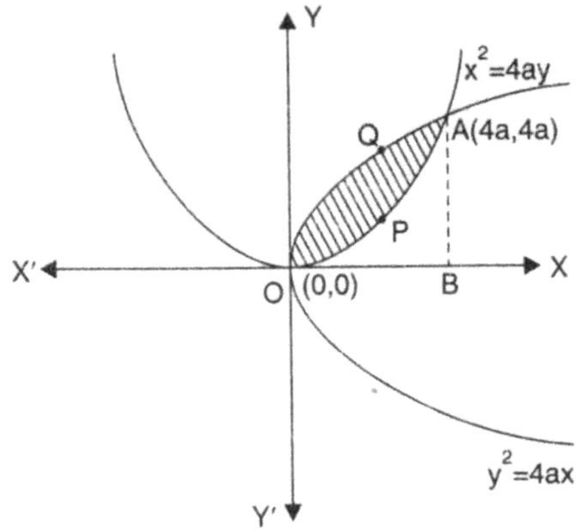

From (2), $\therefore \quad y = \dfrac{x^2}{4a}$

Putting this value of y *in* (1), we get

$$\left(\frac{x^2}{4a}\right)^2 = 4ax$$

$$\Rightarrow \quad \frac{x^4}{16a^2} = 4ax \quad \Rightarrow \quad x^4 - 64a^3 x = 0$$

$$\Rightarrow \quad x(x^3 - 64a^3) = 0 \quad \Rightarrow \quad x = 0, \quad x = 4a$$

For $x = 0$, $\quad y = 0$

For $x = 4a$, $\quad y = 4a$.

$\therefore$ The points of intersection of two curves are (0, 0) and $(4a, 4a)$.

$\therefore$ Required area = Area of the shaded region

= Area OPAQO

= Area OQAB – Area OPAB

$$= \int_0^{4a} (y_{\text{upper curve}} - y_{\text{lower curve}}) \, . \, dx$$

$$= \int_0^{4a} \left(\sqrt{4ax} - \frac{x^2}{4a}\right) dx = 2\sqrt{a} \int_0^{4a} \sqrt{x} \, dx - \frac{1}{4a} \int_0^{4a} x^2 \, dx$$

$$= 2\sqrt{a} \left[\frac{x^{\frac{1}{2}+1}}{\frac{1}{2}+1}\right]_0^{4a} - \frac{1}{4a}\left[\frac{x^3}{3}\right]_0^{4a} = 2\sqrt{a} \, . \, \frac{2}{3}\left[x^{3/2}\right]_0^{4a} - \frac{1}{12a}\left[x^3\right]_0^{4a}$$

$$= \frac{4\sqrt{a}}{3} \left[(4a)^{3/2} - 0\right] - \frac{1}{12a} \left[(4a)^3 - 0\right]$$

$$= \frac{4\sqrt{a}}{3} . \, 8a^{3/2} - \frac{1}{12a} \, . \, 64a^3 = \frac{32a^2}{3} - \frac{16a^2}{3}$$

$$= \frac{16a^2}{3} \text{ sq. units.}$$

Example 11. *Find the area of the region bounded by the parabola $y = x^2$ and the line $y = x$.*

Solution. The given equations of the curve are :

$$y = x^2 \qquad ...(1)$$

$$y = x \qquad ...(2)$$

Equation (1) represents a parabola with vertex at the origin (0, 0) and symmetric about y-axis as it contains only even powers of x.

Equation (2) represents a straight line passing through the origin.

Table of values of x and y satisfying equation (1) is given by

x	0	1	-1	2	-2
y	0	1	1	4	4

Table of values of x and y satisfying equation (2) is given by

x	0	1	2	3
y	0	1	2	3

By joining these points, we obtain a rough sketch of the given curves as shown in the figure.

Now, to find the points of intersection of the given curves, let us solve equations (1) and (2) simultaneously.

From (1) $\quad y = x^2$

$\Rightarrow \quad x = x^2 \quad [\because \ y = x]$

$\Rightarrow \quad x^2 - x = 0 \ \Rightarrow \ x(x-1) = 0$

$\Rightarrow \quad x = 0, x = 1$

For $x = 0$, $\quad y = 0$

For $x = 1$, $\quad y = 1$

$\therefore$ The points of intersection of the given curves are (0, 0) and (1, 1).

$\therefore$ Required area = Area of the shaded region

= Area OABO – Area OABCO

$$= \int_0^1 (y_{\text{upper curve}} - y_{\text{lower curve}}) \, . \, dx$$

$$= \int_0^1 (x - x^2) \, dx = \left[\frac{x^2}{2} - \frac{x^3}{3}\right]_0^1$$

$$= \left[\left(\frac{(1)^2}{2} - \frac{(1)^3}{3}\right) - 0\right] = \frac{1}{2} - \frac{1}{3}$$

$$= \frac{3-2}{6} = \frac{1}{6} \text{ sq. units.}$$

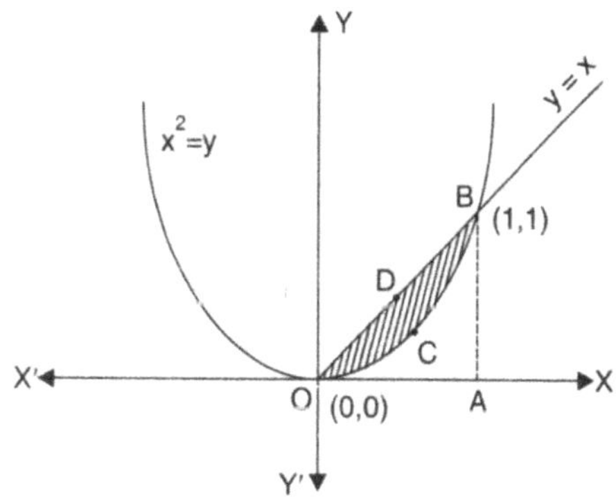

Example 12. *Find the area of the region bounded by the parabola $y = x^2$ and lines $y = |x|$.*

Solution. The given equations of the curves are

$$y = x^2 \quad \text{...(1)}$$

$$y = |x| \quad \text{...(2)}$$

Equation (1), represents a parabola with vertex at the origin (0, 0) and symmetric to y-axis as it contains only even powers of x.

Equation (2), $y = |x|$ *i.e.*, $y = x$ and $y = -x$ represents two straight lines passing through the origin and making an angle of 45° and 135° with the positive direction of x-axis.

The rough sketch of these curves are as shown in the figure.

$\therefore$ Required area = Area of the shaded region

= 2(Area OMANO)

[$\because$ Both the curves are symmetrical about y-axis]

= 2[Area ONAP – Area OMAP]

$$= 2\int_0^1 (y_{\text{upper curve}} - y_{\text{lower curve}})\,.\,dx$$

$$= 2\int_0^1 (x - x^2)\,dx = 2\left[\frac{x^2}{2} - \frac{x^3}{3}\right]_0^1$$

$$= 2\left[\frac{1}{2} - \frac{1}{3}\right] = 2\left(\frac{3-2}{6}\right)$$

$$= \frac{1}{3} \text{ sq. units.}$$

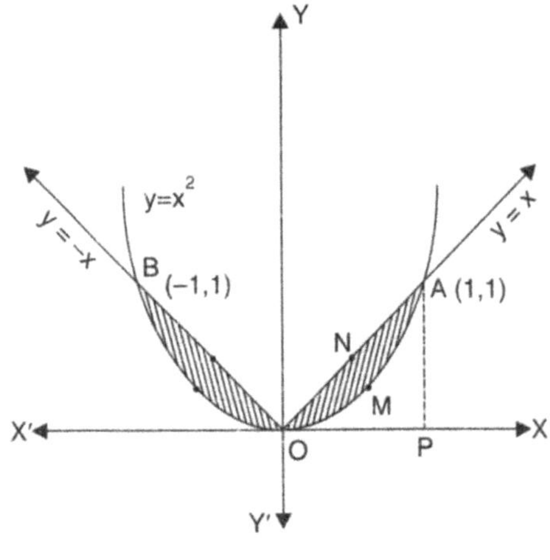

Example 13. *Find the area of the region bounded by the parabola $y^2 = 4ax$ and the line $x = a$.*

Solution. The given equations of the curves are

$$y^2 = 4ax \qquad ...(1)$$

$$x = a \qquad ...(2)$$

Equation (1) represents a parabola with vertex at the origin (0, 0) and symmetric about x-axis as it contains only even powers of y.

Equations (2), $x = a$ represents a straight line parallel to y-axis at a distance a units from it.

Now, to find the points of intersection of these curves put $x = a$ in equation (1), we have

$$y^2 = 4ax$$

$\Rightarrow$ $y^2 = 4a(a) = 4a^2$

$\Rightarrow$ $y = \pm 2a$

For $x = a$, $y = 2a$

For $x = a$, $y = -2a$

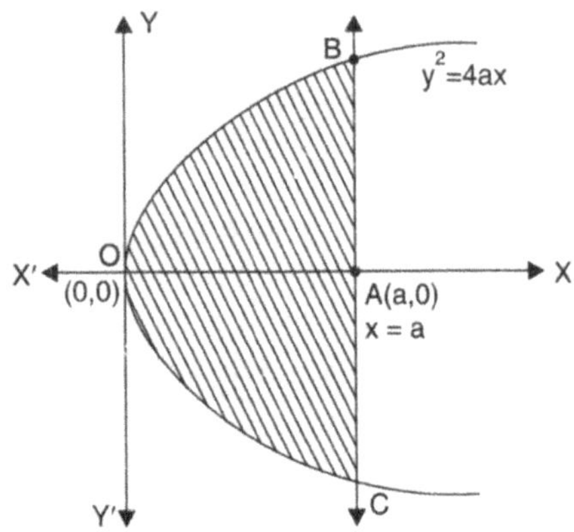

$\therefore$ The points of intersection of these curves are $(a, 2a)$ and $(a, -2a)$.

The rough sketch of these curves are as shown in the figure.

$\therefore$ Required area = Area of the shaded region

= Area OBC = 2(Area OAB)

$$= 2\int_0^a y\,.\,dx = \int_0^a \sqrt{4ax}\,dx = 2\,.\,2\sqrt{a}\int_0^a \sqrt{x}\,dx \qquad [\because\ y = \sqrt{4ax}]$$

$$= 4\sqrt{a}\left[\frac{x^{\frac{1}{2}+1}}{\frac{1}{2}+1}\right]_0^a = 4\sqrt{a}\,.\,\frac{2}{3}\left[x^{3/2}\right]_0^a$$

$$= \frac{8\sqrt{a}}{3}\,[a^{3/2} - 0] = \frac{8}{3}\,a^2 \text{ sq. units.}$$

Example 14. *Find the area of the bounded region bounded by the two parabolas $y = x^2$ and $x = y^2$.*

Solution. The given equations of the curves are

$$y = x^2 \quad ...(1)$$

$$x = y^2 \quad ...(2)$$

Equation (1) $y = x^2$ represents a parabola with vertex at the origin (0, 0) and symmetric about y-axis as it contains only even powers of x.

Equation (2) $x = y^2$ represents a parabola with vertex at the origin (0, 0) and symmetric about x-axis as it contains only even powers of y.

The rough sketch of these curves are as shown in the figure.

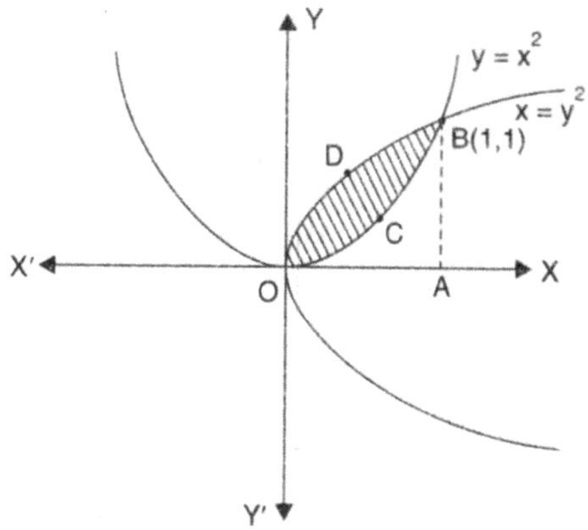

Now, to find the points of intersection of the two curves, let us solve equations (1) and (2).

$\therefore$ From (2) : $x = y^2$

$\Rightarrow$ $y = x^2$

$y = (y^2)^2 = y^4$

$\Rightarrow$ $y^4 - y = 0 \Rightarrow y(y^3 - 1) = 0$

$\Rightarrow$ $y = 0, y = 1$

For $y = 0$, $x = (0)^2 = 0$

For $y = 1$, $x = (1)^2 = 1$

$\therefore$ The points of intersection of the two curves are (0, 0) and (1, 1).

$\therefore$ Required area = Area OCBDO = Area of the shaded region

= Area OABD – Area OABC

$$= \int_0^1 (y_{\text{upper curve}} - y_{\text{lower curve}}) \,.\, dx$$

$$= \int_0^1 (\sqrt{x} - x^2)\, dx$$

$$= \left[\frac{x^{3/2}}{3/2} - \frac{x^3}{3}\right]_0^1 = \left[\frac{2}{3}x^{3/2} - \frac{x^3}{3}\right]_0^1 = \left[\left\{\frac{2}{3}(1)^{3/2} - \frac{(1)^3}{3}\right\} - 0\right]$$

$$= \frac{2}{3} - \frac{1}{3} = \frac{1}{3} \text{ sq. units.}$$

Example 15. *Find the area of the region bounded by the parabola $y^2 = 2x + 1$ and the line $x - y - 1 = 0$.*

Solution. The given equations of the curves are

$$y^2 = 2x + 1$$

$$\Rightarrow \quad y^2 = 2\left(x + \frac{1}{2}\right) \quad ...(1)$$

$$x - y - 1 = 0 \quad ...(2)$$

Equation (1) represents a parabola with vertex at $\left(\frac{1}{2}, 0\right)$ and symmetrical about x-axis as it contains only even powers of y.

Clearly, equation (2) $x - y - 1 = 0$ represents a straight line.

Table of values of x and y satisfying equation (1) is given by

x	0	1	4
y	± 1	± 1.73	± 3

Table of values of x and y satisfying equation (2) is given by

x	1	2	3	0	-1	4
y	0	1	2	-1	-2	3

By joining these points, we obtain a rough sketch of the curves as shown in the figure.

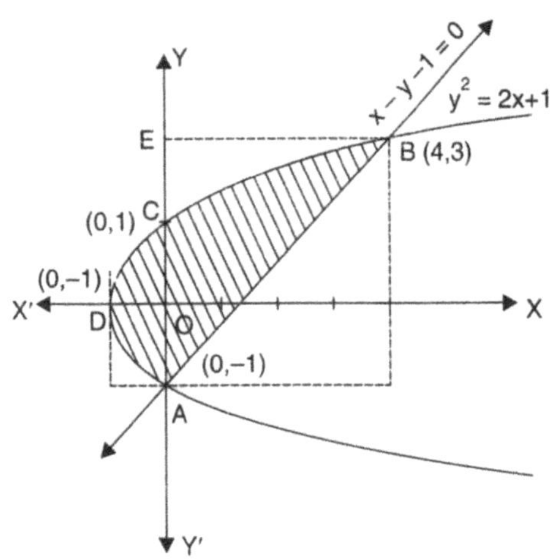

Now, to find the points of intersection of the two curves, let us solve equations (1) and (2).

From (1) $y^2 = 2x + 1$

$$\Rightarrow \quad (x-1)^2 = 2x+1 \qquad \begin{bmatrix} \because & x - y - 1 = 0 \\ \Rightarrow & y = x - 1 \end{bmatrix}$$

$$\Rightarrow \quad x^2 - 2x + 1 = 2x + 1$$

$$\Rightarrow \quad x^2 - 4x = 0$$

$$\Rightarrow \quad x(x-4) = 0 \Rightarrow x = 0, x = 4.$$

For $x = 0$, $y = -1$

For $x = 4$, $y = 3$

$\therefore$ The points of intersection of the two curves are $(0, -1)$ and $(4, 3)$.

$\therefore$ Required area = Area of the shaded region

= Area ABCD

= Area AOCD + Area ABECOA – Area BEC

$$= -\int_{-1}^{1} (x_{\text{parabola}})\, dy + \int_{-1}^{3} (x_{\text{line}})\, dy - \int_{1}^{3} (x_{\text{parabola}})\, dy$$

$$= -\int_{-1}^{1} \left(\frac{y^2-1}{2}\right) dy + \int_{-1}^{3} (y+1)\, dy - \int_{1}^{3} \left(\frac{y^2-1}{2}\right) dy$$

$$= \frac{-1}{2}\left[\frac{y^3}{3} - y\right]_{-1}^{1} + \left[\frac{y^2}{2} + y\right]_{-1}^{3} - \frac{1}{2}\left[\frac{y^3}{3} - y\right]_{1}^{3}$$

$$= \frac{-1}{2}\left[\left\{\frac{(1)^3}{3} - 1\right\} - \left\{\frac{(-1)^3}{3} - (-1)\right\}\right] + \left[\left\{\frac{(3)^2}{2} + 3\right\} - \left\{\frac{(-1)^2}{2} + (-1)\right\}\right]$$

$$- \frac{1}{2}\left[\left\{\frac{(3)^3}{3} - 3\right\} - \left\{\frac{(1)^3}{3} - 1\right\}\right]$$

$$= \frac{-1}{2}\left[\frac{1}{3} - 1 + \frac{1}{3} - 1\right] + \left[\frac{9}{2} + 3 - \frac{1}{2} + 1\right] - \frac{1}{2}\left[9 - 3 - \frac{1}{3} + 1\right]$$

$$= \frac{-1}{2}\left[\frac{2}{3} - 2\right] + [4 + 4] - \frac{1}{2}\left[7 - \frac{1}{3}\right]$$

$$= \frac{-1}{2}\left[\frac{2-6}{3}\right] + 8 - \frac{1}{2}\left[\frac{21-1}{3}\right] = \frac{2}{3} + 8 - \frac{10}{3} = 8 - \frac{8}{3}$$

$$= \frac{24-8}{3} = \frac{16}{3} \text{ sq. units.}$$

Example 16. *Find the area bounded by the curves $y = x$ and $y = x^3$.*

Solution. The given equations of the curves are

$$y = x \quad \ldots(1)$$

$$y = x^3 \quad \ldots(2)$$

Equation (1) $y = x$ represents a straight line passing through the origin and making an angle of 45° with x-axis.

To find the points of intersection of the two curves, let us solve equations (1) and (2) simultaneously.

∴ From (1) and (2), we have

$$x = x^3 \Rightarrow x^3 - x = 0$$

$$\Rightarrow \quad x(x^2 - 1) = 0$$

$$\Rightarrow \quad x = 0, x = \pm 1$$

For $x = 0$, $y = 0$

For $x = -1$, $y = -1$

For $x = 1$, $y = 1$

∴ The points of intersection of the two curves are (0, 0), (– 1, – 1) and (1, 1).

Graph of $y = x^3$ and $y = x$.

Symmetry : The equations (1) and (2) remain unchanged on changing x to $-x$ and y to $-y$.

∴ The curves (1) and (2) are symmetrical in opposite quadrants.

The curves passes through the origin.

Monotonicity : $y = x^3$

$$\Rightarrow \quad \frac{dy}{dx} = 3x^2 \geq 0$$

$$\Rightarrow \quad \frac{dy}{dx} = 0 \Rightarrow 3x^2 = 0 \Rightarrow x = 0$$

$$\Rightarrow \quad \frac{dy}{dx} > 0 \Rightarrow 3x^2 > 0 \quad \text{for } x \neq 0.$$

∴ The curve is strictly increasing for $x \neq 0$.

Extreme values :

$$\frac{d^2y}{dx^2} = 6x$$

$$\Rightarrow \quad \left.\frac{d^2y}{dx^2}\right|_{x=0} = 6(0) = 0.$$

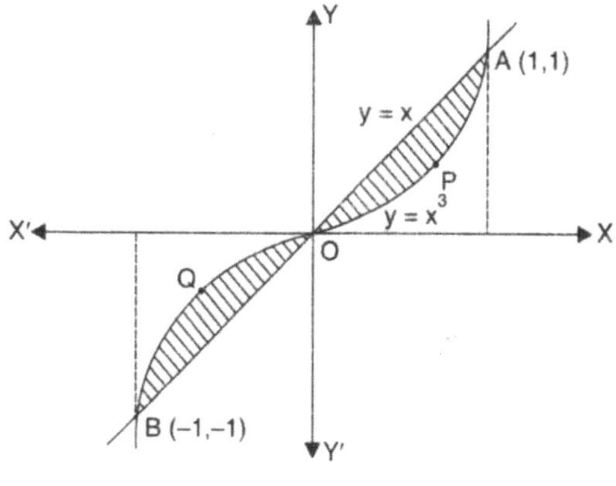

$$\Rightarrow \qquad \frac{d^3y}{dx^3} = 6 \neq 0 \text{ at } x = 0.$$

$\therefore$ $x = 0$ gives a point of inflexion.

Table of values of x and y satisfying equation (1) is given by

x	0	-1	1	-2	2
y	0	-1	1	-8	8

By joining these points we obtain a rough sketch of the curve as shown in the figure.

$\therefore$ Required area = Area of the shaded region

$$= \text{Area BOQB} + \text{Area AOPA}$$

$$= \int_{-1}^{0} [(-y_{\text{line}}) - (-y_{\text{curve}})] \, . \, dx + \int_{0}^{1} [(y_{\text{line}} - y_{\text{curve}})] \, . \, dx$$

$$= \int_{-1}^{0} (-x + x^3) \, dx + \int_{0}^{1} (x - x^3) \, dx$$

$$= \left[-\frac{x^2}{2} + \frac{x^4}{4} \right]_{-1}^{0} + \left[\frac{x^2}{2} - \frac{x^4}{4} \right]_{0}^{1}$$

$$= \left[0 - \left\{ -\frac{(-1)^2}{2} + \frac{(-1)^4}{4} \right\} \right] + \left[\left\{ \frac{(1)^2}{2} - \frac{(1)^4}{4} \right\} - 0 \right]$$

$$= \left[\frac{1}{2} - \frac{1}{4} + \frac{1}{2} - \frac{1}{4} \right] = 1 - \frac{2}{4} = \frac{4-2}{4} = \frac{2}{4}$$

$$= \frac{1}{2} \text{ sq. units.}$$

Example 17. *Draw a rough sketch of the curves $y = \sin x$ and $y = \cos x$ as x varies from 0 to $\frac{\pi}{2}$ and find the area of the region enclosed by the curves and the y-axis.*

Solution. The given equations of the curves are

$$y = \sin x \qquad \text{...(1)}$$

$$y = \cos x \qquad \text{...(2)}$$

Table of values of x and y satisfying equations (1) and (2) are given by

x	0	$\frac{\pi}{6}$	$\frac{\pi}{4}$	$\frac{\pi}{3}$	$\frac{\pi}{2}$
$y = \sin x$	0	$\frac{1}{2} = 0.5$	$\frac{1}{\sqrt{2}} = 0.71$	$\frac{\sqrt{3}}{2} = 0.87$	1
$y = \cos x$	1	$\frac{\sqrt{3}}{2} = 0.87$	$\frac{1}{\sqrt{2}} = 0.71$	$\frac{1}{2} = 0.5$	0

By joining these points with a free hand, we obtain the rough sketch of these curves as shown in the figure.

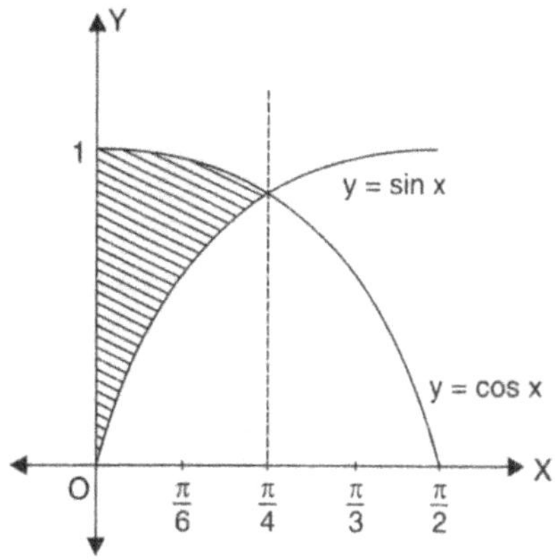

The point of intersection of these curves is $\left(\frac{\pi}{4}, \frac{1}{\sqrt{2}}\right)$.

$\therefore$ Required area = Area of the shaded region

$$= \int_0^{\pi/4} (y_{\text{upper curve}} - y_{\text{lower curve}}) \, . \, dx$$

$$= \int_0^{\pi/4} (\cos x - \sin x)\, dx = \Big[\sin x - (-\cos x)\Big]_0^{\pi/4}$$

$$= \left[\left(\sin\frac{\pi}{4} + \cos\frac{\pi}{4}\right) - (\sin 0^\circ + \cos 0^\circ)\right]$$

$$= \left[\left(\frac{1}{\sqrt{2}} + \frac{1}{\sqrt{2}}\right) - (0+1)\right] = \frac{2}{\sqrt{2}} - 1$$

$= (\sqrt{2} - 1)$ sq. units.

Example 18. *Draw a rough sketch of the curves y = sin x and y = cos x varies from 0 to* $\frac{\pi}{2}$ *and find the area of the region enclosed by them and x-axis.*

Solution. The given equations of the curves are

$$y = \sin x \quad \text{...(1)}$$

$$y = \cos x \quad \text{...(2)}$$

Table of values of x and y satisfying equations (1) and (2) are given by

x	0	$\frac{\pi}{6}$	$\frac{\pi}{4}$	$\frac{\pi}{3}$	$\frac{\pi}{2}$
$y = \sin x$	0	$\frac{1}{2} = 0.5$	$\frac{1}{\sqrt{2}} = 0.71$	$\frac{\sqrt{3}}{2} = 0.87$	1
$y = \cos x$	1	$\frac{\sqrt{3}}{2} = 0.87$	$\frac{1}{\sqrt{2}} = 0.71$	$\frac{1}{2} = 0.5$	0

By joining these points with a free hand, we obtain a rough sketch of these curves as shown in the figure.

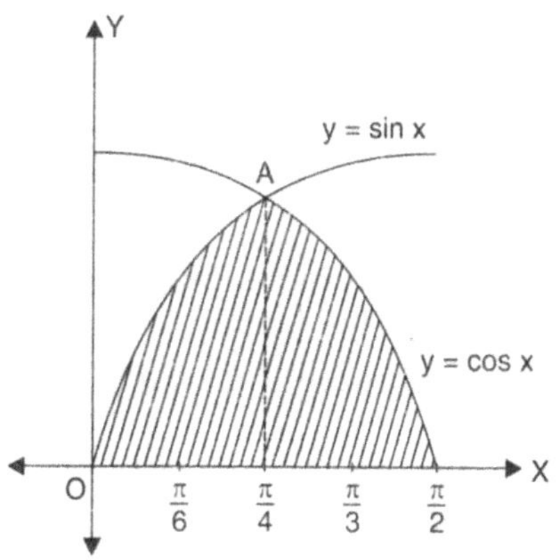

The point of intersection of the two curves is $\left(\frac{\pi}{4}, \frac{1}{\sqrt{2}}\right)$.

$\therefore$ Required area = Area of the shaded region.

$$= \int_0^{\pi/4} \sin x \, dx + \int_{\pi/4}^{\pi/2} \cos x \, dx$$

$$= \Big[-\cos x\Big]_0^{\pi/4} + \Big[\sin x\Big]_{\pi/4}^{\pi/2}$$

$$= -\left[\cos\frac{\pi}{4} - \cos 0°\right] + \left[\sin\frac{\pi}{2} - \sin\frac{\pi}{4}\right]$$

$$= -\left[\frac{1}{\sqrt{2}} - 1\right] + \left[1 - \frac{1}{\sqrt{2}}\right] = -\frac{1}{\sqrt{2}} + 1 + 1 - \frac{1}{\sqrt{2}} = 2 - \frac{2}{\sqrt{2}}$$

$= (2 - \sqrt{2})$ sq. units.

Example 19. *Find the area bounded by the curves*

$$y = 6x - x^2 \text{ and } y = x^2 - 2x.$$

Solution. The given equations of the curves are

$$y = 6x - x^2 \quad ...(1)$$

$$y = x^2 - 2x \quad ...(2)$$

Equation (1) represents a parabola opening downward and intersecting the x-axis at the origin (0, 0) and (0, 6).

Equation (2), $y = x^2 - 2x$ also represents a parabola opening upwards and intersecting the x-axis at the origin (0, 0) and (0, 2).

Now, to find the point of intersection of these curves, let us solve equations (1) and (2) simultaneously.

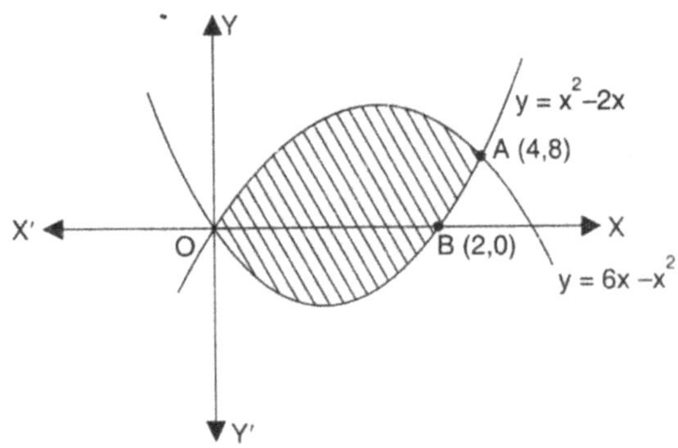

$\therefore$ From (1) and (2), we have

$$y = x^2 - 2x$$

$$\Rightarrow \quad 6x - x^2 = x^2 - 2x$$

$$\Rightarrow \quad 2x^2 - 8x = 0$$

$$\Rightarrow \quad 2x(x - 4) = 0$$

$$x = 0, \quad x = 4$$

For $x = 0$, $y = 0$

For $x = 4$, $y = (4)^2 - 2(4) = 16 - 8 = 8$

$\therefore$ The points of intersection of the two curves are (0, 0) and (4, 8).

$\therefore$ Required area = Area of the shaded region.

$$= \int_0^4 (y_{\text{upper curve}} - y_{\text{lower curve}}) \, . \, dx$$

$$= \int_0^4 [(6x - x^2) - (x^2 - 2x)] \, . \, dx = \int_0^4 (8x - 2x^2) \, dx$$

$$= \left[\frac{8x^2}{2} - \frac{2x^3}{3}\right]_0^4 = \left[\left\{4\,(4)^2 - \frac{2\,(4)^3}{3}\right\} - 0\right] = 64 - \frac{128}{3}$$

$$= \frac{192 - 128}{3} = \frac{64}{3} \text{ sq. units.}$$

Example 20. *Draw a rough sketch of $y^2 = x + 1$ and $y^2 = -x + 1$ and determine the area enclosed by the curves.*

Solution. The equations of the given curves are

$$y^2 = x + 1 \quad ...(1)$$

$$y^2 = -x + 1 \quad ...(2)$$

Equation (1) represents a right handed parabola with vertex at (– 1, 0).

Equation (2) represents a left handed parabola with vertex at (1, 0).

Now, to find the points of intersection of two curves, let us solve equations (1) and (2) simultaneously.

$\therefore \qquad x + 1 = -x + 1$

$\Rightarrow \qquad 2x = 0 \Rightarrow x = 0$

For $x = 0$, $y^2 = 0 + 1 \Rightarrow y^2 = 1 \Rightarrow y = \pm 1$

$\therefore$ The points of intersection of two curves are (0, 1) and (0, – 1).

The rough sketch of the two curves is as shown in the figure.

By symmetry of parabolas :

Area OAB = Area OAD and Area OBC = Area OCD.

$\therefore$ Required area = Area of shaded region

= 2(Area OBC) + 2(Area OAB)

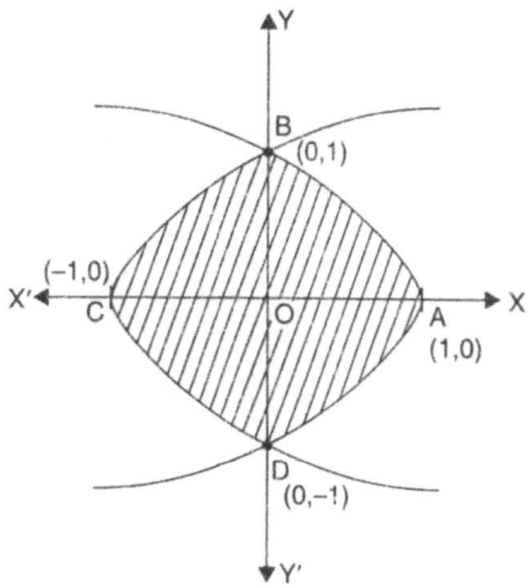

$$= 2\int_{-1}^{0} \sqrt{x+1}\, dx + 2\int_{0}^{1} \sqrt{-x+1}\, dx$$

$$= 2\left[\frac{(x+1)^{\frac{1}{2}+1}}{\frac{1}{2}+1}\right]_{-1}^{0} + 2\left[\frac{(-x+1)^{\frac{1}{2}+1}}{(-1)\left(\frac{1}{2}+1\right)}\right]_{0}^{1}$$

$$= 2.\frac{2}{3}\left[(x+1)^{3/2}\right]_{-1}^{0} - 2.\frac{2}{3}\left[(-x+1)^{3/2}\right]_{0}^{1}$$

$$= \frac{4}{3}[(0+1)^{3/2} - (-1+1)^{3/2}] - \frac{4}{3}[(-1+1)^{3/2} - (0+1)^{3/2}] = \frac{4}{3}(1-0) - \frac{4}{3}(0-1)$$

$$= \frac{4}{3} + \frac{4}{3} = \frac{8}{3} \text{ sq. units.}$$

Example 21. *Find the area of the region common to the two parabolas $y = 2x^2$ and $x^2 = y - 4$.*

Solution. The given equations of the curves are

$$y = 2x^2 \qquad \text{...(1)}$$

$$x^2 = y - 4 \qquad \text{...(2)}$$

Equation (1) represents a parabola opening upward with vertex at the origin (0, 0).

Equation (2) also represents an upward parabola with vertex at (0, 4).

Table of values of x and y satisfying equations (1) and (2) is given by

x	0	– 1	1	2	– 2
$y = 2x^2$	0	2	2	8	8
$y = x^2 + 4$	4	5	5	8	8

By joining these points with a free hand, we obtain a rough sketch of the curves as shown in the figure.

Now, to find the points of intersection of the two curves, let us solve equation (1) and (2), simultaneously.

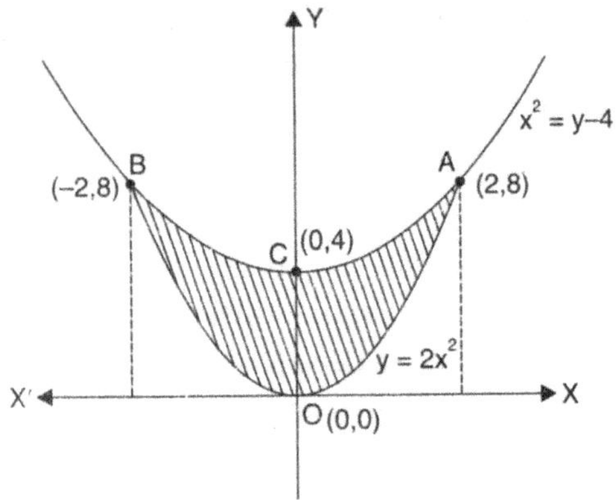

From (2), we have

$$x^2 = y - 4 \qquad [\because \; y = 2x^2]$$

$$\Rightarrow \qquad x^2 = 2x^2 - 4$$

$$\Rightarrow \qquad x^2 = 4 \;\Rightarrow\; x = \pm 2$$

For $x = 2$, $\quad y = 2(2)^2 = 8$

For $x = -2$, $\quad y = 2(-2)^2 = 8$

$\therefore$ The points of intersection of the two curves are (2, 8) and (– 2, 8).

$\therefore$ Required area = Area of the shaded region = Area OACBO

$$= \int_{-2}^{2} (y_{\text{upper curve}} - y_{\text{lower curve}}) \, . \, dx$$

$$= \int_{-2}^{2} (x^2 + 4 - 2x^2) \, dx = \int_{-2}^{2} (4 - x^2) \, dx$$

$$= \left[4x - \frac{x^3}{3}\right]_{-2}^{2} = \left[\left\{4\,(2) - \frac{(2)^3}{3}\right\} - \left\{4\,(-2) - \frac{(-2)^3}{3}\right\}\right]$$

$$= \left[\left(8 - \frac{8}{3}\right) - \left(-8 + \frac{8}{3}\right)\right] = 8 - \frac{8}{3} + 8 - \frac{8}{3} = 16 - \frac{16}{3} = \frac{48 - 16}{3}$$

$$= \frac{32}{3} \text{ sq. units.}$$

Example 22. *Using integration, find the area of the triangle whose sides have the equations $y = 2x + 1$, $y = 3x + 1$ and $x = 4$.*

Solution. The given equations of the curves are

$$y = 2x + 1 \qquad \text{...(1)}$$

$$y = 3x + 1 \qquad \text{...(2)}$$

$$x = 4 \qquad \text{...(3)}$$

Now, to find the points of intersection of these lines *i.e.*, the vertices of the triangle, let us solve these equations.

$\therefore$ From (1) and (2), we have

$$2x + 1 = 3x + 1$$

$$\Rightarrow \qquad x = 0$$

For $x = 0$, $\quad y = 2(0) + 1 = 1$

$\therefore$ (0, 1) is the point of intersection of lines (1) and (2). Now, from (1) and (3), we have

$$y = 2(4) + 1 = 8 + 1 = 9$$

$\therefore$ (4, 9) is the point of intersection of lines (1) and (3).

Now, from (2) and (3), we have

$$y = 3(4) + 1 = 12 + 1 = 13$$

$\therefore$ (4, 13) is the point of intersection of lines (2) and (3).

The rough sketch of these lines is as shown in the figure.

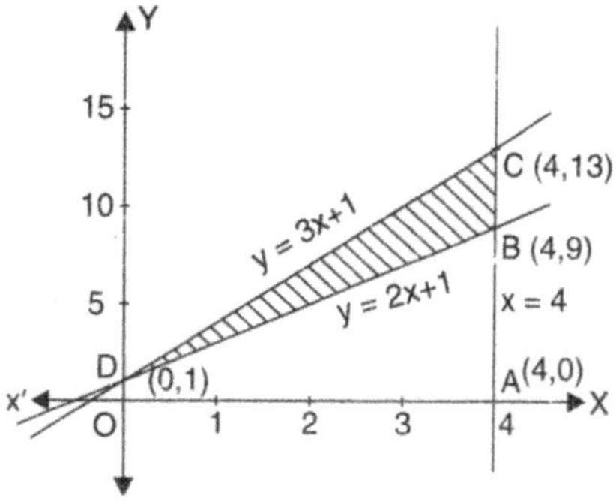

$\therefore$ Required area = Area of the shaded region

= Area ΔBCD

= Area ODCA – Area ODBA

$= \int_0^4 (y_{\text{line DC}}) \cdot dx - \int_0^4 (y_{\text{line DB}}) \cdot dx$

$= \int_0^4 (3x+1)\, dx - \int_0^4 (2x+1)\, dx$

$= \left[\frac{3x^2}{2} + x\right]_0^4 - \left[\frac{2x^2}{2} + x\right]_0^4 = \left[\left(\frac{3}{2}(4)^2 + 4\right) - 0\right] - [((4)^2 + 4) - 0]$

= 24 + 4 – 20 = 8 sq. units.

Example 23. *Compute the area bounded by the lines $x + 2y = 2$, $y - x = 1$ and $2x + y = 7$.*

Solution. The given equations of the lines are

$$x + 2y = 2 \quad \text{...(1)}$$
$$y - x = 1 \quad \text{...(2)}$$
$$2x + y = 7 \quad \text{...(3)}$$

Now, to find the points of intersection of these lines, let us solve equations (1), (2) and (3).

$\therefore$ From equations (1) and (2), we have

$x + 2y = 2$ $\qquad [\because \; y - x = 1 \Rightarrow y = 1 + x]$

$x + 2(1 + x) = 2$

$x + 2 + 2x = 2$

$3x = 0 \Rightarrow x = 0$

For $x = 0 \Rightarrow 0 + 2y = 2 \Rightarrow y = 1$

$\therefore$ (0, 1) is the point of intersection of lines (1) and (2). From equations (2) and (3), we have

$2x + y = 7$ $\qquad [\because \; y - x = 1 \Rightarrow y = 1 + x]$

$\Rightarrow \quad 2x + (1 + x) = 7$

$\Rightarrow \quad 2x + 1 + x = 7$

$\Rightarrow \quad 3x = 6$

$\Rightarrow \quad x = 2$

For $x = 2$, $2(2) + y = 7 \Rightarrow y = 3$

$\therefore$ (2, 3) is the point of intersection of lines (2) and (3).

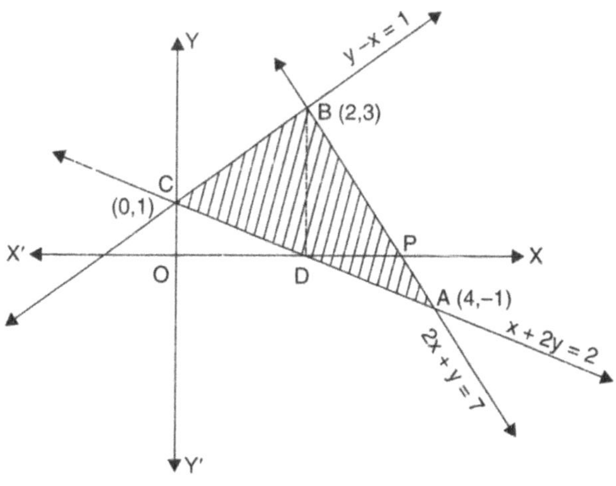

From equations (1) and (3), we have

$2x + y = 7$

$[\because \; x + 2y = 2 \Rightarrow x = (2 - 2y)]$

$2(2 - 2y) + y = 7$

$\Rightarrow \quad 4 - 4y + y = 7$

$\Rightarrow \quad -3y = 3 \Rightarrow y = -1$

For $y = -1$, $2x + (-1) = 7 \Rightarrow x = 4$

$\therefore$ (4, – 1) is the point of intersection of lines (1) and (3). The rough sketch of these lines is as shown in the figure.

$\therefore$ Required area = Area of the shaded region.

= Area BCD + Area BDP + Area ADP

$$= \int_0^2 [y_{\text{line }(2)} - y_{\text{line }(1)}]\,dx + \int_2^{3.5} [y_{\text{line }(3)}]\,.\,dx + \left[\int_2^4 [-\,y_{\text{line }(1)}]\,dx - \int_{3.5}^4 [-\,y_{\text{line }(3)}]\,dx\right]$$

$$= \int_0^2 \left[(x+1) - \frac{2-x}{2}\right] dx + \int_2^{3.5} (7-2x)\,dx + \left[\int_2^4 \frac{x-2}{2}\,dx - \int_{3.5}^4 (2x-7)\,dx\right]$$

$$= \int_0^2 \left(\frac{2x+2-2+x}{2}\right) dx + \int_2^{3.5} (7-2x)\,dx + \frac{1}{2}\int_2^4 (x-2)\,dx - \int_{3.5}^4 (2x-7)\,dx$$

$$= \frac{1}{2}\left[\frac{3x^2}{2}\right]_0^2 + \left[7x - \frac{2x^2}{2}\right]_2^{3.5} + \frac{1}{2}\left[\frac{x^2}{2} - 2x\right]_2^4 - \left[\frac{2x^2}{2} - 7x\right]_{3.5}^4$$

$$= \frac{1}{4}[3(2)^2 - 0] + \left[\left\{7\left(\frac{7}{2}\right) - \left(\frac{7}{2}\right)^2\right\} - \{7(2) - (2)^2\}\right] + \frac{1}{2}\left[\left\{\frac{(4)^2}{2} - 2(4)\right\} - \left\{\frac{(2)^2}{2} - 2(2)\right\}\right]$$

$$- \left[\{(4)^2 - 7(4)\} - \left\{\left(\frac{7}{2}\right)^2 - 7\left(\frac{7}{2}\right)\right\}\right]$$

$$= \frac{12}{4} + \left[\left(\frac{49}{2} - \frac{49}{4}\right) - (14-4)\right] + \frac{1}{2}[(8-8) - (2-4)] - \left[(16-28) - \left(\frac{49}{4} - \frac{49}{2}\right)\right]$$

$$= 3 + \frac{49}{2} - \frac{49}{4} - 10 + 1 + 12 + \frac{49}{4} - \frac{49}{2}$$

= 16 – 10 = 6 sq. units.

Example 24. *Using integration, find the area of the triangle ABC whose vertices have co-ordinates A(2, 5), B(4, 7) and C(6, 2).*

Solution. Let A(2, 5), B(4, 7) and C(6, 2) be the vertices of a triangle. The rough sketch of the triangle is as shown in the figure.

Equation of line AB is given by

$$(y-5) = \frac{7-5}{4-2}(x-2)$$

$\left[\because\right.$ Equation of a line passing through two points (x_1, y_1) and (x_2, y_2) is given by $\left.\left[(y - y_1) = \frac{y_2 - y_1}{x_2 - x_1}(x - x_1)\right]\right]$

$$\Rightarrow \quad y - 5 = \frac{2}{2}(x-2)$$

$$\Rightarrow \quad x - y + 3 = 0 \qquad \text{...(1)}$$

Equation of line BC is given by

$$(y-7) = \frac{2-7}{6-4}(x-4)$$

$$\Rightarrow \quad (y-7) = -\frac{5}{2}(x-4)$$

$$\Rightarrow \quad 2y - 14 = -5x + 20$$

$$\Rightarrow \quad 5x + 2y - 34 = 0 \quad ...(2)$$

Equation of line AC is given by

$$(y-5) = \frac{2-5}{6-2}(x-2)$$

$$\Rightarrow \quad (y-5) = -\frac{3}{4}(x-2)$$

$$\Rightarrow \quad 4y - 20 = -3x + 6$$

$$\Rightarrow \quad 3x + 4y - 26 = 0 \quad ...(3)$$

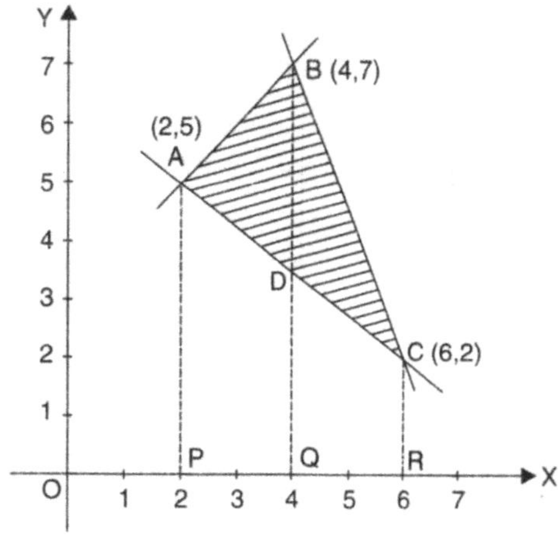

$\therefore$ Required area = Area of the shaded region

= Area of ΔABC

= Area ADB + Area BDC

= [Area APQB – Area APQD] + [Area BQRC – Area DQRC]

$$= \int_2^4 (y_{AB} - y_{AC})\,dx + \int_4^6 (y_{BC} - y_{AC})\,dx$$

$$= \int_2^4 \left[(x+3) - \left(\frac{26-3x}{4}\right)\right].dx + \int_4^6 \left[\left(\frac{34-5x}{2}\right) - \left(\frac{26-3x}{4}\right)\right].dx$$

$$= \frac{1}{4}\int_2^4 (4x + 12 - 26 + 3x)\,dx + \frac{1}{4}\int_4^6 (68 - 10x - 26 + 3x)\,dx$$

$$= \frac{1}{4}\int_2^4 (7x - 14)\,dx + \frac{1}{4}\int_4^6 (42 - 7x)\,dx$$

$$= \frac{1}{4}\left[\frac{7x^2}{2} - 14x\right]_2^4 + \frac{1}{4}\left[42x - \frac{7x^2}{2}\right]_4^6$$

$$= \frac{1}{4}\left[\left\{\frac{7(4)^2}{2} - 14(4)\right\} - \left\{\frac{7(2)^2}{2} - 14(2)\right\}\right] + \frac{1}{4}\left[\left\{42(6) - \frac{7(6)^2}{2}\right\} - \left\{42(4) - \frac{7(4)^2}{2}\right\}\right]$$

$$= \frac{1}{4}[(56 - 56) - (14 - 28)] + \frac{1}{4}[252 - 126) - (168 - 56)]$$

$$= \frac{1}{4}(0 + 14) + \frac{1}{4}(126 - 112) = \frac{14}{4} + \frac{14}{4} = \frac{28}{4}$$

= 7 sq. units.

Example 25. *Using integration, find the area of the triangle ABC whose vertices are A(2, 1), B(3, 4) and C(5, 2).*

Solution. Let A(2, 1), B(3, 4) and C(5, 2) be the vertices of a triangle. The rough sketch of the triangle is as shown in the figure.

Equation of line AB is given by

$$(y-1) = \frac{4-1}{3-2}(x-2)$$

$$\Rightarrow \quad y - 1 = \frac{3}{1}(x - 2)$$

$$\Rightarrow \quad y - 1 = 3x - 6$$

$$\Rightarrow \quad 3x - y - 5 = 0 \quad \ldots(1)$$

$\left[\begin{array}{l} \because \text{ Equation of a line passing through two points } (x_1, y_1) \text{ and } (x_2, y_2) \text{ is given by} \\ (y - y_1) = \dfrac{y_2 - y_1}{x_2 - x_1}(x - x_1) \end{array}\right]$

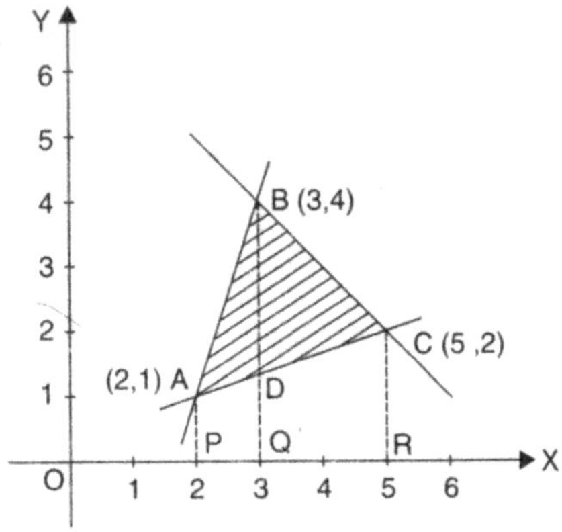

Equation of line BC is given by

$$(y - 4) = \frac{2-4}{5-3}(x - 3)$$

$$\Rightarrow \quad (y - 4) = \frac{-2}{2}(x - 3)$$

$$\Rightarrow \quad y - 4 = -x + 3$$

$$\Rightarrow \quad x + y - 7 = 0 \quad \ldots(2)$$

Equation of line AC is given by

$$(y - 1) = \frac{2-1}{5-2}(x - 2)$$

$$\Rightarrow \quad (y - 1) = \frac{1}{3}(x - 2)$$

$$\Rightarrow \quad 3y - 3 = x - 2$$

$$\Rightarrow \quad x - 3y + 1 = 0 \quad \ldots(3)$$

$\therefore$ Required area = Area of the shaded region

= Area of ΔABC

= Area ABD + Area BDC

= [Area APQB – Area APQD] + [Area BQRC – Area DQRC]

$$= \int_2^3 (y_{AB} - y_{AC}).dx + \int_3^5 (y_{BC} - y_{AC}).dx$$

$$= \int_2^3 \left[(3x - 5) - \left(\frac{x+1}{3}\right)\right].dx + \int_3^5 \left[(7 - x) - \left(\frac{x+1}{3}\right)\right].dx$$

$$= \frac{1}{3}\int_2^3 (9x - 15 - x - 1)\,dx + \frac{1}{3}\int_3^5 (21 - 3x - x - 1)\,dx$$

$$= \frac{1}{3}\int_2^3 (8x - 16).dx + \frac{1}{3}\int_3^5 (20 - 4x).dx$$

$$= \frac{8}{3}\int_2^3 (x - 2).dx + \frac{4}{3}\int_3^5 (5 - x).dx$$

$$= \frac{8}{3}\left[\frac{x^2}{2} - 2x\right]_2^3 + \frac{4}{3}\left[5x - \frac{x^2}{2}\right]_3^5$$

$$= \frac{8}{3}\left[\left\{\frac{(3)^2}{2} - 2(3)\right\} - \left\{\frac{(2)^2}{2} - 2(2)\right\}\right] + \frac{4}{3}\left[\left\{5(5) - \frac{(5)^2}{2}\right\} - \left\{5(3) - \frac{(3)^2}{2}\right\}\right]$$

$$= \frac{8}{3}\left[\left(\frac{9}{2} - 6\right) - (2 - 4)\right] + \frac{4}{3}\left[\left(25 - \frac{25}{2}\right) - \left(15 - \frac{9}{2}\right)\right]$$

$$= \frac{8}{3}\left[\frac{9}{2} - 6 + 2\right] + \frac{4}{3}\left[\frac{25}{2} - 15 + \frac{9}{2}\right] = \frac{8}{3}\left[\frac{9-8}{2}\right] + \frac{4}{3}\left[\frac{34}{2} - 15\right]$$

$$= \frac{8}{3}\left[\frac{1}{2}\right] + \frac{4}{3}\left[\frac{4}{2}\right] = \frac{4}{3} + \frac{8}{3}$$

$$= \frac{12}{3} = 4 \text{ sq. units.}$$

Example 26. *Using integration, find the area of the region bounded by the triangle whose vertices are (1, 0), (2, 2) and (3, 1).*

Solution. Let A(1, 0), B(2, 2) and C(3, 1) be the vertices of a triangle. The rough sketch of the triangle is as shown in the figure.

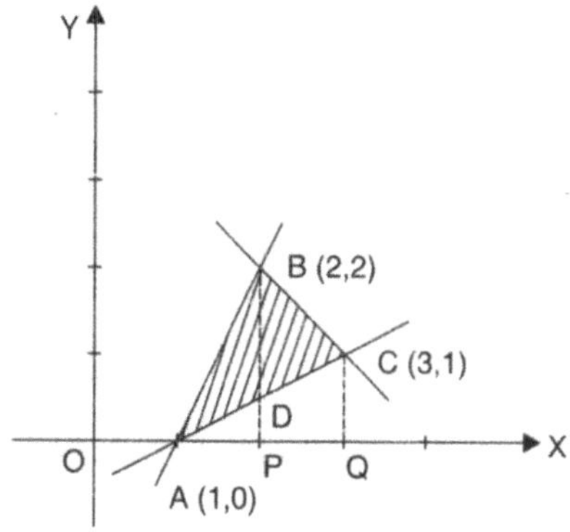

Equation of line AB is given by

$$(y - 0) = \frac{2-0}{2-1}(x - 1)$$

$\Rightarrow \quad y = 2(x - 1)$

$\Rightarrow \quad 2x - y - 2 = 0 \quad \text{...(1)}$

Equation of line BC is given by

$$(y - 1) = \frac{2-1}{2-3}(x - 3)$$

[$\because$ Equation of a line passing through two points (x_1, y_1) and (x_2, y_2) is given by $(y - y_1) = \frac{y_2 - y_1}{x_2 - x_1}(x - x_1)$]

$\Rightarrow \quad (y - 1) = \frac{1}{-1}(x - 3)$

$\Rightarrow \quad y - 1 = -x + 3$

$\Rightarrow \quad x + y - 4 = 0 \quad \text{...(2)}$

Equation of line AC is given by

$$(y - 0) = \frac{1-0}{3-1}(x - 1)$$

$\Rightarrow \quad y = \frac{1}{2}(x - 1)$

$\Rightarrow \quad 2y = x - 1$

$\Rightarrow \quad x - 2y - 1 = 0 \quad \text{...(3)}$

$\therefore$ Required area = Area of the shaded region.

= Area of ΔABC

= Area ABD + Area BDC

= [Area APB – Area APD] + [Area BPQC – Area DPQC]

$$= \int_1^2 (y_{AB} - y_{AC})\, dx + \int_2^3 (y_{BC} - y_{AC})\, dx$$

$$= \int_1^2 \left[(2x-2) - \left(\frac{x-1}{2}\right)\right] dx + \int_2^3 \left[(4-x) - \left(\frac{x-1}{2}\right)\right] dx$$

$$= \frac{1}{2}\int_1^2 (4x - 4 - x + 1)\, dx + \frac{1}{2}\int_2^3 (8 - 2x - x + 1)\, dx$$

$$= \frac{1}{2}\int_1^2 (3x-3)\, dx + \frac{1}{2}\int_2^3 (9-3x)\, dx = \frac{3}{2}\int_1^2 (x-1)\, dx + \frac{3}{2}\int_2^3 (3-x)\, dx$$

$$= \frac{3}{2}\left[\frac{x^2}{2} - x\right]_1^2 + \frac{3}{2}\left[3x - \frac{x^2}{2}\right]_2^3$$

$$= \frac{3}{2}\left[\left\{\frac{(2)^2}{2} - 2\right\} - \left\{\frac{(1)^2}{2} - 1\right\}\right] + \frac{3}{2}\left[\left\{3\,(3) - \frac{(3)^2}{2}\right\} - \left\{3(2) - \frac{(2)^2}{2}\right\}\right]$$

$$= \frac{3}{2}\left[0 + \frac{1}{2}\right] + \frac{3}{2}\left[9 - \frac{9}{2} - 6 + 2\right] = \frac{3}{4} + \frac{15}{2} - \frac{27}{4} = \frac{3+30-27}{4}$$

$$= \frac{6}{4} = \frac{3}{2} \text{ sq. units.}$$

Example 27. *Using integration, find the area of the region bounded by the triangle whose vertices are (– 1, 1), (0, 5) and (3, 2).*

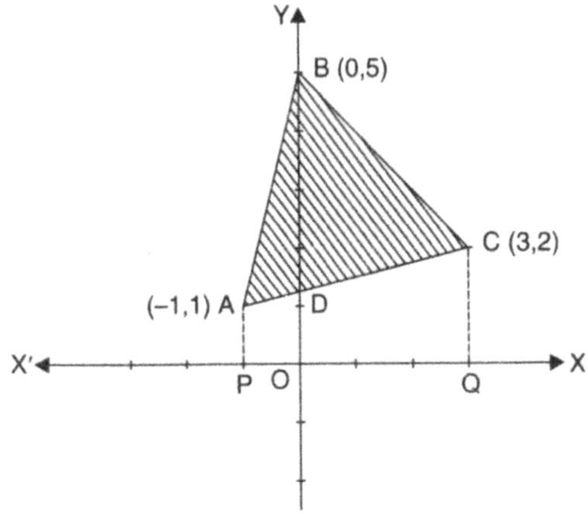

Solution. Let A(– 1, 1), B(0, 5) and C(3, 2) be the vertices of a triangle. The rough sketch of the triangle is shown in the figure.

Equation of line AB is given by

$$(y-1) = \frac{5-1}{0+1}(x+1)$$

$\Rightarrow \quad (y-1) = 4(x+1)$

$\Rightarrow \quad y - 1 = 4x + 4$

$\Rightarrow \quad 4x - y + 5 = 0 \quad \ldots(1)$

Equation of line BC is given by

[∵ Equation of a line passing through two points (x_1, y_1) and (x_2, y_2) is given by $(y - y_1) = \frac{y_2 - y_1}{x_2 - x_1}(x - x_1)$]

$$(y-5) = \frac{2-5}{3-0}(x-0)$$

$\Rightarrow \quad y - 5 = -\frac{3}{3}(x)$

$\Rightarrow \quad y - 5 = -x$

$\Rightarrow \quad x + y - 5 = 0 \quad \ldots(2)$

Equation of line AC is given by

$$(y-1)=\frac{2-1}{3+1}(x+1)$$

$$\Rightarrow \quad (y-1)=\frac{1}{4}(x+1)$$

$$\Rightarrow \quad 4y-4=x+1$$

$$\Rightarrow \quad x-4y+5=0 \qquad \text{...(3)}$$

$\therefore$ Required area = Area of the shaded region

= Area of ΔABC

= Area ABD + Area BDC

= [Area APOB – Area APOD] + [Area BOQC – Area DOQC]

$$=\int_{-1}^{0}(y_{AB}-y_{AC})\,dx+\int_{0}^{3}(y_{BC}-y_{AC})\,dx$$

$$=\int_{-1}^{0}\left[(4x+5)-\left(\frac{x+5}{4}\right)\right]dx+\int_{0}^{3}\left[(5-x)-\left(\frac{x+5}{4}\right)\right]dx$$

$$=\frac{1}{4}\int_{-1}^{0}(16x+20-x-5)\,dx+\frac{1}{4}\int_{0}^{3}(20-4x-x-5)\,dx$$

$$=\frac{1}{4}\int_{-1}^{0}(15x+15)\,dx+\frac{1}{4}\int_{0}^{3}(15-5x)\,dx$$

$$=\frac{15}{4}\int_{-1}^{0}(x+1)\,dx+\frac{5}{4}\int_{0}^{3}(3-x)\,dx$$

$$=\frac{15}{4}\left[\frac{x^2}{2}+x\right]_{-1}^{0}+\frac{5}{4}\left[3x-\frac{x^2}{2}\right]_{0}^{3}$$

$$=\frac{15}{4}\left[0-\left\{\frac{(-1)^2}{2}+(-1)\right\}\right]+\frac{5}{4}\left[\left\{3(3)-\frac{(3)^2}{2}\right\}-0\right]$$

$$=\frac{15}{4}\left[-\left(\frac{1}{2}-1\right)\right]+\frac{5}{4}\left[9-\frac{9}{2}\right]=\frac{15}{8}+\frac{45}{8}=\frac{60}{8}$$

$$=\frac{15}{2}\text{ sq. units.}$$

Example 28. *Using integration, find the area of the triangle whose vertices are (2, 0), (4, 5) and (6, 3).*

Solution. Let A(2, 0), B(4, 5) and C(6, 3) be the vertices of a triangle. The rough sketch of the triangle is as shown in the figure.

Equation of line AB is given by

$$(y-0)=\frac{5-0}{4-2}(x-2)$$

$$\Rightarrow \quad y=\frac{5}{2}(x-2)$$

$$\Rightarrow \quad 2y=5x-10$$

$\Rightarrow \quad 5x - 2y - 10 = 0 \qquad ...(1)$

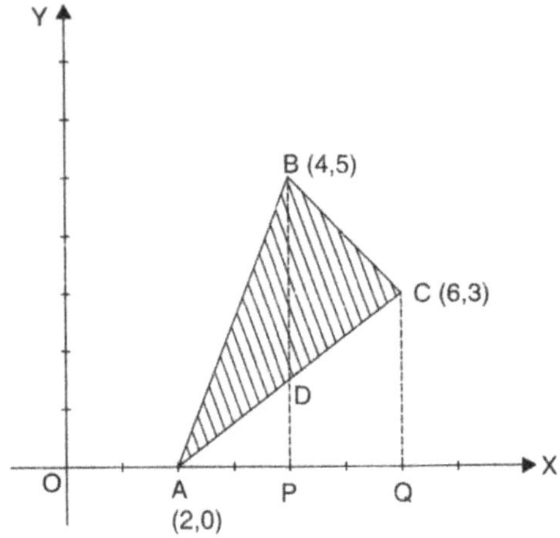

Equation of line BC is given by

$$(y - 5) = \frac{3-5}{6-4}(x - 4)$$

$$\left[\because \text{ Equation of a line passing through two points } (x_1, y_1) \text{ and } (x_2, y_2) \text{ is given by } (y - y_1) = \frac{y_2 - y_1}{x_2 - x_1}(x - x_1)\right]$$

$\Rightarrow \quad (y - 5) = \frac{-2}{2}(x - 4)$

$\Rightarrow \quad y - 5 = -x + 4$

$\Rightarrow \quad x + y - 9 = 0 \qquad ...(2)$

Equation of line AC is given by

$$(y - 0) = \frac{3-0}{6-2}(x - 2)$$

$\Rightarrow \quad y = \frac{3}{4}(x - 2)$

$\Rightarrow \quad 4y = 3x - 6$

$\Rightarrow \quad 3x - 4y - 6 = 0 \qquad ...(3)$

$\therefore$ Required area = Area of the shaded region.

= Area of ΔABC

= Area ABD + Area BDC

= [Area ABP – Area APD] + [Area BPQC – Area DPQC]

$$= \int_2^4 (y_{AB} - y_{AC})\, dx + \int_4^6 (y_{BC} - y_{AC})\, dx$$

$$= \int_2^4 \left[\left(\frac{5x-10}{2}\right) - \left(\frac{3x-6}{4}\right)\right] dx + \int_4^6 \left[(9 - x) - \left(\frac{3x-6}{4}\right)\right] dx$$

$$= \frac{1}{4}\int_2^4 (10x - 20 - 3x + 6)\, dx + \frac{1}{4}\int_4^6 (36 - 4x - 3x + 6)\, dx$$

$$= \frac{1}{4}\int_2^4 (7x - 14)\, dx + \frac{1}{4}\int_4^6 (42 - 7x)\, dx$$

$$= \frac{7}{4}\int_2^4 (x - 2)\, dx + \frac{7}{4}\int_4^6 (6 - x)\, dx$$

$$= \frac{7}{4}\left[\frac{x^2}{2} - 2x\right]_2^4 + \frac{7}{4}\left[6x - \frac{x^2}{2}\right]_4^6$$

$$= \frac{7}{4}\left[\left\{\frac{(4)^2}{2} - 2(4)\right\} - \left\{\frac{(2)^2}{2} - 2(2)\right\}\right] + \frac{7}{4}\left[\left\{6(6) - \frac{(6)^2}{2}\right\} - \left\{6(4) - \frac{(4)^2}{2}\right\}\right]$$

$$= \frac{7}{4}[(8-8)-(2-4)] + \frac{7}{4}\left[\left(36-\frac{36}{2}\right)-(24-8)\right]$$

$$= \frac{7}{4}(0+2) + \frac{7}{4}\left[\frac{36}{2}-16\right] = \frac{7}{2}+\frac{7}{4}\left(\frac{36-32}{2}\right) = \frac{7}{2}+\frac{7}{2}$$

$= 7$ sq. units.

Example 29. *Using integration find the area of the region bounded by the following curves.*

$$y = 1 + |x+1|, \quad x = -2, \quad x = 3, \quad y = 0.$$

Solution. The given equations of the curves are

$$y = 1 + |x+1| \qquad \text{...(1)}$$

$$\Rightarrow \qquad y = \begin{cases} 1+(x+1) & ; \quad x+1>0 \\ 1-(x+1) & ; \quad x+1 \le 0 \end{cases}$$

$$\Rightarrow \qquad y = \begin{cases} x+2 & ; \quad x > -1 \\ -x & ; \quad x \le -0 \end{cases}$$

$x = -2$ is a straight line parallel to y-axis at a distance of 2 units to the left side of y-axis.
$x = 3$ is a straight line parallel to y-axis at a distance of 3 units to the right side of y-axis.
Table of values of x and y satisfying equation $y = x + 2$; $x > -1$ is given by

x	0	1	2	3
y	2	3	4	5

Table of values of x and y satisfying equation. $y = -x$; $x \le -1$ is given by

x	-1	-2
y	1	2

By joining these points we obtain a rough sketch of these curves is as shown in the figure.

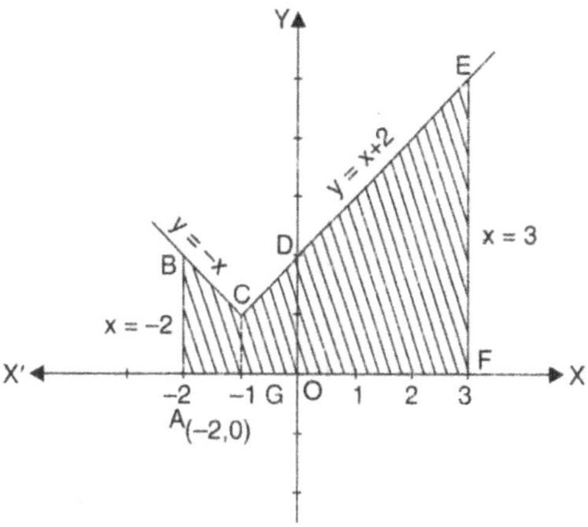

$\therefore$ Required area

= Area of the shaded region.

= Area ABCEF

= Area ABCG + Area GCEF

$$= \int_{-2}^{-1}(-x)\,dx + \int_{-1}^{3}(x+2)\,dx$$

$$= -\left[\frac{x^2}{2}\right]_{-2}^{-1} + \left[\frac{x^2}{2}+2x\right]_{-1}^{3}$$

$$= -\left[\frac{(-1)^2}{2}-\frac{(-2)^2}{2}\right] + \left[\left\{\frac{(3)^2}{2}+2(3)\right\}-\left\{\frac{(-1)^2}{2}+2(-1)\right\}\right]$$

$$= -\left[\frac{1}{2} - 2\right] + \left[\left(\frac{9}{2} + 6\right) - \left(\frac{1}{2} - 2\right)\right]$$

$$= +\frac{3}{2} + \left(\frac{9}{2} + 6 - \frac{1}{2} + 2\right) = \frac{3}{2} + 4 + 8 = \frac{3}{2} + 12$$

$$= \frac{3+24}{2} = \frac{27}{2} \text{ sq. units.}$$

Example 30. *Find the area enclosed by the curves*

$$y = |x - 1| \text{ and } y = 1 - |x - 1|.$$

Solution. The given equations of the curves are

$$y = |x - 1| \quad \text{...(1)}$$

$$\Rightarrow \quad y = \begin{cases} x - 1 & ; \ x - 1 > 0 \\ -(x - 1) & ; \ x - 1 \le 0 \end{cases}$$

$$\Rightarrow \quad y = \begin{cases} x - 1 & ; \ x > 1 \\ 1 - x & ; \ x \le 1 \end{cases} \quad \text{...(2)}$$

$$\Rightarrow \quad y = 1 - |x - 1|$$

$$\Rightarrow \quad y = \begin{cases} 1 - (x - 1) & ; \ x - 1 > 0 \\ 1 + x - 1 & ; \ x - 1 \le 0 \end{cases}$$

$$\Rightarrow \quad y = \begin{cases} 2 - x & ; \ x > 1 \\ x & ; \ x \le 1 \end{cases} \quad \text{...(3)}$$

Now, to find the points of intersection of two curves, let us solve equations (2) and (3) simultaneously.

For $x \ge 1$,

$$y = x - 1 \quad \text{and} \quad y = 2 - x \Rightarrow x = 2 - y$$

$$\Rightarrow \quad y = 2 - y - 1 \Rightarrow 2y = 1 \Rightarrow y = \frac{1}{2}$$

$\therefore$ From equation (1), we have

$$|x - 1| = \frac{1}{2}$$

$$\Rightarrow \quad x - 1 = \pm\frac{1}{2}$$

For $x - 1 = \frac{1}{2}$, $\quad x = \frac{3}{2}$

For $x - 1 = -\frac{1}{2}$, $\quad x = \frac{1}{2}$

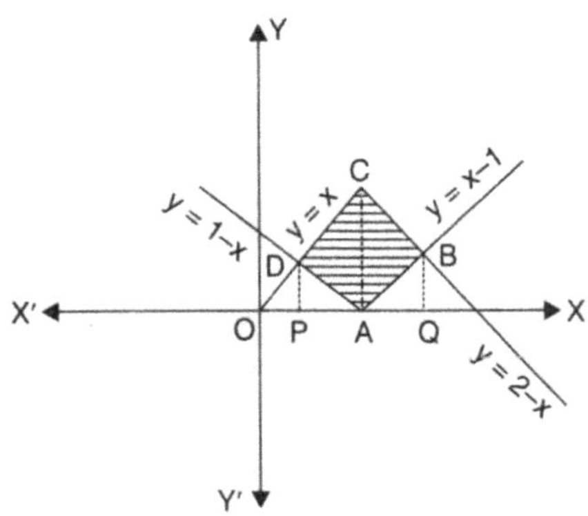

$\therefore$ The points of intersection of two curves are $\left(\frac{3}{2}, \frac{1}{2}\right)$ and $\left(\frac{1}{2}, \frac{1}{2}\right)$.

The rough sketch of these curves is as shown in the figure.

$\therefore$ Required area = Area of the shaded region

$$= \text{Area ABCD}$$
$$= \text{Area DAC + Area CAB}$$
$$= [\text{Area DPAC} - \text{Area DPA}] + [\text{Area CAQB} - \text{Area AQB}]$$
$$= \int_{1/2}^{1} [x-(1-x)]\,dx + \int_{1}^{\frac{3}{2}} [(2-x)-(x-1)]\,dx$$
$$= \int_{1/2}^{1} (2x-1)\,dx + \int_{1}^{\frac{3}{2}} (3-2x)\,dx$$
$$= \left[\frac{2x^2}{2} - x\right]_{\frac{1}{2}}^{1} + \left[3x - \frac{2x^2}{2}\right]_{1}^{\frac{3}{2}}$$
$$= \left[\{(1)^2 - 1\} - \left\{\left(\frac{1}{2}\right)^2 - \left(\frac{1}{2}\right)\right\}\right] + \left[\left\{3\left(\frac{3}{2}\right) - \left(\frac{3}{2}\right)^2\right\} - \{3(1) - (1)^2\}\right]$$
$$= \left[0 - \left(\frac{1}{4} - \frac{1}{2}\right)\right] + \left[\left(\frac{9}{2} - \frac{9}{4}\right) - (3-1)\right]$$
$$= -\left(\frac{1-2}{4}\right) + \left(\frac{18-9}{4}\right) - 2 = \frac{1}{4} + \frac{9}{4} - 2 = \frac{10}{4} - 2 = \frac{10-8}{4}$$
$$= \frac{2}{4} = \frac{1}{2} \text{ sq. units.}$$

Example 31. *Find the area bounded by the curve $y^2 = 4a^2(x-1)$ and the line $x = 1$ and $y = 4a$.*

Solution. The given equations of the curves are

$$y^2 = 4a^2(x-1) \qquad ...(1)$$

$$\Rightarrow \quad (y-0)^2 = 4a^2(x-1)$$

and $\quad x = 1, \quad y = 4a$

Equation (1) represents a parabola with vertex at (1, 0) opening right hand side of x-axis.

$x = 1$ is a straight line parallel to y-axis at a distance of 1 unit to the right side of it.

$y = 4a$ is a straight line parallel to the x-axis at a distance $4a$ units upwards to the x-axis.

The rough sketch of the curves is as shown in the figure.

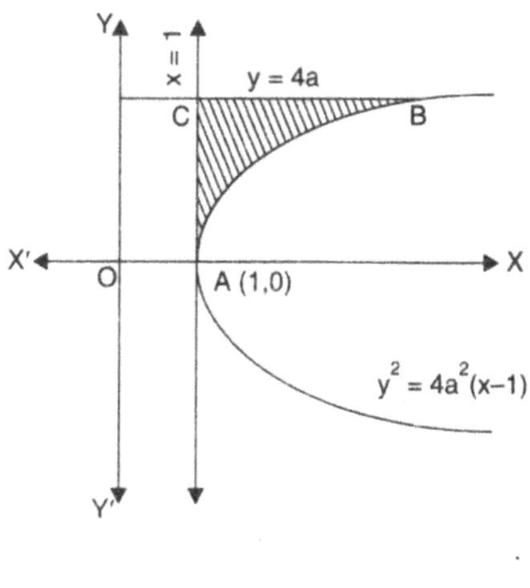

$\therefore$ Required Area = Area of the shaded region

$$= \text{Area ABC}$$
$$= \int_0^{4a} (x-1)\,dy$$

$$\left[\because y^2 = 4a^2(x-1) \Rightarrow \frac{y^2}{4a^2} = (x-1)\right]$$

$$\int_0^{4a} \frac{y^2}{4a^2}.dy = \frac{1}{4a^2}\left[\frac{y^3}{3}\right]_0^{4a}$$

$$= \frac{1}{12a^2}[(4a)^3 - 0] = \frac{1}{12a^2}.64a^3$$

$$= \frac{16a}{3} \text{ sq. units.}$$

Example 32. *Find the area of the region enclosed by the parabola $x^2 = y$ and the line $y = x + 2$ and the x-axis.*

Solution. The given equations of the curves are

$$x^2 = y \quad \text{...(1)}$$

$$y = x + 2 \quad \text{...(2)}$$

Equation (1) represents a parabola with vertex at the origin (0, 0) and symmetric along positive direction of y-axis.

$y = x + 2$ represents a straight line.

Table of values of x and y satisfying equation (1) is given by

x	0	– 1	1	– 2	2
y	0	1	1	4	4

Table of values of x and y satisfying equation (2) is given by

x	0	– 1	1	2
y	2	1	3	4

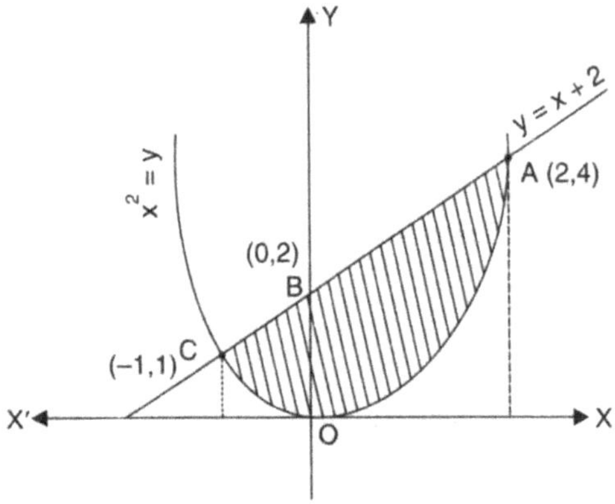

By joining these points with a free hand, we obtain a rough sketch of the two curves as shown in the figure.

Now, to find the points of intersection of the two curves, let us solve equations (1) and (2) simultaneously

$\therefore$ From (1) and (2), we have

$$y = x^2$$

$$\Rightarrow \quad x + 2 = x^2$$

$$\Rightarrow \quad x^2 - x - 2 = 0$$

$$\Rightarrow \quad x^2 - 2x + x - 2 = 0$$

$$\Rightarrow \quad x(x-2) + 1(x-2) = 0$$

$$\Rightarrow \quad (x-2)(x+1) = 0$$

$$\Rightarrow \quad x = -1, \quad x = 2$$

For $x = -1$, $y = (-1)^2 = 1$

For $x = 2$, $y = (2)^2 = 4$

$\therefore$ The points of intersection of the two curves are (– 1, 1) and (2, 4).

$\therefore$ Required area = Area of the shaded region

$$= \text{Area OABCO.}$$

$$= \int_{-1}^{2} (y_{\text{upper curve}} - y_{\text{lower curve}}).\,dx$$

$$= \int_{-1}^{2} [(x+2) - x^2]\,dx = \left[\frac{x^2}{2} + 2x - \frac{x^3}{3}\right]_{-1}^{2}$$

$$= \left[\left\{\frac{(2)^2}{2} + 2(2) - \frac{(2)^3}{3}\right\} - \left\{\frac{(-1)^2}{2} + 2(-1) - \frac{(-1)^3}{3}\right\}\right]$$

$$= \left[\left(2 + 4 - \frac{8}{3}\right) - \left(\frac{1}{2} - 2 + \frac{1}{3}\right)\right] = 6 - \frac{8}{3} - \frac{1}{2} + 2 - \frac{1}{3}$$

$$= 8 - \frac{9}{3} - \frac{1}{2} = 8 - 3 - \frac{1}{2} = 5 - \frac{1}{2} = \frac{9}{2} \text{ sq. units.}$$

Example 33. *Sketch the curves and identify the region bounded by the curves* $x = \frac{1}{2}$, $x = 2$, $y = \log x$ *and* $y = 2^x$. *Find the area of this region.*

Solution. The given equations of the curves are

$$y = 2^x \quad \ldots(1)$$

$$y = \log x \quad \ldots(2)$$

$$x = \frac{1}{2} \quad \ldots(3)$$

$$x = 2 \quad \ldots(4)$$

Equation (1) is an exponential curve and equation (2) is a logarithmic curve.

Since the inverse of an exponential function is a logarithmic function and vice-versa. Therefore, these two curves are on the opposite sides of the line $y = x$.

Thus, the curves $y = 2^x$ and $y = \log x$ do not intersect.

The rough sketch of these curves is as shown in the figure.

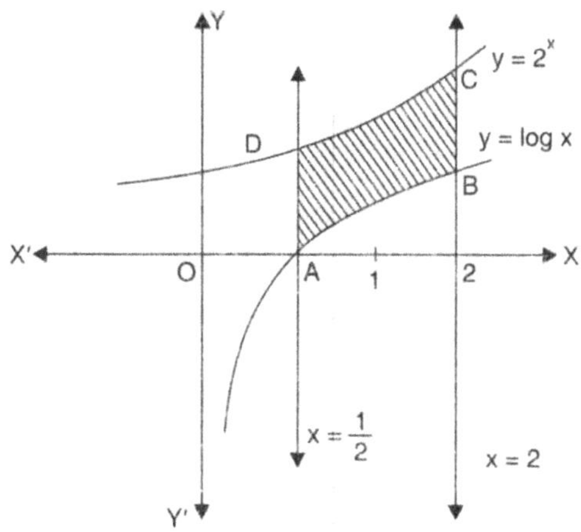

$\therefore$ Required Area

$$= \text{Area of the shaded region.}$$

$$= \int_{1/2}^{2} (y_{\text{upper curve}} - y_{\text{lower curve}}).\,dx$$

$$= \int_{1/2}^{2} (2^x - \log x).\,dx$$

$$= \left[\frac{2^x}{\log 2} - x \log x + x\right]_{1/2}^{2}$$

$$= \left[\left\{\frac{2^2}{\log 2} - 2\log 2 + 2\right\} - \left\{\frac{2^{1/2}}{\log 2} - \frac{1}{2}\log\frac{1}{2} + \frac{1}{2}\right\}\right]$$

$$\left[\begin{array}{l} \because \int \underset{\text{II}}{\text{I}} . \underset{\text{I}}{\log x}\, dx \text{ (Integrating by parts)} \\ = \log x \,.\, x - \int \frac{d}{dx}(\log x)\,.\, x\, dx \\ = x\log x - \int \frac{1}{x}\,.\, x\, dx = x\log x - \int 1\,.\, dx = (x\log x - x) \end{array}\right]$$

$$= \left[\frac{4}{\log 2} - 2\log 2 + 2 - \frac{\sqrt{2}}{\log 2} + \frac{1}{2}\log\frac{1}{2} - \frac{1}{2}\right] = \left[\frac{4-\sqrt{2}}{\log 2} + \frac{3}{2} - 2\log 2 - \frac{1}{2}\log 2\right]$$

$$= \left[\frac{4-\sqrt{2}}{\log 2} - \frac{3}{2} - \frac{5}{2}\log 2\right] \text{ sq. units.}$$

Example 34. *Find the area of the parabola $x^2 = 2 - y$ cut off by the line $x + y = 0$.*

Solution. The given equations of the curves are

$$x^2 = 2 - y$$

$$\Rightarrow \quad x^2 = -(y-2) \qquad \text{...(1)}$$

$$x + y = 0$$

$$\Rightarrow \quad y = -x \qquad \text{...(2)}$$

Equation (1) represents a parabola opening downwards to y-axis with vertex at (0, 2) and meets the x-axis at $x = \pm\sqrt{2}$.

Equation (2), $x + y = 0$ represents a straight line.

Now, to find the points of intersection of two curves, let us solve equations (1) and (2).

$\therefore$ From (1) and (2), we have

$$x^2 = 2 - y \quad \text{and} \quad y = -x$$

$$\Rightarrow \quad x^2 = 2 - (-x)$$

$$\Rightarrow \quad x^2 - x - 2 = 0$$

$$\Rightarrow \quad x^2 - 2x + x - 2 = 0$$

$$\Rightarrow \quad x(x-2) + 1(x-2) = 0$$

$$\Rightarrow \quad (x+1)(x-2) = 0$$

$$\Rightarrow \quad x = -1,\ x = 2$$

For $x = -1$, $y = -(-1) = 1$

For $x = 2$, $y = -(2) = -2$

$\therefore$ The points of intersection of the two curves are $(-1, 1)$ and $(2, -2)$.

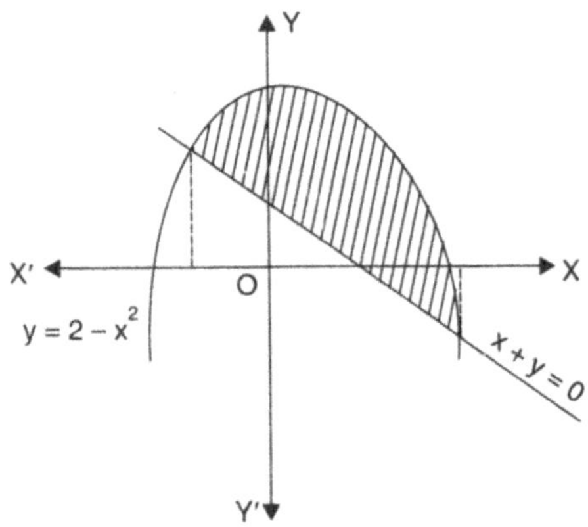

The rough sketch of these curves are as shown in the figure.

$\therefore$ Required area = Area of the shaded region

$$= \int_{-1}^{2} (y_{\text{upper curve}} - y_{\text{lower curve}})\, dx$$

$$= \int_{-1}^{2} [2 - x^2 - (-x)]\, dx = \int_{-1}^{2} (2 - x^2 + x)\, dx$$

$$= \left[2x - \frac{x^3}{3} + \frac{x^2}{2}\right]_{-1}^{2}$$

$$= \left[\left\{2(2) - \frac{(2)^3}{3} + \frac{(2)^2}{2}\right\} - \left\{2(-1) - \frac{(-1)^3}{3} + \frac{(-1)^2}{2}\right\}\right]$$

$$= \left(4 - \frac{8}{3} + 2\right) - \left(-2 + \frac{1}{3} + \frac{1}{2}\right)$$

$$= 6 - \frac{8}{3} + 2 - \frac{1}{3} - \frac{1}{2} = 8 - \frac{9}{3} - \frac{1}{2} = \frac{48 - 18 - 3}{6}$$

$$= \frac{27}{6} = \frac{9}{2} \text{ sq. units.}$$

Example 35. *Sketch the region common to the circle $x^2 + y^2 = 16$ and the parabola $x^2 = 6y$. Also, find the area of the region using integration.*

Solution. The given equations of the curves are

$$x^2 + y^2 = 16 \qquad \text{...(1)}$$

$$x^2 = 6y \qquad \text{...(2)}$$

Equation (1) represents a circle with centre at the origin (0, 0) and radius 4.

Equation (2) represents a parabola opening upwards with vertex at the origin (0, 0).

Now, to find the points of intersection of the two curves, let us solve equations (1) and (2) simultaneously.

$$x^2 + y^2 = 16 \qquad \left[\because x^2 = 6y \Rightarrow y = \frac{x^2}{6}\right]$$

$$\Rightarrow \quad x^2 + \left(\frac{x^2}{6}\right)^2 = 16$$

$$\Rightarrow \quad x^2 + \frac{x^4}{36} = 16$$

$$\Rightarrow \quad x^4 + 36x^2 - 576 = 0$$

$$\Rightarrow \quad x^4 + 48x^2 - 12x^2 - 576 = 0$$

$$\Rightarrow \quad x^2(x^2 + 48) - 12(x^2 + 48) = 0$$

$$\Rightarrow \quad (x^2 + 48)(x^2 - 12) = 0$$

$$x^2 = -48 \quad \text{(Not possible)}$$

$$x^2 = 12 \Rightarrow x = \pm 2\sqrt{3}.$$

For $x = +2\sqrt{3}$, $y = \dfrac{(2\sqrt{3})^2}{6} = \dfrac{12}{6} = 2$

For $x = -2\sqrt{3}$, $y = \dfrac{(-2\sqrt{3})^2}{6} = \dfrac{12}{6} = 2$

$\therefore$ The points of intersection of the two curves are $(2\sqrt{3}, 2)$ and $(-2\sqrt{3}, 2)$.

Now, equation (1) *i.e.,* $x^2 + y^2 = 16$ is symmetrical about both x-axis and y-axis as the equation contains only even powers of x and y.

Equation (2) *i.e.* $x^2 = 6y$ is symmetrical about y-axis as it contains only even powers of x.

$\therefore$ Area bounded between the curves is symmetrical about y-axis.

The rough sketch of the two curves is as shown in the figure.

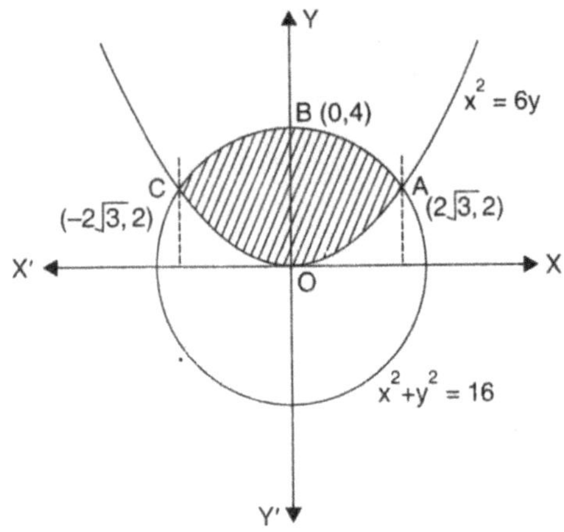

$\therefore$ Required area = Area of the shaded region.

$$= \int_{-2\sqrt{3}}^{2\sqrt{3}} (y_{\text{circle}} - y_{\text{parabola}})\, dx$$

$$= \int_{-2\sqrt{3}}^{2\sqrt{3}} \left(\sqrt{16 - x^2} - \frac{x^2}{6}\right) dx$$

$$= 2\int_{0}^{2\sqrt{3}} \left(\sqrt{16 - x^2} - \frac{x^2}{6}\right) dx \qquad [\because \text{ Integrand is an even function}]$$

$$= 2\left[\frac{x\sqrt{16 - x^2}}{2} + \frac{16}{2}\sin^{-1}\frac{x}{4} - \frac{x^3}{18}\right]_0^{2\sqrt{3}}$$

$$\left[\because \int \sqrt{a^2 - x^2}\, dx = \frac{x\sqrt{a^2 - x^2}}{2} + \frac{a^2}{2}\sin^{-1}\frac{x}{a}\right]$$

$$= 2\left[\left\{\frac{2\sqrt{3}\sqrt{16 - 12}}{2} + 8\sin^{-1}\left(\frac{2\sqrt{3}}{4}\right) - \frac{(2\sqrt{3})^3}{18}\right\} - 0\right]$$

$$= 2\left[2\sqrt{3} + 8\sin^{-1}\frac{\sqrt{3}}{2} - \frac{24\sqrt{3}}{18}\right]$$

$$= 4\sqrt{3} + 16\frac{\pi}{3} - \frac{8}{\sqrt{3}} = \frac{12 - 8}{\sqrt{3}} + \frac{16\pi}{3}$$

$$= \left(\frac{4}{\sqrt{3}} + \frac{16\pi}{3}\right) \text{ sq. units.}$$

Example 36. *Find the area of the circle $4x^2 + 4y^2 = 9$ which is interior to the parabola $y^2 = 4x$.*

Solution. The given equations of the parabola are :

$$4x^2 + 4y^2 = 9$$

$$\Rightarrow \qquad x^2 + y^2 = \frac{9}{4} \qquad \text{...(1)}$$

$$y^2 = 4x \qquad \text{...(2)}$$

Equation (1) represents a circle with centre at the origin (0, 0) and radius equal to $\frac{3}{2}$.

Equation (2), $y^2 = 4x$ represents a parabola whose vertex is at the origin and symmetrical to x-axis as it contains only even powers of y.

Now, to find the points of intersection of two curves, let us solve equations (1) and (2) simultaneously.

$\therefore$ From (1) and (2), we have

$$x^2 + y^2 = \frac{9}{4} \quad \text{and} \quad y^2 = 4x$$

$$\Rightarrow \quad x^2 + 4x = \frac{9}{4}$$

$$\Rightarrow \quad 4x^2 + 16x - 9 = 0$$

$$\Rightarrow \quad 4x^2 + 18x - 2x - 9 = 0$$

$$\Rightarrow \quad 2x(2x + 9) - 1(2x + 9) = 0$$

$$\Rightarrow \quad (2x + 9)(2x - 1) = 0$$

$$\Rightarrow \quad 2x + 9 = 0, \quad 2x - 1 = 0$$

$$\Rightarrow \quad x = -\frac{9}{2}, \quad x = \frac{1}{2}$$

For $x = \frac{1}{2}$, $\quad y^2 = 4\left(\frac{1}{2}\right) \Rightarrow y^2 = 2 \Rightarrow y = \pm\sqrt{2}$

For $x = -\frac{9}{2}$, $\quad y^2 = 4\left(-\frac{9}{2}\right) \Rightarrow y^2 = -18 \Rightarrow y$ has imaginary value.

$\therefore$ The points of intersection of the two curves are $\left(\frac{1}{2}, \sqrt{2}\right)$ and $\left(\frac{1}{2}, -\sqrt{2}\right)$.

The rough sketch of these curves are as shown in the figure.

$\therefore$ Required area

= Area of the shaded region.

= Area OBAC

= 2(Area OBA)

= 2[Area OBD + Area BDA]

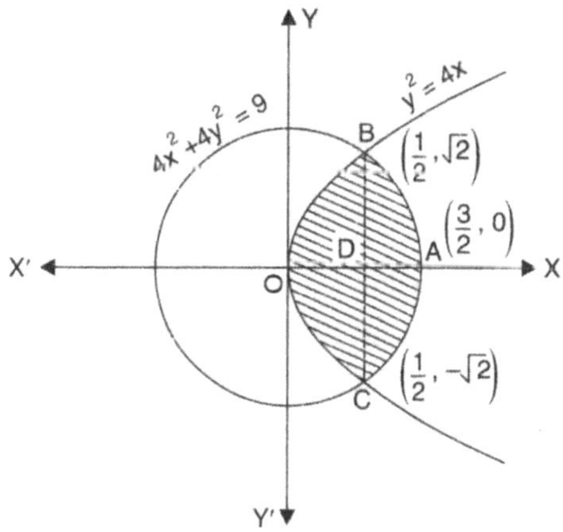

$$= 2\int_0^{1/2} (y_{\text{parabola}})\, dx + 2\int_{1/2}^{3/2} (y_{\text{circle}})\, dx$$

$$= 2\int_0^{1/2} \sqrt{4x}\, dx + 2\int_{1/2}^{3/2} \left(\sqrt{\frac{9}{4} - x^2}\right) dx$$

$$= 4\int_0^{1/2} x^{1/2}\, dx + 2\int_{1/2}^{3/2} \sqrt{\left(\frac{3}{2}\right)^2 - x^2}\, dx$$

$$= 4\left[\frac{x^{3/2}}{\frac{3}{2}}\right]_0^{\frac{1}{2}} + 2\left[\frac{x\sqrt{\frac{9}{4} - x^2}}{2} + \frac{\frac{9}{4}}{2}\sin^{-1}\frac{x}{\frac{3}{2}}\right]_{\frac{1}{2}}^{\frac{3}{2}}$$

$$\left[\because \int \sqrt{a^2 - x^2}\, dx = \frac{x\sqrt{a^2 - x^2}}{2} + \frac{a^2}{2}\sin^{-1}\frac{x}{a}\right]$$

$$= \frac{8}{3}\left[\left(\frac{1}{2}\right)^{\frac{3}{2}} - 0\right] + 2\left[\frac{x\sqrt{9-4x^2}}{4} + \frac{9}{8}\sin^{-1}\frac{2x}{3}\right]_{\frac{1}{2}}^{\frac{3}{2}}$$

$$= \frac{8}{3}\left(\frac{1}{2\sqrt{2}}\right) + \frac{2}{8}\left[2x\sqrt{9-4x^2} + 9\sin^{-1}\frac{2x}{3}\right]_{\frac{1}{2}}^{\frac{3}{2}}$$

$$= \frac{4}{3\sqrt{2}} + \frac{1}{4}\left[\left\{2\left(\frac{3}{2}\right)\sqrt{9-4\left(\frac{3}{2}\right)^2} + 9\sin^{-1}\frac{2\left(\frac{3}{2}\right)}{3}\right\} - \left\{2\left(\frac{1}{2}\right)\sqrt{9-4\left(\frac{1}{2}\right)^2} + 9\sin^{-1}\frac{2\left(\frac{1}{2}\right)}{3}\right\}\right]$$

$$= \frac{4}{3\sqrt{2}} + \frac{1}{4}\left[\left(3\sqrt{9-9} + 9\sin^{-1}1\right) - \left(\sqrt{9-1} + 9\sin^{-1}\frac{1}{3}\right)\right]$$

$$= \frac{4}{3\sqrt{2}} + \frac{1}{4}\left[\left(0 + \frac{9\pi}{2}\right) - \left(\sqrt{8} + 9\sin^{-1}\frac{1}{3}\right)\right]$$

$$= \frac{4}{3\sqrt{2}} + \frac{9\pi}{8} - \frac{2\sqrt{2}}{4} - \frac{9}{4}\sin^{-1}\frac{1}{3} = \frac{4}{3\sqrt{2}} - \frac{1}{\sqrt{2}} + \frac{9\pi}{8} - \frac{9}{4}\sin^{-1}\frac{1}{3}$$

$$= \frac{4-3}{3\sqrt{2}} + \frac{9\pi}{8} - \frac{9}{4}\sin^{-1}\frac{1}{3}$$

$$= \left(\frac{1}{3\sqrt{2}} + \frac{9\pi}{8} - \frac{9}{4}\sin^{-1}\frac{1}{3}\right) \text{ sq. units.}$$

Example 37. *Find the area of the region enclosed by two circles $x^2 + y^2 = 1$ and $(x - 1)^2 + y^2 = 1$.*

Solution. The given equations of the two curves are

$$x^2 + y^2 = 1 \quad \text{...(1)}$$

$$(x-1)^2 + y^2 = 1 \quad \text{...(2)}$$

Equation (1) represents a circle with centre at the origin (0, 0) and radius 1.

Equation (2), also represents a circle with centre at (1, 0) and radius 1.

Since both the curves are symmetrical about x-axis as both the equation contain only even powers of y.

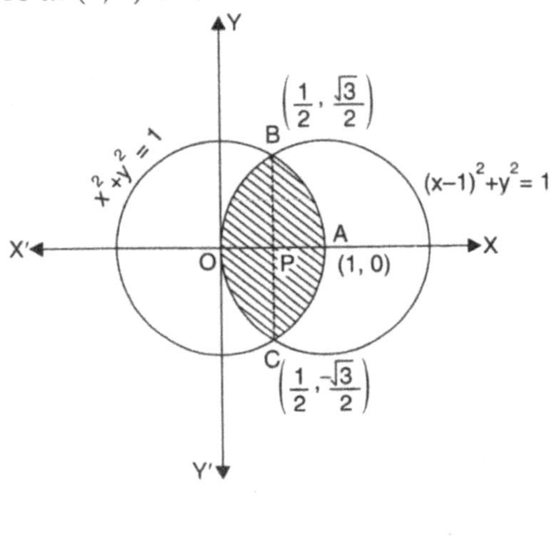

The rough sketch of the two curves are as shown in the figure.

Now, to find the points of intersection of the two curves, let us solve the equations (1) and (2) simultaneously.

$\therefore$ From (1) and (2), $x^2 + y^2 = 1$

$\Rightarrow \quad y^2 = 1 - x^2$

$\Rightarrow \quad (x-1)^2 + y^2 = 1$

$\Rightarrow \quad (x-1)^2 + 1 - x^2 = 1$

$\Rightarrow \quad x^2 + 1 - 2x + 1 - x^2 = 1$

$\Rightarrow \quad -2x = -1$

$\Rightarrow \quad x = \frac{1}{2}$

For $x = \frac{1}{2}$, $y^2 = 1 - \left(\frac{1}{2}\right)^2 = 1 - \frac{1}{4} = \frac{3}{4} \Rightarrow y = \pm\frac{\sqrt{3}}{2}$

$\therefore$ The points of intersection of the two curves are $\left(\frac{1}{2}, \frac{\sqrt{3}}{2}\right)$ and $\left(\frac{1}{2}, -\frac{\sqrt{3}}{2}\right)$.

$\therefore$ Required area = Area of the shaded region
= Area OBAC = 2(Area OAB)
= 2 [Area OBP + Area BAP]

$$= 2\int_0^{1/2} y_{\text{circle (2)}} \, .dx + 2\int_{1/2}^{1} y_{\text{circle (1)}} \, .dx$$

$$= 2\int_0^{1/2} \sqrt{1-(x-1)^2}\,.dx + 2\int_{1/2}^{1} \sqrt{1-x^2}\,.dx$$

$$= 2\left[\frac{(x-1)\sqrt{1-(x-1)^2}}{2} + \frac{1}{2}\sin^{-1}\left(\frac{x-1}{1}\right)\right]_0^{\frac{1}{2}} + 2\left[\frac{x\sqrt{1-x^2}}{2} + \frac{1}{2}\sin^{-1} x\right]_{\frac{1}{2}}^{1}$$

$$\left[\because \int \sqrt{a^2 - x^2}\, dx = \frac{x\sqrt{a^2-x^2}}{2} + \frac{a^2}{2}\sin^{-1}\frac{x}{a}\right]$$

$$= 2\left[\left\{\frac{\left(\frac{1}{2}-1\right)\sqrt{1-\left(\frac{1}{2}-1\right)^2}}{2} + \frac{1}{2}\sin^{-1}\left(\frac{1}{2}-1\right)\right\} - \left\{\frac{(0-1)\sqrt{1-(0-1)^2}}{2} + \frac{1}{2}\sin^{-1}(-1)\right\}\right]$$

$$+ 2\left[\left\{\frac{1\sqrt{1-1}}{2} + \frac{1}{2}\sin^{-1} 1\right\} - \left\{\frac{\frac{1}{2}\sqrt{1-\left(\frac{1}{2}\right)^2}}{2} + \frac{1}{2}\sin^{-1}\frac{1}{2}\right\}\right]$$

$[\because \sin(-\theta) = -\sin\theta]$

$$= 2\left[\left\{\frac{-\sqrt{1-\frac{1}{4}}}{4} + \frac{1}{2}\sin^{-1}\left(-\frac{1}{2}\right)\right\} - \left\{0 - \frac{1}{2}\sin^{-1} 1\right\}\right]$$

$$+ 2\left[\left\{0 + \frac{1}{2}\left(\frac{\pi}{2}\right)\right\} - \left\{\frac{1\sqrt{1-\frac{1}{4}}}{4} + \frac{1}{2}\sin^{-1}\left(\frac{1}{2}\right)\right\}\right]$$

$$= 2\left[-\frac{\sqrt{3}}{8} - \frac{1}{2}\left(\frac{\pi}{6}\right) + \frac{1}{2}\left(\frac{\pi}{2}\right)\right] + 2\left[\frac{\pi}{4} - \frac{\sqrt{3}}{8} - \frac{1}{2}\left(\frac{\pi}{6}\right)\right] = -\frac{\sqrt{3}}{4} - \frac{\pi}{6} + \frac{\pi}{2} + \frac{\pi}{2} - \frac{\sqrt{3}}{4} - \frac{\pi}{6}$$

$$= -\frac{\sqrt{3}}{2} + \pi - \frac{\pi}{3} = \left(\frac{2\pi}{3} - \frac{\sqrt{3}}{2}\right) \text{ sq. units.}$$

Example 38. *Find the area of the smaller part of the circle $x^2 + y^2 = a^2$ cut off by the line $x = \frac{a}{\sqrt{2}}$.*

Solution. The given equations of the curves are

$$x^2 + y^2 = a^2 \quad \text{...(1)}$$

$$x = \frac{a}{\sqrt{2}} \quad \text{...(2)}$$

Equation (1) represents a circle with centre at the origin (0, 0) and radius a.

Equation (2), $x = \frac{a}{\sqrt{2}}$ is a straight line parallel to y-axis at a distance of $\frac{a}{\sqrt{2}}$ in the positive directon of x-axis.

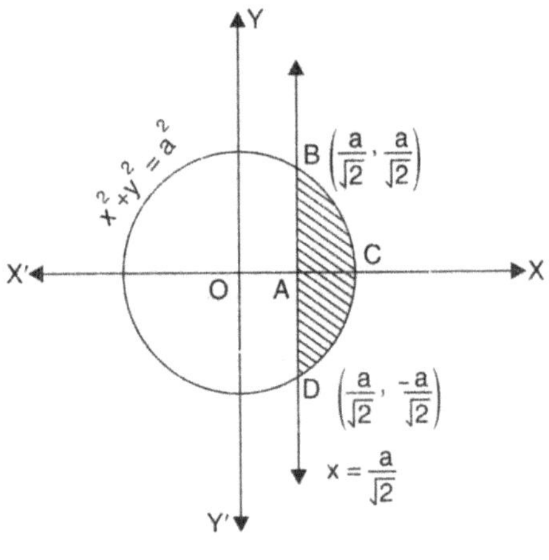

The rough sketch of the curves are as shown in the figure.

Now, to find the points of intersection of the two curves, let us solve equations (1) and (2) simultaneously.

$\therefore$ From (1) and (2), we have

$$x = \frac{a}{\sqrt{2}}$$

$$\Rightarrow \quad x^2 + y^2 = a^2$$

$$\Rightarrow \quad \left(\frac{a}{\sqrt{2}}\right)^2 + y^2 = a^2$$

$$\Rightarrow \quad y^2 = a^2 - \frac{a^2}{2} = \frac{a^2}{2}$$

$$\Rightarrow \quad y = \pm\frac{a}{\sqrt{2}}$$

$\therefore$ The points of intersection of the two curves are $\left(\frac{a}{\sqrt{2}}, \frac{a}{\sqrt{2}}\right)$ and $\left(\frac{a}{\sqrt{2}}, -\frac{a}{\sqrt{2}}\right)$.

$\therefore$ Required area = Area of the shaded region

= Area ABCD

$$= \int_{\frac{a}{\sqrt{2}}}^{a} \sqrt{a^2 - x^2}\,.\,dx$$

$$= \left[\frac{x\sqrt{a^2 - x^2}}{2} + \frac{a^2}{2}\sin^{-1}\frac{x}{a}\right]_{\frac{a}{\sqrt{2}}}^{a} \quad \left[\because \int \sqrt{a^2 - x^2}\,.\,dx = \frac{x\sqrt{a^2 - x^2}}{2} + \frac{a^2}{2}\sin^{-1}\frac{x}{a}\right]$$

$$= \left[\left\{\frac{a\sqrt{a^2-a^2}}{2}+\frac{a^2}{2}\sin^{-1}\frac{a}{a}\right\}-\left\{\frac{\frac{a}{\sqrt{2}}\sqrt{a^2-\frac{a^2}{2}}}{2}+\frac{a^2}{2}\sin^{-1}\frac{\frac{a}{\sqrt{2}}}{a}\right\}\right]$$

$$= \left(0+\frac{a^2}{2}\sin^{-1}1\right)-\left(\frac{a}{2\sqrt{2}}\cdot\frac{a}{\sqrt{2}}+\frac{a^2}{2}\sin^{-1}\frac{1}{\sqrt{2}}\right)$$

$$= \frac{a^2}{2}\cdot\frac{\pi}{2}-\frac{a^2}{4}-\frac{a^2}{2}\left(\frac{\pi}{4}\right)=\frac{a^2\pi}{4}-\frac{a^2}{4}-\frac{a^2\pi}{8}=\frac{a^2\pi}{8}-\frac{a^2}{4}$$

$$= \frac{a^2}{4}\left(\frac{\pi}{2}-1\right) \text{ sq. units.}$$

Example 39. *Find the area of the region between the circles $x^2+y^2=4$ and $(x-2)^2+y^2=4$.*

Solution. The given equations of the curves are

$$x^2+y^2=4 \qquad \text{...(1)}$$

$$(x-2)^2+y^2=4 \qquad \text{...(2)}$$

Equation (1) represents a circle with centre at the origin (0, 0) and radius 2.

Equation (2) also represents a circle with centre (2, 0) and radius 2.

Since both the curves are symmetrical about x-axis, as both the equations have only even powers of y.

The rough sketch of the curves are as shown in the figure.

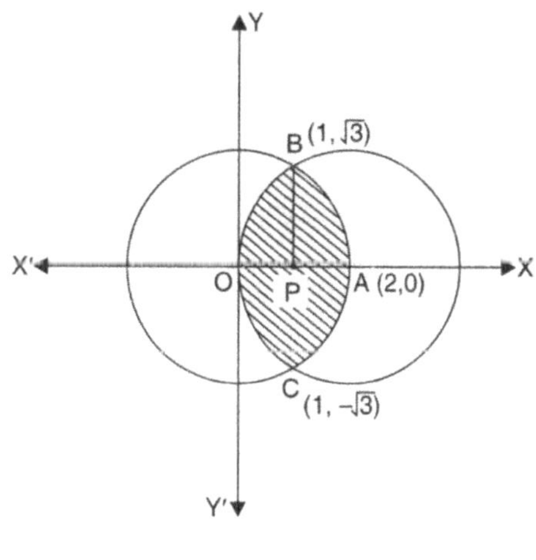

Now, to find the points of intersection of the two curves, let us solve equations (1) and (2) simultaneously.

∴ From (1) and (2), we have

$$x^2+y^2=4$$

$$\Rightarrow \quad y^2=4-x^2$$

$$(x-2)^2+y^2=4$$

$$\Rightarrow \quad x^2+4-4x+4-x^2=4$$

$$\Rightarrow \quad 4x=4 \Rightarrow x=1$$

For $x=1$, $y^2=4-(1)^2=4-1=3$

$$\Rightarrow \quad y=\pm\sqrt{3}$$

∴ The points of intersection of the two curves are $(1, \sqrt{3})$ and $(1, -\sqrt{3})$.

∴ Required area = Area of the shaded region.

= Area OBACO = 2 [Area OBA]

= 2 [Area OBP + Area BAP]

$$= 2\int_0^1 y_{\text{circle (2)}}\cdot dx + 2\int_1^2 y_{\text{circle (1)}}\cdot dx$$

$$= 2\int_0^1 \sqrt{4-(x-2)^2}\, dx + 2\int_1^2 \sqrt{4-x^2}\, dx$$

$$= 2\left[\frac{(x-2)\sqrt{4-(x-2)^2}}{2} + \frac{4}{2}\sin^{-1}\left(\frac{x-2}{2}\right)\right]_0^1 + 2\left[\frac{x\sqrt{4-x^2}}{2} + \frac{4}{2}\sin^{-1}\frac{x}{2}\right]_1^2$$

$$\left[\because \int \sqrt{a^2-x^2}\,.dx = \frac{x\sqrt{a^2-x^2}}{2} + \frac{a^2}{2}\sin^{-1}\frac{x}{a}\right]$$

$$= 2\left[\left\{\frac{(1-2)\sqrt{4-(1-2)^2}}{2} + 2\sin^{-1}\left(\frac{1-2}{2}\right)\right\} - \left\{\frac{(0-2)\sqrt{4-(0-2)^2}}{2} + 2\sin^{-1}\left(\frac{-2}{2}\right)\right\}\right]$$

$$+ 2\left[\left\{\frac{2\sqrt{4-4}}{2} + 2\sin^{-1}\frac{2}{2}\right\} - \left\{\frac{\sqrt{4-1}}{2} + 2\sin^{-1}\frac{1}{2}\right\}\right]$$

$$= 2\left[\left\{\frac{-\sqrt{3}}{2} + 2\sin^{-1}\left(-\frac{1}{2}\right)\right\} - \{0 + 2\sin^{-1}(-1)\}\right] + 2\left[(0 + 2\sin^{-1} 1) - \left(\frac{\sqrt{3}}{2} + 2\sin^{-1}\frac{1}{2}\right)\right]$$

$$[\because \quad \sin(-\theta) = -\sin\theta]$$

$$= -\sqrt{3} - 4\sin^{-1}\left(\frac{1}{2}\right) + 4\sin^{-1} 1 + 4\sin^{-1} 1 - \sqrt{3} - 4\sin^{-1}\left(\frac{1}{2}\right)$$

$$= -2\sqrt{3} - 8\sin^{-1}\left(\frac{1}{2}\right) + 8\sin^{-1} 1$$

$$= -2\sqrt{3} - 8\left(\frac{\pi}{6}\right) + 8\left(\frac{\pi}{2}\right) = -2\sqrt{3} - \frac{4\pi}{3} + 4\pi$$

$$= \left(\frac{8\pi}{3} - 2\sqrt{3}\right) \text{ sq. units.}$$

Example 40. *Calculate the area of the region enclosed between the circles $x^2 + y^2 = 1$ and $\left(x - \frac{1}{2}\right)^2 + y^2 = 1$.*

Solution. The given equations of the curves are

$$x^2 + y^2 = 1 \quad \text{...(1)}$$

$$\left(x - \frac{1}{2}\right)^2 + y^2 = 1 \quad \text{...(2)}$$

Equation (1) represents a circle with centre at the origin (0, 0) and radius unity.

Equation (2) also represents a circle with centre at $\left(\frac{1}{2}, 0\right)$ and radius unity.

Since, both the curves are symmetrical about x-axis, as both the equation contain only even powers of y.

The rough sketch of the curves is as shown in the figure.

Now, to find the points of intersection of the two curves, let us solve equations (1) and (2) simultaneously.

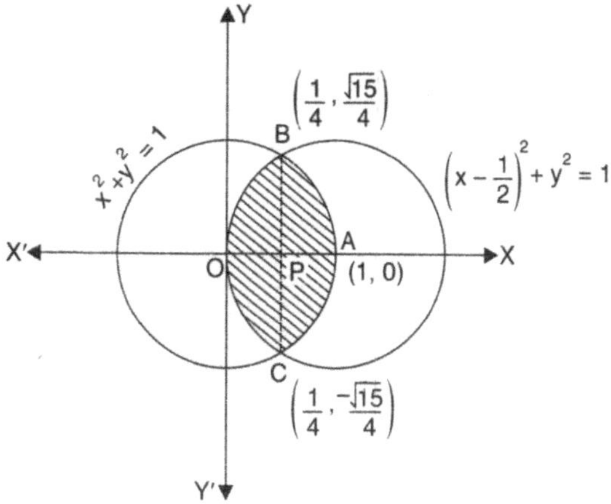

$\therefore$ From (1) and (2), we have

$$x^2 + y^2 = 1$$

$$\Rightarrow \quad y^2 = 1 - x^2$$

$$\left(x - \frac{1}{2}\right)^2 + y^2 = 1$$

$$\Rightarrow \quad x^2 + \frac{1}{4} - x + 1 - x^2 = 1$$

$$\Rightarrow \quad -x + \frac{5}{4} = 1 \Rightarrow x = \frac{1}{4}$$

For $x = \frac{1}{4}$, $y^2 = 1 - x^2$

$$\Rightarrow \quad y^2 = 1 - \left(\frac{1}{4}\right)^2 = 1 - \frac{1}{16} = \frac{15}{16}$$

$$\Rightarrow \quad y = \pm \frac{\sqrt{15}}{4}$$

$\therefore$ The points of intersection of the two curves are $\left(\frac{1}{4}, \frac{\sqrt{15}}{4}\right)$ and $\left(\frac{1}{4}, -\frac{\sqrt{15}}{4}\right)$.

$\therefore$ Required area = Area of the shaded region

= Area OBACO

= 2 Area OBA

= 2[Area OBP + Area BAP]

$$= 2\int_{-1/2}^{1/4} y_{\text{circle (2)}} \cdot dx + 2\int_{1/4}^{1} y_{\text{circle (1)}} \cdot dx$$

$$= 2\int_{-1/2}^{1/4} \sqrt{1 - \left(x - \frac{1}{2}\right)^2} \cdot dx + 2\int_{1/4}^{1} \sqrt{1 - x^2}\, dx$$

$$= 2\left[\frac{\left(x - \frac{1}{2}\right)\sqrt{1 - \left(x - \frac{1}{2}\right)^2}}{2} + \frac{1}{2}\sin^{-1}\left(x - \frac{1}{2}\right)\right]_{-\frac{1}{2}}^{\frac{1}{4}} + 2\left[\frac{x\sqrt{1-x^2}}{1} + \frac{1}{2}\sin^{-1} x\right]_{\frac{1}{4}}^{1}$$

$$\left[\because \int \sqrt{a^2 - x^2}\, dx = \frac{x\sqrt{a^2 - x^2}}{2} + \frac{a^2}{2}\sin^{-1}\frac{x}{a}\right]$$

$$= 2\left[\left\{\frac{\left(\frac{1}{4}-\frac{1}{2}\right)\sqrt{1-\left(\frac{1}{4}-\frac{1}{2}\right)^2}}{2}+\frac{1}{2}\sin^{-1}\left(\frac{1}{4}-\frac{1}{2}\right)\right\}\right.$$

$$\left.-\left\{\frac{\left(-\frac{1}{2}-\frac{1}{2}\right)\sqrt{1-\left(-\frac{1}{2}-\frac{1}{2}\right)^2}}{2}+\frac{1}{2}\sin^{-1}\left(-\frac{1}{2}-\frac{1}{2}\right)\right\}\right]$$

$$+2\left[\left\{\frac{\sqrt{1-x^2}}{2}+\frac{1}{2}\sin^{-1}1\right\}-\left\{\frac{\frac{1}{4}\sqrt{1-\left(\frac{1}{4}\right)^2}}{2}+\frac{1}{2}\sin^{-1}\left(\frac{1}{4}\right)\right\}\right]$$

$$=\left[\left(-\frac{1}{4}\right)\sqrt{1-\frac{1}{4}}+\sin^{-1}\left(-\frac{1}{4}\right)\right]-[(-1)\sqrt{0}+\sin^{-1}(-1)]$$

$$+[1\sqrt{0}+\sin^{-1}(1)]-\left[\frac{1}{4}\cdot\frac{\sqrt{15}}{4}+\sin^{-1}\frac{1}{4}\right] \quad [\because \sin(-\theta)=-\sin\theta]$$

$$=-\frac{\sqrt{3}}{8}-\sin^{-1}\left(\frac{1}{4}\right)+\frac{\pi}{2}+\frac{\pi}{2}-\frac{\sqrt{15}}{16}-\sin^{-1}\frac{1}{4}=-\frac{\sqrt{3}}{8}-\frac{\sqrt{15}}{16}-2\sin^{-1}\frac{1}{4}+\pi$$

$$=\left(\frac{-(2\sqrt{3}+\sqrt{15})}{16}-2\sin^{-1}\frac{1}{4}+\pi\right) \text{ sq. units.}$$

Example 41. *Compute the area of the figure bounded by the straight lines $x = 0$, $x = 2$ and the curves $y = 2^x$, $y = 2x - x^2$.*

Solution. The given equations of the curves are

$$y = 2x - x^2 \quad \ldots(1)$$

$$y = 2^x \quad \ldots(2)$$

$$x = 0 \quad \ldots(3)$$

$$x = 2 \quad \ldots(4)$$

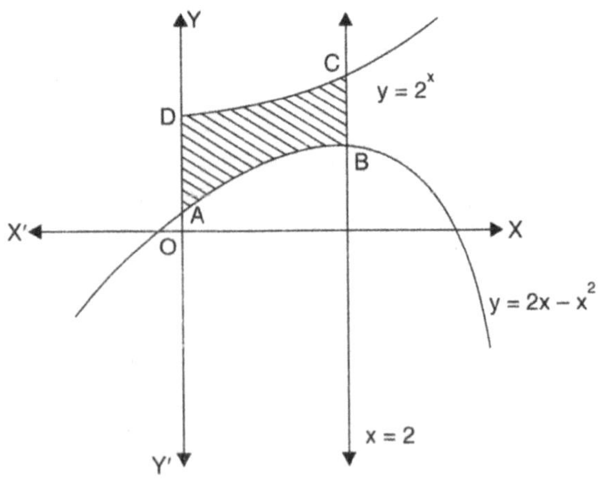

Equation (1) represents a parabola opening downwards having vertex at (1, 1) and cuts the x-axis at points (0, 0) and (2, 0).

Equation (2), $y = 2^x$ shows an exponential curve.

$x = 0$ is the y-axis and $x = 2$ is a straight line parallel to the y-axis at a distance 2 units from it.

The rough sketch of the curves is as shown in the figure.

$\therefore$ Required area

$$= \text{Area of the shaded region}$$

$$= \int_0^2 (y_{\text{upper curve}} - y_{\text{lower curve}}) \, . \, dx$$

$$= \int_0^2 [2^x - (2x - x^2)]. \, dx = \int_0^2 (2^x - 2x + x^2) \, . \, dx$$

$$= \left[\frac{2^x}{\log 2} - \frac{2x^2}{2} + \frac{x^3}{3} \right]_0^2$$

$$= \left[\left\{ \frac{2^2}{\log 2} - (2)^2 + \frac{(2)^3}{3} \right\} - \left\{ \frac{2^0}{\log 2} - 0 + 0 \right\} \right] = \frac{4}{\log 2} - 4 + \frac{8}{3} - \frac{1}{\log 2}$$

$$= \left(\frac{3}{\log 2} - \frac{4}{3} \right) \text{ sq. units.}$$

Example 42. *Find the area of the smaller region bounded by the ellipse* $\frac{x^2}{a^2} + \frac{y^2}{b^2} = 1$ *and the straight line* $\frac{x}{a} + \frac{y}{b} = 1$.

Solution. The given equations of the curves are

$$\frac{x^2}{a^2} + \frac{y^2}{b^2} = 1 \qquad \text{...(1)}$$

$$\frac{x}{a} + \frac{y}{b} = 1 \qquad \text{...(2)}$$

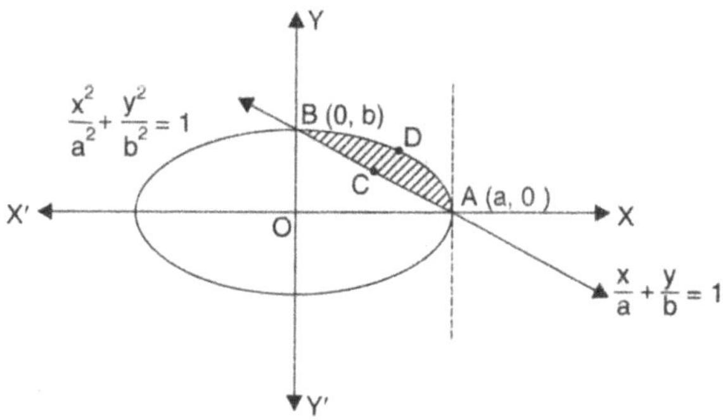

Equation (1) represents an ellipse with centre at the origin, length of major and minor axis are $2a$ and $2b$ respectively.

The ellipse intersects x-axis and y-axis at $(\pm a, 0)$ and $(0, \pm b)$ respectively.

Equation (2) represents a straight line in the intercept form having intercepts a and b on axis.

$\therefore$ Equation (2) intersects x-axis at $(a, 0)$ and y-axis at $(0, b)$.

$\therefore$ Points of intersection of the two curves are $(a, 0)$ and $(0, b)$.

The rough sketch of the curves are as shown in the figure.

$\therefore$ Required area = Area of the shaded region

$$= \text{Area ACBDA}$$

$$= \int_0^a (y_{\text{upper curve}} - y_{\text{lower curve}}) \, . \, dx$$

$$= \int_0^a \left[\frac{b}{a}\sqrt{a^2 - x^2} - \frac{b}{a}(a - x)\right].dx \qquad \left[\begin{aligned} &\because \text{From (1)}: \frac{x^2}{a^2} + \frac{y^2}{b^2} = 1 \Rightarrow \frac{y^2}{b^2} = 1 - \frac{x^2}{a^2} \\ &\Rightarrow y^2 = \frac{b^2}{a^2}(a^2 - x^2) \\ &\Rightarrow y = \pm\frac{b}{a}\sqrt{a^2 - x^2} \end{aligned}\right]$$

$$= \frac{b}{a}\int_0^a \left(\sqrt{a^2 - x^2} - a + x\right) dx$$

$$= \frac{b}{a}\left[\frac{x\sqrt{a^2 - x^2}}{2} + \frac{a^2}{2}\sin^{-1}\frac{x}{a} - ax + \frac{x^2}{2}\right]_0^a$$

$$\left[\because \int \sqrt{a^2 - x^2}\,.dx = \frac{x\sqrt{a^2 - x^2}}{2} + \frac{a^2}{2}\sin^{-1}\frac{x}{a}\right]$$

$$= \frac{b}{a}\left[\left\{\frac{a\sqrt{a^2 - a^2}}{2} + \frac{a^2}{2}\sin^{-1}\frac{a}{a} - a^2 + \frac{a^2}{2}\right\} - \left\{0 + \frac{a^2}{2}\sin^{-1}0 - 0 + 0\right\}\right]$$

$$= \frac{b}{a}\left[\left(0 + \frac{a^2}{2}\sin^{-1}1 - \frac{a^2}{2}\right) - (0)\right] = \frac{b}{a}\left[\frac{a^2}{2}\cdot\left(\frac{\pi}{2}\right) - \frac{a^2}{2}\right]$$

$$= \left(\frac{ab\pi}{4} - \frac{a^2}{2}\right) \text{ sq. units.}$$

Example 43. *Find the area of the smaller region enclosed by the ellipse* $\frac{x^2}{9} + \frac{y^2}{4} = 1$ *and the straight line* $\frac{x}{3} + \frac{y}{2} = 1$.

Solution. The given equations of the curves are

$$\frac{x^2}{9} + \frac{y^2}{4} = 1 \qquad \text{...(1)}$$

$$\frac{x}{3} + \frac{y}{2} = 1 \qquad \text{...(2)}$$

Equation (1) represents an ellipse with centre at the origin, length of major and minor axis are 2(3) = 6 and 2(2) = 4 respectively.

The ellipse intersects x-axis and y-axis at (± 3, 0) and (0, ± 2) respectively.

Equation (2) represents a straight line in the intercept form having intercepts 3 and 2 on the axis.

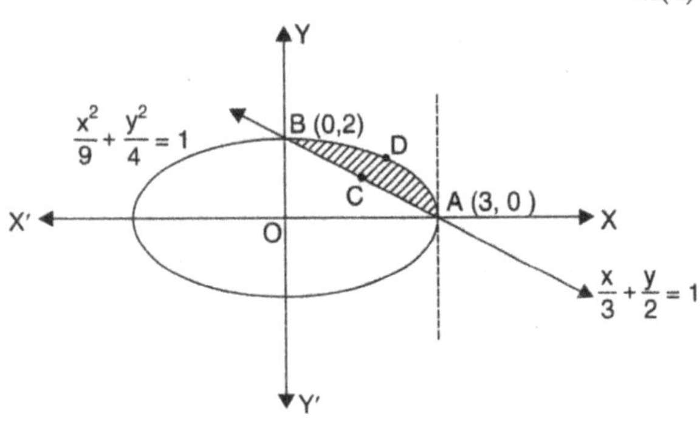

∴ Equation (2) intersects x-axis at (3, 0) and y-axis at (0, 2).

∴ Points of intersection of the two curves are (3, 0) and (0, 2).

The rough sketch of the two curves is as shown in the figure.

∴ Required area = Area of the shaded region = Area BCAD

$$= \int_0^3 (y_{\text{upper curve}} - y_{\text{lower curve}}) \, . \, dx$$

$$\left[\begin{array}{l} \because \text{ From (1), } \dfrac{x^2}{9} + \dfrac{y^2}{4} = 1 \\ \Rightarrow \dfrac{y^2}{4} = 1 - \dfrac{x^2}{9} \\ \Rightarrow y^2 = \dfrac{4}{9}(9 - x^2) \\ \Rightarrow y = \pm \dfrac{2}{3}\sqrt{9 - x^2} \end{array}\right]$$

$$= \int_0^3 \left[\frac{2}{3}\sqrt{9 - x^2} - \frac{6 - 2x}{3}\right] dx$$

$$= \frac{2}{3}\int_0^3 \left[\sqrt{(3)^2 - x^2} - (3 - x)\right] dx$$

$$= \frac{2}{3}\left[\frac{x\sqrt{9 - x^2}}{2} + \frac{9}{2}\sin^{-1}\frac{x}{3} - 3x + \frac{x^2}{2}\right]_0^3$$

$$\left[\because \int \sqrt{a^2 - x^2}\, dx = \frac{x\sqrt{a^2 - x^2}}{2} + \frac{a^2}{2}\sin^{-1}\frac{x}{a}\right]$$

$$= \frac{2}{3}\left[\left\{\frac{3\sqrt{9 - 9}}{2} + \frac{9}{2}\sin^{-1} 1 - 9 + \frac{9}{2}\right\} - \left\{0 + \frac{9}{2}\sin^{-1} 0 - 0\right\}\right]$$

$$= \frac{2}{3}\left[0 + \frac{9}{2}\left(\frac{\pi}{2}\right) - \frac{9}{2} - 0\right] = \left(\frac{3\pi}{2} - 3\right) \text{ sq. units.}$$

Example 44. Find the area of the region bounded by the curves $y = 4x - x^2$ and $y = x^2 - x$ above x-axis.

Solution. The given equations of the curves are

$$y = 4x - x^2 \quad \text{...(1)}$$

$$y = x^2 - x \quad \text{(2)}$$

From equation (1), we have

$$x^2 = 4x - y$$

$$\Rightarrow \quad x^2 - 4x = -y$$

$$\Rightarrow \quad x^2 - 4x + 4 = -y + 4 \quad \text{[Adding 4 on both sides]}$$

$$\Rightarrow \quad (x - 2)^2 = -(y - 4) \quad \text{...(3)}$$

∴ Equation (3) represents a parabola opening downward and having vertex at (2, 4).

Now, from equation (2), we have

$$x^2 - x = y$$

$$\Rightarrow \quad x^2 - x + \frac{1}{4} = y + \frac{1}{4} \qquad \left[\text{Adding } \frac{1}{4} \text{ on both sides}\right]$$

$$\Rightarrow \quad \left(x - \frac{1}{2}\right)^2 = y + \frac{1}{4} \qquad ...(4)$$

$\therefore$ Equation (4) represents a parabola opening upwards and having vertex at $\left(\frac{1}{2}, -\frac{1}{4}\right)$.

The rough sketch of these curves is as shown in the figure.

Now, to find the points of intersection of the two curves, let us solves equations (1) and (2) simultaneously.

$\therefore$ From (1) and (2), we have

$$4x - x^2 = x^2 - x$$

$$\Rightarrow \quad 2x^2 = 5x$$

$$\Rightarrow \quad 2x^2 - 5x = 0$$

$$\Rightarrow \quad x(2x - 5) = 0$$

$$\Rightarrow \quad x = 0, x = \frac{5}{2}$$

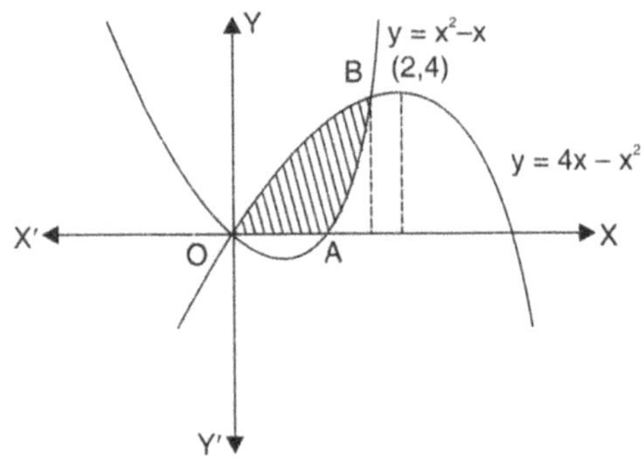

For $x = 0$, $\quad y = 4(0) - (0)^2 = 0$

For $x = \frac{5}{2}$, $\quad y = 4\left(\frac{5}{2}\right) - \left(\frac{5}{2}\right)^2$

$$= 10 - \frac{25}{4} = \frac{40 - 25}{4} = \frac{15}{4}$$

$\therefore$ The points of intersection of the two curves are $(0, 0)$ and $\left(\frac{5}{2}, \frac{15}{4}\right)$.

$\therefore$ Required area = Area of shaded region

= Area of region OAB

$$= \int_0^{5/2} (y_{\text{upper curve}} - y_{\text{lower curve}}).\,dx$$

$$= \int_0^{5/2} [(4x - x^2) - (x^2 - x)].dx$$

$$= \int_0^{5/2} (4x - x^2 - x^2 + x)\,dx = \int_0^{5/2} (5x - 2x^2)\,dx$$

$$= \left[\frac{5x^2}{2} - \frac{2x^3}{3}\right]_0^{5/2} = \frac{1}{6}\left[15x^2 - 4x^3\right]_0^{5/2} = \frac{1}{6}\left[\left\{15\left(\frac{5}{2}\right)^2 - 4\left(\frac{5}{2}\right)^3\right\} - \{0 - 0\}\right]$$

$$= \frac{1}{6}\left[15\left(\frac{25}{4}\right) - 4\left(\frac{125}{8}\right) - 0\right] = \frac{1}{6}\left[\frac{375}{4} - \frac{250}{4}\right] = \frac{1}{6}\left(\frac{125}{4}\right)$$

$$= \frac{125}{24} \text{ sq. units.}$$

Example 45. *Find the area of the region :*

$$\{(x, y) : 0 \le y \le x^2 + 1,\ 0 \le y \le x + 1,\ 0 \le x \le 2\}.$$

Solution. Let $R = \{(x, y) : 0 \le y \le x^2 + 1, 0 \le y \le x + 1, 0 \le x \le 2\}$.

$\Rightarrow \quad R = \{(x, y) : 0 \le y \le x^2 + 1\} \cap \{(x, y) : 0 \le y \le x + 1\} \cap \{(x, y) : 0 \le x \le 2\}$

$\Rightarrow \quad R = R_1 \cap R_2 \cap R_3 \quad$ (say)

where $R_1 = \{(x, y) : 0 \le y \le x^2 + 1\}$

$R_2 = \{(x, y) : 0 \le y \le x + 1\}$

$R_3 = \{(x, y) : 0 \le x \le 2\}$

Now, to find the point of intersection of the curves, let us solve equations $y = x^2 + 1$ and $y = x + 1$.

$\therefore \quad x^2 + 1 = x + 1$

$\Rightarrow \quad x^2 - x = 0$

$\Rightarrow \quad x(x - 1) = 0$

$\Rightarrow \quad x = 0, \quad x = 1$

For $x = 0$, $y = 0 + 1 = 1$

For $x = 1$, $y = 1 + 1 = 2$

$\therefore$ The points of intersection of these curves are (0, 1) and (1, 2).

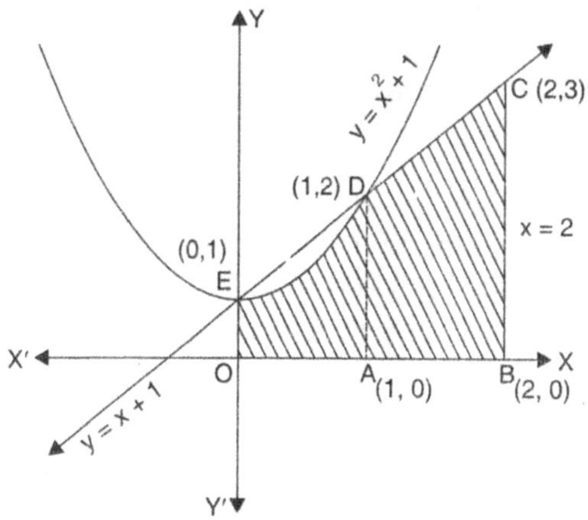

Region R_1 :

We have

$$y = x^2 + 1$$

$$\Rightarrow \quad x^2 = y - 1$$

$$\Rightarrow \quad (x - 0)^2 = (y - 1) \quad \text{...(1)}$$

Clearly, equation (1) represents a parabola opening upwards and having vertex at (0, 1) and axis along positive direction of y-axis.

Region R_1 represents the portion of xy-plane which is on or above x-axis and outside the parabola.

Region R_2 :

We have $\quad y = x + 1 \quad$ ...(2)

Equation (2) represents a straight line passing through (0, 1) and (1, 2).

Region R_2 is the portion of xy-plane which is on or above the x-axis and below the line.

Region R_3 : Clearly, R_3 is the region lying between the lines $x = 0$ *i.e.*, y-axis and $x = 2$.

$\therefore$ Required area = Area of the shaded region

= Area OADE + Area ABCD

$$= \int_0^1 y_{\text{parabola}} \, . dx + \int_1^2 y_{\text{line}} \, . dx$$

$$= \int_0^1 (x^2 + 1) \, dx + \int_1^2 (x + 1) \, dx = \left[\frac{x^3}{3} + x\right]_0^1 + \left[\frac{x^2}{\cdot 2} + x\right]_1^2$$

$$= \left[\left\{\left(\frac{1}{3} + 1\right) - (0 + 0\right\}\right] + \left[\left(\frac{4}{2} + 2\right) - \left(\frac{1}{2} + 1\right)\right] = \frac{4}{3} + 4 - \frac{3}{2} = \frac{8 + 24 - 9}{6}$$

$$= \frac{23}{6} \text{ sq. units.}$$

Example 46. *Make a sketch of the region given below and find its area using integration.*

$$\{(x, y) : 0 \le y \le x^2 + 3,\ 0 \le y \le 2x + 3,\ 0 \le x \le 3\}.$$

Solution. Let $R = \{(x, y) : 0 \le y \le x^2 + 3, 0 \le y \le 2x + 3, 0 \le x \le 3\}$

$\Rightarrow$ $R = \{(x, y) : 0 \le y \le x^2 + 3\} \cap \{(x, y) : 0 \le y \le 2x + 3\} \cap \{(x, y) : 0 \le x \le 3\}$

$\Rightarrow$ $R = R_1 \cap R_2 \cap R_3$ (say).

Now, to find the points of intersection of these curves, let us solve equations $y = x^2 + 3$ and $y = 2x + 3$.

$\therefore$ $x^2 + 3 = 2x + 3$

$\Rightarrow$ $x^2 - 2x = 0$

$\Rightarrow$ $x(x - 2) = 0$

$\Rightarrow$ $x = 0, \quad x = 2$

For $x = 0$, $y = 2(0) + 3 = 3$

For $x = 2$, $y = 2(2) + 3 = 7$

$\therefore$ The points of intersection of these curves are (0, 3) and (2, 7).

Region R_1 :

We have $y = x^2 + 3$

$\Rightarrow$ $x^2 = y - 3$

$\Rightarrow$ $(x - 0)^2 = y - 3$ ...(1)

Equation (1) represents a parabola opening upwards and having vertex at (0, 3), axis along the positive direction of y-axis.

$\therefore$ Region R_1 is the region lying on or above the x-axis and outside the parabola.

Region R_2 :

We have $y = 2x + 3$ ...(2)

Equation (2) represents a straight line intersecting x-axis at $\left(-\frac{3}{2}, 0\right)$ and y-axis at (0, 3).

$\therefore$ Region R_2 is lying on or above x-axis and below the line.

Region R_3 : Region R_3 is lying between the lines $x = 0$ *i.e.*, y-axis and $x = 3$. The rough sketch of these curves is as shown in the figure.

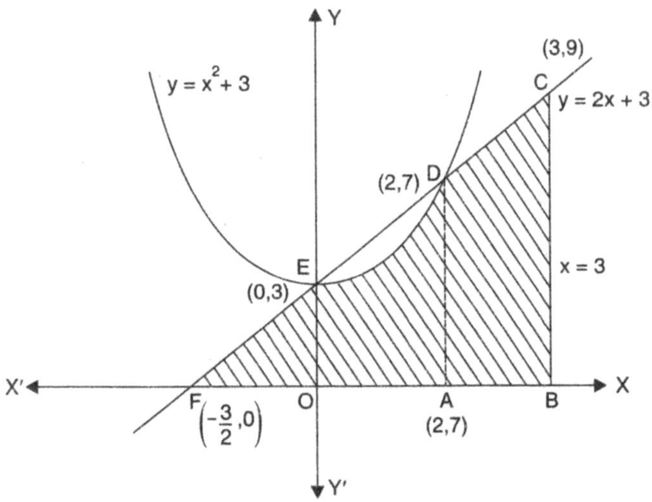

$\therefore$ Required area = Area of the shaded region

= Area OADE + Area ABCD

$$= \int_0^2 (y_{\text{parabola}})\,.\,dx + \int_2^3 (y_{\text{line}})\,.\,dx$$

$$= \int_0^2 (x^2 + 3).\,dx + \int_2^3 (2x + 3)\,dx = \left[\frac{x^3}{3} + 3x\right]_0^2 + \left[\frac{2x^2}{2} + 3x\right]_2^3$$

$$= \left[\left\{\frac{(2)^3}{3} + 3(2)\right\} - \{0+0\}\right] + [\{(3)^2 + 3(3)\} - \{(2)^2 + 3(2)\}]$$

$$= \left(\frac{8}{3} + 6\right) + (9+9) - (4+6) = \frac{8}{3} + 6 + 18 - 10 = \frac{8}{3} + 14 = \frac{8+42}{3}$$

$$= \frac{50}{3} \text{ sq. units.}$$

Example 47. *Find the area of the region :*

$$\{(x, y) : x^2 + y^2 \le 1 \le x + y\}.$$

Solution. Let $R = \{(x, y) : x^2 + y^2 \le 1 \le x + y\}$

$\Rightarrow \quad R = \{(x, y) : x^2 + y^2 \le 1\} \cap \{(x, y) : 1 \le x + y\}$

$\Rightarrow \quad R = R_1 \cap R_2$

where $\quad R_1 = \{(x, y) : x^2 + y^2 \le 1\}$

$\quad R_2 = \{(x, y) : 1 \le x + y\}.$

Now, to find the points of intersection of the two curves let us solve the equations $x^2 + y^2 = 1$ and $x + y = 1$.

$\therefore \quad y = 1 - x$

$\Rightarrow \quad x^2 + (1 - x)^2 = 1$

$\Rightarrow \quad x^2 + 1 + x^2 - 2x = 1$

$\Rightarrow \quad 2x^2 - 2x = 0$

$\Rightarrow \quad 2x(x - 1) = 0$

$\Rightarrow \quad x = 0, x = 1$

For $x = 0$, $\quad y = 1 - 0 = 1$

For $x = 1$, $\quad y = 1 - 1 = 0$

$\therefore$ The points of intersection of the two curves are (0, 1) and (1, 0).

The rough sketch of the two curves is as shown in the figure.

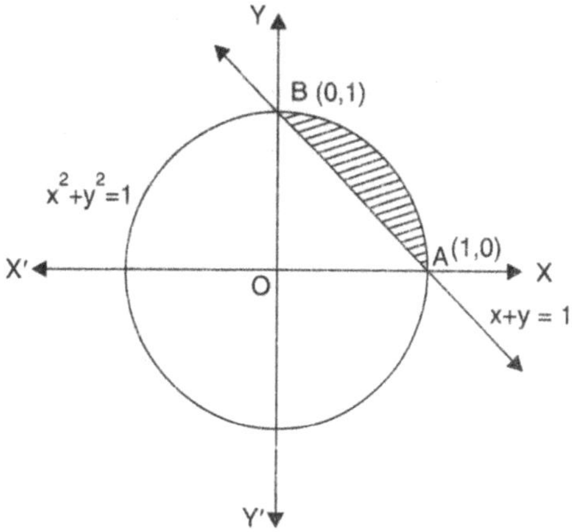

Region R_1 :

We have $\quad x^2 + y^2 = 1 \quad$ ...(1)

Equation (1), represents a circle with centre at the origin (0, 0) and radius 1.

Region R_1 represents the interior of the circle $x^2 + y^2 = 1$.

Region R_2 :

We have $\quad x + y = 1 \quad$ ...(2)

Equation (2), represents a straight line passes through the points (0, 1) and (1, 0). R_2 is the region lying above the line $x + y = 1$.

$\therefore$ Required area = Area of the shaded region.

$$= \int_0^1 (y_{\text{circle}} - y_{\text{line}}) \, . \, dx$$

$$= \int_0^1 [(\sqrt{1 - x^2}) - (1 - x)] \, dx = \int_0^1 \left(\sqrt{1 - x^2} - 1 + x\right) dx$$

$$= \left[\frac{x\sqrt{1-x^2}}{2} + \frac{1}{2}\sin^{-1} x - x + \frac{x^2}{2} \right]_0^1$$

$$\left[\because \int \sqrt{a^2 - x^2}\,.dx = \frac{x\sqrt{a^2-x^2}}{2} + \frac{a^2}{2}\sin^{-1}\frac{x}{a} \right]$$

$$= \left[\left\{ \frac{1\sqrt{1-1}}{2} + \frac{1}{2}\sin^{-1} 1 - 1 + \frac{1}{2} \right\} - \left\{ 0 + \frac{1}{2}\sin^{-1} 0 - 0 + 0 \right\} \right]$$

$$= \left[0 + \frac{1}{2}\left(\frac{\pi}{2}\right) - \frac{1}{2} - 0 \right] = \left(\frac{\pi}{4} - \frac{1}{2} \right) \text{ sq. units.}$$

Example 48. *Find the area of the region :*

$\{(x, y) : x^2 + y^2 \le 2ax,\ y^2 \ge ax,\ x \ge 0,\ y \ge 0\}$.

Solution. Let $R = \{(x, y) : x^2 + y^2 \le 2ax,\ y^2 \ge ax,\ x \ge 0,\ y \ge 0\}$

$\Rightarrow \quad R = \{(x, y) : x^2 + y^2 \le 2ax\} \cap \{(x, y) : y^2 \ge ax\} \cap \{(x, y) : x \ge 0, y \ge 0\}$

$\Rightarrow \quad R = R_1 \cap R_2 \cap R_3$.

where $\quad R_1 = \{(x, y) : x^2 + y^2 \le 2ax\}$

$R_2 = \{(x, y) : y^2 \ge ax\}$

$R_3 = \{(x, y) : x \ge 0, y \ge 0\}$.

Now, to find the points of intersection of these curves, let us solve the equations

$x^2 + y^2 = 2ax \quad$ and $\quad y^2 = ax$

$\Rightarrow \quad x^2 + ax = 2ax$

$\Rightarrow \quad x^2 - ax = 0$

$\Rightarrow \quad x(x - a) = 0$

$\Rightarrow \quad x = 0, \quad x = a$

For $x = 0$; $\quad y = 0$

For $x = a$; $\quad y = a$

$\therefore$ The points of intersection of these curves are (0, 0) *i.e.*, the origin and (a, a).

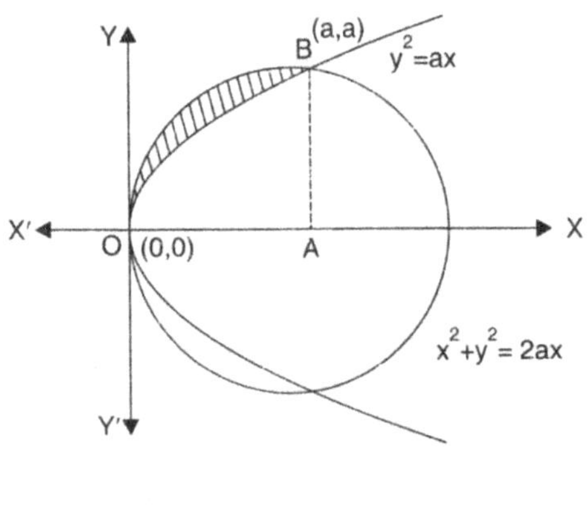

The rough sketch of these curves is as shown in the figure.

Region R_1 :

We have $\quad x^2 + y^2 = 2ax \quad$ ...(1)

$\Rightarrow \quad x^2 - 2ax + y^2 = 0$

Adding a^2 on both sides

$\Rightarrow \quad x^2 - 2ax + a^2 + y^2 = 0 + a^2$

$\Rightarrow \quad (x - a)^2 + (y - 0)^2 = a^2 \quad$ ...(2)

It represents a circle with centre at $(a, 0)$ and radius a.

$\therefore$ Region R_1 is the region inside the circle.

Region R_2 :

We have $\quad y^2 = ax \quad$ ...(3)

Equation (3) represents a parabola with vertex at (0, 0) *i.e.*, at the origin and its axis along x-axis.

$\therefore$ Region R_2 is lying outside the parabola.

Region R_3 :

It is given that $x \geq 0$ and $y \geq 0$.

$\therefore$ Region R_3 is the region in the first quadrant.

$\therefore$ Required area = Area of shaded region

$$= \int_0^a (y_{\text{upper curve}} - y_{\text{lower curve}}) . dx$$

$$= \int_0^a (y_{\text{circle}} - y_{\text{parabola}})\, dx$$

$$= \int_0^a \left(\sqrt{2ax - x^2} - \sqrt{ax}\right) dx = \int_0^a \sqrt{2ax - x^2} . dx - \sqrt{a} \int_0^a \sqrt{x} . dx$$

$$= \int_0^a \sqrt{a^2 - (x-a)^2} . dx - \sqrt{a} \int_0^a x^{1/2}\, dx$$

$$= \left[\frac{(x-a)\sqrt{a^2 - (x-a)^2}}{2} + \frac{a^2}{2} \sin^{-1}\left(\frac{x-a}{a}\right)\right]_0^a - \sqrt{a}\left[\frac{x^{3/2}}{\frac{3}{2}}\right]_0^a$$

$$\left[\because \int \sqrt{a^2 - x^2} . dx = \frac{x\sqrt{a^2 - x^2}}{2} + \frac{a^2}{2} \sin^{-1} \frac{x}{a}\right]$$

$$= \left[\left\{\frac{(a-a)\sqrt{a^2 - (a-a)^2}}{2} + \frac{a^2}{2} \sin^{-1}\left(\frac{a-a}{a}\right)\right\} - \left\{\frac{(0-a)\sqrt{a^2 - (0-a)^2}}{2}\right.\right.$$

$$\left.\left. + \frac{a^2}{2} \sin^{-1}\left(\frac{0-a}{a}\right)\right\}\right] - \frac{2\sqrt{a}}{3} [a^{3/2} - 0]$$

$$= \left[\left\{0 + \frac{a^2}{2} \sin^{-1} (0)\right\} - \left\{0 + \frac{a^2}{2} \sin^{-1} (-1)\right\}\right] - \frac{2a^2}{3}$$

$$= \frac{a^2}{2} (0) + \frac{a^2}{2} \sin^{-1} (1) - \frac{2a^2}{3} = \frac{a^2}{2}\left(\frac{\pi}{2}\right) - \frac{2a^2}{3} = \frac{a^2\pi}{4} - \frac{2a^2}{3}$$

$$= a^2\left(\frac{\pi}{4} - \frac{2}{3}\right) \text{ sq. units.}$$

Example 49. *Find the area enclosed by the region :*

$\{(x, y) : y^2 \leq 5x, 5x^2 + 5y^2 \leq 36\}$.

Solution. Let $R = \{(x, y) : y^2 \leq 5x, 5x^2 + 5y^2 \leq 36\}$

$\Rightarrow \quad R = \{(x, y) : y^2 \leq 5x\} \cap \{(x, y) : 5x^2 + 5y^2 \leq 36\}$

$\Rightarrow \quad R = R_1 \cap R_2$

where $\quad R_1 = \{(x, y) : y^2 \leq 5x\}$

$R_2 = \{(x, y) : 5x^2 + 5y^2 \leq 36\}$

Region R_1 :

We have $y^2 = 5x$ ...(1)

Equation (1) represents a parabola with vertex at origin (0, 0) and x-axis as its axis.

The curve is symmetrical about x-axis. Region R_1 is lying in xy-plane and inside this parabola.

Region R_2 :

We have $5x^2 + 5y^2 = 36$ ...(2)

Equation (2) represents a circle with centre at origin (0, 0) and radius $\frac{6\sqrt{5}}{5}$.

Region R_2 is the region in xy-plane and is inside this circle.

Now, to find the points of intersection of two curves, let us solve the equations (1) and (2) simultaneously:

$$y^2 = 5x$$

$$\Rightarrow \quad 5x^2 + 5(5x) = 36$$

$$\Rightarrow \quad 5x^2 + 25x - 36 = 0$$

$$\Rightarrow \quad x = \frac{-25 \pm \sqrt{625 + 720}}{10}$$

$$\Rightarrow \quad x = \frac{-25 \pm \sqrt{1345}}{10}$$

$\left[\because \text{ For a quadratic equation : } ax^2 + bx + c = 0. \quad x = \frac{-b \pm \sqrt{b^2 - 4ac}}{2a} \right]$

$$\Rightarrow \quad x = \frac{-25 \pm 36.6}{10}$$

$$\Rightarrow \quad x = -\frac{61.6}{10}, \quad x = \frac{11.6}{10}$$

$$\Rightarrow \quad x = -6.16, \quad x = 1.16$$

For $x = -6.16$; $y^2 = 5(-6.16)$

$\Rightarrow$ $y^2 = -30.18$ $\Rightarrow$ y has imaginary values.

For $x = 1.16$; $y^2 = 5(1.16) = 5.8$ $\Rightarrow$ $y = \pm 2.4$

$\therefore$ The points of intersection of two curves are (1.16, 2.4) and (1.16, – 2.4).

The rough sketch of these curves is as shown in the figure :

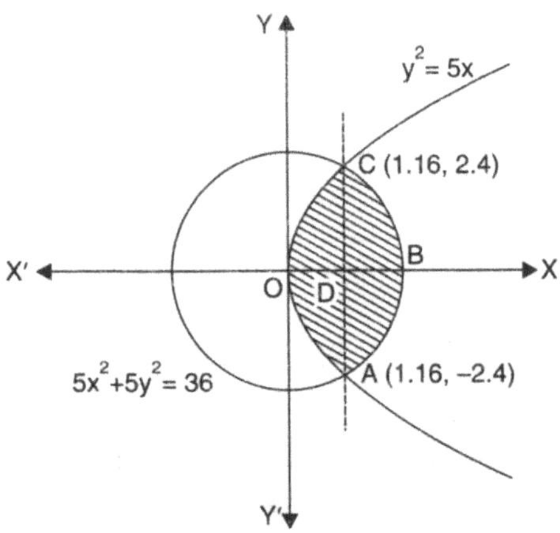

$\therefore$ Required area = Area of the shaded region

= Area OABC

= 2(Area OBC)

= 2[Area OCD + Area DCB]

$$= 2\int_0^{1.16} y_{\text{parabola}} \cdot dx + 2\int_{1.16}^{\frac{6\sqrt{5}}{5}} y_{\text{circle}} \cdot dx$$

$$= 2\int_0^{1.16} \sqrt{5x}\,.dx + \int_{1.16}^{\frac{6\sqrt{5}}{5}} \sqrt{\frac{36}{5} - x^2}\,.dx$$

$$= 2\sqrt{5}\left[\frac{x^{\frac{1}{2}+1}}{\frac{1}{2}+1}\right]_0^{1.16} + 2\left[\frac{x\sqrt{\frac{36}{5} - x^2}}{2} + \frac{\left(\sqrt{\frac{36}{5}}\right)^2}{2}\sin^{-1}\frac{x}{\left(\frac{6}{\sqrt{5}}\right)}\right]_{1.16}^{\frac{6\sqrt{5}}{5}}$$

$$\left[\because \int\sqrt{a^2 - x^2}\,dx = \frac{x\sqrt{a^2 - x^2}}{2} + \frac{a^2}{2}\sin^{-1}\frac{x}{a}\right]$$

$$= 2\sqrt{5}\,.\frac{2}{3}\left[x^{3/2}\right]_0^{1.16} + \left[x\sqrt{\frac{36}{5} - x^2} + \frac{36}{5}\sin^{-1}\frac{\sqrt{5}\,x}{6}\right]_{1.16}^{\frac{6\sqrt{5}}{5}}$$

$$= \frac{4\sqrt{5}}{3}\left[(1.16)^{3/2} - 0\right] + \left[\left\{\frac{6\sqrt{5}}{5}\sqrt{\frac{36}{5} - \left(\frac{6\sqrt{5}}{5}\right)^2} + \frac{36}{5}\sin^{-1}\left(\frac{\sqrt{5}\left(\frac{6\sqrt{5}}{5}\right)}{6}\right)\right\}\right.$$

$$\left. - \left\{1.16\sqrt{\frac{36}{5} - (1.16)^2} + \frac{36}{5}\sin^{-1}\frac{\sqrt{5}\,(1.16)}{6}\right\}\right]$$

[where $z = 1.16$]

$$= \frac{4\sqrt{5}}{3}[z^{3/2}] + \left[\left\{\frac{6\sqrt{5}}{5}\sqrt{\frac{36}{5} - \frac{36}{5}} + \frac{36}{5}\sin^{-1}1\right\} - \left\{z\sqrt{\frac{36}{5} - z^2} + \frac{36}{5}\sin^{-1}\frac{\sqrt{5}\,z}{6}\right\}\right]$$

$$= \frac{4\sqrt{5}\,z^{3/2}}{3} + \frac{36}{5}\left(\frac{\pi}{2}\right) - z\sqrt{\frac{36}{5} - z^2} - \frac{36}{5}\sin^{-1}\frac{\sqrt{5}\,z}{6}$$

$$= \left[\frac{4\sqrt{5}\,z^{3/2}}{3} + \frac{18\pi}{5} - z\sqrt{\frac{36}{5} - z^2} - \frac{36}{5}\sin^{-1}\left(\frac{z\sqrt{5}}{6}\right)\right] \text{ sq. units.}$$

Example 50. *Find the area of the region* $\{(x, y) : y^2 \le 3x,\ 3x^2 + 3y^2 \le 16\}$.

Solution. Let $R = \{(x, y) : y^2 \le 3x,\ 3x^2 + 3y^2 \le 16\}$.

$\Rightarrow$ $R = \{(x, y) : y^2 \le 3x\} \cap \{(x, y) : 3x^2 + 3y^2 \le 16\}$

$\Rightarrow$ $R = R_1 \cap R_2$ (say)

where $R_1 = \{(x, y) : y^2 \le 3x\}$

$R_2 = \{(x, y) : 3x^2 + 3y^2 \le 16\}$.

Region R_1 :

We have $y^2 = 3x$ ...(1)

Equation (1) represents a parabola with vertex at origin (0, 0) and x-axis as its x-axis. Region R_1 is the portion in xy-plane which is on or inside the parabola.

Region R_2 :

We have $3x^2 + 3y^2 = 16$ ...(2)

$$\Rightarrow \quad x^2 + y^2 = \frac{16}{3}$$

$$\Rightarrow \quad x^2 + y^2 = \left(\frac{4}{\sqrt{3}}\right)^2 \qquad \text{...(3)}$$

Equation (2) represents a circle with centre at origin (0, 0) and radius $\left(\frac{4\sqrt{3}}{3}\right)$.

Region R_2 is the portion of xy-plane which is on or inside the circle.

Now, to find the points of intersection of these two curves, let us solve equations (1) and (2) simultaneously.

$$y^2 = 3x$$

$$\Rightarrow \quad 3x^2 + 3(3x) = 16$$

$$\Rightarrow \quad 3x^2 + 9x - 16 = 0$$

$$\Rightarrow \quad x = \frac{-9 \pm \sqrt{81 + 192}}{6}$$

$$\left[\because \text{ For a quadratic equation } ax^2 + bx + c = 0 \quad x = \frac{-b \pm \sqrt{b^2 - 4ac}}{2a}\right]$$

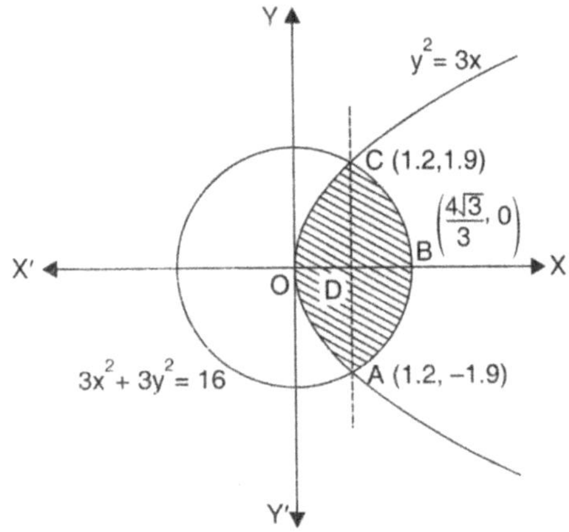

$$\Rightarrow \quad x = \frac{-9 \pm \sqrt{273}}{6}$$

$$\Rightarrow \quad x = \frac{-9 \pm 16.5}{6}$$

$$\Rightarrow \quad x = -\frac{25.5}{6}, \quad x = \frac{7.5}{2}$$

$$\Rightarrow \quad x = -4.2, \quad x = 1.2$$

For $x = -4.2$, $y^2 = 3(-4.2) = -12.6 \Rightarrow y$ has imaginary values.

For $x = 1.2$, $\Rightarrow y^2 = 3(1.2) = 3.6 \Rightarrow y = \pm 1.9$

$\therefore$ The points of intersection of two curves are (1.2, – 1.9) and (1.2, 1.9).

The rough sketch of the two curves is as shown in the figure.

$\therefore$ Required area = Area of the shaded region

= Area OABC

= 2[Area OBC] = 2[Area ODC + Area DCB]

$$= 2\int_0^{1.2} y_{\text{parabola}} \, . dx + 2\int_{1.2}^{\frac{4\sqrt{3}}{3}} y_{\text{circle}} \, . dx$$

$$= 2\int_0^{1.2} \sqrt{3x}\, dx + 2\int_{1.2}^{\frac{4\sqrt{3}}{3}} \sqrt{\frac{16}{3} - x^2} \, . dx$$

$$= 2\sqrt{3}\int_0^{1.2} \sqrt{x}\, dx + 2\int_{1.2}^{\frac{4\sqrt{3}}{3}} \sqrt{\left(\sqrt{\frac{16}{3}}\right)^2 - x^2} \, . dx$$

$$= 2\sqrt{3}\left[\frac{x^{3/2}}{\frac{3}{2}}\right]_0^{1.2} + 2\left[\frac{x\sqrt{\frac{16}{3} - x^2}}{2} + \frac{\frac{16}{3}}{2}\sin^{-1}\left(\frac{x}{\frac{4}{\sqrt{3}}}\right)\right]_{1.2}^{\frac{4\sqrt{3}}{3}}$$

$$\left[\because \int\sqrt{a^2 - x^2}\,dx = \frac{x\sqrt{a^2 - x^2}}{2} + \frac{a^2}{2}\sin^{-1}\frac{x}{a}\right]$$

$$= \frac{4\sqrt{3}}{3}[(1.2)^{3/2} - 0] + \left[\left\{\frac{4\sqrt{3}}{3}\sqrt{\frac{16}{3} - \left(\frac{4\sqrt{3}}{3}\right)^2} + \frac{16}{3}\sin^{-1}\frac{\sqrt{3}}{4}\frac{4\sqrt{3}}{3}\right\}\right.$$

$$\left. - \left\{(1.2)\sqrt{\frac{16}{3} - (1.2)^2} + \frac{16}{3}\sin^{-1}\frac{\sqrt{3}\,(1.2)}{4}\right\}\right]$$

$$= \frac{4\sqrt{3}}{3}(1.31) + \left[\left\{\frac{4\sqrt{3}}{3}\sqrt{\frac{16}{3} - \frac{16}{3}} + \frac{16}{3}\sin^{-1} 1\right\} - \left\{1.2\sqrt{\frac{16}{3} - 1.44} + \frac{16}{3}\sin^{-1}\frac{3\sqrt{3}}{10}\right\}\right]$$

$$= 3.02 + 0 + \frac{16}{3}\left(\frac{\pi}{2}\right) - 2.36 + \frac{16}{3}\sin^{-1}\frac{3\sqrt{3}}{10}$$

$$= \left[0.66 + \frac{8\pi}{3} + \frac{16}{3}\sin^{-1}\frac{3\sqrt{3}}{10}\right] \text{ sq. units.}$$

EXERCISE FOR PRACTICE

1. Find the area bounded by the curve $y = x^3$, x-axis and the ordinates $x = 2$, $x = 4$.
2. Using integration, find the area of the region bounded by the curves $y = 1 + |x + 1|$, $x = -3$, $x = 3$, $y = 0$.
3. Draw a rough sketch of the curve $y = x^2 - 9$ and find the area bounded by the curve, the lines $x = 0$, $x = 2$ and the x-axis.
4. Sketch the region bounded by $y = 2x - x^2$ and x-axis, and find its area using integration.
5. Find the area under the curve $y = (x^2 + 2)^2 + 2x$, between the ordinates $x = 0$, $x = 2$ and the x-axis.
6. Using integration, find the area of the region bounded by the line $3y = 2x + 4$, x-axis and the lines $x = 1$ and $x = 3$.
7. Find the area lying above the x-axis and under the parabola $y = 4x - x^2$.
8. Draw a rough sketch of the graph of the curve $\frac{x^2}{4} + \frac{y^2}{9} = 1$ and evaluate the area of the region under the curve and above the x-axis.
9. Find the area bounded by the curve $y = \cos x$, x-axis and the ordinates $x = 0$ and $x = 2\pi$.
10. Sketch the graph of $y = |x - 5|$. Evaluate $\int_0^1 |x - 5|\,dx$, what does this value of the integral represent on the graph.

11. Draw the rough sketch of $y^2 + 1 = x$, $x \le 2$. Find the area enclosed by the curve and the line $x = 2$.
12. Using integration, find the area of the region bounded between the line $x = 2$ and the parabola $y^2 = 8x$.
13. Find the area of the region common to the parabolas $4y^2 = 9x$ and $3x^2 = 16y$.
14. Find the area of the region enclosed between the circles $x^2 + y^2 = 1$, $(x - 1)^2 + y^2 = 1$.
15. Using integration, find the area of the region bounded by the triangle whose vertices are (2, 1), (3, 4) and (5, 2).
16. Find the area enclosed between two curves $y^2 = 9x$ and $x^2 = 9y$.
17. Find the area common to the circle $x^2 + y^2 = 16a^2$ and the parabola $y^2 = 6ax$.
18. Find the area bounded by the lines
$x + 2y = 2$, $y - x = 1$ and $2x + y = 7$.
19. Compare the areas under the curves
$y = \cos^2 x$ and $y = \sin^2 x$ between $x = 0$ and $x = \pi$.
20. Find the area of the region :
$\{(x, y) : x^2 + y^2 \le 1 \le x + y\}$.
21. Find the area of the region enclosed by the parabola $x^2 = y$ and the line $y = x + 2$.
22. Find the area bounded by the curve $y^2 = 4a^2(x - 1)$ and the line $x = 1$ and $y = 4a$.
23. Find the area enclosed by the parabolas
$y = 5x^2$ and $y = 2x^2 + 9$.
24. Find the area of the region $\{(x, y) : y^2 \le 5x, 5x^2 + 5y^2 \le 36\}$.

Answers

1. 60 sq. units
2. 16 sq. units
3. $\frac{46}{3}$ sq. units
4. $\frac{4}{3}$ sq. units
5. $\frac{436}{15}$ sq. units
6. $\frac{16}{3}$ sq. units
7. $\frac{32}{3}$ sq. units
8. 3π sq. units
9. 4 sq. units
10. $\frac{9}{2}$ sq. units
11. $\frac{4}{3}$ sq. units
12. $\frac{32}{3}$ sq. units
13. 4 sq. units
14. $\frac{4}{3}$ sq. units
15. 4 sq. units
16. 27 sq. units
17. $\frac{4a^2}{3}(4\pi + \sqrt{3})$ sq. units
18. 6 sq. units
19. $\frac{\pi}{2}$ sq. units [Each]
20. $\left(\frac{\pi - 2}{4}\right)$ sq. units
21. $\frac{50}{3}$ sq. units
22. $\frac{16a}{3}$ sq. units
23. $12\sqrt{3}$ sq. units
24. $\left[\frac{4\sqrt{5}}{3} z^{3/2} + \frac{18\pi}{5} - z\sqrt{\frac{36}{5} - z^2} - \frac{36}{5}\sin^{-1}\frac{z\sqrt{15}}{6}\right]$ sq. units

where $z = 1.16$.

5

Differential Equations

5.1. INTRODUCTION

It is a well-accepted fact that differential equations are now employed in modelling almost all natural phenomena, whether of Physics, Chemistry, Biology, Anthropology, Geology etc. or of Social sciences, or even of Economics. In fact, a Mathematical model of any natural phenomenon comprises differential equations. Hence, an in-depth study of differential equations has assumed prime importance in all modern scientific investigations.

5.1.1 Definition : Differential Equation. An equation involving independent variables, dependent variables and their derivatives (differentials) is called a differential equation.

Following are a few examples of differential equations :

(*i*) $\frac{dy}{dx} = x^2 + 1$ (*ii*) $3\frac{dy}{dx} + 5y = 2$

(*iii*) $\frac{dy}{dx} = \sin x$ (*iv*) $6\frac{d^2y}{dx^2} + 4\frac{dy}{dx} + 3y = \cos x$

(*v*) $y = x\frac{dy}{dx} + \sqrt{1+\left(\frac{dy}{dx}\right)^2}$ (*vi*) $xdx + ydy = 0$

(*vii*) $(x + y)^2 \frac{dy}{dx} = 1$ (*viii*) $\left[1+\left(\frac{dy}{dx}\right)^2\right]^3 = 3\left(\frac{d^2y}{dx^2}\right)^2.$

5.2. TYPES OF DIFFERENTIAL EQUATIONS

5.2.1 Ordinary Differential Equation. A differential equation which involves only one independent variable is called an ordinary equation.

All the above cited examples are the examples of ordinary differential equations.

5.2.2 Partial Differential Equation. A differential equation having more than one independent variable is called a partial differential equation.

e.g.,
$$x.\frac{\partial z}{\partial x} + y\frac{\partial z}{\partial y} = 2z$$

$$x.\frac{\partial x}{\partial y} + y.\frac{\partial y}{\partial z} = pz \text{ etc.}$$

In our present course, we shall be considering only ordinary differential equations.

5.3. ORDER AND DEGREE OF DIFFERENTIAL EQUATIONS

5.3.1 Order. The order of a differential equation is the order of the highest order derivative occuring in the equation.

The order of a differential equation is a positive integer.

5.3.2 Degree. The degree of a differential equation is the degree of the highest order differential co-efficient appearing in it, provided the differential co-efficients are made free from radicals and fractions.

e.g., (*i*) $\frac{d^2y}{dx^2} + 3\frac{dy}{dx} + 2y = 0$. Its order is 2 and degree is 1.

(*ii*) $\left[1+\left(\frac{dy}{dx}\right)^2\right]^3 = 2\left(\frac{d^2y}{dx^2}\right)^2$. Its order is 2 and degree is 2.

(*iii*) $\frac{dy}{dx} = \sin x + \cos x$. Its order is 1 and degree is 1.

(*iv*) $\left[1+\left(\frac{dy}{dx}\right)^2\right]^{\frac{3}{2}} = p\left(\frac{d^2y}{dx^2}\right)$

The order of highest order differential co-efficient is 2. So, its order is 2.

Now, to find its degree, first we have to express it as a polynomial in derivatives.

i.e., on squaring both sides, we have

$$\left[1+\left(\frac{dy}{dx}\right)^2\right]^3 - p^2\left(\frac{d^2y}{dx^2}\right)^2 = 0$$

Here, the power of the highest order differential co-efficient is 2.

Therefore, its degree is 2.

5.4. LINEAR AND NON-LINEAR DIFFERENTIAL EQUATIONS

A differential equation is said to be linear, if the dependent variable and its derivatives occur only in the first degree and are not multiplied together.

i.e., if it is expressible in the form

$$A_0\frac{d^ny}{dx^n} + A_1\frac{d^{n-1}y}{dx^{n-1}} + A_2\frac{d^{n-2}y}{dx^{n-2}} + \ldots\ldots A_{n-1}\frac{dy}{dx} + A_ny = B$$

where $A_0, A_1, A_2, \ldots\ldots, A_{n-1}, A_n$ and B are either constants or functions of independent variable x.

In particular, a linear differential equation of order one is of the form

$$A_0 \frac{dy}{dx} + A_1 y = B$$

Note. The degree of a linear differential equation is always one. But, the converse is not true.

e.g., $$x \frac{dy}{dx} + y \cos x = 1$$

and $$\frac{d^2 y}{dx^2} + \frac{5y}{x} = x^2 \sin x$$ are linear differential equations.

The differential equation $y \frac{dy}{dx} - p = x$ is not a linear differential equation, because the dependent variable y and its derivative $\frac{dy}{dx}$ are multiplied together.

A differential equation which is not linear is called a Non-linear Differential Equation.

i.e., A differential equation will be non-linear differential equation if :

(*i*) its degree is more than one.

(*ii*) any of the differential co-efficient has exponent more than one.

(*iii*) exponent of the dependent variable is more than one.

(*iv*) products containing dependent variable and its differential co-efficients are present.

e.g., $$\left(\frac{d^2 y}{dx^2}\right)^2 + 3\left(\frac{dy}{dx}\right)^2 + 5 = 0.$$

Order of highest order derivative *i.e.,* of $\left(\frac{d^2 y}{dx^2}\right)^2$ is 2. Degree of $\left(\frac{d^2 y}{dx^2}\right)$ is 2.

$\therefore$ Order of the equation is 2 and degree of the equation is 2.

This is an example of a non-linear differential equation, because $\left(\frac{d^2 y}{dx^2}\right)$ is multiplied by itself.

5.5. FORMATION OF A DIFFERENTIAL EQUATION

Let there be an equation involving independent variable y, the dependent variable x and 'n' arbitrary constants, to form the differential equation of such family of curves we are to eliminate the 'n' arbitrary constants from the given equation.

This can be achieved by differentiating given equation n times and, we get a differential equation of nth order corresponding to the given equation.

If n be an arbitrary constant and

$$y = n \sin 6x \quad \ldots(1)$$

be an equation.

For different values of n, we get different equations.

If we eliminate the constant n by differentiating the equation (1), then the equation, which we get is called the differential equation of the given equation.

Differentiating equation (1) w.r.t. x, we have

$$\frac{dy}{dx} = 6n \cos 6x \qquad ...(2)$$

$$\Rightarrow \qquad n = \frac{1}{6 \cos 6x} \cdot \frac{dy}{dx}$$

Putting this value of n in equation (1), we get

$$y = \frac{1}{6 \cos 6x} \cdot \frac{dy}{dx} \sin 6x = \frac{1}{6} \tan 6x \frac{dy}{dx}$$

$$\Rightarrow \qquad \frac{dy}{dx} = \frac{6y}{\tan 6x} = 6y \cot 6x$$

which is the required differential equation.

Hence, by eliminating one arbitrary constant, a differential equation of first order is obtained.

Remark. Constant numbers are of two types :

(*i*) Pure constants. (*ii*) Arbitrary constants.

(*i*) **Pure constants :** Pure constants are fixed forever.

e.g., 5, 3, – 2, π etc.

(*ii*) **Arbitrary constants :**

Let $\qquad y = mx \qquad ...(1)$

Clearly, equation (1) represents a straight line passing through the origin.

For different values of m say (1, 2, 3 etc.), we get different straight lines.

i.e., $\qquad y = x, \quad y = 2x \quad$ and $\quad y = 3x.$

passing through the origin.

$\therefore$ Equation (1) represents a family of straight lines for different values of m.

Here, m is an arbitrary constant, x and y are variables.

The differential equation which represents a family of curves is free from arbitrary constants and its order is equal to the number of arbitrary constants in the equation of the family.

Arbitrary constants are also called parameters.

5.6. SOLUTION OF A DIFFERENTIAL EQUATION

The solution of a differential equation is a functional relation between the variables, not involving the differential co-efficients, such that this relation and the derivatives obtained from it must statisfy the given differential equation.

Illustration :

1. *Verify that $y = A \cos x + B \sin x$ is a solution of the differential equation $\frac{d^2y}{dx^2} + y = 0$.*

Solution. Let us consider that

$$y = A \cos x + B \sin x \qquad ...(1)$$

where A and B are arbitrary constants.

Differentiating (1) w.r.t. x, we get

$$\frac{dy}{dx} = -A \sin x + B \cos x \qquad ...(2)$$

Differentiating (2) w.r.t. x, we get

$$\frac{d^2y}{dx^2} = -A\cos x - B\sin x$$

$$= -(A\cos x + B\sin x) = -y \qquad \text{[Using (1)]}$$

$$\Rightarrow \qquad \frac{d^2y}{dx^2} + y = 0 \qquad ...(3)$$

This shows that $y = A\cos x + B\sin x$ satisfies the given differential equation and hence is a solution of given equation.

5.6.1 Types of Solutions. There are three types of solutions :

(*a*) **General Solution.** A solution of a differential equation is called the general solution (or complete solution), if it contains as many arbitrary constants as the order of differential equation.

e.g., $y = A\cos x + B\sin x$, is a general solution of the differential equation

$$\frac{d^2y}{dx^2} + y = 0.$$

because the number of arbitrary constants A and B *i.e.*, 2 is same as the order of differential equation.

(*b*) **Particular Solution.** A solution obtained by giving particular values to arbitrary constants in the general solution of a differential equation is called a particular solution of the differential equation, under consideration.

e.g., $y = 2\cos x + 3\sin x$ is a particular solution of the given differential equation

$$\frac{d^2y}{dx^2} + y = 0.$$

e.g., $y = 7x^4$ is a particular solution of the differential equation $x\frac{dy}{dx} - 4y = 0$.

(*c*) **Singular Solution.** The solution containing no arbitrary constant and which cannot be obtained by giving particular values to arbitrary constants in the general solution is called a singular solution.

e.g., (*i*) $y^2 = 4ax$ is the singular solution of the differential equation $y = x\frac{dy}{dx} + \frac{a}{\frac{dy}{dx}}$.

5.6.2 Note. 1. The finding of a solution of a given differential equation is generally referred to as solving the differential equation or integrating the differential equation. The process involves two steps :

(*a*) First to find the solution of the given differential equation *i.e.*, To find a relation, connecting the independent variable (say) x, and the dependent variable (say) y such that the relation (called the solution or primitive) does not contain any derivative.

(*b*) Second, to derive from the solution, if possible, y in terms of x or x in terms of y.

2. Suppose a solution of a differential equation contains n-arbitrary constants. To eliminate n arbitrary constants, we need $(n + 1)$ equations. The given relation along with n more relations obtained by successively differentiating it n-times provide us with $(n + 1)$ equations. The differential equation thus obtained is clearly of the nth order.

   Hence, a solution of nth order differential equation contain n arbitrary constants.

5.7. INITIAL VALUE PROBLEM

We have seen that a first order differential equation represents a one-parameter family of curves, a second-order differential equation represents a two-parameters family of curves, and so on. In order to specify a particular member of such a family, we need to assign particular value (*s*) to the parameter (*s*) involved.

So we require, besides the differential equation, some other conditions (known as subsidiary conditions) for the specification of the parameter. Usually, these subsidiary conditions are prescribed by assigning values to the unknown function and the requisite number (depending on the number of parameters to be determined) of the derivatives of the unknown function at some point of the domain of its definition. *e.g.*,

(*i*) Let the function $f(x) = 3x + 2$. Satisfies $f'(x) = 3$ and $f(1) = 5$.

Here, the function $f(x) = 3x + A$ is the solution of the differential equation

$$f'(x) = 3$$

where A is a parameter.

(*ii*) Let the function $f(x) = 3x^2 + 3$ satisfies $f'(x) = 6x$ and $f(1) = 6$.

Here, the function $f(x) = 3x + B$ is the solution of the differential equation

$$f'(x) = 6x$$

where B is the parameter.

These subsidiary conditions are generally prescribed of only one point of the domain of definition of the unknown function, these conditions are referred to as the initial conditions.

The differential equation with these initial values or initial conditions is generally known as an initial value problem.

5.8. DIFFERTIAL EQUATIONS OF THE FIRST ORDER AND FIRST DEGREE

5.8.1 Definition. A differential equation of first order and first degree is an equation of the form

$$\frac{dy}{dx} = f(x, y) \quad \text{or} \quad Mdx + Ndy = 0$$

where M, N are functions of x and y.

5.9. METHODS OF SOLVING A FIRST ORDER AND FIRST DEGREE DIFFERENTIAL EQUATION

Differential equations with variable separable :

5.9.1 Definition. If in a differential equation, it is possible to get all the functions of x and dx to one side, and all the functions of y and dy to the other, the variables are said to be separable.

5.9.2 Solution of a Differential Equation by the Method of Variables Separable. Let us consider the differential equation :

$$\frac{dy}{dx} = f(x)\, g(y) \qquad ...(1)$$

where $f(x)$ and $g(y)$ are some functions of x and y respectively.

Now, let us bring the terms containing y on one side and the terms containing x on the other side.

We have $$\frac{dy}{g(y)} = f(x)dx \qquad ...(2) \quad \text{[Variables have been separated]}$$

Now, integrating both sides of equation (2), we get

$$\int \frac{dy}{g(y)} = \int f(x)\, dx + C \qquad ...(3)$$

where C is an arbitrary constant.

This represents the general solution of the differential equation (1).

Special Cases :

Case I. If $g(y) = 1$

$\therefore$ From equation (1), we have

$$\frac{dy}{dx} = f(x)$$

$$\Rightarrow \quad dy = f(x)dx \qquad \text{[Variables are separated]}$$

On integrating, we have :

$$\int dy = \int f(x)\, dx + C$$

$$\Rightarrow \quad y = \int f(x)\, dx + C \qquad ...(4)$$

where C is an arbitrary constant.

Case II. If $f(x) = 1$

$\therefore$ From equation (1), we have

$$\frac{dy}{dx} = g(y)$$

$$\Rightarrow \quad \frac{dy}{g(y)} = dx \qquad \text{[Variables are separated]}$$

On integrating, we have

$$\int \frac{dy}{g(y)} = \int dx + C$$

$$\Rightarrow \quad \int \frac{1}{g(y)}\, dy = x + C \qquad ...(5)$$

where C is an arbitrary constant.

5.9.3 Note. 1. Never forget to add an arbitrary constant on one side (only). A solution without this constant is wrong, for it is not the general solution.

2. The nature of the arbitrary constant depends upon the nature of the problem.

3. The solution of a differential equation must be put in a form as simple as possible.

4. **Remember :**

(i) $\log x + \log y = \log xy$

(ii) $\log x - \log y = \log \dfrac{x}{y}$

(iii) $n \log x = \log x^n$

(iv) $\tan^{-1} x + \tan^{-1} y = \tan^{-1} \dfrac{x+y}{1-xy}$

(v) $\tan^{-1} x - \tan^{-1} y = \tan^{-1} \dfrac{x-y}{1+xy}$.

5.10. EQUATIONS REDUCIBLE TO THE FORM IN WHICH VARIABLES CAN BE SEPARATED

Equations of the form $\dfrac{dy}{dx} = f(ax + by + c)$ can be reduced to the form in which the variables are separable.

Put $\quad ax + by + c = z$

$$\Rightarrow \quad a + b\frac{dy}{dx} = \frac{dz}{dx} \quad \Rightarrow \quad \frac{dy}{dx} = \frac{1}{b}\left(\frac{dz}{dx} - a\right)$$

$\therefore$ The given equation becomes

$$\frac{1}{b}\left(\frac{dz}{dx} - a\right) = f(z) \quad \Rightarrow \quad \frac{dz}{dx} - a = bf(z)$$

$$\Rightarrow \quad \frac{dz}{dx} = a + bf(z) \quad \Rightarrow \quad \frac{dz}{dx} = a + bf(z)$$

$$\Rightarrow \quad \frac{dz}{a + bf(z)} = dx \qquad \text{[Variables are separated]}$$

On integrating both sides, we have

$$\int \frac{1}{a + bf(z)}\, dz = \int dx + C \quad \Rightarrow \quad \int \frac{1}{a + bf(z)}\, dz = x + C$$

where C is an arbitrary constant, and $z = ax + by + C$ which is the required solution.

SOME SOLVED EXAMPLES

Example 1. *Determine the order and degree of the following differential equations. State also, if these are linear or non-linear.*

(i) $\dfrac{d^2 y}{dx^2} = \sqrt{1 + \dfrac{dy}{dx}}$

(ii) $xy\dfrac{dy}{dx} = \dfrac{(1 + y^2)(1 + x + x^2)}{1 + x^2}$

(iii) $\left(\dfrac{d^2 s}{dt^2}\right)^2 + 3\left(\dfrac{ds}{dt}\right)^3 + 8 = 0$

(iv) $y = \dfrac{dy}{dx} + \dfrac{C}{\dfrac{dy}{dx}}$

(*v*) $\frac{d^2y}{dx^2} = 1 + \sqrt{\frac{dy}{dx}}$ (*vi*) $\frac{d^3y}{dx^3} + y = 0$

(*vii*) $y = \frac{dy}{dx} + \sqrt{1 + \left(\frac{dy}{dx}\right)^3}$ (*viii*) $y = x\frac{dy}{dx} + \sqrt{a^2\left(\frac{dy}{dx}\right)^2 + b^2}$

(*ix*) $\frac{1}{x}\frac{d^2y}{dx^2} = e^x$ (*x*) $(xy^2 + x)dx + (y - x^2y)dy = 0.$

Solution. (*i*) The given differential equation is

$$\frac{d^2y}{dx^2} = \sqrt{1 + \frac{dy}{dx}}$$

On squaring both sides, we get

$$\left(\frac{d^2y}{dx^2}\right)^2 = \left(1 + \frac{dy}{dx}\right)$$

In this equation, the order of highest order derivative, *i.e.*, of $\frac{d^2y}{dx^2}$ is 2.

Degree of highest order derivative is 2.

∴ Order and degree of the given differential equation are 2 each.

The given differential equation is Non-linear, because $\frac{d^2y}{dx^2}$ is multiplied by itself.

(*ii*) The given differential equation is

$$xy\frac{dy}{dx} = \frac{(1+y^2)(1+x+x^2)}{(1+x^2)}$$

The order of highest order derivative, *i.e.*, of $\frac{dy}{dx}$ is 1.

Degree of highest order derivative is 1.

∴ Order and degree of the given differential equation are 1 each.

The given differential equation is non-linear, because y and $\frac{dy}{dx}$ are multiplied together.

(*iii*) The given differential equation is

$$\left(\frac{d^2s}{dt^2}\right)^2 + 3\left(\frac{ds}{dt}\right)^3 + 8 = 0$$

The order of highest order derivative *i.e.*, of $\left(\frac{d^2s}{dt^2}\right)$ is 2.

Degree of highest order derivative is 2.

∴ Order and degree of given differential equation are 2 each.

The given differential equation is Non-linear, because $\frac{d^2s}{dt^2}$ is multiplied by itself.

(*iv*) The given differential equation is

$$y = \frac{dy}{dx} + \frac{C}{\frac{dy}{dx}}$$

$$\Rightarrow \qquad y\frac{dy}{dx} = \left(\frac{dy}{dx}\right)^2 + C$$

$$\Rightarrow \qquad \left(\frac{dy}{dx}\right)^2 - y\frac{dy}{dx} + C = 0$$

The order of highest order derivative, *i.e.*, of $\frac{dy}{dx}$ is 1. Degree of highest order derivative is 2.

$\therefore$ The order of given differential equation is 1 and degree is 2.

The given differential equation is Non-linear, because y and $\frac{dy}{dx}$ are multiplied together.

(*v*) The given differential equation is

$$\frac{d^2y}{dx^2} = 1 + \sqrt{\frac{dy}{dx}}$$

$$\Rightarrow \qquad \frac{d^2y}{dx^2} - 1 = \sqrt{\frac{dy}{dx}}$$

On squaring both sides, we get

$$\left(\frac{d^2y}{dx^2}\right)^2 + 1 - 2\frac{d^2y}{dx^2} = \frac{dy}{dx}$$

$$\Rightarrow \qquad \left(\frac{d^2y}{dx^2}\right)^2 - 2\frac{d^2y}{dx^2} - \frac{dy}{dx} + 1 = 0$$

The order of highest order derivative, *i.e.*, of $\frac{d^2y}{dx^2}$ is 2. Degree of highest order derivative is 2.

$\therefore$ Order and degree of the given differential equation is 2.

The given differential equaton is Non-linear, because $\frac{d^2y}{dx^2}$ is multiplied by itself,.

(*vi*) The given differential equation is

$$\frac{d^3y}{dx^3} + y = 0$$

In this equation, the order of highest order derivative *i.e.*, $\frac{d^3y}{dx^3}$ is 3. Degree of highest order derivative is 1.

$\therefore$ Order of given differential equation is 3 and degree is 1.

The given differential equation is linear as the dependent variable and its derivative occur only in first degree and are not multiplied together.

(*vii*) The given differential equation is

$$y = \frac{dy}{dx} + \sqrt{1+\left(\frac{dy}{dx}\right)^3}$$

$$\Rightarrow \qquad \left(y - \frac{dy}{dx}\right) = \sqrt{1+\left(\frac{dy}{dx}\right)^3}$$

On squaring both sides, we get

$$\Rightarrow \qquad \left(y - \frac{dy}{dx}\right)^2 = 1 + \left(\frac{dy}{dx}\right)^3$$

$$\Rightarrow \qquad y^2 + \left(\frac{dy}{dx}\right)^2 - 2y\frac{dy}{dx} = 1 + \left(\frac{dy}{dx}\right)^3$$

$$\Rightarrow \qquad \left(\frac{dy}{dx}\right)^3 - \left(\frac{dy}{dx}\right)^2 + 2y\frac{dy}{dx} - y^2 + 1 = 0$$

In this equation, the order of highest order derivative *i.e.*, $\frac{dy}{dx}$ is 1.

Degree of the highest order derivative is 3.

$\therefore$ The order of given equation is 1 and degree is 3.

The given differential equation is Non-linear, because $\frac{dy}{dx}$ is multiplied by itself.

(*viii*) The given differential equation is

$$y = x\frac{dy}{dx} + \sqrt{a^2\left(\frac{dy}{dx}\right)^2 + b^2}$$

$$\Rightarrow \qquad y - x\frac{dy}{dx} = \sqrt{a^2\left(\frac{dy}{dx}\right)^2 + b^2}$$

On squaring both sides, we get

$$\left(y - x\frac{dy}{dx}\right)^2 = a^2\left(\frac{dy}{dx}\right)^2 + b^2$$

$$\Rightarrow \qquad y^2 + x^2\left(\frac{dy}{dx}\right)^2 - 2xy\frac{dy}{dx} = a^2\left(\frac{dy}{dx}\right)^2 + b^2$$

$$\Rightarrow \qquad (x^2 - a^2)\left(\frac{dy}{dx}\right)^2 - 2xy\frac{dy}{dx} + (y^2 - b^2) = 0$$

In this equation, the order of highest order derivative *i.e.*, $\frac{dy}{dx}$ is 1.

Degree of highest order derivative is 2.

$\therefore$ The order of given differential equation is 1 and degree is 2.

The given differential equation is Non-linear, because $\left(\frac{dy}{dx}\right)$ is multiplied by itself.

(*ix*) The given differential equation is $\frac{1}{x}\frac{d^2y}{dx^2} = e^x$

Order of highest order derivative *i.e.*, of $\frac{d^2y}{dx^2}$ is 2. Degree of highest order derivative is 1.

$\therefore$ Order of given equation is 2 and degree is 1.

The given differential equation is Linear, because dependent variable and its derivative occur only in the first degree and are not multiplied together.

(*x*) The given differential equation is

$$(xy^2 + x)dx + (y - x^2y)dy = 0$$

$$\Rightarrow \quad (xy^2 + x) + (y - x^2y)\frac{dy}{dx} = 0$$

Order of highest order derivative *i.e.*, of $\frac{dy}{dx}$ is 1. Degree of highest order derivative is 1.

$\therefore$ Order of given differential equation is 1 and degree is 1.

The given differential equation is Non-linear, because y and $\frac{dy}{dx}$ are multiplied together.

Example 2. *Find the order and degree of the following differential equations. State also, if they are linear or non-linear.*

(*i*) $\sqrt{1-y^2}\,dx + y\sqrt{1-x^2}\,dy = 0$

(*ii*) $\frac{d^2y}{dx^2} = \cos 3x + \sin 3x$

(*iii*) $y = x\frac{dy}{dx} + \frac{1}{\frac{dy}{dx}}$

(*iv*) $\frac{\left[1+\left(\frac{dy}{dx}\right)^2\right]^{\frac{3}{2}}}{\frac{d^2y}{dx^2}} = p$

(*v*) $\left[1+\left(\frac{dy}{dx}\right)^3\right]^{4/3} = K\frac{d^2y}{dx^2}$

(*vi*) $t^2\frac{d^2s}{dt^2} + st\frac{ds}{dt} = s$

(*vii*) $\frac{d^2y}{dx^2} + y\frac{dy}{dx} - x^2 = 0$

(*viii*) $\left(\frac{ds}{dt}\right)^4 + 3s\frac{d^2s}{dt^2} = 0$

(*ix*) $y + \frac{dy}{dx} = \frac{1}{6}\int y.dx$

(*x*) $x\frac{d^3y}{dx^3} + 6\cos x\frac{d^2y}{dx^2} + y\sin x = \sqrt{1-x^2}$.

Solution. (*i*) The given differential equation is

$$\sqrt{1-y^2}\,dx + y\sqrt{1-x^2}\,dy = 0$$

$$\Rightarrow \qquad \sqrt{1-y^2} + y\sqrt{1-x^2}\,\frac{dy}{dx} = 0$$

Order of the highest order derivative *i.e.*, of $\frac{dy}{dx}$ is 1.

Degree of highest order derivative is 1.

$\therefore$ Order and degree of the given differential equation are 1 each.

The given differential equation is Non-linear, because y and $\frac{dy}{dx}$ are multiplied together.

(*ii*) The given differential equation is

$$\frac{d^2y}{dx^2} = \cos 3x + \sin 3x$$

The order of highest order derivative *i.e.*, of $\frac{d^2y}{dx^2}$ is 2.

Degree of highest order derivative is 1.

$\therefore$ The order of given equation is 2 and degree is 1. The given differential equation is Linear.

(*iii*) The given differential equation is

$$y = x\frac{dy}{dx} + \frac{1}{\frac{dy}{dx}}$$

$$\Rightarrow \qquad y\frac{dy}{dx} = x\left(\frac{dy}{dx}\right)^2 + 1$$

Order of highest order derivative *i.e.*, of $\frac{dy}{dx}$ is 1.

Degree of highest order derivative is 2.

$\therefore$ Order of the given differential equation is 1 and degree is 2.

It is a Non-linear equation, because y and $\frac{dy}{dx}$ are multiplied together.

(*iv*) The given differential equation is

$$\frac{\left[1+\left(\frac{dy}{dx}\right)^2\right]^{3/2}}{\frac{d^2y}{dx^2}} = p$$

$$\Rightarrow \qquad \left[1+\left(\frac{dy}{dx}\right)^2\right]^{3/2} = p\frac{d^2y}{dx^2}$$

On squaring both sides, we get

$$\left[1+\left(\frac{dy}{dx}\right)^2\right]^3 = p^2\left(\frac{d^2y}{dx^2}\right)^2$$

Order of highest order derivative *i.e.*, of $\frac{d^2y}{dx^2}$ is 2. Degree of highest order derivative is 2.

$\therefore$ Order and degree of the given differential equation are 2 each.

It is a Non-linear differential equation, because $\frac{dy}{dx}$ and $\frac{d^2y}{dx^2}$ are multiplied by itself.

(*v*) The given differential equation is

$$\left[1+\left(\frac{dy}{dx}\right)^3\right]^{\frac{4}{3}} = K\,\frac{d^2y}{dx^2}$$

Cubing both sides, we get

$$\left[1+\left(\frac{dy}{dx}\right)^3\right]^4 = K^3\left(\frac{d^2y}{dx^2}\right)^3$$

Order of highest order derivative *i.e.*, $\frac{d^2y}{dx^2}$ is 2. Degree of highest order derivative is 3.

$\therefore$ Order of the given differential equation is 2 and degree is 3.

It is a Non-linear differential equation as $\frac{dy}{dx}$ is multiplied by itself.

(*vi*) The given differential equation is

$$t^2\,\frac{d^2s}{dt^2} + st\,\frac{ds}{dt} = s$$

Order of highest order derivative *i.e.*, of $\frac{d^2s}{dt^2}$ is 2. Degree of highest order derivative is 1.

$\therefore$ Order of the given differential equation is 2 and degree is 1.

The given differential equation is Non-linear, because s and $\frac{ds}{dt}$ are multiplied together.

(*vii*) The given differential equation is

$$\frac{d^2y}{dx^2} + y\,\frac{ds}{dx} - x^2 = 0$$

Order of the highest order derivative *i.e.*, of $\frac{d^2y}{dx^2}$ is 2. Degree of highest order derivative is 1.

$\therefore$ The order of the given differential equation is 2 and degree is 1.

It is a Non-linear differential equation, because y and $\frac{dy}{dx}$ are multiplied together.

(*viii*) The given differential equation is

$$\left(\frac{ds}{dt}\right)^4 + 3s\,\frac{d^2s}{dt^2} = 0$$

Order of the highest order derivative *i.e.*, of $\frac{d^2s}{dt^2}$ is 2. Degree of highest order derivative is 1.

∴ Order of the given differential equation is 2 and degree is 1.

It is a Non-linear differential equation, because $\left(\frac{ds}{dt}\right)$ is multiplied by itself.

(*ix*) The given differential equation is

$$y + \frac{dy}{dx} = \frac{1}{6}\int y.dx$$

Differentiating both sides w.r.t. x, we get

$$\frac{dy}{dx} + \frac{d^2y}{dx^2} = \frac{1}{6}y$$

$$\Rightarrow \qquad 6\frac{d^2y}{dx^2} + 6\frac{dy}{dx} - y = 0$$

In this equation, order of highest order derivative *i.e.*, of $\frac{d^2y}{dx^2}$ is 2.

Degree of highest order derivative is 1.

∴ Order of the given differential equation is 2 and degree is 1.

It is a Linear differential equation as the dependent variable and its derivative occur only in first degree.

(*x*) The given differential equation is

$$x\frac{d^3y}{dx^3} + 6\cos x\frac{d^2y}{dx^2} + y\sin x = \sqrt{1-x^2}$$

Order of the highest order derivative *i.e.*, of $\frac{d^3y}{dx^3}$ is 3. Degree of the highest order derivative is 1.

∴ The order of the given differential equation is 3 and degree is 1.

It is a Linear differential equation, because the dependent variable and its derivative occur only in first degree.

5.11. SOME IMPORTANT RESULTS OF CO-ORDINATE GEOMETRY

(*i*) Equation of any straight line is $y = mx + C$, where m and C are arbitrary constants.

(*ii*) Equation of any straight line passing through the origin is $y = mx$ where m and C are arbitrary constants.

(*iii*) Equation of the circle having centre (h, k) and radius r is

$$(x-h)^2 + (y-k)^2 = r^2$$

(*iv*) Equation of the circle in general form is

$$x^2 + y^2 + 2gx + 2fy + C = 0$$

where Centre $= (-g, -f)$ and Radius $= \sqrt{g^2 + f^2 - C}$

(*v*) Equation of the circle passing through the origin is

$$x^2 + y^2 + 2gx + 2fy = 0$$

where f and g are arbitrary constants.

(*vi*) Equation of circle passing through the origin and having centre on the x-axis is

$$(x-a)^2 + y^2 = a^2$$

where a is an arbitrary constant.

(*vii*) Equation of circle passing through the origin and having centre on the y-axis is

$$x^2 + (y - b)^2 = b^2$$

where b is an arbitrary constant.

(*viii*) Equation of parabola whose axis (*i.e.*, line of symmetry) is parallel to x-axis is

$$(y - k)^2 = 4a(x - h)$$

where a, h and k are arbitrary constants.

(*ix*) Equation of parabola whose axis is parallel to y-axis is

$$(x - h)^2 = 4a(y - k)$$

where a, h and k are arbitrary constants.

(*x*) Equation of ellipse having centre at the origin and axis along co-ordinate axes is

$$\frac{x^2}{a^2} + \frac{y^2}{b^2} = 1$$

where a and b are arbitrary constants.

(*xi*) Equation of the conic whose axes coincide with co-ordinate axes is

$$Ax^2 + By^2 = 1$$

where A and B are arbitrary constants.

Example 3. *Form the differential equations of the following family of curves :*

(*i*) $ay^2 = (x - C)^3$ *; where C is arbitrary constant.*

(*ii*) $y = A \cos mx + B \sin mx$ *; where m is fixed and A, B are arbitrary constants.*

(*iii*) $C(y + C)^2 = x^3$ *; where C is a parameter.*

(*iv*) $y = mx$ *; where m is an arbitrary constant.*

(*v*) $y = A \sin 2x + B \cos 2x$ *; where A and B are arbitrary constants.*

(*vi*) $y = 2(x^2 - 1) + Ce^{-x^2}$ *; where C is an arbitrary constant.*

Solution. (*i*) The given equation is

$$ay^2 = (x - C)^3 \qquad \text{...(1)}$$

Differentiating both sides w.r.t. x, we have

$$2ay \frac{dy}{dx} = 3(x - C)^2 \qquad \text{...(2)}$$

Dividing equation (1) by (2),

$$\frac{ay^2}{2ay \frac{dy}{dx}} = \frac{(x - C)^3}{3(x - C)^2} \quad \Rightarrow \quad \frac{y}{2\frac{dy}{dx}} = \frac{(x - C)}{3}$$

$$\Rightarrow \quad (x - C) = \frac{3y}{2\left(\frac{dy}{dx}\right)}$$

$$\Rightarrow \quad C = x - \frac{3y}{2\frac{dy}{dx}} \qquad \text{...(3)}$$

Substituting this value of C in equation (1), we get

$$ay^2 = \left[x - \left(x - \frac{3y}{2\frac{dy}{dx}}\right)\right]^3 = \left(x - x + \frac{3y}{2\frac{dy}{dx}}\right)^3$$

$$= \left(\frac{3y}{2\frac{dy}{dx}}\right)^3 = \frac{27y^3}{8\left(\frac{dy}{dx}\right)^3}$$

$$\Rightarrow \qquad 8a\left(\frac{dy}{dx}\right)^3 = 27y$$

which is the required differential equation.

(*ii*) The given equation is

$$y = \text{A} \cos mx + \text{B} \sin mx \qquad ...(1)$$

Since the given equation has two arbitrary constants, we shall differentiate it twice and we shall get a differential equation of second order.

Differentiating equation (1) w.r.t. x, we get

$$\frac{dy}{dx} = - m\text{A} \sin mx + m\text{B} \cos mx$$

Differentiating again w.r.t. x, we get

$$\frac{d^2y}{dx^2} = - m^2 \cos mx - m^2 \text{B} \sin mx$$

$$= - m^2 [\text{A} \cos mx + \text{B} \sin mx] \qquad \text{[Using (1)]}$$

$$\Rightarrow \qquad \frac{d^2y}{dx^2} = - m^2 y$$

$$\Rightarrow \qquad \frac{d^2y}{dx^2} + m^2y = 0$$

which is the required differential equation.

(*iii*) The given equation is

$$\text{C}(y + \text{C})^2 = x^3 \qquad ...(1)$$

Since only one parameter is there, so we shall get a differential equation of first order.

Differentiating equation (1) w.r.t. x, we get

$$2\text{C}(y + \text{C})\frac{dy}{dx} = 3x^2 \qquad ...(2)$$

Dividing (1) by (2), we get

$$\frac{\text{C}(y+\text{C})^2}{2\text{C}(y+\text{C})\frac{dy}{dx}} = \frac{x^3}{3x^2} \qquad \Rightarrow \qquad \frac{(y+\text{C})}{2\frac{dy}{dx}} = \frac{x}{3}$$

$$\Rightarrow \qquad y + \text{C} = \frac{2}{3}x\frac{dy}{dx}$$

$$\Rightarrow \qquad \text{C} = \left(\frac{2}{3}\frac{x\,dy}{dx} - y\right) \qquad ...(3)$$

Substituting this value of C in equation (1), we have

$$\left[\frac{2}{3}x\frac{dy}{dx} - y\right]\left[y + \frac{2}{3}x\frac{dy}{dx} - y\right]^2 = x^3$$

$$\Rightarrow \qquad \left(\frac{2}{3}x\frac{dy}{dx} - y\right)\left(\frac{2}{3}x\frac{dy}{dx}\right)^2 = x^3$$

$$\Rightarrow \quad \left(\frac{2}{3}x\frac{dy}{dx} - y\right)\left[\frac{4}{9}x^2\left(\frac{dy}{dx}\right)^2\right] = x^3$$

$$\Rightarrow \quad x^2\left[\frac{8x}{27}\left(\frac{dy}{dx}\right)^3 - \frac{4y}{9}\left(\frac{dy}{dx}\right)^2\right] = x^3$$

$$\Rightarrow \quad \frac{8x}{27}\left(\frac{dy}{dx}\right)^3 - \frac{4y}{9}\left(\frac{dy}{dx}\right)^2 = x$$

which is the required differential equation.

(*iv*) The given equation is

$$y = mx \qquad ...(1)$$

Since the given equation has only one arbitrary constant, we differentiate it once.

Differentiating equation (1) w.r.t. x, we get

$$\frac{dy}{dx} = m \qquad ...(2)$$

Substituting this value of m in equation (1), we have

$$y = \left(\frac{dy}{dx}\right)x \quad \Rightarrow \quad y = x\frac{dy}{dx}$$

which is the required differential equation.

(*v*) The given equation is

$$y = \text{A} \sin 2x + \text{B} \cos 2x \qquad ...(1)$$

Please try yourself.

[**Hint :** See Part (*ii*) of the same example.]

(*vi*) The given equation is

$$y = 2(x^2 - 1) + \text{C}e^{-x^2} \qquad ...(1)$$

Since there is only one arbitrary constant in the given equation, so we differentiate it once and we shall get a differential equation of first order.

Differentiating equation (1) w.r.t. x, we get

$$\frac{dy}{dx} = 2(2x - 0) + \text{C}e^{-x^2}(-2x)$$

$$\Rightarrow \quad \frac{dy}{dx} = 4x - 2\text{C}xe^{-x^2} \qquad ...(2)$$

From equation (1), we have

$$\text{C}e^{-x^2} = y - 2x^2 + 2 \qquad ...(3)$$

Substituting this value of $\text{C}e^{-x^2}$ in equation (2), we have

$$\frac{dy}{dx} = 4x - 2x(y - 2x^2 + 2)$$

$$\Rightarrow \quad \frac{dy}{dx} = 4x - 2xy + 4x^3 - 4x$$

$$\Rightarrow \quad \frac{dy}{dx} = 4x^3 - 2xy$$

$$\Rightarrow \quad \frac{dy}{dx} + 2xy = 4x^3$$

which is the required differential equation.

Example 4. *Find the differential equations of the following families of curves :*

(*i*) $y = C(x - C)^2$; where C is an arbitrary constant.

(*ii*) $y = Ae^{mx} + Be^{nx}$; *where m, n are fixed and A, B are arbitrary constants.*

(*iii*) $y^2 = m(a^2 - x^2)$; *where m and a are arbitrary constants.*

(*iv*) $y^2 - 2ay + x^2 = a^2$; *where a is an arbitrary constant.*

(*v*) $e^x + Ce^y = 1$; *where C is an arbitrary constant.*

(*vi*) $y = e^x (A \cos x + B \sin x)$; *A and B are arbitrary constants.*

Solution. (*i*) The given equation is

$$y = C(x - C)^2 \qquad ...(1)$$

Since there is one arbitrary constant in the given equation, so we differentiate it once and we shall get a differential equation of first order.

Differentiating equation (1) w.r.t. x, we get

$$\frac{dy}{dx} = 2C(x - C) \qquad ...(2)$$

Dividing (1) by (2), we get

$$\frac{y}{\frac{dy}{dx}} = \frac{C(x-C)^2}{2C(x-C)}$$

$$\Rightarrow \quad \frac{y}{\frac{dy}{dx}} = \frac{x-C}{2} \quad \Rightarrow \quad x - C = \frac{2y}{\frac{dy}{dx}}$$

$$\Rightarrow \quad C = \left(x - \frac{2y}{\frac{dy}{dx}}\right) \qquad ...(3)$$

Substituting this value of C in equation (1), we have

$$y = \left(x - \frac{2y}{\frac{dy}{dx}}\right)\left[x - \left(x - \frac{2y}{\frac{dy}{dx}}\right)\right]^2 = \left(x - \frac{2y}{\frac{dy}{dx}}\right)\left[x - x + \frac{2y}{\frac{dy}{dx}}\right]^2$$

$$= \left(x - \frac{2y}{\frac{dy}{dx}}\right)\left(\frac{4y^2}{\left(\frac{dy}{dx}\right)^2}\right) = \frac{4xy^2}{\left(\frac{dy}{dx}\right)^2} - \frac{8y^3}{\left(\frac{dy}{dx}\right)^3}$$

$$\Rightarrow \quad y\left(\frac{dy}{dx}\right)^3 = 4xy^2\left(\frac{dy}{dx}\right) - 8y^3$$

$$\Rightarrow \quad \left(\frac{dy}{dx}\right)^3 = 4y\left(x\frac{dy}{dx} - 2y\right) \qquad \text{[Cancelling } y \text{ throughout]}$$

which is the required differential equation.

(*ii*) The given equation is

$$y = Ae^{mx} + Be^{nx} \qquad ...(1)$$

Since the given equation contains two arbitrary constants A and B ; we shall differentiate it twice and we shall get a differential equation of second order.

Differentiating equation (1) w.r.t. x, we get

$$\frac{dy}{dx} = m\text{A}e^{mx} + n\text{B}e^{nx} \qquad ...(2)$$

Differentiating again w.r.t. x, we get

$$\frac{d^2y}{dx^2} = m^2\text{A}e^{mx} + n^2\text{B}e^{nx} \qquad ...(3)$$

Multiplying (2) by m and subtracting equation (3) from the result, we get

$$m\frac{dy}{dx} - \frac{d^2y}{dx^2} = (mn - n^2)\,\text{B}e^{nx}$$

$$\Rightarrow \quad \frac{1}{n(m-n)}\left[m\frac{dy}{dx} - \frac{d^2y}{dx^2}\right] = \text{B}e^{nx} \qquad ...(4)$$

Multiplying (2) by n subtracting equation (3) from the result, we get

$$n\frac{dy}{dx} - \frac{d^2y}{dx^2} = (mn - m^2)\,\text{A}e^{mx}$$

$$\Rightarrow \quad \frac{1}{m(n-m)}\left[n\frac{dy}{dx} - \frac{d^2y}{dx^2}\right] = \text{A}e^{mx}$$

$$\Rightarrow \quad \frac{1}{m(m-n)}\left[\frac{d^2y}{dx^2} - n\frac{dy}{dx}\right] = \text{A}e^{mx} \qquad ...(5)$$

Substituting the values of $\text{A}e^{mx}$ and $\text{B}e^{nx}$ in equation (1), we get

$$y = \frac{1}{m(m-n)}\left[\frac{d^2y}{dx^2} - n\frac{dy}{dx}\right] + \frac{1}{n(m-n)}\left[m\frac{dy}{dx} - \frac{d^2y}{dx^2}\right]$$

$$= \frac{1}{mn(m-n)}\left[n\frac{d^2y}{dx^2} - n^2\frac{dy}{dx} + m^2\frac{dy}{dx} - m\frac{d^2y}{dx^2}\right]$$

$$= \frac{1}{mn(m-n)}\left[(m^2 - n^2)\frac{dy}{dx} - (m-n)\frac{d^2y}{dx^2}\right]$$

$$= \frac{(m-n)}{mn(m-n)}\left[(m+n)\frac{dy}{dx} - \frac{d^2y}{dx^2}\right]$$

$$\Rightarrow \quad mny = (m+n)\frac{dy}{dx} - \frac{d^2y}{dx^2}$$

$$\Rightarrow \quad \frac{d^2y}{dx^2} - (m+n)\frac{dy}{dx} + mny = 0$$

which is the required differential equation.

(*iii*) The given equation is

$$y^2 = m(a^2 - x^2) \qquad ...(1)$$

Since the given equation contains two arbitrary constants, so we shall differentiate it twice and we shall get a differential equation of second order.

Differentiating equation (1), w.r.t. x, we get

$$2y\frac{dy}{dx} = m(0 - 2x)$$

$$\Rightarrow \qquad y\frac{dy}{dx} = -mx \qquad ...(2)$$

Differentiating again w.r.t. x, we have

$$y\frac{d^2y}{dx^2} + \left(\frac{dy}{dx}\right)^2 = -m \qquad ...(3)$$

Substituting this value of $(-m)$ in equation (2), we have

$$y\frac{dy}{dx} = \left[y\frac{d^2y}{dx^2} + \left(\frac{dy}{dx}\right)^2\right].x = xy\frac{d^2y}{dx^2} + x\left(\frac{dy}{dx}\right)^2$$

$$\Rightarrow \qquad xy\frac{d^2y}{dx^2} + x\left(\frac{dy}{dx}\right)^2 - y\frac{dy}{dx} = 0$$

which is the required differential equation.

(*iv*) The given equation is

$$y^2 - 2ay + x^2 = a^2 \qquad ...(1)$$

Since the given equation has only one arbitrary constant, so we shall get a differential equation of first order.

Differentiating equation (1), w.r.t. x, we get

$$2y\frac{dy}{dx} - 2a\frac{dy}{dx} + 2x = 0$$

$$\Rightarrow \qquad y\frac{dy}{dx} - a\frac{dy}{dx} + x = 0 \qquad \text{[Cancelling 2 throughout]}$$

$$\Rightarrow \qquad a\frac{dy}{dx} = x + y\frac{dy}{dx}$$

$$\Rightarrow \qquad a = \frac{x + y\frac{dy}{dx}}{\frac{dy}{dx}} \qquad ...(2)$$

Substituting this value of a in equation (1), we have

$$y^2 - 2y\left(\frac{x + y\frac{dy}{dx}}{\frac{dy}{dx}}\right) + x^2 = \left(\frac{x + y\frac{dy}{dx}}{\frac{dy}{dx}}\right)^2$$

$$\Rightarrow \qquad y^2 - 2y\left(\frac{x + yy_1}{y_1}\right) + x^2 = \left(\frac{x + yy_1}{y_1}\right)^2 \qquad \left[\text{Writing } \frac{dy}{dx} = y_1\right]$$

$$\Rightarrow \qquad y^2 - \frac{2xy}{y_1} - 2y^2 + x^2 = \frac{x^2 + y^2y_1^2 + 2xyy_1}{y_1^2}$$

$$\Rightarrow \qquad x^2 - y^2 - \frac{2xy}{y_1} = \frac{x^2 + y^2y_1^2 + 2xyy_1}{y_1^2}$$

$$\Rightarrow \qquad \frac{x^2y_1 - y^2y_1 - 2xy}{y_1} = \frac{x^2 + y^2y_1^2 + 2xyy_1}{y_1^2}$$

$$\Rightarrow \qquad y_1(x^2y_1 - y^2y_1 - 2xy) = x^2 + y^2y_1^2 + 2xyy_1$$

$$\Rightarrow \qquad x^2y_1^2 - y^2y_1^2 - 2xyy_1 = x^2 + y^2y_1^2 + 2xyy_1$$

$$\Rightarrow \qquad x^2y_1^{\,2} - 2y^2y_1^{\,2} - 4xyy_1 - x^2 = 0$$

$$\Rightarrow \qquad x^2\left(\frac{dy}{dx}\right)^2 - 2y^2\left(\frac{dy}{dx}\right)^2 - 4xy\frac{dy}{dx} - x^2 = 0$$

which is the required differential equation.

(*v*) The given equation is

$$e^x + Ce^y = 1 \qquad \text{...(1)}$$

Since the given equation contains one arbitrary constant, so we shall get a differential equation of second order.

Differentiating equation (1) w.r.t. x, we have

$$e^x + Ce^y\frac{dy}{dx} = 0 \qquad \Rightarrow \qquad Ce^y\frac{dy}{dx} = -e^x$$

$$\Rightarrow \qquad Ce^y = -\frac{e^x}{\frac{dy}{dx}} \qquad \text{...(2)}$$

Substituting this value of Ce^y in equation (1), we have

$$e^x - \frac{e^x}{\frac{dy}{dx}} = 1 \qquad \Rightarrow \qquad e^x\frac{dy}{dx} - e^x = \frac{dy}{dx}$$

$$\Rightarrow \qquad (e^x - 1)\frac{dy}{dx} = e^x$$

which is the required differential equation

(*vi*) The given equation is

$$y = e^x(A\cos x + B\sin x) \qquad \text{...(1)}$$

Since the given equation involves two arbitrary constants, we shall differentiate it twice and we shall get a differential equation of second order.

Differentiating equation (1), w.r.t. x, we have

$$\frac{dy}{dx} = e^x(-A\sin x + B\cos x) + e^x(A\cos x + B\sin x)$$

$$\Rightarrow \qquad \frac{dy}{dx} = e^x(-A\sin x + B\cos x) + y \qquad \text{...(2)} \quad \text{[Using equation (1)]}$$

Differentiating again w.r.t. x, we have

$$\frac{d^2y}{dx^2} = e^x(-A\cos x - B\sin x) + e^x(-A\sin x + B\cos x) + \frac{dy}{dx}$$

$$= -e^x(A\cos x + B\sin x) + \left(\frac{dy}{dx} - y\right) + \frac{dy}{dx} \qquad \text{[Using equation (2)]}$$

$$= -y + 2\frac{dy}{dx} - y \qquad \text{[Using equation (1)]}$$

$$\Rightarrow \qquad \frac{d^2y}{dx^2} = 2\frac{dy}{dx} - 2y$$

$$\frac{d^2y}{dx^2} - 2\frac{dy}{dx} + 2y = 0$$

which is the required differential equation.

Example 5. *Find the differential equation of the family of all straight lines passing through the origin.*

Solution. The general equation of the family of all straight lines passing through the origin is given by

$$y = mx$$

Please try yourself

[Hint : See Example 3. Part (*iv*)]

Example 6. *Find the differential equation of the following families of curves :*

(*i*) $y = Ae^{Bx}$ *; where A and B are arbitrary constants.*

(*ii*) $y = A\cos(x + B)$ *; where A and B are arbitrary constants.*

(*iii*) $y = Ax + \frac{B}{x}$ *; where A and B are arbitrary constants.*

(*iv*) $y = Ae^{3x} + Be^{5x}$ *; where A and B are arbitrary constants.*

(*v*) $y = ae^{x} + be^{2x} + ce^{3x}$ *; where a, b and c are arbitrary constants.*

(*vi*) $y = a\sin(bx + c)$ *; where b is fixed, and a, c are arbitrary constants.*

Solution. (*i*) The given equation is

$$y = \mathrm{A}e^{\mathrm{B}x} \qquad ...(1)$$

Since the given equation contains two arbitrary constants A and B, so we shall differentiate it twice and we shall get a differential equation of second order.

Differentiating equation (1) w.r.t. x, we have

$$\frac{dy}{dx} = \mathrm{AB}e^{\mathrm{B}x} = \mathrm{B}(\mathrm{A}e^{\mathrm{B}x})$$

$$\Rightarrow \quad \frac{dy}{dx} = \mathrm{B}y \qquad ...(2) \quad \text{[Using equation (1)]}$$

Differentiating again w.r.t. x, we have

$$\frac{d^2y}{dx^2} = \mathrm{B}\frac{dy}{dx} \qquad ...(3)$$

From equation (2), we have

$$\mathrm{B} = \frac{1}{y}\frac{dy}{dx}$$

Substituting this value of B in equation (3), we get

$$\frac{d^2y}{dx^2} = \left(\frac{1}{y}\frac{dy}{dx}\right)\frac{dy}{dx} \quad \Rightarrow \quad y\frac{d^2y}{dx^2} = \left(\frac{dy}{dx}\right)^2$$

which is the required differential equation.

(*ii*) The given equation is

$$y = \mathrm{A}\cos(x + \mathrm{B}) \qquad ...(1)$$

Since the given equation has two arbitrary constants A and B, so we shall differentiate it twice and we shall get a differential equation of order two.

Differentiating equation (1) w.r.t. x, we have

$$\frac{dy}{dx} = -\mathrm{A}\sin(x + \mathrm{B}) \qquad ...(2)$$

Differentiating again w.r.t. x, we have

$$\frac{d^2y}{dx^2} = -\mathrm{A}\cos(x + \mathrm{B})$$

$$\Rightarrow \qquad \frac{d^2y}{dx^2} = -y \qquad \text{[Using equation (1)]}$$

$$\Rightarrow \qquad \frac{d^2y}{dx^2} + y = 0$$

which is the required differential equation.

(*iii*) The given equation is

$$y = Ax + \frac{B}{x} \qquad ...(1)$$

Since the given equation involves two arbitrary constants A and B, so we shall differentiate it two times and we shall obtain a differential equation of second order.

Differentiating equation (1) w.r.t. x, we get

$$\frac{dy}{dx} = A - \frac{B}{x^2} \qquad ...(2)$$

Differentiating again w.r.t. x, we have

$$\frac{d^2y}{dx^2} = \frac{2B}{x^3} \qquad ...(3)$$

$$\Rightarrow \qquad 2B = x^3 \frac{d^2y}{dx^2}$$

$$\Rightarrow \qquad B = \frac{1}{2} x^3 \frac{d^2y}{dx^2} \qquad ...(4)$$

Substituting the value of B in equation (2), we have

$$\frac{dy}{dx} = A - \frac{1}{x^2}\left(\frac{1}{2} x^3 \frac{d^2y}{dx^2}\right) = A - \frac{1}{2} x \frac{d^2y}{dx^2}$$

$$\Rightarrow \qquad A = \frac{1}{2} x \frac{d^2y}{dx^2} + \frac{dy}{dx} \qquad ...(5)$$

Now, substituting the values of A and B in equation (1), we have

$$y = x\left(\frac{1}{2} x \frac{d^2y}{dx^2} + \frac{dy}{dx}\right) + \frac{1}{x}\left(\frac{1}{2} x^3 \frac{d^2y}{dx^2}\right)$$

$$= \frac{x^2}{2}\frac{d^2y}{dx^2} + x\frac{dy}{dx} + \frac{x^2}{2}\frac{d^2y}{dx^2} = x^2\frac{d^2y}{dx^2} + x\frac{dy}{dx}$$

$$\Rightarrow \qquad x^2 \frac{d^2y}{dx^2} + x\frac{dy}{dx} - y = 0$$

which is the required differential equation.

(*iv*) The given equation is

$$y = Ae^{3x} + Be^{5x} \qquad ...(1)$$

Since the given equation involves two arbitrary constants A and B, so we shall differentiate it twice and we shall get a differential equation of order two.

Differentiating equation (1), w.r.t. x, we get

$$\frac{dy}{dx} = 3Ae^{3x} + 5Be^{5x} \qquad ...(2)$$

Differentiating equation (2), w.r.t. x, we get

$$\frac{d^2y}{dx^2} = 9Ae^{3x} + 25Be^{5x} \quad ...(3)$$

Now, multiplying equation (2) by 5 and subtracting equation (3) from the result, we have

$$5\frac{dy}{dx} - \frac{d^2y}{dx^2} = 6Ae^{3x}$$

$$\Rightarrow \quad Ae^{3x} = \frac{1}{6}\left(5\frac{dy}{dx} - \frac{d^2y}{dx^2}\right) \quad ...(4)$$

Multiplying equation (2) by 3 and subtracting the result from equation (3), we have

$$\frac{d^2y}{dx^2} - 3\frac{dy}{dx} = 10Be^{5x}$$

$$\Rightarrow \quad Be^{5x} = \frac{1}{10}\left(\frac{d^2y}{dx^2} - 3\frac{dy}{dx}\right) \quad ...(5)$$

Substituting the values of Ae^{3x} and Be^{5x} from equations (4) and (5) in equation (1), we have

$$y = \frac{1}{6}\left(5\frac{dy}{dx} - \frac{d^2y}{dx^2}\right) + \frac{1}{10}\left(\frac{d^2y}{dx^2} - 3\frac{dy}{dx}\right)$$

$$\Rightarrow \quad y = \frac{5}{6}\frac{dy}{dx} - \frac{1}{6}\frac{d^2y}{dx^2} + \frac{1}{10}\frac{d^2y}{dx^2} - \frac{3}{10}\frac{dy}{dx}$$

$$\Rightarrow \quad y = \left(\frac{5}{6} - \frac{3}{10}\right)\frac{dy}{dx} + \left(\frac{1}{10} - \frac{1}{6}\right)\frac{d^2y}{dx^2}$$

$$\Rightarrow \quad y = \frac{16}{30}\frac{dy}{dx} - \frac{2}{30}\frac{d^2y}{dx^2}$$

$$\Rightarrow \quad 30y = 16\frac{dy}{dx} - 2\frac{d^2y}{dx^2}$$

$$\Rightarrow \quad 2\frac{d^2y}{dx^2} - 16\frac{dy}{dx} + 30y = 0$$

$$\Rightarrow \quad \frac{d^2y}{dx^2} - 8\frac{dy}{dx} + 15y = 0 \quad \text{[Cancelling 2 throughout]}$$

which is the required differential equation.

(v) The given equation is

$$y = ae^x + be^{2x} + ce^{3x} \quad ...(1)$$

Since the given equation contains three arbitrary constants a, b and c, so we shall differentiate it three times and we shall get a differential equation of order three.

Differentiating equation (1) w.r.t. x, we have

$$\frac{dy}{dx} = ae^x + 2be^{2x} + 3ce^{3x} \quad ...(2)$$

Subtracting equation (1) from (2), we have

$$\frac{dy}{dx} - y = be^{2x} + 2ce^{3x} \quad ...(3) \quad \text{[Elimination of 'a']}$$

Differentiating equation (3), w.r.t. x, we have

$$\frac{d^2y}{dx^2} - \frac{dy}{dx} = 2be^{2x} + 6ce^{3x} \qquad ...(4)$$

Now, multiplying equation (3) by 2 and subtract the result from equation (4), we have

$$\frac{d^2y}{dx^2} - 3\frac{dy}{dx} + 2y = 2ce^{3x} \qquad ...(5) \quad \text{[Elimination of 'b']}$$

Differentiating equation (5), w.r.t. x, we have

$$\frac{d^3y}{dx^3} - 3\frac{d^2y}{dx^2} + 2\frac{dy}{dx} = 6ce^{3x} \qquad ...(6)$$

Now, multiplying equation (5) by 3 and subtract the result from equation (6), we have

$$\frac{d^3y}{dx^3} - 6\frac{d^2y}{dx^2} + 11\frac{dy}{dx} - 6y = 0 \qquad \text{[Elimination of 'c']}$$

which is the required differential equation.

(*vi*) The given equation is

$$y = a \sin (bx + c) \qquad ...(1)$$

Since, the given equation contains two arbitrary constants a and c, therefore, we shall differentiate it two times and we shall get a differential equation of second order.

Differentiating equation (1), w.r.t. x, we have

$$\frac{dy}{dx} = ab \cos (bx + c) \qquad ...(2)$$

Differentiating again w.r.t. x, we have

$$\frac{d^2y}{dx^2} = -ab^2 \sin (bx + c)$$

$$\Rightarrow \quad \frac{d^2y}{dx^2} = -b^2y \qquad \text{[Using equation(1)]}$$

$$\Rightarrow \quad \frac{d^2y}{dx^2} + b^2y = 0$$

which is the required differential equation.

Example 7. *Obtain the differential equation of all circles of radius r.*

Solution. The general equation of the family of circles of radius r is given by

$$(x - a)^2 + (y - b)^2 = r^2 \qquad ...(1)$$

where a and b are arbitrary constants.

Since the equation (1) has two arbitrary constants, so we shall differentiate it two times and we shall get a differential equation of order two.

Differentiating equation (1), w.r.t. x, we have

$$2(x - a) + 2(y - b)\frac{dy}{dx} = 0$$

$$\Rightarrow \quad (x - a) + (y - b)\frac{dy}{dx} = 0 \qquad ...(2)$$

Differentiating again w.r.t. x, we have

$$(1 - 0) + (y - b)\frac{d^2y}{dx^2} + \frac{dy}{dx}.\frac{dy}{dx} = 0$$

$$\Rightarrow \quad 1 + (y - b)\frac{d^2y}{dx^2} + \left(\frac{dy}{dx}\right)^2 = 0$$

$$\Rightarrow \quad (y - b) = \frac{-\left[1 + \left(\frac{dy}{dx}\right)^2\right]}{\frac{d^2y}{dx^2}} \qquad ...(3)$$

Substituting this value of $(y - b)$ in equation (2), we have

$$(x - a) - \frac{\left[1 + \left(\frac{dy}{dx}\right)^2\right]}{\frac{d^2y}{dx^2}} \cdot \frac{dy}{dx} = 0$$

$$\Rightarrow \quad (x - a) = \frac{\left[1 + \left(\frac{dy}{dx}\right)^2\right]}{\frac{d^2y}{dx^2}} \cdot \frac{dy}{dx} \qquad ...(4)$$

Now, substituting the values of $(y - b)$ and $(x - a)$ from equations (3) and (4) in equation (1), we have

$$\left[\left(\frac{1 + \left(\frac{dy}{dx}\right)^2}{\frac{d^2y}{dx^2}}\right) \cdot \frac{dy}{dx}\right]^2 + \left[-\left(\frac{1 + \left(\frac{dy}{dx}\right)^2}{\frac{d^2y}{dx^2}}\right)\right]^2 = r^2$$

$$\Rightarrow \quad \left[\frac{(1 + y_1^2)}{y_2} \cdot y_1\right]^2 + \left[-\left(\frac{1 + y_1^2}{y_2}\right)\right]^2 = r^2$$

$$\Rightarrow \quad \frac{(1 + y_1^2)^2 y_1^2}{y_2^2} + \frac{(1 + y_1^2)^2}{y_2^2} = r^2$$

$$\Rightarrow \quad (1 + y_1^2)^2 [y_1^2 + 1] = r^2 y_2^2$$

$$\Rightarrow \quad (1 + y_1^2)^3 = r^2 y_2^2$$

$$\Rightarrow \quad \left[1 + \left(\frac{dy}{dx}\right)^2\right]^3 = r^2 \left[\frac{d^2y}{dx^2}\right]^2$$

which is the required differential equation.

Example 8. *Find the differential equation of all circles passing through origin and having their centres on the axis of x.*

Or

Find the differential equation of the system of circles touching the y-axis at the origin.

Solution. The general equation of the circle is

$$x^2 + y^2 + 2gx + 2fy + c = 0 \qquad ...(1)$$

Since, the circle passes through the origin, *i.e.*, (0, 0)

$$\therefore \quad (0)^2 + (0)^2 + 2g(0) + 2f(0) + c = 0$$

$$\Rightarrow \quad c = 0$$

$\therefore$ Equation of circle reduces to

$$x^2 + y^2 + 2gx + 2fy = 0 \qquad ...(2)$$

If the centre $(-g, -f)$ lies on x-axis.

$$\therefore \qquad f = 0$$

$\therefore$ Equation (2) becomes

$$x^2 + y^2 + 2gx = 0 \qquad ...(3)$$

Differentiating equation (3), w.r.t. x, we have

$$2x + 2y\frac{dy}{dx} + 2g = 0$$

$$\Rightarrow \qquad x + y\frac{dy}{dx} + g = 0 \qquad \text{[Cancelling 2 throughout]}$$

$$\Rightarrow \qquad g = -\left(x + y\frac{dy}{dx}\right) \qquad ...(4)$$

Substituting this value of g in equation (3), we have

$$x^2 + y^2 + 2\left[-\left(x + y\frac{dy}{dx}\right)\right]x = 0$$

$$\Rightarrow \qquad x^2 + y^2 - 2x^2 - 2xy\frac{dy}{dx} = 0$$

$$\Rightarrow \qquad y^2 - x^2 - 2xy\frac{dy}{dx} = 0$$

which is the required differential equation.

Example 9. *Find the differential equation of lines in xy-plane.*

Solution. The general equation of a straight line in xy-plane is given by

$$y = mx + c \qquad ...(1)$$

where m and c are arbitrary constants.

Since the equation (1) has two arbitrary constants, so, we shall differentiate it two times and we shall get a differential equation of second order.

Differentiating equation (1) w.r.t. x, we have

$$\frac{dy}{dx} = m \qquad ...(2)$$

Differentiating again w.r.t. x, we have

$$\frac{d^2y}{dx^2} = 0 \qquad ...(3)$$

which is the required differential equation.

Example 10. *Find the differential equation corresponding to* $y^2 = a(b - x)(b + x)$ *by eliminating parameters a and b.*

Solution. Please try yourself.

$$\left[\textbf{Ans. } x\left(y\frac{d^2y}{dx^2} + \left(\frac{dy}{dx}\right)^2\right) = y\frac{dy}{dx}\right]$$

[**Hint :** $\qquad y^2 = a(b - x)(b + x)$

$$\Rightarrow \qquad y^2 = a(b^2 - x^2)$$

See Example 4 Part (*iii*).]

Example 11. (*a*) *Find the differential equation of all parabolas with latus rectum '4a' and whose axes are parallel to x-axis.*

(*b*) *Find the differential equation of all parabolas whose axes are parallel to y-axis.*

Solution. (*a*) The equation of the parabola with given latus rectum $4a$ and axis parallel to x-axis is given by

$$(y - k)^2 = 4a(x - h) \quad ...(1)$$

where h and k are arbitrary constants.

Since equation (1) has two arbitrary constants, so we shall differentiate it two times and we shall get a differential equation of second order.

Differentiating equation (1) w.r.t. x, we have

$$2(y - k)\frac{dy}{dx} = 4a(1 - 0) \quad \Rightarrow \quad 2(y - k)\frac{dy}{dx} = 4a$$

$$\Rightarrow \quad (y - k)\frac{dy}{dx} = 2a \quad ...(2)$$

Differentiating again w.r.t. x, we have

$$(y - k)\frac{d^2y}{dx^2} + \frac{dy}{dx}.\frac{dy}{dx} = 0 \quad \Rightarrow \quad (y - k)\frac{d^2y}{dx^2} = -\left(\frac{dy}{dx}\right)^2$$

$$\Rightarrow \quad (y - k) = \frac{-\left(\frac{dy}{dx}\right)^2}{\frac{d^2y}{dx^2}} \quad ...(3)$$

Substituting this value of $(y - k)$ in equation (2), we have

$$\frac{-\left(\frac{dy}{dx}\right)^2}{\frac{d^2y}{dx^2}}.\frac{dy}{dx} = 2a \quad \Rightarrow \quad -\left(\frac{dy}{dx}\right)^3 = 2a\frac{d^2y}{dx^2}$$

$$\Rightarrow \quad 2a\frac{d^2y}{dx^2} + \left(\frac{dy}{dx}\right)^3 = 0$$

which is the required differential equation.

(*b*) The general equation of parabola, whose axis is parallel to y-axis is given by

$$(x - h)^2 = 4a(y - k) \quad ...(1)$$

where h, k and a arbitrary constants.

Differentiating equation (1) w.r.t. x, we have

$$2(x - h)(1 - 0) = 4a\left(\frac{dy}{dx} - 0\right)$$

$$\Rightarrow \quad (x - h) = 2a\frac{dy}{dx} \quad ...(2)$$

Differentiating again w.r.t. x, we have

$$(1 - 0) = 2a\frac{d^2y}{dx^2}$$

$$\Rightarrow \quad \frac{d^2y}{dx^2} = \frac{1}{2a} \quad ...(3)$$

Differentiating equation (3), w.r.t. x, we have

$$\frac{d^3 y}{dx^3} = 0$$

which is the required differential equation.

Example 12. *Formulate the differential equations of the family of curves :*

(*i*) $Ax^2 + By^2 = 1$; *where A and B are arbitrary constants.*

(*ii*) $(x^2 - y^2) = c(x^2 + y^2)^2$; *where C is an arbitrary constant.*

(*iii*) $\frac{x}{a} + \frac{y}{b} = 1$; *where a and b are arbitrary constants.*

(*iv*) $y = Ae^{2x} + Be^{-3x}$; *where A and B are arbitrary constants.*

Solution. (*i*) The given equation is

$$Ax^2 + By^2 = 1 \quad ...(1)$$

Since equation (1) contains two arbitrary constants A and B, so we shall differentiate it twice and we shall get a differential equation of order two.

Differentiating equation (1), w.r.t. x, we have

$$2Ax + 2By\frac{dy}{dx} = 0$$

$$\Rightarrow \quad Ax + By\frac{dy}{dx} = 0 \quad ...(2)$$

Differentiating again w.r.t. x, we have

$$A + By\frac{d^2y}{dx^2} + B\frac{dy}{dx}.\frac{dy}{dx} = 0$$

$$\Rightarrow \quad A + By\frac{d^2y}{dx^2} + B\left(\frac{dy}{dx}\right)^2 = 0 \quad ...(3)$$

From equation (2), $\quad Ax = -By\frac{dy}{dx}$

$$\Rightarrow \quad A = -B\frac{y}{x}\frac{dy}{dx} \quad ...(4)$$

Substituting this value of A in equation (3), we have

$$-B\frac{y}{x}\frac{dy}{dx} + By\frac{d^2y}{dx^2} + B\left(\frac{dy}{dx}\right)^2 = 0$$

$$\Rightarrow \quad -\frac{y}{x}\frac{dy}{dx} + y\frac{d^2y}{dx^2} + \left(\frac{dy}{dx}\right)^2 = 0 \quad \text{[Cancelling B throughout]}$$

$$\Rightarrow \quad xy\frac{d^2y}{dx^2} + x\left(\frac{dy}{dx}\right)^2 - y\frac{dy}{dx} = 0$$

which is the required differential equation.

(*ii*) The given equation is

$$x^2 - y^2 = c(x^2 + y^2)^2 \quad ...(1)$$

Differentiating equation (1) w.r.t. x, we have

$$2x - 2y\frac{dy}{dx} = 2c(x^2 + y^2)\left(2x + 2y\frac{dy}{dx}\right)$$

$$\Rightarrow \quad \left(x - y\frac{dy}{dx}\right) = c\,(x^2 + y^2)\left(x + y\frac{dy}{dx}\right) \quad ...(2) \quad \text{[Cancelling 2 throughout]}$$

From equation (1), we have

$$c = \frac{x^2 - y^2}{(x^2 + y^2)^2} \quad ...(3)$$

Substituting this value of c in equation (2), we have

$$\left(x - y\frac{dy}{dx}\right) = \frac{2\,(x^2 - y^2)}{(x^2 + y^2)^2}.(x^2 + y^2)\left(x + y\frac{dy}{dx}\right)$$

$$\Rightarrow \quad (x^2 + y^2)\left(x - y\frac{dy}{dx}\right) = 2\,(x^2 - y^2)\left(x + y\frac{dy}{dx}\right)$$

$$\Rightarrow \quad x^3 + xy^2 - x^2y\frac{dy}{dx} - y^3\frac{dy}{dx} = 2x^3 + 2x^2y\frac{dy}{dx} - 2xy^2 - 2y^3\frac{dy}{dx}$$

$$\Rightarrow \quad (3xy^2 - x^3) = \frac{dy}{dx}(3x^2y - y^3)$$

$$\Rightarrow \quad \frac{dy}{dx} = \frac{3xy^2 - x^3}{3x^2y - y^3}$$

which is the required differential equation.

(*iii*) The given equation is

$$\frac{x}{a} + \frac{y}{b} = 1 \quad ...(1)$$

Differentiating equation (1) w.r.t. x, we have

$$\frac{1}{a} + \frac{1}{b}\frac{dy}{dx} = 0 \quad \Rightarrow \quad \frac{1}{b}\frac{dy}{dx} = -\frac{1}{a}$$

$$\Rightarrow \quad \frac{dy}{dx} = -\frac{b}{a} \quad ...(2)$$

Differentiating equation (2), w.r.t. x, we have

$$\frac{d^2y}{dx^2} = 0$$

which is the required differential equation.

(*iv*) The given equation is

$$y = Ae^{2x} + Be^{-3x} \quad ...(1)$$

Differentiating equation (1) w.r.t. x, we have

$$\frac{dy}{dx} = 2Ae^{2x} - 3Be^{-3x} \quad ...(2)$$

Differentiating equation (2), w.r.t. x again, we have

$$\frac{d^2y}{dx^2} = 4Ae^{2x} + 9Be^{-3x} \quad (3)$$

Adding equations (2) and (3), we have

$$\frac{d^2y}{dx^2} + \frac{dy}{dx} = 6Ae^{2x} + 6Be^{-3x}$$

$$\Rightarrow \quad \frac{d^2y}{dx^2} + \frac{dy}{dx} = 6(Ae^{2x} + Be^{-3x})$$

$$\Rightarrow \quad \frac{d^2y}{dx^2} + \frac{dy}{dx} = 6y \qquad \text{[Using equation (1)]}$$

$$\Rightarrow \quad \frac{d^2y}{dx^2} + \frac{dy}{dx} - 6y = 0$$

which is the required differential equation.

Example 13. *Obtain a differential equation that should be satisfied by the family of concentric circles $x^2 + y^2 = a^2$.*

Solution. The given equation is

$$x^2 + y^2 = a^2 \qquad ...(1)$$

where a is an arbitrary constant.

Since the given equation (1) has only one arbitrary constant, so we shall differentiate it once, and we shall get a differential equation of first order.

Differentiating equation (1) w.r.t. x, we have

$$2x + 2y\frac{dy}{dx} = 0$$

$$\Rightarrow \quad x + y\frac{dy}{dx} = 0$$

which is the required differential equation.

Example 14. *Find the differential equation of all conics whose axes coincides with the axes of co-ordinates.*

Solution. The equation of any conic whose axes coincide with the axes of co-ordinate is given by

$$Ax^2 + By^2 = 1 \qquad ...(1)$$

where A and B are arbitrary constants. [For solution See Example 12 Part (*i*)]

Example 15. *Find the differential equation of all the circles in the first quadrant which touch the co-ordinate axes.*

Solution. The equation of all the circles in the first quadrant which touch the co-ordinate axes is given by

$$(x - a)^2 + (y - a)^2 = a^2 \qquad ...(1)$$

$$\Rightarrow \quad x^2 + a^2 - 2ax + y^2 + a^2 - 2ay = a^2$$

$$\Rightarrow \quad x^2 + y^2 - 2ax - 2ay + a^2 = 0 \qquad ...(2)$$

Differentiating equation (2), w.r.t. x, we have

$$2x + 2y\frac{dy}{dx} - 2a - 2a\frac{dy}{dx} = 0$$

$$\Rightarrow \quad x + y\frac{dy}{dx} - a - a\frac{dy}{dx} = 0 \qquad \text{[Cancelling 2 throughout]}$$

$$\Rightarrow \quad a\left(1 + \frac{dy}{dx}\right) = x + y\frac{dy}{dx}$$

$$\Rightarrow \quad a = \left(\frac{x + y\dfrac{dy}{dx}}{1 + \dfrac{dy}{dx}}\right) \qquad ...(3)$$

Substituting this value of a in equation (1) ; we have

$$\left(x - \frac{x + yy_1}{1 + y_1}\right)^2 + \left(y - \frac{x + yy_1}{1 + y_1}\right)^2 = \left(\frac{x + yy_1}{1 + y_1}\right)^2$$

$$\Rightarrow \quad \left(\frac{x + xy_1 - x - yy_1}{1 + y_1}\right)^2 + \left(\frac{y + yy_1 - x - yy_1}{1 + y_1}\right)^2 = \left(\frac{x + yy_1}{1 + y_1}\right)^2$$

$$\Rightarrow \quad [(x - y)y_1]^2 + (y - x)^2 = (x + yy_1)^2$$

$$\Rightarrow \quad (x - y)^2 y_1^2 + (x - y)^2 = (x + yy_1)^2$$

$$\Rightarrow \quad (x - y)^2(1 + y_1^2) = (x + yy_1)^2$$

$$\Rightarrow \quad (x - y)^2\left[1 + \left(\frac{dy}{dx}\right)^2\right] = \left[x + y\frac{dy}{dx}\right]^2$$

which is the required differential equation.

Example 16. *Find the differential equation of all ellipses centred at the origin.*

Solution. The equation of ellipse with centre at the origin is given by

$$\frac{x^2}{a^2} + \frac{y^2}{b^2} = 1 \quad ...(1)$$

Let us consider that $\quad \frac{1}{a^2} = A, \frac{1}{b^2} = B$

$$\Rightarrow \quad Ax^2 + By^2 = 1 \quad ...(2)$$

[**Hint.** See example 12 part (*i*)] $\quad \left[\textbf{Ans. } xy\frac{d^2y}{dx^2} + x\left(\frac{dy}{dx}\right)^2 - y\frac{dy}{dx} = 0\right]$

Example 17. *Find the differential equation of the family of curves represented by* $c(y + c)^2 = x^3$.

Solution. The given equation is

$$c(y + c)^2 = x^3 \quad ...(1)$$

where c is an arbitrary constant.

Differentiating equation (1) w.r.t. x, we have

$$2c(y + c)\frac{dy}{dx} = 3x^2 \quad ...(2)$$

Dividing equation (1) by (2), we have

$$\frac{c(y + c)^2}{2c(y + c)\frac{dy}{dx}} = \frac{x^3}{3x^2} \quad \Rightarrow \quad \frac{y + c}{2\frac{dy}{dx}} = \frac{x}{3}$$

$$\Rightarrow \quad y + c = \frac{2}{3}x\frac{dy}{dx}$$

$$\Rightarrow \quad c = \left(\frac{2}{3}x\frac{dy}{dx} - y\right) \quad ...(3)$$

Substituting this value of c in equation (2), we have

$$2\left[\frac{2}{3}x\frac{dy}{dx} - y\right]\left[y + \frac{2}{3}x\frac{dy}{dx} - y\right]\frac{dy}{dx} = 3x^2$$

$$\Rightarrow \quad \left(\frac{4}{3}x\frac{dy}{dx}-2y\right)\left[\frac{2}{3}x\left(\frac{dy}{dx}\right)^2\right] = 3x^2$$

$$\Rightarrow \quad \frac{8}{9}x^2\left(\frac{dy}{dx}\right)^3-\frac{4}{3}xy\left(\frac{dy}{dx}\right)^2 = 3x^2$$

$$\Rightarrow \quad 8x^2\left(\frac{dy}{dx}\right)^3-12xy\left(\frac{dy}{dx}\right)^2 = 27x^2$$

$$\Rightarrow \quad 8x\left(\frac{dy}{dx}\right)^3-12y\left(\frac{dy}{dx}\right)^2 = 27x \qquad \text{[Cancelling `}x\text{' throughout]}$$

which is the required differential equation.

Example 18. *Show that the differential equation that represents all parabolas having their axis of symmetry coincident with the axis of x is* $yy_2 + y_1^2 = 0$.

Solution. The general equation of any parabola having its axis of symmetry coincident with the axis of x is given by

$$y^2 = 4a(x-h) \qquad ...(1)$$

where a and h are arbitrary constants.

Since the equation (1) has two arbitrary constants, so we shall differentiate it twice and we shall get a differential equation of second order.

Differentiating equation (1) w.r.t. x, we have

$$2yy_1 = 4a(1-0)$$

$$\Rightarrow \quad yy_1 = 2a \qquad ....(2)$$

Differentiating again w.r.t. x, we have

$$yy_2 + y_1^2 = 0$$

which is the required differential equation.

Example 19. *Form a differential equation of which* $y = e^x(A \cos 2x + B \sin 2x)$ *is a solution.*

Solution. The given equation is

$$y = e^x\,(\text{A} \cos 2x + \text{B} \sin 2x) \qquad ...(1)$$

where A and B are arbitrary constants.

Differentiating equation (1) w.r.t. x, we have

$$\frac{dy}{dx} = e^x(\text{A} \cos 2x + \text{B} \sin 2x) + e^x(-2\text{A} \sin 2x + 2\text{B} \cos 2x)$$

$$\Rightarrow \quad \frac{dy}{dx} = y + e^x(-2\text{A} \sin 2x + 2\text{B} \cos 2x) \qquad ...(2) \quad \text{[Using equation (1)]}$$

Differentiating again w.r.t. x, we have

$$\frac{d^2y}{dx^2} = \frac{dy}{dx} + e^x(-2\text{A} \sin 2x + 2\text{B} \cos 2x) + e^x(-4\text{A} \cos 2x - 4\text{B} \sin 2x)$$

$$\Rightarrow \quad \frac{d^2y}{dx^2} = \frac{dy}{dx} + \left(\frac{dy}{dx} - y\right) - 4e^x(\text{A} \cos 2x + \text{B} \sin 2x) \qquad \text{[Using equation (2)]}$$

$$\Rightarrow \quad \frac{d^2y}{dx^2} = 2\frac{dy}{dx} - y - 4y \qquad \text{[Using equation (1)]}$$

$$\Rightarrow \quad \frac{d^2y}{dx^2} - 2\frac{dy}{dx} + 5y = 0$$

which is the required differential equation.

Example 20. *Assume that a spherical rain drop evaporates at a rate proportional to its surface area. Form a differential equation involving the rate of change of the radius of the rain drop.*

Solution. Let r be the radius of the rain drop at any time t.

Let V be the volume of the rain drop and S be the surface area of the rain drop.

Now, according to the given condition

$$\frac{dv}{dt} \propto S$$

$$\Rightarrow \quad \frac{dv}{dt} = -kS \quad ...(1)$$

where k is the constant of proportionality. (–ve) sign shows that volume is decreasing due to evaporation.

$$\Rightarrow \quad \frac{d}{dt}\left(\frac{4}{3}\pi r^3\right) = -k(4\pi r^2) \quad \left[\begin{array}{l}\because \text{Volume of sphere} = \frac{4}{3}\pi r^3 \\ \text{Surface area of sphere} = 4\pi r^2\end{array}\right]$$

$$\Rightarrow \quad \frac{4}{3}\pi . 3r^2 \frac{dr}{dt} = -4\pi r^2 k$$

$$\Rightarrow \quad \frac{dr}{dt} = -k$$

which is the required differential equation.

Example 21. *Find the differential equation of the family of curves* $xy = Ae^x + Be^{-x} + x^2$.

Solution. The given equation is

$$xy = Ae^x + Be^{-x} + x^2 \quad ...(1)$$

where A and B are arbitrary constants.

Since the given equation contains two arbitrary constants, so we shall differentiate it two times and we shall get a differential equation of second order.

Differentiating equation (1) w.r.t. x, we have

$$y + x\frac{dy}{dx} = Ae^x - Be^{-x} + 2x \quad ...(2)$$

Differentiating again w.r.t. x, we have

$$\frac{dy}{dx} + \frac{dy}{dx} + x\frac{d^2y}{dx^2} = Ae^x + Be^x + 2$$

$$\Rightarrow \quad 2\frac{dy}{dx} + x\frac{d^2y}{dx^2} = xy - x^2 + 2 \quad \text{[Using equation (1)]}$$

$$\Rightarrow \quad x\frac{d^2y}{dx^2} + 2\frac{dy}{dx} - xy + x^2 - 2 = 0$$

which is the required differential equation.

Example 22. *Find the differential equation of all circles in the xy-plane.*

Solution. The general equation of any circle in the xy-plane is given by

$$x^2 + y^2 + 2gx + 2fy + c = 0 \quad ...(1)$$

where g, f and c are arbitrary constants.

Since the given differential equation has three arbitrary constants, so, we shall differentiate it three times and we shall get a differential equation of order three.

Differentiating equation (1) w.r.t. x, we have

$$2x + 2yy_1 + 2g + 2fy_1 = 0$$

$$\Rightarrow \quad x + yy_1 + g + fy_1 = 0 \quad \text{...(2) [Cancelling 2 throughout]}$$

Differentiating again w.r.t. x, we have

$$1 + yy_2 + y_1^2 + fy_2 = 0 \quad \text{...(3)}$$

Differentiating again w.r.t. x, we have

$$0 + yy_3 + y_1y_2 + 2y_1y_2 + fy_3 = 0$$

$$\Rightarrow \quad yy_3 + 3y_1y_2 + fy_3 = 0 \quad \text{...(4)}$$

Now, multiplying equation (3) by y_3 and (4) by y_2 and subtract, we have

$$y_3 + yy_2y_3 + y_1^2y_3 - yy_2\,y_3 - 3y_1y_2^2 = 0$$

$$\Rightarrow \quad (1 + y_1^2)y_3 - 3y_1y_2^2 = 0$$

$$\Rightarrow \quad \left[1+\left(\frac{dy}{dx}\right)^2\right]\frac{d^3y}{dx^3} - 3\frac{dy}{dx}\left(\frac{d^2y}{dx^2}\right)^2 = 0$$

which is the required differential equation.

Example 23. *Eliminate m from* $\sqrt{1-x^2} + \sqrt{1-y^2} = m(x-y)$.

Solution. The given equation is

$$\sqrt{1-x^2} + \sqrt{1-y^2} = m(x-y) \quad \text{...(1)}$$

Putting $x = \sin\theta$, $y = \sin\phi$ in equation (1) ; we have

$$\sqrt{1-\sin^2\theta} + \sqrt{1-\sin^2\phi} = m(\sin\theta - \sin\phi)$$

$$\Rightarrow \quad \sqrt{\cos^2\theta} + \sqrt{\cos^2\phi} = m\left[2\cos\left(\frac{\theta+\phi}{2}\right)\sin\left(\frac{\theta-\phi}{2}\right)\right]$$

$$\left[\begin{array}{l} \because \cos^2 A + \sin^2 A = 1 \\ \sin C - \sin D = 2\cos\dfrac{C+D}{2}\sin\dfrac{C-D}{2} \\ \cos C + \cos D = 2\cos\dfrac{C+D}{2}\cos\dfrac{C-D}{2} \end{array}\right]$$

$$\Rightarrow \quad \cos\theta + \cos\phi = 2m\cos\left(\frac{\theta+\phi}{2}\right)\sin\left(\frac{\theta-\phi}{2}\right)$$

$$\Rightarrow \quad 2\cos\left(\frac{\theta+\phi}{2}\right)\cos\left(\frac{\theta-\phi}{2}\right) = 2m\cos\left(\frac{\theta+\phi}{2}\right)\sin\left(\frac{\theta-\phi}{2}\right)$$

$$\Rightarrow \quad \cos\left(\frac{\theta-\phi}{2}\right) = m\sin\left(\frac{\theta-\phi}{2}\right)$$

$$\Rightarrow \quad m = \cot\left(\frac{\theta-\phi}{2}\right)$$

$$\Rightarrow \quad \frac{\theta-\phi}{2} = \cot^{-1} m$$

$$\Rightarrow \qquad \theta - \phi = 2\cot^{-1} m$$

$$\Rightarrow \qquad \sin^{-1} x - \sin^{-1} y = 2\cot^{-1} m \qquad ...(2) \quad [\because \sin\theta = x, \sin\phi = y]$$

Differentiating equation (2), w.r.t. x, we have

$$\frac{1}{\sqrt{1-x^2}} - \frac{1}{\sqrt{1-y^2}}\frac{dy}{dx} = 0$$

$$\Rightarrow \qquad \frac{1}{\sqrt{1-y^2}}\frac{dy}{dx} = \frac{1}{\sqrt{1-x^2}}$$

$$\Rightarrow \qquad \frac{dy}{dx} = \frac{\sqrt{1-y^2}}{\sqrt{1-x^2}}$$

$$\Rightarrow \qquad \frac{dy}{dx} = \sqrt{\frac{1-y^2}{1-x^2}}$$

which is the required differential equation.

5.12. SOLUTION OF A DIFFERENTIAL EQUATION

In this section we shall discuss the problems based on finding the solution of a differential equation.

SOME SOLVED EXAMPLES

Example 1. *Show that $y = Ax + \frac{B}{x}$, $x \neq 0$ is a solution of the differential equation*

$$x^2\frac{d^2y}{dx^2} + x\frac{dy}{dx} - y = 0.$$

Solution. The given equation is

$$y = Ax + \frac{B}{x}, \quad x \neq 0 \qquad ...(1)$$

Differentiating equation (1) w.r.t. x, we have

$$\frac{dy}{dx} = A - \frac{B}{x^2} \qquad ....(2)$$

Differentiating again w.r.t. x, we have

$$\frac{d^2y}{dx^2} = \frac{2B}{x^3} \qquad ...(3)$$

Now, substituting the values of equations (1), (2) and (3) in the equation

$$x^2\frac{d^2y}{dx^2} + x\frac{dy}{dx} - y = 0$$

$$\Rightarrow \qquad x^2\left(\frac{2B}{x^3}\right) + x\left(A - \frac{B}{x^2}\right) - \left(Ax + \frac{B}{x}\right) = 0$$

$$\Rightarrow \qquad \frac{2B}{x} + Ax - \frac{B}{x} - Ax - \frac{B}{x} = 0$$

$\therefore$ $\qquad y = Ax + \frac{B}{x}$ is a solution of the given differential equation.

Example 2. *Show that $y = 4 \sin 3x$ is a solution of the differential equation $\frac{d^2y}{dx^2} + 9y = 0$.*

Solution. The given equation is

$$y = 4 \sin 3x \qquad ...(1)$$

The given differential equation is

$$\frac{d^2y}{dx^2} + 9y = 0 \qquad ...(2)$$

Differentiating equation (1) w.r.t. x, we have

$$\frac{dy}{dx} = 12 \cos 3x \qquad ...(3)$$

Differentiating again w.r.t. x, we have

$$\frac{d^2y}{dx^2} = -36 \sin 3x \qquad ...(4)$$

Now, substituting the values of y and $\frac{d^2y}{dx^2}$ in the L.H.S. of equation (2), we have

$$\frac{d^2y}{dx^2} + 9y = -36 \sin 3x + 9 (4 \sin 3x) = -36 \sin 3x + 36 \sin 3x = 0$$

$\therefore$ $y = 4 \sin 3x$ is a solution of the given differential equation.

Example 3. *Show that the equation $y = A \cos 2x - B \sin 2x$ is a solution of the differential equation $\frac{d^2y}{dx^2} + 4y = 0$.*

Solution. The given function is

$$y = A \cos 2x - B \sin 2x \qquad ...(1)$$

The given differential equation is

$$\frac{d^2y}{dx^2} + 4y = 0 \qquad ...(2)$$

Differentiating equation (1) w.r.t. x, we have

$$\frac{dy}{dx} = -2A \sin 2x - 2B \cos 2x \qquad ...(3)$$

Differentiating again w.r.t. x, we have

$$\frac{d^2y}{dx^2} = -4A \cos 2x + 4B \sin 2x = -4[A \cos 2x - B \sin 2x]$$

$$\Rightarrow \quad \frac{d^2y}{dx^2} = -4y \qquad \text{[Using equation (1)]}$$

$$\Rightarrow \quad \frac{d^2y}{dx^2} + 4y = 0$$

which is same as the given differential equation.

$\therefore$ $y = A \cos 2x - B \sin 2x$ is a solution of the given differential equation.

Example 4. *Show that $y = a \cos x + b \sin x$ is a solution of the differential equation.*

$$\cos x \frac{dy}{dx} + y \sin x = b.$$

Solution. The given function is

$$y = a \cos x + b \sin x \qquad ...(1)$$

The given differential equation is

$$\cos x \frac{dy}{dx} + y \sin x = b \qquad ...(2)$$

Differentiating equation (1) w.r.t. x, we have

$$\frac{dy}{dx} = -a \sin x + b \cos x \qquad ...(3)$$

Now, substituting the values of y and $\frac{dy}{dx}$ in the L.H.S. of the equation (2), we have

$$\cos x \frac{dy}{dx} + y \sin x = \cos x(-a \sin x + b \cos x) + (a \cos x + b \sin x) \sin x$$

$$= -a \sin x \cos x + b \cos^2 x + a \sin x \cos x + b \sin^2 x$$

$$= b(\cos^2 x + \sin^2 x)$$

$$= b \text{ which is true.} \qquad [\because \cos^2 A + \sin^2 A = 1]$$

$\therefore$ $y = a \cos x + b \sin x$ is a solution of the given differential equation.

Example 5. *Verify that $y = (x^3 - x) \log (cx)$ satisfies the differential equation*

$$(x^3 - x)y_1 - (3x^2 - 1)y = x^5 - 2x^3 + x \quad \textit{for all values of } c.$$

Solution. The given function is

$$y = (x^3 - x) \log (cx) \qquad ...(1)$$

The given differential equation is

$$(x^3 - x)y_1 - (3x^2 - 1)y = x^5 - 2x^3 + x \qquad ...(2)$$

Differentiating equation (1) w.r.t. x, we have

$$y_1 = (x^3 - x) . \frac{1}{cx} . c + (3x^2 - 1) \log (cx)$$

$$\Rightarrow \quad y_1 = (x^2 - 1) + (3x^2 - 1) \log (cx) \qquad ...(3)$$

Now, substituting the values of y and y_1 in the L.H.S. of the equation (2), we have

$$(x^3 - x)y_1 - (3x^2 - 1)y = (x^3 - x)[(x^2 - 1) + (3x^2 - 1) \log cx] - (3x^2 - 1)[(x^3 - x) \log(cx)]$$

$$= (x^2 - 1)(x^3 - x) + (x^3 - x)(3x^2 - 1) \log cx - (x^3 - x)(3x^2 - 1) \log(cx)$$

$$= x^5 - x^3 - x^3 + x = x^5 - 2x^3 + x \quad \text{which is true.}$$

$\therefore$ $y = (x^3 - x) \log(cx)$ satisfies the given differential equation.

Example 6. *Show that the equation $y = ae^{2x} + be^{-x}$ is a solution of the differential equation*

$$\frac{d^2y}{dx^2} - \frac{dy}{dx} - 2y = 0.$$

Solution. The given equation is

$$y = ae^{2x} + be^{-x} \qquad ...(1)$$

The given differential equation is

$$\frac{d^2y}{dx^2} - \frac{dy}{dx} - 2y = 0 \qquad ...(2)$$

Differentiating equation (1) w.r.t. x, we have

$$\frac{dy}{dx} = 2ae^{2x} - be^{-x} \qquad ...(3)$$

Differentiating again w.r.t. x, we have

$$\frac{d^2y}{dx^2} = 4ae^{2x} + be^{-x} \qquad ...(4)$$

Now, substituting the values of y, $\frac{dy}{dx}$ and $\frac{d^2y}{dx^2}$ in the L.H.S. of equation (2), we have

$$\frac{d^2y}{dx^2} - \frac{dy}{dx} - 2y = (4ae^{2x} + be^{-x}) - (2ae^{2x} - be^{-x}) - 2(ae^{2x} + be^{-x})$$

$$= 4ae^{2x} + be^{-x} - 2ae^{2x} + be^{-x} - 2ae^{2x} - 2be^{-x} = 0 \text{ which is true.}$$

$\therefore$ $y = ae^{2x} + be^{-x}$ is a solution of the given differential equation.

Example 7. *Show that* $y = \frac{c}{x} + d$ *is a solution of the differential equation*

$$\frac{d^2y}{dx^2} + \frac{2}{x}\left(\frac{dy}{dx}\right) = 0.$$

Solution. The given function is

$$y = \frac{c}{x} + d \qquad ...(1)$$

The given differential equation is

$$\frac{d^2y}{dx^2} + \frac{2}{x}\left(\frac{dy}{dx}\right) = 0 \qquad ...(2)$$

Differentiating equation (1) w.r.t. x, we have

$$\frac{dy}{dx} = -\frac{c}{x^2} \qquad ...(3)$$

Differentiating again w.r.t. x, we have

$$\frac{d^2y}{dx^2} = \frac{2c}{x^3} \qquad ...(4)$$

Now, substituting the values of $\frac{dy}{dx}$ and $\frac{d^2y}{dx^2}$ in the L.H.S. of equation (2), we have

$$\frac{d^2y}{dx^2} + \frac{2}{x}\left(\frac{dy}{dx}\right) = \frac{2c}{x^3} + \frac{2}{x}\left(-\frac{c}{x^2}\right)$$

$$= \frac{2c}{x^3} - \frac{2c}{x^3} = 0$$

which is true.

$\therefore$ $y = \frac{c}{x} + d$ is a solution of the given differential equation.

Example 8. *Verify that* $y + x + 1 = 0$ *is a solution of the differential equation*

$$(y - x)dy - (y^2 - x^2)dx = 0.$$

Solution. The given function is

$$y + x + 1 = 0 \qquad ...(1)$$

The given differential equation is

$$(y - x)dy - (y^2 - x^2)dx = 0 \qquad ...(2)$$

Differentiating equation (1) w.r.t. x, we have

$$\frac{dy}{dx} + 1 + 0 = 0$$

$$\Rightarrow \quad \frac{dy}{dx} = -1$$

$$\Rightarrow \quad dy = -dx \qquad ...(3)$$

Substituting this value in the L.H.S. of equation (2), we have

$$(y - x)dy - (y^2 - x^2)dx = (y - x)[dy - (y + x)dx]$$

$$= (y - x)[-dx - (-1)dx] \qquad [\because \; y + x + 1 = 0 \Rightarrow (y + x) = -1]$$

$$= (y - x)(-dx + dx) = (y - x).0 = 0 \quad \text{which is true.}$$

$\therefore$ $y + x + 1 = 0$ is a solution of the given differential equation.

Example 9. *Show that the function $y = e^{3x}(A + Bx)$ is a solution of the differential equation*

$$\frac{d^2y}{dx^2} - 6\frac{dy}{dx} + 9y = 0.$$

Solution. The given function is

$$y = e^{3x}(A + Bx) \qquad ...(1)$$

The given differential equation is

$$\frac{d^2y}{dx^2} - 6\frac{dy}{dx} + 9y = 0 \qquad ...(2)$$

Differentiating equation (1) w.r.t. x, we have

$$\frac{dy}{dx} = 3e^{3x}(A + Bx) + Be^{3x} \qquad ...(3)$$

Differentiating again w.r.t. x, we have

$$\frac{d^2y}{dx^2} = 9e^{3x}(A + Bx) + 6Be^{3x} \qquad ...(4)$$

Now, substituting the values of y, $\frac{dy}{dx}$ and $\frac{d^2y}{dx^2}$ in the L.H.S. of equation (2), we have

$$\frac{d^2y}{dx^2} - 6\frac{dy}{dx} + 9y = [9e^{3x}(A + Bx) + 6Be^{3x}] - 6[3e^{3x}(A + Bx) + Be^{3x}] + 9[e^{3x}(A + Bx)]$$

$$= 9Ae^{3x} + 9Bxe^{3x} + 6Be^{3x} - 18Ae^{3x} - 18Bxe^{3x} - 6Be^{3x} + 9Ae^{3x} + 9Bxe^{3x}$$

$$= 18Ae^{3x} - 18Ae^{3x} + 0 = 0 \quad \text{which is true.}$$

$\therefore$ $y = e^{3x}(A + Bx)$ is a solution of the given differential equation.

Example 10. *Show that $y = Ce^{-x}$ is a solution of the differential equation $\frac{dy}{dx} + y = 0$.*

Solution. The given function is

$$y = Ce^{-x} \quad ...(1)$$

The given differential equation is

$$\frac{dy}{dx} + y = 0 \quad ...(2)$$

Differentiating equation (1) w.r.t. x, we have

$$\frac{dy}{dx} = -Ce^{-x}$$

$$\Rightarrow \quad \frac{dy}{dx} = -y \quad \text{[Using equation (1)]}$$

$$\Rightarrow \quad \frac{dy}{dx} + y = 0$$

which is same as the given differential equation

$\therefore$ $y = Ce^{-x}$ is a solution of the differential equation $\frac{dy}{dx} + y = 0$.

Example 11. *Show that $y = a \cos(\log x) + b \sin(\log x)$ is a solution of the differential equation*

$$x^2 \frac{d^2y}{dx^2} + x\frac{dy}{dx} + y = 0.$$

Solution. The given function is

$$y = a \cos(\log x) + b \sin(\log x) \quad ...(1)$$

The given differential equation is

$$x^2 \frac{d^2y}{dx^2} + x\frac{dy}{dx} + y = 0 \quad ...(2)$$

Differentiating equation (1) w.r.t. x, we have

$$\frac{dy}{dx} = -\frac{a \sin(\log x)}{x} + \frac{b \cos(\log x)}{x}$$

$$\Rightarrow \quad x\frac{dy}{dx} = -a \sin(\log x) + b \cos(\log x) \quad ...(3)$$

Differentiating again w.r.t. x, we have

$$x\frac{d^2y}{dx^2} + 1.\frac{dy}{dx} = -\frac{a \cos(\log x)}{x} - \frac{b \sin(\log x)}{x}$$

$$\Rightarrow \quad x^2 \frac{d^2y}{dx^2} + x\frac{dy}{dx} = -[a \cos(\log x) + b \sin(\log x)]$$

$$\Rightarrow \quad x^2 \frac{d^2y}{dx^2} + x\frac{dy}{dx} = -y \quad \text{[Using equation (1)]}$$

$$\Rightarrow \quad x^2 \frac{d^2y}{dx^2} + x\frac{dy}{dx} + y = 0$$

which is same as equation (2).

$\therefore$ $y = a \cos(\log x) + b \sin(\log x)$ is a solution of the given differential equation.

Example 12. *Show that $y = e^{-x} + ax + b$ is a solution of differential equation $e^x \frac{d^2y}{dx^2} = 1$.*

Solution. The given function is

$$y = e^{-x} + ax + b \quad ...(1)$$

The given differential equation is

$$e^x \frac{d^2y}{dx^2} = 1 \quad ...(2)$$

Differentiating equation (1) w.r.t. x, we have

$$\frac{dy}{dx} = -e^{-x} + a \quad ...(3)$$

Differentiating again w.r.t. x, we have

$$\frac{d^2y}{dx^2} = -e^{-x}(-1) = e^{-x} = \frac{1}{e^x}$$

$$\Rightarrow \quad e^x \frac{d^2y}{dx^2} = 1$$

which is same as equation (2).

$\therefore$ $y = e^{-x} + ax + b$ is the solution of the given differential equation.

Example 13. *Show that the differential equation, of which $y = 2(x^2 - 1) + Ce^{-x^2}$ is a solution is $\frac{dy}{dx} + 2xy = 4x^3$.*

Solution. The given equation is

$$y = 2(x^2 - 1) + Ce^{-x^2} \quad ...(1)$$

The given differential equation is

$$\frac{dy}{dx} + 2xy = 4x^3 \quad ....(2)$$

Differentiating equation (1) w.r.t. x, we have

$$\frac{dy}{dx} = 2(2x - 0) + Ce^{-x^2}(-2x)$$

$$\Rightarrow \quad \frac{dy}{dx} = 4x - 2Cxe^{-x^2} \quad ...(3)$$

Now, substituting the values of y and $\frac{dy}{dx}$ in the L.H.S. of equation (2), we have

$$\frac{dy}{dx} + 2xy = 4x - 2Cxe^{-x^2} + 2x[2(x^2 - 1) + Ce^{-x^2}]$$

$$= 4x - 2Cxe^{-x^2} + 4x^3 - 4x + 2Cxe^{-x^2} = 4x^3$$

which is true.

$\therefore$ $y = 2(x^2 - 1) + Ce^{-x^2}$ is a solution of the given differential equation.

Example 14. *Show that $Ax^2 + By^2 = 1$ is a solution of the differential equation*

$$x\left[y\frac{d^2y}{dx^2} + \left(\frac{dy}{dx}\right)^2\right] = y.\frac{dy}{dx}.$$

Solution. The given equation is

$$Ax^2 + By^2 = 1 \quad ...(1)$$

The given differential equation is

$$x\left[y\frac{d^2y}{dx^2} + \left(\frac{dy}{dx}\right)^2\right] = y.\frac{dy}{dx} \quad ...(2)$$

Differentiating equation (1) w.r.t. x, we have

$$2Ax + 2By\frac{dy}{dx} = 0$$

$$\Rightarrow \quad Ax + By\frac{dy}{dx} = 0 \quad ...(3) \quad \text{[Cancelling '2' throughout]}$$

Differentiating again w.r.t. x, we have

$$A + B\left[\left(\frac{dy}{dx}\right)^2 + y\frac{d^2y}{dx^2}\right] = 0$$

$$\Rightarrow \quad A = -B\left[\left(\frac{dy}{dx}\right)^2 + y\frac{d^2y}{dx^2}\right]$$

$$\Rightarrow \quad -\frac{A}{B} = \left(\frac{dy}{dx}\right)^2 + y\frac{d^2y}{dx^2} \quad ...(4)$$

From equation (3), we have

$$Ax = -By\frac{dy}{dx}$$

$$\Rightarrow \quad -\frac{A}{B} = \frac{y}{x}\frac{dy}{dx} \quad ...(5)$$

Equating equations (4) and (5), we have

$$\frac{y}{x}\frac{dy}{dx} = \left(\frac{dy}{dx}\right)^2 + y\frac{d^2y}{dx^2}$$

$$\Rightarrow \quad y\frac{dy}{dx} = x\left(\frac{dy}{dx}\right)^2 + xy\frac{d^2y}{dx^2}$$

$$\Rightarrow \quad xy\frac{d^2y}{dx^2} + x\left(\frac{dy}{dx}\right)^2 - y\frac{dy}{dx} = 0$$

which is same as given equation (2)

$\therefore$ $Ax^2 + By^2 = 1$ is the solution of the given differential equation.

Example 15. *Show that $y = be^x + Ce^{2x}$ is a solution of the differential equation*

$$y_2 - 3y_1 + 2y = 0.$$

Solution. The given function is

$$y = be^x + Ce^{2x} \quad ...(1)$$

The given differential equation is

$$\frac{d^2y}{dx^2} - 3\frac{dy}{dx} + 2y = 0 \quad ...(2)$$

Differentiating equation (1), w.r.t. x, we have

$$\frac{dy}{dx} = be^x + 2Ce^{2x} \qquad ...(3)$$

Differentiating again w.r.t. x, we have

$$\frac{d^2y}{dx^2} = be^x + 4Ce^{2x} \qquad ...(4)$$

Now, substituting the values of y, $\frac{dy}{dx}$ and $\frac{d^2y}{dx^2}$ in the L.H.S. of equation (2), we have

$$\frac{d^2y}{dx^2} - 3\frac{dy}{dx} + 2y = (be^x + 4Ce^{2x}) - 3(be^x + 2Ce^{2x}) + 2(be^x + Ce^{2x})$$

$$= be^x + 4Ce^{2x} - 3be^x - 6Ce^{2x} + 2be^x + 2Ce^{2x}$$

$$= 3bex^e - 3be^x + 6Ce^{2x} - 6Ce^{2x} = 0$$

which is true.

$\therefore$ The equation $y = be^x + Ce^{2x}$ is a solution of the given differential equation.

Example. 16. *Show that $y = Cx + \frac{a}{C}$ and $y^2 = 4ax$ are the solutions of the differential equation $y = x\frac{dy}{dx} + a\frac{dx}{dy}$.*

Solution. The given differential equation is

$$y = x\frac{dy}{dx} + a\frac{dx}{dy} \qquad ...(1)$$

(*i*) The given equation is

$$y = Cx + \frac{a}{C} \qquad ...(2)$$

Differentiating equation (2), w.r.t. x

$$\frac{dy}{dx} = C \qquad ...(3)$$

$$\Rightarrow \qquad \frac{dx}{dy} = \frac{1}{C} \qquad ...(4)$$

Now, substituting these values of $\frac{dy}{dx}$ and $\frac{dx}{dy}$ in equation (1), we have

$$y = xC + \frac{a}{C} \quad \text{which is true.}$$

$$\therefore \qquad y = Cx + \frac{a}{C} \text{ is a solution of } y = x\frac{dy}{dx} + a\frac{dx}{dy}$$

(*ii*) The given equation is

$$y^2 = 4ax \qquad ...(5)$$

Differentiating equation (5), w.r.t. x, we have

$$2y\frac{dy}{dx} = 4a \qquad \Rightarrow \qquad \frac{dy}{dx} = \frac{4a}{2y}$$

$$\Rightarrow \quad \frac{dy}{dx} = \frac{2a}{y} \quad ...(6)$$

$$\Rightarrow \quad \frac{dx}{dy} = \frac{y}{2a} \quad ...(7)$$

Now, substituting these values of $\frac{dy}{dx}$ and $\frac{dx}{dy}$ in equation (1), we have

$$y = x\frac{dy}{dx} + a\frac{dx}{dy} = x\left(\frac{2a}{y}\right) + a\left(\frac{y}{2a}\right)$$

$$= \frac{2ax}{y} + \frac{y}{2} = \frac{4ax + y^2}{2y}$$

$$= \frac{y^2 + y^2}{2y} = \frac{4y^2}{2y} = y \quad \text{[Using equation (5)]}$$

which is true.

$\therefore$ $y^2 = 4ax$ is a solution of the differential equation $y = x\frac{dy}{dx} + a\frac{dx}{dy}$.

Example 17. *Show that $xy = ae^x + be^{-x} + x^2$ is a solution of the differential equation*

$$x\frac{d^2y}{dx^2} + 2\frac{dy}{dx} - xy + x^2 - 2 = 0.$$

Solution. The given equation is

$$xy = ae^x + be^{-x} + x^2 \quad ...(1)$$

The given differential equation is

$$x\frac{d^2y}{dx^2} + 2\frac{dy}{dx} - xy + x^2 - 2 = 0 \quad ...(2)$$

Differentiating equation (1) w.r.t. x, we have

$$x\frac{dy}{dx} + y = ae^x - be^{-x} + 2x \quad ...(3)$$

Differentiating again w.r.t. x, we have

$$x\frac{d^2y}{dx^2} + \frac{dy}{dx} + \frac{dy}{dx} = ae^x + be^{-x} + 2$$

$$\Rightarrow \quad x\frac{d^2y}{dx^2} + 2\frac{dy}{dx} = ae^x + be^{-x} + 2 \quad ...(4)$$

$$\Rightarrow \quad x\frac{d^2y}{dx^2} + 2\frac{dy}{dx} = (xy - x^2) + 2 \quad \text{[Using equation (1)]}$$

$$\Rightarrow \quad x\frac{d^2y}{dx^2} + 2\frac{dy}{dx} - xy + x^2 - 2 = 0$$

which is same as equation (2).

$\therefore$ $xy = ae^x + be^{-x} + x^2$ is a solution of the given differential equation.

Example 18. *Show that $y = x^3 + ax^2 + bx + C$ is a solution of the differential equation*

$$\frac{d^3y}{dx^3} = 6.$$

Solution. The given equation is

$$y = x^3 + ax^2 + bx + C \qquad ...(1)$$

The given differential equation is

$$\frac{d^3y}{dx^3} = 6 \qquad ...(2)$$

Differentiating equation (1) w.r.t. x, we have

$$\frac{dy}{dx} = 3x^2 + 2ax + b \qquad ...(3)$$

Differentiating again w.r.t. x, we have

$$\frac{d^2y}{dx^2} = 6x + 2a \qquad ...(4)$$

Differentiating again w.r.t. x, we have

$$\frac{d^3y}{dx^3} = 6$$

which is same as the given equation (2).

$\therefore \quad y = x^3 + ax^2 + bx + C$ is a solution of the given differential equation $\dfrac{d^3y}{dx^3} = 6$.

Example 19. *Solve the following differential equations :*

(*i*) $\dfrac{dy}{dx} = \dfrac{x}{x^2+1}$ (*ii*) $\dfrac{dy}{dx} = \dfrac{1}{x^2+2x+2}$

(*iii*) $(x^2 + 1)dy = dx$ (*iv*) $\dfrac{dy}{dx} + 2x = e^{3x}$

(*v*) $(x + 2)\dfrac{dy}{dx} = x^2 + 4x - 9$ (*vi*) $\dfrac{dy}{dx} - x\sin^2 x = \dfrac{1}{x\log x}$.

Solution. (*i*) The given equation is

$$\frac{dy}{dx} = \frac{x}{x^2+1} \qquad ...(1)$$

Separating the variables

$$dy = \frac{x}{x^2+1}.dx$$

Integrating both sides, we get

$$\int dy = \int \frac{x}{x^2+1}\,dx$$

$$\Rightarrow \quad y = \frac{1}{2}\int \frac{2x}{x^2+1}\,dx \qquad \text{[Multiply and divided by 2]}$$

$$\Rightarrow \quad y = \frac{1}{2}\log|x^2+1| + C \qquad \left[\because \int \frac{f'(x)}{f(x)}\,dx = \log|f(x)| + C\right]$$

which is the required solution.

(*ii*) The given equation is

$$\frac{dy}{dx} = \frac{1}{x^2+2x+2} \qquad ...(1)$$

Separating the variables,

$$dy = \frac{1}{x^2 + 2x + 2} . dx$$

Integrating both sides, we get

$$\int dy = \int \frac{1}{x^2 + 2x + 2} . dx$$

$$\Rightarrow \quad y = \int \frac{1}{x^2 + 2x + 1 + 2 - 1} . dx$$

$$\left[\begin{array}{l}\text{Add and subtract 1 to the denominator.} \\ \because \left(\frac{1}{2} \text{co-eff. of } x\right)^2 = 1\end{array}\right]$$

$$= \int \frac{1}{(x+1)^2 + (1)^2} . dx \qquad \left[\because \int \frac{1}{x^2 + a^2} . dx = \frac{1}{a} \tan^{-1} \frac{x}{a}\right]$$

$$= \tan^{-1}(x + 1) + C$$

which is the required solution.

(*iii*) The given equation is

$$(x^2 + 1) \frac{dy}{dx} = 1 \qquad ...(1)$$

Separating the variables,

$$\frac{dy}{dx} = \frac{1}{x^2 + 1}$$

$$\Rightarrow \quad dy = \frac{1}{x^2 + 1} . dx$$

Integrating both sides, we get

$$\int dy = \int \frac{1}{x^2 + 1} . dx$$

$$\Rightarrow \quad y = \tan^{-1} x + C \qquad \left[\because \int \frac{1}{x^2 + a^2} . dx = \frac{1}{a} \tan^{-1} \frac{x}{a}\right]$$

which is the required solution.

(*iv*) The given equation is

$$\frac{dy}{dx} + 2x = e^{3x}$$

$$\Rightarrow \quad \frac{dy}{dx} = e^{3x} - 2x$$

Separating the variables, $dy = (e^{3x} - 2x)dx$

Integrating both sides, we get

$$\int dy = \int (e^{3x} - 2x) . dx$$

$$\Rightarrow \qquad y = \frac{e^{3x}}{3} - 2.\frac{x^2}{2} + C$$

$$\Rightarrow \qquad y = \frac{e^{3x}}{3} - x^2 + C$$

which is the required solution.

(*v*) The given equation is

$$(x + 2)\frac{dy}{dx} = (x^2 + 4x - 9)$$

$$\Rightarrow \qquad \frac{dy}{dx} = \frac{x^2 + 4x - 9}{x + 2}$$

Separating the variables $dy = \dfrac{x^2 + 4x - 9}{x + 2}.dx$

Integrating both sides, we get

$$\int dy = \int\left(x + 2 - \frac{13}{x + 2}\right) dx$$

$$\Rightarrow \qquad y = \int (x + 2).dx - 13\int \frac{1}{x + 2}\, dx$$

$$\Rightarrow \qquad y = \frac{x^2}{2} + 2x - 13 \log|x + 2| + C$$

which is the required solution.

(*vi*) The given equation is

$$\frac{dy}{dx} - x\sin^2 x = \frac{1}{x\log x}$$

$$\Rightarrow \qquad \frac{dy}{dx} = \frac{1}{x\log x} + x\sin^2 x$$

$$\Rightarrow \qquad \frac{dy}{dx} = \frac{1}{x\log x} + x\left(\frac{1 - \cos 2x}{2}\right) \qquad \left[\begin{array}{l} \because 1 - \cos 2A = 2\sin^2 A \\ \Rightarrow \dfrac{1 - \cos 2A}{2} = \sin^2 A \end{array}\right]$$

Separating the variables,

$$dy = \left(\frac{1/x}{\log x} + \frac{x}{2} - \frac{x\cos 2x}{2}\right).dx$$

Integrating both sides, we get

$$\int dy = \int\left(\frac{1/x}{\log x} + \frac{x}{2} - \frac{x\cos 2x}{2}\right).dx$$

$$\Rightarrow \qquad y = \int \frac{1/x}{\log x}\, dx + \frac{1}{2}\int x\, dx - \frac{1}{2}\int x\cos 2x\, dx$$

$$\left[\because \int \frac{f'(x)}{f(x)}\, dx = \log|(f(x)|\right]$$

$$= \log|\log x| + \frac{1}{2}.\frac{x^2}{2} - \frac{1}{2}\int \underset{\text{I}}{x} \underset{\text{II}}{\cos 2x}\, dx \qquad \text{[Integrating by parts]}$$

$$= \log|\log x| + \frac{x^2}{4} - \frac{1}{2}\left[x.\int \cos 2x\, dx - \int\left[\frac{d}{dx}(x)\int \cos 2x.dx\right]dx\right]$$

$$= \log|\log x| + \frac{x^2}{4} - \frac{1}{2}\left[x\frac{\sin 2x}{2} - \int \frac{1.\sin 2x}{2}dx\right]$$

$$\Rightarrow \quad y = \log|\log x| + \frac{x^2}{4} - \frac{1}{4}x\sin 2x - \frac{1}{8}\cos 2x + C$$

which is the required solution.

Example 20. *Solve the following differential equations :*

(*i*) $\frac{dy}{dx} = e^{x+y} + x^2e^y$

(*ii*) $\frac{dy}{dx} = e^{x-y} + e^{2\log x - y}$

(*iii*) $y - x\frac{dy}{dx} = a\left(y^2 + \frac{dy}{dx}\right)$

(*iv*) $\frac{dy}{dx} = \tan^2 x$

(*v*) $\frac{dy}{dx} = x^2 + \sin 3x$

(*vi*) $\frac{dy}{dx} = \sin^3 x\cos^2 x + xe^x.$

Solution. (*i*) The given equation is

$$\frac{dy}{dx} = e^{x+y} + x^2e^y = e^x.e^y + x^2e^y$$

$$\Rightarrow \quad \frac{dy}{dx} = e^y(e^x + x^2)$$

Separating the variables,

$$\frac{1}{e^y}dy = (e^x + x^2)dx$$

Integrating both sides, we get

$$\int e^{-y}\, dy = \int (e^x + x^2)\, dx$$

$$\Rightarrow \quad \frac{e^{-y}}{-1} = e^x + \frac{x^3}{3} + C$$

$$\Rightarrow \quad e^x + e^{-y} + \frac{x^3}{3} + C = 0$$

which is the required solution.

(*ii*) The given equation is

$$\frac{dy}{dx} = e^{x-y} + e^{2\log x - y} \qquad ...(1)$$

$$\Rightarrow \quad \frac{dy}{dx} = e^x.e^{-y} + e^{\log x^2}.e^{-y}$$

$$\Rightarrow \quad \frac{dy}{dx} = e^x.e^{-y} + x^2.e^{-y} \qquad [\because \ e^{\log f(x)} = f(x)]$$

$$\Rightarrow \quad \frac{dy}{dx} = e^{-y}(e^x + x^2)$$

Separating the variables,

$$\frac{1}{e^{-y}}\, dy = (e^x + x^2)\, dx$$

Integrating both sides, we get

$$\int e^y dy = \int (e^x + x^2)\, dx$$

$$\Rightarrow \qquad e^y = e^x + \frac{x^3}{3} + C$$

which is the required solution.

(*iii*) The given equation is

$$y - x\frac{dy}{dx} = a\left(y^2 + \frac{dy}{dx}\right) \qquad ...(1)$$

$$\Rightarrow \qquad y - x\frac{dy}{dx} = ay^2 + a\frac{dy}{dx}$$

$$\Rightarrow \qquad (y - ay^2) = (a + x)\frac{dy}{dx}$$

Separating the variables,

$$\frac{1}{a+x}.dx = \frac{1}{y(1-ay)}.dy$$

Integrating both sides, we get

$$\int \frac{1}{a+x}.dx = \int \frac{1}{y(1-ay)}.dy$$

$$\Rightarrow \qquad \int \frac{1}{a+x}.dx = \int \left(\frac{1}{y} + \frac{a}{1-ay}\right).dy \qquad \text{[By partial fractions]}$$

$$\Rightarrow \qquad \log(a+x) = \log y + a\log(1-ay)\left(-\frac{1}{a}\right) + \log C$$

$$\Rightarrow \qquad \log(a+x) = \log y - \log(1-ay) + \log C$$

$$\Rightarrow \qquad \log(a+x) = \log\left(\frac{y}{1-ay}\right) + \log C \qquad \left[\because \log m - \log n = \log\frac{m}{n}\right]$$

$$\Rightarrow \qquad \log(a+x) = \log\left(\frac{Cy}{1-ay}\right) \qquad [\because \log m + \log n = \log mn]$$

$$\Rightarrow \qquad (a+x) = \frac{Cy}{1-ay}$$

$$\Rightarrow \qquad Cy = (a+x)(1-ay)$$

$$\Rightarrow \qquad y = (a+x)(1-ay)C_1 \quad \text{where} \quad C_1 = \frac{1}{C}$$

which is the required solution.

(*iv*) The given equation is

$$\frac{dy}{dx} = \tan^2 x \qquad ...(1)$$

Separating the variables

$$dy = \tan^2 x\, dx$$

Integrating both sides, we get

$$\int dy = \int \tan^2 x . dx \qquad [\because \ \sec^2 A - \tan^2 A = 1]$$

$$\Rightarrow \qquad y = \int (\sec^2 x - 1)\, dx$$

$$\Rightarrow \qquad y = \tan x - x + C$$

which is the required solution.

(*v*) The given equation is

$$\frac{dy}{dx} = x^2 + \sin 3x \qquad ...(1)$$

Separating the variables

$$dy = (x^2 + \sin 3x)dx$$

Integrating both sides, we get

$$\int dy = \int (x^2 + \sin 3x)\, dx$$

$$\Rightarrow \qquad y = \frac{x^3}{3} - \frac{\cos 3x}{3} + C$$

which is the required solution.

(*vi*) The given equation is

$$\frac{dy}{dx} = \sin^3 x \cos^2 x + xe^x \qquad ...(1)$$

Separating the variables

$$dy = (\sin^3 x \cos^2 x + xe^x)dx$$

Integrating both sides, we get

$$\int dy = \int (\sin^3 x \cos^2 x + xe^x)\, dx$$

$$\Rightarrow \qquad y = \int \sin^3 x \cos^2 x\, dx + \int xe^x dx$$

$$\Rightarrow \qquad y = \int \sin^2 x \sin x \cos^2 x\, dx + \int \underset{\text{I}}{x}\, \underset{\text{II}}{e^x}\, dx$$

$$\Rightarrow \qquad y = \int (1 - \cos^2 x) \cos^2 x . \sin x\, dx + x . \int e^x dx - \int 1 . e^x dx$$

$$[\because \ \sin^2 A + \cos^2 A = 1]$$

Put $\cos x = z \Rightarrow -\sin x\, dx = dz \Rightarrow \sin x\, dx = -dz$

$$\therefore \qquad y = -\int (1 - z^2)\, z^2 dz + xe^x - e^x + C$$

$$= \int (z^4 - z^2)\, dz + xe^x - e^x + C$$

$$= \frac{z^5}{5} - \frac{z^3}{3} + xe^x - e^x + C$$

$$= \frac{1}{5} \cos^5 x - \frac{1}{3} \cos^3 x + xe^x - e^x + C \qquad [\because \ z = \cos x]$$

which is the required solution.

Example 21. *Solve the following differential equations*

(i) $(e^x + e^{-x})dy = (e^x - e^{-x})dx$ (ii) $(1 + \cos x)dy = (1 - \cos x)dx$

(iii) $(\sin x + \cos x)dy + (\cos x - \sin x)dx = 0$

(iv) $\frac{1}{x}\frac{dy}{dx} = \tan^{-1} x$

(v) $(x^2 - yx^2)dy + (y^2 + xy^2)dx = 0$ (vi) $(x - y^2 x)dx - (y - x^2y)dy = 0.$

Solution. (i) The given equation is

$$(e^x + e^{-x})dy = (e^x - e^{-x})dx \qquad ...(1)$$

Separating the variables

$$dy = \frac{e^x - e^{-x}}{e^x + e^{-x}}\,dx$$

Integrating both sides, we get

$$\int dy = \int \frac{e^x - e^{-x}}{e^x + e^{-x}}\,dx$$

$$\Rightarrow \quad y = \log|e^x + e^{-x}| + C \qquad \left[\because \int \frac{f'(x)}{f(x)}\,dx = \log|f(x)|\right]$$

which is the required solution.

(ii) The given equation is

$$(1 + \cos x)dy = (1 - \cos x)dx \qquad ...(1)$$

Separating the variables,

$$dy = \frac{1 - \cos x}{1 + \cos x}\,dx \qquad \left[\begin{array}{l} \because\ 1 - \cos 2A = 2\sin^2 A \\ \Rightarrow\ 1 - \cos A = 2\sin^2 \frac{A}{2} \\ \quad 1 + \cos 2A = 2\cos^2 A \\ \Rightarrow\ 1 + \cos A = 2\cos^2 \frac{A}{2} \end{array}\right]$$

$$\Rightarrow \quad dy = \frac{2\sin^2 \frac{x}{2}}{2\cos^2 \frac{x}{2}}\,dx$$

$$\Rightarrow \quad dy = \tan^2 \frac{x}{2}\,dx$$

Integrating both sides, we get

$$\int dy = \int \tan^2 \frac{x}{2}\,dx$$

$$\Rightarrow \quad y = \int \left(\sec^2 \frac{x}{2} - 1\right) dx \qquad [\because\ \sec^2 A - \tan^2 A = 1]$$

$$\Rightarrow \quad y = \frac{\tan \frac{x}{2}}{\frac{1}{2}} - x + C$$

$$\Rightarrow \qquad y = 2\tan\frac{x}{2} - x + C$$

which is the required solution.

(*iii*) The given equation is

$$(\sin x + \cos x)dy + (\cos x - \sin x)dx = 0 \qquad ...(1)$$

$$\Rightarrow \qquad (\sin x + \cos x)dy = -(\cos x - \sin x)dx$$

Separating the variables,

$$dy = \frac{(\sin x - \cos x)}{(\sin x + \cos x)}dx$$

Integrating both sides, we get

$$\int dy = \int \frac{-(\cos x - \sin x)}{\sin x + \cos x}dx \qquad \left[\because \int \frac{f'(x)}{f(x)}dx = \log|f(x)|\right]$$

$$\Rightarrow \qquad y = -\log|\sin x + \cos x| + C$$

$$\Rightarrow \quad y + \log|\sin x + \cos x| + C_1 = 0 \quad \text{where : } C_1 = -C$$

which is the required solution.

(*iv*) The given equation is

$$\frac{1}{x}\frac{dy}{dx} = \tan^{-1}x \qquad ...(1)$$

Separating the variables

$$dy = x\tan^{-1}x.dx$$

Integrating both sides, we get

$$\int dy = \int \underset{II}{x}\,\underset{I}{\tan^{-1}x}\,dx \qquad \text{[Integrating by parts]}$$

$$\Rightarrow \qquad y = \tan^{-1}x.\int x.dx - \int\left\{\frac{d}{dx}(\tan^{-1}x).\int x\,dx\right\}dx$$

$$\Rightarrow \qquad y = \tan^{-1}x.\frac{x^2}{2} - \int\frac{1}{1+x^2}.\frac{x^2}{2}dx$$

$$= \frac{x^2}{2}\tan^{-1}x - \frac{1}{2}\int\frac{x^2}{1+x^2}dx$$

$$= \frac{1}{2}x^2\tan^{-1}x - \frac{1}{2}\int\frac{1+x^2-1}{1+x^2}dx$$

[Add and subtract 1 to the numerator]

$$= \frac{1}{2}x^2\tan^{-1}x - \frac{1}{2}\int\left(1 - \frac{1}{1+x^2}\right)dx$$

$$= \frac{1}{2}x^2\tan^{-1}x - \frac{1}{2}\int 1.dx + \frac{1}{2}\int\frac{1}{1+x^2}dx$$

$$\therefore \qquad y = \frac{1}{2}x^2\tan^{-1}x - \frac{1}{2}x + \frac{1}{2}\tan^{-1}x + C \qquad \left[\because \int\frac{1}{x^2+a^2}dx = \frac{1}{a}\tan^{-1}\frac{x}{a}\right]$$

which is the required solution.

(*v*) The given equation is

$$(x^2 - yx^2)dy + (y^2 + xy^2)dx = 0 \qquad ...(1)$$

$\Rightarrow \quad x^2dy - yx^2dy + y^2dx + xy^2dx = 0$

$\Rightarrow \quad x^2dy - yx^2dy = - y^2dx - xy^2dx$

$\Rightarrow \quad x^2(1 - y)dy = - y^2(1 + x)dx$

Separating the variables,

$$\frac{1-y}{y^2}\,dy = -\frac{1+x}{x^2}\,dx$$

$$\Rightarrow \quad \left(\frac{1}{y^2} - \frac{1}{y}\right)dy = -\left(\frac{1}{x^2} + \frac{1}{x}\right)dx$$

Integrating both sides, we get

$$\int \frac{1}{y^2}\,dy - \int \frac{1}{y}\,dy = -\int \frac{1}{x^2}\,dx - \int \frac{1}{x}\,dx$$

$$\Rightarrow \quad \frac{y^{-1}}{-1} - \log y = -\frac{x^{-1}}{-1} - \log x + C$$

$$\Rightarrow \quad -\frac{1}{y} - \log y = \frac{1}{x} - \log x + C$$

$$\Rightarrow \quad \log x - \log y = \frac{1}{x} + \frac{1}{y} + C$$

$$\Rightarrow \quad \log\left(\frac{x}{y}\right) - \left(\frac{1}{x} + \frac{1}{y}\right) = C \qquad \left[\because \log m - \log n = \log \frac{m}{n}\right]$$

which is the required solution.

(*vi*) The given equation is

$$(x - y^2 x)dx - (y - x^2 y)dy = 0 \qquad ...(1)$$

$\Rightarrow \quad (x - y^2 x)dx = (y - x^2y)dy$

$\Rightarrow \quad (1 - y^2)xdx = (1 - x^2)ydy$

Separating the variables,

$$\frac{x}{1-x^2}\,dx = \frac{y}{1-y^2}\,dy$$

Integrating both sides, we get

$$\int \frac{x}{1-x^2}\,dx = \int \frac{y}{1-y^2}\,dy$$

$$\Rightarrow \quad -\frac{1}{2}\int \frac{-2x}{1-x^2}\,dx = -\frac{1}{2}\int \frac{-2y}{1-y^2}\,dy \qquad \text{[Multiply and divided by } -2\text{]}$$

$$\Rightarrow \quad -\frac{1}{2}\log|1 - x^2| = -\frac{1}{2}\log|1 - y^2| + C$$

$$\Rightarrow \quad \log|(1 - x^2)| = \log|1 - y^2| - 2C_1$$

$$\Rightarrow \quad \log|(1 - x^2)| = \log|1 - y^2| - \log C \qquad \text{[where } 2C_1 = \log C\text{]}$$

$$\Rightarrow \quad \log(1 - x^2) = \log\left(\frac{1-y^2}{C}\right)$$

$$\Rightarrow \quad (1 - x^2) = \frac{1-y^2}{C}$$

$$\Rightarrow \qquad (1 - y^2) = C(1 - x^2)$$

which is the required solution.

Example 22. *Solve the following differential equation :*

(i) $\dfrac{dy}{dx} = \dfrac{1}{\sin^4 x + \cos^4 x}$ (ii) $\dfrac{dy}{dx} = \dfrac{3e^{2x} + 3e^{4x}}{e^x + e^{-x}}$

(iii) $\sin^4 x \dfrac{dy}{dx} = \cos x$ (iv) $y\dfrac{dy}{dx} = 1 + y^2$

(v) $\sqrt{a+x}\dfrac{dy}{dx} + x = 0$ (vi) $\left(\cos x \dfrac{dy}{dx}\right) + \cos 2x = \cos 3x.$

Solution. (i) The given equation is

$$\frac{dy}{dx} = \frac{1}{\sin^4 x + \cos^4 x} \qquad ...(1)$$

Separating the variables

$$dy = \frac{1}{\sin^4 x + \cos^4 x} dx$$

Integrating both sides, we get

$$\int dy = \int \frac{1}{\sin^4 x + \cos^4 x} dx$$

$$\Rightarrow \qquad y = \int \frac{\frac{1}{\cos^4 x}}{\frac{\sin^4 x}{\cos^4 x} + \frac{\cos^4 x}{\cos^4 x}} dx$$

[Dividing numerator and denominatr on R.H.S. by $\cos^4 x$]

$$= \int \frac{\sec^4 x}{\tan^4 x + 1} dx$$

$$= \int \frac{\sec^2 x . \sec^2 x}{(1 + \tan^4 x)} dx = \int \frac{(1 + \tan^2 x)\sec^2 x}{(1 + \tan^4 x)} dx$$

Put $\tan x = z \Rightarrow \sec^2 x dx = dz$

$$\therefore \qquad y = \int \frac{1 + z^2}{1 + z^4} dz$$

$$= \int \frac{\frac{1}{z^2} + 1}{\frac{1}{z^2} + z^2} dz$$

[Dividing numerator and denominator by z^2]

$$= \int \frac{1 + \frac{1}{z^2}}{\left(z - \frac{1}{z}\right)^2 + 2} dz$$

Once again, substitute, $\left(z - \dfrac{1}{z}\right) = t \Rightarrow \left(1 + \dfrac{1}{z^2}\right) dz = dt$

$\therefore$ We have $$y = \int \frac{1}{y^2 + (\sqrt{2})^2}\, dt$$

$$= \frac{1}{\sqrt{2}} \tan^{-1} \frac{t}{\sqrt{2}} + C \qquad \left[\because \int \frac{1}{x^2 + a^2}\, dx = \frac{1}{a} \tan^{-1} \frac{x}{a}\right]$$

$$= \frac{1}{\sqrt{2}} \tan^{-1} \left(\frac{z - \frac{1}{z}}{\sqrt{2}} \right) + C \qquad \left[\because t = \left(z - \frac{1}{z}\right)\right]$$

$$= \frac{1}{\sqrt{2}} \tan^{-1} \left(\frac{z^2 - 1}{\sqrt{2}\, z} \right) + C$$

$$\Rightarrow \qquad y = \frac{1}{\sqrt{2}} \tan^{-1} \left(\frac{\tan^2 x - 1}{\sqrt{2} \tan x} \right) + C \qquad [\because \ z = \tan x]$$

which is the required solution.

(*ii*) The given equation is

$$\frac{dy}{dx} = \frac{3e^{2x} + 3e^{4x}}{e^x + e^{-x}} \qquad ...(1)$$

Separating the variables,

$$dy = \frac{3e^{2x} + 3e^{4x}}{e^x + e^{-x}}\, dx$$

$$= \frac{3e^{2x}(1 + e^{2x})}{e^x + \frac{1}{e^x}}\, dx = \frac{3e^{2x}(1 + e^{2x})}{\left(\frac{e^{2x} + 1}{e^x}\right)}\, dx$$

$$= \frac{3e^{2x}(1 + e^{2x}).e^x}{(1 + e^{2x})}\, dx$$

$$\Rightarrow \qquad dy = 3e^{3x}\, dx$$

Integrating both sides, we get

$$\int dy = 3 \int e^{3x}\, dx$$

$$\Rightarrow \qquad y = 3.\left(\frac{e^{3x}}{3}\right) + C$$

$$\Rightarrow \qquad y = e^{3x} + C$$

which is the required solution.

(*iii*) The given equation is

$$\sin^4 x \frac{dy}{dx} = \cos x$$

Separating the variables,

$$dy = \frac{\cos x}{\sin^4 x}\, dx$$

Integrating both sides, we get

$$\int dy = \int \frac{\cos x}{\sin^4 x}\, dx$$

Put $\sin x = z \Rightarrow \cos x dx = dz$

$$\therefore \quad \int dy = \int \frac{1}{z^4}\, dz$$

$$\Rightarrow \quad y = \frac{z^{-4+1}}{-4+1} + C = -\frac{1}{3z^3} + C$$

$$= -\frac{1}{3\sin^3 x} + C \qquad [\because \ z = \sin x]$$

$$\Rightarrow \quad y = -\frac{1}{3}\operatorname{cosec}^3 x + C$$

which is the required solution.

(*iv*) The given equation is

$$y\frac{dy}{dx} = 1 + y^2$$

Separating the variables,

$$\frac{y}{1+y^2}\, dy = dx$$

Integrating both sides, we get

$$\int \frac{y}{1+y^2}\, dy = \int dx$$

$$\Rightarrow \quad \frac{1}{2}\int \frac{2y}{1+y^2}\, dy = x + C \qquad \text{[Multiply and divided by 2]}$$

$$\Rightarrow \quad \frac{1}{2}\log|1+y^2| = x + C \qquad \left[\because \int \frac{f'(x)}{f(x)}\, dx = \log|f(x)|\right]$$

which is the required solution.

(*v*) The given equation is

$$\sqrt{a+x}\,\frac{dy}{dx} + x = 0$$

$$\Rightarrow \quad \sqrt{a+x}\,\frac{dy}{dx} = -x$$

Separating the variables,

$$dy = -\frac{x}{\sqrt{a+x}}\, dx$$

$$\Rightarrow \quad dy = -\left[\frac{a+x-a}{\sqrt{a+x}}\right] dx \qquad \text{[Add and subtract `}a\text{' to the numerator]}$$

$$\Rightarrow \quad dy = -\left[\sqrt{a+x} - \frac{a}{\sqrt{a+x}}\right] dx$$

Integrating both sides, we get

$$\int dy = -\int\left(\sqrt{a+x} - \frac{a}{\sqrt{a+x}}\right)dx$$

$$\Rightarrow \qquad y = -\int \sqrt{a+x}\,dx + a\int \frac{1}{\sqrt{a+x}}\,dx$$

$$\Rightarrow \qquad y = -\frac{(a+x)^{\frac{1}{2}+1}}{\frac{1}{2}+1} + a\,\frac{(a+x)^{-\frac{1}{2}+1}}{-\frac{1}{2}+1} + C$$

$$\Rightarrow \qquad y = -\frac{(a+x)^{3/2}}{3/2} + \frac{a(a+x)^{1/2}}{1/2} + C$$

$$\Rightarrow \qquad y = -\frac{2}{3}(a+x)^{3/2} + 2a\sqrt{a+x} + C$$

which is the required solution.

(*vi*) The given equation is

$$\cos x\,\frac{dy}{dx} + \cos 2x = \cos 3x \qquad ...(1)$$

$$\Rightarrow \qquad \cos x\,\frac{dy}{dx} = \cos 3x - \cos 2x$$

$$\Rightarrow \qquad \frac{dy}{dx} = \frac{\cos 3x - \cos 2x}{\cos x}$$

$$\Rightarrow \qquad \frac{dy}{dx} = \frac{4\cos^3 x - 3\cos x}{\cos x} - \frac{2\cos^2 x - 1}{\cos x}$$

$$[\because \quad \cos 3A = 4\cos^3 A - 3\cos A,\ \cos 2A = 2\cos^2 A - 1]$$

$$\Rightarrow \qquad \frac{dy}{dx} = 4\cos^2 x - 3 - 2\cos x + \frac{1}{\cos x}$$

$$\Rightarrow \qquad \frac{dy}{dx} = 4\left(\frac{1+\cos 2x}{2}\right) - 3 - 2\cos x + \sec x$$

$$\Rightarrow \qquad \frac{dy}{dx} = (2\cos 2x - 1 - 2\cos x + \sec x)$$

Separating the variables,

$$dy = (2\cos 2x - 1 - 2\cos x + \sec x)dx$$

Integrating both sides, we get

$$\int dy = \int (2\cos 2x - 1 - 2\cos x + \sec x)\,dx$$

$$\Rightarrow \qquad y = \sin 2x - x - 2\sin x + \log|\sec x + \tan x| + C$$

which is the required solution.

Example 23. *Solve the following differential equations :*

(*i*) $xy(y+1)dy = (x^2+1)dx$ (*ii*) $(x^2 - yx^2)dy + (y^2 + x^2y^2)dx = 0$

(*iii*) $\frac{dy}{dx} = x^5 \tan^{-1}(x^3)$ (*iv*) $\sqrt{1-x^6}\,dy = x^2dx$

(*v*) $\frac{dy}{dx} = \cos^3 x \sin^4 x + x\sqrt{2x+1}$ (*vi*) $\frac{dy}{dx} = \log x$.

Solution. (*i*) The given equation is

$$xy(y+1)dy = (x^2+1)dx \quad \text{...(1)}$$

Separating the variables,

$$y(y+1)dy = \left(\frac{x^2+1}{x}\right)dx$$

Integrating both sides, we get

$$\int (y^2+y)dy = \int \left(\frac{x^2}{x}+\frac{1}{x}\right)dx$$

$$\Rightarrow \quad \int (y^2+y)dy = \int \left(x+\frac{1}{x}\right)dx$$

$$\Rightarrow \quad \frac{y^3}{3}+\frac{y^2}{2} = \frac{x^2}{2} + \log|x| + C$$

which is the required solution.

(*ii*) The given equation is

$$(x^2 - yx^2)dy + (y^2 + x^2y^2)dx = 0 \quad \text{...(1)}$$

$$\Rightarrow \quad x^2(1-y)dy = -y^2(1+x^2)dx$$

Separating the variables,

$$\frac{1-y}{y^2}\,dy = -\frac{1+x^2}{x^2}\,dx$$

Integrating both sides, we get

$$\int \left(\frac{1-y}{y^2}\right)dy = -\int \left(\frac{1+x^2}{x^2}\right)dx$$

$$\Rightarrow \quad \int \left(\frac{1}{y^2}-\frac{1}{y}\right)dy = -\int \left(\frac{1}{x^2}+1\right)dx$$

$$\Rightarrow \quad \frac{y^{-2+1}}{-2+1} - \log|y| = -\left[\frac{x^{-2+1}}{-2+1}+x\right] + C$$

$$\Rightarrow \quad -\frac{1}{y} - \log|y| = -\left(-\frac{1}{x}+x\right) + C$$

$$\Rightarrow \quad -\frac{1}{y} - \log|y| = \frac{1}{x} - x + C$$

$$\Rightarrow \quad \frac{1}{x}+\frac{1}{y} + \log|y| - x + C = 0$$

which is the required solution.

(*iii*) The given equation is

$$\frac{dy}{dx} = x^5 \tan^{-1}(x^3) \quad \text{...(1)}$$

Separating the variables

$$dy = x^5 \tan^{-1}(x^3).dx$$

Integrating both sides, we get

$$\int dy = \int x^5 \tan^{-1}(x^3)\, dx$$

$\Rightarrow \qquad y = \int x^2 . x^3 \tan^{-1}(x^3)\, dx$

Put $\qquad x^3 = z \;\Rightarrow\; 3x^2\,dx = dz \;\Rightarrow\; x^2\,dx = \frac{1}{3}dz$

$$\therefore \qquad y = \int z \tan^{-1} z\left(\frac{1}{3}dz\right) = \frac{1}{3}\int \underset{\text{II}}{z} \underset{\text{I}}{\tan^{-1} z}\, dz \qquad \text{(Integrating by parts)}$$

$$= \frac{1}{3}\left[\tan^{-1} z . \frac{z^2}{2} - \int \frac{d}{dx}(\tan^{-1} z). \frac{z^2}{2}\, dz\right]$$

$$= \frac{1}{6}z^2 \tan^{-1} z - \frac{1}{6}\int \frac{1}{1+z^2}. z^2\, dz$$

$$= \frac{1}{6}z^2 \tan^{-1} z - \frac{1}{6}\int \frac{1+z^2-1}{1+z^2}\, dz$$

[Add and subtract 1 to the numerator]

$$= \frac{1}{6}z^2 \tan^{-1} z - \frac{1}{6}\int\left(1 - \frac{1}{1+z^2}\right) dz$$

$$= \frac{1}{6}z^2 \tan^{-1} z - \frac{1}{6}[z - \tan^{-1} z] + C \qquad \left[\because \int \frac{1}{a^2+x^2}\, dx = \frac{1}{a}\tan^{-1}\frac{x}{a}\right]$$

$$= \frac{1}{6}z^2 \tan^{-1} z - \frac{1}{6}z + \frac{1}{6}\tan^{-1} z + C$$

$$= \frac{1}{6}[(x^3)^2 \tan^{-1} x^3 - x^3 + \tan^{-1} x^3] + C \qquad [\because x^3 = z]$$

$$= \frac{1}{6}[x^6 \tan^{-1} x^3 - x^3 + \tan^{-1} x^3] + C$$

which is the required solution.

(*iv*) The given equation is

$$\sqrt{1-x^6}\; dy = x^2\, dx \qquad \ldots(1)$$

Separating the variables,

$$dy = \frac{x^2}{\sqrt{1-x^6}}\, dx$$

Integrating both sides, we get

$$\int dy = \int \frac{x^2}{\sqrt{1-x^6}}\, dx$$

$$\Rightarrow \qquad y = \int \frac{x^2}{\sqrt{1-(x^3)^2}}\, dx$$

Put $\qquad x^3 = z \;\Rightarrow\; 3x^2\,dx = dz \;\Rightarrow\; x^2\,dx = \frac{1}{3}dz$

$$\therefore \qquad y = \int \frac{1}{\sqrt{1-z^2}}\left(\frac{1}{3}dz\right) = \frac{1}{3}\int \frac{1}{\sqrt{1-z^2}}\, dz$$

$$= \frac{1}{3}(\sin^{-1} z) + C \qquad \left[\because \int \frac{1}{\sqrt{a^2 - x^2}}\, dx = \sin^{-1}\frac{x}{a}\right]$$

$$\Rightarrow \qquad y = \frac{1}{3}\sin^{-1} x^3 + C \qquad [\because \ z = x^3]$$

which is the required solution.

(*v*) The given equation is

$$\frac{dy}{dx} = \cos^3 x \sin^4 x + x\sqrt{2x+1} \qquad ...(1)$$

Separating the variables,

$$dy = (\cos^3 x \sin^4 x + x\sqrt{2x+1})\, dx$$

Integrating both sides, we get

$$\int dy = \int (\cos^3 x \sin^4 x + x\sqrt{2x+1})\, dx$$

$$\Rightarrow \qquad y = \int \cos^3 x \sin^4 dx + \int x\sqrt{2x+1}\, dx = I_1 + I_2 \text{ (say)} \qquad ...(2)$$

where $\qquad I_1 = \int \cos^3 x \sin^4 x\, dx \quad \text{and} \quad I_2 = \int x\sqrt{2x+1}\, dx$

Now $\qquad I_1 = \int \cos^3 x \sin^4 x\, dx$

$$= \int \cos^2 x \cos x \sin^4 x\, dx$$

$$= \int (1 - \sin^2 x) \cos x \sin^4 x\, dx \qquad [\because \ \cos^2 A + \sin^2 A = 1]$$

Put $\qquad \sin x = z \ \Rightarrow \ \cos x\, dx = dz$

$$\therefore \qquad I_1 = \int (1 - z^2)\, z^4\, dz = \int (z^4 - z^6)\, dz$$

$$= \frac{z^5}{5} - \frac{z^7}{7} + C_1$$

$$\Rightarrow \qquad I_1 = \frac{1}{5}\sin^5 x - \frac{1}{7}\sin^7 x + C_1 \qquad ...(3) \quad [\because \ z = \sin x]$$

Now $\qquad I_2 = \int x\sqrt{2x+1}\, dx$

Put $\qquad \sqrt{2x+1} = z \ \Rightarrow \ 2x + 1 = z^2 \ \Rightarrow \ x = \left(\frac{z^2 - 1}{2}\right)$

$$\Rightarrow \qquad 2dx = 2z\, dz \ \Rightarrow \ dx = zdz$$

$$\therefore \qquad I_2 = \int \left(\frac{z^2 - 1}{2}\right). z. zdz = \frac{1}{2}\int (z^2 - 1)\, z^2 dz$$

$$= \frac{1}{2}\int (z^4 - z^2)\, dz = \frac{1}{2}\left[\frac{z^5}{5} - \frac{z^3}{3}\right] + C_2 = \frac{1}{10} z^5 - \frac{1}{6} z^3 + C_2$$

$$= \frac{1}{10}(\sqrt{2x+1})^5 - \frac{1}{6}(\sqrt{2x+1})^3 + C_2 \qquad [\because \ z = \sqrt{2x+1}$$

$$\Rightarrow \qquad I_2 = \frac{1}{10}(2x+1)^{\frac{5}{2}} - \frac{1}{6}(2x+1)^{\frac{3}{2}} + C_2 \qquad ...(4)$$

Substituting the values of equations (3) and (4) in equation (2), we have

$$y = I_1 + I_2$$

$$= \frac{1}{5}\sin^5 x - \frac{1}{7}\sin^7 x + \frac{1}{10}(2x+1)^{\frac{5}{2}} - \frac{1}{6}(2x+1)^{\frac{3}{2}} + C_1 + C_2$$

$$= \frac{1}{5}\sin^5 x - \frac{1}{7}\sin^7 x + \frac{1}{10}(2x+1)^{\frac{5}{2}} - \frac{1}{6}(2x+1)^{\frac{3}{2}} + C$$

where $C = C_1 + C_2$

which is the required solution.

(*vi*) The given equation is $\frac{dy}{dx} = \log x$ ...(1)

Separating the variables, $dy = \log x \, dx$

Integrating both sides, we have

$$\int dy = \int \underset{\text{II}}{1} . \underset{\text{I}}{\log x} \, dx$$ [Integrating by parts]

$$\Rightarrow \quad y = \log x . x - \int \frac{d}{dx}(\log x) . x \, dx$$

$$= x \log x - \int \frac{1}{x} . x \, dx = x \log x - \int 1 . dx$$

$$\Rightarrow \quad y = x \log x - x + C$$

which is the required solution.

Example 24. *Solve the following differential equation :*

(*i*) $\frac{dy}{dx} = \frac{1}{y^2 + \sin y}$

(*ii*) $(1 - x^2)dy + xydx = xy^2 \, dx$

(*iii*) $\frac{dy}{dx} = e^{x+y}$

(*iv*) $(1 + x^2)dy = (1 + y^2)dx$

(*v*) $\frac{dy}{dx} + y = 1$

(*vi*) $(1 - y)x \frac{dy}{dx} + (1 + x)y = 0$

(*vii*) $\frac{dy}{dx} = \sec y.$

Solution. (*i*) The given equation is

$$\frac{dy}{dx} = \frac{1}{y^2 + \sin y} \quad \text{...(1)}$$

Separating the variables,

$$(y^2 + \sin y)dy = dx$$

Integrating both sides, we get

$$\int (y^2 + \sin y) \, dy = \int dx \quad \Rightarrow \quad \frac{y^3}{3} + (-\cos y) = x + C$$

$$\Rightarrow \quad \frac{y^3}{3} - \cos y = x + C$$

which is the required solution.

(*ii*) The given equation is

$$(1 - x^2)dy + xydx = xy^2 \, dx \quad \text{...(1)}$$

$$\Rightarrow \quad (1 - x^2)dy = (xy^2 - xy)dx$$

$\Rightarrow \qquad (1 - x^2)dy = xy(y - 1)dx$

Separating the variables,

$$\frac{1}{y(y-1)}\,dy = \frac{x}{1-x^2}\,dx$$

Integrating both sides, we get

$$\int \frac{1}{y(y-1)}\,dy = \int \frac{x}{1-x^2}\,dx$$

$$\Rightarrow \quad \int\left(\frac{1}{y-1} - \frac{1}{y}\right)dy = -\frac{1}{2}\int \frac{-2x}{1-x^2}\,dx \qquad \text{[By partial fractions]}$$

$$\Rightarrow \quad \log|y-1| - \log|y| = -\frac{1}{2}\log|1-x^2| + \log C \qquad \left[\because \int \frac{f'(x)}{f(x)}\,dx = \log|f(x)|\right]$$

$$\Rightarrow \quad \log\left|\frac{y-1}{y}\right| = \frac{1}{2}(2\log C - \log|1-x^2|)$$

$$\Rightarrow \quad 2\log\left|\frac{y-1}{y}\right| = \log\left|\frac{C^2}{1-x^2}\right| \qquad \left[\because \log m - \log n = \log\frac{m}{n}\right]$$

$$\Rightarrow \quad \log\left(\frac{y-1}{y}\right)^2 = \log\left(\frac{C^2}{|1-x^2|}\right) \qquad [\because 2\log C = \log C^2]$$

$$\Rightarrow \quad \frac{(y-1)^2}{y^2} = \frac{C_1}{|1-x^2|} \qquad [\because C^2 = C_1]$$

$$\Rightarrow \quad C_1 y^2 = (y-1)^2\,|1-x^2|$$

which is the required solution.

(*iii*) The given equation is

$$\frac{dy}{dx} = e^{x+y} \qquad \text{...(1)}$$

$$\Rightarrow \quad \frac{dy}{dx} = e^x.e^y$$

Separating the variables

$$\frac{1}{e^y}\,dy = e^x\,dx$$

Integrating both sides, we get

$$\int e^{-y}\,dy = \int e^x\,dx$$

$$\Rightarrow \quad -e^{-y} = e^x + C$$

$$\Rightarrow \quad e^x + e^{-y} + C = 0$$

which is the required solution.

(*iv*) The given equation is

$$(1 + x^2)dy = (1 + y^2)dx \qquad \text{...(1)}$$

Separating the variables,

$$\frac{1}{1+y^2}\,dy = \frac{1}{1+x^2}\,dx$$

Integrating both sides, we get

$$\int \frac{1}{1+y^2}\,dy = \int \frac{1}{1+x^2}\,dx$$

$$\Rightarrow \quad \tan^{-1} y = \tan^{-1} x + \tan^{-1} C \qquad \left[\because \int \frac{1}{a^2+x^2}\,dx = \frac{1}{a}\tan^{-1}\frac{x}{a}\right]$$

$$\Rightarrow \quad \tan^{-1} y - \tan^{-1} x = \tan^{-1} C$$

$$\Rightarrow \quad \tan^{-1}\left(\frac{y-x}{1+xy}\right) = \tan^{-1} C \qquad \left[\because \tan^{-1}x - \tan^{-1}y = \tan^{-1}\left(\frac{x-y}{1+xy}\right)\right]$$

$$\Rightarrow \quad \frac{y-x}{1+xy} = C$$

$$\Rightarrow \quad (y-x) = C(1+xy)$$

which is the required solution.

(*v*) The given equation is

$$\frac{dy}{dx} + y = 1 \qquad \text{...(1)}$$

$$\Rightarrow \quad \frac{dy}{dx} = 1 - y$$

Separating the variables,

$$\frac{1}{1-y}\,dy = dx$$

Integrating both sides, we get

$$\int \frac{1}{1-y}\,dy = \int 1.dx$$

$$\Rightarrow \quad -\log|1-y| = x + C \quad \Rightarrow \quad x + \log|1-y| + C$$

which is the required solution.

(*vi*) The given equation is

$$(1-y)x\frac{dy}{dx} + (1+x)y = 0 \qquad \text{...(1)}$$

$$\Rightarrow \quad (1-y)x\frac{dy}{dx} = -(1+x)y$$

Separating the variables,

$$\frac{(1-y)}{y}\,dy = -\frac{(1+x)}{x}\,dx$$

Integrating both sides, we get

$$\int \frac{1-y}{y}\,dy = -\int \frac{1+x}{x}\,dx$$

$$\Rightarrow \quad \int\left(\frac{1}{y} - 1\right)dy = -\int\left(\frac{1}{x} + 1\right)dx$$

$$\Rightarrow \quad \log|y| - y = -[\log|x| + x] + C$$

$$\Rightarrow \quad \log|y| - y = -\log|x| - x + C$$

$$\Rightarrow \quad \log|x| + \log|y| + x - y = C \qquad [\because \log m + \log n = \log mn]$$

$$\Rightarrow \quad \log|xy| + x - y = C$$

which is the required solution.

(*vii*) The given equation is

$$\frac{dy}{dx} = \sec y \qquad ...(1)$$

Separating the variables

$$\frac{1}{\sec y} dy = dx \qquad \Rightarrow \quad \cos y\, dy = dx$$

Integrating both sides, we get

$$\int \cos y\, dy = \int dx$$

$$\sin y = x + C \qquad \Rightarrow \quad x = \sin y - C$$

which is the required solution.

Example 25. *Solve the following differential equations :*

(*i*) $\frac{dy}{dx} + \frac{1 + \cos 2y}{1 - \cos 2x} = 0$ (*ii*) $(2 - 10y^3)\frac{dy}{dx} + y = 0$

(*iii*) $\frac{dy}{dx} = 1 - x + y - xy$ (*iv*) $(1 + x^2)dy = x(1 + y^2)dx$

(*v*) $\frac{dy}{dx} = \sin^2 y$ (*vi*) $(x + 1)\frac{dy}{dx} = 2xy$

(*vii*) $\frac{dy}{dx} = 1 + x + y + xy.$

Solution. (*i*) The given equation is

$$\frac{dy}{dx} + \frac{1 + \cos 2y}{1 - \cos 2x} = 0 \qquad ...(1)$$

$$\Rightarrow \quad \frac{dy}{dx} = -\frac{1 + \cos 2y}{1 - \cos 2x}$$

Separating the variables,

$$\frac{1}{1 + \cos 2y} dy = -\frac{1}{1 - \cos 2x} dx$$

Integrating both sides, we get

$$\int \frac{1}{1 + \cos 2y} dy = -\int \frac{1}{1 - \cos 2x} dx$$

$$\Rightarrow \quad \int \frac{1}{2\cos^2 y} dy = -\int \frac{1}{2\sin^2 x} dx \quad [\because \ 1 + \cos 2A = 2\cos^2 A,\ 1 - \cos 2A = 2\sin^2 A]$$

$$\Rightarrow \quad \frac{1}{2}\int \sec^2 y\, dy = -\frac{1}{2}\int \text{cosec}^2\, x\, dx$$

$$\Rightarrow \quad \frac{1}{2}\tan y = -\frac{1}{2}(-\cot x) + C$$

$$\Rightarrow \quad \frac{1}{2}[\tan y - \cot x] = C$$

$$\Rightarrow \quad \tan y - \cot x = 2C$$

$$\Rightarrow \quad \tan y - \cot x = C_1, \quad \text{where } C_1 = 2C$$

which is the required solution.

(*ii*) The given equation is

$$(2 - 10y^3)\frac{dy}{dx} + y = 0 \qquad ...(1)$$

$$\Rightarrow \qquad (2 - 10y^3)\frac{dy}{dx} = -y$$

Separating the variables,

$$\frac{2 - 10y^3}{y}\,dy = -\,dx \qquad \Rightarrow \qquad \frac{10y^3 - 2}{y}\,dy = dx$$

Integrating both sides, we get

$$\int \frac{10y^3 - 2}{y}\,dy = \int dx$$

$$\Rightarrow \qquad \int\left(10y^2 - \frac{2}{y}\right)dy = x + C$$

$$\Rightarrow \qquad 10\int y^2\,dy - 2\int \frac{1}{y}\,dy = x + C$$

$$\Rightarrow \qquad \frac{10}{3}y^3 - 2\log|y| = x + C$$

which is the required solution.

(*iii*) The given equation is

$$\frac{dy}{dx} = 1 - x + y - xy \qquad ...(1)$$

$$\Rightarrow \qquad \frac{dy}{dx} = (1 - x) + y(1 - x)$$

$$\Rightarrow \qquad \frac{dy}{dx} = (1 - x)(1 + y)$$

Separating the variables,

$$\frac{1}{1 + y}\,dy = (1 - x)dx$$

Integrating both sides, we get

$$\int \frac{1}{1 + y}\,dy = \int (1 - x)\,dx$$

$$\Rightarrow \qquad \log|1 + y| = x - \frac{x^2}{2} + C$$

which is the required solution.

(*iv*) The given equation is

$$(1 + x^2)dy = x(1 + y^2)dx \qquad ...(1)$$

Separating the variables,

$$\frac{1}{1+y^2}\,dy = \frac{x}{1+x^2}\,dx$$

Integrating both sides, we get

$$\int \frac{1}{1+y^2}\,dy = \int \frac{x}{1+x^2}\,dx$$

$$\Rightarrow \quad \int \frac{1}{1+y^2}\,dy = \frac{1}{2}\int \frac{2x}{1+x^2}\,dx$$ [Multiply and divide the R.H.S. by 2]

$$\Rightarrow \quad \tan^{-1} y = \frac{1}{2}\log|1+x^2| + C$$ $\left[\because \int \frac{1}{a^2+x^2}\,dx = \frac{1}{a}\tan^{-1}\frac{x}{a}; \quad \int \frac{f'(x)}{f(x)}\,dx = \log|f(x)|\right]$

which is the required solution.

(*v*) The given equation is

$$\frac{dy}{dx} = \sin^2 y \qquad ...(1)$$

Separating the variables,

$$\frac{1}{\sin^2 y}\,dy = dx$$

Integrating both sides, we get

$$\int \frac{1}{\sin^2 y}\,dy = \int 1.dx$$

$$\Rightarrow \quad \int \text{cosec}^2 y\,dy = \int dx$$

$$\Rightarrow \quad -\cot y = x + C$$

$$\Rightarrow \quad x + \cot y + C = 0$$

which is the required solution.

(*vi*) The given equation is

$$(x+1)\frac{dy}{dx} = 2xy \qquad ...(1)$$

Separating the variables,

$$\frac{1}{y}\,dy = \frac{2x}{x+1}\,dx$$

Integrating both sides, we get

$$\int \frac{1}{y}\,dy = \int \frac{2x}{x+1}\,dx$$

$$\Rightarrow \quad \int \frac{1}{y}\,dy = 2\int \frac{x}{x+1}\,dx$$

$$\Rightarrow \quad \int \frac{1}{y}\, dy = 2\int \frac{x+1-1}{x+1}\, dx$$

[Add and subtract 1 to the numerator in R.H.S.]

$$\Rightarrow \quad \log|y| = 2\int\left(1-\frac{1}{x+1}\right)dx$$

$$\Rightarrow \quad \log|y| = 2[x - \log|x+1|] + C$$

which is the required solution.

(*vii*) The given equation is

$$\frac{dy}{dx} = 1 + x + y + xy \qquad ...(1)$$

Please try yourself.

[**Hint :** See part (*iii*) of the same example.] $\left[\textbf{Ans. } \log|1+y| = x + \frac{x^2}{2} + C\right]$

Example 26. *Solve the following differential equations*

(*i*) $\sec^2 x \tan y\, dx + \sec^2 y \tan x\, dy = 0$

(*ii*) $(1+x)(1+y^2)dx + (1+y)(1+x^2)dy = 0$

(*iii*) $(y + xy)dx + (x - xy^2)dy = 0$

(*iv*) $(y^3 + 1)(e^x + xe^x)dx - xe^x y^2 dy = 0$

(*v*) $\cos x\,(1 + \cos y)dx - \sin y\,(1 + \sin x)dy = 0$

(*vi*) $\frac{dy}{dx} = y^2 \tan ax$ (*vii*) $\frac{dy}{dx} = (1+x)(1+y^2)$

(*viii*) $x\sqrt{1+y^2}\, dx + y\sqrt{1+x^2}\, dy = 0.$

Solution. (*i*) The given equation is

$$\sec^2 x \tan y\, dx + \sec^2 y \tan x\, dy = 0 \qquad ...(1)$$

Separating the variables,

$$\sec^2 x \tan y\, dx = -\sec^2 y \tan x\, dy$$

$$\Rightarrow \quad \frac{\sec^2 x}{\tan x}\, dx = -\frac{\sec^2 y}{\tan y}\, dy$$

Integrating both sides, we get

$$\int \frac{\sec^2 x}{\tan x}\, dx = -\int \frac{\sec^2 y}{\tan y}\, dy \qquad \left[\because \int \frac{f'(x)}{f(x)}\, dx = \log|f(x)|\right]$$

$$\Rightarrow \quad \log|\tan x| = -\log|\tan y| + \log C$$

$$\Rightarrow \quad \log|\tan x| + \log|\tan y| = \log C \qquad [\because \log m + \log n = \log mn]$$

$$\Rightarrow \quad \log|\tan x \tan y| = \log C$$

$$\Rightarrow \quad |\tan x \tan y| = C$$

which is the required solution.

(*ii*) The given equation is

$$(1+x)(1+y^2)dx + (1+y)(1+x^2)dy = 0 \qquad ...(1)$$

$$\Rightarrow \quad (1+x)(1+y^2)dx = -(1+y)(1+x^2)dy$$

Separating the variables,

$$\frac{1+x}{1+x^2}\,dx = -\frac{(1+y)}{1+y^2}\,dy$$

Integrating both sides, we get

$$\int \frac{1+x}{1+x^2}\,dx = -\int \frac{1+y}{1+y^2}\,dy$$

$$\Rightarrow \quad \int\left(\frac{1}{1+x^2}+\frac{x}{1+x^2}\right)dx = -\int\left(\frac{1}{1+y^2}+\frac{y}{1+y^2}\right)dy$$

$$\Rightarrow \quad \int \frac{1}{1+x^2}\,dx + \int \frac{x}{1+x^2}\,dx = -\int \frac{1}{1+y^2}\,dy - \int \frac{y}{1+y^2}\,dy$$

$$\Rightarrow \quad \int \frac{1}{1+x^2}\,dx + \frac{1}{2}\int \frac{2x}{1+x^2}\,dx = -\int \frac{1}{1+y^2}\,dy - \frac{1}{2}\int \frac{2y}{1+y^2}\,dy$$

$$\Rightarrow \quad \tan^{-1}x + \frac{1}{2}\log|1+x^2| = -\tan^{-1}y - \frac{1}{2}\log|1+y^2| + C$$

$$\left[\because \int \frac{1}{a^2+x^2}\,dx = \frac{1}{a}\tan^{-1}\frac{x}{a} \qquad \int \frac{f'(x)}{f(x)}\,dx = \log|f(x)|\right]$$

$$\Rightarrow \quad \tan^{-1}x + \tan^{-1}y + \frac{1}{2}\log|1+x^2| + \frac{1}{2}\log|1+y^2| = C$$

$$\Rightarrow \quad \tan^{-1}x + \tan^{-1}y + \frac{1}{2}[\log|1+x^2| + \log|1+y^2|] = C$$

$$\Rightarrow \quad \tan^{-1}x + \tan^{-1}y + \frac{1}{2}\log|(1+x^2)(1+y^2)| = C$$

$$[\because \log m + \log n = \log mn]$$

which is the required solution.

(*iii*) The given equation is

$$(y + xy)dx + (x - xy^2)dy = 0$$

$$\Rightarrow \quad y(1+x)dx + x(1-y^2)dy = 0$$

$$\Rightarrow \quad y(1+x)dx = x(y^2-1)dy$$

Separating the variables,

$$\frac{1+x}{x}\,dx = \frac{y^2-1}{y}\,dy$$

Integrating both sides, we get

$$\int\left(\frac{1+x}{x}\right)dx = \int\left(\frac{y^2-1}{y}\right)dy$$

$$\Rightarrow \quad \int\left(\frac{1}{x}+1\right)dx = \int\left(y-\frac{1}{y}\right)dy$$

$$\Rightarrow \quad \log|x| + x = \frac{y^2}{2} - \log|y| + C$$

$$\Rightarrow \quad \log|x| + \log|y| + x = \frac{y^2}{2} + C$$

$$\Rightarrow \quad \log|xy| + x = \frac{y^2}{2} + C \qquad [\because \log m + \log n = \log mn]$$

which is the required solution.

(*iv*) The given equation is

$$(y^3 + 1)(e^x + xe^x)dx - xe^x y^2\, dy = 0 \qquad \ldots(1)$$

$$\Rightarrow \quad (y^3 + 1)(e^x + xe^x)dx = xe^x y^2\, dy$$

Separating the variables,

$$\frac{e^x(1+x)\,dx}{xe^x} = \frac{y^2}{y^3+1}\,dy$$

$$\Rightarrow \quad \frac{1+x}{x}\,dx = \frac{y^2}{y^3+1}\,dy$$

Integrating both sides, we get

$$\int\frac{1+x}{x}\,dx = \int\frac{y^2}{y^3+1}\,dy$$

$$\Rightarrow \quad \int\left(\frac{1}{x}+1\right)dx = \frac{1}{3}\int\frac{3y^2}{y^3+1}\,dy \qquad \text{[Multiply and divide the R.H.S. by 3]}$$

$$\Rightarrow \quad \log|x| + x = \frac{1}{3}\log|y^3+1| + C \qquad \left[\because \int\frac{f'(x)}{f(x)}\,dx = \log|f(x)|\right]$$

which is the required solution.

(*v*) The given equation is

$$\cos x(1 + \cos y)dx - \sin y(1 + \sin x)dy = 0 \qquad \ldots(1)$$

$$\Rightarrow \quad \cos x(1 + \cos y)dx = \sin y(1 + \sin x)dy$$

Separating the variables,

$$\frac{\cos x}{1+\sin x}\,dx = \frac{\sin y}{1+\cos y}\,dy$$

Integrating both sides, we get

$$\int\frac{\cos x}{1+\sin x}\,dx = -\int -\frac{\sin y}{1+\cos y}\,dy \qquad \text{[Note this step]}$$

$$\Rightarrow \quad \log|1+\sin x| = -\log|1+\cos y| + \log C \qquad \left[\because \int\frac{f'(x)}{f(x)}\,dx = \log|f(x)|\right]$$

$$\Rightarrow \quad \log|1+\sin x| + \log|1+\cos y| = \log C$$

$$\Rightarrow \quad \log|(1+\sin x)(1+\cos y)| = \log C \qquad [\because \log m + \log n = \log mn]$$

$$\Rightarrow \quad |(1+\sin x)(1+\cos y)| = C$$

which is the required soltuion.

(*vi*) The given equation is

$$\frac{dy}{dx} = y^2 \tan ax \qquad ...(1)$$

Separting the variables,

$$\frac{1}{y^2}\,dy = \tan ax\,dx$$

Integrating both sides, we get

$$\int \frac{1}{y^2}\,dy = \int \tan ax\,dx$$

$$\Rightarrow \quad \int y^{-2}\,dy = \int \tan ax\,dx$$

$$\Rightarrow \quad -y^{-1} = \log\frac{|\sec ax|}{a} + C$$

$$\Rightarrow \quad -\frac{1}{y} = \frac{1}{a}\log|\sec ax| + C$$

which is the required solution.

(*vii*) The given equation is

$$\frac{dy}{dx} = (1+x)(1+y^2) \qquad ...(1)$$

Separating the variables,

$$\frac{1}{1+y^2}\,dy = (1+x)\,dx$$

Integrating both sides, we get

$$\int \frac{1}{1+y^2}\,dy = \int (1+x)\,dx$$

$$\Rightarrow \quad \tan^{-1} y = x + \frac{x^2}{2} + C \qquad \left[\because \int \frac{1}{a^2+x^2}\,dx = \frac{1}{a}\tan^{-1}\frac{x}{a}\right]$$

which is the required solution.

(*viii*) The given equation is

$$x\sqrt{1+y^2}\,dx + y\sqrt{1+x^2}\,dy = 0 \qquad ...(1)$$

$$\Rightarrow \quad x\sqrt{1+y^2}\,dx = -y\sqrt{1+x^2}\,dy$$

Separating the variables,

$$\frac{x}{\sqrt{1+x^2}}\,dx = -\frac{y}{\sqrt{1+y^2}}\,dy$$

Integrating both sides, we get

$$\int \frac{x}{\sqrt{1+x^2}}\,dx = -\int \frac{y}{\sqrt{1+y^2}}\,dy$$

$$\Rightarrow \quad \frac{1}{2}\int \frac{2x}{\sqrt{1+x^2}}\,dx = -\frac{1}{2}\int \frac{2y}{\sqrt{1+y^2}}\,dy \qquad \text{[Multiply and divided by 2]}$$

$$\Rightarrow \quad \frac{1}{2}\int (1+x^2)^{-\frac{1}{2}}(2x)\,dx = -\frac{1}{2}\int (1+y^2)^{-\frac{1}{2}}(2y)\,dy$$

$$\Rightarrow \quad \frac{1}{2}\,\frac{(1+x^2)^{-\frac{1}{2}+1}}{-\frac{1}{2}+1} = -\frac{1}{2}\,\frac{(1+y^2)^{-\frac{1}{2}+1}}{\left(-\frac{1}{2}+1\right)} + C$$

$$\left[\because \int [f(x)]^n f'(x)\,dx = \frac{[f(x)]^{n+1}}{n+1}\right]$$

$$\Rightarrow \quad (1+x^2)^{1/2} = -(1+y^2)^{1/2} + C$$

$$\Rightarrow \quad \sqrt{1+x^2} + \sqrt{1+y^2} = C$$

which is the required solution.

Example 27. *Solve the following differential equations :*

(i) $3e^x \tan y\,dx + (1-e^x)\sec^2 y\,dy = 0$

(ii) $(x+2)\dfrac{dy}{dx} = 4x^2y$ *(iii)* $\left(y - x\dfrac{dy}{dx}\right)x = y$

(iv) $\dfrac{dy}{dx}\tan y = \sin(x+y) + \sin(x-y)$ *(v)* $\dfrac{dy}{dx} = \dfrac{x(2\log x+1)}{\sin y + y\cos y}$

(vi) $x(1+y^2)dx + y(1+x^2)dy = 0.$

Solution. *(i)* The given equation is

$$3e^x \tan y\,dx + (1-e^x)\sec^2 y\,dx = 0$$

$$\Rightarrow \quad 3e^x \tan y\,\mathbf{dx} = -(1-e^x)\sec^2 y\,dy$$

Separating the variables,

$$\frac{3e^x}{-(1-e^x)}\,dx = \frac{\sec^2 y}{\tan y}\,dy$$

Integrating both sides, we get

$$3\int \frac{e^x}{e^x-1}\,dx = \int \frac{\sec^2 y}{\tan y}\,dy$$

$$\Rightarrow \quad 3\log|e^x-1| = \log|\tan y| + C \qquad \left[\because \int \frac{f'(x)}{f(x)}\,dx = \log|f(x)|\right]$$

$$\Rightarrow \quad \log|(e^x-1)|^3 - \log|\tan y| = C$$

$$\Rightarrow \quad \log\left|\frac{(e^x-1)^3}{\tan y}\right| = C$$

$$\Rightarrow \quad \left|\frac{(e^x-1)^3}{\tan y}\right| = e^C$$

$$\Rightarrow \quad \frac{(e^x-1)^3}{\tan y} = \pm e^C$$

$$\Rightarrow \quad (e^x-1)^3 = C_1 \tan y \qquad [\because \; C_1 = \pm e^C]$$

which is the required solution.

(*ii*) The given equation is

$$(x+2)\frac{dy}{dx} = 4x^2y \qquad ...(1)$$

Separating the variables,

$$\frac{1}{y}\,dy = \frac{4x^2}{x+2}\,dx$$

Integrating both sides, we get

$$\int \frac{1}{y}\,dy = \int \frac{4x^2}{x+2}\,dx$$

$$\Rightarrow \quad \int \frac{1}{y}\,dy = \int \left(4x-8+\frac{16}{x+2}\right)dx$$

$$\Rightarrow \quad \log|y| = 4\int\left[(x-2)+\frac{4}{x+2}\right]dx$$

$\because$

$$\begin{array}{r} x+2\,\overline{)\,4x^2\,}\,(4x-8 \\ \underline{\pm 4x^2 \pm 8x} \\ -8x \\ -8x-16 \\ \underline{+\quad +} \\ 16 \end{array}$$

$$\Rightarrow \quad \log|y| = 4\left[\frac{x^2}{2} - 2x + 4\log|x+2|\right] + C$$

$$\Rightarrow \quad \log|y| = 2x^2 - 8x + 16\log|x+2| + C$$

which is the required solution.

(*iii*) The given equation is

$$\left(y - x\frac{dy}{dx}\right)x = y \qquad ...(1)$$

$$\Rightarrow \quad xy - x^2\frac{dy}{dx} = y$$

$$\Rightarrow \quad xy - y = x^2\frac{dy}{dx}$$

Separating the variables,

$$y(x-1) = x^2\frac{dy}{dx}$$

$$\Rightarrow \qquad \frac{x-1}{x^2}\,dx = \frac{1}{y}\,dy$$

Integrating both sides, we get

$$\int \frac{x-1}{x^2}\,dx = \int \frac{1}{y}\,dy$$

$$\Rightarrow \qquad \int \left(\frac{1}{x} - \frac{1}{x^2}\right) dx = \int \frac{1}{y}\,dy$$

$$\Rightarrow \qquad \int \frac{1}{x}\,dx - \int x^{-2}\,dx = \int \frac{1}{y}\,dy$$

$$\Rightarrow \qquad \log|x| - \frac{x^{-2+1}}{-2+1} = \log|y| + C$$

$$\Rightarrow \qquad \log|x| + \frac{1}{x} = \log|y| + C$$

$$\Rightarrow \qquad \log|x| - \log|y| + \frac{1}{x} = C$$

$$\Rightarrow \qquad \log\left|\frac{x}{y}\right| + \frac{1}{x} + C_1 = 0 \quad \text{where : } C_1 = -C \qquad \left[\because \log m - \log n = \log \frac{m}{n}\right]$$

which is the required solution.

(*iv*) The given equation is

$$\frac{dy}{dx}\tan y = \sin(x+y) + \sin(x-y) \qquad ...(1)$$

$$\left[\because \sin C + \sin D = 2\sin\frac{C+D}{2}\cos\frac{C-D}{2}\right]$$

$$\Rightarrow \qquad \frac{dy}{dx}\tan y = 2\sin\left(\frac{x+y+x-y}{2}\right)\cos\left(\frac{x+y-x+y}{2}\right)$$

$$\Rightarrow \qquad \frac{dy}{dx}\tan y = 2\sin x\cos y$$

Separating the variables,

$$\frac{\tan y}{\cos y}\,dy = 2\sin x\,dx$$

Integrating both sides, we get

$$\int \tan y\,\frac{1}{\cos y}\,dy = 2\int \sin x\,dx$$

$$\Rightarrow \qquad \int \sec y \tan y\,dy = 2\int \sin x\,dx$$

$$\Rightarrow \qquad \sec y = -2\cos x + C$$

$$\Rightarrow \qquad \sec y + 2\cos x = C$$

which is the required solution.

(*v*) The given equation is

$$\frac{dy}{dx} = \frac{x(2\log x + 1)}{\sin y + y\cos y} \qquad ...(1)$$

Separating the variables,

$$(\sin y + y\cos y)dy = x(2\log x + 1)dx$$

Integrating both sides, we get

$$\int (\sin y + y\cos y)\,dy = \int (2x\log x + x)\,dx$$

$$\Rightarrow \quad \int \sin y\,dy + \int \underset{\text{I}}{y}\,\underset{\text{II}}{\cos y}\,dy = 2\int \underset{\text{I}}{x}\,\underset{\text{II}}{\log x}\,dx + \int x\,dx \qquad \text{[Integrating by parts]}$$

$$\Rightarrow \quad -\cos y + y\sin y - \int 1.\sin y\,dy = 2\left[\log x.\frac{x^2}{2} - \int \frac{1}{x}.\frac{x^2}{2}\,dx\right] + \frac{x^2}{2} + C$$

$$\Rightarrow \quad -\cos y + y\sin y - (-\cos y) = x^2\log x - \int x\,dx + \frac{x^2}{2} + C$$

$$\Rightarrow \quad -\cos y + y\sin y + \cos y = x^2\log x - \frac{x^2}{2} + \frac{x^2}{2} + C$$

$$\Rightarrow \quad y\sin y = x^2\log x + C$$

which is the required solution.

(*vi*) The given equation is

$$x(1 + y^2)dx + y(1 + x^2)dy = 0$$

Separating the variables,

$$\frac{x}{1+x^2}\,dx + \frac{y}{1+y^2}\,dy = 0$$

Integrating both sides, we get

$$\int \frac{x}{1+x^2}\,dx + \int \frac{y}{1+y^2}\,dy = C$$

$$\Rightarrow \quad \frac{1}{2}\int \frac{2x}{1+x^2}\,dx + \frac{1}{2}\int \frac{2y}{1+y^2}\,dy = C \qquad \text{[Multiply and divided by 2]}$$

$$\Rightarrow \quad \frac{1}{2}\log|1 + x^2| + \frac{1}{2}\log|1 + y^2| = C \qquad \left[\because \int \frac{f'(x)}{f(x)}\,dx = \log|f(x)|\right]$$

$$\Rightarrow \quad \log|1 + x^2| + \log|1 + y^2| = 2C$$

$$\Rightarrow \quad \log|(1 + x^2)(1 + y^2)| = 2C \qquad [\because \log m + \log n = \log mn]$$

$$\Rightarrow \quad \log|(1 + x^2)(1 + y^2)| = \log C_1 \qquad [\text{where } \log C_1 = 2C]$$

$$|(1 + x^2)(1 + y^2)| = C_1$$

which is the required solution.

Example 28. *Solve the following differential equations :*

(*i*) $(e^y + 1) \cos x\, dx + e^y \sin x\, dy = 0$ (*ii*) $x \cos^2 y\, dx = y \cos^2 x\, dy$

(*iii*) $(1 + x^2)dy = xy\, dx$ (*iv*) $\frac{dy}{dx} = \frac{xe^x \log x + e^x}{x \cos y}$

(*v*) $\frac{dy}{dx} + \sqrt{\frac{1 - y^2}{1 - x^2}} = 0$ (*vi*) $x \log x\, dy - y dx = 0$

(*vii*) $\log \frac{dy}{dx} = ax + by.$

Solution. (*i*) The given equation is

$$(e^y + 1) \cos x\, dx + e^y \sin x\, dy = 0 \qquad ...(1)$$

$$\Rightarrow \qquad (e^y + 1) \cos x\, dx = - e^y \sin x\, dy$$

Separating the variables,

$$\frac{\cos x}{\sin x} dx = - \frac{e^y}{e^y + 1} dy$$

Integrating both sides, we get

$$\int \frac{\cos x}{\sin x} dx = - \int \frac{e^y}{e^y + 1} dy$$

$$\Rightarrow \qquad \log |\sin x| = - \log |e^y + 1| + \log C \qquad \left[\because \int \frac{f'(x)}{f(x)} dx = \log |f(x)|\right]$$

$$\Rightarrow \qquad \log |e^y + 1| + \log |\sin x| = \log C$$

$$\Rightarrow \qquad \log |(e^y + 1) \sin x| = \log C \qquad [\because \log m + \log n = \log mn]$$

$$\Rightarrow \qquad |(e^y + 1) \sin x| = C$$

which is the required solution

(*ii*) The given equation is

$$x \cos^2 y\, dx = y \cos^2 x\, dy \qquad ...(1)$$

Separating the variables,

$$\frac{x}{\cos^2 x} dx = \frac{y}{\cos^2 y} dy$$

$$\Rightarrow \qquad x \sec^2 x\, dx = y \sec^2 y\, dy$$

Integrating both sides, we get

$$\int \underset{\text{I}}{x} \underset{\text{II}}{\sec^2 x}\, dx = \int \underset{\text{I}}{y} \underset{\text{II}}{\sec^2 y}\, dy \qquad \text{[Integrating by parts]}$$

$$\Rightarrow \qquad x. \tan x - \int 1. \tan x. dx = y. \tan y - \int 1. \tan y\, dy + C$$

$$\Rightarrow \qquad x \tan x - (- \log |\cos x|) = y \tan y - (- \log |\cos y|) + C$$

$$\Rightarrow \qquad x \tan x + \log |\cos x| = y \tan y + \log |\cos y| + C$$

which is the required solution.

(*iii*) The given equation is

$$(1 + x^2)dy = xy\,dx \qquad ...(1)$$

Separating the variables,

$$\frac{1}{y}\,dy = \frac{x}{1+x^2}\,dx$$

Integrating both sides, we get

$$\int \frac{1}{y}\,dy = \int \frac{x}{1+x^2}\,dx$$

$$\Rightarrow \quad \log|y| = \frac{1}{2}\int \frac{2x}{1+x^2}\,dx \qquad \text{[Multiply and divide the R.H.S. by 2]}$$

$$\Rightarrow \quad \log|y| = \frac{1}{2}\log|1+x^2| + C \qquad \left[\because \int \frac{f'(x)}{f(x)}\,dx = \log|f(x)|\right]$$

$$\Rightarrow \quad 2\log|y| = \log|1+x^2| + 2C$$

$$\Rightarrow \quad \log y^2 - \log(1+x^2) = 2C$$

$$\Rightarrow \quad \log\frac{y^2}{1+x^2} = 2C$$

$$\Rightarrow \quad \frac{y^2}{1+x^2} = e^{2C}$$

$$\Rightarrow \quad \frac{y^2}{1+x^2} = C_1 \qquad [\because \; C_1 = e^{2C}]$$

$$\Rightarrow \quad y^2 = C_1(1+x^2)$$

which is the required solution.

(*iv*) The given equation is

$$\frac{dy}{dx} = \frac{xe^x\log x + e^x}{x\cos y} \qquad ...(1)$$

Separating the variables,

$$\cos y\,dy = \frac{e^x(x\log x + 1)}{x}\,dx$$

Integrating both sides, we get

$$\int \cos y\,dy = \int e^x\left(\log x + \frac{1}{x}\right)dx$$

$$\Rightarrow \quad \sin y = e^x\log x + C \qquad \left[\because \int e^x[f(x) + f'(x)]dx = e^x f(x)\right]$$

which is the required solution.

(*v*) The given equation is

$$\frac{dy}{dx}+\sqrt{\frac{1-y^2}{1-x^2}}=0 \qquad ...(1)$$

Separating the variables,

$$\frac{dy}{dx}+\frac{\sqrt{1-y^2}}{\sqrt{1-x^2}}=0$$

$$\Rightarrow \qquad \frac{1}{\sqrt{1-y^2}}\,dy+\frac{1}{\sqrt{1-x^2}}\,dx=0$$

Integrating both sides, we get

$$\int\frac{1}{\sqrt{1-y^2}}\,dy+\int\frac{1}{\sqrt{1-x^2}}\,dx=\text{C}$$

$$\Rightarrow \qquad \sin^{-1}y+\sin^{-1}x=\text{C} \qquad \left[\because \int\frac{1}{\sqrt{a^2-x^2}}\,dx=\sin^{-1}\frac{x}{a}\right]$$

which is the required solution.

(*vi*) The given equation is

$$x\log x\,dy-ydx=0 \qquad ...(1)$$

$$\Rightarrow \qquad x\log x\,dy=y\,dx$$

Separating the variables,

$$\frac{1}{y}\,dy=\frac{1}{x\log x}\,dx$$

Integrating both sides, we get

$$\int\frac{1}{y}\,dy=\int\frac{1}{x\log x}\,dx$$

$$\Rightarrow \qquad \log|y|=\int\frac{1/x}{\log x}\,dx$$

$$\Rightarrow \qquad \log|y|=\log|\log x|+\log\text{C} \qquad \left[\because \int\frac{f'(x)}{f(x)}\,dx=\log|f(x)|\right]$$

$$\Rightarrow \quad \log|y|-\log|\log x|=\log\text{C}$$

$$\Rightarrow \qquad \log\left|\frac{y}{\log x}\right|=\log\text{C} \qquad \left[\because \log m-\log n=\log\frac{m}{n}\right]$$

$$\Rightarrow \qquad \frac{y}{\log x}=\text{C}$$

$$\Rightarrow \qquad y=\text{C}\log x$$

which is the required solution.

(*vii*) The given equation is

$$\log \frac{dy}{dx} = ax + by \qquad ...(1)$$

$$\Rightarrow \qquad \frac{dy}{dx} = e^{(ax + by)}$$

$$\Rightarrow \qquad \frac{dy}{dx} = e^{ax} \cdot e^{by}$$

Separating the variables,

$$\frac{1}{e^{by}} dy = e^{ax} dx$$

Integrating both sides, we get

$$\int \frac{1}{e^{by}} dy = \int e^{ax} dx$$

$$\Rightarrow \qquad \int e^{-by} dy = \int e^{ax} dx$$

$$\Rightarrow \qquad \frac{e^{-by}}{-b} = \frac{e^{ax}}{a} + C$$

$$\Rightarrow \qquad \frac{e^{ax}}{a} + \frac{e^{-by}}{b} + C = 0$$

$$\Rightarrow \qquad be^{ax} + ae^{-by} + abC = 0$$

which is the required solution.

Example 29. *Solve the following differential equation*

(*i*) $e^x \sqrt{1 - y^2}\, dx + \frac{y}{x} dy = 0$ (*ii*) $\sin^3 x \frac{dx}{dy} = \sin y$

(*iii*) $\frac{dy}{dx} + \frac{y^2 + y + 1}{x^2 + x + 1} = 0$ (*iv*) $\operatorname{cosec} x \log y\, dy + x^2 y\, dx = 0$

(*v*) $(\cos x \cos y) \frac{dy}{dx} = -\sin x \sin y$ (*vi*) $\sec^2 x \tan y\, dy + \sec^2 y \tan x\, dx = 0.$

Solution. (*i*) The given equation is

$$e^x \sqrt{1 - y^2}\, dx + \frac{y}{x} dy = 0 \qquad ...(1)$$

Separating the variables,

$$e^x \sqrt{1 - y^2}\, dx = -\frac{y}{x} dy$$

$$\Rightarrow \qquad x e^x dx = -\frac{y}{\sqrt{1 - y^2}} dy$$

Integrating both sides, we get

$$\int x e^x \, dx = -\int \frac{y}{\sqrt{1-y^2}} \, dy$$

$$\Rightarrow \quad \int \underset{\text{I}}{x} \underset{\text{II}}{e^x} \, dx = \frac{1}{2}\int -\frac{2y}{\sqrt{1-y^2}} \, dy$$ [Multiply and divide the R.H.S. by 2]

(Integrating by parts)

$$\Rightarrow \quad x.e^x - \int 1.e^x \, dx = \frac{1}{2}\int (1-y^2)^{-1/2} (-2y) \, dy + C$$

$$\left[\because \int [f(x)]^n f'(x) \, dx = \frac{[f(x)]^{n+1}}{n+1}\right]$$

$$\Rightarrow \quad xe^x - e^x = \frac{1}{2} \frac{(1-y^2)^{-\frac{1}{2}+1}}{-\frac{1}{2}+1} + C$$

$$\Rightarrow \quad e^x(x-1) = \sqrt{1-y^2} + C$$

which is the required solution.

(*ii*) The given equation is

$$\sin^3 x \frac{dx}{dy} = \sin y \qquad \text{...(1)}$$

Separating the variables

$$\sin^3 x \, dx = \sin y \, dy$$

Integrating both sides, we get

$$\int \sin^3 x \, dx = \int \sin y \, dy$$ $$\left[\begin{array}{l}\because \sin 3A = 3 \sin A - 4 \sin^3 A \\ \Rightarrow \sin^3 A = \dfrac{3 \sin A - \sin 3A}{4}\end{array}\right]$$

$$\Rightarrow \quad \int \frac{3 \sin x - \sin 3x}{4} \, dx = \int \sin y \, dy$$

$$\Rightarrow \quad \frac{3}{4}\int \sin x \, dx - \frac{1}{4}\int \sin 3x \, dx = \int \sin y \, dy$$

$$\Rightarrow \quad -\frac{3}{4}\cos x + \frac{1}{4}\frac{\cos 3x}{3} = -\cos y + C$$

$$\Rightarrow \quad \cos y - \frac{3}{4}\cos x + \frac{1}{12}\cos 3x = C$$

which is the required solution.

(*iii*) The given equation is

$$\frac{dy}{dx} + \frac{y^2 + y + 1}{x^2 + x + 1} = 0 \qquad \text{...(1)}$$

$$\Rightarrow \qquad \frac{dy}{dx} = -\frac{y^2+y+1}{x^2+x+1}$$

Separating the variables,

$$\frac{1}{y^2+y+1}\,dy = -\frac{1}{x^2+x+1}\,dx$$

Integrating both sides, we get

$$\int \frac{1}{y^2+y+1}\,dy = -\int \frac{1}{x^2+x+1}\,dx$$

$$\Rightarrow \qquad \int \frac{1}{y^2+y+\frac{1}{4}+1-\frac{1}{4}}\,dy = -\int \frac{1}{x^2+x+\frac{1}{4}+1-\frac{1}{4}}\,dx$$

[Now, completing the squares on both sides. Add and subtract $\left(\frac{1}{2}\text{ co-eff. of } x\right)^2$]

$$\Rightarrow \qquad \int \frac{1}{\left(y+\frac{1}{2}\right)^2+\frac{3}{4}}\,dy = -\int \frac{1}{\left(x+\frac{1}{2}\right)^2+\frac{3}{4}}\,dx$$

$$\Rightarrow \qquad \int \frac{1}{(y+1/2)^2+\left(\frac{\sqrt{3}}{2}\right)^2}\,dy = -\int \frac{1}{(x+1/2)^2+\left(\frac{\sqrt{3}}{2}\right)^2}\,dx.$$

$$\Rightarrow \qquad \frac{1}{\frac{\sqrt{3}}{2}}\tan^{-1}\left(\frac{y+1/2}{\frac{\sqrt{3}}{2}}\right) = -\frac{1}{\frac{\sqrt{3}}{2}}\tan^{-1}\left(\frac{x+1/2}{\frac{\sqrt{3}}{2}}\right)+C$$

$$\left[\because \int \frac{1}{x^2+a^2}\,dx = \frac{1}{a}\tan^{-1}\frac{x}{a}\right]$$

$$\Rightarrow \qquad \frac{2}{\sqrt{3}}\tan^{-1}\left(\frac{2y+1}{\sqrt{3}}\right) = -\frac{2}{\sqrt{3}}\tan^{-1}\left(\frac{2x+1}{\sqrt{3}}\right)+C$$

$$\Rightarrow \qquad \frac{2}{\sqrt{3}}\left[\tan^{-1}\left(\frac{2y+1}{\sqrt{3}}\right)+\tan^{-1}\left(\frac{2x+1}{\sqrt{3}}\right)\right] = C$$

$$\Rightarrow \qquad \frac{2}{\sqrt{3}}\left[\tan^{-1}\left\{\frac{\frac{2y+1}{\sqrt{3}}+\frac{2x+1}{\sqrt{3}}}{1-\left(\frac{2y+1}{\sqrt{3}}\right)\left(\frac{2x+1}{\sqrt{3}}\right)}\right\}\right] = C \qquad \left[\because \tan^{-1}a+\tan^{-1}b = \tan^{-1}\left(\frac{a+b}{1-ab}\right)\right]$$

$$\Rightarrow \quad \frac{2}{\sqrt{3}}\left[\tan^{-1}\frac{(2x+2y+2)/\sqrt{3}}{\dfrac{3-(4xy+2x+2y+1)}{3}}\right]=C$$

$$\Rightarrow \quad \left[\tan^{-1}\frac{\sqrt{3}\,(2x+2y+2)}{(2-4xy-2x-2y)}\right]=\frac{\sqrt{3}}{2}C$$

$$\Rightarrow \quad \tan^{-1}\frac{\sqrt{3}\,(2x+2y+2)}{2\,(1-2xy-x-y)}=\tan^{-1}C_1 \qquad \left[\text{where } \frac{\sqrt{3}}{2}C=\tan^{-1}C_1 = \text{another constant.}\right]$$

$$\Rightarrow \quad \frac{\sqrt{3}\,(x+y+1)}{(1-2xy-x-y)}=C_1$$

$$\Rightarrow \quad \sqrt{3}\,(x+y+1)=C_1(1-2xy-x-y)$$

which is the required solution.

(*iv*) The given equation is

$$\text{cosec } x \log y\, dy + x^2 y\, dx = 0 \qquad ...(1)$$

Separating the variables

$$\text{cosec } x \log y\, dy = -x^2 y\, dx$$

$$\Rightarrow \quad \frac{1}{y}\log y\, dy = -\frac{x^2}{\text{cosec } x}dx$$

$$\Rightarrow \quad \frac{1}{y}\log y\, dy = -x^2 \sin x\, dx$$

Integrating both sides, we get

$$\int\frac{1}{y}\log y\, dy = -\int \underset{\text{I}}{x^2}\, \underset{\text{II}}{\sin x}\, dx \qquad \text{[Integrating by parts on R.H.S.]}$$

$$\Rightarrow \quad \frac{(\log y)^2}{2} = -\left[x^2.(-\cos x)-\int 2x.(-\cos x)\,dx\right]+C$$

$$\left[\because \int [f(x)]^n f'(x)\,dx = \frac{[f(x)]^{n+1}}{n+1}\right]$$

$$\Rightarrow \quad \frac{1}{2}(\log y)^2 = x^2\cos x - 2\int \underset{\text{I}}{x}\,\underset{\text{II}}{\cos x}\,dx + C$$

(Integrating by parts again)

$$\Rightarrow \quad \frac{1}{2}(\log y)^2 = x^2\cos x - 2\left[x\sin x-\int 1.\sin x\,dx\right]+C$$

$$\Rightarrow \quad \frac{1}{2}(\log y)^2 = x^2\cos x - 2x\sin x + 2\int \sin x\,dx + C$$

$$\Rightarrow \quad (\log y)^2 = 2\,[x^2\cos x - 2x\sin x - 2\cos x + C]$$

$$\Rightarrow \qquad (\log y)^2 = 2[(x^2 - 2)\cos x - 2x \sin x + C]$$

which is the required solution.

(*v*) The given equation is

$$(\cos x \cos y)\frac{dy}{dx} = -\sin x \sin y \qquad \text{...(1)}$$

Sepaerating the variables,

$$\cos x \cos y\, dy = -\sin x \sin y\, dx$$

$$\Rightarrow \qquad \frac{\cos y}{\sin y}\, dy = -\frac{\sin x}{\cos x}\, dx$$

Integrating both sides, we get

$$\int \frac{\cos y}{\sin y}\, dy = \int \frac{-\sin x}{\cos x}\, dx$$

$$\Rightarrow \qquad \log|\sin y| = \log|\cos x| + \log C \qquad \left[\because \int \frac{f'(x)}{f(x)}\, dx = \log|f(x)|\right]$$

$$\Rightarrow \qquad \log|\sin y| - \log|\cos x| = \log C$$

$$\Rightarrow \qquad \log\left|\frac{\sin y}{\cos x}\right| = \log C \qquad \left[\because \log m - \log n = \log\frac{m}{n}\right]$$

$$\Rightarrow \qquad \frac{\sin y}{\cos x} = C$$

$$\Rightarrow \qquad \sin y = C \cos x$$

which is the required solution.

(*vi*) The given equation is

$$\sec^2 x \tan y\, dy + \sec^2 y \tan x\, dx = 0 \qquad \text{...(1)}$$

$$\Rightarrow \qquad \sec^2 x \tan y\, dy = -\sec^2 y \tan x\, dx$$

Separating the variables,

$$\frac{\tan y}{\sec^2 y}\, dy = -\frac{\tan x}{\sec^2 x}\, dx$$

Integrating both sides, we get

$$\int \frac{\tan y}{\sec^2 y}\, dy = -\int \frac{\tan x}{\sec^2 x}\, dx$$

$$\Rightarrow \qquad \int \frac{\frac{\sin y}{\cos y}}{1/\cos^2 y}\, dy = -\int \frac{\frac{\sin x}{\cos x}}{1/\cos^2 x}\, dx$$

$$\Rightarrow \qquad \int \sin y \cos y\, dy = -\int \sin x \cos x\, dx$$

$$\Rightarrow \qquad \frac{1}{2}\int 2\sin y \cos y\, dy = -\frac{1}{2}\int 2 \sin x \cos x\, dx \qquad \text{[Multiply and divided by 2]}$$

$$\Rightarrow \quad \frac{1}{2}\int \sin 2y\, dy = -\frac{1}{2}\int \sin 2x\, dx \qquad [\because \ 2 \sin A \cos A = \sin 2A]$$

$$\Rightarrow \quad \frac{1}{2}\left[-\frac{\cos 2y}{2}\right] = -\frac{1}{2}\left[-\frac{\cos 2x}{2}\right] + C$$

$$\Rightarrow \quad -\frac{1}{4}\cos 2y = +\frac{1}{4}\cos 2x + C$$

$$\Rightarrow \quad \frac{1}{4}\cos 2x + \frac{1}{4}\cos 2y + C = 0$$

$$\Rightarrow \quad \cos 2x + \cos 2y + C_1 = 0 \quad \text{where } C_1 = 4C$$

which is the required solution.

Example 30. *Solve the following differential equations :*

(*i*) $y - x\dfrac{dy}{dx} = 3\left(1 + x^2\dfrac{dy}{dx}\right)$ (*ii*) $(e^x + 1)ydy = (y + 1)e^x\, dx$

(*iii*) $(1 + e^x)\sec^2 y\dfrac{dy}{dx} - e^x \tan y = 0$ (*iv*) $(1 - x^2)(1 - y)\, dx = xy(1 + y)\, dy$

(*v*) $a(xdy + 2ydx) = xydy$ (*vi*) $x(e^{2y} - 1)dy + (x^2 - 1)e^y\, dx = 0$

(*vii*) $\dfrac{dy}{dx} + x^2 = x^2 e^{3y}$ (*viii*) $x^2\dfrac{dy}{dx} + y = 1.$

Solution. (*i*) The given equation is

$$y - x\frac{dy}{dx} = 3\left(1 + x^2\frac{dy}{dx}\right) \qquad ...(1)$$

$$\Rightarrow \quad y - x\frac{dy}{dx} = 3 + 3x^2\frac{dy}{dx}$$

$$\Rightarrow \quad (y - 3) = (3x^2 + x)\frac{dy}{dx}$$

Separating the variables,

$$\frac{1}{3x^2 + x}\, dx = \frac{1}{y - 3}\, dy$$

Integrating both sides, we get

$$\int \frac{1}{3x^2 + x}\, dx = \int \frac{1}{y - 3}\, dy$$

$$\Rightarrow \quad \int\left(\frac{1}{x} - \frac{3}{3x + 1}\right) dx = \int \frac{1}{y - 3}\, dy \qquad \text{[By using partial fractions]}$$

$$\Rightarrow \quad \int \frac{1}{x}\, dx - 3\int \frac{1}{3x + 1}\, dx = \int \frac{1}{y - 3}\, dy$$

$$\Rightarrow \quad \log x - 3\frac{1}{3}\log(3x+1) = \log(y-3) + \log C$$

$$\Rightarrow \quad \log x - \log(3x+1) = \log(y-3) + \log C$$

$$\Rightarrow \quad \log\left(\frac{x}{3x+1}\right) = \log C(y-3) \qquad [\because \ \log m + \log n = \log mn]$$

$$\Rightarrow \quad \frac{x}{3x+1} = C(y-3)$$

$$\Rightarrow \quad x = C(y-3)(3x+1)$$

which is the required solution.

(*ii*) The given equation is

$$(e^x + 1)y\,dy = (y+1)e^x\,dx \qquad \ldots(1)$$

Separating the variables,

$$\frac{y}{y+1}\,dy = \frac{e^x}{e^x+1}\,dx$$

Integrating both sides, we get

$$\int \frac{y}{y+1}\,dy = \int \frac{e^x}{e^x+1}\,dx$$

$$\Rightarrow \quad \int \frac{y+1-1}{y+1}\,dy = \int \frac{e^x}{e^x+1}\,dx$$

$$\Rightarrow \quad \int \left(1 - \frac{1}{y+1}\right)dy = \int \frac{e^x}{e^x+1}\,dx$$

$$\Rightarrow \quad y - \log|y+1| = \log|e^x+1| + \log C \qquad \left[\because \ \int \frac{f'(x)}{f(x)}\,dx = \log|f(x)|\right]$$

$$\Rightarrow \quad y = \log|y+1| + \log|e^x+1| + \log C$$

$$[\because \ \log u + \log v + \log w = \log uvw]$$

$$\Rightarrow \quad y = \log|(y+1)(e^x+1)C|$$

$$\Rightarrow \quad C(e^x+1)(y+1) = e^y$$

which is the required solution.

(*iii*) The given equation is

$$(1+e^x)\sec^2 y\,\frac{dy}{dx} - e^x \tan y = 0 \qquad \ldots(1)$$

$$\Rightarrow \quad (1+e^x)\sec^2 y\,\frac{dy}{dx} = e^x \tan y$$

Separating the variables

$$\frac{\sec^2 y}{\tan y}\,dy = \frac{e^x}{1+e^x}\,dx$$

Integrating both sides, we get

$$\int \frac{\sec^2 y}{\tan y}\,dy = \int \frac{e^x}{1+e^x}\,dx \qquad \left[\because \ \int \frac{f'(x)}{f(x)}\,dx = \log|f(x)|\right]$$

$\Rightarrow \quad \log |\tan y| = \log |1 + e^x| + \log C$

$\Rightarrow \quad \log |\tan y| = \log |(1 + e^x)C|$ $\quad [\because \log m + \log n = \log mn]$

$\Rightarrow \quad \tan y = C(1 + e^x)$

which is the required solution.

(*iv*) The given equation is

$$(1 - x^2)(1 - y)\, dx = xy(1 + y)\, dy \qquad ...(1)$$

Separating the variables,

$$\frac{1-x^2}{x}\, dx = \frac{y(1+y)}{1-y}\, dy$$

$$\Rightarrow \quad \frac{1-x^2}{x}\, dx = \frac{y+y^2}{1-y}\, dy$$

$$\Rightarrow \quad \left(\frac{1}{x} - x\right) dx = \left(-y - 2 + \frac{2}{1-y}\right) dy$$

$1 - y\,\big)\,\overline{y^2 + y}\,\big(\, -y - 2$
$y^2 - y$
$- \quad +$
$2y$
$-2y \mp 2$
2

Integrating both sides, we get

$$\int \left(\frac{1}{x} - x\right) dx = \int \left(-y - 2 + \frac{2}{1-y}\right) dy$$

$$\Rightarrow \quad \int \frac{1}{x}\, dx - \int x\, dx = -\int y\, dy - 2\int 1.\, dy + 2\int \frac{1}{1-y}\, dy$$

$$\Rightarrow \quad \log |x| - \frac{x^2}{2} = \frac{-y^2}{2} - 2y - 2\log |(1-y)| + C$$

$$\Rightarrow \quad \frac{x^2}{2} - \frac{y^2}{2} - 2y - 2\log |1 - y| - \log |x| + C$$

which is the required solution.

(*v*) The given equation is

$$a(x\, dy + 2y\, dx) = xy\, dy \qquad ...(1)$$

Separating the variables,

$$ax\, dy + 2ay\, dx = xy\, dy$$

$$\Rightarrow \quad (ax - xy)\, dy + 2ay\, dx = 0$$

$$\Rightarrow \quad x(a - y)\, dy + 2ay\, dx = 0$$

$$\Rightarrow \quad 2ay\, dx = x(y - a)\, dy$$

$$\Rightarrow \quad \frac{2a}{x}\, dx = \left(\frac{y-a}{y}\right) dy$$

Integrating both sides, we get

$$\int \frac{2a}{x}\, dx = \int \left(\frac{y-a}{y}\right) dy$$

$$\Rightarrow \quad 2a \int \frac{1}{x}\, dx = \int \left(1 - \frac{a}{y}\right) dy$$

$$\Rightarrow \quad 2a \log |x| = y - a \log |y| + C$$

$$\Rightarrow \qquad y = 2a \log|x| + a \log|y| - C$$

which is the required solution.

(*vi*) The given equation is

$$x(e^{2y} - 1)\,dy + (x^2 - 1)e^y\,dx = 0 \qquad ...(1)$$

$$\Rightarrow \qquad x(e^{2y} - 1)\,dy = (1 - x^2)e^y\,dx$$

Separating the variables,

$$\frac{e^{2y} - 1}{e^y}\,dy = \frac{1 - x^2}{x}\,dx$$

$$\Rightarrow \qquad (e^y - e^{-y})\,dy = \left(\frac{1}{x} - x\right)dx$$

Integrating both sides, we get

$$\int (e^y - e^{-y})\,dy = \int \left(\frac{1}{x} - x\right)dx$$

$$\Rightarrow \qquad e^y - \frac{e^{-y}}{-1} = \log|x| - \frac{x^2}{2} + C$$

$$\Rightarrow \qquad (e^y + e^{-y}) = \log|x| - \frac{x^2}{2} + C$$

which is the required solution.

(*vii*) The given equation is

$$\frac{dy}{dx} + x^2 = x^2 e^{3y} \qquad ...(1)$$

$$\Rightarrow \qquad \frac{dy}{dx} = x^2 e^{3y} - x^2$$

$$\Rightarrow \qquad \frac{dy}{dx} = x^2(e^{3y} - 1)$$

Separating the variables,

$$\left(\frac{1}{e^{3y} - 1}\right)dy = x^2\,dx$$

Dividing the numerator and denominator on the L.H.S. by e^{3y}, we have

$$\frac{e^{-3y}}{1 - e^{-3y}}\,dy = x^2\,dx$$

Integrating both sides, we get

$$\int \frac{e^{-3y}}{1 - e^{-3y}}\,dy = \int x^2\,dx$$

$$\Rightarrow \qquad \frac{1}{3}\int \frac{3e^{-3y}}{1 - e^{-3y}}\,dy = \frac{x^3}{3} + C \qquad \text{[Multiply and divided by 3]}$$

$$\Rightarrow \qquad \frac{1}{3}\log|1 - e^{-3y}| = \frac{x^3}{3} + C \qquad \left[\because \int \frac{f'(x)}{f(x)}\,dx = \log|f(x)|\right]$$

$$\Rightarrow \qquad \log |1 - e^{-3y}| = x^3 + 3C$$

which is the required solution.

(*viii*) The given equation is

$$x^2 \frac{dy}{dx} + y = 1 \qquad ...(1)$$

$$\Rightarrow \qquad x^2 \frac{dy}{dx} = (1 - y)$$

Separating the variables,

$$\frac{1}{1-y}\, dy = \frac{1}{x^2}\, dx$$

Integrating both sides, we get

$$\int \frac{1}{1-y}\, dy = \int x^{-2}\, dx$$

$$\Rightarrow \qquad -\log |1 - y| = \frac{x^{-2+1}}{-2+1} + C = \frac{-1}{x} + C$$

$$\Rightarrow \qquad \log |1 - y| = \frac{1}{x} + C_1 \qquad \text{where } C_1 = -C$$

which is the required solution.

Example 31. *Solve the following differential equation :*

(*i*) $y\, dx + (1 + x^2) \tan^{-1} x\, dy = 0$ (*ii*) $\sqrt{1 + x^2 + y^2 + x^2 y^2} + xy \frac{dy}{dx} = 0$

(*iii*) $xy \frac{dy}{dx} = \frac{(1 + y^2)(1 + x + x^2)}{1 + x^2}$.

Solution. (*i*) The given equation is

$$y\, dx + (1 + x^2) \tan^{-1} x\, dy = 0 \qquad ...(1)$$

$$\Rightarrow \qquad y\, dx = -(1 + x^2) \tan^{-1} x\, dy$$

Separating the variables,

$$\frac{1}{(1 + x^2) \tan^{-1} x}\, dx = \frac{-1}{y}\, dy$$

Integrating both sides, we get

$$\int \frac{1}{(1 + x^2) \tan^{-1} x}\, dx = -\int \frac{1}{y}\, dy$$

$$\Rightarrow \qquad \int \frac{\frac{1}{1 + x^2}}{\tan^{-1} x}\, dx = -\int \frac{1}{y}\, dy$$

$$\Rightarrow \qquad \log |\tan^{-1} x| = -\log |y| + \log C \qquad \left[\because \int \frac{f'(x)}{f(x)}\, dx = \log |f(x)|\right]$$

$$\Rightarrow \qquad \log |\tan^{-1} x| + \log |y| = \log C$$

$$\Rightarrow \qquad \log |y \tan^{-1} x| = \log C \qquad [\because \log m + \log n = \log mn]$$

$$\Rightarrow \qquad |\, y \tan^{-1} x \,| = C$$

which is the required solution.

(*ii*) The given equation is

$$\sqrt{1 + x^2 + y^2 + x^2 y^2} + xy\frac{dy}{dx} = 0 \qquad \text{...(1)}$$

$$\Rightarrow \qquad \sqrt{(1 + x^2) + y^2(1 + x^2)} = -\,xy\frac{dy}{dx}$$

$$\Rightarrow \qquad \sqrt{(1 + x^2)(1 + y^2)} = -\,xy\frac{dy}{dx}$$

$$\Rightarrow \qquad \sqrt{1 + x^2}\,\sqrt{1 + y^2} = -\,xy\frac{dy}{dx}$$

Separating the variables,

$$\frac{\sqrt{1 + x^2}}{x}\,dx = -\frac{y}{\sqrt{1 + y^2}}\,dy$$

Integrating both sides, we get

$$\int \frac{\sqrt{1 + x^2}}{x}\,dx = -\int \frac{y}{\sqrt{1 + y^2}}\,dy + C$$

$$\Rightarrow \qquad \int \frac{\sqrt{1 + x^2}}{x}\,dx + \int \frac{y}{\sqrt{1 + y^2}}\,dy = C \qquad \text{...(2)}$$

$$\Rightarrow \qquad I_1 + I_2 = C \qquad \text{...(3)}$$

Now $\qquad I_1 = \int \frac{\sqrt{1 + x^2}}{x}\,dx$

Put $\sqrt{1 + x^2} = z$

$\Rightarrow \quad 1 + x^2 = z^2 \qquad \Rightarrow \quad x^2 = z^2 - 1$

$\Rightarrow \quad 2x\,dx = 2z\,dz \qquad \Rightarrow \quad x\,dx = z\,dz$

$$\therefore \qquad I_1 = \int \frac{z}{z^2 - 1}(z\,dz) = \int \frac{z^2}{z^2 - 1}\,dz$$

$$= \int \frac{z^2 - 1 + 1}{z^2 - 1}\,dz = \int \left(1 + \frac{1}{z^2 - 1}\right)dz \qquad \text{[Add and subtract 1 to the numerator]}$$

$$= \int 1 \,.\, dz + \int \frac{1}{z^2 - (1)^2}\,dz$$

$$= z + \frac{1}{2}\log\left|\frac{z - 1}{z + 1}\right| + C_1 \qquad \left[\because \int \frac{1}{x^2 - a^2}\,dx + \frac{1}{2a}\log\left|\frac{x - a}{x + a}\right|\right]$$

$$\Rightarrow \qquad I_1 = \sqrt{1 + x^2} + \frac{1}{2}\log\left|\frac{\sqrt{1 + x^2} - 1}{\sqrt{1 + x^2} + 1}\right| + C_1 \qquad \text{...(4)} \quad [\because \; z = \sqrt{1 + x^2}\,]$$

and $$I_2 = \int \frac{y}{\sqrt{1+y^2}}\, dy$$

$$= \frac{1}{2}\int \frac{2y}{\sqrt{1+y^2}}\, dy \qquad \text{[Multiply and divided by 2]}$$

$$= \frac{1}{2}\int (1+y^2)^{-1/2}.(2y)\, dy$$

$$= \frac{1}{2}\left[\frac{(1+y^2)^{\frac{-1}{2}+1}}{\frac{-1}{2}+1}\right] + C_2 \qquad \left[\because \int [f(x)]^n f'(x)\, dx = \frac{[f(x)]^{n+1}}{n+1}\right]$$

$$\Rightarrow \quad I_2 = \sqrt{1+y^2} + C_2 \qquad ...(5)$$

$\therefore$ From equation (3),

$$I_1 + I_2 = C$$

$$\Rightarrow \quad \sqrt{1+x^2} + \frac{1}{2}\log\left|\frac{\sqrt{1+x^2}-1}{\sqrt{1+x^2}+1}\right| + C_1 + \sqrt{1+y^2} + C_2 = C$$

[Using equations (4) and (5)]

$$\Rightarrow \quad \sqrt{1+x^2} + \sqrt{1+y^2} + \frac{1}{2}\log\left|\frac{\sqrt{1+x^2}-1}{\sqrt{1+x^2}+1}\right| = C - C_1 - C_2$$

$$\Rightarrow \quad \sqrt{1+x^2} + \sqrt{1+y^2} + \frac{1}{2}\log\left|\frac{\sqrt{1+x^2}-1}{\sqrt{1+x^2}+1}\right| = C_0 \text{ where } C_0 = C - C_1 - C_2$$

which is the required solution.

(*iii*) The given equation is

$$xy\frac{dy}{dx} = \frac{(1+y^2)(1+x+x^2)}{(1+x^2)} \qquad ...(1)$$

Separating the variables,

$$\frac{y}{1+y^2}\, dy = \frac{(1+x+x^2)}{x(1+x^2)}\, dx$$

$$\Rightarrow \quad \frac{y}{1+y^2}\, dy = \left[\frac{1+x^2}{x(1+x^2)} + \frac{x}{x(1+x^2)}\right] dx$$

$$\Rightarrow \quad \frac{y}{1+y^2}\, dy = \left(\frac{1}{x} + \frac{1}{1+x^2}\right) dx$$

Integrating both sides, we get

$$\int \frac{y}{1+y^2}\, dy = \int \left(\frac{1}{x} + \frac{1}{1+x^2}\right) dx$$

$$\Rightarrow \quad \frac{1}{2}\int \frac{2y}{1+y^2}\, dy = \int \frac{1}{x}\, dx + \int \frac{1}{1+x^2}\, dx$$

$$\Rightarrow \quad \frac{1}{2} \log | 1 + y^2 | = \log | x | + \tan^{-1} x + C \qquad \left[\because \int \frac{f'(x)}{f(x)}\, dx = \log | f(x) | \text{ and } \int \frac{1}{a^2+x^2}\, dx = \frac{1}{a}\tan^{-1}\frac{x}{a}\right]$$

which is the required solution.

Example 32. *Solve the following initial value problems :*

(i) $\frac{dy}{dx} + 2y^2 = 0$, *when* $y(1) = 1$.

(ii) If $\frac{dy}{dx} = e^{x+y}$ *and it is given that for* $x = 1$, $y = 1$; *find* y *when* $x = -1$.

(iii) $\frac{dy}{dx} = 2e^x y^3$; *when* $y(0) = \frac{1}{2}$

(iv) $\frac{dy}{dx} = \sec y$; *when* $y(0) = 0$

(v) $(x - 1) \frac{dy}{dx} = 2xy$; *when* $y(2) = 1$

(vi) $e^x \frac{dy}{dx} = 3y^2$; *when* $y(0) = \frac{1}{2}$.

Solution. (*i*) The given equation is

$$\frac{dy}{dx} + 2y^2 = 0 \qquad ...(1)$$

Separating the variables,

$$\frac{dy}{dx} = -2y^2 \quad \Rightarrow \quad \frac{-1}{2y^2}\, dy = dx$$

Integrating both sides, we get

$$\frac{-1}{2}\int \frac{1}{y^2}\, dy = \int 1 . dx \quad \Rightarrow \quad \frac{-1}{2}\left[\frac{y^{-2+1}}{-2+1}\right] = x + C$$

$$\Rightarrow \quad \frac{1}{2y} = x + C \qquad ...(2)$$

It is given that $y(1) = 1$

$\Rightarrow$ $y = 1$ when $x = 1$

Putting these values in equation (2), we have

$$\frac{1}{2(1)} = 1 + C \quad \Rightarrow \quad C = \frac{1}{2} - 1 \quad \Rightarrow \quad C = \frac{-1}{2}$$

Putting this value of C in equation (2), we have

$$\frac{1}{2y} = x - \frac{1}{2} \quad \Rightarrow \quad \frac{1}{2y} = \frac{2x-1}{2}$$

$$\Rightarrow \qquad 2y = \frac{2}{2x-1} \quad \Rightarrow \quad y = \frac{1}{2x-1}.$$

(*ii*) The given equation is

$$\frac{dy}{dx} = e^{x+y} \qquad ...(1)$$

$$\Rightarrow \qquad \frac{dy}{dx} = e^x \cdot e^y$$

Separating the variables,

$$\frac{1}{e^y} dy = e^x dx$$

Integrating both sides, we get

$$\int e^{-y} dy = \int e^x dx$$

$$\Rightarrow \qquad -e^{-y} = e^x + C \qquad ...(2)$$

It is given that $x = 1, y = 1$

$\therefore$ From equation (2), we have

$$-e^{-1} = e^1 + C$$

$$\Rightarrow \qquad \frac{-1}{e} - e = C \quad \Rightarrow \quad C = -\left(\frac{1}{e} + e\right) = -\left(\frac{1+e^2}{e}\right)$$

Putting this value of C in equation (2), we have

$$-e^{-y} = e^x - \left(\frac{1+e^2}{e}\right)$$

Now, when $x = 1$,

$$-e^{-y} = e^{-1} - \left(\frac{1+e^2}{e}\right)$$

$$= \frac{1}{e} - \frac{1+e^2}{e} = \frac{1-1-e^2}{e} = \frac{-e^2}{e} = -e$$

$$\Rightarrow \qquad -e^{-y} = -e \quad \Rightarrow \quad e^{-y} = e^1 \quad \Rightarrow \quad -y = 1 \quad \Rightarrow \quad y = -1.$$

(*iii*) The given equation is

$$\frac{dy}{dx} = 2e^x y^3 \qquad ...(1)$$

Separating the variables,

$$\frac{1}{y^3} dy = 2e^x dx$$

Integrating both sides, we get

$$\int \frac{1}{y^3} dy = \int 2e^x dx$$

$$\Rightarrow \qquad \int y^{-3} dy = 2\int e^x dx \quad \Rightarrow \quad \frac{-1}{2y^2} = 2e^x + C \qquad ...(2)$$

It is given that $y(0) = \frac{1}{2}$ *i.e.*, $y = \frac{1}{2}$ at $x = 0$

$\therefore$ From equation (2), we have

$$\Rightarrow \quad \frac{-1}{2\left(\frac{1}{2}\right)^2} = 2e^0 + C \qquad [\because \ e^0 = 1]$$

$$\Rightarrow \quad \frac{-1}{1/2} = 2 + C \ \Rightarrow \ -2 - 2 = C \ \Rightarrow \ C = -4.$$

Putting this value of C in equation (2), we have

$$\frac{-1}{2y^2} = 2e^x - 4 \quad \Rightarrow \quad \frac{-1}{2y^2} = 2(e^x - 2)$$

$$\Rightarrow \quad 4y^2(e^x - 2) + 1 = 0.$$

(*iv*) The given equation is

$$\frac{dy}{dx} = \sec y \qquad ...(1)$$

Separating the variables,

$$\frac{1}{\sec y}\, dy = dx$$

Integrating both sides, we get

$$\int \cos y\, dy = \int 1 \, . \, dx$$

$$\sin y = x + C \qquad ...(2)$$

It is given that $y(0) = 0$ *i.e.*, $y = 0$ at $x = 0$

$\therefore$ From equation (2), we have

$$\sin 0 = 0 + C \ \Rightarrow \ C = 0$$

Putting this value of C in equation (2), we have $\sin y = x$.

(*v*) The given equation is

$$(x - 1)\frac{dy}{dx} = 2xy \qquad ...(1)$$

Separating the variables,

$$\frac{1}{y}\, dy = 2\frac{x}{x-1}\, dx$$

Integrating both sides, we get

$$\int \frac{1}{y}\, dy = 2\int \frac{x}{x-1}\, dx$$

$$\Rightarrow \quad \log|y| = 2\int \frac{x-1+1}{x-1}\, dx \qquad \text{[Add and subtract 1 to the numerator]}$$

$$\Rightarrow \quad \log|y| = 2\int \left(1 + \frac{1}{x-1}\right) dx$$

$$\Rightarrow \quad \log|y| = 2\,[x + \log|x - 1|] + C$$

$\Rightarrow \qquad \log |y| = 2x + 2\log|x-1| + C$

$\Rightarrow \qquad \log |y| = 2x + \log|x-1|^2 + C \qquad [\because m \log n = \log n^m]$

$\Rightarrow \quad \log|y| - \log|(x-1)^2| - 2x = C \qquad ...(2)$

It is given that $y(2) = 1$ *i.e.,* $y = 1$ at $x = 2$.

$\therefore$ From equation (2) ; we have

$$\log|1| - \log|(2-1)^2| - 2(2) = C$$

$\Rightarrow \qquad 0 - 0 - 4 = C \quad \Rightarrow \quad C = -4$

Putting this value of C in equation (2), we have

$$\log|y| - 2x - \log|(x-1)^2| = -4$$

$\Rightarrow \quad \log|y| - \log|(x-1)^2| = 2x - 4.$

(*vi*) The given equation is

$$e^x \frac{dy}{dx} = 3y^2 \qquad ...(1)$$

Separating the variables, we have

$$\frac{1}{y^2}\, dy = \frac{3}{e^x}\, dx$$

Integrating both sides, we get

$$\int y^{-2}\, dy = 3\int e^{-x}\, dx$$

$\Rightarrow \qquad \dfrac{-1}{y} = -3e^{-x} + C = \dfrac{-3}{e^x} + C$

$\Rightarrow \qquad \dfrac{3}{e^x} = \dfrac{1}{y} + C \qquad ...(2)$

It is given that $y(0) = \dfrac{1}{2}$ *i.e.,* $y = \dfrac{1}{2}$ at $x = 0$.

$\therefore$ From equation (2) ; we have

$$\frac{3}{e^0} = \frac{1}{1/2} + C \qquad [\because e^0 = 1]$$

$\Rightarrow \qquad 3 = 2 + C \quad \Rightarrow \quad C = 1$

Putting this value of C in equation (2), we have

$$\frac{3}{e^x} = \frac{1}{y} + 1$$

$\Rightarrow \qquad 3y = e^x + ye^x.$

Example 33. *Solve the following initial value problems*

(*i*) $(1 + e^{2x})\, dy + (1 + y^2)\, e^x\, dx = 0$; *when* $x = 0, y = 1$

(*ii*) $(y - yx)\, dx + (x + xy)\, dy = 0$; *when* $x = 1, y = 1$

(*iii*) $\dfrac{dy}{dx} = y \cot 2x$; *when* $y\left(\dfrac{\pi}{4}\right) = 2$

(*iv*) $2x \dfrac{dy}{dx} = 3y$; *when* $y(1) = 4$

(*v*) $(e^x + e^{-x})\, dy = (e^x - e^{-x})\, dx$; *when* $y(0) = 0$

(*vi*) $\cos y\, dy + \cos x \sin y\, dx = 0$; *when* $y\left(\dfrac{\pi}{2}\right) = \dfrac{\pi}{2}$.

Solution. (*i*) The given equation is

$$(1 + e^{2x})\, dy + (1 + y^2)\, e^x\, dx = 0 \quad ...(1)$$

$$\Rightarrow \quad (1 + e^{2x})\, dy = -(1 + y^2)\, e^x\, dx$$

Separating the variables,

$$\frac{1}{1+y^2}\, dy = \frac{-e^x}{1+e^{2x}}\, dx$$

Integrating both sides, we get

$$\int \frac{1}{1+y^2}\, dy = -\int \frac{e^x}{1+e^{2x}}\, dx$$

Put $e^x = z \Rightarrow e^x\, dx = dz$

$$\therefore \quad \int \frac{1}{1+y^2}\, dy = -\int \frac{1}{1+z^2}\, dz \qquad [\because\ e^{2x} = (e^x)^2]$$

$$\Rightarrow \quad \tan^{-1} y = -\tan^{-1} z + C \qquad \left[\because \int \frac{1}{a^2+x^2}\, dx = \frac{1}{a} \tan^{-1} \frac{x}{a}\right]$$

$$\Rightarrow \quad \tan^{-1} y + \tan^{-1} e^x = C \quad ...(2) \quad [\because\ z = e^x]$$

It is given that $x = 0, y = 1$

$\therefore$ From equation (2), we have

$$\tan^{-1}(1) + \tan^{-1}(e^0) = C \qquad [\because\ e^0 = 1]$$

$$\Rightarrow \quad \frac{\pi}{4} + \frac{\pi}{4} = C \Rightarrow C = \frac{\pi}{2}$$

Putting this value of C in equation (2), we have

$$\tan^{-1} y + \tan^{-1} e^x = \frac{\pi}{2}$$

$$\Rightarrow \quad \tan^{-1}\left(\frac{y+e^x}{1-ye^x}\right) = \frac{\pi}{2} \qquad \left[\because\ \tan^{-1} a + \tan^{-1} b = \tan^{-1}\left(\frac{a+b}{1-ab}\right)\right]$$

$$\Rightarrow \quad \frac{y+e^x}{1-ye^x} = \tan\left(\frac{\pi}{2}\right) = \infty = \frac{1}{0}$$

$$\Rightarrow \quad 1 - ye^x = 0 \Rightarrow ye^x = 1 \Rightarrow y = \frac{1}{e^x}.$$

(*ii*) The given equation is

$$(y - yx)\, dx + (x + xy)\, dy = 0 \quad ...(1)$$

$$\Rightarrow \quad y(1 - x)\, dx + x(1 + y)\, dy = 0$$

Separating the variables,

$$\left(\frac{1-x}{x}\right) dx = -\left(\frac{1+y}{y}\right) dy$$

Integrating both sides, we get

$$\int \left(\frac{1-x}{x}\right) dx = -\int \left(\frac{1+y}{y}\right) dy$$

$$\Rightarrow \quad \int\left(\frac{1}{x}-1\right)dx = -\int\left(\frac{1}{y}+1\right)dy$$

$$\Rightarrow \quad \log|x| - x = -\log|y| - y + C$$

$$\Rightarrow \quad \log|x| + \log|y| = x - y + C$$

$$\Rightarrow \quad \log|xy| = x - y + C \qquad \text{...(2)} \quad [\because \ \log m + \log n = \log mn]$$

Now, when $x = 1$, $y = 1$.

$\therefore$ From equation (2), we have

$$\log 1 = 1 - 1 + C \ \Rightarrow \ C = 0 \qquad [\because \ \log 1 = 0]$$

Putting this value of C in equation (2), we have

$$\log|xy| = x - y.$$

(*iii*) The given equation is

$$\frac{dy}{dx} = y \cot 2x \qquad \text{...(1)}$$

Separating the variables

$$\frac{1}{y}\,dy = \cot 2x\,dx$$

Integrating both sides, we get

$$\int \frac{1}{y}\,dy = \int \cot 2x\,dx$$

$$\Rightarrow \quad \log|y| = \frac{\log|\sin 2x|}{2} + C \qquad \text{...(2)}$$

Now, it is given that

$$y\left(\frac{\pi}{4}\right) = 2 \quad i.e., \quad x = \frac{\pi}{4},\ y = 2$$

$\therefore$ From equation (2), we have

$$\log|2| = \frac{\log\left|\sin\frac{\pi}{2}\right|}{2} + C \qquad [\because \ \log 1 = 0]$$

$$\Rightarrow \quad \log 2 = \frac{\log 1}{2} + C \ \Rightarrow \ C = \log 2$$

Putting this value of C in equation (2), we get

$$\log|y| = \frac{\log|\sin 2x|}{2} + \log 2$$

$$\Rightarrow \quad \log|y| = \frac{1}{2}\log|\sin 2x| + \log 2$$

$$\Rightarrow \quad \log|y| = \log|\sin 2x|^{1/2} + \log 2$$

$$\Rightarrow \quad \log|y| = \log|(2\sqrt{\sin 2x})| \qquad [\because \ \log m + \log n = mn]$$

$$\Rightarrow \quad |y| = |2\sqrt{\sin 2x}| \ \Rightarrow \ y^2 = 4\sin 2x.$$

(*iv*) The given equation is

$$2x \frac{dy}{dx} = 3y \qquad ...(1)$$

Separating the variables,

$$\frac{1}{3y}\, dy = \frac{1}{2x}\, dx$$

Integrating both sides, we get

$$\int \frac{1}{y}\, dy = \frac{3}{2} \int \frac{1}{x}\, dx$$

$$\Rightarrow \qquad \log | y | = \frac{3}{2} \log | x | + C \qquad ...(2)$$

Now, it is given that $y(1) = 4$. *i.e.*, $x = 1$, $y = 4$

$\therefore$ From equation (2), we have

$$\log 4 = \frac{3}{2} \log 1 + C \qquad [\because \log 1 = 0]$$

$$\Rightarrow \qquad C = \log 4$$

Putting this value in equation (2), we have

$$\log | y | = \frac{3}{2} \log | x | + \log 4$$

$$\Rightarrow \qquad \log | y | = \log | x |^{3/2} + \log 4 \qquad [\because m \log n = \log n^m]$$

$$\Rightarrow \qquad \log | y | = \log 4 | x |^{3/2} \qquad [\because \log m + \log n = \log mn]$$

$$\Rightarrow \qquad y = 4x^{3/2}.$$

(*v*) The given equation is

$$(e^x + e^{-x})\, dy = (e^x - e^{-x})\, dx \qquad ...(1)$$

Separating the variables,

$$dy = \frac{e^x - e^{-x}}{e^x + e^{-x}}\, dx$$

Integrating both sides, we get

$$\int dy = \int \frac{e^x - e^{-x}}{e^x + e^{-x}}\, dx \qquad ...(2) \qquad \left[\because \int \frac{f'(x)}{f(x)}\, dx = \log | f(x) |\right]$$

$$\Rightarrow \qquad y = \log | e^x + e^{-x} | + C$$

Now, it is given that

$$y(0) = 0 \quad i.e., \quad x = 0, \quad y = 0$$

$\therefore$ From equation (2), we have

$$0 = \log | e^0 + e^0 | + C$$

$$\Rightarrow \qquad 0 = \log 2 + C \quad \Rightarrow \quad C = -\log 2$$

Putting this value of C in equation (2), we get

$$y = \log | e^x + e^{-x} | - \log 2$$

$$\Rightarrow \qquad y = \log \frac{| e^x + e^{-x} |}{2}. \qquad \left[\because \log m - \log n = \log \frac{m}{n}\right]$$

(*vi*) The given equation is

$$\cos y\, dy + \cos x \sin y\, dx = 0 \quad \text{...(1)}$$

$$\Rightarrow \quad \cos y\, dy = -\cos x \sin y\, dx$$

Separating the variables,

$$\frac{\cos y}{\sin y}\, dy = -\cos x\, dx \quad \Rightarrow \quad \cot y\, dy = -\cos x\, dx$$

Integrating both sides, we get

$$\int \cot y\, dy = -\int \cos x\, dx$$

$$\Rightarrow \quad \log |\sin y| = -\sin x + C \quad \text{...(2)}$$

Now, it is given that

$$y = \frac{\pi}{2} \text{ at } x = \frac{\pi}{2}.$$

$\therefore$ From equation (2), we have

$$\log \left| \sin \frac{\pi}{2} \right| = -\sin \frac{\pi}{2} + C$$

$$\Rightarrow \quad \log 1 = -1 + C \quad [\because \log 1 = 0]$$

$$\Rightarrow \quad 0 = -1 + C \quad \Rightarrow \quad C = 1$$

Putting this value of C in equation (2), we have

$$\log |\sin y| + \sin x = 1.$$

Example 34. *Solve the following differential equations :*

(*i*) *$\log\left(\frac{dy}{dx}\right) = 2x + y$, given that $y = 0$ when $x = 0$.*

(*ii*) *$(1 + x^2)\frac{dy}{dx} + (1 + y^2) = 0$, given that $y = 1$ when $x = 0$.*

(*iii*) *$\log\left(\frac{dy}{dx}\right) = 3x + 4y$, given that $y = 0$ when $x = 0$.*

(*iv*) *$x(x\, dy - y\, dx) = y\, dx$, when $y(1) = 1$.*

(*v*) *$(1 + y^2)(1 + \log x)\, dx + x\, dy = 0$, given that $y = 1$, when $x = 1$.*

(*vi*) *$\frac{dy}{dx} = e^{-y} \cos x$, when $y(0) = 0$.*

(*vii*) *$e^{\frac{dy}{dx}} = x + 1$, given that $y = 5$ when $x = 0$.*

Solution. (*i*) The given equation is

$$\log\left(\frac{dy}{dx}\right) = 2x + y \quad \text{...(1)}$$

$$\Rightarrow \quad \frac{dy}{dx} = e^{2x+y} \quad \Rightarrow \quad \frac{dy}{dx} = e^{2x} \,.\, e^{y}$$

Separating the variables,

$$\frac{1}{e^{y}}\, dy = e^{2x} \,.\, dx$$

Integrating both sides, we get

$$\int e^{-y}\, dy = \int e^{2x}\, dx$$

$$\Rightarrow \quad -e^{-y} = \frac{e^{2x}}{2} + C \qquad ...(2)$$

Now, it is given that $y = 0$ when $x = 0$

$\therefore$ From equation (2), we have

$$-e^0 = \frac{1}{2} e^0 + C$$

$$\Rightarrow \quad -1 = \frac{1}{2} + C \Rightarrow C = \frac{-3}{2}$$

Putting this value of C in equation (2), we get

$$-e^{-y} = \frac{1}{2} e^{2x} - \frac{3}{2}$$

$$\Rightarrow \quad -2e^{-y} = e^{2x} - 3 \Rightarrow e^{2x} + 2e^{-y} - 3 = 0.$$

(*ii*) The given equation is

$$(1 + x^2) \frac{dy}{dx} + (1 + y^2) = 0 \qquad ...(1)$$

$$\Rightarrow \quad (1 + x^2) \frac{dy}{dx} = -(1 + y^2)$$

Separating the variables,

$$\frac{1}{1+y^2} dy = -\frac{1}{1+x^2} dx$$

Integrating both sides, we get

$$\int \frac{1}{1+y^2} dy = -\int \frac{1}{1+x^2} dx$$

$$\Rightarrow \quad \tan^{-1} y = -\tan^{-1} x + C \qquad \left[\because \int \frac{1}{a^2+x^2} dx = \frac{1}{a} \tan^{-1} \frac{x}{a}\right]$$

$$\Rightarrow \quad \tan^{-1} y + \tan^{-1} x = C \qquad ...(2)$$

Now, it is given that $y = 1$ when $x = 0$

$\therefore$ From equation (2), we have

$$\tan^{-1} 1 + \tan^{-1} 0 = C$$

$$\Rightarrow \quad \frac{\pi}{4} + 0 = C \Rightarrow C = \frac{\pi}{4}$$

Putting this value of C in equation (2), we have

$$\tan^{-1} x + \tan^{-1} y = \frac{\pi}{4}$$

$$\Rightarrow \quad \tan^{-1}\left(\frac{x+y}{1-xy}\right) = \frac{\pi}{4} \qquad \left[\because \tan^{-1} a + \tan^{-1} b = \tan^{-1}\left(\frac{a+b}{1-ab}\right)\right]$$

$$\Rightarrow \quad \frac{x+y}{1-xy} = \tan\frac{\pi}{4} = 1 \Rightarrow x + y = 1 - xy.$$

(*iii*) The given equation is

$$\log\left(\frac{dy}{dx}\right) = 3x + 4y \qquad ...(1)$$

Please try yourself.

[**Hint.** See part (*i*) of the same example.] [**Ans.** $4e^{3x} + 3e^{-4y} = 7$.]

(*iv*) The given equation is

$$x(xdy - ydx) = ydx \qquad ...(1)$$

$$\Rightarrow \quad x^2dy - xy\,dx = ydx$$

Separating the variables,

$$x^2dy = (xy + y)\,dx \Rightarrow x^2dy = y(x+1)\,dx \Rightarrow \frac{1}{y}\,dy = \frac{x+1}{x^2}\,dx$$

Integrating both sides, we get

$$\int \frac{1}{y}\,dy = \int \frac{x+1}{x^2}\,dx$$

$$\Rightarrow \quad \int \frac{1}{y}\,dy = \int \left(\frac{1}{x} + \frac{1}{x^2}\right) dx$$

$$\Rightarrow \quad \log|y| = \int \frac{1}{x}\,dx + \int x^{-2}\,dx = \log|x| - \frac{1}{x} + C$$

$$\Rightarrow \quad \log|y| - \log|x| + \frac{1}{x} = C$$

$$\Rightarrow \quad \log\frac{|y|}{|x|} + \frac{1}{x} = C \qquad \left[\because \log m - \log n = \log\frac{m}{n}\right]$$

$$\Rightarrow \quad \log\left|\frac{y}{x}\right| + \frac{1}{x} = C \qquad ...(2)$$

Now, it is given that $y(1) = 1$ *i.e.*, $y = 1$ when $x = 1$

$\therefore$ From equation (2), we have

$$\log|1| + 1 = C$$

$$\Rightarrow \quad C = 1 \qquad [\because \log 1 = 0]$$

Putting this value of C in equation (2), we have

$$\log\left|\frac{y}{x}\right| + \frac{1}{x} = 1$$

$$\Rightarrow \quad \log\left|\frac{y}{x}\right| = \left(1 - \frac{1}{x}\right) \Rightarrow \left|\frac{y}{x}\right| = e^{\left(1-\frac{1}{x}\right)}$$

$$\Rightarrow \quad |y| = |x|\,e^{\left(1-\frac{1}{x}\right)}.$$

(*v*) The given equation is

$$(1 + y^2)(1 + \log x)\,dx + xdy = 0 \qquad ...(1)$$

$$\Rightarrow \quad (1 + y^2)(1 + \log x)\,dx = -\,xdy$$

Separating the variables,

$$\left(\frac{1+\log x}{x}\right)dx = \frac{-1}{1+y^2}\,dy$$

Integrating both sides, we get

$$\int (1+\log x)\left(\frac{1}{x}\right)dx = -\int \frac{1}{1+y^2}\,dy$$

$$\Rightarrow \quad \frac{(1+\log x)^2}{2} = -\tan^{-1} y + C \qquad \left[\because \int [f(x)]^n f'(x)\,dx = \frac{[f(x)]^{n+1}}{n+1}\right]$$

$$\left[\because \int \frac{1}{a^2+x^2}\,dx = \frac{1}{a}\tan^{-1}\frac{x}{a}\right]$$

$$\Rightarrow \quad \frac{1}{2}(1+\log x)^2 + \tan^{-1} y = C \qquad ...(2)$$

Now, it is given that $y = 1$ when $x = 1$

$\therefore$ From equation (2), we have

$$\frac{1}{2}(1+\log 1)^2 + \tan^{-1} 1 = C$$

$$\Rightarrow \quad \frac{1}{2}(1+0)^2 + \frac{\pi}{4} = C \quad \Rightarrow \quad C = \left(\frac{1}{2}+\frac{\pi}{4}\right)$$

Putting this value of C in equation (2), we have

$$\frac{1}{2}(1+\log x)^2 + \tan^{-1} y = \frac{1}{2}+\frac{\pi}{4}$$

$$\Rightarrow \quad 2(1+\log x)^2 + 4\tan^{-1} y = 2+\pi.$$

(*vi*) The given equation is

$$\frac{dy}{dx} = e^{-y}\cos x \qquad ...(1)$$

Separating the variables,

$$\frac{1}{e^{-y}}\,dy = \cos x\,dx$$

Integrating both sides, we get

$$\int e^y\,dy = \int \cos x\,dx$$

$$\Rightarrow \quad e^y = \sin x + C \qquad ...(2)$$

Now, it is given that $y(0) = 0$ *i.e.*, $y = 0$, when $x = 0$

$\therefore$ From equation (2), we have

$$\Rightarrow \quad e^0 = \sin 0 + C \quad \Rightarrow \quad C = 1$$

$\therefore$ Putting this value of C in equation (2), we get

$$e^y = \sin x + 1.$$

(*vii*) The given equation is

$$e^{\frac{dy}{dx}} = x + 1 \qquad ...(1)$$

Taking log on both sides, we get

$$\log e^{\frac{dy}{dx}} = \log (x + 1) \qquad [\because \log e^{f(x)} = f(x)]$$

$$\Rightarrow \quad \frac{dy}{dx} = \log (x + 1)$$

Separating the variables,

$$dy = \log (x + 1)\, dx$$

Integrating both sides, we get

$$\int dy = \int \underset{\text{II}}{1} . \underset{\text{I}}{\log} (x + 1)\, dx \qquad \text{(Integrating by parts)}$$

$$\Rightarrow \quad y = \log (x + 1) . x - \int \frac{1}{x+1} . x\, dx + C$$

$$\Rightarrow \quad y = x \log (x + 1) - \int \frac{x}{x+1} dx + C$$

$$\Rightarrow \quad y = x \log (x + 1) - \int \frac{x+1-1}{x+1} dx + C$$

[Add and subtract 1 to the numerator.]

$$\Rightarrow \quad y = x \log (x + 1) - \int \left(1 - \frac{1}{x+1}\right) dx + C$$

$$\Rightarrow \quad y = x \log (x + 1) - x + \log | x + 1 | + C \qquad ...(2)$$

Now, it is given that $y = 5$ at $x = 0$

$\therefore$ From equation (2), we have

$$5 = 0 \log (0 + 1) - 0 + \log | 0 + 1 | + C$$

$$\Rightarrow \quad 5 = 0 - 0 + 0 + C \Rightarrow C = 5$$

Putting this value of C in equation (2), we get

$$y = x \log (x + 1) - x + \log | x + 1 | + 5.$$

Example 35. *The slope of tangent at any point of a curve is λ times the slope of the straight line joining the point of contact to the origin. Formulate the differential equation representing the problem and hence find the equation of the curve.*

Solution. Let the equation of the curve is

$$y = f(x)$$

Let P(x, y) be any point on the curve.

$\therefore$ Slope of the tangent at point P(x, y) $= \dfrac{dy}{dx}$

Now, slope of any line passing through the origin and the point P(x, y) is given by

$$\frac{y-0}{x-0} = \frac{y}{x}.$$

$\therefore$ According to the question,

$$\frac{dy}{dx} = \lambda \frac{y}{x} \qquad ...(1)$$

Separating the variables,

$$\frac{dy}{y} = \lambda \frac{1}{x} dx$$

Integrating both sides, we get

$$\int \frac{1}{y} dy = \lambda \int \frac{1}{x} dx$$

$$\Rightarrow \quad \log |y| = \lambda \log |x| + C$$

$$\Rightarrow \quad \log |y| = \log |x|^{\lambda} + C$$

$$\Rightarrow \log |y| - \log |x^{\lambda}| = C \Rightarrow \log \left|\frac{y}{x^{\lambda}}\right| = C \qquad \left[\because \log m - \log n = \log \frac{m}{n}\right]$$

$$\Rightarrow \quad \left|\frac{y}{x^{\lambda}}\right| = e^{C} \Rightarrow \frac{y}{x^{\lambda}} = \pm e^{C}$$

$$\Rightarrow \quad y = x^{\lambda}C_1 \qquad [\text{where } C_1 = \pm e^{C}]$$

which is the required equation of curve.

Example 36. *The slope of the tangent at a point P(x, y) on the curve is* $\left(-\frac{y+3}{x+2}\right)$. *If the curve passes through the origin, find the equation of the curve.*

Solution. Let the slope of the tangent at any point (x, y) on the curve $= \frac{dy}{dx}$.

$$\therefore \quad \frac{dy}{dx} = -\frac{y+3}{x+2} \qquad ...(1)$$

Separating the variables,

$$\frac{1}{y+3} dy = -\frac{1}{x+2} dx$$

Integrating both sides, we get

$$\int \frac{1}{y+3} dy = -\int \frac{1}{x+2} dx$$

$$\Rightarrow \quad \log |y+3| = -\log |x+2| + C \qquad ...(2)$$

$$\Rightarrow \quad \log |y+3| + \log |x+2| = C$$

$$\Rightarrow \quad \log |(y+3)(x+2)| = C \qquad ...(3) \quad [\because \log m + \log n = \log mn]$$

Since, the curve passes through the origin.

$$\therefore \quad x = 0, \quad y = 0.$$

$\therefore$ Equation (3), becomes

$$\log |(0+3)(0+2)| = C \Rightarrow \log 6 = C$$

Now, putting this value of C in equation (3), we get

$$\log |(x+2)(y+3)| = \log 6$$

$$\Rightarrow \quad (x+2)(y+3) = 6$$

$$\Rightarrow \quad xy + 3x + 2y + 6 = 6$$

$$\Rightarrow \quad xy + 3x + 2y = 0$$

which is the required equation of curve.

Example 37. *The temperature T of a cooling object drops at a rate which is proportional to the difference T – S, where S is the constant temperature of the surrounding medium.*

Thus, $$\frac{dT}{dt} = -k(T - S)$$

where k > 0 is a constant and t is time. Solve the differential equation, if it is given that

$$T(0) = 150.$$

Solution. The given equation is

$$\frac{dT}{dt} = -k(T - S) \quad ...(1)$$

Separating the variables,

$$\frac{1}{(T - S)}\, dT = -k\,dt$$

Integrating both sides, we get

$$\int \frac{1}{(T - S)}\, dT = -k \int 1 . dt$$

$$\Rightarrow \quad \log (T - S) = -kt + c \quad ...(2) \quad \left[\begin{array}{l}\because \text{ S is constant,} \\ \Rightarrow \text{ (T – S) is (+ ve)}\end{array}\right]$$

Now, it is given that

$$T(0) = 150$$

i.e., at $t = 0$, T = 150.

∴ Equation (2) becomes

$$\log (150 - S) = -k . 0 + C$$

$$\Rightarrow \quad C = \log (150 - S)$$

Putting this value of C in equation (2), we get

$$\log (T - S) = -kt + \log (150 - S)$$

$$\Rightarrow \quad \log (T - S) - \log (150 - S) = -kt$$

$$\Rightarrow \quad \log \frac{(T - S)}{(150 - S)} = -kt \quad \left[\because \log m - \log n = \log \frac{m}{n}\right]$$

$$\Rightarrow \quad \frac{T - S}{150 - S} = e^{-kt}$$

which is the required solution.

Example 38. *A population grows at the rate of 2% per year. How long does it take for the population to double.*

Solution. Let the initial population $= P_0$

Population after t years $= P.$

∴ Rate of increase of population $= \frac{dP}{dt}$

∴ According to given condition,

$$\frac{dP}{dt} = 2\% \text{ of } P = \frac{2P}{100}$$

$$\Rightarrow \qquad \frac{dP}{dt} = \frac{P}{50} \qquad ...(1)$$

Separating the variables,

$$\frac{1}{P}\, dP = \frac{1}{50}\, dt$$

Integrating both sides, we get

$$\int \frac{1}{P}\, dP = \frac{1}{50} \int 1 . dt$$

$$\Rightarrow \qquad \log P = \frac{1}{50} t + C \qquad ...(2)$$

Now, when $t = 0$, $P = P_0$

$\therefore$ Equation (2) becomes

$$\log P_0 = 0 + C \Rightarrow C = \log P_0$$

Putting this value of C in equation (2), we get

$$\log P = \frac{1}{50} t + \log P_0$$

$$\Rightarrow \qquad \log P - \log P_0 = \frac{1}{50} t$$

$$\Rightarrow \qquad \log \left(\frac{P}{P_0}\right) = \frac{1}{50} t \qquad \left[\because \log m - \log n = \log \frac{m}{n}\right]$$

$$\Rightarrow \qquad t = 50 \log \left(\frac{P}{P_0}\right) \qquad ...(3)$$

When, the population becomes double, then

$$P = 2\, P_0. \Rightarrow \left(\frac{P}{P_0}\right) = 2.$$

$\therefore$ From equation (3), we have

$$t = 50 \log 2$$

$$\Rightarrow \qquad t = 50\ (0.3010) \qquad [\because \log 2 = 0.\ 3010]$$

$$\Rightarrow \qquad t = 15.05 \approx 15 \text{ years.}$$

Example 39. *In a condenser discharging electricity, the voltage V satisfies the equation,*

$$k \frac{dV}{dt} + V = 0$$

where k is a constant and t is the time measured in seconds.

Given $k = 50$, find the time 't' in which V decreases to one-tenth of its original value.

Solution. The given equation is

$$k \frac{dV}{dt} + V = 0 \quad \Rightarrow \quad k \frac{dV}{dt} = -V \qquad ...(1)$$

Separating the variables,

$$\frac{1}{V}\, dV = -\frac{1}{k}\, dt$$

Integrating both sides, we get

$$\int \frac{1}{V}\, dV = -\int \frac{1}{k}\, dt$$

$$\Rightarrow \qquad \log V = -\frac{1}{k}t + C \qquad \text{...(2)} \quad [\because \ V > 0]$$

Now, when $t = 0$, let $V = V_0$

$\therefore$ Equation (2), becomes

$$\log V_0 = 0 + C \quad \Rightarrow \quad C = \log V_0$$

Putting this value of C in equation (2), we have

$$\log V = -\frac{1}{k}t + \log V_0$$

$$\Rightarrow \qquad \log V - \log V_0 = -\frac{1}{k}t$$

$$\Rightarrow \qquad \log \frac{V}{V_0} = -\frac{1}{k}t \qquad \text{...(3)} \quad \left[\because \ \log m - \log n = \log \frac{m}{n}\right]$$

It is given that, V decreases to $\frac{1}{10}$th of its original value and $k = 50$.

$\therefore$ Putting $V = \frac{V_0}{10} \Rightarrow \left(\frac{V}{V_0} = \frac{1}{10}\right)$ and $k = 50$ in equation (3), we have

$$\log\left(\frac{1}{10}\right) = -\frac{1}{50}t \quad \Rightarrow \quad -\log\left(\frac{1}{10}\right) = \frac{1}{50}t$$

$$\Rightarrow \qquad \log 10 = \frac{1}{50}t \quad \Rightarrow \quad t = 50 \log 10 \text{ seconds.}$$

Example 40. *The line normal to a given curve at each point (x, y) on the curve passes through the point $(2, 0)$. If the curve contains the point $(2, 3)$. Find its equation.*

Solution. The normal at (x, y) passes through the point $(2, 0)$.

$$\therefore \quad \text{Slope of normal} = \frac{y-0}{x-2} = \frac{y}{x-2} \qquad \text{...(1)}$$

Since, the slope of tangent at $(x, y) = \frac{dy}{dx}$

$$\therefore \quad \text{Slope of normal} = -\frac{1}{\frac{dy}{dx}} = -\frac{dx}{dy} \qquad \text{...(2)}$$

$\therefore$ From (1) and (2),

$$-\frac{dx}{dy} = \frac{y}{x-2} \qquad \text{...(3)}$$

Separating the variables,

$$-(x-2)\, dx = y dy$$

Integrating both sides, we get

$$\int (2-x)\, dx = \int y\, dy$$

$$\Rightarrow \quad 2x - \frac{x^2}{2} = \frac{y^2}{2} + C \quad \Rightarrow \quad 4x - x^2 = y^2 + 2C$$

$$\Rightarrow \quad x^2 + y^2 - 4x = -2C \qquad ...(4)$$

Since the curve passes through the point (2, 3)

i.e., $x = 2, \quad y = 3$

$\therefore$ Equation (3), becomes

$$(2)^2 + (3)^2 - 4(2) = -2C \quad \Rightarrow \quad 4 + 9 - 8 = -2C$$

$$\Rightarrow \quad 5 = -2C \quad \Rightarrow \quad C = -\frac{5}{2}$$

Putting this value of C in equation (4), we get

$$x^2 + y^2 - 4x = 5$$

which is the required equation of the curve.

Example 41. *A wet porous substance in the open air loses its moisture at a rate proportional to the moisture contents. If a sheet hung in the wind loses half its moisture during the first hour, when will it have lost 90%, weather conditions remains the same.*

Solution. Let initially the moisture content = m_0 and moisture content after time $t = m$.

$\therefore$ Rate of loss of moisture $= \frac{dm}{dt}$

Now, according to given condition,

$$\frac{dm}{dt} \propto m$$

$$\Rightarrow \quad \frac{dm}{dt} = km \qquad ...(1)$$

where $k \to$ constant of proportionality.

Separating the variables

$$\frac{1}{m}\, dm = k dt$$

Integrating both sides, we get

$$\int \frac{1}{m}\, dm = k \int 1 \,.\, dt$$

$$\Rightarrow \quad \log m = kt + C \qquad ...(2) \quad [\because \ m > 0]$$

Now, initially, at $t = 0$, $m = m_0$

$\therefore$ Equation (2) becomes

$$\log m_0 = k(0) + C \quad \Rightarrow \quad C = \log m_0$$

Putting this value of C in equation (2), we have

$$\log m = kt + \log (m_0)$$

$$\Rightarrow \quad \log m - \log m_0 = kt \qquad \left[\because \ \log m - \log n = \log \frac{m}{n}\right]$$

$$\Rightarrow \quad \log \left(\frac{m}{m_0}\right) = kt \qquad ...(3)$$

After 1 hour, *i.e.*, $t = 1$

$$m = \frac{m_0}{2} \quad \Rightarrow \quad \frac{m}{m_0} = \frac{1}{2}$$

$\therefore$ Equation (3), becomes

$$\log\left(\frac{1}{2}\right) = k(1) \quad \Rightarrow \quad k = \log(2^{-1})$$

$$\Rightarrow \qquad k = -\log 2 \qquad \text{...(4)}$$

Putting this value of k in equation (3), we have

$$\log \frac{m}{m_0} = -t \log 2 \qquad \text{...(5)}$$

Now, when the sheet have lost 90% of moisture.

$$\therefore \qquad m = \frac{m_0}{10} \quad \Rightarrow \quad \frac{m}{m_0} = \frac{1}{10}$$

$\therefore$ Equation (5), becomes

$$\log\left(\frac{1}{10}\right) = -t \log 2 \quad \Rightarrow \quad -\log 10 = -t \log 2$$

$$\Rightarrow \qquad t = \frac{\log 10}{\log 2} \text{ hours.}$$

Example 42. *The surface area of a baloon being inflated changes at a constant rate. If initially, its radius is 3 units and after 2 seconds it is 5 units, find the radius after 1 second.*

Solution. Let the surface area of a baloon after t second = S.

$$\therefore \qquad S = 4\pi r^2 \qquad \text{[Baloon being spherical]}$$

$$\Rightarrow \qquad \frac{dS}{dt} = 8\pi r \frac{dr}{dt} \qquad \text{...(1)}$$

It is given that, $\frac{dS}{dt}$ = constant = k(say).

$\therefore$ Equation (1) becomes

$$8\pi r \frac{dr}{dt} = k$$

Separating the variables,

$$8\pi r\, dr = k dt$$

Integrating both sides, we get

$$\int 8\pi r\, dr = \int k\, dt$$

$$\Rightarrow \qquad 8\pi \frac{r^2}{2} = kt + C$$

$$\Rightarrow \qquad 4\pi r^2 = kt + C \qquad \text{...(2)}$$

Now, at $t = 0$, $r = 3$ units and at $t = 2$, $r = 5$ units.

$\therefore$ From equation (2), at $t = 0$, $r = 3$

$$4\pi (3)^2 = k(0) + C \Rightarrow C = 36\pi$$

at $t = 2, \quad r = 5$

$$4\pi(5)^2 = 2k + 36\pi \Rightarrow 100\pi - 36\pi = 2k \Rightarrow 64\pi = 2k \Rightarrow k = 32\pi$$

Putting these values of C and k in equation (2), we have

$$4\pi r^2 = 32\pi t + 36\pi$$

$$\Rightarrow \quad r^2 = 8t + 9$$

$$\Rightarrow \quad r = \sqrt{8t+9}.$$

Example 43. *The equation of electromotive forces for an electric circuit containing resistance and self-inductance is*

$$E = Ri + L\frac{di}{dt}$$

where $E \rightarrow$ *electromotive force given to the circuit,*

$R \rightarrow$ *resistance*

$L \rightarrow$ *co-efficient of inductance.*

Find the current 'i' at time t, when

(*i*) $E = 0$ (*ii*) E = *a non-zero constant.*

Solution. The given differential equation is

$$E = Ri + \frac{L\,di}{dt} \qquad ...(1)$$

$$\Rightarrow \quad \frac{E - Ri}{L} = \frac{di}{dt}$$

Separating the variables,

$$dt = \frac{L}{E - Ri}\,di$$

Integrating both sides, we get

$$\int 1.\,dt = \int \frac{L}{E - Ri}\,di$$

$$\Rightarrow \quad t = \frac{L \log|E - Ri|}{-R} + C$$

$$\Rightarrow \quad \frac{L}{R}\log|E - Ri| = (C - t)$$

$$\Rightarrow \quad \log|E - Ri| = \frac{(C - t)R}{L}$$

$$\Rightarrow \quad (E - Ri) = e^{\frac{(C-t)R}{L}} \qquad [\text{Let } E > Ri]$$

$$\Rightarrow \quad Ri = E - e^{\left(\frac{CR}{L} - \frac{tR}{L}\right)}$$

$$\Rightarrow \quad i = \frac{E}{R} - \frac{1}{R}e^{\frac{CR}{L}}.e^{\frac{-R}{L}t}$$

$$\Rightarrow \quad i = \frac{E}{R} + C_1 e^{\frac{-R}{L}t} \qquad ...(2) \quad \left[\text{where } C_1 = \frac{-1}{R}e^{\frac{CR}{L}}\right]$$

(*i*) Now, when E = 0, equation (2) becomes

$$i = \frac{0}{R} + C_1 e^{\frac{-R}{L}t} \Rightarrow i = C_1 e^{\frac{-R}{L}t}$$

(*ii*) When E ≠ 0, equation (2) becomes

$$i = \frac{E}{R} + C_1 e^{\frac{-R}{L}t}.$$

Example 44. *Solve the following differential equations*

(*i*) $(x - y)^2 \frac{dy}{dx} = a^2$ (*ii*) $\frac{dy}{dx} = (x + y)^2$

(*iii*) $\frac{dy}{dx} = (4x + y + 1)^2$ (*iv*) $\frac{dy}{dx} = 1 + e^{x-y}$

(*v*) $\frac{dy}{dx} = \cot^2(x + y)$ (*vi*) $\sin^{-1}\left(\frac{dy}{dx}\right) = x + y$

(*vii*) $\cos(x + y)\, dy = dx$; *when* $y(0) = 0$ (*viii*) $\frac{dy}{dx} + 1 = e^{x+y}$

Solution. (*i*) The given equation is

$$(x - y)^2 \frac{dy}{dx} = a^2 \qquad ...(1)$$

Put $x - y = z \Rightarrow 1 - \frac{dy}{dx} = \frac{dz}{dx} \Rightarrow \left(1 - \frac{dz}{dx}\right) = \frac{dy}{dx}$

∴ Equation (1) becomes

$$z^2\left(1 - \frac{dz}{dx}\right) = a^2 \Rightarrow 1 - \frac{dz}{dx} = \frac{a^2}{z^2}$$

$$\Rightarrow \frac{dz}{dx} = 1 - \frac{a^2}{z^2} \Rightarrow \frac{dz}{dx} = \left(\frac{z^2 - a^2}{z^2}\right)$$

Separating the variables,

$$\left(\frac{z^2}{z^2 - a^2}\right) dz = dx$$

Integrating both sides, we get

$$\int \frac{z^2}{z^2 - a^2}\, dz = \int 1.dx$$

$$\Rightarrow \int \frac{z^2 - a^2 + a^2}{z^2 - a^2}\, dz = \int 1.dx$$

[∵ Add and subtract a^2 to the numerator on L.H.S]

$$\Rightarrow \int \left(1 + \frac{a^2}{z^2 - a^2}\right) dz = x + C$$

$$\Rightarrow \quad \int 1.dz + a^2 \int \frac{1}{z^2 - a^2}\, dz = x + C$$

$$\Rightarrow \quad z + a^2 \frac{1}{2a} \log \left| \frac{z-a}{z+a} \right| = x + C \qquad \left[\because \int \frac{1}{x^2 - a^2}\, dx = \frac{1}{2a} \log \left| \frac{x-a}{x+a} \right|\right]$$

$$\Rightarrow \quad x - y + \frac{a}{2} \log \left| \frac{x-y-a}{x-y+a} \right| = x + C \qquad [\because \ z = x - y]$$

$$\Rightarrow \quad \frac{a}{2} \log \left| \frac{x-y-a}{x-y+a} \right| = y + C$$

which is the required solution.

(*ii*) The given equation is

$$\frac{dy}{dx} = (x + y)^2 \qquad ...(1)$$

Put $x + y = z \Rightarrow 1 + \frac{dy}{dx} = \frac{dz}{dx} \Rightarrow \frac{dy}{dx} = \left(\frac{dz}{dx} - 1\right)$

$\therefore$ Equation (1) becomes

$$\left(\frac{dz}{dx} - 1\right) = z^2 \Rightarrow \frac{dz}{dx} = 1 + z^2$$

Separating the variables,

$$\frac{1}{1+z^2}\, dz = 1.dx$$

Integrating both sides, we get

$$\int \frac{1}{1+z^2}\, dz = \int 1.dx$$

$$\Rightarrow \quad \tan^{-1} z = x + C \qquad \left[\because \int \frac{1}{a^2 + x^2}\, dx = \frac{1}{a} \tan^{-1} \frac{x}{a}\right]$$

$$\Rightarrow \quad \tan^{-1} (x + y) = x + C \qquad (\because \ z = x + y)$$

$$\Rightarrow \quad x + y = \tan (x + C)$$

which is the required solution.

(*iii*) The given equation is

$$\frac{dy}{dx} = (4x + y + 1)^2 \qquad ...(1)$$

Put $(4x + y + 1) = z \Rightarrow 4 + \frac{dy}{dx} = \frac{dz}{dx} \Rightarrow \frac{dy}{dx} = \left(\frac{dz}{dx} - 4\right)$

$\therefore$ Equation (1) becomes

$$\frac{dz}{dx} - 4 = z^2 \Rightarrow \frac{dz}{dx} = z^2 + 4$$

Separating the variables,

$$\frac{1}{z^2+4}\,dz = dx$$

Integrating both sides, we get

$$\int \frac{1}{z^2+4}\,dz = \int dx \qquad \left[\because \int \frac{1}{a^2+x^2}\,dx = \frac{1}{a}\tan^{-1}\frac{x}{a}\right]$$

$$\Rightarrow \quad \frac{1}{2}\tan^{-1}\frac{z}{2} = x + C$$

$$\Rightarrow \quad \frac{1}{2}\tan^{-1}\frac{(4x+y+1)}{z} = x + C \qquad [\because \; z = (4x+y+1)]$$

which is the required solution.

(*iv*) The given equation is

$$\frac{dy}{dx} = 1 + e^{x-y} \qquad \text{...(1)}$$

Put $x - y = z \Rightarrow 1 - \dfrac{dy}{dx} = \dfrac{dz}{dx} \Rightarrow \dfrac{dy}{dx} = \left(1 - \dfrac{dz}{dx}\right)$

$\therefore$ Equation (1) becomes

$$1 - \frac{dz}{dx} = 1 + e^z \Rightarrow -\frac{dz}{dx} = e^z$$

Separating the variables,

$$-\frac{1}{e^z}\,dz = dx$$

Integrating both sides, we get

$$-\int \frac{1}{e^z}\,dz = \int 1.\,dx$$

$$\Rightarrow \quad -\int e^{-z}\,dz = x + C$$

$$\Rightarrow \quad e^{-z} = x + C$$

$$\Rightarrow \quad e^{-(x-y)} = x + C \qquad [\because \; z = (x-y)]$$

$$\Rightarrow \quad e^{y-x} = x + C$$

which the required solution.

(*v*) The given equation is

$$\frac{dy}{dx} = \cot^2(x+y) \qquad \text{...(1)}$$

Put $x + y = z \Rightarrow 1 + \dfrac{dy}{dx} = \dfrac{dz}{dx} \Rightarrow \dfrac{dy}{dx} = \left(\dfrac{dz}{dx} - 1\right)$

$\therefore$ Equation (1) becomes

$$\left(\frac{dz}{dx} - 1\right) = \cot^2 z$$

$$\Rightarrow \quad \frac{dz}{dx} = 1 + \cot^2 z$$

$$\Rightarrow \quad \frac{dz}{dx} = \text{cosec}^2 z \qquad [\because \ \text{cosec}^2 A - \cot^2 A = 1]$$

Separating the variables,

$$\frac{1}{\text{cosec}^2 z} dz = dx \Rightarrow \sin^2 z \, dz = dx$$

Integrating both sides, we get

$$\int \sin^2 z \, dz = \int 1.dx$$

$$\Rightarrow \quad \int \left(\frac{1 - \cos 2z}{2} \right) dz = x + C$$

$$\left[\because \ 1 - \cos 2A = 2 \sin^2 A \Rightarrow \frac{1 - \cos 2A}{2} = \sin^2 A \right]$$

$$\Rightarrow \quad \frac{1}{2} \int (1 - \cos 2z) \, dz = x + C$$

$$\Rightarrow \quad \frac{1}{2} \left[z - \frac{\sin 2z}{2} \right] = x + C$$

$$\Rightarrow \quad \frac{1}{2} z - \frac{1}{4} \sin 2z = x + C$$

$$\Rightarrow \quad 2z - \sin 2z = 4x + 4C$$

$$\Rightarrow \quad 2(x + y) - \sin 2(x + y) = 4x + 4C \qquad [\because \ z = x + y]$$

$$\Rightarrow \quad 2x + 2y - 4x - \sin 2(x + y) = C_1$$

$$\Rightarrow \quad 2y - 2x - \sin 2(x + y) = C_1$$

where $C_1 = 4C$

which is the required solution.

(*vi*) The given equation is

$$\sin^{-1} \frac{dy}{dx} = (x + y)$$

$$\Rightarrow \quad \frac{dy}{dx} = \sin (x + y) \qquad \text{...(1)}$$

Put $x + y = z \Rightarrow 1 + \frac{dy}{dx} = \frac{dz}{dx} \Rightarrow \frac{dy}{dx} = \left(\frac{dz}{dx} - 1 \right)$

$\therefore$ Equation (1) becomes

$$\left(\frac{dz}{dx} - 1 \right) = \sin z$$

$$\Rightarrow \quad \frac{dz}{dx} = (1 + \sin z)$$

Separating the variables,

$$\frac{1}{1+\sin z}\,dz = dx$$

Integrating both sides, we get

$$\int \frac{1}{1+\sin z}\,dz = \int dx$$

$$\Rightarrow \quad \int \left(\frac{1}{1+\sin z} \times \frac{1-\sin z}{1-\sin z}\right) dz = x \div C \qquad \text{[Rationalization]}$$

$$\Rightarrow \quad \int \frac{(1-\sin z)}{(1+\sin z)(1-\sin z)}\,dz = x + C$$

$$\Rightarrow \quad \int \frac{1-\sin z}{1-\sin^2 z}\,dz = x + C$$

$$\Rightarrow \quad \int \left(\frac{1-\sin z}{\cos^2 z}\right) dz = x + C \qquad [\because \ \sin^2 A + \cos^2 A = 1]$$

$$\Rightarrow \quad \int \left(\frac{1}{\cos^2 z} - \frac{\sin z}{\cos^2 z}\right) dz = x + C$$

$$\Rightarrow \quad \int \sec^2 z\,dz - \int \sec z \tan z\,dz = x + C$$

$$\Rightarrow \quad \tan z - \sec z = x + C \qquad [\because \ z = x + y]$$

$$\Rightarrow \quad \tan (x+y) - \sec (x+y) = x + C$$

which is the required solution.

(*vii*) The given equation is

$$\cos (x+y)\,dy = dx$$

$$\Rightarrow \quad \frac{dy}{dx} = \frac{1}{\cos (x+y)} \qquad ...(1)$$

Put $x + y = z \quad \Rightarrow \quad 1 + \frac{dy}{dx} = \frac{dz}{dx} \quad \Rightarrow \quad \frac{dy}{dx} = \frac{dz}{dx} - 1$

$\therefore$ Equation (1) becomes

$$\frac{dz}{dx} - 1 = \frac{1}{\cos z}$$

$$\Rightarrow \quad \frac{dz}{dx} = 1 + \frac{1}{\cos z}$$

$$\Rightarrow \quad \frac{dz}{dx} = \frac{1+\cos z}{\cos z}$$

Separating the variables,

$$\left(\frac{\cos z}{1+\cos z}\right) dz = dx$$

Integrating both sides, we get

$$\int \frac{\cos z}{1+\cos z}\, dz = \int 1.dx$$

$$\Rightarrow \quad \int \frac{\cos z\,(1-\cos z)}{(1+\cos z)(1-\cos z)}\, dz = x + C \qquad \text{[Rationalization]}$$

$$\Rightarrow \quad \int \frac{\cos z - \cos^2 z}{1-\cos^2 z}\, dz = x + C$$

$$\Rightarrow \quad \int \frac{\cos z - \cos^2 z}{\sin^2 z}\, dz = x + C \qquad [\because \ \sin^2 A + \cos^2 A = 1]$$

$$\Rightarrow \quad \int \left(\frac{\cos z}{\sin^2 z} - \frac{\cos^2 z}{\sin^2 z}\right) dz = x + C$$

$$\Rightarrow \quad \int \cot z \operatorname{cosec} z\, dz - \int \cot^2 z\, dz = x + C \qquad [\because \ \operatorname{cosec}^2 A - \cot^2 A = 1]$$

$$\Rightarrow \int \cot z \operatorname{cosec} z\, dz - \int (\operatorname{cosec}^2 z - 1)\, dz = x + C$$

$$\Rightarrow \quad -\operatorname{cosec} z + \cot z + z = x + C \qquad [\because \ z = x + y]$$

$$\Rightarrow \quad -\operatorname{cosec}(x+y) + \cot(x+y) + x + y) = x + C$$

$$\Rightarrow \quad y - \operatorname{cosec}(x+y) + \cot(x+y) = C \qquad ...(2)$$

$$\Rightarrow \quad y - \frac{1}{\sin(x+y)} + \frac{\cos(x+y)}{\sin(x+y)} = C$$

$$\Rightarrow \quad y + \left[\frac{1-\cos(x+y)}{\sin(x+y)}\right] = C$$

$$\Rightarrow \quad y + \left[\frac{2\sin^2\left(\frac{x+y}{2}\right)}{2\sin\left(\frac{x+y}{2}\right)\cos\left(\frac{x+y}{2}\right)}\right] = C \qquad \left[\begin{array}{l}\because \ 1-\cos 2A = 2\sin^2 A \\ \Rightarrow \ 1-\cos A = 2\sin^2 \frac{A}{2} \\ \text{and} \\ \sin 2A = 2\sin A\cos A \\ \Rightarrow \ \sin A = 2\sin\frac{A}{2}\cos\frac{A}{2}\end{array}\right]$$

$$\Rightarrow \quad y + \tan\left(\frac{x+y}{2}\right) = C \qquad ...(3)$$

Now, it is given that $y(0) = 0$ *i.e.*, $y = 0$ at $x = 0$

$\therefore$ Equation (3) becomes

$$0 + \tan 0 = C \Rightarrow C = 0$$

Putting this value of C in equation (3), we have

$$y + \tan\left(\frac{x+y}{2}\right) = 0$$

which is the required solution.

(*viii*) The given equation is

$$\frac{dy}{dx} + 1 = e^{x+y} \qquad ...(1)$$

Please try yourself

[**Hint.** See part (*iv*) of the same example.] [**Ans.** $1 + (x + c)\, e^{x+y} = 0$]

Example 45. *Solve the following differential equations :*

(*i*) $(x + y)^2 \dfrac{dy}{dx} = a^2$

(*ii*) $\dfrac{dy}{dx} = (3x + 2y + 1)^2$

(*iii*) $(x + y + 1)\dfrac{dy}{dx} = 1$

(*iv*) $\dfrac{dy}{dx} = \dfrac{2}{x + 2y - 3}$

(*v*) $\dfrac{dy}{dx} = \sin (x + y) + \cos (x + y)$

(*vi*) $(x + y)(dx - dy) = dx + dy$

(*vii*) $\dfrac{dy}{dx} = \cos (x + y).$

Solution. (*i*) The given equation is

$$(x + y)^2 \frac{dy}{dx} = a^2 \qquad ...(1)$$

Put $x + y = z \;\Rightarrow\; 1 + \dfrac{dy}{dx} = \dfrac{dz}{dx} \;\Rightarrow\; \dfrac{dy}{dx} = \dfrac{dz}{dx} - 1$

$\therefore$ Equation (1) becomes

$$z^2\left(\frac{dz}{dx} - 1\right) = a^2 \;\Rightarrow\; \frac{dz}{dx} - 1 = \frac{a^2}{z^2}$$

$$\Rightarrow \qquad \frac{dz}{dx} = 1 + \frac{a^2}{z^2} = \frac{z^2 + a^2}{z^2}$$

Separating the variables,

$$\frac{z^2}{z^2 + a^2}\, dz = dx$$

Integrating both sides, we get

$$\int \frac{z^2}{z^2 + a^2}\, dz = \int 1. dx$$

$$\Rightarrow \qquad \int \left(\frac{z^2 + a^2 - a^2}{z^2 + a^2}\right) dz = x + C$$

[Add and subtract a^2 to the numerator on the L.H.S.]

$$\Rightarrow \qquad \int \left(1 - \frac{a^2}{z^2 + a^2}\right) dz = x + C$$

$$\Rightarrow \qquad \int 1. dz - a^2 \int \frac{1}{z^2 + a^2}\, dz = x + C$$

$$\Rightarrow \quad z - a^2 \frac{1}{a} \tan^{-1} \frac{z}{a} = x + C \qquad \left[\because \int \frac{1}{a^2 + x^2}\, dx = \frac{1}{a} \tan^{-1} \frac{x}{a}\right]$$

$$\Rightarrow \quad x + y - a \tan^{-1}\left(\frac{x+y}{a}\right) = x + C \qquad [\because \; z = x + y]$$

$$\Rightarrow \quad y = a \tan^{-1}\left(\frac{x+y}{a}\right) + C$$

which is the required solution.

(*ii*) The given equation is

$$\frac{dy}{dx} = (3x + 2y + 1)^2 \qquad \ldots(1)$$

Put $\quad 3x + 2y + 1 = z \quad \Rightarrow \quad 3 + 2\frac{dy}{dx} = \frac{dz}{dx}$

$$\Rightarrow \quad 2\frac{dy}{dx} = \frac{dz}{dx} - 3 \quad \Rightarrow \quad \frac{dy}{dx} = \frac{1}{2}\left(\frac{dz}{dx} - 3\right)$$

$\therefore$ Equation (1) becomes

$$\frac{1}{2}\left(\frac{dz}{dx} - 3\right) = z^2$$

$$\Rightarrow \quad \frac{dz}{dx} = 2z^2 + 3$$

Separating the variables,

$$\frac{1}{2z^2 + 3}\, dz = dx$$

Integrating both sides, we get

$$\int \frac{1}{2z^2 + 3}\, dz = \int 1 . dx$$

$$\Rightarrow \quad \frac{1}{2}\int \frac{1}{z^2 + \frac{3}{2}}\, dz = x + C$$

$$\Rightarrow \quad \frac{1}{2}\int \frac{1}{z^2 + \left(\sqrt{\frac{3}{2}}\right)^2}\, dz = x + C \qquad \left[\because \int \frac{1}{a^2 + x^2}\, dx = \frac{1}{a} \tan^{-1} \frac{x}{a}\right]$$

$$\Rightarrow \quad \frac{1}{2} \cdot \frac{1}{\sqrt{\frac{3}{2}}} \tan^{-1} \frac{z}{\sqrt{\frac{3}{2}}} = x + C$$

$$\Rightarrow \quad \frac{1}{2} \cdot \frac{\sqrt{2}}{\sqrt{3}} \tan^{-1} \frac{\sqrt{2}\, z}{\sqrt{3}} = x + C \qquad [\because \; z = 3x + 2y + 1]$$

$$\Rightarrow \qquad \frac{1}{\sqrt{6}} \tan^{-1} \frac{\sqrt{2}(3x+2y+1)}{\sqrt{3}} = x + C$$

$$\Rightarrow \qquad \tan^{-1} \frac{\sqrt{2}(3x+2y+1)}{\sqrt{3}} = \sqrt{6}\, x + \sqrt{6}\, C \qquad \text{[where : } C_1 = \sqrt{6}\, C]$$

$$\Rightarrow \qquad \frac{\sqrt{2}(3x+2y+1)}{\sqrt{3}} = \tan(\sqrt{6}\, x + C_1)$$

$$\Rightarrow \qquad \sqrt{2}(3x+2y+1) = \sqrt{3}\tan(\sqrt{6}\, x + C_1)$$

which is the required solution.

(*iii*) The given equation is

$$(x+y+1)\frac{dy}{dx} = 1 \qquad \text{...(1)}$$

$$\Rightarrow \qquad \frac{dy}{dx} = \frac{1}{x+y+1}$$

Put $x + y + 1 = z \Rightarrow 1 + \frac{dy}{dx} = \frac{dz}{dx} \Rightarrow \frac{dy}{dx} = \left(\frac{dz}{dx} - 1\right)$

$\therefore$ Equation (1) becomes

$$z\left(\frac{dz}{dx} - 1\right) = 1 \qquad \Rightarrow \qquad \frac{dz}{dx} - 1 = \frac{1}{z}$$

$$\Rightarrow \qquad \frac{dz}{dx} = 1 + \frac{1}{z} = \frac{z+1}{z}$$

Separating the variables,

$$\frac{z}{z+1}\, dz = dx$$

Integrating both sides, we get

$$\int \frac{z}{z+1}\, dz = \int dx$$

$$\Rightarrow \qquad \int \frac{z+1-1}{z+1}\, dz = x + C$$

[Add and subtract 1 to the numerator on L.H.S.]

$$\Rightarrow \qquad \int \left(1 - \frac{1}{z+1}\right) dz = x + C$$

$$\Rightarrow \qquad z - \log|z+1| = x + C \qquad \text{...(2)}$$

$$\Rightarrow \qquad (x+y+1) - \log|x+y+1+1| = x + C \qquad [\because \; z = x + y + 1]$$

$$\Rightarrow \qquad x + y - \log|x+y+2| = x + C - 1$$

$$\Rightarrow \qquad y - \log|x+y+2| = C_1 \qquad \text{where } C_1 = (C-1)$$

which is the required solution.

(*iv*) The given equation is

$$\frac{dy}{dx} = \frac{2}{x+2y-3} \qquad ...(1)$$

Put $x + 2y - 3 = z \quad \Rightarrow \quad 1 + 2\frac{dy}{dx} = \frac{dz}{dx} \quad \Rightarrow \quad \frac{dy}{dx} = \frac{1}{2}\left(\frac{dz}{dx} - 1\right)$

$\therefore$ Equation (1) becomes

$$\frac{1}{2}\left(\frac{dz}{dx} - 1\right) = \frac{2}{z} \quad \Rightarrow \quad \frac{dz}{dx} - 1 = \frac{4}{z}$$

$$\Rightarrow \quad \frac{dz}{dx} = 1 + \frac{4}{z} = \frac{z+4}{z}$$

Separating the variables,

$$\frac{z}{z+4}\, dz = dx$$

Integrating both sides, we get

$$\int \frac{z}{z+4}\, dz = \int dx$$

$$\Rightarrow \quad \int \frac{z+4-4}{z+4}\, dz = \int dx \quad \text{[Add and subtract 4 to the numerator on L.H.S.]}$$

$$\Rightarrow \quad \int \left(1 - \frac{4}{z+4}\right) dz = x + C$$

$$\Rightarrow \quad z - 4 \log | z + 4 | = x + C$$

$$\Rightarrow \quad x + 2y - 3 - 4 \log | x + 2y - 3 + 4 | = x + C \qquad [\because \ z = x + 2y - 3]$$

$$\Rightarrow \quad 2y - 4 \log | x + 2y + 1 | = C + 3$$

$$\Rightarrow \quad 2y - 4 \log | x + 2y + 1 | = C_1 \qquad [\text{where } C_1 = C + 3]$$

which is the required solution.

(*v*) The given equation is

$$\frac{dy}{dx} = \sin (x + y) + \cos (x + y) \qquad ...(1)$$

Put $(x + y) = z \quad \Rightarrow \quad 1 + \frac{dy}{dx} = \frac{dz}{dx} \quad \Rightarrow \quad \frac{dy}{dx} = \left(\frac{dz}{dx} - 1\right)$

$\therefore$ Equation (1) becomes

$$\left(\frac{dz}{dx} - 1\right) = \sin z + \cos z$$

$$\Rightarrow \quad \frac{dz}{dx} = (1 + \sin z + \cos z)$$

Separating the variables,

$$\frac{1}{1+\cos z+\sin z}\,dz = dx$$

$$\Rightarrow \qquad \frac{1}{2\cos^2\frac{z}{2}+2\sin\frac{z}{2}\cos\frac{z}{2}}\,dz = dx$$

$$\left[\begin{array}{l} \because\ 1+\cos 2A = 2\cos^2 A \\ \Rightarrow\ 1+\cos A = 2\cos^2\frac{A}{2} \\ \text{and } \sin 2A = 2\sin A\cos A \\ \Rightarrow\ \sin A = 2\sin\frac{A}{2}\cos\frac{A}{2} \end{array}\right]$$

$$\Rightarrow \qquad \frac{\sec^2\frac{z}{2}}{2\left(1+\tan\frac{z}{2}\right)}\,dz = dx$$

$$\left[\text{Dividing the numerator and denominator by } \cos^2\frac{z}{2}\right]$$

Integrating both sides, we get

$$\int\frac{\frac{1}{2}\sec^2\frac{z}{2}}{\left(1+\tan\frac{z}{2}\right)}\,dz = \int 1.dx$$

$$\Rightarrow \qquad \log\left|1+\tan\frac{z}{2}\right| = x + C \qquad \left[\because\ \int\frac{f'(x)}{f(x)}\,dx = \log|f(x)|\right]$$

$$\Rightarrow \qquad \log\left|1+\tan\left(\frac{x+y}{2}\right)\right| = x + C \qquad [\because\ z = x+y]$$

which is the required solution.

(*vi*) The given equation is

$$(x+y)(dx-dy) = dx+dy$$

$$\Rightarrow \qquad (x+y)\,dx-(x+y)\,dy = dx+dy$$

$$\Rightarrow \qquad (x+y-1)\,dx = (x+y+1)\,dy$$

$$\Rightarrow \qquad \frac{dy}{dx} = \frac{x+y-1}{x+y+1} \qquad \ldots(1)$$

Put $x+y=z \Rightarrow \quad 1+\frac{dy}{dx} = \frac{dz}{dx} \Rightarrow \frac{dy}{dx} = \left(\frac{dz}{dx}-1\right)$

$\therefore$ Equation (1) becomes

$$\frac{dz}{dx}-1 = \frac{z-1}{z+1}$$

$$\Rightarrow \qquad \frac{dz}{dx} = \frac{z-1}{z+1} + 1 = \frac{z-1+z+1}{z+1} = \frac{2z}{z+1}$$

Separating the variables,

$$\frac{z+1}{z}\, dz = 2\, dx$$

Integrating both sides, we get

$$\int \frac{z+1}{z}\, dz = 2 \int 1.dx$$

$$\Rightarrow \qquad \int \left(1 + \frac{1}{z}\right) dz = 2(x + C)$$

$$\Rightarrow \qquad z + \log|z| = 2x + 2C \qquad [\because \ z = x + y]$$

$$\Rightarrow \qquad x + y + \log|x + y| = 2x + C_1 \qquad [\text{where } C_1 = 2C]$$

$$\Rightarrow \qquad \log|x + y| = (x - y) + C_1$$

which is the required solution.

(*vii*) The given equation is

$$\frac{dy}{dx} = \cos(x + y) \qquad ...(1)$$

Put $x + y = z \Rightarrow 1 + \frac{dy}{dx} = \frac{dz}{dx} \Rightarrow \frac{dy}{dx} = \left(\frac{dz}{dx} - 1\right)$

$\therefore$ Equation (1) becomes

$$\left(\frac{dz}{dx} - 1\right) = \cos z$$

$$\Rightarrow \qquad \frac{dz}{dx} = (1 + \cos z)$$

Separating the variables,

$$\frac{1}{1+\cos z}\, dz = dx$$

$$\Rightarrow \qquad \frac{1}{2\cos^2 \frac{z}{2}}\, dz = dx \qquad \left[\begin{array}{l} \because \ 1 + \cos 2A = 2\cos^2 A \\ \Rightarrow \ 1 + \cos A = 2\cos^2 \frac{A}{2} \end{array}\right]$$

$$\Rightarrow \qquad \frac{1}{2} \sec^2 \frac{z}{2}\, dz = dx$$

Integrating both sides, we get

$$\frac{1}{2} \int \sec^2 \frac{z}{2}\, dz = \int 1.dx$$

$$\Rightarrow \qquad \frac{1}{2}\frac{\tan z/2}{\frac{1}{2}} = x + C \quad \Rightarrow \quad \tan\frac{z}{2} = x + C$$

$$\Rightarrow \qquad \tan\left(\frac{x+y}{2}\right) = x + C \qquad [\because \; z = x + y]$$

which is the required solution.

Example 46. *Solve the following differential equations*

(*i*) $\dfrac{dy}{dx} = \dfrac{x + 2y - 1}{x + 2y + 1}$

(*ii*) $\dfrac{dy}{dx} = \dfrac{x + y - 3}{x + y + 3}$, *given that* $y = 1$ *when* $x = 0$

(*iii*) $\dfrac{dy}{dx} = \dfrac{1}{(x + y + 1)^2}$, *given that* $y = 0$, *when* $x = -1$

Solution. (*i*) The given equation is

$$\frac{dy}{dx} = \frac{x + 2y - 1}{x + 2y + 1} \qquad ...(1)$$

Put $x + 2y = z \;\Rightarrow\; 1 + 2\dfrac{dy}{dx} = \dfrac{dz}{dx} \;\Rightarrow\; \dfrac{dy}{dx} = \dfrac{1}{2}\left(\dfrac{dz}{dx} - 1\right)$

$\therefore$ Equation (1) becomes

$$\frac{1}{2}\left(\frac{dz}{dx} - 1\right) = \frac{z - 1}{z + 1}$$

$$\Rightarrow \qquad \frac{dz}{dx} - 1 = \frac{2z - 2}{z + 1}$$

$$\Rightarrow \qquad \frac{dz}{dx} = \frac{2z - 2}{z + 1} + 1 = \frac{2z - 2 + z + 1}{z + 1} = \frac{3z - 1}{z + 1}$$

Separating the variables,

$$\frac{z + 1}{3z - 1}\,dz = dx$$

Integrating both sides, we get

$$\int \frac{z + 1}{3z - 1}\,dz = \int 1.\,dx$$

$$\Rightarrow \qquad \frac{1}{3}\int \frac{3z + 3}{3z - 1}\,dz = \int 1.\,dx \qquad \text{[Multiply and divided by 3]}$$

$$\Rightarrow \qquad \frac{1}{3}\int \frac{(3z - 1) + 4}{3z - 1}\,dz = x + C$$

$$\Rightarrow \qquad \frac{1}{3}\int\left(1+\frac{4}{3z-1}\right) dz = x + C$$

$$\Rightarrow \qquad \frac{1}{3}\int 1.\,dz + \frac{4}{3}\int \frac{1}{3z-1}\, dz = x + C$$

$$\Rightarrow \qquad \frac{1}{3}\, z + \frac{4}{3}\, \frac{\log|3z-1|}{3} = x + C$$

$$\Rightarrow \qquad z + \frac{4}{3}\log\, |\, 3z - 1\, | = 3x + 3C$$

$$\Rightarrow \qquad 3z + 4 \log\, |\, 3z - 1\, | = 9x + 9C$$

$$\Rightarrow \; 3(x + 2y) + 4 \log\, |\, 3(x + 2y) - 1\, | = 9x + C_1 \qquad [\because\; z = x + 2y]$$

$$\Rightarrow \qquad 3x + 6y + 4 \log\, |\, 3x + 6y - 1\, | = 9x + C_1 \qquad [\text{where } C_1 = 9C]$$

$$\Rightarrow \qquad 6y - 6x + 4 \log\, |\, 3x + 6y - 1\, | = C_1$$

$$\Rightarrow \qquad 6(y - x) + 4 \log\, |\, 3x + 6y - 1\, | = C_1$$

which is the required solution.

(*ii*) The given equation is

$$\frac{dy}{dx} = \frac{x + y - 3}{x + y + 3} \qquad ...(1)$$

Put $x + y = z \;\Rightarrow\; 1 + \dfrac{dy}{dx} = \dfrac{dz}{dx} \;\Rightarrow\; \dfrac{dy}{dx} = \left(\dfrac{dz}{dx} - 1\right)$

$\therefore$ Equation (1) becomes

$$\left(\frac{dz}{dx} - 1\right) = \frac{z-3}{z+3}$$

$$\Rightarrow \qquad \frac{dz}{dx} = \frac{z-3}{z+3} + 1 = \frac{z - 3 + z + 3}{z+3} = \frac{2z}{z+3}$$

Separating the variables,

$$\frac{z+3}{z}\, dz = 2dx$$

Integrating both sides, we get

$$\int\left(\frac{z+3}{z}\right) dz = 2\int 1.\, dx$$

$$\Rightarrow \qquad \int\left(1+\frac{3}{z}\right) dz = 2(x + C)$$

$$\Rightarrow \qquad z + 3 \log\, |\, z\, | = 2x + 2C$$

$$\Rightarrow \qquad x + y + 3 \log\, |\, x + y\, | = 2x + 2C \qquad [\because\; z = x + y]$$

$$\Rightarrow \qquad y - x + 3 \log\, |\, x + y\, | = 2C \qquad ...(2)$$

Now, it is given that $y = 1$ when $x = 0$

$\therefore$ Equation (2) becomes

$$1 - 0 + 3 \log | 0 + 1 | = 2C \Rightarrow 2C = 1$$

Putting this value of 2C in equation (2), we get

$$(y - x) + 3 \log | x + y | = 1$$

which is the required solution.

(*iii*) The given equation is

$$\frac{dy}{dx} = \frac{1}{(x + y + 1)^2} \quad ...(1)$$

Put $x + y + 1 = z \quad \Rightarrow \quad 1 + \frac{dy}{dx} = \frac{dz}{dx} \quad \Rightarrow \quad \frac{dy}{dx} = \left(\frac{dz}{dx} - 1\right)$

$\therefore$ Equation (1) becomes

$$\left(\frac{dz}{dx} - 1\right) = \frac{1}{z^2}$$

$$\Rightarrow \quad \frac{dz}{dx} = \frac{1}{z^2} + 1 = \frac{1 + z^2}{z^2}$$

Separating the variables,

$$\frac{z^2}{1 + z^2}\, dz = dx$$

Integrating both sides, we get

$$\int \frac{z^2}{z^2 + 1}\, dz = \int 1.\, dx$$

$$\Rightarrow \quad \int \frac{z^2 + 1 - 1}{z^2 + 1}\, dz = x + C$$

[Add and subtract 1 to the numerator on L.H.S.]

$$\Rightarrow \quad \int \left(1 - \frac{1}{z^2 + 1}\right) dz = x + C$$

$$\Rightarrow \quad \int 1.\, dz - \int \frac{1}{z^2 + 1}\, dz = x + C$$

$$\Rightarrow \quad z - \tan^{-1} z = x + C \qquad \left[\because \int \frac{1}{a^2 + x^2}\, dx = \frac{1}{a} \tan^{-1} \frac{x}{a}\right]$$

$$\Rightarrow \quad x + y + 1 - \tan^{-1} (x + y + 1) = x + C \qquad [\because z = x + y + 1]$$

$$\Rightarrow \quad y + 1 - \tan^{-1} (x + y + 1) = C \quad ...(2)$$

Now, it is given that $y = 0$ when $x = -1$

$\therefore$ Equation (2) becomes

$$0 + 1 - \tan^{-1} (-1 + 0 + 1) = C \Rightarrow 1 - 0 = C \Rightarrow C = 1$$

Putting this value of C in equation (2), we get

$$y + 1 - \tan^{-1}(x + y + 1) = 1$$

$$\Rightarrow \quad y = \tan^{-1}(x + y + 1)$$

$$\Rightarrow \quad (x + y + 1) = \tan y$$

which is the required solution.

EXERCISE FOR PRACTICE

1. Form the differential equation corresponding to $y^2 = a(b - x^2)$ by eliminating a and b.
2. Obtain a differential equation that should be satisfied by the family of concentric circles
$$x^2 + y^2 = a^2.$$
3. Find the differential equation of the system of circles which passes through the origin and having centres on the x-axis.
4. Form the differential equation corresponding to $(x - a)^2 + (y - b)^2 = r^2$ by eliminating a and b.
5. Find the differential equation of all the parabolas with latus rectum '$4a$' and whose axis are parallel to x-axis.
6. Find the differential equation of all circles in the xy-plane.
7. Show that $y = a\cos(\log x) + b\sin(\log x)$ is a solution of $x^2\dfrac{d^2y}{dx^2} + x\dfrac{dy}{dx} + y = 0$.
8. Show that $y = x^3 + ax^2 + bx + c$ is a solution of $\dfrac{d^3y}{dx^3} = 6$.
9. Verify that $y^2 = 4ax$ is a solution of the differential equation $y = x\dfrac{dy}{dx} + a\dfrac{dx}{dy}$.
10. Show that the function $y = A\cos 2x - B\sin 2x$ is a solution of the differential equation
$$\frac{d^2y}{dx^2} + 4y = 0.$$
11. Show that the differential equation of which $y = 2(x^2 - 1) + ce^{-x^2}$ is a solution is $\dfrac{dy}{dx} + 2xy = 4x^3$.

Solve the following differential equations :

12. $y + x\dfrac{dy}{dx} = a\left(y^2 + \dfrac{dy}{dx}\right)$
13. $x\sqrt{1-y^2}\,dx + y\sqrt{1-x^2}\,dy = 0$
14. $4e^x \tan y\, dx + 3(1 + e^x)\sec^2 y\, dy = 0$
15. $\cos x\dfrac{dy}{dx} - \cos 2x = \cos 3x$
16. $\dfrac{dy}{dx} = (x + y)^2$
17. $\dfrac{dy}{dx} = \cos(x + y)$
18. $x\cos y\, dy = (xe^x \log x + e^x)dx$
19. $(e^y + 1)\cos x\, dx + e^y \sin x\, dy = 0$
20. $(1 - x^2)dy + xy\, dx = xy^2\, dx$
21. $(y + xy)dx + (x - xy^2)dy = 0$
22. $(x + y)(dx - dy) = dx + dy.$

Answers

1. $y\dfrac{dy}{dx} = x\left[y\dfrac{d^2y}{dx^2} + \left(\dfrac{dy}{dx}\right)^2\right]$

2. $x + y\dfrac{dy}{dx} = 0$

3. $2xy\dfrac{dy}{dx} + x^2 - y^2 = 0$

4. $(1 + p^2)^3 = r^2\left(\dfrac{d^2y}{dx^2}\right)^2$, where $p = \dfrac{dy}{dx}$

5. $2ay_2 + y_1{}^3 = 0$

6. $\left[1 + \left(\dfrac{dy}{dx}\right)^2\right]\dfrac{d^3y}{dx^3} - 3\dfrac{dy}{dx}\left(\dfrac{d^2y}{dx^2}\right)^2 = 0$

12. $(ay - 1) = cy(x - a)$

13. $\sqrt{1 - x^2} + \sqrt{1 - y^2} = C$

14. $4 \log (1 + e^x) + 3 \log | \tan y | = C$

15. $y = \sin 2x - x + 2 \sin x + \log | \sec x + \tan x) + C$

16. $(x + y) = \tan (x + C)$

17. $\tan\left(\dfrac{x + y}{2}\right) = x + C$

18. $\sin y = e^x \log x + C$

19. $\sin x(e^y + 1) = C$

20. $\log | y - 1 | - \log | y | = -\dfrac{1}{2} \log | 1 - x^2 | + C$

21. $\log x + x + \log y - \dfrac{1}{2}y^2 = C$

22. $\dfrac{1}{2}(y - x) + \dfrac{1}{2} \log | x + y | = C.$

6

Homogeneous Differential Equations

6.1. HOMOGENEOUS FUNCTION

A function $f(x, y)$ is said to be homogeneous of the nth degree in x and y if it can be put in the form

$$x^n f\left(\frac{y}{x}\right).$$

6.2. HOMOGENEOUS EQUATION

A differential equation in x and y is said to be homogeneous if it can be put in the form $\frac{dy}{dx} = \frac{f_1(x, y)}{f_2(x, y)}$, where $f_1(x, y)$ and $f_2(x, y)$ are homogeneous functions of the same degree in x and y.

6.3. METHOD TO SOLVE A HOMOGENEOUS DIFFERENTIAL EQUATION

To solve the equation $\frac{dy}{dx} = \frac{f_1(x, y)}{f_2(x, y)}$, where $f_1(x, y)$ and $f_2(x, y)$ are homogeneous functions of the same degree in x and y.

The given equation is

$$\frac{dy}{dx} = \frac{f_1(x, y)}{f_2(x, y)} \qquad ...(1)$$

Let $\quad f_1(x, y) = x^n \phi_1\left(\frac{y}{x}\right)$ and $f_2(x, y) = x^n \phi_2\left(\frac{y}{x}\right)$.

[$\because$ $f_1(x, y)$ and $f_2(x, y)$ are homogeneous functions of the same degree in x and y.]

$\therefore$ Equation (1) becomes

$$\frac{dy}{dx} = \frac{x^n \phi_1\left(\frac{y}{x}\right)}{x^n \phi_2\left(\frac{y}{x}\right)} = \frac{\phi_1\left(\frac{y}{x}\right)}{\phi_2\left(\frac{y}{x}\right)}$$

$$\Rightarrow \quad \frac{dy}{dx} = f\left(\frac{y}{x}\right) \text{ (say)} \qquad ...(2)$$

Now, put $y = vx$

$$\Rightarrow \quad \frac{dy}{dx} = v + x\frac{dv}{dx}$$

$\therefore$ From equation (2), we have

$$v + x\frac{dv}{dx} = f\left(\frac{vx}{x}\right)$$

$$\Rightarrow \quad x\frac{dv}{dx} = f(v) - v$$

Separating the variables,

$$\frac{1}{f(v) - v}\,dv = \frac{1}{x}\,dx \qquad ...(3)$$

which can be easily integrated. In the solution by putting $v = \frac{y}{x}$, we get the required solution.

Remark. A function $f(x, y)$ of x and y is said to be homogeneous function if

$$f(\lambda x, \lambda y) = \lambda^n f(x, y).$$

Then, number n is called the degree of the homogeneous function $f(x, y)$.

6.4. EQUATIONS REDUCIBLE TO HOMOGENEOUS

To solve the equation $\frac{dy}{dx} = \frac{ax + by + c}{a_1x + b_1y + c_1}$ when $\frac{a}{a_1} \neq \frac{b}{b_1}$.

The given equation is

$$\frac{dy}{dx} = \frac{ax + by + c}{a_1x + b_1y + c_1} \qquad ...(1)$$

Put $x = X + h$ and $y = Y + k$ where X, Y are new variables and h, k are constants.

$$\Rightarrow \quad dx = dX \quad \text{and} \quad dy = dY.$$

$\therefore$ Equation (1) becomes

$$\frac{dY}{dX} = \frac{a(X + h) + b(Y + k) + c}{a_1(X + h) + b_1(Y + k) + c_1}$$

$$\Rightarrow \quad \frac{dY}{dX} = \frac{(aX + bY) + (ah + bk + c)}{(a_1X + b_1Y) + (a_1h + b_1k + c_1)} \qquad ...(2)$$

The constants h and k are to be chosen such that

$$ah + bk + c = 0 \qquad ...(3)$$

$$a_1h + bk_1 + c_1 = 0 \qquad ...(4)$$

Solving equations (3) and (4),

$$\frac{h}{bc_1 - b_1c} = \frac{k}{ca_1 - c_1a} = \frac{1}{ab_1 - a_1b}$$

$$\Rightarrow \quad h = \frac{bc_1 - b_1c}{ab_1 - a_1b} \quad \text{and} \quad k = \frac{ca_1 - c_1a}{ab_1 - a_1b} \qquad ...(5)$$

Now, equation (2) becomes

$$\frac{dY}{dX} = \frac{aX + bY}{a_1X + b_1Y} \quad ...(6)$$

This is a homogeneous equation in X and Y and can be solved by putting $Y = vX$.

In the final solution, by putting $X = x - h$ and $Y = y - k$, we get the required solution.

Note. If $\frac{a}{a_1} = \frac{b}{b_1}$ *i.e.,* $ab_1 - a_1 b = 0$.

Then, h and k both become infinite. Hence the method fails.

6.5. LINEAR DIFFERENTIAL EQUATIONS

6.5.1 Definition. A linear differential equation is that in which the dependent variable and its differential co-efficient occur only in the first degree and are not multiplied together.

Thus, the standard form of a linear differential equation of the first order is

$$\frac{dy}{dx} + Py = Q$$

where P and Q are functions of x or constants.

6.6. SOLUTION OF LINEAR DIFFERENTIAL EQUATIONS OF FIRST ORDER

To solve the equation $\frac{dy}{dx} + Py = Q$, where P and Q are functions of x only.

The given equation is

$$\frac{dy}{dx} + Py = Q \quad ...(1)$$

Multiplying both sides by $e^{\int P\,dx}$, we get

$$\frac{dy}{dx} \cdot e^{\int P\,dx} + Py \cdot e^{\int P\,dx} = Q \cdot e^{\int P\,dx}$$

$$\because \quad \frac{d}{dx}(y \cdot e^{\int P\,.\,dx}) = \left[\frac{dy}{dx} \cdot e^{\int P\,.\,dx} + y \cdot e^{\int P\,.\,dx} \cdot \frac{d}{dx}\left(\int P\,.\,dx\right)\right]$$

$$= \left[\frac{dy}{dx} \cdot e^{\int P\,.\,dx} + Py \cdot e^{\int P\,.\,dx}\right]$$

$$\Rightarrow \quad \frac{d}{dx}[y \cdot e^{\int P\,.\,dx}] = Q \cdot e^{\int P\,.\,dx} \quad ...(2)$$

Integrating both sides, we get

$$y \cdot e^{\int P\,.\,dx} = \int Q \cdot e^{\int P\,.\,dx} \cdot dx + c \quad ...(3)$$

which is the required solution of the given linear differential equation.

Remarks. (*i*) The factor $e^{\int P\,dx}$, on multiplying by which the L.H.S. of the differential equation becomes the differential co-efficient of some function of x and y, is called an integrating factor of the differential equation, and is denoted as I.F.

i.e., $\text{I.F.} = e^{\int P\,dx}$

(*ii*) The solution of the linear differential equation

$$\frac{dy}{dx} + Py = Q$$

where P and Q are functions of x only, is given as

$$y \, . \, (\text{I.F.}) = \int Q(\text{I.F.}) \, dx + c.$$

(*iii*) A linear differential equation of the type

$$\frac{dx}{dy} + Px = Q$$

where P and Q are functions of y only has a solution of the type

$$x(\text{I.F.}) = \int Q(\text{I.F.}) \, dy + c.$$

(*iv*) In evaluating integrating factor (I.F.), the result $e^{\log f(x)} = f(x)$ is frequently used.

(*v*) The co-efficient of $\frac{dy}{dx}$, if not unity, must be made unity by dividing throughout by it.

6.7. BERNOULLI'S EQUATION

Consider the differential equation

$$\frac{dy}{dx} + Py = Qy^n \qquad ...(1)$$

where P and Q are functions of x only and $n \neq 0, 1$.

Equation (1) is known as Bernoulli's Equation.

Dividing both sides of equation (1) by y^n, to make the R.H.S. a function of x only.

$$\Rightarrow \qquad y^{-n} \frac{dy}{dx} + Py^{1-n} = Q \qquad ...(2)$$

Put $\qquad y^{1-n} = z \;\Rightarrow\; (1-n)y^{-n} \frac{dy}{dx} = \frac{dz}{dx} \;\Rightarrow\; y^{-n} \frac{dy}{dx} = \frac{1}{(1-n)} \cdot \frac{dz}{dx}.$

$\therefore$ Equation (2) becomes

$$\frac{1}{1-n} \cdot \frac{dz}{dx} + Pz = Q$$

$$\Rightarrow \qquad \frac{dz}{dx} + (1-n) \, Pz = (1-n)Q \qquad ...(3)$$

Equation (3), is a linear differential equation

Integrating factor $\qquad (\text{I.F.}) = e^{\int P(1-n) \, dx}$.

$\therefore$ The solution of equation (3) is of the form

$$ze^{\int P(1-n) \, dx} = \int Q(1-n) \, e^{\int P \, . \, (1-n) \, dx} \, . \, dx + c$$

In this equation putting $z = y^{1-n}$, we get the required solution.

6.8. SOLUTION OF DIFFERENTIAL EQUATIONS OF SECOND ORDER $\left(\text{i.e., } \frac{d^2y}{dx^2} = f(x)\right)$

The given differential equation is

$$\frac{d^2 y}{dx^2} = f(x) \qquad ..(1)$$

Integrating both sides w.r.t. x, we get

$$\frac{dy}{dx} = \int f(x)\, dx + c_1 \qquad ...(2)$$

Integrating again w.r.t. x, we get

$$y = \int\left(\int f(x)\, dx + c_1\right).\, dx + c_2$$

which is the general solution of the given differential equation.

SOME SOLVED EXAMPLES

Example 1. *Solve the following differential equations :*

(i) $2xy \frac{dy}{dx} = x^2 + 3y^2$ (ii) $(x^2 - y^2)\, dx + 2xy\, dy = 0$

(iii) $x \frac{dy}{dx} = x + y$ (iv) $(x^2 + y^2)\, dx + 2xy\, dy = 0$

(v) $x^2 \frac{dy}{dx} = x^2 + xy + y^2$ (vi) $\frac{dy}{dx} = \frac{y}{x} + \tan \frac{y}{x}$.

Solution. (*i*) The given equation is

$$2xy \frac{dy}{dx} = x^2 + 3y^2 \Rightarrow \frac{dy}{dx} = \frac{x^2 + 3y^2}{2xy} \qquad ...(1)$$

$$\Rightarrow \qquad \frac{dy}{dx} = \frac{x^2\left(1 + \frac{3y^2}{x^2}\right)}{2x \, . \, x\left(\frac{y}{x}\right)} = \frac{1 + 3\left(\frac{y}{x}\right)^2}{2\left(\frac{y}{x}\right)} \qquad ...(2)$$

which is a homogeneous differential equation.

$\therefore$ Put $y = vx \Rightarrow \frac{dy}{dx} = v + x \frac{dv}{dx}$

$\therefore$ Equation (1) becomes

$$v + x \frac{dv}{dx} = \frac{x^2 + 3v^2x^2}{2x(vx)} = \frac{x^2(1 + 3v^2)}{x^2(2v)} = \frac{1 + 3v^2}{2v}$$

$$\Rightarrow \qquad x \frac{dv}{dx} = \frac{1 + 3v^2}{2v} - \frac{v}{1} = \frac{1 + 3v^2 - 2v^2}{2v}$$

$$\Rightarrow \qquad x\frac{dv}{dx}=\frac{1+v^2}{2v}$$

Separating the variables,

$$\frac{2v}{1+v^2}\,dv=\frac{1}{x}\,dx$$

Integrating both sides, we get

$$\int\frac{2v}{1+v^2}\,dv=\int\frac{1}{x}\,dx$$

$$\Rightarrow \qquad \log|1+v^2|=\log|x|+\log c \qquad \left[\because \int\frac{f'(x)}{f(x)}\,dx=\log|f(x)|\right]$$

$$\Rightarrow \qquad \log|1+v^2|-\log|x|=\log c$$

$$\Rightarrow \qquad \log\left|\frac{1+v^2}{x}\right|=\log c \qquad \left[\because \log m-\log n=\log\frac{m}{n}\right]$$

$$\Rightarrow \qquad \frac{1+v^2}{x}=c \;\Rightarrow\; 1+\frac{y^2}{x^2}=cx \qquad \left[\because y=vx \;\Rightarrow\; v=\left(\frac{y}{x}\right)\right]$$

$$\Rightarrow \qquad \frac{x^2+y^2}{x^2}=cx \;\Rightarrow\; x^2+y^2=cx^3$$

which is the required solution.

(*ii*) The given equation is

$$(x^2-y^2)\,dx+2xy\,dy=0$$

$$\Rightarrow \qquad 2xy\,dy=-(x^2-y^2)\,dx$$

$$\Rightarrow \qquad \frac{dy}{dx}=\frac{-(x^2-y^2)}{2xy} \qquad \ldots(1)$$

$$=\frac{-x^2\left(1-\frac{y^2}{x^2}\right)}{2x\,.\,x\left(\frac{y}{x}\right)}=\frac{-\left[1-\left(\frac{y}{x}\right)^2\right]}{2\left(\frac{y}{x}\right)}$$

which a homogeneous differential equation.

$\therefore$ Put $\qquad y=vx \;\Rightarrow\; \dfrac{dy}{dx}=v+x\dfrac{dv}{dx}$

$\therefore$ Equation (1) becomes

$$v+x\frac{dv}{dx}=\frac{-(x^2-v^2x^2)}{2x(vx)}=\frac{-x^2(1-v^2)}{2x^2v}=\frac{-(1-v^2)}{2v}$$

$$\Rightarrow \qquad x\frac{dv}{dx}=\frac{-(1-v^2)}{2v}-\frac{v}{1}=\frac{-1+v^2-2v^2}{2v}$$

$$\Rightarrow \qquad x\frac{dv}{dx}=-\left(\frac{v^2+1}{2v}\right)$$

Separating the variables,

$$\frac{2v}{1+v^2}\,dv = \frac{-1}{x}\,dx$$

Integrating both sides, we get

$$\int \frac{2v}{1+v^2}\,dv = -\int \frac{1}{x}\,dx$$

$$\Rightarrow \quad \log|1+v^2| = -\log|x| + \log c \qquad \left[\because \int \frac{f'(x)}{f(x)}\,dx = \log|f(x)|\right]$$

$$\Rightarrow \quad \log|1+v^2| + \log|x| = \log c$$

$$\Rightarrow \quad \log|x(1+v^2)| = \log c \qquad [\because \log m + \log n = \log mn]$$

$$\Rightarrow \quad x\left(1+\frac{y^2}{x^2}\right) = c \qquad \left[\because y = vx \Rightarrow v = \left(\frac{y}{x}\right)\right]$$

$$\Rightarrow \quad \frac{x(x^2+y^2)}{x^2} = c \Rightarrow x^2 + y^2 = cx$$

which is the required solution.

(*iii*) The given equation is

$$x\frac{dy}{dx} = x + y \qquad \text{...(1)}$$

$$\Rightarrow \quad \frac{dy}{dx} = \frac{x+y}{x} = \frac{x\left(1+\frac{y}{x}\right)}{x} = \left(1+\frac{y}{x}\right)$$

which is a homogeneous differential equation.

$\therefore$ Put $y = vx \Rightarrow \dfrac{dy}{dx} = v + x\dfrac{dv}{dx}$

$\therefore$ Equation (1) becomes

$$v + x\frac{dv}{dx} = \frac{x+vx}{x} = \frac{x(1+v)}{x} = 1 + v$$

$$\Rightarrow \quad x\frac{dv}{dx} = (1+v) - v \Rightarrow \frac{dv}{dx} = \frac{1}{x}$$

Separating the variables,

$$dv = \frac{1}{x}\,dx$$

Integrating both sides, we get

$$\int dv = \int \frac{1}{x}\,dx$$

$$\Rightarrow \quad v = \log|x| + c$$

$$\Rightarrow \quad \frac{y}{x} = \log|x| + c \qquad \left[\because y = vx \Rightarrow v = \left(\frac{y}{x}\right)\right]$$

$$\Rightarrow \quad y = x\log|x| + cx$$

which is the required solution.

(*iv*) The given equation is

$$(x^2 + y^2)\, dx + 2xy\, dy = 0$$

$$\Rightarrow \quad 2xy\, dy = -(x^2 + y^2)\, dx$$

$$\frac{dy}{dx} = \frac{-(x^2 + y^2)}{2xy} \qquad ...(1)$$

$$= \frac{-x^2\left(1 + \frac{y^2}{x^2}\right)}{2x \, . \, x\left(\frac{y}{x}\right)} = \frac{-\left[1 + \left(\frac{y}{x}\right)^2\right]}{2\left(\frac{y}{x}\right)}$$

which is a homogeneous differential equation.

$\therefore$ Put $y = vx \Rightarrow \frac{dy}{dx} = v + x\frac{dv}{dx}$

$\therefore$ Equation (1) becomes

$$v + x\frac{dv}{dx} = \frac{-(x^2 + v^2x^2)}{2x(vx)} = \frac{-x^2(1 + v^2)}{2vx^2}$$

$$\Rightarrow \quad x\frac{dv}{dx} = \frac{-(1 + v^2)}{2v} - v = \frac{-1 - v^2 - 2v^2}{2v} = \frac{-1 - 3v^2}{2v}$$

$$\Rightarrow \quad x\frac{dv}{dx} = \frac{-(1 + 3v^2)}{2v}.$$

Separating the variables,

$$\frac{2v}{1 + 3v^2}\, dv = \frac{-1}{x}\, dx$$

Integrating both sides, we get

$$\int \frac{2v}{1 + 3v^2}\, dv = -\int \frac{1}{x}\, dx$$

$$\Rightarrow \quad \frac{1}{3}\int \frac{6v}{1 + 3v^2}\, dv = -\int \frac{1}{x}\, dx \qquad \text{[Multiply and divided by 3]}$$

$$\Rightarrow \quad \frac{1}{3}\log|1 + 3v^2| = -\log|x| + \log c \qquad \left[\because \int \frac{f'(x)}{f(x)}\, dx = \log|f(x)|\right]$$

$$\Rightarrow \quad \log|1 + 3v^2| = -3\log|x| + 3\log c$$

$$\Rightarrow \quad \log\left|1 + \frac{3y^2}{x^2}\right| = \log|x^{-3}| + \log c^3 \qquad \left[\because y = vx \Rightarrow v = \left(\frac{y}{x}\right)\right]$$

$$\Rightarrow \quad \log\left|\frac{x^2 + 3y^2}{x^2}\right| = \log|x^{-3}| + \log c_1 \qquad [\text{where } c_1 = c^3]$$

$$\Rightarrow \quad \log\left|\frac{x^2 + 3y^2}{x^2}\right| = \log|cx^{-3}| \qquad [\because \log m + \log n = \log mn]$$

$$\Rightarrow \qquad \frac{x^2+3y^2}{x^2} = cx^{-3} \Rightarrow x^2 + 3y^2 = cx^{-1}$$

$$\Rightarrow \qquad x^2 + 3y^2 = \frac{c}{x} \Rightarrow x(x^2 + 3y^2) = c$$

which is the required solution.

(*v*) The given equation is

$$x^2 \frac{dy}{dx} = x^2 + xy + y^2 \Rightarrow \frac{dy}{dx} = \frac{x^2 + xy + y^2}{x^2} \qquad ...(1)$$

$$= \frac{x^2\left[1+\left(\frac{y}{x}\right)+\frac{y^2}{x^2}\right]}{x^2} = \left[1+\left(\frac{y}{x}\right)+\left(\frac{y}{x}\right)^2\right]$$

which is a homogeneous differential equation.

$\therefore$ Put $y = vx \Rightarrow \dfrac{dy}{dx} = v + x\dfrac{dv}{dx}$

$\therefore$ Equation (1) becomes

$$v + x\frac{dv}{dx} = \frac{x^2 + x(vx) + v^2x^2}{x^2}$$

$$\Rightarrow \qquad v + x\frac{dv}{dx} = \frac{x^2(1+v+v^2)}{x^2} = (1 + v + v^2)$$

$$\Rightarrow \qquad x\frac{dv}{dx} = (1 + v + v^2) - v$$

$$\Rightarrow \qquad x\frac{dv}{dx} = 1 + v^2$$

Separating the variables,

$$\frac{1}{1+v^2}\,dv = \frac{1}{x}\,dx$$

Integrating both sides, we get

$$\int\frac{1}{1+v^2}\,dv = \int\frac{1}{x}\,dx$$

$$\Rightarrow \qquad \tan^{-1} v = \log |x| + c \qquad \left[\because \int\frac{1}{a^2+x^2}\,dx = \frac{1}{a}\tan^{-1}\frac{x}{a}\right]$$

$$\Rightarrow \qquad \tan^{-1}\left(\frac{y}{x}\right) = \log |x| + c \qquad \left[\because y = vx \Rightarrow v = \left(\frac{y}{x}\right)\right]$$

which is the required solution.

(*vi*) The given equation is

$$\frac{dy}{dx} = \frac{y}{x} + \tan\left(\frac{y}{x}\right) \qquad ...(1)$$

which is a homogeneous differential equation.

$\therefore$ Put $y = vx \Rightarrow \dfrac{dy}{dx} = v + x\dfrac{dv}{dx}$

$\therefore$ Equation (1) becomes

$$v + x\frac{dv}{dx} = \frac{vx}{x} + \tan\left(\frac{vx}{x}\right)$$

$$\Rightarrow \quad v + x\frac{dv}{dx} = v + \tan v \qquad \Rightarrow \quad x\frac{dv}{dx} = \tan v$$

Separating the variables,

$$\frac{1}{\tan v}\, dv = \frac{1}{x}\, dx$$

Integrating both sides, we get

$$\int \frac{1}{\tan v}\, dv = \int \frac{1}{x}\, dx$$

$$\Rightarrow \quad \int \cot v\, dv = \log |x| + \log c$$

$$\Rightarrow \quad \log |\sin v| = \log |x| + \log c$$

$$\Rightarrow \quad \log \left|\sin\left(\frac{y}{x}\right)\right| = \log |cx| \qquad \left[\because\ y = vx \Rightarrow v = \left(\frac{y}{x}\right),\ \log m + \log n = \log mn\right]$$

$$\Rightarrow \quad \sin\left(\frac{y}{x}\right) = cx$$

which is the required solution.

Example 2. *Solve the following differential equations :*

(*i*) $(x - y)\dfrac{dy}{dx} = x + 2y$ (*ii*) $x^2y\, dx - (x^3 + y^3)\, dy = 0$

(*iii*) $(x^2 + y^2)\dfrac{dy}{dx} = xy$ (*iv*) $xy\dfrac{dy}{dx} = x^2 - y^2$

(*v*) $(3xy + y^2)\, dx + (x^2 + xy)\, dy = 0$ (*vi*) $(x + y)\, dy + (x - y)\, dx = 0$

(*vii*) $(x + 2y)\, dx - (2x - y)\, dy = 0$ (*viii*) $x\dfrac{dy}{dx} - y + x\sin\left(\dfrac{y}{x}\right) = 0.$

Solution. (*i*) The given equation is

$$(x - y)\frac{dy}{dx} = x + 2y$$

$$\Rightarrow \quad \frac{dy}{dx} = \frac{x + 2y}{x - y} \qquad ...(1)$$

$$\Rightarrow \quad \frac{dy}{dx} = \frac{x\left[1 + 2\left(\frac{y}{x}\right)\right]}{x\left[1 - \left(\frac{y}{x}\right)\right]} = \frac{1 + 2\left(\frac{y}{x}\right)}{1 - \left(\frac{y}{x}\right)}$$

which is a homogeneous differential equation.

$\therefore$ Put $y = vx \Rightarrow \dfrac{dy}{dx} = v + x\dfrac{dv}{dx}$

$\therefore$ Equation (1) becomes

$$v + x\frac{dv}{dx} = \frac{x + 2vx}{x - vx} = \frac{1+2v}{1-v}$$

$$\Rightarrow \quad x\frac{dv}{dx} = \frac{1+2v}{1-v} - \frac{v}{1} = \frac{1+2v-v+v^2}{1-v}$$

$$\Rightarrow \quad x\frac{dv}{dx} = \frac{1+v+v^2}{1-v}$$

Separating the variables,

$$\frac{1-v}{1+v+v^2}\,dv = \frac{1}{x}\,dx$$

Integrating both sides, we get

$$\int \frac{1-v}{v^2+v+1}\,dv = \int \frac{1}{x}\,dx$$

$$\Rightarrow \quad \int\left(\frac{1}{v^2+v+1} - \frac{v}{v^2+v+1}\right)dv = \int\frac{1}{x}\,dv$$

$$\Rightarrow \quad \int\frac{1}{v^2+v+1}\,dv - \int\frac{v}{v^2+v+1}\,dv = \log|x| + c$$

[Note this step]

$$\Rightarrow \quad \int\frac{1}{v^2+v+\frac{1}{4}+1-\frac{1}{4}}\,dv - \frac{1}{2}\int\frac{2v}{v^2+v+1}\,dv = \log|x| + c$$

$$\Rightarrow \quad \int\frac{1}{\left(v+\frac{1}{2}\right)^2+\frac{3}{4}}\,dv - \frac{1}{2}\int\frac{2v+1-1}{v^2+v+1}\,dv = \log|x| + c$$

$$\Rightarrow \int\frac{1}{\left(v+\frac{1}{2}\right)^2+\left(\frac{\sqrt{3}}{2}\right)^2}\,dv - \frac{1}{2}\int\frac{2v+1}{v^2+v+1}\,dv + \frac{1}{2}\int\frac{1}{v^2+v+1}\,dv = \log|x| + c$$

$$\Rightarrow \int\frac{1}{\left(v+\frac{1}{2}\right)^2+\left(\frac{\sqrt{3}}{2}\right)^2}\,dv - \frac{1}{2}\int\frac{2v+1}{v^2+v+1}\,dv + \frac{1}{2}\int\frac{1}{\left(v+\frac{1}{2}\right)^2+\left(\frac{\sqrt{3}}{2}\right)^2}\,dv = \log|x| + c$$

[Note this step]

$$\Rightarrow \quad \frac{3}{2}\int\frac{1}{\left(v+\frac{1}{2}\right)^2+\left(\frac{\sqrt{3}}{2}\right)^2}\,dv - \frac{1}{2}\int\frac{2v+1}{v^2+v+1}\,dv = \log|x| + c$$

$$\Rightarrow \quad \frac{3}{2}\cdot\frac{1}{\sqrt{3}/2}\tan^{-1}\left(\frac{v+1/2}{\sqrt{3}/2}\right)-\frac{1}{2}\log|v^2+v+1| = \log|x|+c$$

$$\left[\because \int \frac{1}{x^2+a^2}\,dx = \frac{1}{a}\tan^{-1}\frac{x}{a} \qquad \int \frac{f'(x)}{f(x)}\,dx = \log|f(x)|\right]$$

$$\Rightarrow \quad \frac{3}{2}\cdot\frac{2}{\sqrt{3}}\tan^{-1}\left(\frac{2v+1}{\sqrt{3}}\right)-\frac{1}{2}\log|v^2+v+1| = \log|x|+c \qquad [\because \; y = vx]$$

$$\Rightarrow \quad 2\sqrt{3}\tan^{-1}\left(\frac{\frac{2y}{x}+1}{\sqrt{3}}\right)-\log\left|\frac{y^2}{x^2}+\frac{y}{x}+1\right| = 2\log|x|+2c$$

$$\Rightarrow \quad 2\sqrt{3}\tan^{-1}\left(\frac{2y+x}{x\sqrt{3}}\right)-\log\left|\frac{y^2+xy+x^2}{x^2}\right| = \log|x^2|+c_1 \text{ [where } c_1 = 2c]$$

$$\Rightarrow \quad 2\sqrt{3}\tan^{-1}\left(\frac{2y+x}{x\sqrt{3}}\right)-\log|y^2+xy+x^2|+\log|x^2| = \log|x^2|+c_1$$

$$\left[\because \log\left(\frac{m}{n}\right) = \log m - \log n\right]$$

$$\Rightarrow \quad 2\sqrt{3}\tan^{-1}\left(\frac{2y+x}{x\sqrt{3}}\right)-\log|y^2+xy+x^2| = c_1$$

which is the required solution.

(*ii*) The given equation is

$x^2y\,dx-(x^3+y^3)\,dy = 0 \;\Rightarrow\; x^2y\,dx = (x^3+y^3)\,dy$

$$\Rightarrow \quad \frac{dy}{dx} = \frac{x^2y}{x^3+y^3} \qquad \ldots(1)$$

$$= \frac{x^2\cdot x\left(\frac{y}{x}\right)}{x^3\left[1+\left(\frac{y}{x}\right)^3\right]} = \frac{\left(\frac{y}{x}\right)}{\left[1+\left(\frac{y}{x}\right)^3\right]}$$

which is a homogeneous differential equation.

$\therefore$ Put $y = vx \;\Rightarrow\; \dfrac{dy}{dx} = v + x\dfrac{dv}{dx}$

$\therefore$ Equation (1) becomes

$$v+\frac{x\,dv}{dx} = \frac{x^2(vx)}{x^3+v^3x^3} = \frac{x^3v}{x^3(1+v^3)} = \frac{v}{1+v^3}$$

$$\Rightarrow \quad x\frac{dv}{dx} = \frac{v}{1+v^3} - \frac{v}{1} = \frac{v - v(1+v^3)}{1+v^3} = \frac{v - v - v^4}{1+v^3}$$

$$\Rightarrow \quad x\frac{dv}{dx} = -\frac{v^4}{1+v^3}.$$

Separating the variables,

$$\frac{1+v^3}{v^4}\,dv = -\frac{1}{x}\,dx$$

Integrating both sides, we get

$$\int \frac{1+v^3}{v^4}\,dv = -\int \frac{1}{x}\,dx$$

$$\Rightarrow \quad \int \left(\frac{1}{v^4} + \frac{1}{v}\right) dv = -\int \frac{1}{x}\,dx$$

$$\Rightarrow \quad \int v^{-4}\,dv + \int \frac{1}{v}\,dv = -\int \frac{1}{x}\,dx$$

$$\Rightarrow \quad \frac{v^{-3}}{-3} + \log|v| = -\log|x| + c$$

$$\Rightarrow \quad \frac{-1}{3v^3} + \log|v| + \log|x| = c$$

$$\Rightarrow \quad \frac{-1}{3v^3} + \log|vx| = c \qquad [\because \log m + \log n = \log m \,.\, n]$$

$$\Rightarrow \quad -\frac{x^3}{3y^3} + \log\left|\frac{y}{x}\,.\,x\right| = c \qquad \left[\because y = vx \Rightarrow v = \left(\frac{y}{x}\right)\right]$$

$$\Rightarrow \quad -\frac{x^3}{3y^3} + \log|y| = c$$

which is the required solution.

(*iii*) The given equation is

$$(x^2+y^2)\frac{dy}{dx} = xy \Rightarrow \frac{dy}{dx} = \frac{xy}{x^2+y^2} \qquad ...(1)$$

$$\Rightarrow \quad \frac{dy}{dx} = \frac{x\,.\,x\left(\frac{y}{x}\right)}{x^2\left(1+\frac{y^2}{x^2}\right)} = \frac{\left(\frac{y}{x}\right)}{\left[1+\left(\frac{y}{x}\right)^2\right]}$$

which is a homogeneous differential equation.

$$\therefore \quad \text{Put } y = vx \Rightarrow \frac{dy}{dx} = v + x\frac{dv}{dx}$$

∴ Equation (1) becomes

$$v + x\frac{dv}{dx} = \frac{x\,vx}{x^2 + v^2x^2} = \frac{x^2v}{x^2(1+v^2)} = \frac{v}{1+v^2}$$

$$\Rightarrow \qquad x\frac{dv}{dx} = \frac{v}{1+v^2} - v = \frac{v - v(1+v^2)}{1+v^2}$$

$$\Rightarrow \qquad x\frac{dv}{dx} = \frac{v - v - v^3}{1+v^2} = \frac{-v^3}{1+v^2}$$

Separating the variables,

$$\frac{1+v^2}{v^3}\,dv = -\frac{1}{x}\,dx$$

Integrating both sides, we get

$$\int\left(\frac{1+v^2}{v^3}\right)dv = -\int\frac{1}{x}\,dx$$

$$\Rightarrow \qquad \int\left(\frac{1}{v^3}+\frac{1}{v}\right)dv = -\int\frac{1}{x}\,dx$$

$$\Rightarrow \qquad \int v^{-3}\,dv + \int\frac{1}{v}\,dv = -\log|x| + c$$

$$\Rightarrow \qquad \frac{v^{-2}}{-2} + \log|v| = -\log|x| + c$$

$$\Rightarrow \qquad -\frac{1}{2v^2} + \log|v| = -\log|x| + c$$

$$\Rightarrow \qquad -\frac{x^2}{2y^2} + \log\left|\frac{y}{x}\right| = -\log|x| + c \qquad \left[\because\ y = vx \Rightarrow v = \left(\frac{y}{x}\right)\right]$$

$$\Rightarrow \qquad -\frac{x^2}{2y^2} + \log|y| - \log|x| = -\log|x| + c \qquad \left[\because\ \log\left(\frac{m}{n}\right) = \log m - \log n\right]$$

$$\Rightarrow \qquad -\frac{x^2}{2y^2} + \log|y| = c$$

which is the required solution.

(*iv*) The given equation is

$$xy\frac{dy}{dx} = x^2 - y^2 \quad \Rightarrow \quad \frac{dy}{dx} = \frac{x^2 - y^2}{xy} \qquad ...(1)$$

$$\Rightarrow \qquad \frac{dy}{dx} = \frac{x^2\left(1 - \frac{y^2}{x^2}\right)}{x\,.\,x\left(\frac{y}{x}\right)} = \frac{\left[1 - \left(\frac{y}{x}\right)^2\right]}{\left(\frac{y}{x}\right)}$$

which is a homogeneous differential equation.

$\therefore$ Put $y = vx \Rightarrow \dfrac{dy}{dx} = v + x\dfrac{dv}{dx}$

$\therefore$ Equation (1) becomes

$$v + x\frac{dv}{dx} = \frac{x^2 - v^2x^2}{x(vx)} = \frac{x^2(1-v^2)}{x^2v} = \frac{1-v^2}{v}$$

$$\Rightarrow \quad x\frac{dv}{dx} = \frac{1-v^2}{v} - v = \frac{1-v^2-v^2}{v}$$

$$\Rightarrow \quad x\frac{dv}{dx} = \frac{1-2v^2}{v}$$

Separating the variables,

$$\frac{v}{1-2v^2}\,dv = \frac{1}{x}\,dx$$

Integrating both sides, we get

$$\int \frac{v}{1-2v^2}\,dv = \int \frac{1}{x}\,dx$$

$$\Rightarrow \quad -\frac{1}{4}\int \frac{-4v}{1-2v^2}\,dv = \int \frac{1}{x}\,dx \qquad \text{[Multiply and divided by } -4]$$

$$\Rightarrow \quad -\frac{1}{4}\log|1-2v^2| = \log|x| + \log c \qquad \left[\because \int \frac{f'(x)}{f(x)}\,dx = \log|f(x)|\right]$$

$$\Rightarrow \quad \log|(1-2v^2)^{-1/4}| = \log|cx| \qquad [\because \log m + \log n = \log mn]$$

$$\Rightarrow \quad (1-2v^2)^{-1/4} = cx$$

$$\Rightarrow \quad 1-2v^2 = (cx)^{-4}$$

$$\Rightarrow \quad 1 - \frac{2y^2}{x^2} = \frac{1}{c^4x^4} \qquad \left[\because y = vx \Rightarrow v = \left(\frac{y}{x}\right)\right]$$

$$\Rightarrow \quad \frac{x^2-2y^2}{x^2} = \frac{1}{c^4x^4} \Rightarrow x^2(x^2-2y^2) = \frac{1}{c^4}$$

$$\Rightarrow \quad x^2(x^2-2y^2) = c_1 \qquad \left[\text{where } c_1 = \frac{1}{c^4}\right]$$

which is the required solution.

(v) The given equation is $(3xy + y^2)\,dx + (x^2 + xy)\,dy = 0$

$$\Rightarrow \quad (x^2 + xy)\,dy = -(3xy + y^2)\,dx$$

$$\Rightarrow \quad \frac{dy}{dx} = \frac{-(3xy+y^2)}{(x^2+xy)} \qquad \text{...(1)}$$

$$\Rightarrow \quad \frac{dy}{dx} = \frac{-\left[3\left(\frac{y}{x}\right) + \left(\frac{y}{x}\right)^2\right]}{\left[1+\left(\frac{y}{x}\right)\right]} \qquad \left[\begin{array}{l}\text{Dividing the numerator and}\\ \text{denominator on R.H.S. by } x^2\end{array}\right]$$

which is a homogeneous differential equation.

$\therefore$ Put $y = vx \Rightarrow \dfrac{dy}{dx} = v + x\dfrac{dv}{dx}$

$\therefore$ Equation (1) becomes

$$v + x\frac{dv}{dx} = -\frac{3x(vx) + v^2x^2}{x^2 + x(vx)} = -\frac{x^2(3v + v^2)}{x^2(1+v)}$$

$$\Rightarrow \quad x\frac{dv}{dx} = -\frac{3v + v^2}{1+v} - v = -\frac{3v + v^2 + v(1+v)}{1+v}$$

$$\Rightarrow \quad x\frac{dv}{dx} = -\frac{3v + v^2 + v + v^2}{1+v} = -\frac{2v^2 + 4v}{1+v}$$

$$\Rightarrow \quad x\frac{dv}{dx} = -\frac{2v(v+2)}{1+v}$$

Separating the variables,

$$\frac{1+v}{v(v+2)}dv = -\frac{2}{x}dx$$

Integrating both sides, we get

$$\int\frac{1+v}{v(v+2)}dv = -2\int\frac{1}{x}dx$$

$$\Rightarrow \quad \frac{1}{2}\int\frac{2+2v}{v(v+2)}dv = -2\int\frac{1}{x}dx \qquad \text{[Multiply and divided by 2]}$$

$$\Rightarrow \quad \frac{1}{2}\int\frac{(2+v)+v}{v(v+2)}.\,dv = -2\int\frac{1}{x}dx$$

$$\Rightarrow \quad \frac{1}{2}\int\left(\frac{1}{v} + \frac{1}{v+2}\right)dv = -2\int\frac{1}{x}dx$$

$$\Rightarrow \quad \log|v| + \log|v+2| = -4\log|x| + \log c$$

$$\Rightarrow \quad \log|v(v+2)| = \log|x^{-4}| + \log c \qquad [\because \log m + \log n = \log mn]$$

$$\Rightarrow \quad \log|v(v+2)| = \log|cx^{-4}|$$

$$\Rightarrow \quad v(v+2) = \frac{c}{x^4}$$

$$\Rightarrow \quad \frac{y}{x}\left(\frac{y}{x} + 2\right) = \frac{c}{x^4} \qquad \left[\because\ y = vx \Rightarrow v = \left(\frac{y}{x}\right)\right]$$

$$\Rightarrow \quad \frac{y^2}{x^2} + \frac{2y}{x} = \frac{c}{x^4}$$

$$\Rightarrow \quad y^2 + 2xy = \frac{c}{x^2}$$

$$\Rightarrow \quad x^2y^2 + 2x^3y = c$$

which is the required solution.

(*vi*) The given equation is

$$(x + y)\,dy + (x - y)\,dx = 0 \Rightarrow (x + y)\,dy = (y - x)\,dx$$

$$\Rightarrow \quad \frac{dy}{dx} = \frac{y-x}{y+x} \qquad ...(1)$$

$$\Rightarrow \quad \frac{dy}{dx} = \frac{\left[\left(\frac{y}{x}\right)-1\right]}{\left[\left(\frac{y}{x}\right)+1\right]} \qquad \begin{bmatrix}\text{Dividing the numerator and}\\ \text{denominator on R.H.S. by } x\end{bmatrix}$$

which is a homogeneous differential equation.

$\therefore$ Put $y = vx \Rightarrow \dfrac{dy}{dx} = v + x\dfrac{dv}{dx}$

$\therefore$ Equation (1) becomes

$$v + x\frac{dv}{dx} = \frac{vx-x}{vx+x} = \frac{x(v-1)}{x(v+1)} = \frac{v-1}{v+1}$$

$$\therefore \quad x\frac{dv}{dx} = \frac{v-1}{v+1} - v = \frac{v-1-v(v+1)}{v+1}$$

$$\Rightarrow \quad x\frac{dv}{dx} = \frac{v-1-v^2-v}{v+1} = \frac{-(1+v^2)}{v+1}$$

Separating the variables,

$$\frac{1+v}{1+v^2}\,dv = -\frac{1}{x}\,dx$$

Integrating both sides, we get

$$\int \frac{1+v}{1+v^2}\,dv = -\int \frac{1}{x}\,dx$$

$$\Rightarrow \quad \int \left(\frac{1}{1+v^2} + \frac{v}{1+v^2}\right) dv = -\int \frac{1}{x}\,dx$$

$$\Rightarrow \quad \int \frac{1}{1+v^2}\,dv + \frac{1}{2}\int \frac{2v}{1+v^2}\,dv = -\int \frac{1}{x}\,dx$$

$$\Rightarrow \quad \tan^{-1} v + \frac{1}{2}\log|1+v^2| = -\log|x| + c$$

$$\left[\because \int \frac{1}{a^2+x^2}\,dx = \frac{1}{a}\tan^{-1}\frac{x}{a} \quad \int \frac{f'(x)}{f(x)}\,dx = \log|f(x)|\right]$$

$$\Rightarrow \quad \tan^{-1}\left(\frac{y}{x}\right) + \frac{1}{2}\log\left|1+\frac{y^2}{x^2}\right| = -\log|x| + c \qquad \left[\because y = vx \Rightarrow v = \left(\frac{y}{x}\right)\right]$$

$$\Rightarrow \quad \tan^{-1}\left(\frac{y}{x}\right) + \frac{1}{2}\log\left|\frac{x^2+y^2}{x^2}\right| + \log|x| = c$$

which is the required solution.

(*vii*) The given equation is

$(x + 2y)\, dx - (2x - y)\, dy = 0 \quad \Rightarrow \quad (x + 2y)\, dx = (2x - y)\, dy$

$$\Rightarrow \qquad \frac{dy}{dx} = \frac{x+2y}{2x-y} \qquad ...(1)$$

$$\Rightarrow \qquad \frac{dy}{dx} = \frac{1+2\left(\frac{y}{x}\right)}{2-\left(\frac{y}{x}\right)} \qquad ...(2) \qquad \left[\begin{array}{l}\text{Dividing the numerator and}\\ \text{denominator on R.H.S. by } x\end{array}\right]$$

which is a homogeneous differential equation.

$\therefore$ Put $y = vx \;\Rightarrow\; v = \left(\frac{y}{x}\right). \;\Rightarrow\; \frac{dy}{dx} = v + x\,\frac{dv}{dx}$

$\therefore$ Equation (2) becomes

$$v + x\,\frac{dv}{dx} = \frac{1+2v}{2-v}$$

$$\Rightarrow \qquad x\,\frac{dv}{dx} = \frac{1+2v}{2-v} - v = \frac{1+2v-2v+v^2}{2-v} = \frac{1+v^2}{2-v}$$

Separating the variables,

$$\frac{2-v}{1+v^2}\, dv = \frac{1}{x}\, dx$$

Integrating both sides, we get

$$\int \left(\frac{2-v}{1+v^2}\right) dv = \int \frac{1}{x}\, dx$$

$$\Rightarrow \qquad \int \left(\frac{2}{1+v^2} - \frac{v}{1+v^2}\right) dv = \int \frac{1}{x}\, dx$$

$$\Rightarrow \qquad 2\int \frac{1}{1+v^2}\, dv - \frac{1}{2}\int \frac{2v}{1+v^2}\, dv = \int \frac{1}{x}\, dx$$

$$\Rightarrow \qquad 2\tan^{-1} v - \frac{1}{2}\log|1+v^2| = \log|x| + \log c$$

$$\left[\because \int \frac{1}{a^2+x^2}\, dx = \frac{1}{a}\tan^{-1}\frac{x}{a} \qquad \int \frac{f'(x)}{f(x)}\, dx = \log|f(x)|\right]$$

$$\Rightarrow \qquad 2\tan^{-1}\left(\frac{y}{x}\right) - \frac{1}{2}\log\left|1+\frac{y^2}{x^2}\right| = \log|x| + \log c \qquad \left[\because y = vx \Rightarrow v = \left(\frac{y}{x}\right)\right]$$

$$\Rightarrow \qquad 2\tan^{-1}\left(\frac{y}{x}\right) - \log\left|\frac{x^2+y^2}{x^2}\right|^{1/2} = \log|x| + \log c$$

$$\Rightarrow \quad 2\tan^{-1}\left(\frac{y}{x}\right) - \log|x^2+y^2|^{1/2} + \log|x^2|^{1/2} = \log|x| + \log c$$

$$\Rightarrow \quad 2\tan^{-1}\left(\frac{y}{x}\right) - \log|x^2+y^2|^{1/2} = \log c$$

$$\Rightarrow \quad 2\tan^{-1}\left(\frac{y}{x}\right) = \log c + \log|x^2+y^2|^{1/2}$$

$$\Rightarrow \quad 2\tan^{-1}\left(\frac{y}{x}\right) = \log|c(x^2+y^2)^{1/2}| \qquad [\because \log m + \log n = \log m.n]$$

$$\Rightarrow \quad c(x^2+y^2)^{1/2} = e^{2\tan^{-1}\left(\frac{y}{x}\right)}$$

$$\Rightarrow \quad (x^2+y^2)^{1/2} = c_1\, e^{2\tan^{-1}\left(\frac{y}{x}\right)} \qquad \left[\text{where } c_1 = \frac{1}{c}\right]$$

which is the required solution.

(*viii*) The given equation is

$$x\frac{dy}{dx} - y + x\sin\left(\frac{y}{x}\right) = 0 \quad \Rightarrow \quad x\frac{dy}{dx} = y - x\sin\left(\frac{y}{x}\right).$$

$$\Rightarrow \quad \frac{dy}{dx} = \frac{y}{x} - \sin\left(\frac{y}{x}\right) \qquad ...(1)$$

which is a homogeneous differential equation.

$\therefore$ Put $y = vx \Rightarrow v = \left(\frac{y}{x}\right) \Rightarrow \frac{dy}{dx} = v + x\frac{dv}{dx}$

$\therefore$ Equation (1) becomes

$$v + x\frac{dv}{dx} = v - \sin v$$

$$\Rightarrow \quad x\frac{dv}{dx} = -\sin v$$

Separating the variables,

$$\frac{1}{\sin v}\,dv = -\frac{1}{x}\,dx$$

Integrating both sides, we get

$$\int \frac{1}{\sin v}\,dv = -\int \frac{1}{x}\,dx$$

$$\Rightarrow \quad -\int \operatorname{cosec} v\,dv = \int \frac{1}{x}\,dx$$

$$\Rightarrow \quad \log|\operatorname{cosec} v + \cot v| = \log|x| + \log c$$

$$\Rightarrow \quad \log|\operatorname{cosec} v + \cot v| = \log|cx| \qquad [\because \log m + \log n = \log mn]$$

$$\Rightarrow \quad \operatorname{cosec} v + \cot v = cx$$

$$\Rightarrow \quad \frac{1}{\sin v} + \frac{\cos v}{\sin v} = cx$$

$$\Rightarrow \quad 1 + \cos v = cx \sin v$$

$$\Rightarrow \quad 1 + \cos\left(\frac{y}{x}\right) = cx \sin\left(\frac{y}{x}\right) \qquad \left[\because\ y = vx \Rightarrow v = \left(\frac{y}{x}\right)\right]$$

which is the required solution.

Example 3. *Solve the following differential equations :*

(*i*) $x\dfrac{dy}{dx} - y = \sqrt{x^2 + y^2}$ (*ii*) $y^2 + x^2\dfrac{dy}{dx} = xy\dfrac{dy}{dx}$

(*iii*) $(x^2 - y^2)\,dx + 2xy\,dy = 0$, *given that* $y = 1$ *when* $x = 1$.

(*iv*) $x^2\dfrac{dy}{dx} = 2xy + y^2$ (*v*) $(x^3 - 3xy^2)\,dx = (y^3 - 3x^2y)\,dy$

(*vi*) $(2x^2y + y^3)\,dx + (xy^2 - 3x^3)\,dy = 0$.

Solution. (*i*) The given equation is

$$x\frac{dy}{dx} - y = \sqrt{x^2 + y^2} \Rightarrow x\frac{dy}{dx} = \sqrt{x^2 + y^2} + y$$

$$\Rightarrow \quad \frac{dy}{dx} = \frac{\sqrt{x^2 + y^2} + y}{x} \qquad \text{...(1)}$$

$$\Rightarrow \quad \frac{dy}{dx} = \frac{\sqrt{1 + \left(\frac{y}{x}\right)^2} + \left(\frac{y}{x}\right)}{1}$$

$$\text{...(2)} \qquad \left[\begin{array}{l}\text{Dividing the numerator and}\\ \text{denominator on the R.H.S. by } x\end{array}\right]$$

which is a homogeneous differential equation.

$\therefore$ Put $y = vx \Rightarrow v = \left(\dfrac{y}{x}\right) \Rightarrow \dfrac{dy}{dx} = v + x\dfrac{dv}{dx}$

$\therefore$ Equation (2) becomes

$$v + x\frac{dv}{dx} = \sqrt{1 + v^2} + v$$

$$\Rightarrow \quad x\frac{dv}{dx} = \sqrt{1 + v^2}$$

Separating the variables,

$$\frac{1}{\sqrt{1 + v^2}}\,dv = \frac{1}{x}\,dx$$

Integrating both sides, we get

$$\int \frac{1}{\sqrt{1 + v^2}}\,dv = \int \frac{1}{x}\,dx$$

$$\Rightarrow \quad \log|\,v + \sqrt{1 + v^2}\,| = \log|\,x\,| + \log c \qquad \left[\because\ \int \frac{1}{\sqrt{x^2 + a^2}}\,dx = \log|\,x + \sqrt{x^2 + a^2}\,|\right]$$

$$\Rightarrow \quad \log | v + \sqrt{1+v^2} | = \log | cx | \qquad [\because \log m + \log n = \log mn]$$

$$\Rightarrow \quad v + \sqrt{1+v^2} = cx$$

$$\Rightarrow \quad \frac{y}{x} + \sqrt{1+\frac{y^2}{x^2}} = cx \qquad \left[\because y = vx \Rightarrow v = \left(\frac{y}{x}\right)\right]$$

$$\Rightarrow \quad \frac{y}{x} + \frac{\sqrt{x^2+y^2}}{x} = cx \Rightarrow y + \sqrt{x^2+y^2} = cx^2$$

which is the required solution.

(*ii*) The given equation is

$$y^2 + x^2 \frac{dy}{dx} = xy \frac{dy}{dx}$$

$$\Rightarrow \quad (xy - x^2) \frac{dy}{dx} = y^2$$

$$\Rightarrow \quad \frac{dy}{dx} = \frac{y^2}{xy - x^2} \qquad ...(1)$$

$$\Rightarrow \quad \frac{dy}{dx} = \frac{\left(\frac{y}{x}\right)^2}{\left(\frac{y}{x} - 1\right)} \qquad ...(2) \qquad \left[\begin{array}{l}\text{Dividing the numerator and} \\ \text{denominator on R.H.S. by } x^2\end{array}\right]$$

which is a homogeneous differential equation

$\therefore$ Put $\quad y = vx \Rightarrow v = \left(\frac{y}{x}\right) \Rightarrow \frac{dy}{dx} = v + x\frac{dv}{dx}$

$\therefore$ Equation (2) becomes

$$v + x\frac{dv}{dx} = \frac{v^2}{v-1}$$

$$\Rightarrow \quad x\frac{dv}{dx} = \frac{v^2}{v-1} - v = \frac{v^2 - v^2 + v}{v-1} = \frac{v}{v-1}$$

Separating the variables,

$$\frac{v-1}{v} dv = \frac{1}{x} dx$$

Integrating both sides, we get

$$\int \frac{v-1}{v} dv = \int \frac{1}{x} dx \Rightarrow \int \left(1 - \frac{1}{v}\right) dv = \int \frac{1}{x} dx$$

$$\Rightarrow \quad v - \log | v | = \log | x | + c$$

$$\Rightarrow \quad \frac{y}{x} - \log\left|\frac{y}{x}\right| = \log | x | + c \qquad \left[\because y = vx \Rightarrow v = \left(\frac{y}{x}\right)\right]$$

$$\Rightarrow \quad \frac{y}{x} - \log | y | + \log | x | = \log | x | + c \qquad \left[\because \log\left(\frac{m}{n}\right) = \log m - \log n\right]$$

$$\Rightarrow \quad \frac{y}{x} - \log|y| = c \quad \Rightarrow \quad \frac{y}{x} = \log|y| + c$$

which is the required solution.

(*iii*) The given equation is $(x^2 - y^2)\,dx + 2xy\,dy = 0$

Please try yourself.

[**Hint.** See example 1, part (*ii*)]. [**Ans.** $x^2 + y^2 = 2x$]

(*iv*) The given equation is

$$x^2 \frac{dy}{dx} = 2xy + y^2$$

$$\Rightarrow \quad \frac{dy}{dx} = \frac{2xy + y^2}{x^2} \quad \text{...(1)}$$

$$\Rightarrow \quad \frac{dy}{dx} = \frac{2\left(\frac{y}{x}\right) + \left(\frac{y}{x}\right)^2}{1} \quad \text{...(2)} \quad \left[\begin{array}{l}\text{Dividing the numerator and} \\ \text{denominator on R.H.S. by } x^2\end{array}\right]$$

which is a homogeneous differential equation.

$\therefore$ Put $\quad y = vx \Rightarrow v = \left(\frac{y}{x}\right) \Rightarrow \frac{dy}{dx} = v + x\frac{dv}{dx}$

$\therefore$ Equation (2) becomes

$$v + x\frac{dv}{dx} = 2v + v^2 \quad \Rightarrow \quad x\frac{dv}{dx} = v + v^2$$

Separating the variables,

$$\frac{1}{v(1+v)}\,dv = \frac{1}{x}\,dx$$

Integrating both sides, we get

$$\int \frac{1}{v(1+v)}\,dv = \int \frac{1}{x}\,dx$$

$$\Rightarrow \quad \int \left(\frac{1}{v} - \frac{1}{1+v}\right) dv = \int \frac{1}{x}\,dx \quad \text{[By partial fractions]}$$

$$\Rightarrow \quad \log|v| - \log|1+v| = \log|x| + \log c$$

$$\Rightarrow \quad \log\left|\frac{v}{1+v}\right| = \log|cx| \quad \left[\begin{array}{l}\because \log m + \log n = \log(m\,.\,n) \\ \log m - \log n = \log\left(\frac{m}{n}\right)\end{array}\right]$$

$$\Rightarrow \quad \frac{v}{1+v} = cx$$

$$\Rightarrow \quad \frac{\frac{y}{x}}{1 + \frac{y}{x}} = cx \quad \left[\because y = vx \Rightarrow v = \left(\frac{y}{x}\right)\right]$$

$$\Rightarrow \quad \frac{y}{x+y} = cx \quad \Rightarrow \quad y = cx(x+y)$$

which is the required solution.

(*v*) The given equation is

$$(x^3 - 3xy^2)\, dx = (y^3 - 3x^2y)\, dy$$

$$\Rightarrow \quad \frac{dy}{dx} = \frac{x^3 - 3xy^2}{y^3 - 3x^2 y} \quad ...(1)$$

$$\Rightarrow \quad \frac{dy}{dx} = \frac{1 - 3\left(\frac{y}{x}\right)^2}{\left(\frac{y}{x}\right)^3 - 3\left(\frac{y}{x}\right)} \quad ...(2) \quad \left[\begin{array}{l}\text{Dividing the numerator and} \\ \text{denominator on R.H.S. by } x^3\end{array}\right]$$

which is a homogeneous differential equation.

$\therefore$ Put $y = vx \Rightarrow v = \left(\frac{y}{x}\right) \Rightarrow \frac{dy}{dx} = v + x\frac{dv}{dx}$

$\therefore$ Equation (2) becomes

$$v + x\frac{dv}{dx} = \frac{1 - 3v^2}{v^3 - 3v}$$

$$\Rightarrow \quad x\frac{dv}{dx} = \frac{1 - 3v^2}{v^3 - 3v} - v = \frac{1 - 3v^2 - v^4 + 3v^2}{v^3 - 3v}$$

$$\Rightarrow \quad x\frac{dv}{dx} = \frac{1 - v^4}{v^3 - 3v}$$

Separating the variables,

$$\frac{v^3 - 3v}{1 - v^4}\, dv = \frac{1}{x}\, dx$$

Integrating both sides, we get

$$\int \frac{v^3 - 3v}{1 - v^4}\, dv = \int \frac{1}{x}\, dx$$

$$\Rightarrow \quad \int \frac{v^3}{1 - v^4}\, dv - 3\int \frac{v}{1 - v^4}\, dv = \int \frac{1}{x}\, dx$$

$$\Rightarrow \quad -\frac{1}{4}\int \frac{-4v^3}{1 - v^4}\, dv - \frac{3}{2}\int \frac{2v}{1 - (v^2)^2}\, dv = \int \frac{1}{x}\, dx$$

$$\Rightarrow \quad -\frac{1}{4}\int \frac{-4v^3}{1 - v^4}\, dv - \frac{3}{2}\int \frac{1}{1 - z^2}\, dz = \int \frac{1}{x}\, dx \quad [\text{where } z = v^2]$$

$$\Rightarrow \quad -\frac{1}{4}\log|1 - v^4| - \frac{3}{2}\cdot\frac{1}{2}\log\left|\frac{1 + z}{1 - z}\right| = \log|x| + \log c$$

$$\left[\begin{array}{l}\because \int \frac{f'(x)}{f(x)}\, dx = \log|f(x)| \\ \int \frac{1}{a^2 - x^2}\, dx = \frac{1}{2a}\log\left|\frac{a + x}{a - x}\right|\end{array}\right]$$

$$\Rightarrow \quad -\frac{1}{4}\log|1-v^4|-\frac{3}{4}\log\left|\frac{1+v^2}{1-v^2}\right| = \log|cx|$$

$$\Rightarrow \quad \log|1-v^4| + 3\log\left|\frac{1+v^2}{1-v^2}\right| = -4\log|cx|$$

$$\Rightarrow \quad \log|1-v^4| + \log\left|\frac{(1+v^2)^3}{(1-v^2)^3}\right| = \log(|cx|^{-4}) \qquad [\because \log m + \log n = \log(mn)]$$

$$\Rightarrow \quad \log\left|(1-v^4)\cdot\frac{(1+v^2)^3}{(1-v^2)^3}\right| = \log\left|\frac{1}{c^4x^4}\right|$$

$$\Rightarrow \quad \frac{(1-v^2)(1+v^2)(1+v^2)^3}{(1-v^2)^3} = \frac{1}{c^4x^4}$$

$$\Rightarrow \quad \frac{(1+v^2)^4}{(1-v^2)^2} = \frac{1}{c^4x^4} \qquad \left[\because y = vx \Rightarrow v = \left(\frac{y}{x}\right)\right]$$

$$\Rightarrow \quad c^4x^4\left(1+\frac{y^2}{x^2}\right)^4 = \left(1-\frac{y^2}{x^2}\right)^2$$

$$\Rightarrow \quad c^4x^4\left(\frac{x^2+y^2}{x^2}\right)^4 = \left(\frac{x^2-y^2}{x^2}\right)^2$$

$$\Rightarrow \quad c^4x^4\,\frac{(x^2+y^2)^4}{x^8} = \frac{(x^2-y^2)^2}{x^4}$$

$$\Rightarrow \quad c_1(x^2+y^2)^4 = (x^2-y^2)^2 \qquad [\text{where } c_1 = c^4]$$

which is the required solution.

(*vi*) The given equation is

$$(2x^2y + y^3)\,dx + (xy^2 - 3x^3)\,dy = 0$$

$$\Rightarrow \quad (xy^2 - 3x^3)\,dy = -(2x^2y + y^3)$$

$$\Rightarrow \quad \frac{dy}{dx} = \frac{2x^2y + y^3}{3x^3 - xy^2} \qquad \text{...(1)}$$

$$\Rightarrow \quad \frac{dy}{dx} = \frac{2\left(\frac{y}{x}\right)+\left(\frac{y}{x}\right)^3}{3-\left(\frac{y}{x}\right)^2} \qquad \text{...(2)}$$

[Dividing the numerator and denominator on R.H.S. by x^3]

which is a homogeneous differential equation.

$\therefore$ Put $y = vx \Rightarrow v = \left(\frac{y}{x}\right) \Rightarrow \frac{dy}{dx} = v + x\frac{dv}{dx}$

$\therefore$ Equation (2) becomes

$$v + x\frac{dv}{dx} = \frac{2v + v^3}{3 - v^2} \Rightarrow x\frac{dv}{dx} = \frac{2v + v^3}{3 - v^2} - v = \frac{2v + v^3 - 3v + v^3}{3 - v^2}$$

$$\Rightarrow \qquad x\frac{dv}{dx} = \frac{2v^3 - v}{3 - v^2}$$

Separating the variables,

$$\frac{3 - v^2}{2v^3 - v}\,dv = \frac{1}{x}\,dx$$

Integrating both sides, we get

$$\int \frac{3 - v^2}{v(2v^2 - 1)}\,dv = \int \frac{1}{x}\,dx \qquad \text{...(3)}$$

Now
$$\frac{3 - v^2}{v(2v^2 - 1)} = \frac{A}{v} + \frac{Bv + C}{2v^2 - 1} \qquad \text{...(4)}$$

$$\Rightarrow \qquad 3 - v^2 = A(2v^2 - 1) + (Bv + C)v$$

Put $v = 0$, we get : $3 = A(0 - 1) + 0 \Rightarrow A = -3$.

Comparing co-efficients of v^2 on both sides, we get

$$-1 = 2A + B \Rightarrow -1 = 2(-3) + B \Rightarrow B = 6 - 1 \Rightarrow B = 5.$$

Comparing co-efficients of v on both sides, we get

$$0 = C \Rightarrow C = 0$$

Putting these values of A, B and C in equation (4), we get

$$\frac{3 - v^2}{v(2v^2 - 1)} = \frac{-3}{v} + \frac{5v}{2v^2 - 1} \qquad \text{...(5)}$$

$\therefore$ From equations (3) and (5), we get

$$\int \left(\frac{-3}{v} + \frac{5v}{2v^2 - 1}\right) dv = \int \frac{1}{x}\,dx$$

$$\Rightarrow \qquad -3\int \frac{1}{v}\,dv + \frac{5}{4}\int \frac{4v}{2v^2 - 1}\,dv = \int \frac{1}{x}\,dx$$

$$\Rightarrow \qquad -3\log|v| + \frac{5}{4}\log|2v^2 - 1| = \log|x| + \log c \qquad \left[\because \int \frac{f'(x)}{f(x)} = \log|f(x)|\right]$$

$$\Rightarrow \quad -3\log|v| + \frac{5}{4}\log|2v^2 - 1| = \log|cx| \qquad [\log m + \log n = \log(m.n)]$$

$$\Rightarrow \quad -12\log|v| + 5\log|2v^2 - 1| = 4\log|cx|$$

$$\Rightarrow \quad \log|(2v^2 - 1)^5| - \log|v^{12}| = \log|c^4x^4|$$

$$\Rightarrow \qquad \log\left|\frac{(2v^2 - 1)^5}{v^{12}}\right| = \log|c^4x^4|$$

$$\Rightarrow \quad \frac{\left(\frac{2y^2}{x^2} - 1\right)}{\left(\frac{y^{12}}{x^{12}}\right)} = c^4 x^4 \qquad \left[\because \; y = vx \Rightarrow v = \frac{y}{x}\right]$$

$$\Rightarrow \quad \frac{(2y^2 - x^2)x^{12}}{x^2 y^{12}} = c^4 x^4$$

$$\Rightarrow \quad (2y^2 - x^2)x^{12} = c^4 x^6 y^{12}$$

$$\Rightarrow \quad (2y^2 - x^2)x^6 = c^4 y^{12}$$

which is the required solution.

Example 4. *Solve the following differential equations :*

(*i*) $x\frac{dy}{dx} + \frac{y^2}{x} = y$ (*ii*) $x^2\, dy + y(x + y)\, dx = 0$

(*iii*) $x\frac{dy}{dx} = y(\log y - \log x + 1)$ (*iv*) $x\frac{dy}{dx} = y - \cos^2\left(\frac{y}{x}\right)$

(*v*) $y^2\, dx - (xy + x^2)\, dy = 0$

(*vi*) $(x + y)\, dy + (x - y)\, dx = 0$, *given that* $y = 1$ *when* $x = 1$

(*vii*) $\frac{dy}{dx} = \frac{y}{x} + \frac{x}{y}$.

Solution. (*i*) The given equation is

$$x\frac{dy}{dx} + \frac{y^2}{x} = y$$

$$\Rightarrow \quad x\frac{dy}{dx} = y - \frac{y^2}{x} = \frac{xy - y^2}{x}$$

$$\Rightarrow \quad \frac{dy}{dx} = \frac{xy - y^2}{x^2} \qquad ...(1)$$

$$\Rightarrow \quad \frac{dy}{dx} = \frac{(y/x) - (y/x)^2}{1} \qquad ...(2) \qquad \left[\text{Dividing the numerator and denominator on R.H.S. by } x^2\right]$$

which is a homogeneous differential equation.

$\therefore$ Put $y = vx \Rightarrow v = \left(\frac{y}{x}\right) \Rightarrow \frac{dy}{dx} = v + x\frac{dv}{dx}$

$\therefore$ Equation (2) becomes

$$v + x\frac{dv}{dx} = v - v^2 \Rightarrow x\frac{dv}{dx} = -v^2$$

Separating the variables,

$$\frac{1}{v^2}\, dv = -\frac{1}{x}\, dx$$

Integrating both sides, we get

$$\int \frac{1}{v^2}\, dv = -\int \frac{1}{x}\, dx$$

$$\Rightarrow \quad \frac{v^{-1}}{-1} = -\log|x| + c$$

$$\Rightarrow \quad -\frac{1}{v} = -\log|x| + c$$

$$\Rightarrow \quad \frac{-x}{y} + \log|x| = c \qquad \left[\because\ y = vx \Rightarrow v = \left(\frac{y}{x}\right)\right]$$

which is the required solution.

(*ii*) The given equation is $x^2\,dy + y(x+y)\,dx = 0$

$$\Rightarrow \quad x^2\,dy = -y(x+y)\,dx$$

$$\Rightarrow \quad \frac{dy}{dx} = \frac{-y(x+y)}{x^2} \qquad \ldots(1)$$

$$\Rightarrow \quad \frac{dy}{dx} = \frac{-\left(\frac{y}{x}\right)\left(1+\frac{y}{x}\right)}{1} \qquad \ldots(2) \qquad \left[\begin{array}{l}\text{Dividing the numerator and}\\ \text{denominator on R.H.S. by } x^2\end{array}\right]$$

which is a homogeneous differential equation.

$\therefore$ Put $y = vx \Rightarrow v = \left(\frac{y}{x}\right) \Rightarrow \frac{dy}{dx} = v + x\frac{dv}{dx}$

$\therefore$ Equation (2) becomes

$$v + x\frac{dv}{dx} = -v(1+v)$$

$$\Rightarrow \quad x\frac{dv}{dx} = -v - v^2 - v \quad \Rightarrow \quad x\frac{dv}{dx} = -(v^2 + 2v)$$

Separating the variables,

$$\frac{1}{v^2+2v}\,dv = \frac{-1}{x}\,dx$$

Integrating both sides, we get

$$\int \frac{1}{v(v+2)}\,dv = -\int \frac{1}{x}\,dx$$

$$\Rightarrow \quad \frac{1}{2}\int \left[\frac{1}{v} - \frac{1}{v+2}\right]dv = -\int \frac{1}{x}\,dx \qquad \text{[By partial fractions]}$$

$$\Rightarrow \quad \frac{1}{2}[\log|v| - \log|v+2|] = -\log|x| + \log c$$

$$\Rightarrow \quad \log\left|\frac{v}{v+2}\right| = -2\log|x| + \log c$$

$$\Rightarrow \quad \log\left|\frac{v}{v+2}\right| = \log c - \log|x^2|$$

$$\Rightarrow \quad \log\left|\frac{v}{v+2}\right| = \log\left|\frac{c}{x^2}\right|$$

$$\Rightarrow \quad \frac{v}{v+2} = \frac{c}{x^2} \Rightarrow \frac{y}{2x+y} = \frac{c}{x^2} \qquad \left[\because\ y = vx \Rightarrow v = \left(\frac{y}{x}\right)\right]$$

$$\Rightarrow \quad x^2y = c(2x + y)$$

which is the required solution.

(*iii*) The given equation is

$$x\frac{dy}{dx} = y(\log y - \log x + 1)$$

$$\Rightarrow \quad \frac{dy}{dx} = \frac{y}{x}\left[\log\left(\frac{y}{x}\right) + 1\right] \qquad ...(1)$$

which is a homogeneous differential equation.

$\therefore$ Put $y = vx \Rightarrow v = \left(\frac{y}{x}\right) \Rightarrow \frac{dy}{dx} = v + x\frac{dv}{dx}$

$\therefore$ Equation (1) becomes

$$v + x\frac{dv}{dx} = v(\log v + 1) = v\log v + v$$

$$\Rightarrow \quad x\frac{dv}{dx} = v\log v$$

Separating the variables,

$$\frac{1}{v\log v}dv = \frac{1}{x}dx$$

Integrating both sides, we get

$$\int \frac{1/v}{\log v}dv = \int \frac{1}{x}dx$$

$$\Rightarrow \quad \log|\log v| = \log|x| + \log c \qquad \left[\because\ \int \frac{f'(x)}{f(x)}dx = \log|f(x)|\right]$$

$$\Rightarrow \quad \log|\log v| = \log|cx| \qquad [\because\ \log m + \log n = \log(mn)]$$

$$\Rightarrow \quad \log v = cx$$

$$\Rightarrow \quad \log\left(\frac{y}{x}\right) = cx \qquad \left[\because\ y = vx \Rightarrow v = \frac{y}{x}\right]$$

$$\Rightarrow \quad \frac{y}{x} = e^{cx}$$

$$\Rightarrow \quad y = xe^{cx}$$

which is the required solution.

(*iv*) The given equation is

$$x\frac{dy}{dx} = y - x\cos^2\left(\frac{y}{x}\right) \Rightarrow \frac{dy}{dx} = \frac{y}{x} - \cos^2\left(\frac{y}{x}\right) \qquad ...(1)$$

which is a homogeneous differential equation

$\therefore$ Put $y = vx \Rightarrow v = \left(\frac{y}{x}\right) \Rightarrow \frac{dy}{dx} = v + x\frac{dv}{dx}$

∴ Equation (1) becomes

$$v + x\frac{dv}{dx} = v - \cos^2 v$$

$$\Rightarrow \qquad x\frac{dv}{dx} = -\cos^2 v$$

Separating the variables,

$$\frac{1}{\cos^2 v}\,dv = \frac{-1}{x}\,dx$$

Integrating both sides, we get

$$\int \sec^2 v\,dv = -\int \frac{1}{x}\,dx$$

$$\Rightarrow \qquad \tan v = -\log|x| + c$$

$$\Rightarrow \qquad \tan\left(\frac{y}{x}\right) = -\log|x| + c \qquad \left[\because\ y = vx \Rightarrow v = \left(\frac{y}{x}\right)\right]$$

which is the required solution.

(v) The given equation is

$$y^2\,dx - (xy + x^2)\,dy = 0 \Rightarrow y^2\,dx = (xy + x^2)\,dy$$

$$\Rightarrow \qquad \frac{dy}{dx} = \frac{y^2}{xy + x^2} \qquad ...(1)$$

$$\Rightarrow \qquad \frac{dy}{dx} = \frac{\left(\frac{y}{x}\right)^2}{\left(\frac{y}{x}\right) + 1} \qquad ...(2) \qquad \left[\begin{array}{l}\text{Dividing the numerator and}\\ \text{denominator on R.H.S. by } x^2\end{array}\right]$$

which is a homogeneous differential equation.

∴ Put $y = vx \Rightarrow v = \left(\frac{y}{x}\right) \Rightarrow \frac{dy}{dx} = v + x\frac{dv}{dx}$

∴ Equation (2) becomes

$$v + x\frac{dv}{dx} = \frac{v^2}{v+1}$$

$$\Rightarrow \qquad x\frac{dv}{dx} = \frac{v^2}{v+1} - v = \frac{v^2 - v^2 - v}{v+1} = \frac{-v}{v+1}$$

Separating the variables,

$$\frac{v+1}{v}\,dv = \frac{-1}{x}\,dx$$

Integrating both sides, we get

$$\int\left(\frac{v+1}{v}\right)dv = -\int\frac{1}{x}\,dx$$

$$\Rightarrow \qquad \int\left(1 + \frac{1}{v}\right)dv = -\int\frac{1}{x}\,dx$$

$$\Rightarrow \quad v + \log |v| = -\log|x| + \log c$$

$$\Rightarrow \quad \frac{y}{x} + \log\left|\frac{y}{x}\right| = -\log|x| + \log c \qquad \left[\because \ y = vx \Rightarrow v = \left(\frac{y}{x}\right)\right]$$

$$\Rightarrow \quad \frac{y}{x} + \log|y| - \log|x| = -\log|x| + \log c \qquad \left[\because \ \log\left(\frac{m}{n}\right) = \log m - \log n\right]$$

$$\Rightarrow \quad \frac{y}{x} + \log|y| = \log c \quad \Rightarrow \quad \frac{y}{x} = \log c - \log|y|$$

$$\Rightarrow \quad \frac{y}{x} = \log\left|\frac{c}{y}\right| \quad \Rightarrow \quad \frac{c}{y} = e^{y/x}$$

$$\Rightarrow \quad c = ye^{y/x}$$

which is the required solution.

(*vi*) The given equation is $(x + y)\, dy + (x - y)\, dx = 0$ ...(1)

Please try yourself.

[**Hint.** See example 2, part (*vi*).] [**Ans.** $\log(x^2 + y^2) + 2\tan^{-1}\left(\frac{y}{x}\right) = \frac{\pi}{2} + \log 2$.]

(*vii*) The given equation is

$$\frac{dy}{dx} = \frac{y}{x} + \frac{x}{y} \quad \Rightarrow \quad \frac{dy}{dx} = \frac{y^2 + x^2}{xy} \qquad ...(1)$$

$$\Rightarrow \quad \frac{dy}{dx} = \frac{\left(\frac{y}{x}\right)^2 + 1}{\left(\frac{y}{x}\right)} \qquad ...(2) \qquad \left[\begin{array}{l}\text{Dividing the numerator and} \\ \text{denominator on R.H.S. by } x^2\end{array}\right]$$

which is a homogeneous differential equation.

$\therefore$ Put $y = vx \Rightarrow v = \left(\frac{y}{x}\right) \Rightarrow \frac{dy}{dx} = v + x\frac{dv}{dx}$

$\therefore$ Equation (2) becomes

$$v + x\frac{dv}{dx} = \frac{v^2 + 1}{v}$$

$$\Rightarrow \quad x\frac{dv}{dx} = \frac{v^2 + 1}{v} - v = \frac{v^2 + 1 - v^2}{v} = \frac{1}{v}$$

Separating the variables,

$$v\, dv = \frac{1}{x}\, dx$$

Integrating both sides, we get

$$\int v\, dv = \int \frac{1}{x}\, dx$$

$$\Rightarrow \quad \frac{v^2}{2} = \log|x| + c$$

$$\Rightarrow \qquad \frac{y^2}{2x^2} = \log |x| + c \qquad \left[\because \; y = vx \Rightarrow v = \left(\frac{y}{x}\right)\right]$$

which is the required solution.

Example 5. *Solve the following differential equations :*

(i) $(x^2 + xy)\, dy = (x^2 + y^2)\, dx$

(ii) $y\left[x \cos\left(\frac{y}{x}\right) + y \sin\left(\frac{y}{x}\right)\right] - x\left[y \sin\left(\frac{y}{x}\right) - x \cos\left(\frac{y}{x}\right)\right]\frac{dy}{dx} = 0$

(iii) $2ye^{x/y}\, dx + (y - 2xe^{x/y})\, dy = 0$ (iv) $3x^2 \frac{dy}{dx} = 3xy + y^2$

(v) $\frac{dy}{dx} = \frac{\sqrt{x^2 - y^2} + y}{x}$ (vi) $x^2 \frac{dy}{dx} = \frac{y(x + y)}{2}$.

Solution. (i) The given equation is

$$(x^2 + xy)\, dy = (x^2 + y^2)\, dx \Rightarrow \frac{dy}{dx} = \frac{x^2 + y^2}{x^2 + xy} \qquad ...(1)$$

$$\Rightarrow \qquad \frac{dy}{dx} = \frac{1 + \left(\frac{y}{x}\right)^2}{1 + \left(\frac{y}{x}\right)} \qquad ...(2) \qquad \left[\begin{array}{l}\text{Dividing the numerator and} \\ \text{denominator on R.H.S. by } x^2\end{array}\right]$$

which is a homogeneous differential equation.

$$\therefore \quad \text{Put } y = vx \Rightarrow v = \left(\frac{y}{x}\right) \Rightarrow \frac{dy}{dx} = v + x\frac{dv}{dx}$$

$\therefore$ Equation (2) becomes

$$v + x\frac{dv}{dx} = \frac{1 + v^2}{1 + v}$$

$$\Rightarrow \qquad x\frac{dv}{dx} = \frac{1 + v^2}{1 + v} - v = \frac{1 + v^2 - v - v^2}{1 + v}$$

$$\Rightarrow \qquad x\frac{dv}{dx} = \frac{1 - v}{1 + v}$$

Separating the variables,

$$\frac{1 + v}{1 - v}\, dv = \frac{1}{x}\, dx$$

Integrating both sides, we get

$$\int \frac{1 + v}{1 - v}\, dv = \int \frac{1}{x}\, dx$$

$$\Rightarrow \qquad \int \left(\frac{2 - (1 - v)}{1 - v}\right) dv = \int \frac{1}{x}\, dx$$

$$\Rightarrow \qquad \int \frac{2}{1 - v}\, dv - \int 1 \, . \, dv = \int \frac{1}{x}\, dx$$

$$\Rightarrow \quad -2\log|1-v| - v = \log|x| + c$$

$$\Rightarrow \quad \log|(1-v)^{-2}| - \log|x| = v + c$$

$$\Rightarrow \quad \log\left|\frac{1}{(1-v)^2 x}\right| = v + c \qquad \left[\because \log m - \log n = \log\left(\frac{m}{n}\right)\right]$$

$$\Rightarrow \quad \log\left|\frac{1}{\left(1-\frac{y}{x}\right)^2 x}\right| = \frac{y}{x} + c \qquad \left[\because y = vx \Rightarrow v = \left(\frac{y}{x}\right)\right]$$

$$\Rightarrow \quad \log\left|\frac{x}{(x-y)^2}\right| = \frac{y}{x} + c$$

which is the required solution.

(*ii*) The given equation is

$$y\left[x\cos\frac{y}{x} + y\sin\frac{y}{x}\right] - x\left[y\sin\frac{y}{x} - x\cos\frac{y}{x}\right]\frac{dy}{dx} = 0$$

$$\Rightarrow \quad \frac{dy}{dx} = \frac{y\left[x\cos\left(\frac{y}{x}\right) + y\sin\left(\frac{y}{x}\right)\right]}{x\left[y\sin\left(\frac{y}{x}\right) - x\cos\left(\frac{y}{x}\right)\right]}$$

$$\Rightarrow \quad \frac{dy}{dx} = \frac{\frac{y}{x}\left[\cos\left(\frac{y}{x}\right) + \frac{y}{x}\sin\left(\frac{y}{x}\right)\right]}{\left[\frac{y}{x}\sin\left(\frac{y}{x}\right) - \cos\left(\frac{y}{x}\right)\right]} \qquad \ldots(1)$$

which is a homogeneous differential equation.

$\therefore$ Put $y = vx \Rightarrow v = \left(\frac{y}{x}\right) \Rightarrow \frac{dy}{dx} = v + x\frac{dv}{dx}$

$\therefore$ Equation (1) becomes

$$v + x\frac{dv}{dx} = \frac{v(\cos v + v\sin v)}{(v\sin v - \cos v)}$$

$$\Rightarrow \quad x\frac{dv}{dx} = \frac{v\cos v + v^2\sin v}{v\sin v - \cos v} - v = \frac{v\cos v + v^2\sin v - v^2\sin v + v\cos v}{v\sin v - \cos v}$$

$$\Rightarrow \quad x\frac{dv}{dx} = \frac{2v\cos v}{v\sin v - \cos v}$$

Separating the variables, $\frac{v\sin v - \cos v}{v\cos v}\,dv = \frac{2}{x}\,dx$

Integrating both sides, we get

$$\int\left(\frac{v\sin v}{v\cos v} - \frac{\cos v}{v\cos v}\right)dv = 2\int\frac{1}{x}\,dx$$

$$\Rightarrow \quad \int \frac{\sin v}{\cos v}\, dv - \int \frac{1}{v}\, dv = 2\int \frac{1}{x}\, dx$$

$$\Rightarrow \quad -\log|\cos v| - \log|v| = 2\log|x| + \log c \qquad \left[\because \int \frac{f'(x)}{f(x)}\, dx = \log|f(x)|\right]$$

$$\Rightarrow \quad \log|\cos v| + \log|v| = -\log|x^2| - \log c$$

$$\Rightarrow \quad \log|\cos v| + \log|v| + \log|x^2| = -\log c$$

$$\Rightarrow \quad \log|(v\cos v)\, x^2| = \log c^{-1} \qquad [\because \log m + \log n + \log l = \log(mnl)]$$

$$\Rightarrow \quad x^2 v \cos v = \frac{1}{c}$$

$$\Rightarrow \quad x^2 \cdot \frac{y}{x} \cos\left(\frac{y}{x}\right) = c_1 \qquad \left[\because y = vx \Rightarrow v = \left(\frac{y}{x}\right)\right]$$

$$\Rightarrow \quad xy \cos\left(\frac{y}{x}\right) = c_1 \qquad \left[\text{where } c_1 = \frac{1}{c}\right]$$

which is the required solution.

(*iii*) The given equation is

$$2ye^{x/y}\, dx + (y - 2xe^{x/y})\, dy = 0 \quad \Rightarrow \quad \frac{dx}{dy} = \frac{2xe^{x/y} - y}{2ye^{x/y}} \qquad ...(1)$$

$$\Rightarrow \quad \frac{dx}{dy} = \left[\frac{2\left(\frac{x}{y}\right)e^{x/y} - 1}{2e^{x/y}}\right] \qquad ...(2) \quad \left[\begin{array}{l}\text{Dividing the numerator and}\\ \text{denominator on R.H.S. by } x^2\end{array}\right]$$

which is a homogeneous differential equation.

$\therefore$ Put $x = vy \Rightarrow v = \left(\frac{x}{y}\right)$

$$\Rightarrow \quad \frac{dx}{dy} = v + y\frac{dv}{dy}$$

$\therefore$ Equation (2) becomes

$$v + y\frac{dv}{dy} = \frac{2ve^{v} - 1}{2e^{v}}$$

$$\Rightarrow \quad y\frac{dv}{dy} = \frac{2ve^{v} - 1}{2e^{v}} - v = \frac{2ve^{v} - 1 - 2ve^{v}}{2e^{v}}$$

$$\Rightarrow \quad y\frac{dv}{dy} = \frac{-1}{2e^{v}}.$$

Separating the variables,

$$2e^{v}\, dv = \frac{-1}{y}\, dy$$

Integrating both sides, we get

$$2\int e^v\,dv = -\int \frac{1}{y}\,dv$$

$$\Rightarrow \quad 2e^v = -\log|y| + \log c$$

$$\Rightarrow \quad 2e^v = \log\left|\frac{c}{y}\right| \qquad \left[\because \log m - \log n = \log\left(\frac{m}{n}\right)\right]$$

$$\Rightarrow \quad 2e^{x/y} = \log\left|\frac{c}{y}\right| \qquad \left[\because x = vy \Rightarrow v = \frac{x}{y}\right]$$

which is the required solution.

(*iv*) The given equation is

$$3x^2\frac{dy}{dx} = 3xy + y^2 \Rightarrow \frac{dy}{dx} = \frac{3xy + y^2}{3x^2} \qquad \text{...(1)}$$

$$\Rightarrow \quad \frac{dy}{dx} = \frac{3\left(\frac{y}{x}\right) + \left(\frac{y}{x}\right)^2}{3} \qquad \text{...(2)} \qquad \left[\begin{array}{l}\text{Dividing the numerator and}\\ \text{denominator on R.H.S. by } x^2\end{array}\right]$$

which is a homogeneous differential equation.

$\therefore$ Put $y = vx \Rightarrow v = \left(\frac{y}{x}\right) \Rightarrow \frac{dy}{dx} = v + x\frac{dv}{dx}$

$\therefore$ Equation (2) becomes

$$v + x\frac{dv}{dx} = \frac{3v + v^2}{3}$$

$$\Rightarrow \quad x\frac{dv}{dx} = \frac{3v + v^2}{3} - v = \frac{3v + v^2 - 3v}{3}$$

$$\Rightarrow \quad x\frac{dv}{dx} = \frac{v^2}{3}.$$

Separating the variables,

$$\frac{1}{v^2}\,dv = \frac{1}{3}\,x\,dx$$

Integrating both sides, we get

$$\int v^{-2}\,dv = \frac{1}{3}\int \frac{1}{x}\,dx \Rightarrow -v^{-1} = \frac{1}{3}\log|x| + c$$

$$\Rightarrow \quad \frac{-1}{v} = \frac{1}{3}\log|x| + c$$

$$\Rightarrow \quad \frac{-x}{y} = \frac{1}{3}\log|x| + c \qquad \left[\because y = vx \Rightarrow v = \left(\frac{y}{x}\right)\right]$$

which is the required solution.

(*v*) The given equation is

$$\frac{dy}{dx} = \frac{\sqrt{x^2 - y^2} + y}{x} \qquad \text{...(1)}$$

$$\Rightarrow \qquad \frac{dy}{dx} = \frac{\sqrt{1-\left(\frac{y}{x}\right)^2} + \left(\frac{y}{x}\right)}{1} \qquad ...(2) \qquad \left[\begin{array}{l}\text{Dividing the numerator and}\\ \text{denominator on R.H.S. by } x^2\end{array}\right]$$

which is a homogeneous differential equation.

$\therefore$ Put $y = vx \Rightarrow v = \left(\frac{y}{x}\right) \Rightarrow \frac{dy}{dx} = v + x\frac{dv}{dx}$

$\therefore$ Equation (2) becomes

$$v + x\frac{dv}{dx} = \frac{\sqrt{1-v^2} + v}{1} \qquad \Rightarrow \quad x\frac{dv}{dx} = \sqrt{1-v^2}$$

Separating the variables,

$$\frac{1}{\sqrt{1-v^2}}\,dv = \frac{1}{x}\,dx$$

Integrating both sides, we get

$$\int \frac{1}{\sqrt{1-v^2}}\,dv = \int \frac{1}{x}\,dx$$

$$\Rightarrow \qquad \sin^{-1} v = \log|x| + c$$

$$\Rightarrow \qquad \sin^{-1}\left(\frac{y}{x}\right) = \log|x| + c \qquad \left[\because\ y = vx \Rightarrow v = \left(\frac{y}{x}\right)\right]$$

which is the required solution.

(*vi*) The given equation is

$$x^2\frac{dy}{dx} = \frac{y(x+y)}{2} \quad \Rightarrow \quad \frac{dy}{dx} = \frac{y(x+y)}{2x^2} \qquad ...(1)$$

$$\Rightarrow \qquad \frac{dy}{dx} = \frac{\left(\frac{y}{x}\right)\left[1+\left(\frac{y}{x}\right)\right]}{2} \qquad ...(2) \qquad \left[\begin{array}{l}\text{Dividing the numerator and}\\ \text{denominator on R.H.S. by } x^2\end{array}\right]$$

which is a homogeneous differential equation.

$\therefore$ Put $y = vx \Rightarrow v = \left(\frac{y}{x}\right) \Rightarrow \frac{dy}{dx} = v + x\frac{dv}{dx}$

$\therefore$ Equation (2) becomes

$$v + x\frac{dv}{dx} = \frac{v(1+v)}{2}$$

$$\Rightarrow \qquad x\frac{dv}{dx} = \frac{v+v^2}{2} - v = \frac{v + v^2 - 2v}{2} \quad \Rightarrow \quad x\frac{dv}{dx} = \frac{v^2 - v}{2}.$$

Separating the variables,

$$\frac{2}{v(v-1)}\,dv = \frac{1}{x}\,dx$$

Integrating both sides, we get

$$2\int \frac{1}{v(v-1)}\,dv = \int \frac{1}{x}\,dx$$

$$\Rightarrow \quad 2\int \left(\frac{1}{v-1} - \frac{1}{v}\right) dv = \int \frac{1}{x}\,dx \qquad \text{[By partial fractions]}$$

$$\Rightarrow \quad 2\,[\log|v-1| - \log|v|] = \log|x| + \log c$$

$$\Rightarrow \quad 2\log\left|\frac{v-1}{v}\right| = \log|cx| \qquad [\because \ \log m + \log n = \log(m.n)]$$

$$\Rightarrow \quad \log\left|\left(\frac{v-1}{v}\right)^2\right| = \log|cx| \quad \Rightarrow \quad \frac{(v-1)^2}{v^2} = cx$$

$$\Rightarrow \quad \frac{\left(\frac{y}{x}-1\right)^2}{\frac{y^2}{x^2}} = cx \qquad \left[\because \ y = vx \ \Rightarrow \ v = \left(\frac{y}{x}\right)\right]$$

$$\Rightarrow \quad \frac{(y-x)^2}{y^2} = cx$$

$$\Rightarrow \quad (y-x)^2 = cxy^2$$

which is the required solution.

Example 6. *Solve the following differential equations :*

(*i*) $x\dfrac{dy}{dx}\sin\left(\dfrac{y}{x}\right) + x - y\sin\left(\dfrac{y}{x}\right) = 0\ ;\ y(1) = \dfrac{\pi}{2}.$

(*ii*) $(1 + e^{x/y})\,dx + e^{x/y}\left(1 - \dfrac{x}{y}\right)dy = 0$

(*iii*) $(x^2y - 2xy^2)\,dx - (x^3 - 3x^2y)\,dy = 0$

(*iv*) $(3x^2 + 2xy - y^2)\,dx = (3y^2 + 2xy - x^2)\,dy$

(*v*) $(xe^{y/x} + y)\,dx = x\,dy\ ;\ y(1) = 1.$

Solution. (*i*) The given equation is

$$x\frac{dy}{dx}\sin\left(\frac{y}{x}\right) + x - y\sin\left(\frac{y}{x}\right) = 0$$

$$\Rightarrow \quad x\frac{dy}{dx}\sin\left(\frac{y}{x}\right) = y\sin\left(\frac{y}{x}\right) - x$$

$$\Rightarrow \quad \frac{dy}{dx} = \frac{y\sin\left(\frac{y}{x}\right) - x}{x\sin\left(\frac{y}{x}\right)} \qquad \text{...(1)}$$

$$\Rightarrow \quad \frac{dy}{dx} = \frac{\left[\left(\frac{y}{x}\right)\sin\left(\frac{y}{x}\right) - 1\right]}{\sin\left(\frac{y}{x}\right)} \quad ...(2) \quad \left[\begin{array}{l}\text{Dividing the numerator and} \\ \text{denominator on R.H.S. by } x^2\end{array}\right]$$

which is a homogeneous differential equation.

$\therefore$ Put $y = vx \quad \Rightarrow \quad v = \left(\frac{y}{x}\right) \quad \Rightarrow \quad \frac{dy}{dx} = v + x\frac{dv}{dx}$

$\therefore$ Equation (2) becomes

$$v + x\frac{dv}{dx} = \frac{v \sin v - 1}{\sin v}$$

$$\Rightarrow \quad x\frac{dv}{dx} = \frac{v \sin v - 1}{\sin v} - v = \frac{v \sin v - 1 - v \sin v}{\sin v}$$

$$\Rightarrow \quad x\frac{dv}{dx} = \frac{-1}{\sin v}$$

Separating the variables,

$$\sin v \, dv = \frac{-1}{x} dx$$

Integrating both sides, we get

$$-\int \sin v \, dv = \int \frac{1}{x} dx$$

$$\Rightarrow \quad \cos v = \log |x| + c \qquad \left[\because \ y = vx \ \Rightarrow \ v = \left(\frac{y}{x}\right)\right]$$

$$\Rightarrow \quad \cos\left(\frac{y}{x}\right) = \log |x| + c \qquad ...(3)$$

Now, it is given that $y(1) = \frac{\pi}{2}$ *i.e.*, $y = \frac{\pi}{2}$ at $x = 1$

$\therefore$ Equation (3) becomes

$$\cos\left(\frac{\frac{\pi}{2}}{1}\right) = \log |1| + c \quad \Rightarrow \quad 0 = 0 + c \quad \Rightarrow \quad c = 0$$

Putting this value of c in equation (3), we have

$$\cos\left(\frac{y}{x}\right) = \log |x|$$

which is the required solution.

(*ii*) The given equation is $(1 + e^{x/y})\, dx + e^{x/y}\left(1 - \frac{x}{y}\right) dy = 0$

$$\Rightarrow \quad (1 + e^{x/y})\, dx = -e^{x/y}\left(1 - \frac{x}{y}\right) dy$$

$$\Rightarrow \qquad \frac{dx}{dy} = \frac{e^{x/y}\left[\frac{x}{y} - 1\right]}{(1 + e^{x/y})} \qquad \text{...(1)}$$

which is a homogeneous differential equation.

$\therefore$ Put $x = vy \Rightarrow v = \left(\frac{x}{y}\right) \Rightarrow \frac{dx}{dy} = v + y\frac{dv}{dy}$

$\therefore$ Equation (1) becomes

$$v + y\frac{dv}{dy} = \frac{e^v(v-1)}{1+e^v} \quad \Rightarrow \quad y\frac{dv}{dy} = \frac{ve^v - e^v}{1+e^v} - v$$

$$\Rightarrow \qquad y\frac{dv}{dy} = \frac{ve^v - e^v - v - ve^v}{1+e^v}$$

$$\Rightarrow \qquad y\frac{dv}{dy} = -\frac{(v+e^v)}{1+e^v}.$$

Separating the variables,

$$\frac{1+e^v}{v+e^v}\,dv = \frac{-1}{y}\,dy$$

Integrating both sides, we get

$$\int \frac{1+e^v}{v+e^v}\,dv = -\int \frac{1}{y}\,dy \qquad \left[\because \int \frac{f'(x)}{f(x)}\,dx = \log|f(x)|\right]$$

$$\Rightarrow \qquad \log|v + e^v| = -\log|y| + \log c \qquad \left[\because \log m - \log n = \log\left(\frac{m}{n}\right)\right]$$

$$\Rightarrow \qquad \log|v + e^v| = \log\left|\frac{c}{y}\right|$$

$$\Rightarrow \qquad v + e^v = \frac{c}{y} \qquad \left[\because x = vy \Rightarrow v = \left(\frac{x}{y}\right)\right]$$

$$\Rightarrow \qquad \frac{x}{y} + e^{x/y} = \frac{c}{y} \quad \Rightarrow \quad x + ye^{x/y} = c$$

which is the required solution.

(*iii*) The given equation is $(x^2y - 2xy^2)\,dx - (x^3 - 3x^2y)\,dy = 0$

$$\Rightarrow \qquad (x^2y - 2xy^2)\,dx = (x^3 - 3x^2y)\,dy$$

$$\Rightarrow \qquad \frac{dy}{dx} = \frac{x^2y - 2xy^2}{x^3 - 3x^2y} \qquad \text{...(1)}$$

$$\Rightarrow \qquad \frac{dy}{dx} = \frac{\left(\frac{y}{x}\right) - 2\left(\frac{y}{x}\right)^2}{1 - 3\left(\frac{y}{x}\right)} \qquad \text{...(2)} \quad \left[\begin{array}{l}\text{Dividing the numerator and}\\ \text{denominator on R.H.S. by } x^3\end{array}\right]$$

which is a homogeneous differential equation.

$\therefore$ Put $y = vx \Rightarrow v = \left(\frac{y}{x}\right) \Rightarrow \frac{dy}{dx} = v + x\frac{dv}{dx}$

$\therefore$ Equation (2) becomes

$$v + x\frac{dv}{dx} = \frac{v - 2v^2}{1 - 3v}$$

$$\Rightarrow \quad x\frac{dv}{dx} = \frac{v - 2v^2}{1 - 3v} - v = \frac{v - 2v^2 - v + 3v^2}{1 - 3v}$$

$$\Rightarrow \quad x\frac{dv}{dx} = \frac{v^2}{1 - 3v}$$

Separating the variables, $\frac{1 - 3v}{v^2}\, dv = \frac{1}{x}\, dx$

Integrating both sides, we get

$$\int\left(\frac{1}{v^2} - \frac{3}{v}\right) dv = \int\frac{1}{x}\, dx$$

$$\Rightarrow \quad \int v^{-2}\, dv - 3\int\frac{1}{v}\, dv = \int\frac{1}{x}\, dx \Rightarrow -v^{-1} - 3\log|v| = \log|x| + \log c$$

$$\Rightarrow \quad \frac{-1}{v} = 3\log|v| + \log|x| + \log c$$

$$\Rightarrow \quad \frac{-1}{v} = \log(v^3 xc) \qquad [\because \log l + \log m + \log n = \log(l.m.n)]$$

$$\Rightarrow \quad v^3 xc = e^{-1/v}$$

$$\Rightarrow \quad \frac{y^3}{x^3} \cdot x \cdot c = e^{-x/y} \qquad \left[\because y = vx \Rightarrow v = \frac{y}{x}\right]$$

$$\Rightarrow \quad cy^3 = x^2 e^{-x/y} \Rightarrow x^2 = cy^3 e^{x/y}$$

which is the required solution.

(*iv*) The given equation is

$$(3x^2 + 2xy - y^2)\, dx = (3y^2 + 2xy - x^2)\, dy$$

$$\Rightarrow \quad \frac{dy}{dx} = \frac{3x^2 + 2xy - y^2}{3y^2 + 2xy - x^2} \qquad ...(1)$$

$$\Rightarrow \quad \frac{dy}{dx} = \frac{3 + 2\left(\frac{y}{x}\right) - \left(\frac{y}{x}\right)^2}{3\left(\frac{y}{x}\right)^2 + 2\left(\frac{y}{x}\right) - 1} \qquad ...(2) \qquad \left[\begin{array}{l}\text{Dividing the numerator and}\\ \text{denominator on R.H.S. by } x^2\end{array}\right]$$

which is a homogeneous differential equation.

$\therefore$ Put $y = vx \Rightarrow v = \left(\frac{y}{x}\right) \Rightarrow \frac{dy}{dx} = v + x\frac{dv}{dx}$

$\therefore$ Equation (2) becomes

$$v + x\frac{dv}{dx} = \frac{3 + 2v - v^2}{3v^2 + 2v - 1}$$

$$\Rightarrow \quad x\frac{dv}{dx} = \frac{3 + 2v - v^2}{3v^2 + 2v - 1} - v = \frac{3 + 2v - v^2 - 3v^3 - 2v^2 + v}{3v^2 + 2v - 1}$$

$$\Rightarrow \quad x\frac{dv}{dx} = \frac{-3v^3 - 3v^2 + 3v + 3}{3v^2 + 2v - 1}$$

$$\Rightarrow \quad x\frac{dv}{dx} = \frac{-3(v^3 + v^2 - v - 1}{3v^2 + 2v - 1}$$

Separating the variables,

$$\frac{3v^2 + 2v - 1}{v^3 + v^2 - v - 1}\, dv = \frac{-3}{x}\, dx$$

Integrating both sides, we get

$$\int \frac{3v^2 + 2v - 1}{v^3 + v^2 - v - 1}\, dv = -3\int \frac{1}{x}\, dx$$

$$\Rightarrow \quad \log|v^3 + v^2 - v - 1| = -3\log|x| + \log c \qquad \left[\because \int \frac{f'(x)}{f(x)}\, dx = \log|f(x)|\right]$$

$$\Rightarrow \quad \log|v^3 + v^2 - v - 1| = \log c - \log|x^3|$$

$$\Rightarrow \quad \log|v^3 + v^2 - v - 1| = \log\left|\frac{c}{x^3}\right| \qquad \left[\because \log m - \log n = \log\left(\frac{m}{n}\right)\right]$$

$$\Rightarrow \quad v^3 + v^2 - v - 1 = \frac{c}{x^3}$$

$$\Rightarrow \quad \frac{y^3}{x^3} + \frac{y^2}{x^2} - \frac{y}{x} - 1 = \frac{c}{x^3} \qquad \left[\because y = vx \Rightarrow v = \left(\frac{y}{x}\right)\right]$$

$$\Rightarrow \quad y^3 + xy^2 - x^2y - x^3 = c \quad \Rightarrow \quad y^2(y + x) - x^2(y + x) = c$$

$$\Rightarrow \quad (y + x)(y^2 - x^2) = c \quad \Rightarrow \quad (y + x)(y + x)(y - x) = c$$

$$\Rightarrow \quad (y + x)^2\,(y - x) = c$$

which is the required solution.

(v) The given equation is

$$(xe^{y/x} + y)\, dx = x\, dy$$

$$\Rightarrow \quad \frac{dy}{dx} = \left(\frac{xe^{y/x} + y}{x}\right) \quad \Rightarrow \quad \frac{dy}{dx} = e^{y/x} + \left(\frac{y}{x}\right) \qquad ...(1)$$

which is a homogeneous differential equation.

$$\therefore \text{ Put } y = vx \quad \Rightarrow \quad v = \left(\frac{y}{x}\right) \quad \Rightarrow \quad \frac{dy}{dx} = v + x\frac{dv}{dx}$$

$\therefore$ Equation (1) becomes

$$v + x\frac{dv}{dx} = e^v + v \quad \Rightarrow \quad x\frac{dv}{dx} = e^v$$

Separating the variables,

$$\frac{1}{e^v}\,dv = \frac{1}{x}\,dx$$

Integrating both sides, we get

$$\int \frac{1}{e^v}\,dv = \int \frac{1}{x}\,dx$$

$$\Rightarrow \quad -e^{-v} = \log|x| + c \qquad \left[\because\ y = vx \Rightarrow v = \left(\frac{y}{x}\right)\right]$$

$$\Rightarrow \quad -e^{-y/x} = \log|x| + c \qquad \ldots(2)$$

Now, it is given that $y(1) = 1$ *i.e.*, $y = 1$ at $x = 1$.

$$\Rightarrow \quad -e^{-1} = \log 1 + c$$

$$\Rightarrow \quad \frac{-1}{e} = 0 + c \quad \Rightarrow \quad c = \frac{-1}{e}$$

Putting this value of c in equation (2), we have

$$-e^{-y/x} = \log|x| + (-e^{-1})$$

$$\Rightarrow \quad -e^{-y/x} = \log|x| - e^{-1}$$

$$\Rightarrow \quad e^{-y/x} = \frac{1}{e} - \log|x|$$

which is the required solution.

Example 7. *Find the general solution of the following differential equations :*

(*i*) $\dfrac{dy}{dx} + \dfrac{y}{x} = x^2$ (*ii*) $\dfrac{dy}{dx} = \dfrac{x+y+1}{x+1}$

(*iii*) $\dfrac{dy}{dx} + 2y = 6e^x$ (*iv*) $2y' + 4y = x^2 - x$

(*v*) $x\dfrac{dy}{dx} - 3y = -2mx$ (*vi*) $4\dfrac{dy}{dx} + 8y = 5e^{-3x}$.

Solution. (*i*) The given equation is

$$\frac{dy}{dx} + \frac{y}{x} = x^2 \qquad \ldots(1)$$

Comparing it with $\dfrac{dy}{dx} + \text{P}y = \text{Q}$

where $\text{P} = \dfrac{1}{x}$ and $\text{Q} = x^2$

$$\therefore \quad \text{I.F.} = e^{\int \text{P}\,.\,dx} = e^{\int \frac{1}{x}dx} = e^{\log|x|} = x$$

$\therefore$ The required solution is given by

$$y(\text{I.F.}) = \int \text{Q}\,.\,(\text{I.F.})\,dx + c$$

$\Rightarrow \qquad yx = \int x^2 \cdot x\, dx + c = \int x^3 \cdot dx + c$

$\Rightarrow \qquad xy = \dfrac{x^4}{4} + c.$

(*ii*) The given equation is

$$\frac{dy}{dx} = \frac{x+y+1}{x+1} = \frac{(x+1)+y}{x+1}$$

$\Rightarrow \qquad \dfrac{dy}{dx} = 1 + \dfrac{y}{x+1} \quad \Rightarrow \quad \dfrac{dy}{dx} - \dfrac{1}{x+1}\, y = 1 \qquad$ ...(1)

Comparing it with $\dfrac{dy}{dx} + \text{P}y = \text{Q}$

where $\qquad \text{P} = \dfrac{-1}{x+1}$ and $\text{Q} = 1$

$\therefore \qquad \text{I.F.} = e^{\int \text{P}\, dx} = e^{-\int \frac{1}{x+1} dx} = e^{-\log|x+1|}$

$$= e^{\log|x+1|^{-1}} = (x+1)^{-1} = \frac{1}{x+1}$$

$\therefore$ The required solution is given by

$$y(\text{I.F.}) = \int \text{Q}\,(\text{I.F.})\, dx + c$$

$\Rightarrow \qquad y \cdot \dfrac{1}{x+1} = \int \dfrac{1}{x+1}\, dx + c$

$\Rightarrow \qquad \dfrac{y}{x+1} = \log|x+1| + c.$

(*iii*) The given equation is

$$\frac{dy}{dx} + 2y = 6e^x \qquad \text{...(1)}$$

Comparing it with $\dfrac{dy}{dx} + \text{P}y = \text{Q}$

where $\qquad \text{P} = 2$ and $\text{Q} = 6e^x$

$\therefore \qquad \text{I.F.} = e^{\int \text{P}\, .\, dx} = e^{\int 2\, .\, dx} = e^{2x}$

$\therefore$ The required solution is given by

$$y(\text{I.F.}) = \int \text{Q}\,(\text{I.F.})\, dx + c$$

$\Rightarrow \qquad ye^{2x} = \int 6e^x \cdot e^{2x}\, dx + c \qquad \Rightarrow \quad ye^{2x} = 6\int e^{3x}\, dx + c$

$\Rightarrow \qquad ye^{2x} = 6 \cdot \dfrac{e^{3x}}{3} + c \qquad \Rightarrow \quad ye^{2x} = 2e^{3x} + c$

$\Rightarrow \qquad y = 2e^x + ce^{-2x}.$

(*iv*) The given equation is

$$2y' + 4y = x^2 - x$$

$$\Rightarrow \qquad \frac{dy}{dx} + 2y = \left(\frac{x^2 - x}{2}\right) \qquad ...(1)$$

Comparing it with $\frac{dy}{dx} + Py = Q$

where $P = 2$ and $Q = \frac{1}{2}(x^2 - x)$

$\therefore$ $\text{I.F.} = e^{\int P\,.\,dx} = e^{\int 2\,.\,dx} = e^{2x}$

$\therefore$ The required solution is given by

$$y(\text{I.F.}) = \int Q(\text{I.F.})\,dx + c$$

$$\Rightarrow \qquad ye^{2x} = \int \frac{1}{2}(x^2 - x)\,e^{2x}\,dx + c$$

$$= \frac{1}{2}\int \underset{\text{I}}{(x^2 - x)}\,\underset{\text{II}}{e^{2x}}\,dx + c \qquad \text{[Integrating by parts]}$$

$$= \frac{1}{2}\left[(x^2 - x)\,.\,\frac{e^{2x}}{2} - \int (2x - 1)\,.\,\frac{e^{2x}}{2}\,dx\right] + c$$

$$= \frac{1}{4}(x^2 - x)\,e^{2x} - \frac{1}{4}\int \underset{\text{I}}{(2x - 1)}\,\underset{\text{II}}{e^{2x}}\,dx + c$$

[Integrating again by parts]

$$= \frac{1}{4}(x^2 - x)e^{2x} - \frac{1}{4}\left[(2x - 1)\,.\,\frac{e^{2x}}{2} - \int 2\,.\,\frac{e^{2x}}{2}\,dx\right] + c$$

$$= \frac{1}{4}(x^2 - x)e^{2x} - \frac{1}{8}(2x - 1)e^{2x} + \frac{1}{4}\int e^{2x}\,dx + c$$

$$= \frac{1}{4}(x^2 - x)e^{2x} - \frac{1}{8}(2x - 1)e^{2x} + \frac{1}{4}\,.\,\frac{e^{2x}}{2} + c$$

$$= \frac{1}{8}e^{2x}[2x^2 - 2x - 2x + 1 + 1] + c$$

$$= \frac{1}{8}e^{2x}(2x^2 - 4x + 2) + c$$

$$\Rightarrow \qquad ye^{2x} = \frac{1}{4}e^{2x}(x^2 - 2x + 1) + c$$

$$\Rightarrow \qquad y = \frac{1}{4}(x - 1)^2 + ce^{-2x}. \qquad \text{[Dividing throughout by } e^{2x}\text{]}$$

(*v*) The given equation is

$$x\frac{dy}{dx} - 3y = -2mx \quad \Rightarrow \quad \frac{dy}{dx} - \frac{3}{x}y = -2m \qquad ...(1)$$

Comparing it with $\frac{dy}{dx} + Py = Q$

where $$P = -\frac{3}{x} \quad \text{and} \quad Q = -2m$$

$\therefore$ $$\text{I.F.} = e^{\int P.dx} = e^{\int \frac{-3}{x} dx} = e^{-3\int \frac{1}{x} dx} = e^{-3\log|x|}$$

$$= e^{\log|x^{-3}|} = x^{-3} = \frac{1}{x^3}$$

$\therefore$ The required solution is given by

$$y\,.\,(\text{I.F.}) = \int Q\,.\,(\text{I.F.})\,dx + c$$

$\Rightarrow$ $$y\,.\,\frac{1}{x^3} = \int -2m\left(\frac{1}{x^3}\right) dx + c$$

$\Rightarrow$ $$\frac{y}{x^3} = -2m\int x^{-3}\,dx + c$$

$\Rightarrow$ $$\frac{y}{x^3} = -2m\left(\frac{x^{-3+1}}{-3+1}\right) + c = \frac{m}{x^2} + c$$

$\Rightarrow$ $$y = mx + cx^3.$$

(*vi*) The given equation is

$$4\frac{dy}{dx} + 8y = 5e^{-3x} \quad \Rightarrow \quad \frac{dy}{dx} + 2y = \frac{5}{4}e^{-3x} \qquad ...(1)$$

Comparing it with $\frac{dy}{dx} + Py = Q$

where $$P = 2 \quad \text{and} \quad Q = \frac{5}{4}e^{-3x}$$

$\therefore$ $$\text{I.F.} = e^{\int P\,.\,dx} = e^{\int 2\,.\,dx} = e^{2x}$$

$\therefore$ The required solution is given by

$$y\,.\,(\text{I.F.}) = \int Q\,.\,(\text{I.F.})\,dx + c$$

$\Rightarrow$ $$ye^{2x} = \int \frac{5}{4}e^{-3x}\,.\,e^{2x}\,dx + c = \frac{5}{4}\int e^{-x}\,dx + c$$

$\Rightarrow$ $$ye^{2x} = \frac{-5}{4}e^{-x} + c$$

$\Rightarrow$ $$y = \frac{-5}{4}e^{-3x} + ce^{-2x}.$$ [Dividing throughout by e^{2x}]

Example 8. *Solve the following differential equations :*

(*i*) $\frac{dy}{dx} + \frac{y}{2x} = 3x^2$ (*ii*) $\frac{dy}{dx} + my = e^{ax}$

(*iii*) $x\frac{dy}{dx} - y = (x-1)e^x$ (*iv*) $\frac{dy}{dx} + y = \sin x + \cos x$

(*v*) $(1+x^2)\frac{dy}{dx} + 2xy = 4x^2$ (*vi*) $\frac{dy}{dx} + y\sec x = \tan x.$

Solution. (*i*) The given equation is

$$\frac{dy}{dx} + \frac{y}{2x} = 3x^2 \qquad ...(1)$$

Comparing it with $\frac{dy}{dx} + Py = Q$

where $P = \frac{1}{2x}$ and $Q = 3x^2$

$\therefore$ $$\text{I.F.} = e^{\int P \,.\, dx} = e^{\int \frac{1}{2x} dx} = e^{\frac{1}{2}\int \frac{1}{x} dx} = e^{\frac{1}{2}\log|x|} = e^{\log|x^{1/2}|} = x^{1/2}$$

$\therefore$ The required solution is given by

$$y \,.\, (\text{I.F.}) = \int Q \,.\, (\text{I.F.})\, dx + c$$

$\Rightarrow$ $$yx^{1/2} = \int 3x^2 \,.\, x^{1/2}\, dx + c = 3\int x^{5/2}\, dx + c$$

$$= \frac{3x^{\frac{5}{2}+1}}{\frac{5}{2}+1} + c = 3 \,.\, \frac{2}{7} x^{7/2} + c$$

$\Rightarrow$ $$y\sqrt{x} = \frac{6}{7} x^{7/2} + c.$$

(*ii*) The given equation is $\frac{dy}{dx} + my = e^{ax}$ ...(1)

Comparing it with $\frac{dy}{dx} + Py = Q$

where $P = m$ and $Q = e^{ax}$

$\therefore$ $$\text{I.F.} = e^{\int P \,.\, dx} = e^{\int m \,.\, dx} = e^{mx}$$

$\therefore$ The required solution is given by

$$y \,.\, (\text{I.F.}) = \int Q \,.\, (\text{I.F.})\, dx + c$$

$\Rightarrow$ $$ye^{mx} = \int e^{ax} \,.\, e^{mx}\, dx + c = \int e^{(a+m)x} \,.\, dx + c$$

$\Rightarrow$ $$ye^{mx} = \frac{e^{(a+m)x}}{a+m} + c$$

$\Rightarrow$ $$y = \frac{e^{ax+mx}}{a+m} \,.\, e^{-mx} + ce^{-mx}$$ [Dividing throughout by e^{mx}]

$\Rightarrow$ $$y = \frac{e^{ax}}{a+m} + ce^{-mx}.$$

(*iii*) The given equation is

$$x\frac{dy}{dx} - y = (x-1)e^x \;\Rightarrow\; \frac{dy}{dx} - \frac{y}{x} = \left(\frac{x-1}{x}\right)e^x \qquad ...(1)$$

Comparing it with $\frac{dy}{dx} + \text{P}y = \text{Q}$

where $\text{P} = \frac{-1}{x}$ and $\text{Q} = \left(\frac{x-1}{x}\right)e^x$

$\therefore$ $\text{I.F.} = e^{\int \text{P}\,.\,dx} = e^{-\int \frac{1}{x} dx} = e^{-\log|x|} = e^{\log|x^{-1}|} = x^{-1} = \frac{1}{x}.$

$\therefore$ The required solution is given by

$$y\,.\,(\text{I.F.}) = \int \text{Q}\,.\,(\text{I.F.})\,dx + c$$

$$\Rightarrow \quad y\left(\frac{1}{x}\right) = \int \left(\frac{x-1}{x}\right) e^x \,.\, \frac{1}{x}\,dx + c$$

$$\Rightarrow \quad \frac{y}{x} = \int e^x \left(\frac{x-1}{x^2}\right) dx + c = \int e^x \left[\frac{1}{x} + \left(\frac{-1}{x^2}\right)\right] dx + c$$

$$\Rightarrow \quad \frac{y}{x} = e^x \,.\, \frac{1}{x} + c \qquad \left[\because \int e^x [f(x) + f'(x)]\,dx = e^x f(x)\right]$$

$$\Rightarrow \quad y = e^x + cx.$$

(*iv*) The given equation is

$$\frac{dy}{dx} + y = \sin x + \cos x \qquad ...(1)$$

Comparing it with $\frac{dy}{dx} + \text{P}y = \text{Q}$

where $\text{P} = 1$ and $\text{Q} = (\sin x + \cos x)$

$\therefore$ $\text{I.F.} = e^{\int \text{P}.\,dx} = e^{\int 1\,.\,dx} = e^x$

$\therefore$ The required solution is given by

$$y\,.\,(\text{I.F.}) = \int \text{Q}\,.\,(\text{I.F.})\,dx + c$$

$$\Rightarrow \quad ye^x = \int (\sin x + \cos x)\,e^x\,dx + c \qquad \left[\because \int e^x [f(x) + f'(x)]\,dx = e^x f(x)\right]$$

$$\Rightarrow \quad ye^x = e^x \sin x + c$$

$$\Rightarrow \quad y = \sin x + ce^{-x}. \qquad \text{[Dividing throughout by } e^x\text{]}$$

(*v*) The given equation is

$$(1 + x^2)\frac{dy}{dx} + 2xy = 4x^2$$

$$\Rightarrow \quad \frac{dy}{dx} + \frac{2x}{1+x^2}\,y = \frac{4x^2}{1+x^2} \qquad ...(1)$$

Comparing it with $\frac{dy}{dx} + \text{P}y = \text{Q}$

where $\text{P} = \frac{2x}{1+x^2}$ and $\text{Q} = \frac{4x^2}{1+x^2}$

$$\therefore \qquad \text{I.F.} = e^{\int P\,.\,dx} = e^{\int \frac{2x}{1+x^2} dx} = e^{\log|1+x^2|} = (1 + x^2)$$

$$\left[\because \int \frac{f'(x)}{f(x)}\,dx = \log|f(x)|\right]$$

$\therefore$ The required solution is given by

$$y\,(\text{I.F.}) = \int Q\,.\,(\text{I.F})\,dx + c$$

$$\Rightarrow \qquad y\,(1 + x^2) = \int \frac{4x^2}{(1+x^2)}\,.\,(1 + x^2)\,dx + c$$

$$\Rightarrow \qquad y(1 + x^2) = 4\int x^2\,dx + c$$

$$\Rightarrow \qquad y(1 + x^2) = \frac{4}{3}x^3 + c.$$

(*vi*) The given equation is

$$\frac{dy}{dx} + y \sec x = \tan x \qquad ...(1)$$

Comparing it with $\dfrac{dy}{dx} + \text{P}y = \text{Q}$

where $\text{P} = \sec x$ and $\text{Q} = \tan x$

$$\therefore \qquad \text{I.F.} = e^{\int P\,.\,dx} = e^{\int \sec x\,dx} = e^{\log|\sec x + \tan x|} = (\sec x + \tan x)$$

$\therefore$ The required solution is given by

$$y\,.\,(\text{I.F.}) = \int Q\,.\,(\text{I.F.})\,dx + c$$

$$\Rightarrow \qquad y(\sec x + \tan x) = \int \tan x\,(\sec x + \tan x)\,dx + c$$

$$= \int \sec x \tan x\,dx + \int \tan^2 x\,dx + c$$

$$= \int \sec x\ \tan x\,dx + \int (\sec^2 x - 1)\,.\,dx + c$$

$$= \int \sec x \tan x\,dx + \int \sec^2 x\,dx - \int 1\,.\,dx + c$$

$$\Rightarrow \qquad y(\sec x + \tan x) = \sec x + \tan x - x + c.$$

Example 9. *Solve the following differential equations :*

(*i*) $\dfrac{dy}{dx} + y = (2 - x)$ (*ii*) $\dfrac{dy}{dx} + 2xy = 4x^3$

(*iii*) $\dfrac{dy}{dx} = y \tan x - 2 \sin x$ (*iv*) $\dfrac{dy}{dx} - 2y = \cos 3x$

(*v*) $x \log x \dfrac{dy}{dx} + y = \dfrac{2}{x} \log x$ (*vi*) $\cos^2 x \dfrac{dy}{dx} + y = \tan x.$

Solution. (*i*) The given equation is

$$\frac{dy}{dx} + y = (2 - x) \qquad ...(1)$$

Comparing it with $\frac{dy}{dx} + Py = Q$

where $P = 1$ and $Q = (2 - x)$.

$\therefore$ $$\text{I.F.} = e^{\int P.dx} = e^{\int 1.dx} = e^x$$

$\therefore$ The required solution is given by

$$y \,.\, (\text{I.F.}) = \int Q \,.\, (\text{I.F.})\, dx + c$$

$$\Rightarrow \quad y\, e^x = \int \underset{\text{I}}{(2 - x)}\, \underset{\text{II}}{e^{+x}}\, dx + c \qquad \text{[Integrating by parts]}$$

$$= (2 - x) \,.\, e^x - \int (-1)\, e^x\, dx + c$$

$$= (2 - x)\, e^x + e^x + c = (2 - x)\, e^x + e^x + c$$

$$\Rightarrow \quad y = (2 - x) + 1 + c\, e^{-x} \qquad \text{[Dividing throughout by } e^x]$$

$$\Rightarrow \quad y = 3 - x + c\, e^{-x}.$$

(*ii*) The given equation is

$$\frac{dy}{dx} + 2xy = 4x^3 \qquad \text{...(1)}$$

Comparing it with $\frac{dy}{dx} + Py = Q$

where $P = 2x$ and $Q = 4x^3$

$\therefore$ $$\text{I.F.} = e^{\int P.dx} = e^{\int 2x.dx} = e^{2\frac{x^2}{2}} = e^{x^2}$$

$\therefore$ The required solution is given by

$$y(\text{I.F.}) = \int Q\, (\text{I.F.}) \,.\, dx + c$$

$$\Rightarrow \quad ye^{x^2} = \int 4x^3 \,.\, e^{x^2}\, dx + c$$

$$\Rightarrow \quad ye^{x^2} = 4 \int x^3 \,.\, e^{x^2}\, dx + c = 4 \int x^2 \,.\, e^{x^2} \,.\, x dx + c$$

Put $x^2 = z \Rightarrow 2x\, dx = dz \Rightarrow x\, dx = \frac{1}{2} dz$

$\therefore$ $$y\, e^{x^2} = \frac{4}{2} \int z e^z\, dz + c$$

$$= 2 \int \underset{\text{I}}{z}\, \underset{\text{II}}{e^z}\, dz + c \qquad \text{[Integrating by parts]}$$

$$= 2 \left[z \,.\, e^z - \int 1 \,.\, e^z\, dz \right] + c = 2z\, e^z - 2\, e^z + c$$

$$= 2\, e^z(z - 1) + c$$

$$= 2\, e^{x^2} (x^2 - 1) + c \qquad [\because \; z = x^2]$$

$$\Rightarrow \quad y\, e^{x^2} = 2\, e^{x^2} (x^2 - 1) + c$$

$$\Rightarrow \quad y = 2(x^2 - 1) + c\, e^{-x^2}. \qquad \text{[Dividing throughout by } e^{x^2}]$$

(*iii*) The given equation is

$$\frac{dy}{dx} = y \tan x - 2 \sin x$$

$$\Rightarrow \quad \frac{dy}{dx} - y(\tan x) = -2 \sin x \qquad ...(1)$$

Comparing it with $\frac{dy}{dx} + Py = Q$

where $\quad P = -\tan x \quad$ and $\quad Q = -2 \sin x$

$$\therefore \quad \text{I.F.} = e^{\int P.dx} = e^{-\int \tan x\, dx} = e^{-(-\log|\cos x|)} = e^{\log|(\cos x)|} = \cos x$$

$\therefore$ The required solution is given by

$$y.(\text{I. F.}) = \int Q.(\text{I.F.})\, dx + c$$

$$\Rightarrow \quad y \cos x = \int -2 \sin x \cos x\, dx + c$$

$$\Rightarrow \quad y \cos x = -\int \sin 2x\, dx + c$$

$$\Rightarrow \quad y \cos x = -\frac{(-\cos 2x)}{2} + c$$

$$\Rightarrow \quad y \cos x = \frac{1}{2} \cos 2x + c.$$

(*iv*) The given equation is

$$\frac{dy}{dx} - 2y = \cos 3x \qquad ...(1)$$

Comparing it with $\frac{dy}{dx} + Py = Q$

where $\quad P = -2 \quad$ and $\quad Q = \cos 3x$

$$\therefore \quad \text{I.F.} = e^{\int P.dx} = e^{\int -2.dx} = e^{-2\int 1.dx} = e^{-2x}$$

$\therefore$ The required solution is given by

$$y(\text{I.F.}) = \int Q.(\text{I.F.})\, dx + c$$

$$\Rightarrow \quad y\, e^{-2x} = \int \cos 3x . e^{-2x}\, dx + c$$

$$\Rightarrow \quad y\, e^{-2x} = \int e^{-2x} . \cos 3x\, dx + c \qquad ...(2)$$

Let $\quad I = \int \underset{\text{I}}{e^{-2x}} . \underset{\text{II}}{\cos 3x}\, dx \qquad$ (Integrating by parts)

$$= e^{-2x} . \frac{\sin 3x}{3} - \int -2e^{-2x} . \frac{\sin 3x}{3}\, dx$$

$$= \frac{1}{3} e^{-2x} \sin 3x + \frac{2}{3} \int \underset{\text{I}}{e^{-2x}}\, \underset{\text{II}}{\sin 3x}\, dx \qquad \text{(Integrating again by parts)}$$

$$= \frac{1}{3} e^{-2x} \sin 3x + \frac{2}{3}\left[e^{-2x} . \left(\frac{-\cos 3x}{3}\right) - \int -2e^{-2x} . \left(\frac{-\cos 3x}{3}\right) dx\right]$$

$$= \frac{1}{3} e^{-2x} \sin 3x - \frac{2}{9} e^{-2x} \cos 3x - \frac{4}{9} \int e^{-2x} \cos 3x \, dx$$

$$\therefore \quad I = \frac{1}{9} e^{-2x} [3 \sin 3x - 2 \cos 3x] - \frac{4}{9} I$$

$$\Rightarrow \quad I + \frac{4}{9} I = \frac{1}{9} e^{-2x} [3 \sin 3x - 2 \cos 3x]$$

$$\Rightarrow \quad \frac{13}{9} I = \frac{1}{9} e^{-2x} [3 \sin 3x - 2 \cos 3x]$$

$$\Rightarrow \quad I = \frac{1}{13} e^{-2x} [3 \sin 3x - 2 \cos 3x] \qquad ...(3)$$

Putting this value of equation (3) in (2), we have

$$y e^{-2x} = \frac{1}{13} e^{-2x} [3 \sin 3x - 2 \cos 3x] + c \qquad [\because \text{ Dividing throughout by } e^{-3x}]$$

$$\Rightarrow \quad y = \frac{1}{13} [3 \sin 3x - 2 \cos 3x] + c\, e^{-2x}.$$

(v) The given equation is

$$x \log x \frac{dy}{dx} + y = \frac{2}{x} \log x$$

$$\Rightarrow \quad \frac{dy}{dx} + \frac{1}{x \log x} y = \frac{2 \log x}{x^2 \log x}$$

$$\Rightarrow \quad \frac{dy}{dx} + \frac{1}{x \log x} y = \frac{2}{x^2} \qquad ...(1)$$

Comparing it with $\frac{dy}{dx} + Py = Q$

where $\qquad P = \frac{1}{x \log x} \quad$ and $\quad Q = \frac{2}{x^2}$

$$\therefore \quad \text{I.F.} = e^{\int P.dx} = e^{\int \frac{1}{x \log x} dx} = e^{\int \frac{1/x}{\log x} dx} \qquad \left[\because \int \frac{f'(x)}{f(x)} dx = \log |f(x)|\right]$$

$$= e^{\log |\log x|} = \log x$$

$\therefore$ The required solution is given by

$$y(\text{I.F.}) = \int Q.(\text{I.F.})\, dx + c$$

$$\Rightarrow \quad y \log x = \int \frac{2}{x^2} . \log x \, dx + c = 2 \int \underset{\text{I}}{x^{-2}} \underset{\text{II}}{\log x} \, dx + c \qquad \text{[Integrating by parts]}$$

$$= 2 \left[\log x . \left(\frac{x^{-1}}{-1} \right) - \int \frac{1}{x} \left(\frac{x^{-1}}{-1} \right) dx \right] + c$$

$$= \frac{-2}{x} \log x + 2 \int x^{-2} \, dx + c = \frac{-2 \log x}{x} + \frac{2x^{-1}}{-1} + c$$

$$\Rightarrow \quad y \log x = \frac{-2 \log x}{x} - \frac{2}{x} + c$$

$$\Rightarrow \quad xy \log x = -2 \log x - 2 + cx.$$

(*vi*) The given equation is

$$\cos^2 x \frac{dy}{dx} + y = \tan x \qquad ...(1)$$

$$\Rightarrow \quad \frac{dy}{dx} + \frac{1}{\cos^2 x} y = \frac{1}{\cos^2 x} . \tan x$$

$$\Rightarrow \quad \frac{dy}{dx} + (\sec^2 x)y = \sec^2 x \tan x$$

Comparing it with $\frac{dy}{dx} + \text{P}y = \text{Q}$

where $\text{P} = \sec^2 x$ and $\text{Q} = \sec^2 x \tan x$

$$\therefore \quad \text{I.F.} = e^{\int \text{P}.dx} = e^{\int \sec^2 x\, dx} = e^{\tan x}$$

$\therefore$ The required solution is given by

$$y . (\text{I.F.}) = \int \text{Q}\,(\text{I.F.})\, dx + c$$

$$\Rightarrow \quad y\, e^{\tan x} = \int (\sec^2 x \tan x) . e^{\tan x}\, dx + c$$

$$= \int \tan x . e^{\tan x} . \sec^2 x\, dx + c$$

Put $\tan x = z \Rightarrow \sec^2 x\, dx = dz$

$$\therefore \quad ye^{\tan x} = \int \underset{\text{I}}{z}\, \underset{\text{II}}{e^z}\, dz + c \qquad \text{(Integrating by parts)}$$

$$= \left[z . e^z - \int 1 . e^z\, dz\right] + c = z\, e^z - \int e^z\, dz + c$$

$$= z\, e^z - e^z + c$$

$$\Rightarrow \quad y\, e^{\tan x} = e^{\tan x}(\tan x - 1) + c \qquad [\because \ z = \tan x]$$

$$\Rightarrow \quad y = (\tan x - 1) + c\, e^{-\tan x}. \qquad [\text{Dividing throughout by } e^{\tan x}]$$

Example 10. *Solve the following differential equations :*

(*i*) $\sec x \frac{dy}{dx} = y + \sin x$ (*ii*) $\frac{dy}{dx} + y \tan x = \sec x$

(*iii*) $x \log x \frac{dy}{dx} + y = 2 \log x$ (*iv*) $\frac{dy}{dx} + y \cot x = 2 \cos x$

(*v*) $(1 + x^2) \frac{dy}{dx} + 2xy = \cos x$ (*vi*) $(1 - x^2) \frac{dy}{dx} + xy = ax.$

Solution. (*i*) The given equation is

$$\sec x \frac{dy}{dx} = y + \sin x$$

$$\Rightarrow \quad \frac{dy}{dx} - \frac{1}{\sec x} y = \frac{\sin x}{\sec x}$$

$$\Rightarrow \quad \frac{dy}{dx} - (\cos x)\, y = \sin x \cos x \qquad ...(1)$$

Comparing it with $\frac{dy}{dx} + Py = Q$

where $P = -\cos x$ and $Q = \sin x \cos x$

$\therefore$ $\text{I.F.} = e^{\int P.dx} = e^{-\int \cos x\, dx} = e^{-\sin x}$

$\therefore$ The required solution is given by

$$y.(\text{I.F.}) = \int Q\,(\text{I.F.})\, dx + c$$

$$\Rightarrow \quad y\, e^{-\sin x} = \int \sin x \cos x \,.\, e^{-\sin x}\, dx + c$$

Put $\sin x = z \Rightarrow \cos x\, dx = dz$

$$\therefore \quad y\, e^{-\sin x} = \int \underset{\text{I}}{z}\, \underset{\text{II}}{e^{-z}}\, dz + c \qquad \text{(Integrating by parts)}$$

$$= z \,.\, \frac{e^{-z}}{-1} - \int 1 . \frac{e^{-z}}{-1} . dz + c = -z e^{-z} + \int e^{-z}\, dz + c$$

$$= -z e^{-z} - e^{-z} + c = -e^{-z}(z+1) + c$$

$$\Rightarrow \quad y\, e^{-\sin x} = -e^{-\sin x}(\sin x + 1) + c \qquad [\because \; z = \sin x]$$

$$\Rightarrow \quad y = -(\sin x + 1) + c\, e^{\sin x}. \qquad [\text{Dividing throughout by } e^{-\sin x}]$$

(*ii*) The given equation is

$$\frac{dy}{dx} + y \tan x = \sec x \qquad ...(1)$$

Comparing it with $\frac{dy}{dx} + Py = Q$

where $P = \tan x$ and $Q = \sec x$

$\therefore$ $\text{I.F.} = e^{\int P.dx} = e^{\int \tan x\, dx} = e^{\log|\sec x|} = \sec x$

$\therefore$ The required solution is given by

$$y\,.\,(\text{I.F.}) = \int Q\,(\text{I.F.})\, dx + c$$

$$\Rightarrow \quad y \sec x = \int \sec x \,.\, \sec x\, dx + c$$

$$\Rightarrow \quad y \sec x = \int \sec^2 x\, dx + c$$

$$\Rightarrow \quad y \sec x = \tan x + c$$

$$\Rightarrow \quad y = \frac{\tan x}{\sec x} + \frac{c}{\sec x}$$

$$\Rightarrow \quad y = \sin x + c \cos x.$$

(*iii*) The given equation is

$$x \log x \frac{dy}{dx} + y = 2 \log x$$

$$\Rightarrow \quad \frac{dy}{dx} + \frac{1}{x \log x} . y = \frac{2 \log x}{x \log x} \quad \Rightarrow \quad \frac{dy}{dx} + \frac{1}{x \log x} . y = \frac{2}{x} \qquad ...(1)$$

Comparing it with $\dfrac{dy}{dx} + \text{P}y = \text{Q}$

where $$\text{P} = \frac{1}{x \log x} \quad \text{and} \quad \text{Q} = \frac{2}{x}$$

$\therefore$ $$\text{I.F.} = e^{\int \text{P}.dx} = e^{\int \frac{1}{x \log x} dx} = e^{\int \frac{1/x}{\log x} dx} \qquad \left[\because \int \frac{f'(x)}{f(x)} dx = \log |f(x)|\right]$$

$$= e^{\log |\log x|} = \log x$$

$\therefore$ The required solution is given by

$$y \,.\, (\text{I.F.}) = \int \text{Q} \,.\, (\text{I.F.})\, dx + c$$

$$\Rightarrow \quad y \log x = \int \frac{2}{x} \log x \, dx + c$$

$$\Rightarrow \quad y \log x = 2 \int \frac{1}{x} \log x \, dx + c$$

$$\Rightarrow \quad y \log x = 2 \,.\, \frac{(\log x)^2}{2} + c \qquad \left[\because \int [f(x)]^n f'(x)\, dx = \frac{[f(x)]^{n+1}}{n+1}\right]$$

$$\Rightarrow \quad y \log x = (\log x)^2 + c\,.$$

(*iv*) The given equation is

$$\frac{dy}{dx} + y \cot x = 2 \cos x \qquad \text{...(1)}$$

Comparing it with $\dfrac{dy}{dx} + \text{P}y = \text{Q}$

where $$\text{P} = \cot x \quad \text{and} \quad \text{Q} = 2 \cos x$$

$\therefore$ $$\text{I.F.} = e^{\int \text{P}.dx} = e^{\int \cot x \, dx} = e^{\log |\sin x|} = \sin x$$

$\therefore$ The required solution is given by

$$y \,.\, (\text{I.F.}) = \int \text{Q} \,.\, (\text{I.F.})\, dx + c$$

$$\Rightarrow \quad y \sin x = \int 2 \cos x \,.\, \sin x \, dx + c = \int \sin 2x \, dx + c$$

$$\Rightarrow \quad y \sin x = -\frac{\cos 2x}{2} + c\,.$$

(*v*) The given equation is

$$(1 + x^2) \frac{dy}{dx} + 2xy = \cos x$$

$$\Rightarrow \quad \frac{dy}{dx} + \frac{2x}{1+x^2} y = \frac{1}{1+x^2} \,.\, \cos x$$

Comparing it with $\dfrac{dy}{dx} + \text{P}y = \text{Q}$

where $$\text{P} = \frac{2x}{1+x^2} \quad \text{and} \quad \text{Q} = \frac{1}{1+x^2} \,.\, \cos x$$

$\therefore$ $$\text{I.F.} = e^{\int P.dx} = e^{\int \frac{2x}{1+x^2} dx} = e^{\log|(1+x^2)|} = 1 + x^2 \qquad \left[\because \int \frac{f'(x)}{f(x)} dx = \log|f(x)|\right]$$

$\therefore$ The required solution is given by

$$y\,(\text{I.F.}) = \int Q\,(\text{I.F.})\,dx + c$$

$$\Rightarrow \quad y(1 + x^2) = \int \frac{1}{(1+x^2)} \cos x \,.\, (1 + x^2)\, dx + c = \int \cos x\, dx + c$$

$$\Rightarrow \quad y(1 + x^2) = \sin x + c.$$

(*vi*) The given equation is

$$(1 - x^2) \frac{dy}{dx} + xy = ax$$

$$\Rightarrow \quad \frac{dy}{dx} + \frac{x}{1-x^2} y = \frac{ax}{1-x^2} \qquad \text{...(1)}$$

Comparing it with $\frac{dy}{dx} + Py = Q$

where $$P = \frac{x}{1-x^2} \quad \text{and} \quad Q = \frac{ax}{1-x^2}$$

$\therefore$ $$\text{I.F.} = e^{\int P.dx} = e^{\int \frac{x}{1-x^2} dx} = e^{-\frac{1}{2}\int \frac{-2x}{1-x^2} dx} = e^{-\frac{1}{2}\log|1-x^2|} = e^{\log(1-x^2)^{-1/2}}$$

$$= e^{\log\left(\frac{1}{\sqrt{1-x^2}}\right)} = \frac{1}{\sqrt{1-x^2}}$$

$\therefore$ The required solution is given by

$$y\,.\,(\text{I.F.}) = \int Q\,(\text{I.F.})\,dx + c$$

$$\Rightarrow \quad y\,.\,\frac{1}{\sqrt{1-x^2}} = \int \frac{ax}{(1-x^2)} \cdot \frac{1}{\sqrt{1-x^2}}\,dx + c = \int \frac{ax}{(1-x^2)^{3/2}}\,dx + c$$

$$= -\frac{a}{2}\int (-2x)(1-x^2)^{-\frac{3}{2}}\,dx + c \qquad \left[\because \int [f(x)]^n\,[f'(x)]\,dx = \frac{[f(x)]^{n+1}}{n+1}\right]$$

$$= -\frac{a}{2}\left[\frac{(1-x^2)^{-\frac{3}{2}+1}}{-\frac{3}{2}+1}\right] + c = -\frac{a}{2}\left[\frac{(1-x^2)^{-1/2}}{-\frac{1}{2}}\right] + c$$

$$\Rightarrow \quad \frac{y}{\sqrt{1-x^2}} = \frac{a}{\sqrt{1-x^2}} + c$$

$$\Rightarrow \quad y = a + c\sqrt{1-x^2}\,.$$

Example 11. *Solve the following differential equations :*

(*i*) $\frac{dy}{dx} + y \cos x = \sin 2x$ (*ii*) $\frac{dy}{dx} + \frac{1}{1+x^2} y = \frac{\tan^{-1} x}{1+x^2}$

(*iii*) $\frac{dy}{dx} + y = \frac{1+\sin x}{1+\cos x}$ (*iv*) $\frac{dy}{dx} + \frac{2y}{x} = \frac{1}{x^2}$

(*v*) $(x+1)\frac{dy}{dx} - ny = e^x (x+1)^{n+1}$ (*vi*) $\frac{dy}{dx} + \frac{4x}{x^2+1} y = \frac{1}{(x^2+1)^3}$.

Solution. (*i*) The given equation is

$$\frac{dy}{dx} + y \cos x = \sin 2x$$

Comparing it with $\frac{dy}{dx} + Py = Q$

where $P = \cos x$ and $Q = \sin 2x$

$\therefore$ $\text{I.F.} = e^{\int P.dx} = e^{\int \cos x\, dx} = e^{\sin x}$

$\therefore$ The required solution is given by

$$y . (\text{I.F.}) = \int Q . (\text{I.F.})\, dx + c$$

$$\Rightarrow \quad y e^{\sin x} = \int \sin 2x . e^{\sin x}\, dx = \int 2 \sin x \cos x\, e^{\sin x}\, dx$$

Put $\sin x = z \Rightarrow \cos x\, dx = dz$

$$\therefore \quad ye^{\sin x} = 2 \int \underset{\text{I}}{z}\, \underset{\text{II}}{e^z}\, dz \qquad \text{(Integrating by parts)}$$

$$= 2\left[z\, e^z - \int 1 . e^z\, dz\right] + c = 2\left[ze^z - \int e^z\, dz\right] + c$$

$$= 2\,[z\, e^z - e^z] + c$$

$$\Rightarrow \quad y\, e^{\sin x} = 2\, e^{\sin x}\,[\sin x - 1] + c \qquad [\because \ z = \sin x]$$

$$\Rightarrow \quad y = 2(\sin x - 1) + ce^{-\sin x}. \qquad \text{[Dividing throughout by } e^{\sin x}]$$

(*ii*) The given equation is

$$\frac{dy}{dx} + \frac{1}{1+x^2} y = \frac{\tan^{-1} x}{1+x^2} \qquad ...(1)$$

Comparing it with $\frac{dy}{dx}$ $Py = Q$

where $P = \frac{1}{1+x^2}$ and $Q = \frac{\tan^{-1} x}{1+x^2}$

$\therefore$ $\text{I.F.} = e^{\int P.dx} = e^{\int \frac{1}{1+x^2} dx} = e^{\tan^{-1} x}$

$\therefore$ The required solution is given by

$$y . (\text{I.F.}) = \int Q\, (\text{I.F.})\, dx + c$$

$$\Rightarrow \quad y\, e^{\tan^{-1} x} = \int \frac{\tan^{-1} x}{1+x^2} . e^{\tan^{-1} x}\, dx + c$$

Put $\tan^{-1} x = z \Rightarrow \frac{1}{1+x^2} dx = dz$

$\therefore \quad y e^{\tan^{-1} x} = \int \underset{\text{I}}{z} \, \underset{\text{II}}{e^z} \, dz + c$ (Integrating by parts)

$$= z \cdot e^z - \int 1 \cdot e^z \, dz + c = z e^z - \int e^z \, dz + c$$

$$= z e^z - e^z + c = e^z (z - 1) + c$$

$\Rightarrow \quad y e^{\tan^{-1} x} = e^{\tan^{-1} x} (\tan^{-1} x - 1) + c$ [$\because \; z = \tan^{-1} x$]

$\Rightarrow \quad y = (\tan^{-1} x - 1) + c \, e^{-\tan^{-1} x}.$ [Dividing throughout by $e^{\tan^{-1} x}$]

(*iii*) The given equation is

$$\frac{dy}{dx} + y = \frac{1 + \sin x}{1 + \cos x} \quad \text{...(1)}$$

Comparing it with $\frac{dy}{dx} + \text{P}y = \text{Q}$

where $\text{P} = 1$ and $\text{Q} = \frac{1 + \sin x}{1 + \cos x}$

$\therefore \quad \text{I.F.} = e^{\int \text{P}.dx} = e^{\int 1.dx} = e^x$

$\therefore$ The required solution is given by

$$y \cdot (\text{I.F.}) = \int \text{Q} \cdot (\text{I.F.}) \, dx + c$$

$\Rightarrow \quad y e^x = \int \frac{1 + \sin x}{1 + \cos x} \cdot e^x \, dx + c$

$$= \int \frac{1 + 2 \sin \frac{x}{2} \cdot \cos \frac{x}{2}}{2 \cos^2 \frac{x}{2}} e^x \, dx + c \qquad \left[\begin{array}{l} \because \; \sin 2\text{A} = 2 \sin \text{A} \cos \text{A} \\ \Rightarrow \; \sin \text{A} = 2 \sin \frac{\text{A}}{2} \cos \frac{\text{A}}{2} \\ \because \; 1 + \cos 2\text{A} = 2 \cos^2 \text{A} \\ \Rightarrow \; 1 + \cos \text{A} = 2 \cos^2 \frac{\text{A}}{2} \end{array}\right]$$

$$= \int \left(\frac{1}{2} \sec^2 \frac{x}{2} + \tan \frac{x}{2} \right) e^x \, dx + c$$

$$= \int e^x \left(\tan \frac{x}{2} + \frac{1}{2} \sec^2 \frac{x}{2} \right) dx + c$$

$\Rightarrow \quad y e^x = e^x \tan \frac{x}{2} + c$ $\left[\because \; \int [f(x) + f'(x)] \cdot e^x \, dx = e^x f(x)\right]$

$\Rightarrow \quad y = \tan \frac{x}{2} + c \, e^{-x}.$ [Dividing throughout by e^x]

(*iv*) The given equation is

$$\frac{dy}{dx} + \frac{2}{x} y = \frac{1}{x^2} \quad \text{...(1)}$$

Comparing it with $\dfrac{dy}{dx} + \text{P}y = \text{Q}$

where $\text{P} = \dfrac{2}{x}$ and $\text{Q} = \dfrac{1}{x^2}$

$\therefore$ $\text{I.F.} = e^{\int \text{P}.dx} = e^{\int \frac{2}{x} dx} = e^{2 \log x} = e^{\log x^2} = x^2$

$\therefore$ The required solution is given by

$$y(\text{I.F.}) = \int \text{Q (I.F.)}\, dx + c$$

$$\Rightarrow \quad yx^2 = \int \frac{1}{x^2} . x^2\, dx + c$$

$$\Rightarrow \quad x^2 y = \int 1 . dx + c$$

$$\Rightarrow \quad x^2 y = x + c.$$

(*v*) The given equation is

$$(x + 1) \frac{dy}{dx} - ny = e^x (x + 1)^{n+1}$$

$$\Rightarrow \quad \frac{dy}{dx} - \frac{n}{x+1} y = \frac{e^x (x+1)^{n+1}}{x+1} \quad \Rightarrow \quad \frac{dy}{dx} - \frac{n}{x+1} y = e^x (x+1)^n$$

Comparing it with $\dfrac{dy}{dx} + \text{P}y = \text{Q}$

where $\text{P} = -\dfrac{n}{x+1}$ and $\text{Q} = e^x (x + 1)^n$

$\therefore$ $\text{I.F.} = e^{\int \text{P}.dx} = e^{-n \int \frac{1}{x+1} dx} = e^{-n \log (x+1)} = e^{\log (x+1)^{-n}}$

$$= (x + 1)^{-n} = \frac{1}{(x+1)^n}$$

$\therefore$ The required solution is given by

$$y . (\text{I.F.}) = \int \text{Q (I.F.)}\, dx + c$$

$$\Rightarrow \quad y . \frac{1}{(x+1)^n} = \int e^x (x+1)^n . \frac{1}{(x+1)^n}\, dx + c$$

$$\Rightarrow \quad \frac{y}{(x+1)^n} = \int e^x\, dx + c \quad \Rightarrow \quad \frac{y}{(x+1)^n} = e^x + c$$

$$\Rightarrow \quad y = (x + 1)^n (e^x + c).$$

(*vi*) The given equation is

$$\frac{dy}{dx} + \frac{4x}{x^2+1} y = \frac{1}{(x^2+1)^3} \qquad ...(1)$$

Comparing it with $\frac{dy}{dx} + Py = Q$

where $$P = \frac{4x}{x^2+1} \quad \text{and} \quad Q = \frac{1}{(x^2+1)^3}$$

$\therefore$ $$\text{I.F.} = e^{\int P\,.\,dx} = e^{2\int \frac{2x}{x^2+1}\,.\,dx} = e^{2\log(x^2+1)} = e^{\log(x^2+1)^2} = (x^2+1)^2$$

$\therefore$ The required solution is given by

$$y(\text{I.F.}) = \int Q(\text{I.F.})\,.\,dx + c$$

$$\Rightarrow \quad y(x^2+1)^2 = \int \frac{1}{(x^2+1)^3}\,.\,(x^2+1)^2\,dx + c = \int \frac{1}{(x^2+1)}\,dx + c$$

$$\Rightarrow \quad y(x^2+1)^2 = \tan^{-1} x + c.$$

Example 12. *Solve the following differential equations :*

(*i*) $\frac{dy}{dx} + y \tan x = x^2 \cos^2 x$ (*ii*) $(x^2-1)\frac{dy}{dx} + 2xy = \frac{1}{x^2-1}$

(*iii*) $\frac{dy}{dx} + 2y \tan x = \sin x$; *when* $y\left(\frac{\pi}{3}\right) = 0$

(*iv*) $\sin x \frac{dy}{dx} + y \cos x = 2 \sin^2 x \cos x$ (*v*) $\frac{dy}{dx} + \frac{y}{x} = \cos x + \frac{\sin x}{x}$

(*vi*) $\frac{dy}{dx} + y \tan x = 2x + x^2 \tan x.$

Solution. (*i*) The given equation is

$$\frac{dy}{dx} + y \tan x = x^2 \cos^2 x \qquad ...(1)$$

Comparing it with $\frac{dy}{dx} + Py = Q$

where $$P = \tan x \quad \text{and} \quad Q = x^2 \cos^2 x$$

$\therefore$ $$\text{I.F.} = e^{\int P\,.\,dx} = e^{\int \tan x\,dx} = e^{\log(\sec x)} = \sec x$$

$\therefore$ The required solution is given by

$$y(\text{I.F.}) = \int Q\,.\,(\text{I.F.})\,dx + c$$

$$\Rightarrow \quad y \sec x = \int x^2 \cos^2 x\,.\sec x\,dx + c$$

$$\Rightarrow \quad y \sec x = \int x^2 \cos^2 x\,.\,\frac{1}{\cos x}\,dx + c$$

$$= \int \underset{\text{I}}{x^2}\,\underset{\text{II}}{\cos x}\,dx + c \qquad \text{(Integrating by parts)}$$

$$= x^2\,.\,\sin x - \int 2x\,.\sin x\,dx + c$$

$$= x^2 \sin x - 2\int \underset{\text{I}}{x}\,.\,\underset{\text{II}}{\sin x}\,dx + c \qquad \text{(Integrating again by parts)}$$

$$= x^2 \sin x - 2\left[x.(-\cos x) - \int 1.(-\cos x)\, dx\right] + c$$

$$= x^2 \sin x + 2x \cos x - 2\int \cos x\, dx + c$$

$\Rightarrow \quad y \sec x = x^2 \sin x + 2x \cos x - 2 \sin x + c.$

(*ii*) The given equation is

$$(x^2 - 1)\frac{dy}{dx} + 2xy = \frac{1}{x^2 - 1}$$

$$\Rightarrow \quad \frac{dy}{dx} + \frac{2x}{x^2 - 1}y = \frac{1}{(x^2 - 1)^2} \quad \text{...(1)}$$

Comparing it with $\frac{dy}{dx} + \text{P}y = \text{Q}$

where $\quad \text{P} = \frac{2x}{x^2 - 1}$ and $\text{Q} = \frac{1}{(x^2 - 1)^2}$

$\therefore \quad \text{I.F.} = e^{\int \text{P}\,.\,dx} = e^{\int \frac{2x}{x^2 - 1}dx} = e^{\log (x^2 - 1)} = (x^2 - 1)$

$\therefore$ The required solution is given by

$$y\,.\,(\text{I.F.}) = \int \text{Q}\,.\,(\text{I.F.})\, dx + c$$

$$\Rightarrow \quad y(x^2 - 1) = \int \frac{1}{(x^2 - 1)^2}\,.\,(x^2 - 1)\, dx + c$$

$$\Rightarrow \quad y(x^2 - 1) = \int \frac{1}{x^2 - 1}\, dx + c$$

$$\Rightarrow \quad y(x^2 - 1) = \frac{1}{2}\log\left|\frac{x - 1}{x + 1}\right| + c.$$

(*iii*) The given equation is

$$\frac{dy}{dx} + 2y \tan x = \sin x \quad \text{...(1)}$$

Comparing it with $\frac{dy}{dx} + \text{P}y = \text{Q}$

where $\quad \text{P} = 2 \tan x$ and $\text{Q} = \sin x$

$\therefore \quad \text{I.F.} = e^{\int \text{P}\,.\,dx} = e^{2\int \tan x} = e^{2 \log \sec x} = e^{\log (\sec^2 x)} = \sec^2 x$

$\therefore$ The required solution is given by

$$y\,.\,(\text{I.F.}) = \int \text{Q}\,.\,(\text{I.F.})\, dx + c$$

$$\Rightarrow \quad y \sec^2 x = \int \sin x\,.\,\sec^2 x\, dx + c$$

$$= \int \sin x \cdot \frac{1}{\cos^2 x}\, dx + c = \int \frac{\sin x}{\cos x} \cdot \frac{1}{\cos x}\, dx + c$$

$$= \int \sec x \tan x \, dx + c$$

$$\Rightarrow \quad y \sec^2 x = \sec x + c$$

$$\Rightarrow \quad \frac{y}{\cos^2 x} = \frac{1}{\cos x} + c \quad \Rightarrow \quad y = \cos x + c \cos^2 x \quad \text{...(2)}$$

Now, it is given that $y\left(\frac{\pi}{3}\right) = 0$ *i.e.*, $y = 0$ at $x = \frac{\pi}{3}$.

$\therefore$ Equation (2) becomes

$$0 = \cos \frac{\pi}{3} + c \cos^2 \frac{\pi}{3}$$

$$\Rightarrow \quad 0 = \frac{1}{2} + c\left(\frac{1}{2}\right)^2 \quad \Rightarrow \quad \frac{1}{4} c = \frac{-1}{2} \quad \Rightarrow \quad c = -2$$

Putting this value of c in equation (2), we get

$$y = \cos x - 2 \cos^2 x.$$

(*iv*) The given equation is

$$\sin x \frac{dy}{dx} + y \cos x = 2 \sin^2 x \cos x$$

$$\Rightarrow \quad \frac{dy}{dx} + y \frac{\cos x}{\sin x} = \frac{2 \sin^2 x \cos x}{\sin x}$$

$$\Rightarrow \quad \frac{dy}{dx} + y \cot x = 2 \sin x \cos x$$

Comparing it with $\frac{dy}{dx} + \mathrm{P}y = \mathrm{Q}$

where $\mathrm{P} = \cot x$ and $\mathrm{Q} = 2 \sin x \cos x$

$$\therefore \quad \text{I.F.} = e^{\int \mathrm{P}\, .\, dx} = e^{\int \cot x\, dx} = e^{\log (\sin x)} = \sin x$$

$\therefore$ The required solution is given by

$$y \,.\, (\text{I.F.}) = \int \mathrm{Q} \,.\, (\text{I.F.})\, dx + c$$

$$\Rightarrow \quad y \sin x = \int 2 \sin x \cos x \,.\, \sin x \, dx + c$$

$$\Rightarrow \quad y \sin x = 2 \int \sin^2 x \cos x \, dx + c \quad \left[\because \int [f(x)]^n f'(x)\, dx = \frac{[f(x)]^{n+1}}{n+1}\right]$$

$$\Rightarrow \quad y \sin x = \frac{2}{3} \sin^3 x + c.$$

(*v*) The given equation is

$$\frac{dy}{dx} + \frac{y}{x} = \cos x + \frac{\sin x}{x} \qquad ...(1)$$

Comparing it with $\dfrac{dy}{dx} + \text{P}y = \text{Q}$

where $\text{P} = \dfrac{1}{x}$ and $\text{Q} = \cos x + \dfrac{\sin x}{x}$

$$\therefore \qquad \text{I.F.} = e^{\int \text{P}.dx} = e^{\int \frac{1}{x} dx} = e^{\log x} = x$$

$\therefore$ The required solution is given by

$$y\,.\,(\text{I.F.}) = \int \text{Q}\,.\,(\text{I.F.})\,dx + c$$

$$\Rightarrow \qquad yx = \int x\left[\cos x + \frac{\sin x}{x}\right] dx + c$$

$$\Rightarrow \qquad xy = \int \underset{\text{I}}{x} \underset{\text{II}}{\cos x}\,dx + \int \sin x\,dx + c \qquad \text{(Integrating by parts)}$$

$$\Rightarrow \qquad xy = x\,.\,\sin x - \int 1\,.\,\sin x\,dx + \int \sin x\,dx + c$$

$$\Rightarrow \qquad xy = x\sin x - \int \sin x\,dx + \int \sin x\,dx + c$$

$$\Rightarrow \qquad y = \sin x + \frac{c}{x}.$$

(*vi*) The given equation is

$$\frac{dy}{dx} + y\tan x = 2x + x^2 \tan x \qquad ...(1)$$

Comparing it with $\dfrac{dy}{dx} + \text{P}y = \text{Q}$

where $\text{P} = \tan x$ and $\text{Q} = (2x + x^2 \tan x)$

$$\therefore \qquad \text{I.F.} = e^{\int \text{P}.dx} = e^{\int \tan x\,.\,dx} = e^{\log(\sec x)} = \sec x$$

$\therefore$ The required solution is given by

$$y\,.\,(\text{I.F.}) = \int \text{Q}\,.\,(\text{I.F.})\,dx + c$$

$$\Rightarrow \qquad y\sec x = \int (2x + x^2 \tan x)\,.\,\sec x\,dx + c$$

$$= \int 2x\sec x\,dx + \int \underset{\text{I}}{x^2} \tan \underset{\text{II}}{x \sec x}\,dx + c \qquad \text{[Integrating by parts]}$$

$$= \int 2x\sec x\,dx + x^2\sec x - \int 2x\,.\,\sec x\,dx + c$$

$$\Rightarrow \qquad y\sec x = x^2\sec x + c$$

$$\Rightarrow \qquad y = x^2 + c\cos x. \qquad \text{[Dividing throughout by } \sec x\text{]}$$

Example 13. *Solve the following differential equations :*

(*i*) $(x^2 - 1)\dfrac{dy}{dx} + 2(x + 2)y = 2(x + 1)$ (*ii*) $\dfrac{dy}{dx} + y = \dfrac{1 + x\log x}{x}$

(*iii*) $(1 + x^3)\dfrac{dy}{dx} + 6x^2y = (1 + x^2)$ (*iv*) $\sin 2x\dfrac{dy}{dx} = y + \tan x$

(*v*) $(x^2 + 1)\dfrac{dy}{dx} + 2xy = \sqrt{x^2 + 4}$ (*vi*) $x(x - 1)\dfrac{dy}{dx} - (x - 2)y = x^3(2x - 1)$.

Solution. (*i*) The given equation is

$$(x^2 - 1)\frac{dy}{dx} + 2(x + 2)y = 2(x + 1)$$

$$\Rightarrow \quad \frac{dy}{dx} + \frac{2(x+2)}{(x^2-1)}y = \frac{2(x+1)}{x^2-1}$$

$$\Rightarrow \quad \frac{dy}{dx} + \frac{2(x+2)}{(x^2-1)}y = \frac{2(x+1)}{(x+1)(x-1)}$$

$$\Rightarrow \quad \frac{dy}{dx} + \frac{2(x+2)}{x^2-1}y = \frac{2}{x-1} \qquad ...(1)$$

Comparing it with $\dfrac{dy}{dx} + Py = Q$

where $P = \dfrac{2(x+2)}{x^2-1}$ and $Q = \dfrac{2}{x-1}$

$$\therefore \quad \text{I.F.} = e^{\int P.dx} = e^{\int \frac{2(x+2)}{x^2-1}dx} = e^{\int \frac{2x+4}{(x-1)(x+1)}dx}$$

$$= e^{\int \left(\frac{3}{x-1} - \frac{1}{x+1}\right)dx} \qquad \text{[By partial fractions]}$$

$$= e^{3\int \frac{1}{x-1}dx - \int \frac{1}{x+1}dx} = e^{3\log(x-1) - \log(x+1)}$$

$$= e^{\log(x-1)^3 - \log(x+1)} = e^{\log\left[\frac{(x-1)^3}{x+1}\right]} = \frac{(x-1)^3}{x+1}$$

$\therefore$ The required solution is given by

$$y(\text{I.F.}) = \int Q.(\text{I.F.})\,dx + c$$

$$\Rightarrow \quad \frac{y(x-1)^3}{x+1} = \int \frac{2}{(x-1)}\cdot\frac{(x-1)^3}{(x+1)}dx + c$$

$$\Rightarrow \quad \frac{y(x-1)^3}{x+1} = \int \frac{2(x-1)^2}{x+1}dx + c = 2\int \frac{x^2-2x+1}{x+1}dx + c \quad \text{[By actual division]}$$

$$= 2\int\left(x - 3 + \frac{4}{x+1}\right)dx + c = 2\left[\frac{x^2}{2} - 3x + 4\log(x+1)\right] + c$$

$$\Rightarrow \qquad \frac{y(x-1)^3}{(x+1)} = x^2 - 6x + 8\log(x+1) + c.$$

(*ii*) The given equation is

$$\frac{dy}{dx} + y = \frac{1 + x\log x}{x} \qquad ...(1)$$

Comparing it with $\frac{dy}{dx} + Py = Q$

where $\qquad P = 1 \quad \text{and} \quad Q = \frac{1 + x\log x}{x}$

$\therefore \qquad \text{I.F.} = e^{\int P\,.\,dx} = e^{\int 1\,.\,dx} = e^x$

$\therefore$ The required solution is given by

$$y\,.\,(\text{I.F.}) = \int Q\,.\,(\text{I.F.})\,dx + c$$

$$\Rightarrow \qquad ye^x = \int \left[\frac{1 + x\log x}{x}\right].\,e^x\,dx + c$$

$$= \int \underset{\text{II}}{1/x}\ \underset{\text{I}}{e^x}\,dx + \int e^x \log x\,dx + c \qquad \text{[Integrating by parts]}$$

$$= e^x \log x - \int e^x \log x\,dx + \int e^x \log x + c$$

$$\Rightarrow \qquad ye^x = e^x \log x + c$$

$$\Rightarrow \qquad y = \log x + ce^{-x}. \qquad \text{[Dividing throughout by } e^x\text{]}$$

(*iii*) The given equation is

$$(1 + x^3)\frac{dy}{dx} + 6x^2 y = (1 + x^2)$$

$$\Rightarrow \qquad \frac{dy}{dx} + \frac{6x^2}{1+x^3}y = \frac{1+x^2}{1+x^3} \qquad ...(1)$$

Comparing it with $\frac{dy}{dx} + Py = Q$

where $\qquad P = \frac{6x^2}{1+x^3} \quad \text{and} \quad Q = \frac{1+x^2}{1+x^3}$

$$\therefore \qquad \text{I.F.} = e^{\int P\,.\,dx} = e^{\int \frac{6x^2}{1+x^3}dx} = e^{2\int \frac{3x^2}{1+x^3}dx}$$

$$= e^{2\log(1+x^3)} = e^{\log(1+x^3)^2} = (1 + x^3)^2$$

$\therefore$ The required solution is given by

$$y\,.\,(\text{I.F.}) = \int Q\,.\,(\text{I.F.})\,dx + c$$

$$\Rightarrow \qquad y(1 + x^3)^2 = \int \frac{(1+x^2)}{(1+x^3)}(1 + x^3)^2\,dx + c$$

$$= \int (1+x^2)(1+x^3)\, dx + c = \int (1+x^2+x^3+x^5)\, dx + c$$

$$\Rightarrow \quad y(1+x^3)^2 = x + \frac{x^3}{3} + \frac{x^4}{4} + \frac{x^6}{6} + c.$$

(*iv*) The given equation is

$$\sin 2x \frac{dy}{dx} = y + \tan x$$

$$\Rightarrow \quad \frac{dy}{dx} - \frac{1}{\sin 2x} y = \frac{\tan x}{\sin 2x}$$

$$\Rightarrow \quad \frac{dy}{dx} - (\text{cosec } 2x)y = \frac{\sin x}{\cos x \,.\, (2 \sin x \cos x)}$$

$$\Rightarrow \quad \frac{dy}{dx} - (\text{cosec } 2x)y = \frac{1}{2} \sec^2 x \qquad \ldots(1)$$

Comparing it with $\frac{dy}{dx} + \text{P}y = \text{Q}$

where $\text{P} = -\text{cosec } 2x$ and $\text{Q} = \frac{1}{2} \sec^2 x$

$$\therefore \quad \text{I.F.} = e^{\int \text{P}dx} = e^{-\int \text{cosec } 2x \,.\, dx}$$

$$= e^{-\frac{1}{2} \log (\tan x)}$$

$$= e^{\log (\tan x)^{-1/2}} = (\tan x)^{-1/2} = \frac{1}{\sqrt{\tan x}}$$

$\therefore$ The required solution is given by

$$y \,.\, (\text{I.F.}) = \int \text{Q} \,.\, (\text{I.F.}) \,.\, dx + c$$

$$\Rightarrow \quad y \,.\, \frac{1}{\sqrt{\tan x}} = \int \frac{1}{2} \sec^2 x \,.\, \frac{1}{\sqrt{\tan x}}\, dx + c$$

$$\Rightarrow \quad \frac{y}{\sqrt{\tan x}} = \frac{1}{2} \int (\tan x)^{-1/2} \,.\, \sec^2 x\, dx + c$$

$$\Rightarrow \quad \frac{y}{\sqrt{\tan x}} = \frac{1}{2} \left[\frac{(\tan x)^{\frac{-1}{2}+1}}{\frac{-1}{2}+1} \right] + c \qquad \left[\because \int [f(x)]^n \,.\, f'(x)\, dx = \frac{[f(x)]^{n+1}}{n+1} \right]$$

$$\Rightarrow \quad \frac{y}{\sqrt{\tan x}} = \sqrt{\tan x} + c \qquad \Rightarrow \quad y = \tan x + c\sqrt{\tan x}.$$

(*v*) The given equation is

$$(x^2+1) \frac{dy}{dx} + 2xy = \sqrt{x^2+4} \qquad \Rightarrow \quad \frac{dy}{dx} + \frac{2x}{x^2+1} y = \frac{\sqrt{x^2+4}}{x^2+1} \qquad \ldots(1)$$

Comparing it with $\frac{dy}{dx} + \text{P}y = \text{Q}$

where $$\text{P} = \frac{2x}{x^2+1} \quad \text{and} \quad \text{Q} = \frac{\sqrt{x^2+4}}{x^2+1}$$

$$\therefore \quad \text{I.F.} = e^{\int \text{P}\,.\,dx} = e^{\int \frac{2x}{x^2+1}dx} = e^{\log(x^2+1)} = (x^2+1)$$

$\therefore$ The required solution is given by

$$y\,.\,(\text{I.F.}) = \int \text{Q}\,.\,(\text{I.F.})\,dx + c$$

$$\Rightarrow \quad y(x^2+1) = \int \frac{\sqrt{x^2+4}}{(x^2+1)}\,.\,(x^2+1)\,dx + c$$

$$\Rightarrow \quad y(x^2+1) = \int \sqrt{x^2+4}\,dx + c$$

$$\left[\because \int \sqrt{x^2+a^2}\,dx = \frac{x\sqrt{x^2+a^2}}{2} + \frac{a^2}{2}\log|x+\sqrt{x^2+a^2}|\right]$$

$$\Rightarrow \quad y(x^2+1) = \frac{x}{2}\sqrt{x^2+4} + \frac{4}{2}\log|x+\sqrt{x^2+4}| + c$$

$$\Rightarrow \quad y(x^2+1) = x\sqrt{x^2+4} + 2\log|x+\sqrt{x^2+4}| + c.$$

(*vi*) The given equation is

$$x(x-1)\frac{dy}{dx} - (x-2)y = x^3(2x-1)$$

$$\Rightarrow \quad \frac{dy}{dx} - \frac{(x-2)}{x(x-1)}y = \frac{x^3(2x-1)}{x(x-1)} \qquad ...(1)$$

Comparing it with $\frac{dy}{dx} + \text{P}y = \text{Q}$

where $$\text{P} = \frac{-(x-2)}{x(x-1)} \quad \text{and} \quad \text{Q} = \frac{x^2(2x-1)}{x-1}$$

$$\therefore \quad \text{I.F.} = e^{\int \text{P}dx} = e^{-\int \frac{x-2}{x(x-1)}dx} = e^{-\int\left(\frac{2}{x}-\frac{1}{x-1}\right)dx}$$

$$= e^{-[2\log x - \log(x-1)]} = e^{-[\log x^2 - \log(x-1)]}$$

$$= e^{-\log\left(\frac{x^2}{x-1}\right)} = e^{\log\left(\frac{x^2}{x-1}\right)^{-1}} = \left(\frac{x^2}{x-1}\right)^{-1} = \frac{x-1}{x^2}$$

$\therefore$ The required solution is given by

$$y(\text{I.F.}) = \int \text{Q(I.F.)} \, . \, dx + c$$

$$\Rightarrow \quad \frac{y(x-1)}{x^2} = \int \frac{x^2(2x-1)}{(x-1)} \cdot \frac{(x-1)}{x^2} \, dx + c$$

$$= \int (2x-1) \, dx + c = \frac{2x^2}{2} - x + c$$

$$\Rightarrow \quad \frac{y(x-1)}{x^2} = x^2 - x + c$$

$$\Rightarrow \quad y(x-1) = x^2 \, (x^2 - x + c).$$

Example 14. *Solve the following differential equations :*

(*i*) $\dfrac{dy}{dx} + x \sin 2y = x^3 \cos^2 y$ (*ii*) $\dfrac{dy}{dx} = -\dfrac{x + y\cos x}{1 + \sin x}$

(*iii*) $x(x-1)\dfrac{dy}{dx} - y = x^2 (x-1)^2$ (*iv*) $(x \log x - y^2)\, dx + 2xy \, dy = 0$

(*v*) $\dfrac{dy}{dx} + \dfrac{2}{x} y = \sin x$ (*vi*) $\left(\dfrac{e^{-2\sqrt{x}}}{\sqrt{x}} - \dfrac{y}{\sqrt{x}}\right)\dfrac{dx}{dy} = 1.$

Solution. (*i*) The given equation is

$$\frac{dy}{dx} + x \sin 2y = x^3 \cos^2 y$$

$$\Rightarrow \quad \frac{1}{\cos^2 y}\frac{dy}{dx} + \frac{x \sin 2y}{\cos^2 y} = x^3$$

$$\Rightarrow \quad \sec^2 y \frac{dy}{dx} + \frac{x \, 2 \sin y \cos y}{\cos^2 y} = x^3$$

$$\Rightarrow \quad \sec^2 y \frac{dy}{dx} + 2x \tan y = x^3 \qquad \text{...(1)}$$

Put $\tan y = z \Rightarrow \sec^2 y \dfrac{dy}{dx} = \dfrac{dz}{dx}$

$\therefore$ Equation (1) becomes

$$\frac{dz}{dx} + 2xz = x^3 \qquad \text{...(2)}$$

Comparing it with $\dfrac{dz}{dx} + \text{P}z = \text{Q}$

where $\text{P} = 2x$ and $\text{Q} = x^3$

$$\therefore \quad \text{I.F.} = e^{\int \text{P}dx} = e^{\int 2x\, dx} = e^{2\left(\frac{x^2}{2}\right)} = e^{x^2}$$

$\therefore$ The required solution is given by

$$z\,.\,(\text{I.F.}) = \int Q\,(\text{I.F.})\,dx + c$$

$$\Rightarrow \quad ze^{x^2} = \int x^3\,.\,e^{x^2}\,dx + c = \int x^2\,.\,e^{x^2}\,.\,x\,dx + c$$

Put $x^2 = t$

$\Rightarrow \quad 2x\,dx = dt \Rightarrow \quad x\,dx = \frac{1}{2}\,dt$

$$= \frac{1}{2}\int \underset{\text{I}}{t}\,\underset{\text{II}}{e^t}\,dt + c \qquad \text{[Integrating by parts]}$$

$$= \frac{1}{2}\left[t\,e^t - \int 1\,.\,e^t\,dt\right] + c = \frac{1}{2}\left[t\,e^t - \int e^t\,dt\right] + c$$

$$= \frac{1}{2}\,[t\,e^t - e^t] + c = \frac{e^t}{2}\,[t-1] + c \qquad [\because \; t = x^2]$$

$$\Rightarrow \quad z\,e^{x^2} = \frac{1}{2}\,e^{x^2}\,(x^2-1) + c \qquad [\because \; z = \tan y]$$

$$\Rightarrow \quad \tan y\; e^{x^2} = \frac{1}{2}\,e^{x^2}\,(x^2-1) + c$$

$$\Rightarrow \quad \tan y = \frac{1}{2}(x^2-1) + ce^{-x^2}.$$

(*ii*) The given equation is

$$\frac{dy}{dx} = -\,\frac{x + y\cos x}{1+\sin x}$$

$$\Rightarrow \quad \frac{dy}{dx} + \frac{\cos x}{1+\sin x}\,y = -\,\frac{x}{1+\sin x} \qquad \text{...(1)}$$

Comparing it with $\frac{dy}{dx} + \text{P}y = \text{Q}$

where $\quad \text{P} = \frac{\cos x}{1+\sin x}, \quad \text{Q} = \frac{-\,x}{1+\sin x}$

$$\therefore \quad \text{I.F.} = e^{\int \text{P}.dx} = e^{\int \frac{\cos x}{1+\sin x}dx} = e^{\log(1+\sin x)} = (1+\sin x).$$

$\therefore$ The required solution is given by

$$y.\,(\text{I.F.}) = \int \text{Q}\,.\,(\text{I.F.})\,dx + c$$

$$\Rightarrow \quad y(1+\sin x) = \int \frac{-\,x}{(1+\sin x)}\,(1+\sin x)\,dx + c$$

$$\Rightarrow \quad y(1+\sin x) = -\int x\,dx + c$$

$$\Rightarrow \quad y(1+\sin x) = \frac{-\,x^2}{2} + c$$

$$\Rightarrow \qquad y = \frac{2c - x^2}{2(1 + \sin x)}.$$

(*iii*) The given equation is

$$x(x-1)\frac{dy}{dx} - y = x^2(x-1)^2$$

$$\Rightarrow \qquad \frac{dy}{dx} - \frac{1}{x(x-1)}y = \frac{x^2(x-1)^2}{x(x-1)}$$

$$\Rightarrow \qquad \frac{dy}{dx} - \frac{1}{x(x-1)}y = x(x-1) \qquad \text{...(1)}$$

Comparing it with $\frac{dy}{dx} + \text{P}y = \text{Q}$

where $\text{P} = \frac{-1}{x(x-1)}$ and $\text{Q} = x(x-1)$

$$\therefore \qquad \text{I.F.} = e^{\int \text{P}.dx} = e^{-\int \frac{1}{x(x-1)}dx} = e^{-\int\left(\frac{1}{x-1} - \frac{1}{x}\right)dx}$$

$$= e^{-[\log(x-1) - \log x]} = e^{-\log\left(\frac{x-1}{x}\right)}$$

$$= e^{\log\left(\frac{x-1}{x}\right)^{-1}} = \left(\frac{x-1}{x}\right)^{-1} = \frac{x}{x-1}$$

$\therefore$ The required solution is given by

$$y\,(\text{I.F.}) = \int \text{Q}\,(\text{I.F.})\,dx + c$$

$$\Rightarrow \qquad y\,.\,\frac{x}{x-1} = \int x(x-1)\,.\,\frac{x}{(x-1)}\,dx + c$$

$$\Rightarrow \qquad \frac{xy}{x-1} = \int x^2\,dx + c = \frac{x^3}{3} + c$$

$$\Rightarrow \qquad xy = (x-1)\left(\frac{x^3}{3} + c\right).$$

(*iv*) The given equation is

$$(x \log x - y^2)\,dx + 2xy\,dy = 0$$

$$\Rightarrow \qquad 2xy\,dy = (y^2 - x\log x)\,dx$$

$$\Rightarrow \qquad 2y\frac{dy}{dx} = \frac{y^2 - x\log x}{x}$$

$$\Rightarrow \qquad 2y\frac{dy}{dx} = \frac{y^2}{x} - \log x$$

$$\Rightarrow \qquad 2y\frac{dy}{dx} - \frac{y^2}{x} = -\log x \qquad \text{...(1)}$$

Put $y^2 = z \Rightarrow 2y\dfrac{dy}{dx} = \dfrac{dz}{dx}$

$\therefore$ Equation (1) becomes

$$\frac{dz}{dx} - \frac{1}{x}z = -\log x \qquad ...(2)$$

Comparing it with $\dfrac{dz}{dx} + Pz = Q$

where $\qquad P = -\dfrac{1}{x}$ and $Q = -\log x$

$\therefore \qquad \text{I.F.} = e^{\int P.dx} = e^{-\int \frac{1}{x}dx} = e^{-\log x} = e^{\log x^{-1}} = x^{-1} = \dfrac{1}{x}$

$\therefore$ The required solution is given by

$$z.(\text{I.F.}) = \int Q.(\text{I.F.})\,dx + c$$

$\Rightarrow \qquad z.\dfrac{1}{x} = \displaystyle\int -\log x.\left(\frac{1}{x}\right)dx + c$

$\Rightarrow \qquad \dfrac{z}{x} = -\displaystyle\int (\log x).\left(\frac{1}{x}\right)dx + c \qquad \left[\because \displaystyle\int [f(x)]^n f'(x)\,dx = \frac{[f(x)]^{n+1}}{n+1}\right]$

$\Rightarrow \qquad \dfrac{z}{x} = -\dfrac{(\log x)^2}{2} + c \qquad [\because z = y^2]$

$\Rightarrow \qquad y^2 = -\dfrac{x}{2}(\log x)^2 + c.$

(v) The given equation is

$$\frac{dy}{dx} + \frac{2}{x}y = \sin x \qquad ...(1)$$

Comparing it with $\dfrac{dy}{dx} + Py = Q$

where $\qquad P = \dfrac{2}{x}$ and $Q = \sin x$

$\therefore \qquad \text{I.F.} = e^{\int P.dx} = e^{2\int \frac{1}{x}dx} = e^{2\log x} = e^{\log x^2} = x^2$

$\therefore$ The required solution is given by

$$y.(\text{I.F.}) = \int Q.(\text{I.F.})\,dx + c$$

$\Rightarrow \qquad yx^2 = \displaystyle\int (\sin x)\,x^2\,dx + c$

$\Rightarrow \qquad x^2y = \displaystyle\int \underset{\text{I}}{x^2}\,\underset{\text{II}}{\sin x}\,dx + c \qquad$ (Integrating by parts)

$\Rightarrow \qquad x^2y = x^2.(-\cos x) - \displaystyle\int 2x.(-\cos x)\,dx + c$

$\qquad\qquad = -x^2\cos x + 2\displaystyle\int \underset{\text{I}}{x}\,\underset{\text{II}}{\cos x}\,dx + c \qquad$ [Integrating again by parts]

$$= -x^2 \cos x + 2\left[x \cdot \sin x - \int 1 \cdot \sin x\, dx\right] + c$$

$$= -x^2 \cos x + 2x \sin x - 2\int \sin x\, dx + c$$

$$\Rightarrow \qquad x^2 y = -x^2 \cos x + 2x \sin x + 2 \cos x + c.$$

(*vi*) The given equation is $\left(\dfrac{e^{-2\sqrt{x}}}{\sqrt{x}} - \dfrac{y}{\sqrt{x}}\right)\dfrac{dx}{dy} = 1$

$$\Rightarrow \qquad \frac{dy}{dx} = \frac{e^{-2\sqrt{x}}}{\sqrt{x}} - \frac{y}{\sqrt{x}} \qquad \Rightarrow \qquad \frac{dy}{dx} + \frac{1}{\sqrt{x}}\, y = \frac{e^{-2\sqrt{x}}}{\sqrt{x}} \qquad \text{...(1)}$$

Comparing it with $\dfrac{dy}{dx} + \text{P}y = \text{Q}$

where $\qquad \text{P} = \dfrac{1}{\sqrt{x}}$ and $\text{Q} = \dfrac{e^{-2\sqrt{x}}}{\sqrt{x}}$

$$\therefore \qquad \text{I.F.} = e^{\int \text{P}.dx} = e^{\int \frac{1}{\sqrt{x}} dx} = e^{\int x^{-1/2}.dx} = e^{\frac{x^{-\frac{1}{2}+1}}{-\frac{1}{2}+1}} = e^{2\sqrt{x}}$$

$\therefore$ The required solution is given by

$$y \cdot (\text{I.F.}) = \int \text{Q} \cdot (\text{I.F.})\, dx + c$$

$$\Rightarrow \qquad y\, e^{2\sqrt{x}} = \int \frac{e^{-2\sqrt{x}}}{\sqrt{x}} \cdot e^{2\sqrt{x}}\, dx + c = \int x^{-1/2}\, dx + c$$

$$\Rightarrow \qquad y\, e^{2\sqrt{x}} = 2\sqrt{x} + c.$$

Example 15. *Solve the following differential equations :*

(*i*) $y\, e^y\, dx = (y^3 + 2x\, e^y)\, dy\ ;\ y(0) = 1$

(*ii*) $\sqrt{1 - y^2}\, dx = (\sin^{-1} y - x)\, dy\ ;\ y(0) = 0$

(*iii*) $(1 + y^2)\, dx + (x - e^{-\tan^{-1} y})\, dy = 0\ ;\ y(0) = 0$

(*iv*) $y' + 2y = e^{-2x} \sin x\ ;\ y(0) = 0$

(*v*) $(dx + x\, dy) = e^{-y} \sec^2 y\, dy$

(*vi*) $y\, dx + (x - y^3)\, dy = 0.$

Solution. (*i*) The given equation is

$$y\, e^y\, dx = (y^3 + 2x\, e^y)\, dy$$

$$\Rightarrow \qquad \frac{dx}{dy} - \frac{2x}{y} = y^2 e^{-y} \qquad \text{...(1)}$$

Comparing it with $\dfrac{dx}{dy} + \text{P}x = \text{Q}$

where $\qquad \text{P} = -\dfrac{2}{y}$ and $\text{Q} = y^2 e^{-y}$

$$\therefore \qquad \text{I.F.} = e^{\int \text{P}.dy} = e^{-2\int \frac{1}{y} dy} = e^{-2 \log y} = e^{\log y^{-2}} = y^{-2} = \frac{1}{y^2}$$

$\therefore$ The required solution is given by

$$x \,.\, (\text{I.F.}) = \int \text{Q} \,.\, (\text{I. F.}) \,.\, dy + c$$

$$\Rightarrow \quad x \,.\, \frac{1}{y^2} = \int y^2 \,.\, e^{-y} \,.\, \frac{1}{y^2}\, dy + c = \int e^{-y}\, dy + c$$

$$\Rightarrow \quad \frac{x}{y^2} = -\, e^{-y} + c \qquad ...(2)$$

Now, it is given that

$$y(0) = 1 \quad \textit{i.e.,} \quad x = 0, \quad y = 1$$

$\therefore$ Equation (2) becomes

$$0 = -\, e^{-1} + c \quad \Rightarrow \quad c = e^{-1}$$

Putting this value of c in equation (2), we have

$$\frac{x}{y^2} = -\, e^{-y} + e^{-1}$$

$$\Rightarrow \quad x = y^2(e^{-1} - e^{-y}).$$

(*ii*) The given equation is

$$\sqrt{1 - y^2}\;\; dx = (\sin^{-1} y - x)\, dy$$

$$\Rightarrow \quad \frac{dx}{dy} + \frac{1}{\sqrt{1 - y^2}} \,.\, x = \frac{\sin^{-1} y}{\sqrt{1 - y^2}} \qquad ...(1)$$

Comparing it with $\dfrac{dx}{dy} + \text{P}x = \text{Q}$

where $$\text{P} = \frac{1}{\sqrt{1 - y^2}} \quad \text{and} \quad \text{Q} = \frac{\sin^{-1} y}{\sqrt{1 - y^2}}$$

$$\therefore \quad \text{I.F.} = e^{\int \text{P} . dx} = e^{\int \frac{1}{\sqrt{1 - y^2}} dy} = e^{\sin^{-1} y}$$

$\therefore$ The required solution is given by

$$x \,.\, (\text{I.F.}) = \int \text{Q} \,.\, (\text{I.F.})\, dy + c$$

$$\Rightarrow \quad x e^{\sin^{-1} y} = \int \frac{\sin^{-1} y}{\sqrt{1 - y^2}} \cdot e^{\sin^{-1} y}\, dy + c$$

| Put $\sin^{-1} y = z$
 $\Rightarrow \dfrac{1}{\sqrt{1 - y^2}}\, dy = dz$

$$= \int \underset{\text{I}}{z}\, \underset{\text{II}}{e^z}\, dz + c \qquad \text{(Integrating by parts)}$$

$$= \left[z \,.\, e^z - \int 1 \,.\, e^z\, dz \right] + c = ze^z - e^z + c$$

$$= e^z\,(z - 1) + c \qquad [\because \; z = \sin^{-1} y]$$

$$\Rightarrow \quad xe^{\sin^{-1} y} = e^{\sin^{-1} y} (\sin^{-1} y - 1) + c \quad ...(2)$$

Now, It is given that

$$y(0) = 0 \quad i.e., \quad x = 0, y = 0$$

$\therefore$ Equation (2) becomes

$$0 = e^{\sin^{-1} 0} (\sin^{-1} 0 - 1) + c$$

$$\Rightarrow \quad 0 = 1(0 - 1) + c \quad \Rightarrow \quad c = 1$$

$\therefore$ Putting this value of c in equation (2), we have

$$xe^{\sin^{-1} y} = e^{\sin^{-1} y} (\sin^{-1} y - 1) + 1$$

$$\Rightarrow \quad xe^{\sin^{-1} y} = \sin^{-1} y \, e^{\sin^{-1} y} - e^{\sin^{-1} y} + 1$$

$$\Rightarrow \quad e^{\sin^{-1} y} (x + 1 - \sin^{-1} y) = 1 \quad \Rightarrow \quad (x + 1 - \sin^{-1} y) = e^{-\sin^{-1} y}.$$

(*iii*) The given equation is

$$(1 + y^2)\, dx + (x - e^{-\tan^{-1} y})\, dy = 0$$

$$\Rightarrow \quad (1 + y^2) \frac{dx}{dy} + x = e^{-\tan^{-1} y}$$

$$\Rightarrow \quad \frac{dx}{dy} + \frac{1}{1+y^2} \cdot x = \frac{e^{-\tan^{-1} y}}{1+y^2} \quad ...(1)$$

Comparing it with $\dfrac{dx}{dy} + \text{P}x = \text{Q}$

where $$\text{P} = \frac{1}{1+y^2} \quad \text{and} \quad \text{Q} = \frac{e^{-\tan^{-1} y}}{1+y^2}$$

$$\therefore \quad \text{I.F.} = e^{\int \text{P} \cdot dy} = e^{\int \frac{1}{1+y^2} dy} = e^{\tan^{-1} y}$$

$\therefore$ The required solution is given by

$$x \,.\, (\text{I.F.}) = \int \text{Q} \,.\, (\text{I.F.})\, dy + c$$

$$\Rightarrow \quad xe^{\tan^{-1} y} = \int \frac{e^{-\tan^{-1} y}}{1+y^2} \cdot e^{\tan^{-1} y}\, dy + c$$

$$\Rightarrow \quad xe^{\tan^{-1} y} = \int \frac{1}{1+y^2}\, dy + c$$

$$\Rightarrow \quad xe^{\tan^{-1} y} = \tan^{-1} y + c \quad ...(2)$$

Now, It is given that $y(0) = 0 \quad i.e., \quad x = 0, y = 0$

$\therefore$ Equation (2) becomes

$$0 = \tan^{-1} 0 + c \quad \Rightarrow \quad c = 0$$

$\therefore$ Putting this value of c in equation (2), we have

$$xe^{\tan^{-1} y} = \tan^{-1} y.$$

(*iv*) The given equation is

$$y' + 2y = e^{-2x} \sin x$$

$$\Rightarrow \quad \frac{dy}{dx} + 2y = e^{-2x} \sin x \qquad ...(1)$$

Comparing it with $\frac{dy}{dx} + Py = Q$

where $\quad P = 2 \quad \text{and} \quad Q = e^{-2x} \sin x$

$\therefore \quad \text{I.F.} = e^{\int P dx} = e^{\int 2\, dx} = e^{2x}$

$\therefore$ The required solution is given by

$$y\,.\,(\text{I.F.}) = \int Q\,.\,(\text{I.F.})\, dx + c$$

$$\Rightarrow \quad ye^{2x} = \int e^{-2x} \sin x\,.\,e^{2x}\, dx + c$$

$$\Rightarrow \quad ye^{2x} = \int \sin x\,.\,dx + c \quad \Rightarrow \quad ye^{2x} = -\cos x + c \qquad ...(2)$$

Now, It is given that $y(0) = 0$ *i.e.*, $x = 0, y = 0$

$\therefore$ Equation (2) becomes

$$0 = -\cos 0 + c \quad \Rightarrow \quad 0 = -1 + c \quad \Rightarrow \quad c = 1$$

Putting this value of c in equation (2), we have

$$\Rightarrow \quad ye^{2x} = 1 - \cos x.$$

(*v*) The given equation is

$$dx + x\, dy = e^{-y} \sec^2 y\, dy$$

$$\Rightarrow \quad dx = e^{-y} \sec^2 y\, dy - x\, dy$$

$$\Rightarrow \quad dx = (e^{-y} \sec^2 y - x)\, dy$$

$$\Rightarrow \quad \frac{dx}{dy} = e^{-y} \sec^2 y - x \quad \Rightarrow \quad \frac{dx}{dy} + x = e^{-y} \sec^2 y \qquad ...(1)$$

Comparing it with $\frac{dx}{dy} + Px = Q$.

where $\quad P = 1 \quad \text{and} \quad Q = e^{-y} \sec^2 y$

$\therefore \quad \text{I.F.} = e^{\int P\,.\,dy} = e^{\int 1\,.\,dy} = e^{y}$

$\therefore$ The required solution is given by

$$x\,.\,(\text{I.F.}) = \int Q\,(\text{I.F.})\, dy + c$$

$$\Rightarrow \quad xe^{y} = \int e^{-y} \sec^2 y\,.\,e^{y}\, dy + c$$

$$\Rightarrow \quad xe^{y} = \int \sec^2 y\, dy + c \quad \Rightarrow \quad xe^{y} = \tan y + c.$$

(*vi*) The given equation is $y\, dx + (x - y^3)\, dy = 0$

$$\Rightarrow \quad \frac{dx}{dy} + \frac{x}{y} = y^2 \qquad ...(1)$$

Comparing it with $\frac{dx}{dy} + Px = Q$

where $$\mathrm{P} = \frac{1}{y} \quad \text{and} \quad \mathrm{Q} = y^2$$

$$\therefore \quad \text{I.F.} = e^{\int \mathrm{P}\,.\,dy} = e^{\int \frac{1}{y} dy} = e^{\log y} = y$$

$\therefore$ The required solution is given by

$$x\,.\,(\text{I.F.}) = \int \mathrm{Q}\,.\,(\text{I.F.})\,dy + c$$

$$\Rightarrow \quad xy = \int y^2\,.\,y\,dy + c = \int y^3\,dy + c$$

$$\Rightarrow \quad xy = \frac{y^4}{4} + c \quad \Rightarrow \quad 4xy = y^4 + 4c.$$

Example 16. *Solve the following differential equations :*

(*i*) $\dfrac{d^2y}{dx^2} = xe^x$ (*ii*) $\dfrac{d^2y}{dx^2} = sec^2x + xe^x$

(*iii*) $\dfrac{d^2y}{dx^2} = sin\ 2x$ (*iv*) $\dfrac{d^2y}{dx^2} = 2x^3 + 6x^2 - x + 8$

(*v*) $\dfrac{d^2y}{dx^2} = x\cos x$ (*vi*) $\dfrac{d^2y}{dx^2} = \sin x + \cos 2x.$

Solution. (*i*) The given equation is

$$\frac{d^2y}{dx^2} = xe^x \qquad ...(1)$$

Integrating both sides w.r.t. x, we get

$$\frac{dy}{dx} = \int \underset{\text{I}}{x}\,\underset{\text{II}}{e^x}\,dx \qquad \text{[Integrating by parts]}$$

$$= xe^x - \int 1\,.\,e^x\,dx + c_1 = xe^x - \int e^x\,dx + c_1$$

$$\Rightarrow \quad \frac{dy}{dx} = xe^x - e^x + c_1$$

Integrating again w.r.t. x, we get

$$y = \int (xe^x - e^x + c_1)\,.\,dx + c_2$$

$$= \int \underset{\text{I}}{x}\,\underset{\text{II}}{e^x}\,dx - \int e^x\,dx + c_1 \int 1\,.\,dx + c_2 \qquad \text{[Integrating by parts]}$$

$$= x\,.\,e^x - \int 1\,.\,e^x\,dx - e^x + c_1x + c_2$$

$$\therefore \quad y = xe^x - e^x - e^x + c_1x + c_2 = xe^x - 2e^x + c_1x + c_2$$

which is the required solution.

(*ii*) The given equation is

$$\frac{d^2y}{dx^2} = \sec^2 x + xe^x \qquad ...(1)$$

Integrating both sides w.r.t. x, we get

$$\frac{dy}{dx} = \int (\sec^2 x + xe^x)\, dx + c_1$$

$$= \int \sec^2 x\, dx + \int \underset{\text{I}}{x}\, \underset{\text{II}}{e^x}\, dx + c_1 \qquad \text{[Integrating by parts]}$$

$$= \tan x + xe^x - \int 1 . e^x\, dx + c_1$$

$$\Rightarrow \quad \frac{dy}{dx} = \tan x + xe^x - e^x + c_1 \qquad ...(2)$$

Integrating again w.r.t. x, we get

$$y = \int (\tan x + xe^x - e^x + c_1)\, dx + c_2$$

$$= \int \tan x\, dx + \int xe^x\, dx - \int e^x\, dx + c_1 \int dx + c_2$$

$$= -\log |\cos x| + xe^x - \int 1 . e^x\, dx - e^x + c_1 x + c_2$$

$$= -\log |\cos x| + xe^x - e^x - e^x + c_1 x + c_2$$

$$\therefore \quad y = -\log |\cos x| + e^x (x - 2) + c_1 x + c_2$$

which is the required solution.

(*iii*) The given equation is

$$\frac{d^2 y}{dx^2} = \sin 2x \qquad ...(1)$$

Integrating both sides w.r.t. x, we get

$$\frac{dy}{dx} = \int \sin 2x\, dx + c_1$$

$$\Rightarrow \quad \frac{dy}{dx} = -\frac{1}{2} \cos 2x + c_1 \qquad ...(2)$$

Integrating again w.r.t x, we get

$$\Rightarrow \quad y = \int \left(-\frac{1}{2} \cos 2x + c_1\right) dx + c_2 = -\frac{1}{2} \int \cos 2x\, dx + c_1 \int 1 . dx + c_2$$

$$= -\frac{1}{4} \sin 2x + c_1 x + c_2$$

which is the required solution.

(*iv*) The given equation is

$$\frac{d^2 y}{dx^2} = 2x^3 + 6x^2 - x + 8 \qquad ...(1)$$

Integrating both sides w.r.t. x, we get

$$\frac{dy}{dx} = \int (2x^3 + 6x^2 - x + 8)\, dx + c_1$$

$$\Rightarrow \quad \frac{dy}{dx} = \frac{2x^4}{4} + \frac{6x^3}{3} - \frac{x^2}{2} + 8x + c_1$$

$$\Rightarrow \quad \frac{dy}{dx} = \frac{1}{2}x^4 + 2x^3 - \frac{1}{2}x^2 + 8x + c_1 \qquad ...(2)$$

Integrating again w.r.t. x, we get

$$y = \int\left(\frac{1}{2}x^4 + 2x^3 - \frac{1}{2}x^2 + 8x + c_1\right) dx + c_2$$

$$= \frac{1}{2}.\frac{x^5}{5} + \frac{2x^4}{4} - \frac{1}{2}.\frac{x^3}{3} + 8.\frac{x^2}{2} + c_1x + c_2$$

$$\therefore \quad y = \frac{1}{10}x^5 + \frac{1}{2}x^4 - \frac{1}{6}x^3 + 4x^2 + c_1x + c_2$$

which is the required solution.

(*v*) The given equation is

$$\frac{d^2y}{dx^2} = x \cos x \qquad ...(1)$$

Integrating both sides w.r.t. x, we get

$$\frac{dy}{dx} = \int \underset{\text{I}}{x} \underset{\text{II}}{\cos x}\, dx \qquad \text{[Integrating by parts]}$$

$$\Rightarrow \quad \frac{dy}{dx} = x\,.\,(\sin x) - \int 1 . \sin x\, dx + c_1$$

$$\Rightarrow \quad \frac{dy}{dx} = x \sin x - \int \sin x\, dx + c_1 = x \sin x - (-\cos x) + c_1$$

$$\Rightarrow \quad \frac{dy}{dx} = x \sin x + \cos x + c_1 \qquad ...(2)$$

Integrating again w.r.t. x, we get

$$y = \int (x \sin x + \cos x + c_1)\, dx + c_2$$

$$= \int \underset{\text{I}}{x} \underset{\text{II}}{\sin x}\, dx + \int \cos x\, dx + c_1 \int 1 . dx + c_2 \qquad \text{[Integrating by parts]}$$

$$= x\,.\,(-\cos x) - \int 1 . (-\cos x)\, dx + \sin x + c_1x + c_2$$

$$= -x \cos x + \int \cos x\, dx + \sin x + c_1x + c_2$$

$$= -x \cos x + \sin x + \sin x + c_1x + c_2$$

$$\Rightarrow \quad y = -x \cos x + 2 \sin x + c_1x + c_2$$

which is the required solution.

(*vi*) The given equation is

$$\frac{d^2y}{dx^2} = \sin x + \cos 2x \qquad ...(1)$$

Integrating both sides w.r.t. x, we get

$$\Rightarrow \quad \frac{dy}{dx} = \int (\sin x + \cos 2x)\, dx + c_1$$

$$\Rightarrow \quad \frac{dy}{dx} = \int \sin x\, dx + \int \cos 2x\, dx + c_1$$

$$\Rightarrow \quad \frac{dy}{dx} = -\cos x + \frac{1}{2} \sin 2x + c_1 \qquad ...(2)$$

Integrating again w.r.t. x, we get

$$y = \int \left(-\cos + \frac{1}{2} \sin 2x + c_1\right) dx + c_2$$

$$\Rightarrow \quad y = -\sin x + \frac{1}{4} (-\cos 2x) + c_1 x + c_2$$

$$\Rightarrow \quad y = -\sin x - \frac{1}{4} \cos 2x + c_1 x + c_2$$

which is the required solution.

Example 17. *Solve the following differential equations :*

(*i*) $\dfrac{d^2y}{dx^2} = sec^2 x$ (*ii*) $\dfrac{d^2y}{dx^2} = x \sin x$

(*iii*) $\dfrac{d^2y}{dx^2} = x + \cos x$ (*iv*) $\dfrac{d^2y}{dx^2} = \log x + x^6 + 3x^2$

(*v*) $\dfrac{d^2y}{dx^2} = e^x \left(\dfrac{1}{x} - \dfrac{2}{x^2} + \dfrac{2}{x^3}\right)$ (*vi*) $\dfrac{d^2y}{dx^2} = \log x.$

Solution. (*i*) The given equation is

$$\frac{d^2y}{dx^2} = \sec^2 x \qquad ...(1)$$

Integrating both sides w.r.t. x , we get

$$\frac{dy}{dx} = \int \sec^2 x\, dx + c_1$$

$$\Rightarrow \quad \frac{dy}{dx} = \tan x + c_1 \qquad ...(2)$$

Integrating again w.r.t. x, we get

$$y = \int (\tan x + c_1)\, dx + c_2$$

$$\Rightarrow \quad y = \log |\sec x| + c_1 x + c_2$$

which is the required solution.

(*ii*) The given equation is

$$\frac{d^2y}{dx^2} = x \sin x \qquad ...(1)$$

Integrating both sides w.r.t. x, we get

$$\frac{dy}{dx} = \int \underset{\text{I}}{x} \underset{\text{II}}{\sin x}\, dx + c_1 \qquad \text{[Integrating by parts]}$$

$$\Rightarrow \quad \frac{dy}{dx} = x.(-\cos x) - \int 1.(-\cos x).dx + c_1 = -x\cos x + \int \cos x\, dx + c_1$$

$$\Rightarrow \quad \frac{dy}{dx} = -x\cos x + \sin x + c_1 \qquad ...(2)$$

Integrating again both sides w.r.t. x, we get

$$y = \int(-x\cos x + \sin x + c_1)\, dx + c_2$$

$$= -\int \underset{\text{I}}{x} \underset{\text{II}}{\cos x}\, dx + \int \sin x\, dx + c_1 \int 1.\, dx + c_2 \qquad \text{[Integrating by parts]}$$

$$= -\left[x.\sin x - \int 1.\sin x\, dx\right] + (-\cos x) + c_1 x + c_2$$

$$= -x\sin x + \int \sin x\, dx - \cos x + c_1 x + c_2$$

$$= -x\sin x - \cos x - \cos x + c_1 x + c_2$$

$$= -x\sin x - 2\cos x + c_1 x + c_2$$

which is the required solution.

(*iii*) The given equation is

$$\frac{d^2y}{dx^2} = x + \cos x \qquad ...(1)$$

Integrating both sides w.r.t. x, we get

$$\frac{dy}{dx} = \int(x + \cos x)\, dx + c_1$$

$$\Rightarrow \quad \frac{dy}{dx} = \frac{x^2}{2} + \sin x + c_1 \qquad ...(2)$$

Integrating again w.r.t. x, we get

$$y = \int\left(\frac{x^2}{2} + \sin x + c_1\right) dx + c_2$$

$$\Rightarrow \quad y = \frac{x^3}{6} - \cos x + c_1 x + c_2$$

which is the required solution.

(*iv*) The given equation is

$$\frac{d^2y}{dx^2} = \log x + x^6 + 3x^2 \qquad ...(1)$$

Integrating both sides w.r.t. x, we get

$$\frac{dy}{dx} = \int(\log x + x^6 + 3x^2)\, dx + c_1$$

$$= \int \underset{\text{II}}{1} . \underset{\text{I}}{\log x}\, dx + \int x^6\, dx + \int 3x^2\, dx + c_1 \qquad \text{[Integrating by parts]}$$

$$= \log x \,.\, x - \int \frac{1}{x} \,.\, x\, dx + \frac{x^7}{7} + \frac{3x^3}{3} + c_1$$

$$= x \log x - \int 1 \,.\, dx + \frac{1}{7} x^7 + x^3 + c_1$$

$$\Rightarrow \qquad \frac{dy}{dx} = x \log x - x + \frac{1}{7} x^7 + x^3 + c_1 \qquad ...(2)$$

Integrating again w.r.t. x, we get

$$y = \int \left(x \log x - x + \frac{1}{7} x^7 + x^3 + c_1 \right) dx + c_2$$

$$= \int \underset{\text{II}}{x} \underset{\text{I}}{\log x}\, dx - \int x\, dx + \frac{1}{7} \int x^7\, dx + \int x^3\, dx + c_1 \int 1 \,.\, dx + c_2$$

[Integrating by parts]

$$= \log x \,.\, \frac{x^2}{2} - \int \frac{1}{x} \,.\, \frac{x^2}{2}\, dx - \frac{x^2}{2} + \frac{1}{7} \,.\, \frac{x^8}{8} + \frac{x^4}{4} + c_1 x + c_2$$

$$= \frac{x^2}{2} \log x - \frac{1}{2} \int x\, dx - \frac{x^2}{2} + \frac{1}{56} x^8 + \frac{x^4}{4} + c_1 x + c_2$$

$$= \frac{x^2}{2} \log x - \frac{1}{4} x^2 - \frac{x^2}{2} + \frac{1}{56} x^8 + \frac{1}{4} x^4 + c_1 x + c_2$$

$$\Rightarrow \qquad y = \frac{x^2}{2} \log x - \frac{3x^2}{4} + \frac{x^8}{56} + \frac{x^4}{4} + c_1 x + c_2$$

which is the required solution.

(*v*) The given equation is

$$\frac{d^2 y}{dx^2} = e^x \left(\frac{1}{x} - \frac{2}{x^2} + \frac{2}{x^3} \right)$$

$$\Rightarrow \qquad \frac{d^2 y}{dx^2} = e^x \left(\frac{1}{x} - \frac{1}{x^2} - \frac{1}{x^2} + \frac{2}{x^3} \right) = e^x \left[\left(\frac{1}{x} - \frac{1}{x^2} \right) + \left(-\frac{1}{x^2} + \frac{2}{x^3} \right) \right]$$

$$\Rightarrow \qquad \frac{d^2 y}{dx^2} = e^x \left(\frac{1}{x} - \frac{1}{x^2} \right) + e^x \left(-\frac{1}{x^2} + \frac{2}{x^3} \right)$$

Integrating both sides w.r.t. x, we get

$$\frac{dy}{dx} = \int e^x \left(\frac{1}{x} - \frac{1}{x^2} \right) dx + \int e^x \left(-\frac{1}{x^2} + \frac{2}{x^3} \right) dx + c_1$$

$$\left[\because \int e^x \,[f(x) + f'(x)]\, dx = e^x f(x) \right]$$

$$\Rightarrow \qquad \frac{dy}{dx} = e^x \cdot \frac{1}{x} + e^x \cdot \left(-\frac{1}{x^2}\right) + c_1$$

$$\Rightarrow \qquad \frac{dy}{dx} = e^x \left(\frac{1}{x} - \frac{1}{x^2}\right) + c_1 \qquad \text{...(2)}$$

Integrating again w.r.t. x, we get

$$y = \int \left[e^x \left(\frac{1}{x} - \frac{1}{x^2}\right) + c_1 \right] dx + c_2$$

$$\Rightarrow \qquad y = \int e^x \left(\frac{1}{x} - \frac{1}{x^2}\right) dx + \int c_1 \, dx + c_2$$

$$\Rightarrow \qquad y = e^x \cdot \frac{1}{x} + c_1 x + c_2$$

which is the required solution.

(*vi*) The given equation is

$$\frac{d^2y}{dx^2} = \log x$$

Please try yourself. $\left[\textbf{Ans. } y = \frac{x^2 \log x}{2} - \frac{3x^2}{4} + c_1 x + c_2. \right]$

Example 18. *Solve the following differential equations :*

(*i*) $\dfrac{d^2y}{dx^2} = x^2 \cos x$ (*ii*) $\dfrac{d^2y}{dx^2} = 1 + 2x + 3x^2 + \ldots + (n + 1)\, x^n$

(*iii*) $\dfrac{d^2y}{dx^2} = x^2 \sin x$

(*iv*) $\dfrac{d^2y}{dx^2} = x + \sin x$ *; given that :* $y = 0,\ \dfrac{dy}{dx} = -1$ *when* $x = 0$.

(*v*) $x \dfrac{d^2y}{dx^2} = 1$ *; given that :* $y = 1,\ \dfrac{dy}{dx} = 0,$ *when* $x = 1$.

(*vi*) $\dfrac{d^2y}{dx^2} = \sin^{-1} x$ (*vii*) $\dfrac{d^2y}{dx^2} = \sin x + e^x$ *;* $y(0) = 3,\ y'(0) = 4.$

Solution. (*i*) The given equation is

$$\frac{d^2y}{dx^2} = x^2 \cos x \qquad \text{...(1)}$$

Integrating both sides w.r.t. x, we get

$$\frac{dy}{dx} = \int \underset{\text{I}}{x^2} \underset{\text{II}}{\cos x}\, dx + c_1 \qquad \text{[Integrating by parts]}$$

$$= x^2 \cdot \sin x - \int 2x \cdot \sin x \, dx + c_1$$

$$= x^2 \sin x - 2 \int \underset{\text{I}}{x} \underset{\text{II}}{\sin x}\, dx + c_1 \qquad \text{[Integrating by parts again]}$$

$$= x^2 \sin x - 2 \left[x \,.\, (-\cos x) - \int 1 \,.\, (-\cos x)\, dx \right] + c_1$$

$$= x^2 \sin x + 2x \cos x - 2 \int \cos x\, dx + c_1$$

$$\Rightarrow \qquad \frac{dy}{dx} = x^2 \sin x + 2x \cos x - 2 \sin x + c_1 \qquad \text{...(2)}$$

Integrating again w.r.t. x, we get

$$y = \int (x^2 \sin x + 2x \cos x - 2 \sin x + c_1)\, dx + c_2$$

$$= \int \underset{\text{I}}{x^2} \underset{\text{II}}{\sin x}\, dx + 2 \int \underset{\text{I}}{x} \underset{\text{II}}{\cos x}\, dx - 2 \int \sin x\, dx + c_1 \int 1 \,.\, dx + c_2$$

$$= x^2 (-\cos x) - \int 2x \,.\, (-\cos x)\, dx + 2 \left[x \,.\, \sin x - \int 1 \,.\, \sin x\, dx \right] - 2(-\cos x) + c_1 x + c_2$$

$$= -x^2 \cos x + 2 \int \underset{\text{I}}{x} \underset{\text{II}}{\cos x}\, dx + 2x \sin x - 2 \int \sin x\, dx + 2 \cos x + c_1 x + c_2$$

$$= -x^2 \cos x + 2 \left[x \,.\, \sin x - \int 1 \,.\, \sin x\, dx \right] + 2x \sin x - 2(-\cos x) + 2 \cos x + c_1 x + c_2$$

$$= -x^2 \cos x + 2x \sin x - 2 \int \sin x\, dx + 2x \sin x + 2 \cos x + 2 \cos x + c_1 x + c_2$$

$$= -x^2 \cos x + 4x \sin x - 2(-\cos x) + 4 \cos x + c_1 x + c_2$$

$$= -x^2 \cos x + 4x \sin x + 2 \cos x + 4 \cos x + c_1 x + c_2$$

$$\Rightarrow \quad y = (6 - x^2) \cos x + 4x \sin x + c_1 x + c_2$$

which is the required solution.

(*ii*) The given equation is

$$\frac{d^2 y}{dx^2} = 1 + 2x + 3x^2 + \ldots + (n+1)\, x^n \qquad \text{...(1)}$$

Integrating both sides w.r.t. x, we get

$$\frac{dy}{dx} = \int [1 + 2x + 3x^2 + \ldots + (n+1)\, x^n]\, dx + c_1$$

$$\Rightarrow \qquad \frac{dy}{dx} = x + \frac{2x^2}{2} + \frac{3x^3}{3} + \ldots + (n+1) \frac{x^{n+1}}{n+1} + c_1$$

$$\Rightarrow \qquad \frac{dy}{dx} = x + x^2 + x^3 + \ldots + x^{n+1} + c_1 \qquad \text{...(2)}$$

Integrating again w.r.t. x, we get

$$y = \int (x + x^2 + x^3 + \ldots x^{n+1} + c_1)\, dx + c_2$$

$$= \frac{x^2}{2} + \frac{x^3}{3} + \frac{x^4}{4} + \ldots + \frac{x^{n+2}}{n+2} + c_1 x + c_2$$

$$\Rightarrow \qquad y = \frac{1}{2}x^2 + \frac{1}{3}x^3 + \frac{1}{4}x^4 + \ldots + \frac{1}{n+2}x^{n+2} + c_1x + c_2$$

which is the required solution.

(*iii*) The given equation is

$$\frac{d^2y}{dx^2} = x^2 \sin x \qquad \ldots(1)$$

Integrating both sides w.r.t. x, we get

$$\frac{dy}{dx} = \int \underset{\text{I}}{x^2} \underset{\text{II}}{\sin x}\, dx + c_1 \qquad \text{[Integrating by parts]}$$

$$= x^2 . (-\cos x) - \int 2x . (-\cos x)\, dx + c_1$$

$$= -x^2 \cos x + 2 \int \underset{\text{I}}{x} \underset{\text{II}}{\cos x}\, dx + c_1 \qquad \text{[Integrating by parts]}$$

$$= -x^2 \cos x + 2\left[x . \sin x - \int 1 . \sin x\, dx\right] + c_1$$

$$= -x^2 \cos x + 2x \sin x - 2\int \sin x\, dx + c_1$$

$$= -x^2 \cos x + 2x \sin x - 2\cos x + c_1$$

$$\Rightarrow \qquad \frac{dy}{dx} = (2 - x^2) \cos x + 2x \sin x + c_1 \qquad \ldots(2)$$

Integrating again w.r.t. x, we get

$$y = \int\left[(2 - x^2) \cos x + 2x \sin x + c_1\right] dx + c_2$$

$$= \int \underset{\text{I}}{(2 - x^2)} \underset{\text{II}}{\cos x}\, dx + 2\int x \sin x\, dx + c_1 \int 1 . dx + c_2 \qquad \text{[Integrating by parts]}$$

$$= (2 - x^2) \sin x - \int (-2x) \sin x\, dx + 2\int x \sin x\, dx + c_1x + c_2$$

$$= (2 - x^2) \sin x + 2\int x \sin x\, dx + 2\int x \sin x\, dx + c_1x + c_2$$

$$= (2 - x^2) \sin x + 4\int \underset{\text{I}}{x} \underset{\text{II}}{\sin x}\, dx + c_1x + c_2 \qquad \text{[Integrating by parts]}$$

$$= (2 - x^2) \sin x + 4\left[x . (-\cos x) - \int 1 . (-\cos x)\, dx\right] + c_1x + c_2$$

$$= (2 - x^2) \sin x - 4x \cos x + 4\int \cos x\, dx + c_1x + c_2$$

$$= (2 - x^2) \sin x - 4x \cos x + 4 \sin x + c_1x + c_2$$

$$\Rightarrow \qquad y = (6 - x^2) \sin x - 4x \cos x + c_1x + c_2$$

which is the required solution.

(*iv*) The given equation is

$$\frac{d^2y}{dx^2} = x + \sin x \qquad \ldots(1)$$

Integrating both sides w.r.t. x, we get

$$\frac{dy}{dx} = \int (x + \sin x)\, dx + c_1 = \frac{x^2}{x} + (-\cos x) + c_1$$

$$\Rightarrow \quad \frac{dy}{dx} = \frac{1}{2}x^2 - \cos x + c_1 \quad ...(2)$$

It is given that $\frac{dy}{dx} = -1$ when $x = 0$

$\therefore$ Equation (2) becomes

$$-1 = 0 - \cos 0 + c_1 \Rightarrow -1 = -1 + c_1 \Rightarrow c_1 = 0$$

Putting this value of c_1 in equation (2), we have

$$\frac{dy}{dx} = \frac{1}{2}x^2 - \cos x \quad ...(3)$$

Integrating again w.r.t. x, we have

$$y = \int \left(\frac{1}{2}x^2 - \cos x\right) dx + c_2$$

$$\Rightarrow \quad y = \frac{x^3}{6} - \sin x + c_2 \quad ...(4)$$

It is given that $y = 0$ when $x = 0$.

$\therefore$ Equation (4) becomes

$$0 = 0 - \sin 0 + c_2 \Rightarrow c_2 = 0$$

Putting this value of c_2 in equation (4), we have

$$y = \frac{x^3}{6} - \sin x.$$

which is the required solution.

(v) The given equation is

$$x\frac{d^2y}{dx^2} = 1 \Rightarrow \frac{d^2y}{dx^2} = \frac{1}{x} \quad ...(1)$$

Integrating both sides w.r.t. x, we have

$$\frac{dy}{dx} = \int \frac{1}{x}\, dx + c_1$$

$$\Rightarrow \quad \frac{dy}{dx} = \log x + c_1 \quad ...(2)$$

It is given that $\frac{dy}{dx} = 0$ when $x = 1$

$\therefore$ Equation (2) becomes

$$0 = \log 1 + c_1 \Rightarrow c_1 = 0$$

Putting this value of c_1 in equation (2), we have

$$\frac{dy}{dx} = \log x \quad ...(3)$$

Integrating both sides w.r.t. x, we get

$$y = \int \underset{II}{1} . \underset{I}{\log x} . dx + c_2 \qquad \text{[Integrating by parts]}$$

$$y = \log x . x - \int \frac{1}{x} . x\, dx + c_2 = x \log x - \int 1 . dx + c_2$$

$$\Rightarrow \qquad y = x \log x - x + c_2 \qquad ...(4)$$

It is given that $y = 1$ when $x = 1$

Equation (4) becomes

$$1 = \log 1 - 1 + c_2 \quad \Rightarrow \quad c_2 = 2$$

Putting this value of c_2 in equation (4), we have

$$y = x \log x - x + 2$$

$$\Rightarrow \qquad y = x(\log x - 1) + 2.$$

which is the required solution.

(*vi*) The given equation is

$$\frac{d^2 y}{dx^2} = \sin^{-1} x \qquad ...(1)$$

Integrating both sides w.r.t. x, we have

$$\frac{dy}{dx} = \int \underset{II}{1} . \underset{I}{\sin^{-1} x}\, dx + c_1 \qquad \text{[Integrating by parts]}$$

$$= \sin^{-1} x . x - \int \frac{1}{\sqrt{1-x^2}} . x\, dx + c_1$$

$$= x \sin^{-1} x - \int \frac{x}{\sqrt{1-x^2}}\, dx + c_1 = x \sin^{-1} x + \frac{1}{2} \int \frac{-2x}{\sqrt{1-x^2}}\, dx + c_1$$

$$= x \sin^{-1} x + \frac{1}{2} \int (1-x^2)^{-1/2} (-2x) . dx + c_1$$

$$= x \sin^{-1} x + \frac{1}{2} \frac{(1-x^2)^{-\frac{1}{2}+1}}{-\frac{1}{2}+1} + c_1$$

$$\left[\because \int [f(x)]^n . f'(x)\, dx = \frac{[f(x)]^{n+1}}{n+1}\right]$$

$$\frac{dy}{dx} = x \sin^{-1} x + \sqrt{1-x^2} + c_1 \qquad ...(2)$$

Integrating both sides w.r.t. x, we get

$$y = \int (x \sin^{-1} x + \sqrt{1-x^2} + c_1)\, dx + c_2$$

$$= \int \underset{\text{II}}{x} \underset{\text{I}}{\sin^{-1} x}\, dx + \int \sqrt{1-x^2}\, dx + c_1 \int 1 . dx + c_2 \qquad \text{[Integrating by parts]}$$

$$= \sin^{-1} x . \frac{x^2}{2} - \int \frac{1}{\sqrt{1-x^2}} . \frac{x^2}{2}\, dx + \frac{x}{2}\sqrt{1-x^2} + \frac{1}{2}\sin^{-1} x + c_1 x + c_2$$

$$\left[\because \int \sqrt{a^2 - x^2}\, dx = \frac{x}{2}\sqrt{a^2 - x^2} + \frac{a^2}{2}\sin^{-1}\frac{x}{a}\right]$$

$$= \frac{x^2}{2}\sin^{-1} x + \frac{1}{2}\int \frac{-x^2}{\sqrt{1-x^2}}\, dx + \frac{x}{2}\sqrt{1-x^2} + \frac{1}{2}\sin^{-1} x + c_1 x + c_2$$

$$= \frac{x^2}{2}\sin^{-1} x + \frac{1}{2}\int \frac{1-x^2-1}{\sqrt{1-x^2}}\, dx + \frac{x}{2}\sqrt{1-x^2} + \frac{1}{2}\sin^{-1} x + c_1 x + c_2$$

[Add and subtract 1]

$$= \left(\frac{x^2+1}{2}\right)\sin^{-1} x + \frac{1}{2}\int \left(\frac{1-x^2}{\sqrt{1-x^2}} - \frac{1}{\sqrt{1-x^2}}\right) dx + \frac{x}{2}\sqrt{1-x^2} + c_1 x + c_2$$

$$= \left(\frac{x^2+1}{2}\right)\sin^{-1} x + \frac{1}{2}\int \sqrt{1-x^2}\, dx - \frac{1}{2}\int \frac{1}{\sqrt{1-x^2}}\, dx + \frac{x}{2}\sqrt{1-x^2} + c_1 x + c_2$$

$$= \left(\frac{x^2+1}{2}\right)\sin^{-1} x + \frac{x}{4}\sqrt{1-x^2} + \frac{1}{4}\sin^{-1} x - \frac{1}{2}\sin^{-1} x + \frac{x}{2}\sqrt{1-x^2} + c_1 x + c_2$$

$$\left[\because \int \frac{1}{\sqrt{a^2 - x^2}}\, dx = \sin^{-1}\frac{x}{a}\right]$$

$$= \left(\frac{x^2+1}{2} + \frac{1}{4} - \frac{1}{2}\right)\sin^{-1} x + \left(\frac{x}{4} + \frac{x}{2}\right)\sqrt{1-x^2} + c_1 x + c_2$$

$$= \left(\frac{2x^2+2+1-2}{4}\right)\sin^{-1} x + \frac{3x}{4}\sqrt{1-x^2} + c_1 x + c_2$$

$$\therefore \qquad y = \frac{2x^2+1}{4}\sin^{-1} x + \frac{3x\sqrt{1-x^2}}{4} + c_1 x + c_2$$

which is the required solution.

(*vii*) The given equation is

$$\frac{d^2 y}{dx^2} = \sin x + e^x \qquad ...(1)$$

Integrating both sides w.r.t. x, we get

$$\frac{dy}{dx} = \int (\sin x + e^x)\, dx + c_1$$

$$\Rightarrow \qquad \frac{dy}{dx} = -\cos x + e^x + c_1 \qquad \text{...(2)}$$

It is given that $\frac{dy}{dx} = 4$ when $x = 0$

$\therefore$ Equation (2) becomes

$$4 = -\cos 0 + e^0 + c_1 \Rightarrow 4 = -1 + 1 + c_1 \Rightarrow c_1 = 4.$$

Putting this value of c_1 in equation (2), we have

$$\frac{dy}{dx} = -\cos x + e^x + 4 \qquad \text{...(3)}$$

Integrating again w.r.t. x, we have

$$y = \int (-\cos x + e^x + 4)\, dx + c_2$$

$$\Rightarrow \qquad y = -\sin x + e^x + 4x + c_2 \qquad \text{...(4)}$$

It is given that $y = 3$ when $x = 0$.

$\therefore$ Equation (4) becomes

$$3 = -\sin 0 + e^0 + 0 + c_2 \Rightarrow 3 = 0 + 1 + 0 + c_2 \Rightarrow c_2 = 2$$

Putting this value of c_2 in equation (4), we have

$$y = -\sin x + e^x + 4x + 2$$

which is the required solution.

Example 19. *(i) Solve the differential equation* $\sin^{-1}\left(\frac{d^2y}{dx^2} - 1\right) = x$ *and hence find the particular solution if*

$$y = 0, \ \frac{dy}{dx} = 0 \text{ when } x = 0.$$

(ii) If $x = \frac{1}{n}\cos^{-1}\left(\frac{d^2y}{dx^2}\right)$ *and if* $\frac{dy}{dx} = 0$ *and* $y = 0$ *when* $x = \pi$, *find the particular solution of the differential equation.*

Solution. (*i*) The given equation is

$$\sin^{-1}\left(\frac{d^2y}{dx^2} - 1\right) = x.$$

$$\Rightarrow \qquad \frac{d^2y}{dx^2} - 1 = \sin x \qquad \Rightarrow \qquad \frac{d^2y}{dx^2} = 1 + \sin x \qquad \text{...(1)}$$

Integrating both sides, we get

$$\frac{dy}{dx} = \int (1 + \sin x)\, dx + c_1$$

$$\Rightarrow \qquad \frac{dy}{dx} = x - \cos x + c_1 \qquad ...(2)$$

It is given that $\frac{dy}{dx} = 0$ when $x = 0$

$\therefore$ Equation (2) becomes

$$0 = 0 - \cos 0 + c_1 \Rightarrow 0 = -1 + c_1 \Rightarrow c_1 = 1$$

Putting this value of c_1 in equation (2), we have

$$\frac{dy}{dx} = x - \cos x + 1 \qquad ...(3)$$

Integrating again both sides w.r.t. x, we get

$$y = \int (x - \cos x + 1)\, dx + c_2$$

$$\Rightarrow \qquad y = \frac{x^2}{2} - \sin x + x + c_2 \qquad ...(4)$$

It is given that $y = 0$ when $x = 0$

$\therefore$ Equation (4) becomes

$$0 = 0 - \sin 0 + 0 + c_2 \Rightarrow c_2 = 0$$

Putting this value of c_2 in equation (4), we get

$$y = \frac{x^2}{2} - \sin x + x$$

which is the required solution.

(*ii*) The given equation is

$$x = \frac{1}{n} \cos^{-1}\left(\frac{d^2y}{dx^2}\right)$$

$$\Rightarrow \qquad \cos^{-1}\left(\frac{d^2y}{dx^2}\right) = nx \qquad \Rightarrow \qquad \frac{d^2y}{dx^2} = \cos nx \qquad ...(1)$$

Integrating both sides w.r.t. x, we get

$$\frac{dy}{dx} = \int \cos nx\, dx + c_1$$

$$\Rightarrow \qquad \frac{dy}{dx} = + \frac{\sin nx}{n} + c_1 \qquad ...(2)$$

It is given that $\frac{dy}{dx} = 0$ when $x = \pi$. $\qquad [\because \sin n\pi = 0]$

$\therefore$ Equation (2) becomes

$$0 = \frac{\sin n\pi}{n} + c_1 \Rightarrow c_1 = 0$$

Putting this value of c_1 in equation (2), we have

$$\frac{dy}{dx} = \frac{1}{n} \sin nx \qquad ...(3)$$

Integrating both sides w.r.t. x, we get

$$y = \frac{1}{n}\int \sin nx\, dx + c_2$$

$$\Rightarrow \qquad y = \frac{1}{n}\left(-\frac{\cos nx}{n}\right) + c_2 \qquad ...(4)$$

It is given that $y = 0$ when $x = \pi$

$\therefore$ Equation (4) becomes

$$0 = -\frac{1}{n^2}\cos n\pi + c_2 \Rightarrow 0 = -\frac{1}{n^2}(-1)^n + c_2$$

$$[\because \cos n\pi = (-1)^n]$$

$$\Rightarrow \qquad c_2 = \frac{(-1)^n}{n^2}$$

Putting this value of c_2 in equation (4), we have

$$y = -\frac{1}{n^2}\cos nx + \frac{(-1)^n}{n^2}$$

which is the required solution.

Example 20. *Solve the following differential equation*

$$\frac{d^2y}{dx^2} + \cos x = x \textit{ given that } y = 1 \textit{ and } \frac{dy}{dx} = 0, \textit{ when } x = 0.$$

Solution. The given equation is

$$\frac{d^2y}{dx^2} + \cos x = x \Rightarrow \frac{d^2y}{dx^2} = x - \cos x \qquad ...(1)$$

Integrating both sides w.r.t. x, we get

$$\frac{dy}{dx} = \int (x - \cos x)\, dx + c_1$$

$$\Rightarrow \qquad \frac{dy}{dx} = \frac{x^2}{2} - \sin x + c_1 \qquad ...(2)$$

It is given that $\frac{dy}{dx} = 0$ when $x = 0$

$\therefore$ Equation (2), becomes

$$0 = 0 - \sin 0 + c_1 \Rightarrow c_1 = 0$$

Putting this value of c_1 in equation (2), we have

$$\frac{dy}{dx} = \frac{x^2}{2} - \sin x \qquad ...(3)$$

Integrating again w.r.t. x, we have

$$y = \int\left(\frac{x^2}{2} - \sin x\right) dx + c_2 = \frac{x^3}{6} - (-\cos x) + c_2$$

$$\Rightarrow \qquad y = \frac{x^3}{6} + \cos x + c_2 \qquad ...(4)$$

It is given that $y = 1$ when $x = 0$

$\therefore$ Equation (4) becomes

$$1 = 0 + \cos 0 + c_2 \Rightarrow 1 = 1 + c_2 \Rightarrow c_2 = 0$$

Putting this value of c_2 in equation (4), we have

$$y = \frac{x^3}{6} + \cos x$$

which is the required solution.

EXERCISE FOR PRACTICE

Solve the following differential equations :

1. $(x^2 - y^2)\,dx + 2xy\,dy = 0.$

2. $x\dfrac{dy}{dx} = x + y,\ x \neq 0.$

3. $x^2\dfrac{dy}{dx} = x^2 - 2y^2 + xy.$

4. $x^2\dfrac{dy}{dx} = y(x + y).$

5. $\dfrac{dy}{dx} = \dfrac{y - x}{y + x}.$

6. $xy\dfrac{dy}{dx} = x^2 - y^2.$

7. $\dfrac{dy}{dx} = \dfrac{x}{y} + \tan\left(\dfrac{y}{x}\right).$

8. $x\dfrac{dy}{dx} - y = \sqrt{x^2 + y^2}.$

9. $3x^2\,dy = (3xy + y^2)\,dx.$

10. $(2x^2 + y^2)\,dx + (2xy + 3y^2)\,dy = 0.$

11. $y^2\,dx + (x^2 - xy + y^2)\,dy = 0.$

12. $x\dfrac{dy}{dx} + \dfrac{y^2}{x} = y.$

13. $xy\log\left(\dfrac{x}{y}\right)dx + \left\{y^2 - x^2\log\left(\dfrac{x}{y}\right)\right\}dy = 0.$

14. $(x^2 + 3xy + y^2)\,dx = x^2\,dy.$

15. $\dfrac{dy}{dx} = y\tan x - 2\sin x.$

16. $(1 + x^2)\dfrac{dy}{dx} + y = \tan^{-1} x.$

17. $\dfrac{dy}{dx} + y\tan x = 2x + x^2\tan x.$

18. $(\sin x)\dfrac{dy}{dx} + y\cos x = 2\sin^2 x\cos x.$

19. $\dfrac{dy}{dx} + 2y = \sin x.$

20. $dx + x\,dy = e^{-y}\sec^2 y\,dy.$

21. $\dfrac{d^2y}{dx^2} = xe^x + 1.$

22. $\dfrac{d^2y}{dx^2} = x^3\sin x.$

23. $\dfrac{d^2y}{dx^2} = \dfrac{1}{x} + 6$, given that $\dfrac{dy}{dx} = 0$ and $y = 0$ when $x = 1$.

24. $\dfrac{d^2y}{dx^2} + \sin x = x$, given that $\dfrac{dy}{dx} = 1$ and $y = 1$ when $x = 0$.

Answers

1. $x = c(x^2 + y^2)$

2. $y = x \log |x| + cx,\ x \neq 0$

3. $\frac{1}{2\sqrt{2}} \log \left| \frac{x + \sqrt{2}y}{x - \sqrt{2}y} \right| = \log |x| + c$

4. $y \log |x| + x + cy = 0$

5. $\tan^{-1}\left(\frac{y}{x}\right) + \frac{1}{2} \log (x^2 + y^2) = c$

6. $x^2 (x^2 - 2y^2) = c$

7. $x = c \sin \left(\frac{y}{x}\right)$

8. $y + \sqrt{x^2 + y^2} = cx^2$

9. $-\frac{3x}{y} = \log |x| + c$

10. $3y^3 + 3xy^2 + 2x^3 = c$

11. $y = c\, e^{\tan^{-1}(y/x)}$

12. $\log |x| = \frac{x}{y} + c$

13. $\frac{x^2}{y^2}\left\{\log\left(\frac{x}{y}\right) - \frac{1}{2}\right\} + \log y^2 = c$

14. $\frac{x}{x + y} + \log x + c = 0$

15. $y \cos x = \frac{\cos 2x}{2} + c$

16. $y = \tan^{-1} x - 1 + c\, e^{-\tan^{-1} x}$

17. $y = x^2 + c \cos x.$

18. $y \sin x = \frac{2}{3} \sin^3 x + c$

19. $y = ce^{-2x} + \frac{1}{5} (2 \sin x - \cos x)$

20. $xe^y = \tan y + c$

21. $y = e^x (x - 2) + \frac{x^2}{2} + c_1 x + c_2$

22. $y = (-x^3 + 18x) \sin x + (24 - 6x^2) \cos x + c_1 x + c_2$

23. $y = x \log x + 3x^2 - 7x + 4$

24. $y = \frac{x^3}{6} + \sin x + 1.$